Breakdown in Traffic Networks

Boris S. Kerner

Breakdown in Traffic Networks

Fundamentals of Transportation Science

Springer

Boris S. Kerner
Physics of Transport and Traffic
University Duisburg-Essen
Duisburg, Germany

ISBN 978-3-662-54471-6 ISBN 978-3-662-54473-0 (eBook)
DOI 10.1007/978-3-662-54473-0

Library of Congress Control Number: 2017941004

Printed on acid-free paper

This Springer imprint is published by Springer Nature
The registered company is Springer-Verlag GmbH Germany
The registered company address is: Heidelberger Platz 3, 14197 Berlin, Germany

Preface

Traffic breakdown is a transition from free flow to congested traffic in a traffic or transportation network. Traffic breakdown occurs usually at network bottlenecks.

Users of traffic and transportation networks expect that through the application of traffic control, dynamic traffic assignment, cooperative driving systems, as well as other intelligent transportation systems (ITS), traffic breakdown in a network can be prevented. This is because congested traffic resulting from traffic breakdown causes a considerable increase in travel time, fuel consumption, CO_2 emission, as well as other travel costs. Therefore, any traffic and transportation theory applied for the development of automatic driving vehicles and reliable methods of dynamic traffic assignment and control should be consistent with empirical features of traffic breakdown at a network bottleneck.

The most important empirical feature of traffic breakdown in a network is the nucleation nature of traffic breakdown found in real field traffic data. The term *nucleation nature* of traffic breakdown means that traffic breakdown occurs in a metastable free flow. The metastability of free flow is as follows. There can be many speed (density, flow rate) disturbances in free flow. Amplitudes of the disturbances can be very different. When a disturbance occurs randomly whose amplitude is larger than a critical one, then traffic breakdown occurs. Such a disturbance resulting in the breakdown is called *nucleus* for the breakdown. Otherwise, if the disturbance amplitude is smaller than the critical one, the disturbance decays; as a result, no traffic breakdown occurs.

The nucleation nature of traffic breakdown at a network bottleneck is an empirical fundamental of transportation science.

Unfortunately, generally accepted classical traffic and transportation theories, which have had a great impact on the understanding of many empirical traffic phenomena, have nevertheless failed by their applications in the real world. Even several decades of a very intensive effort to improve and validate network optimization and control models based on the classical traffic and transportation theories have had no success. Indeed, there can be no examples found where online implementations of the network optimization models based on these classical traffic and transportation theories could reduce congestion in real traffic and transportation networks. In particular, in the book, we will show that applications of the classical approaches for dynamic traffic assignment in traffic networks, which are related to the state of the art in traffic and transportation research, deteriorate the traffic system while provoking heavy traffic congestion in urban networks.

This failure of classical traffic and transportation theories can be explained as follows: The classical traffic and transportation theories are not consistent with the nucleation nature of traffic breakdown. The nucleation nature of empirical traffic breakdown was understood only during last 20 years. In contrast, the generally accepted fundamentals and methodologies of traffic and transportation theory were introduced in the 1950s–1960s. Thus, the scientists whose ideas led to the classical traffic and transportation theories could not know the empirical nucleation nature of traffic breakdown. Respectively, the author introduced the three-phase traffic theory and the breakdown minimization (BM) principle.

The three-phase theory is the theoretical fundamental of transportation science that explains the empirical nucleation nature of traffic breakdown.

The BM principle is the theoretical fundamental of transportation science that permits to maximize the network throughput preventing traffic breakdown in the network.

The classical traffic and transportation theories that failed by their applications in the real world are currently the methodologies of teaching programs in most universities and the subject of publications in most transportation research journals and scientific conferences. Consequently, some of book's objectives are as follows:

1. We prove the empirical nucleation nature of traffic breakdown in networks.
2. We discuss the origin of the failure of classical traffic and transportation theories.
3. We show that the three-phase theory is incommensurable with the classical traffic theories.
4. We explain that standard dynamic traffic assignment that is the state of the art in transportation research provokes heavy traffic congestion in networks.
5. We show that applications of the BM principle result in the maximization of the network throughput while ensuring that no breakdown can occur in the network.

Stuttgart, Germany
December 2016

Boris S. Kerner

Acknowledgments

I thank Michael Schreckenberg, Dietrich Wolf, Florian Knorr, and Gerhard Hermanns of the University Duisburg-Essen as well as my former colleagues at the Daimler Company, in particular, Hubert Rehborn, Peter Häussermann, Harald Brunini, Ralf-Guido Herrtwich, Ralf Lamberti, Peter Konhäuser, Martin Schilke, Matthias Schulze, Peter Ebel, Viktor Friesen, Mario Aleksić, Jochen Palmer, Micha Koller, Peter Hemmerle, Gerhard Nöcker, Winfried Kronjäger, and Andreas Hiller for their support, discussions, and advice.

I thank Rui Jiang for many useful discussions and comments to the book. I thank Craig Davis for many useful comments to the three-phase theory used in this book. I thank also Jun-fang Tian for fruitful discussions of the book title.

Particular thanks are to Sergey Klenov, Hubert Rehborn, and Micha Koller who have read the book and made many useful comments.

I thank our partners for their support in the projects "UR:BAN—Urban Space: User oriented assistance systems and network management" and "MEC-View—Object detection for automated driving based on Mobile Edge Computing," funded by the German Federal Ministry of Economic Affairs and Energy.

I thank authorities of the State of Hessen (Germany) for providing real field traffic data measured on German freeways as well as Ralf-Peter Schäfer, Oliver Kannenberg, Stefan Lorkowski, and Nikolaus Witte for providing anonymized GPS probe data of the TomTom Company. I thank also Heiko Böhme, Timo Finke, and Nicolas Gath for providing real field detector data measured in Düsseldorf (Germany).

I am grateful to Sergey Klenov for his help with numerical simulations and the preparation of illustrations for the book.

Finally, I thank my wife, Tatiana Kerner, for her help and understanding.

Contents

Acronyms and Symbols

F	Free traffic flow (free flow traffic phase)
S	Synchronized flow phase of congested traffic
J	Wide moving jam phase of congested traffic
Line J	Characteristic line in the flow–density plane representing a steady propagation of the downstream front of a wide moving jam. The slope of the line J is determined by the mean velocity of the downstream jam front
F→S transition	Phase transition from the free flow phase (F) to the synchronized flow phase (S) (traffic breakdown at a highway bottleneck)
S→J transition	Phase transition from the synchronized flow phase (S) to the wide moving jam phase (J)
S→J instability	The classical traffic flow instability occurring in synchronized flow that leads to a growing speed wave of the local decrease in the speed in synchronized flow. The development of the S→J instability causes an S→J transition, i.e., the formation of a wide moving jam
S→F transition	Phase transition from the synchronized flow phase (S) to the free flow phase (F)
S→F instability	Instability of synchronized flow introduced in the three-phase theory that leads to a growing speed wave of the local increase in the speed in synchronized flow. The development of the S→F instability causes an S→F transition
F→S→J transitions	Sequence of an F→S transition with the following S→J transition
F→S→F transitions	Sequence of an F→S transition with the following S→F transition. A sequence of F→S→F transitions introduced in the three-phase theory results in a random time delay of traffic breakdown (F→S transition) at a highway bottleneck

F→J transition	Phase transition from the free flow phase (F) to the wide moving jam phase (J) that occurs in some classical traffic flow models
SP	Synchronized flow traffic pattern
LSP	Localized SP
WSP	Widening SP
MSP	Moving SP
GP	General congested traffic pattern
EP	Expanded traffic congested pattern
ITS	Intelligent transportation systems
V2V	Vehicle-to-vehicle communication
V2X	Vehicle-to-vehicle communication and/or vehicle-to-infrastructure communication
GM	General Motors
GW	Green wave in city traffic
RW	Red wave in city traffic: a hypothetical case, when all vehicles arrive the signal during the red signal phase *only*, i.e., the arrival flow rate is equal to zero during the green and yellow signal phases
ANCONA	**A**utomatic o**n**-ramp **c**ontrol of c**on**gested p**a**tterns (ANCONA) is on-ramp metering based on a congested pattern control approach
ACC	Adaptive cruise control
GPS	"Global Position System": satellite navigation system
DOST	Dissolving oversaturated traffic at traffic signal in city traffic
UE	Wardrop's user equilibrium
SO	Wardrop's system optimum
BM	Breakdown minimization (BM) principle for dynamic traffic assignment and control in traffic and transportation networks
CA	Cellular automaton
x	Road location
t	Time
v	Vehicle speed [km/h] or [m/s]
$v_{\rm free}(g)$	A dependence of the speed in free flow on the space gap between vehicles
$v_{\rm free}$	Vehicle speed in free flow under assumption that the free flow speed does not depend on the space gap
v_ℓ	Speed of the preceding vehicle [km/h] or [m/s]
$\Delta v = v_\ell - v$	Speed difference between the speed of the preceding vehicle and the vehicle speed [km/h] or [m/s]
$a(t)$	Vehicle acceleration $[{\rm m/s^2}]$
g	Space gap between vehicles [m] that is also called as net distance or space headway
$\tau^{\rm (net)}$	Time headway between vehicles [s] that is also called as net time gap or net time distance

d	Vehicle length [m]. The vehicle length includes the mean space gap between vehicles that are in a standstill within a wide moving jam or within a vehicle queue at traffic signal
q	Flow rate [vehicles/h]
ρ	Vehicle density [vehicles/km]
q_{in}	The flow rate in free flow on the main road upstream of a bottleneck
$\bar{q}_{\text{in}}$	The average arrival flow rate at traffic signal in city traffic
q_{on}	The on-ramp inflow rate at an on-ramp bottleneck
q_{sum}	The flow rate at a network bottleneck under free flow conditions
N	A number of bottlenecks in a traffic or transportation network, $N > 1$
α_k	The set of control parameters of network bottleneck k, where $k = 1, 2, \ldots, N$
$\mathbf{R}_k$	A matrix of percentages of vehicles with different vehicle (and/or driver) characteristics related to a network bottleneck k, where $k = 1, 2, \ldots, N$
M	A number of network links for which link inflow rates can be adjusted in a traffic or transportation network, $M > 1$
q_{m}	The inflow rate q_m for a network link m that can be adjusted in a traffic or transportation network, where $m = 1, 2, \ldots, M$
C	Capacity of free flow at a network bottleneck
C_{min}	Minimum capacity of free flow at a network bottleneck in the three-phase theory
$C_{\text{min}}^{(k)}$	Minimum capacity of free flow at bottleneck k in a traffic or transportation network, where $k = 1, 2, \ldots, N$
C_{cl}	Classical capacity of traffic signal in city traffic
$q_{\text{th}}^{(\text{B})}$	Threshold flow rate for spontaneous traffic breakdown at a network bottleneck
$q_{\text{th}}^{(\text{B},k)}$	Threshold flow rate for spontaneous traffic breakdown at network bottleneck k, where $k = 1, 2, \ldots, N$
$q_{\text{th}}^{(\text{DOST})}$	Threshold flow rate for spontaneous occurrence of dissolving over-saturated traffic (DOST) at traffic signal in city traffic
$q_{\text{th}}^{(\text{DOST},k)}$	Threshold flow rate for spontaneous occurrence of dissolving over-saturated traffic (DOST) at network bottleneck k due to traffic signal
C_{max}	Maximum capacity of free flow at a network bottleneck in the three-phase theory
$C_{\text{max}}^{(k)}$	Maximum capacity of free flow at bottleneck k in a traffic or transportation network, where $k = 1, 2, \ldots, N$

$P^{(\mathrm{B})}$	Probability of spontaneous traffic breakdown at a bottleneck during a given time interval in the three-phase theory
T_{ob}	A time interval for observing of free flow
T_{av}	Averaging time interval for traffic variables
$P^{(\mathrm{B},k)}$	Probability of spontaneous traffic breakdown at network bottleneck k during a given time interval, where $k = 1, 2, \ldots, N$
$P_{\mathrm{C}}^{(\mathrm{B})}$	Probability that free flow remains at a bottleneck during a given time interval
N_{r}	The number of simulation realizations (runs) used for the calculation of the probability of traffic breakdown
n_{r}	The number of simulation realizations (runs) in which traffic breakdown has been found at a bottleneck
$T^{(\mathrm{B})}$	A random time delay of traffic breakdown (F→S transition) at a network bottleneck in the three-phase theory
$T^{(\mathrm{B,\ mean})}$	The mean time delay of traffic breakdown (F→S transition) at a network bottleneck
$T_{\mathrm{ob}}^{(\mathrm{SJ})}$	A time interval for observing of synchronized flow
$T_{\mathrm{FJ}}^{(\mathrm{B})}$	A random time delay of F→J transition at a bottleneck in traffic flow models of the GM model class
$T_{\mathrm{SJ}}^{(\mathrm{B})}$	A random time delay of S→J transition in synchronized flow
$P_{\mathrm{SJ}}^{(\mathrm{B})}$	Probability of S→J transition in synchronized flow during a given time interval $T_{\mathrm{ob}}^{(\mathrm{SJ})}$
$P_{\mathrm{FJ}}^{(\mathrm{B})}$	Probability of F→J transition at a bottleneck during a given time interval in traffic flow models of the GM model class
P_{net}	Probability of the occurrence of traffic breakdown during a given time interval in a traffic or transportation network
$P_{\mathrm{net}}^{(\mathrm{min})}$	The minimum value of probability of the occurrence of traffic breakdown during a given time interval in a traffic or transportation network resulting from the application of the BM principle
$P_{\mathrm{C,\ net}}$	Probability that during a given time interval traffic breakdown does *not* occur in a traffic or transportation network
Q	The total network inflow rate
C_{net}	Network capacity
$q_i^{(\mathrm{o})}(t)$	Network inflow rates at the network boundaries, $i = 1, 2, \ldots, I$
$q_j^{(\mathrm{d})}(t)$	Network outflow rates at the network boundaries, $j = 1, 2, \ldots, J$
O_i–D_j	Origin-destination pair of a network
$q_{ij}^{(\mathrm{o})}(t)$	Origin-destination matrix of a network
$\mathscr{S}_{ij}$	A set of paths (routes) through a network for O_i–D_j pair

$T_{ij}^{(s)}$	The travel time on path (route) s for O_i–D_j pair under free flow conditions at the small enough network inflow rate ($Q \to 0$)
$\mathscr{A}_{ij}$	The set of alternative network routes for O_i–D_j pair related to a constrain "alternative network routes (paths)" used by the applications of the BM principle, $\mathscr{A}_{ij} \subset \mathscr{S}_{ij}$
Λ_{ij}	A threshold difference between route travel times in a network for O_i–D_j pair used in the constrain "alternative network routes (paths)"
$\tau_{\text{del}}^{(\text{acc})}$	The mean time delay in vehicle acceleration
$\tau_{\text{del}}^{(\text{acc})}(0)$	The mean time delay in vehicle acceleration at the vehicle speed that is equal to zero (at the downstream front of a wide moving jam or a moving queue at the signal)
$\rho_{\max}$	The vehicle density within a wide moving jam or within a (moving) vehicle queue at the signal in city traffic
v_{g}	The mean velocity of the downstream front of a wide moving jam
q_{out}	The flow rate in free flow formed in the outflow of a wide moving jam
q_{sat}	The saturated flow rate in free flow formed in the outflow of a moving vehicle queue at traffic signal
$\rho_{\min}$	The vehicle density in free flow formed in the outflow of a wide moving jam
q_0	The maximum flow rate at the maximum point of a theoretical fundamental diagram
$q_{\max}^{(\text{free, emp})}$	The maximum flow rate in empirical free flow
$q_{\text{cr}}^{(\text{J})}$	The critical flow rate at which free flow on a homogeneous road becomes unstable with respect to the classical traffic flow instability of the GM model class
$\rho_{\text{cr}}^{(\text{J})}$	The critical vehicle density at which free flow on a homogeneous road becomes unstable with respect to the classical traffic flow instability of the GM model class
$q_{\text{cr}}^{(\text{B})}$	The critical flow rate at which free flow at a highway bottleneck becomes unstable with respect to the classical traffic flow instability of the GM model class
q_{slow}	The flow rate of slow vehicles in empirical free flow
Δq_{wave}	A difference between the flow rate within a wave propagating downstream in empirical free flow and the average flow rate
Δv_{wave}	A difference between the vehicle speed within a wave propagating downstream in empirical free flow and the average vehicle speed

$\Delta\psi_{\text{wave}}$	A dimensionless characteristic of the share of slow vehicles within a wave propagating downstream in empirical free flow
q_{off}	The flow rate of vehicles leaving the main road to off-ramp at an off-ramp bottleneck
η_{off}	The percentage of vehicles leaving the main road to off-ramp at an off-ramp bottleneck, $\eta_{\text{off}} = (q_{\text{off}}/q_{\text{in}})100\%$
$v^{(\text{B})}_{\text{free}}$	The average vehicle speed within a permanent speed disturbance localized at a highway bottleneck in a qualitative three-phase theory
$v^{(\text{B})}_{\text{cr, FS}}$	The average vehicle speed within a critical nucleus required for traffic breakdown (F→S transition) at a highway bottleneck in a qualitative three-phase theory
$\Delta v^{(\text{B})}_{\text{cr, FS}}$	The amplitude of the critical nucleus required for traffic breakdown (F→S transition) at a highway bottleneck in a qualitative three-phase theory, $\Delta v^{(\text{B})}_{\text{cr, FS}} = v^{(\text{B})}_{\text{free}} - v^{(\text{B})}_{\text{cr, FS}}$
$v_{\text{cr, SJ}}$	The average speed within the critical nucleus required for the emergence of a wide moving jam in synchronized flow (S→J transition)
$\Delta v_{\text{cr, SJ}}$	The amplitude of the critical nucleus required for the emergence of a wide moving jam in synchronized flow (S→J transition)
$\tau^{(\text{ACC})}_{\text{d}}$	A desired net time gap (desired time headway) of the ACC vehicle
g_{safe}	A safe space gap between vehicles
v_{safe}	A safe vehicle speed
τ_{safe}	A safe time headway (safe net time gap) between vehicles
G	A synchronization space gap between vehicles
τ_{G}	A synchronization time headway (synchronization net time gap) between vehicles
τ	A time step in traffic flow models with a discrete time
$\tau^{(\text{gross})}$	A gross time gap between two vehicles
$\tau^{(\text{gross})}_{\text{sat}}$	A saturated gross time headway between two vehicles that is the mean gross time headway in free flow formed by the discharge from a moving queue during the green phase of traffic signal
T_{G}	A duration of the green phase of traffic signal
T_{R}	A duration of the rot phase of traffic signal
T_{Y}	A duration of the yellow phase of traffic signal
ϑ	A duration of the cycle of traffic signal, $\vartheta = T_{\text{G}} + T_{\text{Y}} + T_{\text{R}}$
δt	A lost time at traffic signal
$T^{(\text{eff})}_{\text{G}}$	A duration of the effective green phase of traffic signal, $T^{(\text{eff})}_{\text{G}} = \vartheta - T_{\text{R}} - \delta t$
q_{TS}	The flow rate of traffic passing the signal

$\bar{q}_{\mathrm{TS}}$	The average flow rate of traffic passing the signal
x_{TS}	Location of traffic signal
q_{GW}	The flow rate within an GW (green wave) in city traffic
q_{turn}	The rate of turning-in traffic at traffic signal
$q_{\mathrm{out}}^{(\mathrm{MSP})}$	The outflow rate from an MSP (moving synchronized flow pattern)
$v_{\mathrm{down}}^{(\mathrm{MSP})}$	The mean velocity of the downstream front of the MSP
$\tau_{\mathrm{out}}^{(\mathrm{MSP})}$	The mean time headway (mean net time gap) between vehicles that discharge from the MSP
ΔT_{b}	A time interval between the end of the red signal phase and the beginning of a GW
ΔT_{e}	A time interval between the end of the GW and the beginning of the next red signal phase
T_{GW}	A duration of the undisturbed GW
$\Delta T_{\mathrm{b}}^{(\mathrm{ideal})}$	A time interval between the end of the red signal phase and the beginning of the undisturbed GW
$\Delta T_{\mathrm{e}}^{(\mathrm{ideal})}$	A time interval between the end of the undisturbed GW and the beginning of the next red signal phase
$\gamma^{(\mathrm{ACC})}$	The percentage of automatic driving vehicles in a mixed traffic flow in which automatic driving and manual driving vehicles are randomly distributed
$r = \mathrm{rand}(0, 1)$	A random number uniformly distributed between 0 and 1

Chapter 1
Introduction—The Reason for Paradigm Shift in Transportation Science

Vehicular traffic in traffic and transportation networks occurs in space and time. Therefore, to understand vehicular traffic, real field traffic data measured in space and time should be understood. For this reason, the empirical basis of this book is a *spatiotemporal* analysis of empirical, i.e., real field traffic data. This analysis allows us to reveal empirical spatiotemporal traffic phenomena occurring in traffic and transportation networks. The empirical spatiotemporal traffic phenomena are the basis for the development of traffic control and optimization methods, for dynamic traffic assignment in traffic and transportation networks as well as for the development of a diverse variety of intelligent transportation systems (ITS).[1] Examples of ITS include driver assistant systems in vehicles, vehicle-to-vehicle and/or vehicle-to-infrastructure communication (called as V2X-communication), automatic driving vehicles, individual and collective ITS for dynamic traffic assignment, control, and management of traffic and transportation networks.

There is a huge variety of empirical studies of traffic flow, traffic flow models and theories, theoretical methods for dynamic traffic assignment, traffic control and optimization in traffic and transportation networks developed by several generations of researchers (see, e.g., [11, 13–15, 19–21, 25, 26, 59, 61, 66, 75, 76, 80–86, 89–91, 94–98, 100, 105–110, 113, 115–117, 121–123, 139, 172, 174–177, 208–210, 224–227, 231, 232, 242, 244, 247–249, 251, 253, 266–273, 285–289, 294, 295, 306, 307, 319, 329, 334, 340, 341, 343, 346, 354, 367–369, 371–373, 376, 379, 385, 416, 418, 419, 423–426, 431, 434–436, 438, 439, 448, 454, 456–458, 470, 474, 476, 488, 489, 491, 495, 496, 502, 504, 508–512, 514, 527, 528, 536–538, 647–652, 654, 655, 657, 668, 670, 682, 684, 693, 695, 697, 729, 738–740, 757,

[1] In other words, we do not consider travel decision behavior, which determines traffic demand. Models of travel decision behavior as well as traffic and transportation demand can be found, for example, in the book by Ben-Akiva and Lerman [74] as well as in reviews by Axhausen and Gärling [29] and by Goulias [338]; see also papers [33, 169] and references there. A challenge for transportation demand research following from results of this book is formulated in item 10 of Sect. 15.5.

B.S. Kerner, *Breakdown in Traffic Networks*, DOI 10.1007/978-3-662-54473-0_1

765, 767, 769, 771, 774, 776, 777, 780–783, 787–789, 792, 793, 797, 800, 801, 806–809, 817–819, 823–829, 831, 832, 846, 850, 851, 854, 857, 861, 863, 864, 870–873, 875, 881, 882, 886, 888, 897, 898, 900, 901, 910, 913–916, 918, 920, 932, 954, 955, 957, 959, 963, 982, 995, 1005, 1007–1009, 1018, 1022, 1035, 1038, 1041, 1044, 1057, 1063, 1075, 1077, 1078, 1102, 1104, 1107, 1109, 1111–1113, 1128–1130, 1135, 1142, 1171, 1172, 1181, 1185, 1206] as well as references in reviews and conference proceedings [16, 22, 70, 73, 103, 135, 141, 146, 166, 173, 187, 243, 258, 259, 290, 305, 314, 318, 355, 389, 394, 404, 429, 430, 450, 453, 661, 662, 665, 689, 694, 696, 725, 727, 768, 770, 773, 796, 799, 804, 845, 856, 884, 894, 923, 938, 956, 964, 976, 978–980, 983–985, 1003, 1010, 1012, 1025, 1064, 1089, 1136, 1139, 1143, 1144]).

In this book, rather than making a review of this diverse variety of traffic and transportation theories, models, and methods, we focus on a detailed analysis of empirical traffic data and the consequences of this empirical analysis for traffic and transportation theories. The term *empirical traffic data* ("empirical data" for short) is a synonym of the term *real field traffic data*. The real field traffic data is solely obtained through measurements in real world traffic based on road detectors, video-cameras, global navigation satellite system (GNSS)[2] in vehicles, and other methods for the measurement of traffic variables[3] (vehicle speed, flow rate, density, space gaps and time headway between vehicles, etc.).

1.1 Definitions of Free and Congested Traffic in Empirical Data

Free traffic flow (free flow for short) is usually observed, when the vehicle density in traffic is not large. At small enough vehicle density, interactions between vehicles in free flow are negligible. Therefore, vehicles have an opportunity to move with their desired maximum speeds (if this speed is not restricted by road conditions or traffic regulations). At large enough vehicle density, rather than free flow, congested traffic is observed. There are a huge number of publications devoted to empirical and theoretical studies of congested traffic (see, e.g., [3, 5, 16, 22, 87, 141, 146, 243, 258, 259, 290, 305, 314, 318, 351, 355, 394, 404, 429, 430, 450, 453, 535, 590, 661, 662, 665, 674, 675, 689, 694, 725, 727, 758–761, 768, 773, 794–796, 798, 799, 804, 845, 856, 867–869, 884, 923, 938, 956, 964, 976, 978–980, 983–985, 1010, 1012, 1025, 1055, 1064, 1089, 1114, 1136, 1139, 1143, 1144] and references there).

[2] It is often called also as Globe Position System (GPS).

[3] Definitions of traffic variables, traffic parameters, and traffic patterns can be found in Sec. 2.1 of the book [572].

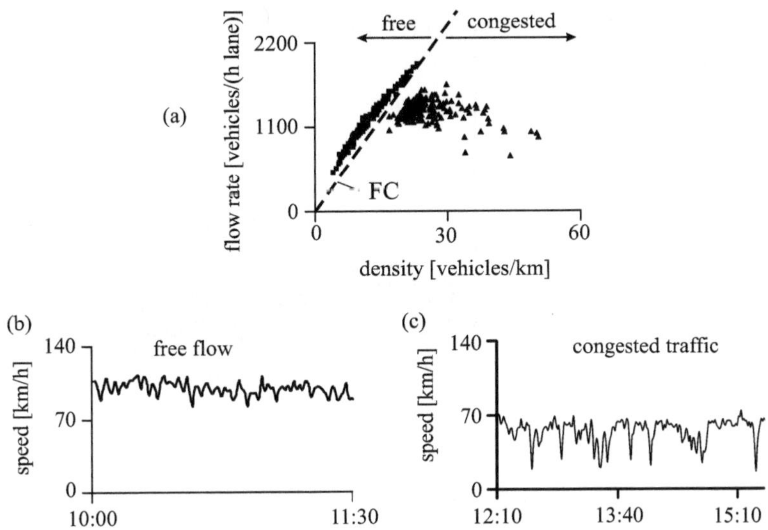

Fig. 1.1 Empirical example of the definitions of free flow and congested traffic [662, 804]: (**a**) Empirical traffic data in the flow–density plane measured at a single road location. The slope of the line *FC* in (**a**) determines a minimum average speed that is possible in free flow. (**b**, **c**) Speed as a time-function for the data in (**a**), respectively, for free flow (**b**) (data points left of the line *FC* in (**a**)) and for congested traffic (**c**) (data points right of the line *FC* in (**a**)). Adapted from [572]

We will use a definition of congested traffic used by many traffic engineers (e.g., [662, 804]):

- Congested traffic is defined as a state of traffic in which the average speed is lower than the minimum average speed that is still possible in free flow[4] (Fig. 1.1a) [662, 804].

Congested traffic occurs due to traffic breakdown in an initial free flow.

- Traffic breakdown is a transition from free flow to congested traffic [662, 804].

In congested traffic, a great variety of congested traffic patterns have been observed (see, e.g., [252, 254–256, 394, 556, 662, 845, 856, 983–985, 1089, 1093, 1094]). A *congested traffic pattern* (congested pattern for short) is defined as follows. A congested traffic pattern is a spatiotemporal traffic pattern within which there is congested traffic. The congested pattern is separated from free flow by the downstream and upstream fronts: At the downstream front, vehicles accelerate from a lower speed within the pattern to a higher speed in free flow downstream; at the

[4]In empirical observations, to determine this minimum average speed in free flow, traffic variables (e.g., the average vehicle speed and flow rate), which are measured during a long enough time interval in a neighborhood of a bottleneck, are usually studied in the flow–density (Fig. 1.1a) or speed–flow planes (e.g., [662, 804]).

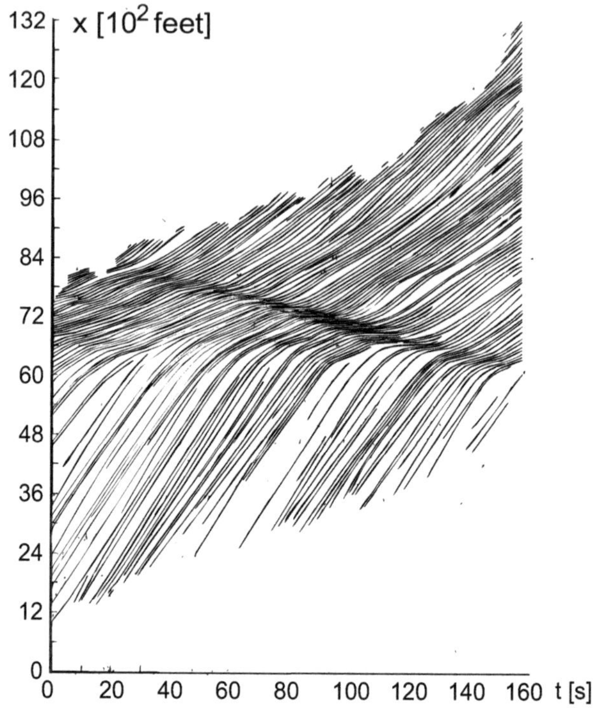

Fig. 1.2 Empirical moving jam: Traffic dynamics derived from aerial photography. Adapted from Treiterer [1093]

upstream front, vehicles decelerate from a free flow speed to a lower speed within the congested pattern.

In particular, one of the congested traffic patterns is a moving traffic jam [252, 254–256, 394, 556, 662, 845, 856, 983–985, 1089, 1093, 1094]. A *moving traffic jam* (moving jam for short) is defined as follows (Fig. 1.2): A moving jam is a localized congested traffic pattern that moves upstream in traffic flow. Within the moving jam the average vehicle speed is very low (sometimes as low as zero), and the density is very high. The moving jam is spatially restricted by the downstream jam front and upstream jam front. Within the downstream jam front vehicles accelerate from low speed states within the jam to higher speeds in traffic flow downstream of the moving jam. Within the upstream jam front vehicles must slow down to the speed within the jam. Both jam fronts move upstream. Within the jam fronts the vehicle speed, flow rate, and density vary abruptly. Moving jams have been studied empirically by many authors, in particular, in classic empirical works by Edie et al. [252, 254–256], Treiterer et al. [1092–1095] (Fig. 1.2), and Koshi et al. [662–664]. In accordance with a theory of moving jam propagation developed firstly in 1994 [616], in a study of real field traffic data it has been found [622] that a moving jam can exhibit characteristic parameters that do not depend on initial conditions. In particular, the moving jam can propagate on a highway while maintaining the mean velocity of the downstream front of the jam (see Sect. 8.3.1).

As already mentioned, traffic occurs in space and time, i.e., traffic phenomena are *spatiotemporal* phenomena. This means that traffic phenomena *cannot* be sufficient understood, if traffic data is measured at a single road location only as presented in Fig. 1.1. An example of the empirical spatiotemporal traffic phenomena is the propagation of a moving jam shown in Fig. 1.2.

Traffic measurements made simultaneously in space and time are needed for the understanding of real traffic phenomena.

For this reason, below in the book only such traffic measurements are the empirical basis for a study of spatiotemporal traffic phenomena.

1.2 Bottlenecks

In empirical data, traffic breakdown occurs usually at a network bottleneck (see, e.g., [70, 73, 103, 135, 166, 173, 187, 258, 259, 305, 318, 355, 394, 429, 430, 661, 662, 727, 768, 796, 804, 845, 856, 884, 923, 938, 956, 978, 980, 1089, 1136, 1139] and references there).

Two main types of bottlenecks can be distinguished: Road bottlenecks fixed at some road locations and moving bottlenecks. Road bottlenecks are caused for example by on- and off-ramps, road gradients, road-works, a decrease in the number of road lines (in the flow direction), traffic signal, etc.

A moving bottleneck that has probably firstly been predicted by Gazis and Herman [320] as well as by Newell [885, 893] is caused by a slow vehicle(s) moving in traffic flow. While approaching the slow vehicle, vehicles try to pass the moving bottleneck. An usual assumption used in simulations of moving bottlenecks [203, 204, 262, 263, 320, 599, 710, 712, 830, 885, 893] is that to pass the slow vehicle (moving bottleneck) moving in the right lane, passenger vehicles moving in the right lane change to the left lane. After passing the slow vehicle, most of these passenger vehicles change back to the right lane. As a result of this passing, a speed disturbance should occur in a neighborhood of the moving bottleneck. In contrast with a road bottleneck, which leads to a speed disturbance that is fixed at a bottleneck location, a speed disturbance caused by a moving bottleneck should move at the speed of the moving bottleneck, i.e., at the speed of the slow vehicle [203, 204, 262, 263, 320, 599, 710, 712, 830, 885, 893, 895].

1.3 Definitions of Synchronized Flow and Wide Moving Jam Phases in Empirical Data for Congested Traffic

From many empirical studies of real field traffic data measured in different countries over many years of traffic flow observations it has been found that in congested traffic two qualitatively different traffic phases should be distinguished: Synchronized flow (S) and wide moving jam (J) [556, 572]. Therefore, in the three-phase traffic flow theory introduced by the author (three-phase theory for short) there are three phases [556, 572][5]:

1. Free flow (F).
2. Synchronized flow (S).
3. Wide moving jam (J).

In the three-phase theory, the traffic phases *wide moving jam* and *synchronized flow* in congested traffic are defined, respectively, through the use of the definitions [J] and [S] as follows (Fig. 1.3) [539–553, 555, 556, 571–574, 578, 584, 586]:

> [J] A wide moving jam is a moving jam that maintains the mean velocity of the downstream front of the jam, even as it propagates through other different traffic states of free flow and synchronized flow or highway bottlenecks. This is a characteristic feature of the wide moving jam phase.
> [S] In contrast to the wide moving jam traffic phase, the downstream front of the synchronized flow phase does *not* maintain the mean velocity of the downstream front. In particular, the downstream front of synchronized flow is often *fixed* at a bottleneck. In other words, the synchronized flow phase does not show the characteristic feature [J] of the wide moving jam phase.

In the synchronized flow phase (S), the average speed is smaller than the average speed in the free flow phase (F), whereas the flow rate can be as large as the flow rate in the phase F (Fig. 1.4a). In contrast with the phase S, within the wide moving jam phase (J) both the speed and flow rate are very small, sometimes as small as zero (Fig. 1.4b,c).[6]

The traffic phases *synchronized flow* and *wide moving jam* in congested traffic can be distinguished in microscopic traffic data (single-vehicle data) through a microscopic criterion of the phases that is consistent with the macroscopic definitions

---

[5]Based on the differentiation of the synchronized flow phase and the wide moving jam phase in the three-phase theory, ASDA and FOTO methods for the reconstruction of spatiotemporal features of traffic congestion have been introduced and developed [589, 591, 627–634, 636, 921, 922, 983–985]. Based on ASDA and FOTO models, Rehborn, Aleksić, Haug, Palmer, and other colleagues of the Daimler Company have developed software-tools that are up to now used in on-line installations in traffic control centers. A detailed consideration of ASDA and FOTO models is out of scope of this book. For results of ASDA and FOTO applications, see a review by Rehborn and Klenov [983] as well as the book [556].

[6]Explanations of the terms "synchronized flow" and "wide moving jam" can be found in Sec. 2.4.4 of the book [572].

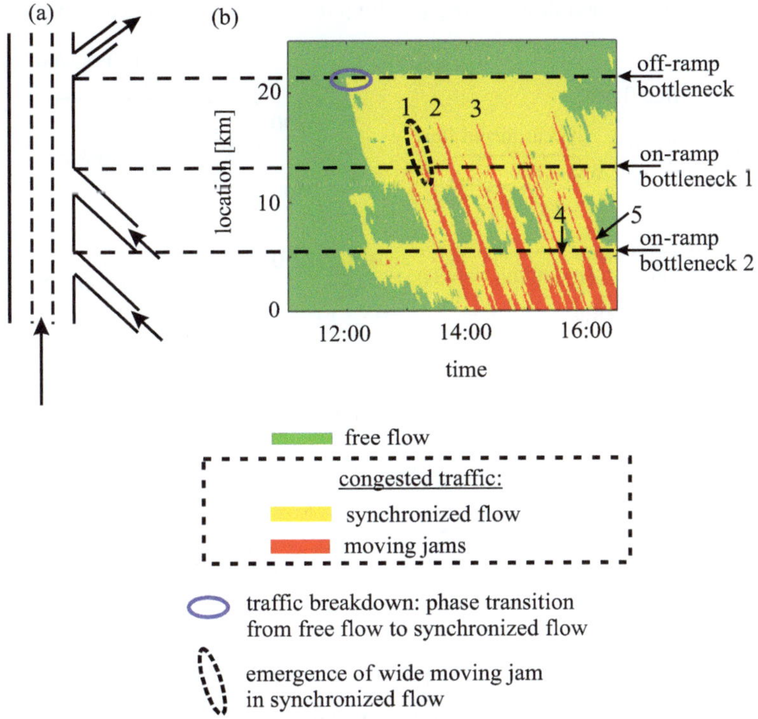

Fig. 1.3 Empirical example of a fragment of complex congested traffic pattern measured on three-lane highway A5-North in Germany on March 23, 2001: (**a**) Sketch of highway section with three highway bottlenecks (off-ramp bottleneck and two on-ramp bottlenecks). (**b**) Empirical speed data presented in space and time with averaging method described in Sec. C.2 of [636]. Adapted from [556]

[J] and [S] [606–608]. The microscopic criterion of the phases *synchronized flow* and *wide moving jam* in congested traffic has been discussed in Sec. 2.6 of the book [572].

The downstream front of synchronized flow separates synchronized flow upstream from free flow downstream. Within the downstream front of synchronized flow vehicles accelerate from lower speeds in synchronized flow upstream of the front to higher speeds in free flow downstream of the front.

The definitions [J] and [S] of the phases J and S in congested traffic are illustrated with empirical data presented in Fig. 1.3b as follows. In Fig. 1.3b, there are many moving jams propagating upstream of an off-ramp bottleneck. Some of the moving jams (jams 1–5 marked in Fig. 1.3b) propagate through two neighborhood on-ramp bottlenecks (labeled by "on-ramp bottleneck 1" and "on-ramp bottleneck 2" in Fig. 1.3b) while maintaining the velocity of the downstream jam front. Thus, in accordance with the definition [J], moving jams 1–5 belong to the wide moving jam phase in congested traffic.

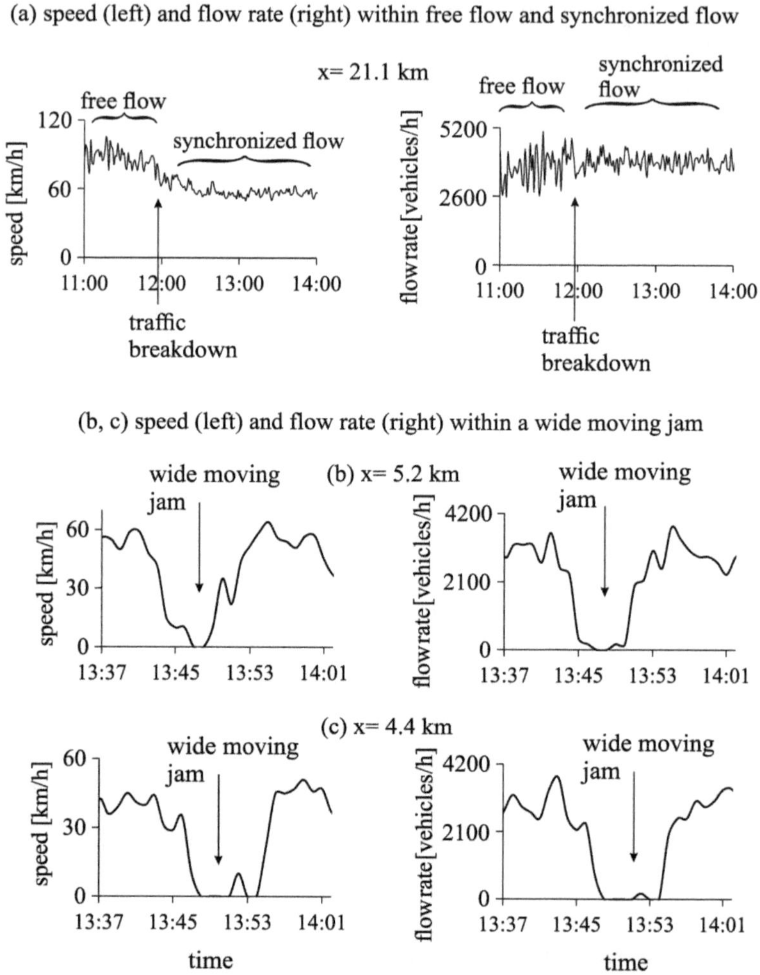

Fig. 1.4 Empirical features of synchronized flow (**a**) and wide moving jams (**b**, **c**) related to empirical data in Fig. 1.3: (**a**) Dependence of speed and total flow rate across the highway at the effective location of the off-ramp bottleneck $x = 21.1$ km at which traffic breakdown (F→S transition) occurs. (**b**, **c**) Dependence of speed and total flow rate across the highway at two road locations $x = 5.2$ km and 4.4 km within a wide moving jam labeled by "1" in Fig. 1.3b; locations $x = 5.2$ km and 4.4 km within the wide moving jam in (**b**, **c**) are about 15 km upstream of the location of traffic breakdown $x = 21.1$ km in (**a**). Adapted from [556]

In Fig. 1.3b, we can also see congested traffic whose downstream front is fixed at the off-ramp bottleneck (dashed line related to the off-ramp bottleneck in Fig. 1.3). Thus, congested traffic resulting from the breakdown satisfies the definition [S] for the synchronized flow phase.

1.4 Traffic Breakdown

Traffic breakdown at a highway bottleneck is a local phase transition from free flow (F) to congested traffic whose downstream front is usually fixed at the bottleneck location. In real traffic data, a typical duration of traffic breakdown at a highway bottleneck is equal to or less than 1 min [258, 259, 661, 804]. During this short time interval, instead of an initial free flow at the bottleneck a state of congested traffic occurs in a neighborhood of the bottleneck (this can already been seen in Figs. 1.3b and 1.4a). The empirical result that the downstream front of traffic congestion resulting from traffic breakdown is fixed at the bottleneck is a *common result* of all empirical observations of traffic breakdown at any highway bottleneck [6, 38–47, 111, 114, 118, 142, 145, 147, 162, 241, 258–260, 327, 356, 358–362, 364–366, 461, 539–553, 555, 556, 571–574, 578, 584, 586, 622–626, 661, 752, 803–805, 947–953, 996, 1017]. As above-mentioned, in the three-phase theory, such congested traffic is called synchronized flow (S) (Figs. 1.3b and 1.4a) [556, 572]. In other words, using the terminology of the three-phase theory, traffic breakdown is a transition from free flow to synchronized flow (called F→S transition) [556, 572]. However, it should be emphasized that before basic features of the F→S transition are not discussed (a brief discussion will be made in next Sect. 1.5 and a detailed discussion of the basic features the F→S transition will be performed in Chaps. 3 and 5), the term *synchronized flow* is nothing more as only the definition of congested traffic whose downstream front is fixed at the bottleneck.

Traffic breakdown at a highway bottleneck is a phase transition from free flow to synchronized flow (F→S transition): Traffic breakdown at the highway bottleneck and the F→S transition are synonyms.

1.5 Empirical Phase Transitions in Traffic Flow

The definition of the traffic phases made in the three-phase theory (Sect. 1.3) leads to the explanation of the nature of phase transitions observed in real field traffic data [539–542]. Based on a study of real field traffic data, it has been found that both a phase transition from free flow to synchronized flow (F→S transition) and a phase transition from synchronized flow to a wide moving jam (called S→J transition) exhibit the nucleation nature [539–542]. These empirical findings have led to the development of the three-phase theory [539–542].

The empirical nucleation nature of the F→S and S→J transitions is as follows [539–542, 556]:

(i) The phase transition from free flow to synchronized flow (F→S transition) occurs in a metastable state of free flow at a highway bottleneck. To understand the sense of the term *metastable state of free flow with respect to the F→S transition at the bottleneck*, we consider a local speed disturbance in free flow that causes a local speed decrease at the bottleneck. The disturbance is localized at the bottleneck (see Sects. 3.4.4 and 5.3). We use the term "the amplitude of the disturbance": The larger the disturbance amplitude, the smaller the free flow speed within the disturbance in comparison with the free flow speed outside the disturbance. If the amplitude of the disturbance is small enough, then no F→S transition occurs at the bottleneck. However, when the amplitude of a local speed disturbance in the metastable free flow is equal to or exceeds a critical amplitude, the F→S transition does occur. A local speed disturbance occurring in the metastable free flow that leads to the F→S transition is called a nucleus for the F→S transition. Respectively, a local speed disturbance in the metastable free flow with a critical amplitude required for traffic breakdown (F→S transition) is called a critical nucleus. In other words, in real traffic there is the metastability of free flow with respect to the F→S transition at the bottleneck. As above mentioned, the term *traffic breakdown at a highway bottleneck* is a synonym of the term *F→S transition at the bottleneck*. Therefore, the term *a nucleus for F→S transition at the bottleneck* is a synonym of the term *a nucleus for traffic breakdown at the bottleneck*. Detailed empirical and theoretical studies of the nucleation nature of traffic breakdown (F→S transition) at highway bottlenecks will been made in Chaps. 3 and 5, respectively.

(ii) The phase transition from synchronized flow to a wide moving jam(s) (S→J transition) occurs in a metastable state of synchronized flow with respect to the S→J transition. We consider a local speed disturbance in synchronized flow that causes a local speed decrease. The disturbance propagates upstream in synchronized flow. When the amplitude of the local speed disturbance in this metastable synchronized flow is small enough, the local disturbance decays. However, when the amplitude of a local speed disturbance in the metastable synchronized flow is equal to or exceeds a critical amplitude, the S→J transition does occur. A local speed disturbance occurring in the metastable synchronized flow that leads to the S→J transition is called a nucleus for the S→J transition.

These empirical results (Figs. 1.3, 1.5, and 1.6) lead to the following conclusion of the three-phase theory [539]:

- Wide moving jams emerge spontaneously in free flow due to a sequence of two phase transitions: Firstly, the F→S transition occurs at the bottleneck leading to synchronized flow upstream of the bottleneck. Synchronized flow propagates upstream of the bottleneck. Within the emergent synchronized flow, at some distance upstream of the bottleneck moving jams emerge spontaneously (S→J transition) (Figs. 1.3b, 1.5b, and 1.6). This sequence of the phase transitions is called the F→S→J transitions (a more detailed consideration of F→S→J transitions is made in Sect. 8.4).

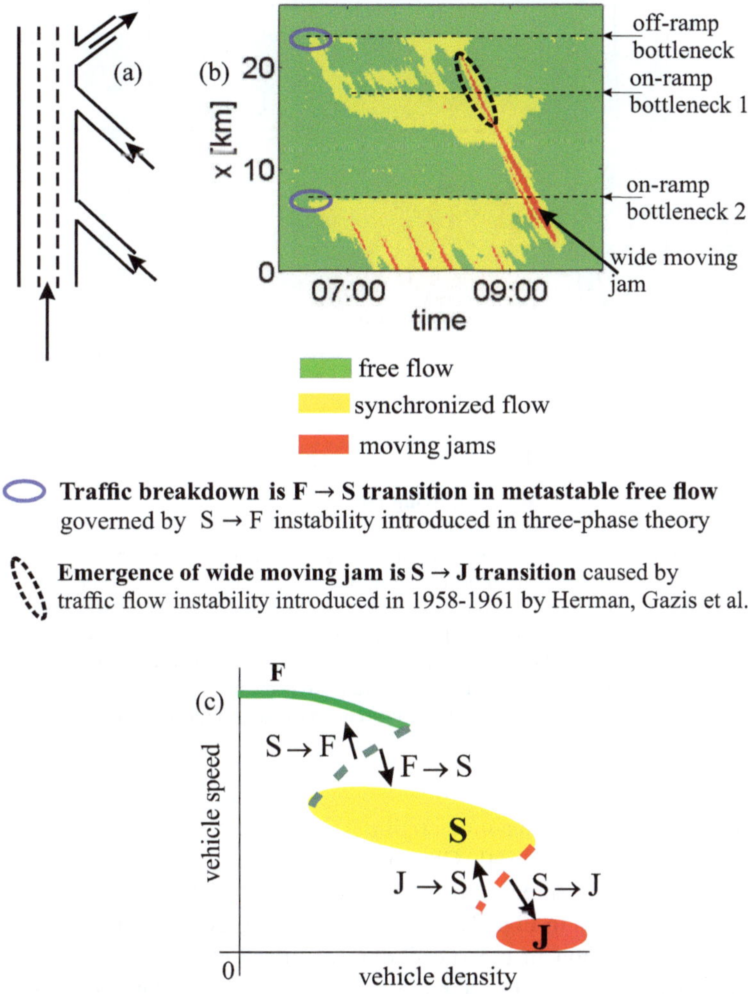

Fig. 1.5 Empirical example of phase transitions in traffic flow illustrating the existence of two traffic flow instabilities in three-phase theory [539]: (**a**) Sketch of section of three-lane highway in Germany with three bottlenecks adapted from Fig. 1.3a. (**b**) Speed data measured on April 20, 1998 on freeway A5-South in Germany with road detectors installed along road section in (**a**); data is presented in space and time with averaging method described in Sec. C.2 of [636]. (**c**) Hypothesis of three-phase theory about features of phase transitions in traffic flow: Qualitative 2Z-characteristic for phase transitions [556, 594]. F—free flow phase, S—synchronized flow phase, J—wide moving jam phase

Both the F→S transition (traffic breakdown) and the S→J transition that occur in space and time in real traffic (Figs. 1.3b, 1.5b, and 1.6) can also be presented in the speed–density, speed–flow-rate, and flow–density planes by double Z-

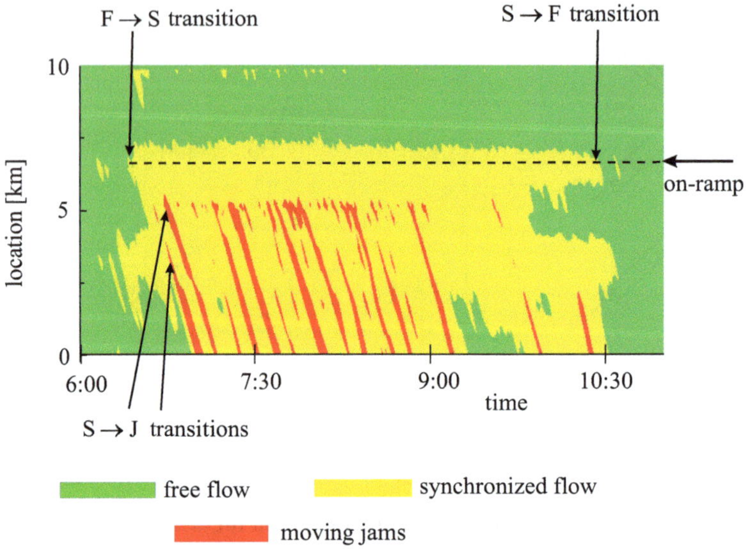

Fig. 1.6 Empirical example of F→S→J transitions in real traffic [539, 548]: Speed in space and time within a congested pattern occurring at on-ramp bottleneck (this bottleneck is on-ramp bottleneck 2 in Fig. 1.5a); 1-min averaged data measured on March 23, 1998 on freeway A5-South in Germany with road detectors installed along three-lane road section; data is presented in space and time with averaging method described in Sec. C.2 of [636]. Arrows F→S and S→F mark symbolically, respectively, the F→S and S→F transitions at the location of on-ramp bottleneck, arrows S→J mark symbolically S→J transitions related to the emergence of two first wide moving jams within synchronized flow

characteristics for phase transitions (called 2Z-characteristic for phase transitions) (Figs. 1.5c and 1.7) [556, 594]. Obviously, there are also return S→F and J→S transitions between the three traffic phases (Figs. 1.6 and 1.7). Respectively, arrows F→S and S→F shown in Figs. 1.5c and 1.7 mark, respectively, the F→S transition and the S→F transition in the speed–density, speed–flow-rate, and flow–density planes associated with real F→S and S→F transitions occurring in space and time (Figs. 1.3b, 1.5b, and 1.6).

1.6 Empirical Fundamental of Transportation Science

As emphasized in Sect. 1.5, based on a study of real field traffic data it has been found that empirical traffic breakdown (F→S transition) at a highway bottleneck exhibits the nucleation nature [539–542].

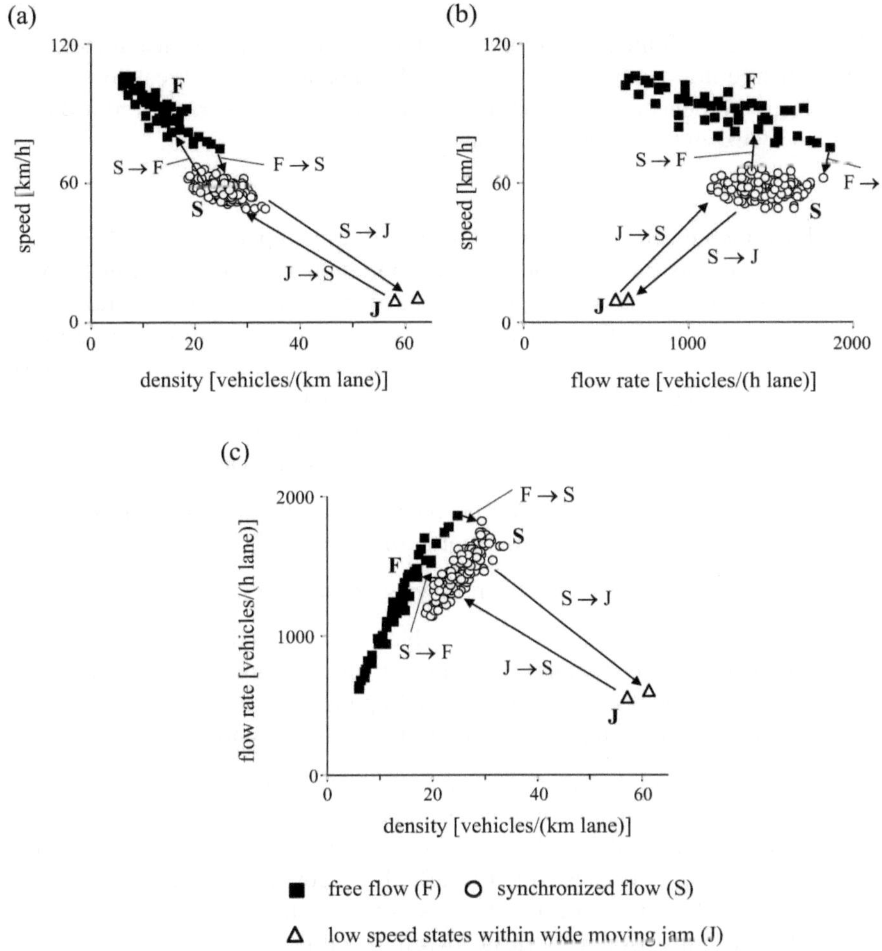

Fig. 1.7 Empirical 2Z-characteristics for phase transitions [556, 594]: Presentation of phase transitions shown in Fig. 1.6 in the speed–density (**a**), speed–flow-rate (**b**) and flow–density planes (**c**) (points J are related to the second of the wide moving jams in Fig. 1.6)

Traffic breakdown at a highway bottleneck is a phase transition from free flow to synchronized flow (F→S transition) in metastable free flow at the bottleneck: In real traffic, there is the metastability of free flow with respect to an F→S transition at the bottleneck. This is the basic empirical feature of traffic breakdown at the bottleneck.

In accordance with the terminology of natural science, the empirical metastability of free flow with respect to the F→S transition (traffic breakdown) at a highway bottleneck means that traffic breakdown (F→S transition) at the bottleneck exhibits the nucleation nature.

As above-mentioned, vehicular traffic is a spatiotemporal process because it occurs in space and time. Traffic and transportation networks are usually very complex. Therefore, it is not surprising that in empirical studies of traffic data measured in vehicular traffic a diverse variety of empirical spatiotemporal traffic phenomena have been discovered. Obviously, each of the traffic and transportation theories and models can explain some real traffic phenomena and each of the models exhibits a limited region of the applicability for the explanation of real traffic and transportation phenomena. Therefore, the following question arises:

- Whether is there an empirical traffic phenomenon that can be considered an fundamental empirical basis of transportation science?

This phenomenon should satisfied the following requirement:

- A traffic and/or transportation theory, which is a theoretical basis for reliable dynamic management, control, assignment, and organization of traffic and transportation networks as well as other ITS-applications must be consistent with the set of fundamental empirical features of this phenomenon.

Users of traffic and transportation networks would expect that through the use of traffic control, dynamic traffic assignment and other methods of dynamic optimization traffic breakdown can be prevented, i.e., free flow can be maintained in the network. This is because due to traffic breakdown congested traffic occurs in which travel time, fuel consumption as well as other travel costs increased considerably in comparison with travel costs in free flow.

Usually, after traffic breakdown has occurred at a network bottleneck the upstream front of the congested pattern propagates upstream. Over time moving jams emerge often within the pattern (Figs. 1.3b, 1.5b, and 1.6). In this case, the mean flow rate and the average speed within congested traffic can become considerably smaller than, respectively, the flow rate and the average speed in free flow upstream of the congested pattern (results of a theoretical analysis of congested traffic patterns will be presented in Sect. 8.4).[7]

- Congested traffic, in which the flow rate reduction through congestion is large enough, can be considered *heavy traffic congestion*.

[7] A more detailed analysis of empirical features of congested traffic patterns than that shown in Figs. 1.3b, 1.5b, and 1.6 can be found in the book [556].

To dissolve heavy congested traffic in urban networks, for example, through the reduction of the flow rate in free flow upstream of a congested traffic pattern, a great limitation of incoming traffic into a congested part of the network through the use of ITS is required. This can lead to heavy traffic congestion in other parts of the network.

Thus, after heavy traffic congestion has already developed in the network (for example, as shown in Figs. 1.3b, 1.5b, and 1.6), it is very difficult with the use of ITS to dissolve heavy traffic congestion or even to limit the upstream propagation of traffic congestion in the network (Sect. 8.5). This leads to a well-known conclusion that rather than controlling of heavy traffic congestion in the network, ITS should prevent either traffic breakdown or limit the development of traffic congestion in the network (see Chap. 11 and Sect. 12.6).

This explains why the understanding of empirical features of real traffic breakdown has the fundamental priority for the development of reliable ITS. Therefore, any traffic and transportation theory, which is claimed to be a basis for the development of reliable ITS-applications, in particular, for automatic driving (called also as self-driving or automated driving), cooperative driving, methods and strategies for traffic control, dynamic traffic assignment as well as dynamic network optimization should be consistent with the fundamental empirical features of traffic breakdown at a highway bottleneck. As explained above, the nucleation nature of real traffic breakdown at highway bottlenecks is the fundamental empirical feature of the breakdown. Consequently, we can also make the following conclusion.

The understanding of empirical nucleation nature of real traffic breakdown at a highway bottleneck has the fundamental priority for the development of reliable ITS.

The nucleation nature of empirical traffic breakdown (F→S transition) at a highway bottleneck can be considered an empirical fundamental of transportation science.

Therefore, rather than features of traffic congested patterns resulting from traffic breakdown, in this book we analyze the impact of the nucleation nature of real traffic breakdown on traffic and transportation theories, their applications for the development of ITS like dynamic traffic assignment and control in traffic and transportation networks. Respectively, traffic flow models and theories, theoretical methods for dynamic traffic assignment, traffic control and optimization, which contradict this fundamental empirical basis of transportation science, fail by their applications in the real world.

This explains why in this book, the criticism of generally accepted classical traffic and transportation theories is **solely** based on an analysis whether the classical traffic and transportation theories are consistent with the empirical nucleation nature of traffic breakdown at a highway bottleneck or not.

To make this critical analysis as clear as possible, in this book we do not perform a detailed consideration of empirical features of congested traffic resulting from traffic breakdown.

- A detailed consideration of empirical features of congested traffic resulting from traffic breakdown is out of scope of this book.[8]

This limitation of this book can be explained as follows: The nucleation empirical nature of traffic breakdown at a road bottleneck is an empirical fundamental of transportation science. If a traffic flow model cannot explain this fundamental empirical basis of transportation science, then the model cannot be applied for the development of reliable systems for automatic driving, traffic control, dynamic traffic assignment, network optimization as well as other ITS-applications in traffic and transportation networks.

Traffic and transportation theories, which are not consistent with the nucleation nature of traffic breakdown (F→S transition) at highway bottlenecks, cannot be applied for the development of reliable traffic control, dynamic traffic assignment as well as other reliable ITS-applications in traffic and transportation networks.

However, a question can arise: Why traffic breakdown can be considered independent of resulting congested patterns? The possibility of such a consideration follows from a huge number of empirical studies of real measured traffic data (see, e.g., [6, 38–47, 111, 114, 118, 142, 145, 147, 162, 241, 258–260, 327, 356, 358–362, 364–366, 461, 661, 752, 803–805, 947–953, 996, 1017]). Indeed, the phase transition "traffic breakdown" and a spatiotemporal evolution of a congested

[8]The exceptions are a brief consideration of the empirical and theoretical characteristic parameters of moving jam propagation, empirical features of phase transitions in congested traffic as well as features of moving synchronized flow patterns (MSP) and moving queues at traffic signal in city traffic. The consideration of these empirical and theoretical features of congested traffic are required for a more clear discussion of the subjects "classical traffic flow theories versus the three-phase theory" made in Chap. 8 as well as "time-delayed traffic breakdown at traffic signal in city traffic" considered in Chap. 9. A detailed consideration of empirical features and a theory of spatiotemporal congested traffic patterns occurring at highway bottlenecks can be found in the book [556] as well as in [570].

traffic pattern resulting from the breakdown exhibit qualitative different empirical spatiotemporal behavior as well as crucially different scales in time and space:

(i) *Different spatiotemporal behavior*. Traffic breakdown at a highway bottleneck is a phase transition from free flow to congested traffic that occurs within a *motionless* region of traffic flow in a neighborhood of the bottleneck. In other words, the localized traffic flow region within which the speed drop occurs during traffic breakdown is fixed at the bottleneck (Figs. 1.3 and 1.4a). In contrast, the phase transition leading to the emergence of a moving jam in congested traffic occurs within a region of traffic flow that *moves* usually propagating upstream along the road stretch about 2–6 km until the moving jam has been developed (Figs. 1.3 and 1.4b,c).
(ii) *Time scale*. As above-mentioned, the characteristic time of traffic breakdown is about 1 min or less. In contrast, a congested pattern develops in space and time during a considerably longer time interval (10–120 min or longer). In particular, moving traffic jams develop in congested traffic often during about 10–30 min (Figs. 1.3 and 1.4b,c).
(iii) *Space scale*. Traffic breakdown occurs in a small neighborhood of the bottleneck within a traffic flow region that is about 500 m long. In contrast, the congested pattern development occurs on a very long road stretch (3–30 km or longer) or even several roads of the network and usually far away upstream of the bottleneck (Figs. 1.3 and 1.4b,c).

1.7 The Origin of Failure of Classical Traffic and Transportation Theories

1.7.1 Nature of Stochastic Highway Capacity

Traffic breakdown at a highway bottleneck limits highway capacity. This means that the origin of traffic breakdown at the bottleneck determines the physical nature of highway capacity [6, 38–47, 111, 112, 114, 118, 142, 145, 147, 162, 200, 241, 258–260, 327, 342, 356, 358–362, 364–366, 461, 661, 752, 803–805, 947–953, 996, 1017].

Beginning from the classical work by Greenshields [342], there have been a huge number of studies of traffic breakdown and highway capacity (see references in the book by May [804]). These works include many studies made by many scientific groups, in particular, by May et al. [250, 803, 805, 1134], by Hall et al. [6, 356–366], by Persaud et al. [365, 947–953], by Elefteriadou et al. [258–260, 661, 752], by Brilon et al. [111, 112, 114, 118, 327], by Banks et al. [38–47], by Cassidy et al. [137–140, 142, 1038] and many other traffic research groups (see, e.g., [55, 56, 134, 145, 162, 165, 240, 241, 461, 467, 468, 475, 485, 931, 986, 987, 996, 1017, 1027, 1030, 1049] and references there). These works have made a significant success in the understanding of some empirical features of traffic breakdown and highway capacity. In particular, in 1995 Elefteriadou et al. [260] found that traffic breakdown exhibits a probabilistic (stochastic) nature. In Chap. 2, we will consider main achievements of these empirical investigations in more details.

The empirical investigations of traffic breakdown have led to the following definition of highway capacity (see, e.g., [6, 38–47, 111, 112, 114, 118, 142, 145, 147, 162, 241, 258–260, 327, 356, 358–362, 364–366, 461, 661, 752, 803–805, 947–953, 996, 1017] and references there):

- At any given time instant, there is a particular value of highway capacity of free flow at a bottleneck.

This definition is also valid for stochastic highway capacity: Because highway capacity is a stochastic value, we do not know and we cannot measure the value of highway capacity of free flow at a bottleneck at a given time instant. However, in accordance with the above understanding of highway capacity, it is assumed that at *any given time instant* there is a particular value of stochastic highway capacity of free flow at a bottleneck (see, e.g., [111, 112, 114, 118, 258, 259, 327, 661]). Because this understanding of highway capacity is accepted in the transportation research community (see, e.g., the book by Elefteriadou [258] as well as papers [259, 661]), we call it as a classical understanding of stochastic highway capacity (or stochastic highway capacity in the classical theory). One of the consequences of the classical understanding of highway capacity is a well-known term "capacity drop" (see, e.g., [40, 258, 358]) (Sect. 4.11.2).

As above-mentioned, the empirical nucleation nature of traffic breakdown at a highway bottleneck is an empirical fundamental of transportation science. However, as we will show in Sect. 4.10 of this book, the classical understanding of stochastic highway capacity (see, e.g., [6, 38–47, 111, 112, 114, 118, 124, 142, 145, 147, 162, 200, 241, 258–260, 327, 356, 358–362, 364–366, 461, 477, 661, 752, 803–805, 947–953, 996, 1017] and references there) contradicts this empirical fundamental of transportation science:

- The classical understanding of stochastic highway capacity is inconsistent with the empirical nucleation nature of traffic breakdown at a highway bottleneck (see Sects. 4.10 and 8.11).

1.7.2 Description of Traffic Breakdown with Classical Traffic Flow Models

There can be distinguished two qualitatively different classical traffic flow models for the description of traffic breakdown:

1. The Lighthill-Whitham-Richards (LWR) model [735, 994] that explains traffic breakdown through a fundamental diagram of traffic flow.

2. The General Motors (GM) model introduced by Herman, Gazis, Montroll, Potts, Rothery, and Chandler [152, 321, 322, 420] that explains traffic breakdown through a traffic flow instability associated with a driver reaction time.

1.7.2.1 About Applications of LWR-Model

The LWR model is the theoretical basis for a huge number of theoretical approaches for the modeling of traffic phenomena in traffic networks, like Daganzo's cell transmission model (CTM) [183, 184], so-called N-curves [143, 736, 830], and a macroscopic fundamental diagram (MFD) (MFD is also called network fundamental diagram (NFD)) (see Sect. 4.11.3) [196, 202, 205, 316, 330, 786, 790, 791] (see also references in [246, 666, 713, 1149, 1168]). The Daganzo's CTM is currently used in a huge number of theoretical studies of ITS-applications (e.g., [30, 186, 189, 198, 202, 206, 233, 335, 520, 746, 747, 749, 816, 1101, 1207]). However, the LWR model as well as all its applications and further developments (see, e.g., [93, 143, 157, 183–185, 190, 191, 193–197, 199, 245, 246, 330, 345, 415, 433, 666, 698, 699, 701–706, 708, 711, 713–715, 730, 736, 763, 830, 862, 889–892, 965, 970, 1042, 1043, 1062, 1145, 1149, 1168, 1177, 1186, 1199]) are inconsistent with the nucleation nature of real traffic breakdown at road bottlenecks.

The LWR model is also the theoretical basis for a huge number of methods of traffic control (e.g., [924–929]). A prominent example is ALINEA on-ramp metering by Papageorgiou et al. [924–929].

However, as we will show in this book, the traffic flow models in the frameworks of the LWR model as well as their applications for the development and analysis of ITS contradict the empirical nucleation nature of traffic breakdown at the bottleneck (Sect. 4.3.3):

- The LWR model as well as its applications, including Daganzo's cell-transmission model, the MFD (Sect. 4.11.3), and Papageorgiou's ALINEA on-ramp metering method (Sect. 4.9.2) are inconsistent with the empirical nucleation nature of traffic breakdown at a highway bottleneck.

1.7.2.2 About Traffic Flow Models of General Motors (GM) Model Class

The classical traffic flow instability of Herman, Gazis, Montroll, Potts, Rothery, and Chandler [152, 321, 322, 420] has been incorporated in a huge number of traffic flow models that can be considered belonging to the General Motors (GM) class of traffic-flow models.

There is a huge number works devoted to studies of traffic flow models of the GM motor class and their applications for different traffic flow simulations (e.g., [9, 28, 34–37, 52, 53, 77, 79, 119, 120, 125, 150, 151, 160, 211, 212, 221, 222, 257, 261, 265, 284, 293, 323–325, 332, 333, 347–350, 386–388, 390–403, 405–

412, 417, 421, 432, 439–447, 449, 455, 486, 487, 494, 495, 498, 505, 512, 515–519, 521, 524, 533, 609, 615, 617–621, 671–673, 678–681, 683, 687, 707, 709, 720–723, 734, 758–760, 762, 766, 833–844, 847–849, 852, 853, 855, 859, 876, 877, 899, 906, 909, 911, 912, 933, 935, 939–946, 972, 974, 977, 993, 1006, 1011, 1013–1015, 1023, 1024, 1034, 1036, 1047, 1050, 1056, 1060, 1061, 1072, 1079, 1080, 1087, 1089, 1098, 1103, 1106, 1108, 1110, 1119, 1120, 1137, 1139–1141, 1160, 1163, 1173–1176, 1178, 1179, 1182, 1187–1189, 1198, 1203, 1205] and references in reviews by Newell [878], Brackstone and McDonald [103], Leutzbach [727], Chowdhury et al. [166], Gartner et al. (eds.) [305], Helbing [394], Hoogendoorn and Bovy [443], Gazis [318], Nagatani [845], Nagel et al. [856], Shvetsov [1032], Maerivoet and De Moor [768], Bellomo et al. [73], Piccoli and Tosin [956], Schadschneider et al. [1010], Treiber and Kesting [1089], Elefteriadou [258], and by Saifuzzaman and Zheng [1004]). Some of the most prominent traffic models of the GM model class are as follows:

- GM car-following model of Herman, Gazis et al. [152, 321, 322, 420].
- Newell's optimal velocity (OV) microscopic model [876, 899].
- Gipps's microscopic model [332, 333].
- Payne's macroscopic model [933–935].
- Wiedemann's microscopic model [1139]
- Whitham's microscopic model [1137]
- The Nagel-Schreckenberg (NaSch) cellular automaton (CA) model [53, 855].
- Bando et al. OV model [35–37].
- A lattice model by Nagatani [833, 834].
- Treiber's intelligent driving model (IDM) [394, 1087, 1089].
- Krauß's microscopic stochastic model [671, 673].
- The Aw-Rascle macroscopic model [28].
- A full velocity difference OV model by Jiang et al. [517].

The empirical nucleation nature of traffic breakdown at a highway bottleneck is an empirical fundamental of transportation science (Sect. 1.6). However, as we will show in this book, the traffic flow models in the framework of the GM model class contradict this empirical fundamental of transportation science:

- Traffic flow models of the GM model class are inconsistent with the empirical nucleation nature of traffic breakdown at a highway bottleneck (Sects. 4.7 and 4.8).

1.7.2.3 Common Critical Statements to Classical Traffic Flow Models and Their Applications

We can formulate the following common critical statements to the LWR model or/and traffic flow models of the GM model class.

None of classical traffic-flow models and theories incorporates the metastability of free flow with respect to an F→S transition at a highway bottleneck.

The origin for the failure of the classical traffic flow models and theories by their ITS applications in the real world is that the classical models and theories cannot explain the nucleation nature of empirical traffic breakdown at highway bottlenecks.

1.7.2.4 About Achievements of Classical Traffic Flow Models

In spite of the above criticism on classical traffic flow models, many ideas of the classical traffic flow theories are very important (for more details, see Sect. 4.5). Therefore, these classical ideas have been used in the three-phase theory. In particular, the following achievements of the classical traffic flow theories have been used in the three-phase theory:

(i) The classical traffic flow instability of Herman, Gazis, Montroll, Potts, Rothery, and Chandler [152, 321, 322, 420]. In the three-phase theory, the classical traffic flow instability is responsible for the spontaneous emergence of moving jams in synchronized flow (S→J transition) (see Hypothesis 9, item (ii) of Sect. 1.9).
(ii) Line J and characteristic parameters of wide moving jam propagation found out by Kerner and Konhäuser [616] (Sect. 4.4.3).
(iii) A diverse variety of driver behavioral assumptions in different driving situations introduced in the classical traffic flow models (Sect. 4.5.2).

These and other achievements of the classical traffic flow models will be considered in more details in Sects. 4.3.2, 4.4.3, 4.5, and 8.3 of this book.

1.7.3 Deterioration of Traffic System Through Standard Dynamic Traffic Assignment in Networks

Under "standard" dynamic traffic assignment in a traffic or transportation network, we understand dynamic traffic assignment whose objective is the minimization of travel times and/or other travel costs in a traffic or transportation network. Standard dynamic traffic assignment that is the state-of-the-art in transportation research is usually based on the classical Wardrop's user equilibrium (UE) and system optimum

(SO) (called also Wardrop's equilibria or Wardrop's principles) [1127] (see, e.g., references in [70, 281, 772, 978, 980, 1028]).

There has been made a great progress in the development of different mathematical methods for standard dynamic traffic assignment by many scientific groups, in particular, by Daganzo and Sheffi [207, 1028], Merchant and Nemhauser [812, 813], Daganzo et al. [179–182, 188, 192, 198, 337], Bell et al. [60, 63–65, 67–69, 71, 148, 149, 1154–1157], Friesz et al. [275, 277–280, 282, 283, 375, 1051, 1076, 1138], Mahmassani et al. [2, 163, 469, 772, 775, 778, 784, 785, 820, 936, 937, 1099, 1183, 1202, 1208] as well as in works of many other researchers (see, e.g., [10, 18, 24, 51, 101, 132, 133, 136, 161, 167, 236, 276, 326, 353, 427, 428, 451, 463, 464, 473, 478, 479, 481, 483, 499, 522, 529–531, 534, 685, 686, 690–692, 726, 744, 750, 751, 754–756, 764, 858, 865, 908, 930, 938, 966, 978, 980, 981, 1026, 1029, 1033, 1059, 1152, 1158, 1167, 1169, 1180, 1184, 1191, 1192, 1194, 1195, 1207], conference proceedings [16, 141, 146, 450, 689, 694, 725, 773, 1064] and references in reviews [70, 772, 978, 980, 1028]). Many mathematical results have been derived within a steady-state analysis of traffic and transportation networks for which standard *stochastic* dynamic traffic assignment has been developed (see, e.g., the book by Sheffi [1028]).[9]

However, as we have shown in [587], independent of the mathematical approach used for standard dynamic traffic assignment, the minimization of travel times in the network leads to a considerable increase in the probability of traffic breakdown (F→S transition) on some of the network links. The increase in the breakdown probability results in the deterioration of the performance of the traffic system.

A real traffic network consists of alternative routes with very different lengths. At a small enough total network inflow rate, we can distinguish routes with short travel times ("short routes") and routes with longer travel times ("long routes") in the network. Following [587, 588], in Chap. 13 we will show that when the total network inflow rate becomes large enough, the minimization of travel times in the network with the use of standard dynamic traffic assignment (see, e.g., [2, 10, 18, 24, 27, 51, 60, 63–65, 67–71, 101, 132, 148, 149, 163, 167, 179–182, 188, 192, 198, 207, 236, 275–280, 282, 283, 326, 337, 375, 427, 428, 451, 463, 469, 473, 479, 499, 522, 530, 531, 534, 685, 686, 690–692, 744, 751, 756, 772, 775, 778, 784, 785, 812, 813, 820, 865, 930, 936–938, 967, 968, 978, 980, 981, 988, 1026, 1028, 1029, 1033, 1051, 1076, 1099, 1138, 1152, 1154–1158, 1180, 1183, 1195, 1202, 1207, 1208] and conference proceedings [16, 141, 146, 450, 689, 694, 725, 773, 1064]) leads to considerably larger increases in the flow rates on short routes in comparison with increases in the flow rates on long routes of the network. We will show that this

[9] In many cases, the total network inflow rate $Q(t)$ changes in a traffic or transportation network over time very slowly in comparison with any characteristic times of dynamic traffic effects at network bottlenecks under free flow conditions in the network. For this reason, as it is usually assumed in classical theories of traffic and transportation networks, at any given Q free flow distribution in the network can be considered as a steady state (steady-state analysis of traffic and transportation networks) [1028]. This means that it is assumed that $Q = Q_{\text{out}}$, where $Q_{\text{out}}(t)$ is the total network outflow rate.

effect provokes heavy traffic congestion in the network. Heavy traffic congestion can occur already at the total network inflow rate that is smaller than a network capacity (the network capacity is a physical measure of a traffic or transportation network that will be explained in Sect. 11.3).

Thus, the wish of humans to use shortest routes of a network contradicts fundamentally another wish to drive under free flow conditions in the network. Therefore, as we will explain in Chap. 13 of this book, the use of the classical Wardrop's equilibria, which are based on the wish of humans to use shortest routes in the network, results in the occurrence of heavy traffic congestion in urban networks.

1.7.4 Failure of Applications of Intelligent Transportation Systems (ITS) Based on Classical Traffic Theories

As above-mentioned, the classical traffic and transportation theories have made a great impact on the understanding of many traffic phenomena. However, network optimization approaches based on these theories have failed by their applications in the real world. Even several decades of a very intensive effort to improve and validate network optimization models have had no success. Indeed, there can be found no examples where on-line implementations of traffic control, standard dynamic traffic assignment, and the network optimization models based on the classical traffic theories could reduce congestion in real traffic and transportation networks.

As emphasized above, traffic flow models and traffic control methods can be used for reliable ITS-applications only when they are consistent with the nucleation nature of traffic breakdown (F→S transition) at highway bottlenecks (Sect. 1.6). However, as emphasized above, the classical understanding of stochastic highway capacity (Sect. 1.7.1), traffic flow models in the framework of the LWR-theory and traffic flow models of the GM model class (Sect. 1.7.2) cannot show the nucleation nature of traffic breakdown (F→S transition) at highway bottlenecks. This explains why applications of the classical traffic flow models for the development of different traffic simulation tools for the analysis of ITS as well as a variety of diverse ITS applications in which dynamic traffic assignment and control are used together with traffic flow simulations in the frameworks of the LWR model and/or traffic flow models of the GM model class *cannot* be used for reliable and valid analysis of ITS performance in the real world (see examples of such ITS applications based on classical traffic theories and models, e.g., in [1, 7, 8, 12, 17, 23, 31, 32, 48–50, 54, 57, 58, 72, 78, 88, 92, 102, 104, 126–131, 144, 153, 154, 156, 158, 159, 164, 168, 170, 178, 230, 237–239, 264, 274, 331, 336, 344, 352, 374, 377, 381–384, 437, 459, 460, 462, 471, 472, 480, 482, 484, 532, 637–646, 676, 688, 718, 728, 733, 737, 743, 748, 753, 779, 802, 822, 860, 902, 903, 905, 907, 917, 924–929, 960, 971, 973–975, 1000–1002, 1021, 1031, 1037, 1039, 1043, 1045, 1048, 1052–

1054, 1058, 1065, 1081, 1083, 1096, 1097, 1100, 1105, 1115–1118, 1121, 1123–1125, 1133, 1146, 1148, 1159, 1170, 1193, 1196, 1197, 1201, 1204]).

Applications of standard dynamic traffic assignment related to the state-of-the-art in traffic and transportation research (see, e.g., [2, 10, 18, 24, 27, 51, 60, 63–65, 67–71, 101, 132, 148, 149, 155, 163, 167, 179–182, 188, 192, 198, 207, 235, 236, 275–280, 282, 283, 296, 299, 308, 326, 328, 337, 375, 427, 428, 451, 463, 469, 473, 479, 499, 522, 530, 531, 534, 685, 686, 690–692, 732, 744, 745, 751, 756, 772, 775, 778, 784, 785, 812, 813, 820, 865, 904, 930, 936–938, 967, 968, 978, 980, 981, 988, 1016, 1026, 1028, 1029, 1033, 1051, 1076, 1099, 1138, 1151, 1152, 1154–1158, 1164, 1180, 1183, 1200, 1202, 1207, 1208] and conference proceedings [16, 141, 146, 450, 689, 694, 725, 773, 1064]) deteriorate the traffic system while provoking heavy traffic congestion in urban networks (Chap. 13).

Traffic engineers that work with real installations for traffic control know this fact: There can be found no examples where on-line implementations of the network optimization models based on the classical traffic theories could reduce congestion in real traffic and transportation networks. It is understandable that traffic researchers would like to publish positive results of applications of their traffic control approaches. This can explain why it is impossible to find publications in which the failure of traffic applications based on the classical traffic and transportation theories in real traffic installations is clearly admitted. However, in 1995 when the author gave a talk "traffic moving jam without obvious reason" at the traffic engineering consulting firm Heusch/Boesefeldt GmbH in Aachen (Germany), the failure of traffic applications of the classical theories in real traffic installations was openly deplored by the engineers. This encounter with the traffic engineers of the Heusch/Boesefeldt GmbH has affected the author to begin with a study of real field traffic data (see references in a review by Rehborn and Klenov [983]), rather than to continue simulations of classical traffic flow models (Sect. 4.4.3).

While working during many years in many engineering projects in the field "traffic" at the Daimler Company, the author recognized that no noticeable progress was done in the field of applications of classical traffic and transportation theories for the reduction of traffic congestion in real traffic and transportation networks. The reason for this failure of classical traffic and transportation theories has been explained above: The classical traffic and transportation theories and models cannot explain or they ignore the nucleation nature of traffic breakdown at network bottlenecks observed in real field traffic data. The result of the study of real field traffic data made between 1995 and 2016 at the Daimler Company by colleagues together with the author (see references in the review by Rehborn and Klenov [983]) is this book.

1.8 Classical Ideas of Transportation Science and Nucleation Nature of Empirical Traffic Breakdown

As emphasized in Sect. 1.7, the classical traffic and transportation theories and models are not consistent with the nucleation nature of empirical traffic breakdown at a highway bottleneck. This is due to the fact that the nucleation nature of empirical traffic breakdown have been understood only during last 20 years [539–542, 556, 598].

In contrast, the classical theoretical works, in particular, made by Wardrop [1127], Lighthill, Whitham, and Richards [735, 994], Herman, Gazis, Montroll, Potts, Rothery, and Chandler [152, 321, 322, 420], Newell [876, 878], Kometani and Sasaki [658–660], Prigogine [962], Reuschel [990–992], and Pipes [958] that are the basis for the generally accepted fundamentals and methodologies of traffic and transportation theory have been introduced in the 1950s–1960s. These and other scientists whose ideas led to the classical fundamentals and methodologies of traffic and transportation theory could not know the nucleation nature of real traffic breakdown at road bottlenecks.

1.9 Three-Phase Traffic Theory

The nucleation nature of traffic breakdown (metastability of free flow with respect to an F→S transition) at highway bottlenecks was understood by the author in 1996–2002 based on an analysis of a number of real field traffic data measured over many days (and years) on several German highways [539–552, 622–626].[10] To explain the empirical nucleation nature of traffic breakdown at highway bottlenecks the author introduced the three-phase traffic theory [539–552, 556].

The main reason for the three-phase theory is the explanation of the empirical nucleation nature of traffic breakdown at highway bottlenecks.

[10]The author was fortunate to work in 1994–1995 for the traffic engineering consulting firm Heusch/Boesefeldt GmbH in Aachen (Germany), where he met many good traffic engineers. Here he had his first opportunity to begin a study of real field traffic data measured on many highways in Germany. The author thanks Heinz Heusch, Jochen Boesefeldt, Hubert Rehborn, Heribert Kirschfink, Ulrich Uerlings, and Thomas Scheiderer, for their help in the preparation of empirical data required for the data analysis, critical discussion of earlier traffic flow theories, and their fruitful cooperation. In particular, the author thanks Hubert Rehborn for his help in the discussion of the analysis of real field traffic data presented in [539–553, 555, 622–626].

The three-phase theory consists of several hypotheses resulting from a spatiotemporal analysis of real field traffic data measured during many years of traffic observations on different highways in a variety of countries.

Main hypotheses of the three-phase theory are as follows [539–552, 556]:

Hypothesis 1: Three traffic phases. There are three traffic phases: (i) Free flow (F). (ii) Synchronized flow (S). (iii) Wide moving jam (J) (Sect. 1.3).
Hypothesis 2: Nucleation nature of traffic breakdown. There is the metastability of free flow with respect to an F→S transition at a highway bottleneck. The nucleation nature of traffic breakdown is as follows: Traffic breakdown is the F→S transition in this metastable free flow at the bottleneck (see detailed explanations in Chaps. 3 and 5).
Hypothesis 3: Z-characteristic for traffic breakdown. The nucleation nature of traffic breakdown is described by a Z-characteristic for the F→S transition (traffic breakdown) and a return phase transition from synchronized flow to free flow (S→F transitions). The Z-characteristic is also called the Z-characteristic for traffic breakdown (see brief explanations of the Z-characteristic for traffic breakdown in Sec 1.12 and detailed explanations in Chap. 5).
Hypothesis 4: Driver behaviors resulting in traffic breakdown. The nucleation nature of traffic breakdown (F→S transition) at a highway bottleneck is explained by a spatiotemporal competition between driver speed adaptation and driver over-acceleration (see explanations in Chap. 5).
Hypothesis 5: Discontinuity of over-acceleration. A density function of the probability of driver over-acceleration exhibits a discontinuity: It is a Z-shaped function of the vehicle density. The Z-shaped density function of the probability of driver over-acceleration is caused by a driver time delay in over-acceleration. The discontinuity in the probability of driver over-acceleration leads to an instability of synchronized flow resulting in a growing wave of a local increase in the speed in synchronized flow. This instability of synchronized flow is called the S→F instability. The S→F instability causes an S→F transition. The S→F instability governs traffic breakdown (F→S transition) at highway bottlenecks (see Sects. 5.12 and 5.13).
Hypothesis 6: Infinite number of highway capacities at any time instant. At any time instant, there are the infinite number of highway capacities of free flow at a highway bottleneck. The highway capacities are between a minimum highway capacity and a maximum highway capacity (see explanations in Chap. 5).
Hypothesis 7: Sequence of F→S→J transitions. A sequence of the F→S→J transitions governs moving jam emergence in traffic flow: Rather than in free flow, wide moving jams (J) emerge spontaneously in synchronized flow (S→J transition). Both the S→J transition and a return J→S transition exhibit the nucleation nature that can be described by a Z-characteristic for S→J and J→S

transitions (see brief explanations in Sec 1.12 and a more detailed explanation in Chap. 8).

Hypothesis 8: Double Z-characteristic for phase transitions. The Z-characteristic for traffic breakdown (hypothesis 3) together with the Z-characteristic for S→J and J→S transitions (hypothesis 7) lead to a double Z-characteristic (2Z-characteristic) for phase transitions in traffic flow (see brief explanations in Sec 1.12 and detailed explanations in Chap. 8).

Hypothesis 9: Two qualitatively different instabilities in synchronized flow. The 2Z-characteristic for phase transitions is associated with the existence of two qualitatively different traffic flow instabilities in synchronized flow (see Sects. 5.12, 5.13, 8.2, 8.3.3 and 8.9):

(i) *The S→F instability*: A synchronized flow instability introduced in the three-phase theory that results in a growing wave of a local **increase** in the speed in synchronized flow. The S→F instability is caused by a driver time delay in over-acceleration (hypothesis 5 above). The S→F instability is responsible for the existence of the metastability of free flow with respect to an F→S transition and for the Z-characteristic for traffic breakdown. The S→F instability will be considered in Sects. 5.12 and 5.13.

(ii) *The S→J instability*: A synchronized flow instability resulting in a growing wave of a local **decrease** in the speed in synchronized flow. The S→J instability leads to an S→J transition. The S→J instability is associated with the classical traffic flow instability of Herman, Gazis, Montroll, Potts, Rothery, and Chandler [152, 321, 322, 420] that is caused by driver over-deceleration effect (driver reaction time). A brief discussion of the S→J instability will be done in Sect. 8.3.3.[11]

Hypothesis 10: Probabilities of F→S and F→J transitions. There is a range of the flow rate at a highway bottleneck within which free flow at the bottleneck is metastable both with respect to an F→S transition *and* with respect to an F→J transition. However, at any flow rate within this flow rate range a nucleus required for the F→S transition is considerably smaller than a nucleus required for the F→J transition (see Sect. 8.9). For this reason, the probability of the F→S transition is considerably larger than the probability of the F→J transition. This explains why real traffic breakdown at a highway bottleneck is the F→S transition.

Hypothesis 11: 2D-states of synchronized flow. Steady states of synchronized flow cover a two-dimensional (2D) region in the flow–density plane (as well as 2D-regions in all other possible presentations of steady states of traffic flow; see explanations in Sect. 5.9.1).[12]

[11]A detailed consideration of the S→J instability is out of the scope of the book.

[12]Clearly, already due to real flow fluctuations "pure" steady states of traffic flow do not exist. Therefore, steady states of synchronized flow cannot be observed in real field data: The fluctuations destroy 2D-steady states of synchronized flow transforming them into dynamic 2D-

Hypothesis 12: Line *J* and S→J transitions. The slope of the line *J* in the flow–density plane is equal to the mean velocity of the downstream front of a wide moving jam (see explanations of the line *J* in Sect. 4.5.1). The mean velocity of the downstream front of the wide moving jam is a characteristic parameter of the jam propagation: The mean velocity of the downstream front of the wide moving jam remains independent of the jam propagation through different states of free flow and synchronized flow as well as through different highway bottlenecks. The line *J* divides 2D-states of synchronized flow into two different classes (Sect. 8.3.3):

- Synchronized flow states below the line *J* are stable with respect to S→J transitions.
- Synchronized flow states that lie on or above the line *J* are metastable with respect to S→J transitions. In the metastable synchronized flow, the S→J instability can occur leading to the S→J transition.
- All synchronized traffic states (the infinite number of the states) that lie on the line *J* in the flow–density plane are threshold flow rates for the emergence of moving jams.

A brief consideration of the history of the three-phase theory can be found in the review by Rehborn and Klenov [983] as well as papers by Rehborn, Klenov et al. [378, 921, 984, 985].

---

states of synchronized flow. However, in the book [556] (see, e.g., Secs. 4.3.4 and 6.3.3 of [556]) we have explained that real dynamic 2D-states of synchronized flow, at least when they are slightly non-homogeneous in space and time, should exhibit almost the same features as those postulated in the three-phase theory for hypothetical 2D-steady states of synchronized flow [539–552, 555, 556]. To make a proof of the most of the features of 2D-states of synchronized flow postulated in the three-phase theory [539–552, 555, 556], microscopic (single-vehicle) data should be measured. Only recently some microscopic (single-vehicle) measurements could be made (see, e.g., [492, 526, 1166]).

In particular, microscopic experimental results of Jiang et al. [492, 497, 526] are in agreement with the hypothesis of the three-phase theory about 2D-states of synchronized flow. Based on a statistical analysis of a microscopic real field traffic data measured in Singapore with a video camera, Yang et al. [1166] have found a region of nearly zero acceleration in the data that proves the empirical existence of the synchronization space gap and the speed adaptation effect within 2D-states of synchronized flow as stated in the three-phase theory [556, 572] (see explanations of the term *synchronization space gap* in Sect. 5.9). Recall that the focus of this book is an analysis of the breakdown in traffic networks. Therefore, a detailed consideration of the hypothesis about the 2D-states of synchronized flow (see Chap. 6 of the book [556]) is out the scope of this book. Respectively, in Sect. 5.9.1 we consider only some of the features of the 2D-states of synchronized flow [539–552, 555, 556] needed for the understanding of the physical mechanism of traffic breakdown at network bottlenecks.

The fundamental empirical basis of transportation science—the empirical nucleation nature of traffic breakdown at a highway bottleneck,—is also the empirical fundamental of the three-phase theory.

The three-phase theory explains the empirical nucleation nature of traffic breakdown at highway bottlenecks.

We will show in the book that the three-phase theory should be considered the theoretical fundamental for future reliable traffic control, dynamic traffic assignment, and optimization of traffic and transportation networks.

1.10 Infinite Number of Stochastic Highway Capacities in Three-Phase Theory

The empirical nucleation nature of real traffic breakdown (F→S transition) at highway bottlenecks leads to the assumption of the three-phase theory that at any time instant there are the infinite number of highway capacities [555, 556, 572]. Indeed, in accordance with empirical nucleation nature of traffic breakdown, there should be a range of the flow rate q_{sum} in free flow at a highway bottleneck within which traffic breakdown can be induced in free flow at the bottleneck. Therefore, within this flow rate range free flow is in a metastable state with respect to an F→S transition at the bottleneck. Empirical observations show that this range of the flow rate is *limited*: When the flow rate q_{sum} in free flow at the bottleneck is smaller than some minimum highway capacity C_{min} no traffic breakdown can be induced at the bottleneck. On contrary, when the flow rate q_{sum} in free flow is equal to or larger than some maximum highway capacity C_{max}, traffic breakdown should occur at the bottleneck during a given time interval with probability $P^{(\text{B})} = 1$.

For these reasons, in the three-phase theory it is assumed that the metastability of free flow with respect to an F→S transition at a bottleneck is realized under the following conditions [556, 572]:

$$C_{\text{min}} \le q_{\text{sum}} < C_{\text{max}}. \tag{1.1}$$

It is assumed in the three-phase theory that when the flow rate q_{sum} satisfies conditions (1.1), traffic breakdown can be induced at the bottleneck. This explains why highway capacity of free flow at a bottleneck is defined in the three-phase theory through the empirical evidence of empirical induced traffic breakdown as follows:

- *At any time instant*, there are the infinite number of the flow rates q_{sum} in free flow at a bottleneck at which traffic breakdown can be induced at the bottleneck. These flow rates are the infinite number of the capacities of free flow at the bottleneck. The range of these capacities of free flow at the bottleneck is limited by the minimum highway capacity C_{min} and the maximum highway capacity C_{max}.
- The existence of an infinite number of highway capacities *at any time instant* means that highway capacity is stochastic. This explains empirical stochastic highway capacity in the framework of the three-phase theory.

Recently, the theoretical conclusion that at any time instant there are the infinite number of road capacities have been generalized for a city bottleneck due to traffic signal [576, 579, 580].

A detailed consideration of the infinite number of capacities of free flow at highway bottlenecks will be done in Chap. 5. A detailed consideration of the infinite number of capacities of traffic signal in city traffic will be done in Chap. 9.

1.11 Breakdown Minimization (BM) Principle

The minimization of travel times and other travel costs in traffic and transportation networks, which is performed with standard dynamic traffic assignment in the networks (Sect. 1.7.3), deteriorates the performance of the traffic system (Chap. 13). For this reason, in 2011 the author introduced a breakdown minimization principle (BM principle) for the optimization of traffic and transportation networks [575, 576]. The BM principle is consistent with the empirical nucleation nature of traffic breakdown at a bottleneck. The BM principle is defined as follows:

- The BM principle states that the optimum of a traffic network with N bottlenecks is reached, when dynamic traffic assignment, optimization and/or control are performed in the network in such a way that the probability for occurrence of traffic breakdown in at least one of the network bottlenecks during a given time interval reaches the minimum possible value.
- The BM principle is equivalent to the maximization of the probability that during the given time interval traffic breakdown occurs at none of the network bottlenecks.

A detailed consideration of the BM principle will be made in Chaps. 10–12.

As we have mentioned in Sect. 1.7.4, in the book we will show that applications of the classical Wardrop's equilibria for standard dynamic traffic assignment and control related to the state-of-the-art in traffic and transportation research [16, 141, 146, 450, 689, 694, 725, 773, 1064] deteriorate the traffic system while provoking heavy traffic congestion in urban networks. Contrarily to these classical approaches, we will also show in this book that the application of the BM principle for dynamic traffic assignment can reduce or even avoid heavy traffic congestion in urban networks.

We will show in the book that the BM principle should be considered the theoretical fundamental for future reliable dynamic traffic assignment and optimization of traffic and transportation networks.

1.12 Mathematical Three-Phase Traffic Flow Models and ITS-Applications of Three-Phase Theory

Rather than a mathematical model of traffic flow, the three-phase theory is a *qualitative theory* that consists of several hypotheses [539–552, 555, 556, 571–574, 578, 584, 586] (Sect. 1.9). A mathematical model of traffic flow can incorporate only one or a few of the hypotheses of the three-phase theory. It can be expected that a diverse variety of different mathematical models of traffic flow can be developed that incorporate some of the hypotheses of the three-phase theory.

It must be emphasized that not any traffic flow model that incorporate some of the hypotheses of the three-phase theory can be considered a *three-phase traffic flow model*. This is because the empirical fundamental of the three-phase theory is the nucleation nature of traffic breakdown (F$\rightarrow$S transition) at a highway bottleneck (Sect. 1.9). Therefore, any three-phase traffic flow model must show a Z-characteristic for traffic breakdown (Fig. 1.8). In Fig. 1.8b, we show simulations of a Z-characteristic for traffic breakdown at an on-ramp bottleneck with the Kerner-Klenov stochastic microscopic three-phase traffic flow model [594]. The Z-characteristic is built by metastable states of the traffic phases F and S (solid curves F and S, respectively) together with a curve for critical nuclei for the F$\rightarrow$S transition (a dashed curve in Fig. 1.8b between the metastable states of the phases F and S). The Z-characteristic for traffic breakdown will be considered in more details in Sect. 5.4.2.

The Z-characteristic for traffic breakdown shows the flow rate range between the minimum highway capacity $C_{\rm min}$ and the maximum highway capacity $C_{\rm max}$ of free flow at the bottleneck within which traffic breakdown can occur (Fig. 1.8b). When free flow is initially at the bottleneck, then the flow rate in this free flow downstream of the bottleneck is equal to $q_{\rm sum} = q_{\rm in} + q_{\rm on}$. Therefore, the infinite number of these flow rates (1.1) are highway capacities of free flow at the bottleneck. It should be emphasized that the Z-characteristic for traffic breakdown (Fig. 1.8b) is a macroscopic characteristic of traffic breakdown at the bottleneck. A microscopic

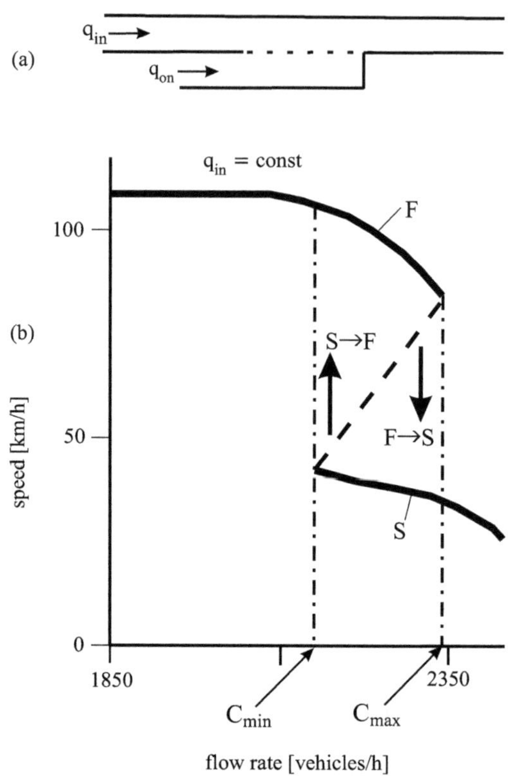

Fig. 1.8 Simulations of Z-characteristic for traffic breakdown with the Kerner-Klenov stochastic microscopic three-phase model: (**a**) Sketch of on-ramp bottleneck on a single-lane road. (**b**) Z-characteristic as function of the flow rate $q_{in} + q_{on}$ at the on-ramp bottleneck when the on-ramp inflow rate q_{on} is a variable and the flow rate upstream of the on-ramp bottleneck q_{in} is a given value (q_{in} = 1850 vehicles/h). F—free flow phase, S—synchronized flow phase. Arrow F→S illustrates traffic breakdown (F→S transition); arrow S→F illustrates symbolically a return S→F transition. C_{min} is the minimum highway capacity, C_{max} is the maximum highway capacity. For the simplification of the consideration of the Z-characteristic, in (**b**) the curve S is related to an averaging of different speeds for a given q_{on} within real 2D-states of synchronized flow (S) following from simulations with the Kerner-Klenov model. Adapted from [594]

theory of traffic breakdown at a bottleneck derived only recently [583] will be considered in Sects. 5.12 and 5.13 of this book.

The Kerner-Klenov model introduced in 2002 [593] was the first mathematical three-phase traffic flow model in the framework of the three-phase theory that can show this Z-characteristic (Fig. 1.8b). Some months later, Kerner, Klenov, and Wolf developed a three-phase cellular automaton (CA) traffic flow model (KKW CA model) [613]. Based on the KKW CA model, the KKS (Kerner-Klenov-Schreckenberg) three-phase CA traffic flow model [610] and the KKSW (Kerner-Klenov-Schreckenberg-Wolf) three-phase CA traffic flow model [605, 611]

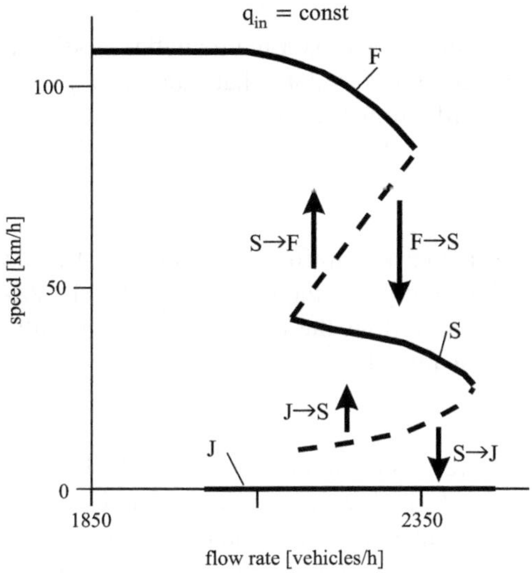

Fig. 1.9 Simulation of 2Z-characteristic for phase transitions with the Kerner-Klenov stochastic microscopic three-phase model: 2Z-characteristic for phase transitions as function of the flow rate $q_{\mathrm{in}} + q_{\mathrm{on}}$ at the on-ramp bottleneck when the on-ramp inflow rate q_{on} is a variable and the flow rate upstream of the on-ramp bottleneck q_{in} is a given value ($q_{\mathrm{in}} = 1850$ vehicles/h). The first Z-characteristic between the phases F and S is the same one as that shown in Fig. 1.8b. The second Z-characteristic between the phases S and J will be discussed in Sect. 8.3. F—free flow phase, S—synchronized flow phase, J—wide moving jam phase. Arrows F→S, S→F, S→J, and J→S illustrate symbolically the related phase transitions between the three traffic phase. Adapted from [594]

have been developed for a more detailed description of empirical features of real traffic.

In the three-phase theory, there are the free flow, synchronized flow, and wide moving jam phases. Therefore, additionally to the Z-characteristic for traffic breakdown (Fig. 1.8b), a three-phase traffic flow model must show a second Z-characteristic for the S→J and J→S transitions. A presentation of the F→S and S→F transitions together with the S→J and J→S transitions leads to a 2Z-characteristic for phase transitions (Fig. 1.9). The theoretical 2Z-characteristic for phase transitions (Fig. 1.9) is in accordance with empirical features of phase transitions in real traffic flow (Sect. 1.5). In more details, the 2Z-characteristic for phase transitions as the basic requirement for a three-phase traffic flow model will be considered in Sect. 8.9.

These explanations lead to the following definition of the term *three-phase traffic flow model.*

A three-phase traffic flow model is a mathematical traffic flow model in the framework of the three-phase theory that can show a 2Z-characteristic of phase transitions in traffic flow (Fig. 1.9).

This definition means that if a traffic flow model that incorporates some of the hypotheses of the three-phase theory cannot show the 2Z-characteristic of phase transitions in traffic flow, the model cannot generally be considered a three-phase traffic flow model.

This explains the above statement that not any traffic flow model that incorporate some of the hypotheses of the three-phase theory can be considered a three-phase traffic flow model. In particular, the use of some of the hypotheses of the three-phase theory in a traffic flow model does not necessarily mean that the model becomes a three-phase traffic flow model: If this traffic flow model cannot show the 2Z-characteristic for the phase transitions, the model cannot be considered a three-phase traffic flow model. In particular, there can be three-phase traffic flow models that do not incorporate 2D-states of traffic flow (hypothesis 10 of Sect. 1.9). Indeed, if these traffic flow models can show a 2Z-characteristic for the phase transitions, the traffic flow models can nevertheless be considered three-phase traffic flow models (see examples of such three-phase traffic flow models in Sec. 12.1 of the book [572]).

The Kerner-Klenov stochastic three-phase traffic flow model has further been developed for different applications in [554, 558–565, 569, 570, 577, 579, 580, 582, 583, 594–597, 599–603, 606–608], in particular, to simulate on-ramp metering [558–561, 563–565, 569], speed limit control [562], dynamic traffic assignment [575, 587, 588], traffic at heavy bottlenecks [570] and at moving bottlenecks [599], features of heterogeneous traffic flow consisting of different vehicles and drivers [595], features of the complexity of spatiotemporal traffic congestion observed in real measured traffic data [577], the diagram of congested traffic patterns at highway bottlenecks in the framework of the three-phase theory [594], jam warning methods [635, 636], vehicle-to-vehicle (V2V) communication [582, 600–603], the ACC (adaptive cruise control) performance [554], traffic breakdown at signals in city traffic [576, 579, 580, 612], over-saturated city traffic [592, 604], vehicle fuel consumption in traffic networks [413, 414, 422] based on a cumulative vehicle acceleration [581].

The three-phase theory allows us to develop a number of reliable ITS that can improve traffic and increase traffic safety. To these ITS-applications belong automatic driving (called also self-driving or automated driving) vehicles based on the three-phase theory [557, 566–568, 572] that will be discussed in Sect. 7.2, traffic control methods like "congested pattern control approach" and the associated ANCONA on-ramp metering [558, 559, 563–565] that will be discussed in

Sects. 12.6 and 13.5 as well as cooperative driving that will be briefly discussed in Sect. 7.3. Other ITS-applications of the three-phase theory can be found in the books [556, 572] and the review [582].

Over time several scientific groups have developed and studied new traffic flow models as well as ITS-applications (e.g., [4, 99, 171, 213–220, 223, 228, 229, 291, 292, 315, 378, 380, 452, 484, 490, 492, 493, 500, 501, 503, 506, 507, 513, 523, 525, 526, 653, 656, 667, 669, 700, 716, 717, 719, 724, 731, 811, 866, 961, 969, 989, 1066–1071, 1073, 1074, 1122, 1126, 1147, 1150, 1153, 1161, 1162, 1190]) that incorporate some of the hypotheses of the three-phase theory [539–543, 556, 572].

In particular, new traffic flow models that incorporate some of the hypotheses of the three-phase theory have been introduced in the works by Jiang, Wu, Gao et al. [291, 503], Davis [213, 215], Lee, Kim, Schreckenberg et al. [716], Schreckenberg, Schadschneider, Knorr et al. [667, 961], as well as Tian, Treiber, Jia, Ma, Jiang et al. [1067–1071].

Through the use of the Kerner-Klenov stochastic three-phase traffic flow model as well as other traffic flow models that incorporate hypotheses of the three-phase theory, Davis has derived a number of novel results related to ITS applications, in particular, for wirelessly connected ACC-vehicles [229], for predicting travel time to limit congestion [220], for realizing Wardrop equilibria with real-time traffic information [219], for traffic control at highway bottlenecks [218], and for on-ramp metering near the transition to the synchronized flow phase [215]. Davis [214] was one of the first who studied the effect of ACC-vehicles on traffic flow with a traffic flow model that incorporates some of the hypotheses of the three-phase theory. Davis [217] was also one of the first who applied hypotheses of the three-phase theory [539–543, 556] for simulations of the cooperative merging at an on-ramp bottleneck to study the prevention of the formation of synchronized flow at the bottleneck.

As mentioned in Sect. 1.7, none of classical traffic-flow theories and models (see, e.g., reviews [73, 103, 166, 173, 187, 258, 305, 318, 355, 394, 429, 430, 727, 768, 796, 804, 845, 856, 884, 923, 956, 978, 1089, 1136]) incorporates the metastability of free flow with respect to an F→S transition at a bottleneck.

- To distinguish three-phase traffic flow models from classical traffic flow theories and models (in particular, the LWR model and traffic flow models of the GM model class as well as any combinations of the classical theories and models), the classical traffic flow theories and models can be considered "two-phase traffic flow models".
- Thus, in the book the term *a two-phase traffic flow model* is a synonym of the terms *a classical traffic flow model* and *a classical traffic flow theory*.

1.13 Criticism of Three-Phase Traffic Theory

The main criticism on the three-phase theory made during last years, in particular in the works by Daganzo et al. [201] and by Helbing, Treiber et al. [412, 1019, 1020, 1091] can be formulated as follows:

- The three-phase theory is not needed because generally accepted traffic flow theories also succeed in simulating of the most essential empirical features of traffic flow described by the three-phase theory.

The authors of these works believe [201, 412, 1019, 1020, 1091] that the properly tuned two-phase traffic flow models, based on extensive simulations can show the same spatiotemporal traffic features as three-phase traffic flow models.

To prove their statements, Helbing, Treiber et al. [412, 1019, 1020, 1091] are arguing that the well-defined empirical features of the spatiotemporal congested patterns, as well as empirical scattering of the flow-density plot, can be captured by two-phase traffic flow models. In particular, two-phase traffic flow models (see a discussion in Sect. 4.8) can indeed show a spatiotemporal congested pattern that looks similar to the empirical spatiotemporal congested pattern presented in Fig. 1.6.

However, rather than empirical features of the spatiotemporal congested patterns and empirical scattering of the flow-density plot, the metastability of free flow at a highway bottleneck with respect to an F→S transition is needed to use as the basic criterion for a comparison of the classical traffic flow models [201, 412, 1019, 1020, 1091] with three-phase traffic flow models [556].

Indeed, the main reason of the three-phase theory is the explanation of traffic breakdown through the metastability of free flow at a bottleneck with respect to an F→S transition (Sects. 1.9 and 1.12). Such a comparison has *not* been made either by Daganzo et al. [201] or by Helbing, Treiber et al. [412, 1019, 1020, 1091]. The reason for the invalid criticism of the three-phase theory made by Daganzo et al. [201] and Helbing, Treiber et al. [412, 1019, 1020, 1091] is that none of the traffic flow models of Daganzo et al. [201] and none of the traffic flow models of Helbing, Treiber et al. [412, 1019, 1020, 1091] is able to show the empirical fundamental of transportation science—the metastability of free flow at a highway bottleneck with respect to an F→S transition.

As we will see in Chap. 4, this critical conclusion is valid for both the traffic flow models of Daganzo, Helbing, Treiber et al. [201, 412, 1019, 1020, 1091] and for any other two-phase traffic flow models that are generally accepted as the state-of-the-art in traffic and transportation research (see, e.g., reviews, books, and conference proceedings [16, 22, 70, 73, 103, 135, 141, 146, 166, 173, 187, 243, 258, 290, 305, 314, 318, 355, 394, 404, 429, 430, 443, 450, 453, 665, 689, 694, 725, 727, 768, 773,

796, 799, 804, 845, 856, 878, 884, 938, 956, 976, 978–980, 1004, 1010, 1012, 1025, 1064, 1089, 1136, 1139, 1143, 1144]).

Thus, in contrast with the three-phase theory, none of the classical traffic theories and two-phase traffic flow models can show the metastability of free flow with respect to an F→S transition at a highway bottleneck. Nevertheless, the following question can arise:

- Why does the metastability of free flow with respect to the F→S transition at the bottleneck, which is the reason for the three-phase theory, exhibit the *fundamental importance* for transportation science?

We have already answered this question in Sect. 1.6: The metastability of free flow with respect to the F→S transition at the bottleneck is needed to explain the empirical nucleation nature of traffic breakdown that is the empirical fundamental of transportation science. For this reason, traffic and transportation theories, which are not consistent the nucleation nature of traffic breakdown (F→S transition) at highway bottlenecks, cannot be applied for the development of reliable traffic control, dynamic traffic assignment, and organization as well as other ITS-applications in traffic and transportation networks.

This explains why *neither* empirical features of spatiotemporal congested patterns, *nor* the empirical scattering of the flow-density plot should be considered as the basic empirical traffic phenomena for a comparison of two-phase and three-phase traffic flow models as made by Daganzo, Helbing, Treiber et al. [201, 412, 1019, 1020, 1091, 1162]. This is because the fundamental difference between two-phase and three-phase traffic flow models is associated with the explanation of the empirical nature of traffic breakdown at a highway bottleneck:

- *None* of the two-phase traffic flow models can simulate and explain the empirical metastability of free flow at a highway bottleneck with respect to an F→S transition as observed in real field traffic data.

This critical statement will be considered in more details in Chaps. 4 and 8.

> The main reason of the author's criticism of the classical traffic flow models is that the models and associated classical traffic flow theories are invalid to explain the empirical metastability of free flow with respect to an F→S transition at a highway bottleneck. This empirical metastability of free flow is the reason for the nucleation nature of real traffic breakdown at the bottleneck. The empirical nucleation nature of traffic breakdown at the bottleneck is an empirical fundamental of transportation science.

As mentioned in Sect. 1.12, the metastability of free flow at a highway bottleneck with respect to an F→S transition is described by the Z-characteristic for traffic breakdown (Fig. 1.8b). Synchronized flow related to this Z-characteristic is metastable with respect to a return S→F transition. In its turn, the synchronized flow is also metastable with respect to an S→J transition. The 2Z-characteristic for phase transitions (Fig. 1.9) illustrates these two types of the metastability of synchronized flow (with respect to the S→F transition and with respect to the S→J transition) predicted in the three-phase theory. For this reason, a three-phase traffic flow model should show the 2Z-characteristic for phase transitions.

As above-mentioned, Daganzo, Helbing, Treiber et al. [201, 412, 1019, 1020, 1091, 1162] have not considered this fundamental empirical feature of traffic breakdown for a comparison of two-phase traffic flow models with three-phase traffic flow models. For this reason, Daganzo, Helbing, Treiber et al. [201, 412, 1019, 1020, 1091, 1162, 1165, 1166] have made the following *incorrect* conclusions:

(i) Two-phase traffic flow models can simulate most of the empirical features of phase transitions in traffic flow.
(ii) Two-phase traffic flow models can simulate most of the empirical features of synchronized flow.
(iii) Two-phase traffic flow models are approximations of the three-phase traffic flow models.

These and other incorrect conclusions of Daganzo, Helbing, Treiber et al. [412, 1019, 1020, 1091, 1162, 1165, 1166] result from the ignoring of the basic feature of synchronized flow: The existence of *two types* of the metastability of synchronized flow (with respect to the S→F transition and with respect to the S→J transition) resulting in the 2Z-characteristic for phase transitions in traffic flow (Fig. 1.9).

None of the two-phase traffic flow models can show the 2Z-characteristic for phase transitions in traffic flow that is the basic requirement for a three-phase traffic flow model (Fig. 1.9). For this reason, none of the two-phase traffic flow models can also describe empirical features of synchronized flow that are responsible for the nucleation nature of traffic breakdown.

The critical conclusion that none of the two-phase traffic flow models can simulate and explain the empirical metastability of free flow with respect to an F→S transition at a highway bottleneck as well as the 2Z-characteristic for phase transitions is related to all generally accepted traffic flow theories and models (see, e.g., reviews [73, 103, 166, 173, 187, 243, 258, 305, 318, 355, 370, 394, 412, 727, 768, 796, 799, 845, 856, 884, 923, 956, 976, 979, 1010, 1089, 1136, 1139, 1143, 1162] and references there). These models include the models of Daganzo, Helbing, Treiber et al. [201, 394, 412, 1019, 1020, 1082–1086, 1088–1091, 1162, 1165, 1166].

The explanation of the empirical metastability of free flow with respect to an F→S transition at a highway bottleneck is the main reason for the three-phase theory [556, 572]. Therefore, the statement "the three-phase theory is not needed" is invalid.

1.14 Incommensurability of Three-Phase Traffic Theory and Classical Traffic Theories

Due to the criticism of classical traffic flow theories made in this book, a question arises:

- May some of the classical traffic-flow theories be relatively easily adjusted to take into account the empirical evidence of the metastability of free flow with respect to an F→S transition at a highway bottleneck?

The explanation of traffic breakdown at a highway bottleneck by an F→S transition that occurs due to the metastability of free flow with respect to this F→S transition at the bottleneck is the basic assumption of the three-phase theory [539–553, 555, 556, 571–574, 578, 584, 586]. None of the classical traffic-flow theories (see, for example, reviews and conference proceedings [16, 22, 70, 73, 103, 135, 141, 146, 166, 173, 187, 243, 258, 259, 281, 290, 305, 314, 318, 355, 394, 404, 429, 430, 450, 453, 661, 662, 665, 689, 694, 725, 727, 768, 772, 773, 796, 799, 804, 845, 856, 884, 923, 938, 956, 964, 976, 978–980, 983–985, 1010, 1012, 1025, 1028, 1064, 1088, 1089, 1136, 1139, 1143, 1144] as well as references there) incorporates the metastability of free flow with respect to the F→S transition at the bottleneck (Chap. 4).

For this reason, the classical traffic-flow models cannot describe the empirical metastability of free flow with respect to the F→S transition at the bottleneck. However, the metastability of free flow with respect to the F→S transition at the bottleneck is the empirical evidence of the nucleation nature of traffic breakdown [614]. Reasons why the classical traffic-flow models cannot describe the metastability of free flow with respect to the F→S transition at the bottleneck will be explained in Chap. 8.

The metastability of free flow with respect to the F→S transition at the bottleneck has no sense for the classical traffic and transportation theories. The three-phase theory has been introduced for the explanation of this free flow metastability.

In accordance with the classical book by Kuhn [677], this shows the incommensurability of the three-phase theory and the classical traffic flow theories. Therefore, the three-phase theory is incommensurable with all classical traffic flow theories [611]. This statement will be explained in details in Chap. 8 of this book. The term "incommensurable" has been introduced by Kuhn [677] to explain a paradigm shift in a scientific field. This explains the title of this Chap. 1.

The existence of the metastability of free flow with respect to an F→S transition at a highway bottleneck does not result from the stochastic nature of traffic: Even if there were no stochastic processes in vehicular traffic, the metastability of free flow with respect to the F→S transition at the bottleneck does exist. For this reason, stochastic approaches to traffic control (see, e.g., references in reviews and conference proceedings [16, 22, 70, 73, 103, 135, 141, 146, 166, 173, 187, 243, 258, 259, 281, 290, 305, 314, 318, 355, 394, 404, 429, 430, 450, 453, 661, 662, 665, 689, 694, 725, 727, 768, 772, 773, 796, 799, 804, 845, 856, 884, 923, 938, 956, 964, 976, 978–980, 983–985, 1010, 1012, 1025, 1028, 1064, 1088, 1089, 1136, 1139, 1143, 1144]), which do not assume a possibility of the metastability of free flow with respect to the F→S transition at the bottleneck, cannot resolve the above-discussed problem of the inconsistency of the classical traffic theories with the empirical nucleation nature of real traffic breakdown.

However, the stochastic nature of traffic influences crucially on the probability of random transitions between the phases F and S (see a discussion of the influence of fluctuations in traffic flow on random transitions between the phases F and S in Sects. 5.6 and 11.5.1). At a given flow rate, this probability can change in several orders of magnitude when stochastic characteristics of traffic change.

1.15 Objectives of the Book

There are a huge number of traffic flow models, models for traffic control, dynamic traffic assignment and optimization in traffic and transportation networks. However, as stressed above, the classical traffic and transportation theories have failed by their applications in the real world. There are a number of books, reviews and books of conference proceedings in which mathematical descriptions of these classical traffic models have been considered in detailed [16, 22, 70, 73, 103, 135, 141, 146, 166, 173, 187, 243, 258, 259, 290, 305, 314, 318, 355, 394, 404, 429, 430, 450, 453, 661, 665, 689, 694, 725, 727, 768, 773, 796, 799, 804, 845, 856, 884, 923, 938, 956, 976, 978–980, 1010, 1012, 1025, 1064, 1088, 1089, 1136, 1139, 1143, 1144]. For these reasons, in this book we will *not* consider the mathematical descriptions of these classical traffic models. Instead, to explain why the classical traffic and transportation theories have failed by their applications in the real world, we present results of their calculations and compare them with the empirical fundamental of transportation science—the empirical nucleation nature of traffic breakdown at highway bottlenecks. In particular, we show that applications of the classical Wardrop's equilibria for standard dynamic traffic assignment in traffic and transportation networks related to the state-of-the-art in traffic and transportation research (see, e.g., [2, 10, 18, 24, 27, 51, 60, 63–65, 67–71, 101, 132, 148, 149, 163, 167, 179–182, 188, 192, 198, 207, 236, 275–280, 282, 283, 326, 337, 375, 427, 428, 451, 463, 469, 473, 479, 499, 522, 530, 531, 534, 685, 686, 690–692, 744, 751, 756, 772, 775, 778, 784, 785, 812, 813, 820, 865, 930, 936–938, 967, 968, 978, 980, 981, 988, 1026, 1028, 1029, 1033, 1051, 1076, 1099, 1138, 1152, 1154–1158, 1180, 1183, 1195, 1202, 1207, 1208] and conference proceedings [16, 141, 146, 450, 689, 694, 725, 773, 1064]) deteriorate basically the traffic system while provoking heavy traffic congestion in urban networks.

We will show that in contrast with standard dynamic traffic assignment, the application of the BM principle for dynamic traffic assignment can lead to the maintenance of free flow in the network during almost the whole day time, including a large interval of rush hour.

In the book, we show also that contrarily to the classical traffic flow theories, the three-phase theory incorporates the empirical nucleation nature of traffic breakdown at highway bottlenecks. Therefore, in this book we will discuss how the three-phase theory explains this empirical fundamental of transportation science. We consider requirements following from the empirical nucleation nature of traffic breakdown at highway bottlenecks for the development of future three-phase traffic flow models, models for traffic control, dynamic traffic assignment and optimization in traffic and transportation networks.

The main objective of this book is a detailed consideration of the effect of the empirical nucleation nature of traffic breakdown at network bottlenecks on traffic and transportation theories, future automatic driving, traffic control, dynamic traffic assignment, and optimization in traffic and transportation networks. This explain the title of the book.

Other objectives of this book are as follows:

1. We explain why the empirical nucleation nature of traffic breakdown at a highway bottleneck should be considered *the empirical fundamental* for any traffic and transportation theory.
2. We show why generally accepted traffic and transportation theories that are the state-of-the-art in traffic and transportation research (see, for example, reviews [70, 73, 103, 135, 166, 173, 187, 243, 258, 259, 305, 317, 318, 355, 394, 429, 430, 661, 727, 768, 796, 799, 804, 845, 856, 884, 923, 938, 956, 964, 976, 978–980, 1010, 1088, 1089, 1136, 1139, 1143]) lead to approaches for control, dynamic traffic assignment, and optimization of traffic and transportation networks which *fail* by their applications in the real world.
3. We explain why the three-phase theory [539–553, 555, 556, 571–574, 578, 584–586] should be the theoretical fundamental for reliable traffic control, dynamic traffic assignment, and optimization of traffic and transportation networks.

4. We show that a reliable traffic control, dynamic traffic assignment, and optimization of traffic and transportation networks can be achieved through *the minimization of the probability* of traffic breakdown in a network based on the BM principle that is consistent with the empirical fundamental of transportation science [575, 576].
5. We explain why the three-phase theory should be the theoretical fundamental for the development of future automatic driving.
6. We show that the use of the BM principle results in the maximization of the network throughput while ensuring free flow conditions in the whole traffic or transportation network [587, 588].
7. Based on a famous theory by Kuhn [677] about a paradigm shift in the science, we proof *the incommensurability* of the three-phase theory with classical traffic theories.
8. We present a theory of time-delayed traffic breakdown at traffic signal in city traffic introduced in 2011 [576]. As shown in [579], in contrast with the classical model of traffic at the signal (e.g., [62, 234, 297, 298, 300–304, 309–313, 339, 465, 466, 741, 742, 810, 814, 815, 821, 874, 879, 880, 883, 887, 896, 919, 997–999, 1040, 1046, 1131, 1132]), the model of random time-delayed traffic breakdown at traffic signal [576] predicts that under-saturated traffic at the signal can exist during a long time interval even when the average arrival flow rate in under-saturated traffic is larger than the classical signal capacity. A theory of city traffic based on the model of random time-delayed traffic breakdown at traffic signal in city traffic allows us to find conditions at which the classical theory of traffic at the signal can be considered a *special case* of the three-phase theory.

1.16 Book's Structure

The book consists of fifteen chapters. In Chap. 2, we discuss achievements of empirical investigations of traffic breakdown at highway bottlenecks and highway capacity, which have made a great impact on the understanding of traffic flow phenomena.

In Chap. 3, based on a study of real field traffic data, we prove the empirical nucleation nature of traffic breakdown (F→S transition) at highway bottlenecks.

As we will see in Chap. 4, none of the classical traffic flow theories incorporates the metastability of free flow with respect to the F→S transition at a highway bottleneck. For this reason, all classical traffic flow theories failed for a reliable dynamic control and optimization of traffic and transportation networks. Nevertheless, we explain that there are also great achievements of these classical theories and models in the understanding of many other traffic phenomena as well as in studies of the driver behavior in vehicular traffic. These achievements of the classical models are also used in the three-phase theory.

In Chap. 5, we show how the three-phase theory explains the nucleation nature of traffic breakdown at highway bottlenecks.

An application of this theory for the analysis of the probability of traffic breakdown in a mixed traffic flow consisting of manual driving and automatic driving vehicles is the subject of Chap. 6. A discussion of automatic driving based on the three-phase theory is presented in Chap. 7.

Reasons for the incomprehensibility of the three-phase theory with the classical traffic flow theories are discussed in Chap. 8.

In Chap. 9, we present a theory of time-delayed traffic breakdown (a time-delayed transition from under- to over-saturated city traffic) at traffic signal in city traffic [576, 579]. We have found that both two-phase traffic flow models of the GM model class and three-phase traffic flow models can describe the time-delayed traffic breakdown at the signal. This is because the signal introduces an extremely large disturbance in traffic flow while forcing all vehicles come to a stop at the signal during the red signal phase. Due to this large disturbance in traffic flow, the existence of the synchronized flow phase in the three-phase theory effects on *quantitative* characteristics of time-delayed traffic breakdown at the signal only: For the description of time-delayed traffic breakdown at the signal, the two-phase models of the GM model class can indeed be considered some approximations of the three-phase models. However, we have also found that the classical model of the transition from under- to over-saturated traffic at the signal can be considered a special case of the theory of time-delayed traffic breakdown at the signal only in a non-realistic case, when all vehicles arrive the signal during the red signal phase.

Chaps. 10–12 are devoted to the BM principle and its applications for dynamic traffic assignment and control in traffic and transportation networks. In particular, we show that the BM principle allows us to maximize the network throughput at which free flow conditions are ensured in the whole network.

A critical comparison of possible applications of the BM principle versus applications of the classical Wardrop's equilibria for dynamic traffic assignment in traffic networks is the subject of Chap. 13. We show that applications of the classical Wardrop's equilibria for standard dynamic traffic assignment in traffic and transportation networks deteriorate basically traffic system. This explains the failure of standard dynamic traffic assignment in the real world applications.

Based on results of Chaps. 10–13, in Chap. 14 we make a qualitative discussion of future dynamic traffic assignment and control in traffic and transportation networks. Conclusions and outlook are formulated in Chap. 15.

In Appendices A and B, we present the Kerner-Klenov stochastic microscopic three-phase traffic flow model (Appendix A) and the KKSW (Kerner-Klenov-Schreckenberg-Wolf) cellular automaton (CA) three-phase traffic flow model (Appendix B) whose numerical simulations are discussed in the main text of the book.

References

1. M. Abdel-Aty, J. Dilmore, A. Dhindsa, Accid. Anal. Prevent. **38**, 335–345 (2006)
2. K. Abdelghany, D. Valdes, A. Abdelfatah, H.S. Mahmassani, Transp. Res. Rec. **1667**, 67–76 (1999)
3. S.M. Abuelenin, A.Y. Abul-Magd, in *Proc. of 2014 International Conference on Connected Vehicles and Expo (ICCVE)* (IEEE, 2014), pp. 391–395
4. S.M. Abuelenin, A.Y. Abul-Magd, IEEE Trans. ITS **16**, 2543–2550 (2015)
5. A.Y. Abul-Magd, Phys. Rev. E **76**, 057101 (2007)
6. J.A. Acha-Daza, F.L. Hall, Transp. Res. B **28**, 235–250 (1994)
7. M. van Aerde, H. Rakha, INTEGRATION ©Release 2.30 for Windows: User's Guide - Volume I: Fundamental Model Features (M. Van Aerde & Assoc., Ltd., Blacksburg, 2007)
8. M. van Aerde, H. Rakha, INTEGRATION ©Release 2.30 for Windows: User's Guide - Volume II: Advanced Model Features, (M. Van Aerde & Assoc., Ltd., Blacksburg, 2007)
9. K. Aghabayk, M. Sarvi, W. Young, J. Adv. Transp. **50**, 1432–1446 (2016)
10. S.D. Ahipas, R. Meskarian, T.L. Magnanti, K. Natarajan, Transp. Res. B **81**, 333–354 (2015)
11. E. Alberti, G. Belli, Transpn. Res. **12**, 33–42 (1978)
12. P. Allaby, B. Hellinga, M. Bullock, IEEE Trans. Intell. Transp. Syst. **8**, 671–680 (2006)
13. B.L. Allen, G.F. Newell, Transp. Sci. **10**, 227–242 (1976)
14. B.L. Allen, G.F. Newell, Transp. Sci. **10**, 243–268 (1976)
15. R.E. Allsop, Transp. Sci. **6**, 260–285 (1972)
16. R.E. Allsop, M.G.H. Bell, B.G. Heydecker (eds.), *Transportation and Traffic Theory 2007* (Elsevier Science, Amsterdam, 2007)
17. J. Al-Obaedi, S. Yousif, IEEE Trans. ITS **13**, 296–306 (2012)
18. M. Amirgholy, E.J. Gonzales, Transp. Res. B **92**, 234–252 (2016)
19. F.C. Andrews, Transp. Res. **4**, 367–377 (1970)
20. F.C. Andrews, Transp. Res. **4**, 359–366 (1970)
21. F.C. Andrews, Transp. Res. **7**, 223–232 (1973)
22. C. Appert-Rolland, F. Chevoir, P. Gondret, S. Lassarre, J.-P. Lebacque, M. Schreckenberg (eds.), *Traffic and Granular Flow'07* (Springer, Heidelberg, 2009)
23. M. Ardelt, C. Coester, N. Kaempchen, IEEE Trans. ITS **13**, 1576–1585 (2012)
24. D. Atmani, J.-P. Lebacque, N. Bhouri, H. Haj-Salem, Transp. Res. Procedia **3**, 895–904 (2014)
25. P. Athol, Highway Res. Rec. **72**, 58–87 (1965)
26. P. Athol, Highway Res. Rec. **72**, 137–155 (1965)
27. A. Aungsuyanon, D. Boyce, B. Ran, in *Proc. TRB 92nd Annual Meeting*, Paper # 13-1423 (TRB, Washington DC, 2013)
28. A. Aw, M. Rascle, SIAM J. Appl. Math. **60**, 916–938 (2000)
29. K.W. Axhausen, T. Gärling, Transp. Rev. **12**, 323–341 (1992)
30. H.M.A. Aziz, S.V. Ukkusuri, Transp. Res. Rec. **2311**, 73–84 (2012)
31. C. Backfrieder, G. Ostermayer, C.F. Mecklenbräuker, IEEE Trans. ITS **18**, 349–363 (2017)
32. M. Bahram, Z. Ghandeharioun, P. Zahn, M. Baur, W. Huber, F. Busch, *17th IEEE Intern. Conference on ITS (ITSC) 2014*, pp. 1752–1757
33. M. Balmer, K.W. Axhausen, K. Nagel, Transp. Res. Rec. **1985**, 125–134 (2006)
34. M. Bando, K. Hasebe, K. Nakanishi, A. Nakayama, Phys. Rev. E **58**, 5429–5435 (1998)
35. M. Bando, K. Hasebe, A. Nakayama, A. Shibata, Y. Sugiyama, Jpn. J. Appl. Math. **11**, 203–223 (1994)
36. M. Bando, K. Hasebe, A. Nakayama, A. Shibata, Y. Sugiyama, Phys. Rev. E **51**, 1035–1042 (1995)
37. M. Bando, K. Hasebe, A. Nakayama, A. Shibata, Y. Sugiyama, J. Phys. I France **5**, 1389–1399 (1995)
38. J.H. Banks, Transp. Res. Rec. **1225**, 53–60 (1989)
39. J.H. Banks, Transp. Res. Rec. **1287**, 20–28 (1990)

40. J.H. Banks, "Evaluation of the Two-Capacity Phenomenon as a Basis for Ramp Metering", *Final Report, Civil Engineering Report Series 9002* (San Diego State University, San Diego, CA, 1990)
41. J.H. Banks, Transp. Res. Rec. **1320**, 234–241 (1991)
42. J.H. Banks, Transp. Res. Rec. **1320**, 83–90 (1991)
43. J.H. Banks, Transp. Res. Rec. **1394**, 17–25 (1993)
44. J.H. Banks, Transp. Res. Rec. **1510**, 1–10 (1995)
45. J.H. Banks, Transp. Res. Rec. **1678**, 128–134 (1999)
46. J.H. Banks, Transp. Res. Rec. **1802**, 225–232 (2002)
47. J.H. Banks, Transp. Res. Rec. **2099**, 14–21 (2009)
48. J. Barceló, in *Int. Sym. on Transport Simulation* (Yokohama, 2002)
49. J. Barceló, J. Casas, Oper. Res. Comp. Sci. Interf. Ser. **31**, 57–98 (2005)
50. J. Barceló, E. Codina, J. Casas, J.L. Ferrer, D. Garcia, J. Intell. Robotic Syst. **41**, 173–203 (2005)
51. H. Bar-Gera, Transp. Sci. **36**, 398–417 (2002)
52. R. Barlović, T. Huisinga, A. Schadschneider, M. Schreckenberg, Phys. Rev. E **66**, 046113 (2002)
53. R. Barlović, L. Santen, A. Schadschneider, M. Schreckenberg, Eur. Phys. J. B **5**, 793–800 (1998)
54. L.D. Baskar, B. De Schutter, H. Hellendoorn, IEEE Trans. ITS **13**, 838–847 (2012)
55. S. Bassan, A. Faghri, A. Polus, Civil Eng. Environ. Syst. **24**, 261–274 (2007)
56. S. Bassan, A. Polus, A. Faghri, Transp. Res. Rec. **1965**, 40–47 (2006)
57. M. Behrisch, L. Bieker, J. Erdmann, D. Krajzewicz, "SUMO - Simulation of Urban MObility: An Overview", in *SIMUL 2011, The Third Inter. Conf. on Adv. in Sys. Simulation* (2011)
58. N. Bekiaris-Liberis, C. Roncoli, M. Papageorgiou, IEEE Trans. ITS **17**, 3484–3497 (2016)
59. M.G.H. Bell, Transp. Sci. **17**, 198–217 (1983)
60. M.G.H. Bell, Transp. Res. B **25**, 13–22 (1991)
61. M.G.H. Bell, Transp. Res. B **25**, 115–125 (1991)
62. M.G.H. Bell, Transp. Res. A **26**, 303–313 (1992)
63. M.G.H. Bell, Transp. Res. B **29**, 125–137 (1995)
64. M.G.H. Bell, Transp. Res. B **29**, 287–295 (1995)
65. M.G.H. Bell, Transp. Res. B **34**, 533–545 (2000)
66. M.G. Bell, D. Brookes, Transp. Res. C **1**, 43–55 (1993)
67. M.G.H. Bell, Ch. Cassir, in *2nd Int. Conf. on Traffic and Transportation Studies (ICTTS), Beijing, China*, ed. by K.C. Wang, G. Xiao, J. Ji (American Society of Civil Engineers, USA, 2000), pp. 9–16
68. M.G.H. Bell, Ch. Cassir, *Reliability of Transport Networks* (Research Studies Press, Baldock, Hertfordshire, 2000)
69. M.G.H. Bell, Ch. Cassir, Transp. Res. B **36**, 671–681 (2002)
70. M.G.H. Bell, Y. Iida, *Transportation Network Analysis* (Wiley, Hoboken, 1997)
71. M.G.H. Bell, C.M. Shielda, F. Busch, G. Kruse, Transp. Res. C **5**, 197–210 (1997)
72. T. Bellemans, B. De Schutter, B. De Moor, Contr. Eng. Pract. **14**, 757–767 (2006)
73. N. Bellomo, V. Coscia, M. Delitala, Math. Mod. Meth. App. Sci. **12**, 1801–1843 (2002)
74. M.E. Ben-Akiva, S.R. Lerman *Discrete Choice Analysis: Theory and Application to Travel Demand* (The Massachussets Institude of Thechnology, 1985)
75. C.R. Bennett, C.M. Dunn, Transp. Res. Rec. **1510**, 70–75 (1995)
76. M.J. Beckmann, C.B. McGuire, C.B. Winsten, *Studies in the Economics of Transportation* (Yale Univ. Press, New Haven, 1956)
77. P. Berg, A. Mason, A.W. Woods, Phys. Rev. E **61**, 1056–1066 (2000)
78. V.A.C. van den Berg, E.T. Verhoef, Transp. Res. B **94**, 43–60 (2016)
79. P. Berg, A.W. Woods, Phys. Rev. E **64**, 035602(R) (2001)
80. R.L. Bertini, J. Intell. Transp. Syst. **9**, 35–45 (2005)
81. R.L. Bertini, M.J. Cassidy, Transp. Res. A **36**, 683–697 (2002)

82. R.L. Bertini, S. Boice, K. Bogenberger, Transp. Res. Rec. **1978**, 149–159 (2006)
83. R.L. Bertini, S. Hansen, K. Bogenberger, Transp. Res. Rec. **1934**, 97–107 (2005)
84. R.L. Bertini, M.T. Leal, J. Transp. Eng. **131**, 397–407 (2005)
85. R.L. Bertini, Sh. Malik, Transp. Res. Rec. **1867**, 25–35 (2004)
86. R.L. Bertini, A. Myton, Transp. Res. Rec. **1925**, 48–57 (2005)
87. C. Bian, C. Yuan, W. Kuang, D. Wu, Math. Problems Eng. **2016**, 1693729 (2016)
88. G.N. Bifulco, L. Pariota, F. Simonelli, R.D. Pace, Transp. Res. C **29**, 156–170 (2013)
89. O. Biham, A.A. Middleton, D. Levine, Phys. Rev. A **46**, R6124 (1992)
90. M. Blank, J. Stat. Phys. **111**, 903–930 (2003)
91. M. Blank, J. Stat. Phys. **120**, 627–658 (2005)
92. L. Bloomberg, J. Dale, Transp. Res. Rec. **1727**, 52–60 (2000)
93. R. Boel, L. Mihaylova, Transp. Res. B **40**, 319–334 (2006)
94. K. Bogenberger, (Ph.D. Thesis, UC, Berkeley University, Berkeley, CA, 2001) (UC Berkeley Transportation Library, PATH Record Number 24483)
95. K. Bogenberger, H. Keller, in *Proceedings of the 34th Annual Hawaii Int. Conference on System Sciences, 2001*, DOI: 10.1109/HICSS.2001.926333
96. K. Bogenberger, H. Keller, S. Vukanovic, in *Proc. of Intelligent Transportation Systems 2001*, DOI: 10.1109/ITSC.2001.948636
97. K. Bogenberger, A.D. May, *Working Paper UCB-ITS-PWP-99-19, California Partners for Advanced Transit and Highways (PATH)* (Institute of Transportation Studies (UCB), UC Berkeley, 1999)
98. P. Böhnke, E. Pfannerstill, ITE J. **56**, 41–45 (1986)
99. R. Borsche, M. Kimathi, A. Klar, Comp. Math. Appl. **64**, 2939–2953 (2012)
100. P.H.L. Bovy (ed.), *Motorway Analysis: New Methodologies and Recent Empirical Findings* (Delft University Press, Delft, 1998)
101. D. Boyce, B. Ralevic-Dekic, H. Bar-Gera, J. Transp. Eng. **130**, 49–55 (2004)
102. S.D. Boyles, in *Proc. TRB 92nd Annual Meeting*, Paper # 13-4519 (TRB, Washington DC, 2013)
103. M. Brackstone, M. McDonald, Transp. Res. F **2**, 181 (1998)
104. M. Brackstone, P. Sykes, P. Vortisch, Traffic Eng. Contr. **55**, 165–171 (2013)
105. D. Branston, Transp. Sci. **10**, 125–148 (1976)
106. L. Breiman, R.L. Lawrence, Transp. Res. **7**, 77–105 (1973)
107. L. Breiman, R.L. Lawrence, Transp. Res. **11**, 177–182 (1977)
108. L. Breiman, R.L. Lawrence, D. Goodwin, B. Bailey, Transp. Res. **11**, 221–228 (1977)
109. W. Brilon, *Intersections Without Traffic Signals II* (Springer, Berlin, 1991)
110. W. Brilon, Transp. Res. Rec. **2483**, 57–65 (2015)
111. W. Brilon, J. Geistefeldt, M. Regler, in *Traffic and Transportation Theory*, ed. by H.S. Mahmassani (Elsevier Science, Amsterdam, 2005), pp. 125–144
112. W. Brilon, J. Geistefeld, H. Zurlinden, Transp. Res. Rec. **2027**, 1–8 (2007)
113. W. Brilon, M. Grossmann, in *Schriftenreihe Forschung Straßenbau und Straßenverkehrstechnik*, Heft 596 (Bonn, Germany, 1991)
114. W. Brilon, M. Regler, J. Geistefeld, Straßenverkehrstechnik, Heft **3**, 136 (2005)
115. W. Brilon, B. Stuwe, Transp. Res. Rec. **1398**, 61–67 (1993)
116. W. Brilon, N. Wu, Traffic Eng. Contr. **31**, 623 (1990)
117. W. Brilon, N. Wu, in *Traffic and Mobility*, ed. by W. Brilon, F. Huber, M. Schreckenberg, H. Wallentowitz (Springer, Berlin 1999), pp.163
118. W. Brilon, H. Zurlinden, Straßenverkehrstechnik, Heft **4**, 164 (2004)
119. E. Brockfeld, R.D. Kühne, A. Skabardonis, P. Wagner, Trans. Res. Rec. **1852**, 124–129 (2003)
120. E. Brockfeld, R.D. Kühne, P. Wagner, in *Proceeding of the Transportation Research Board 84th Annual Meeting*, TRB Paper No. 05-2152 (TRB, Washington, DC, 2005)
121. D.J. Buckley, Transp. Sci. **2**, 107–133 (1968)
122. J.M. Bunker, R.J. Troutbeck, Transp. Res. Rec. **1510**, 19–25 (1995)
123. A.P. Buslaev, A.V. Gasnikov, M.V. Yashina, Int. J. Comput. Math. **89**, 409–432 (2012)

124. S.C. Calvert, H. Taale, S.P. Hoogendoorn, J. Adv. Transp. **50**, 570–588 (2016)
125. J.L. Cao, Z.K. Shi, Int. J. Mod. Phys. C **26**, 1550121 (2015)
126. M. Carey, P. Humphreys, M. McHugh, R. McIvor, Transp. Res. B **65**, 90–104 (2014)
127. R.C. Carlson, I. Papamichail, M, Papageorgiou, Transp. Res. C **46**, 209–221 (2014)
128. R.C. Carlson, I. Papamichail, M. Papageorgiou, A. Messmer, Transp. Sci. **44**, 238–253 (2010)
129. R.C. Carlson, I. Papamichail, M. Papageorgiou, A. Messmer, Transp. Res. C **18**, 193–212 (2010)
130. R.C. Carlson, I. Papamichail, M. Papageorgiou, IEEE Trans. Intell. Transp. Syst. **12**, 1261–1276 (2011)
131. R.C. Carlson, I. Papamichail, M. Papageorgiou, J. Intell. Transp. Syst. **17**, 268–281 (2013)
132. M. Carey, P. Humphreys, M. McHugh, R. McIvor, Transp. Res. B **65**, 90–104 (2014)
133. M. Carey, D. Watling, Transp. Res. B. **46**, 634–648 (2012)
134. G. Carofiglio, V. Gehlen, D. Perino, in *2011 IEEE Intern. Conf. on Communications (ICC)*, pp. 1–6 (2011)
135. E. Cascetta, *Transportation Systems Engineering: Theory and Methods* (Kluwer Academic, Dordrecht, 2001)
136. E. Cascetta, M. Gallo, B. Montella, Ann. Oper. Res. **144**, 301–328 (2006)
137. M.J. Cassidy, in *2001 IEEE Inter. Transp. Sys. Proc.* (IEEE, Oakland, USA, 2001), pp. 513–535
138. M.J. Cassidy, S. Ahn, Transp. Res. Rec. **1934**, 140–147 (2005)
139. M.J. Cassidy, R.L. Bertini, Transp. Res. B **33**, 25–42 (1999)
140. M.J. Cassidy, B. Coifman, Transp. Res. Rec. **1591**, 1–6 (1997)
141. M.J. Cassidy, A. Skabardonis (eds.), *Papers selected for the 19th Int. Sym. on Transp. and Traffic Theory*. Procedia Soc. Behav. Sci. **17**, 1–716 (2011)
142. M.J. Cassidy, A. Skabardonis, A.D. May, Transp. Res. Rec. **1225**, 61–72 (1989)
143. M.J. Cassidy, J.R. Windover, Transp. Res. Rec. **1484**, 73–79 (1995)
144. A.G. Castro, A. Monzon, in *Proc. of the 13th International Conference on Reliability and Statistics in Transportation and Communication (RelStat13)*, Riga, Latvia, October 2013, pp. 117–127
145. A. Ceder, *Investigation of Two-Regime Traffic Flow Models at the Micro- and Macroscopic Levels*, (Ph.D. Thesis, UC, Berkeley University, Berkeley, CA, 1975)
146. A. Ceder (ed.), *Transportation and Traffic Theory* (Elsevier Science, Oxford, 1999)
147. A. Ceder, A.D. May, Transp. Res. Rec. **567**, 1–15 (1976)
148. H. Ceylan, M.G.H. Bell, J. Adv. Transp. **38**, 291–321 (2004)
149. H. Ceylan, M.G.H. Bell, Transp. Res. B **39**, 169–185 (2005)
150. P. Chakroborty, S. Kikuchi, Transp. Res. C **7**, 209–235 (1999)
151. E. Chamberlayne, H. Rakha, D. Bish, Transp. Lett. **4**, 227–242 (2012)
152. R.E. Chandler, R. Herman, E.W. Montroll, Oper. Res. **6**, 165–184 (1958)
153. D. Chen, S. Ahn, A. Hegyi, Transp. Res. B **70**, 340–358 (2014)
154. L.-W. Chen, T.-Y. Hu, IEEE Trans. ITS **13**, 1266–1276 (2012)
155. B.Y. Chen, W.H.K. Lam, Q. Li, J. Adv. Transp. **50**, 1413–1431 (2016)
156. X. Chen, L. Li, Z. Li, IEEE Trans. ITS **13**, 1705–1716 (2012)
157. J. Chen, Z. Shi, Y. Hu, in *Computational Science ICCS 2007: 7th International Conference*, ed. by Y. Shi, G.D. van Albada, J. Dongarra, P.M.A. Sloot. Part I. (Springer, Berlin, Heidelberg, 2007), pp. 17–24
158. B. Chen, W. Tong, W. Zhang, X. Sun, B. Wang, EPL **97**, 14001 (2012)
159. B. Chen, W. Tong, W. Zhang, X. Sun, B. Wang, Comput. Phys. Comm. **183**, 2081–2088 (2012)
160. Q. Chen, Y. Wang, J. Adv. Transp. **50**, 949–966 (2016)
161. X. Chen, C. Xiong, X. He, Z. Zhu, L. Zhang, Transp. Res. C **63**, 71–95 (2016)
162. H.C. Chin, A.D. May, Transp. Res. Rec. **1320**, 75–82 (1991)
163. Y.-C. Chiu, H.S. Mahmassani, Transp. Res. Rec. **1783**, 89–97 (2002)
164. A.H.F. Chow, in [16] (2007), pp. 301–326

165. A.H.F. Chow, X.-Y. Lu, T.Z. Qiu, in *IFAC Proceedings Volumes*, vol. 42, pp. 472–477 (2009)
166. D. Chowdhury, L. Santen, A. Schadschneider, Phys. Rep. **329**, 199 (2000)
167. B.-D. Chung, H.-J. Cho, T.L. Friesz, H. Huang, T. Yao, Netw. Spat. Econ. **14**, 183–207 (2014)
168. B.D. Chung, T. Yao, B. Zhang, Netw. Spat. Econ. **12**, 167–181 (2012)
169. F. Ciari, N. Schuessler, K.W. Axhausen, Int. J. Sustainable Transp. **7**, 70–84 (2013)
170. E. Cipriani, S. Gori, M. Petrelli, Transp. Res. C **20**, 3–14 (2012)
171. R.M. Colombo, SIAM J. Appl. Math. **63**, 708–721 (2002)
172. K.G. Courage, Highway Res. Rec. **279**, 118–120 (1969)
173. M. Cremer, *Der Verkehrsfluss auf Schnellstrassen* (Springer, Berlin, 1979)
174. M. Cremer, J. Ludwig, Math. Comput. Simul. **28**, 297–303 (1986)
175. M. Cremer, F. Meissner, in *Modeling and Simulation*, ed. by A. Pave (Ghent, Soc. Comput. Simul. Int., 1993), pp. 515–519
176. M. Cremer, M. Papageorgiou, Automatica **17**, 837–843 (1981)
177. W.B. Cronje, Transp. Res. Rec. **905**, 93–95 (1983)
178. L. D'Acierno, M. Gallo, B. Montella, Eur. J. Oper. Res. **217**, 459–469 (2012)
179. C.F. Daganzo, Transp. Res. **11**, 385–389 (1977)
180. C.F. Daganzo, Transp. Res. **11**, 433–437 (1977)
181. C.F. Daganzo, Transp. Res. **11**, 439–441 (1977)
182. C.F. Daganzo, Transp. Res. B **14**, 229–239 (1980)
183. C.F. Daganzo, Transp. Res. B **28**, 269–287 (1994)
184. C.F. Daganzo, Transp. Res. B **29**, 79–93 (1995)
185. C.F. Daganzo, Transp. Res. B **29**, 261–276 (1995)
186. C.F. Daganzo, Transp. Res. B **31**, 83–102 (1997)
187. C.F. Daganzo, *Fundamentals of Transportation and Traffic Operations* (Elsevier Science, New York, 1997)
188. C.F. Daganzo, Transp. Sci. **32**, 3–11 (1998)
189. C.F. Daganzo, in [146], pp. 81–104
190. C.F. Daganzo, Transp. Res. B **36**, 131–158 (2002)
191. C.F. Daganzo, Transp. Res. B **36**, 159–169 (2002)
192. C.F. Daganzo, Transp. Res. B **36**, 665–668 (2002)
193. C.F. Daganzo, Transp. Res. B **39**, 934–950 (2005)
194. C.F. Daganzo, Transp. Res. B **39**, 187–196 (2005)
195. C.F. Daganzo, Transp. Res. B **40**, 396–403 (2006)
196. C.F. Daganzo, Transp. Res. B **41**, 49–62 (2007)
197. C.F. Daganzo, Transp. Res. B **45**, 782–788 (2011)
198. C.F. Daganzo, Transp. Res. B **55**, 98–117 (2013)
199. C.F. Daganzo, Transp. Res. B **69**, 50–59 (2014)
200. C.F. Daganzo, M.J. Cassidy, Transp. Res. B **42**, 861–872 (2008)
201. C.F. Daganzo, M.J. Cassidy, R.L. Bertini, Transp. Res. A **33**, 365–379 (1999)
202. C.F. Daganzo, N. Geroliminis, Transp. Res. B **42**, 771–781 (2008)
203. C.F. Daganzo, J.A. Laval, *Report UCB-ITS-RR-93-7* (Institute of Transportation Studies, University of California, Berkeley, 2003)
204. C.F. Daganzo, J.A. Laval, Transp. Res. B **39**, 31–46 (2005)
205. C.F. Daganzo, L.J. Lehe, Transp. Res. B **90**, 56–69 (2016)
206. C.F. Daganzo, W.-H. Lin, J.M. Del Castillo, Transp. Res. B **31**, 103–125 (1997)
207. C.F. Daganzo, Y. Sheffi, Transp. Sci. **11**, 253–274 (1977)
208. R. Damrath, M. Rose, in *Forschung Straßenbau und Straßenverkehrstechnik*, Heft 854 (Bundesministerium für Verkehr, Bau- und Wohnungswesen, Bonn, 2002)
209. A. Daou, Oper. Res. **12**, 360–361 (1964)
210. J.N. Darroch, Ann. Math. Stat. **35**, 380–388 (1964)
211. L.C. Davis, Phys. Rev. E **66**, 038101 (2002)
212. L.C. Davis, Physica A **319**, 557–567 (2003)
213. L.C. Davis, Phys. Rev. E **69**, 016108 (2004)

214. L.C. Davis, Phys. Rev. E. **69**, 066110 (2004)
215. L.C. Davis, Physica A **368**, 541–550 (2006)
216. L.C. Davis, Physica A **379**, 274–290 (2007)
217. L.C. Davis, Physica A **361**, 606–618 (2006)
218. L.C. Davis, Physica A **387**, 6395–6410 (2008)
219. L.C. Davis, Physica A **388**, 4459–4474 (2009)
220. L.C. Davis, Physica A **389**, 3588–3599 (2010)
221. L.C. Davis, Physica A **390**, 943–950 (2011)
222. L.C. Davis, Physica A **390**, 3624 (2011)
223. L.C. Davis, Physica A **391**, 1679 (2012)
224. L.C. Davis, Phys. Lett. A **376**, 2658–2662 (2012)
225. L.C. Davis, Physica A **392**, 3755–3764 (2013)
226. L.C. Davis, Physica A **392**, 3798–3805 (2013)
227. L.C. Davis, Physica A **395**, 580 (2014)
228. L.C. Davis, Physica A **405**, 128–139 (2014)
229. L.C. Davis, Physica A **451**, 320–332 (2016)
230. A.I. Delis, I.K. Nikolos, M. Papageorgiou, Comput. Math. Appl. **70**, 1921–1947 (2015)
231. D.S. Dendrinos, Transp. Res. **12**, 191–194 (1978)
232. T. Dijker, P.H.L. Bovy, R.G.M.M. Vermijs, Transp. Res. Rec. **1644**, 20–28 (1998)
233. J. White, S. Garside, J. Whitaker, in *World Conference on Transportation Research* (Antwerp, Belgium, 1998)
234. F. Dion, H. Rakha, Y.S. Kang, Transp. Res. B **38**, 99–122 (2004)
235. K. Doan, S.V. Ukkusuri, Transp. Res. B **46**, 1218–1238 (2012)
236. K. Doan, S.V Ukkusuri, Transport. Res. C **51**, 41–65 (2015)
237. C. Dong, X. Ma, B. Wang, Phys. Lett. A **374**, 1326–1331 (2010)
238. C. Dong, X. Ma, B. Wang, Sci. China-Infor. Sci. **53**, 2265–2271 (2010)
239. C. Dong, X. Ma, G. Wang, B. Wang, X. Sun, Physica A **389**, 3274–3281 (2010)
240. J. Dong, H. Mahmassani, Transp. Res. Rec. **2124**, 203–212 (2009)
241. J. Drake, J. Shofer, A.D. May, Highway Res. Rec. **154**, 53–87 (1967)
242. D.R. Drew, Highway Res. Rec. **99**, 47–58 (1965)
243. D.R. Drew, *Traffic Flow Theory and Control* (McGraw Hill, New York, 1968)
244. D.R. Drew, C.J. Keese, Highway Res. Rec. **99**, 1–47 (1965)
245. L. Du, S. Gong, L. Wang, X.-Y. Li, Transp. Res. C **73**, 30–48 (2016)
246. J. Du, H. Rakha, V.V. Gayah, Transp. Res. C **66**, 136–149 (2016)
247. R.E. Dudash, A.G.R. Bullen, Transp. Res. Rec. **905**, 115–117 (1983)
248. C.L. Dudek, C.J. Messer, Transp. Res. Rec. **469**, 1–15 (1973)
249. C.L. Dudek, S.H. Richards, Transp. Res. Rec. **869**, 14–18 (1982)
250. S.M. Easa, A.D. May, Transp. Res. Rec. **772**, 24–37 (1980)
251. L.C. Edie, Oper. Res. **2**, 107–138 (1954)
252. L.C. Edie, Oper. Res. **9**, 66–77 (1961)
253. L.C. Edie, E. Baverez, in *Third Int. Sym. on the Theory of Traffic Flow*, 1995, pp. 26–37
254. L.C. Edie, R.S. Foote, in *Highway Research Board Proceedings* (HRB, National Research Council, Washington, D.C., 1958), vol. 37, pp. 334–344
255. L.C. Edie, R.S. Foote, in *Highway Research Board Proceedings* (HRB, National Research Council, Washington, D.C., 1960), vol. 39, pp. 492–505
256. L.C. Edie, P. Herman, T.N. Lam, Transp. Sci. **14**, 55–76 (1980)
257. B. Eisenblätter, L. Santen, A. Schadschneider, M. Schreckenberg, Phys. Rev. E **57**, 1309 (1998)
258. L. Elefteriadou, *An Introduction to Traffic Flow Theory*, Springer Optimization and Its Applications, vol. 84 (Springer, Berlin, 2014)
259. L. Elefteriadou, A. Kondyli, W. Brilon, F.L. Hall, B. Persaud, S. Washburn, J. Transp. Eng. **140**, 04014003 (2014)
260. L. Elefteriadou, R.P. Roess, W.R. McShane, Transp. Res. Rec. **1484**, 80–89 (1995)
261. J. Esser, M. Schreckenberg, Int. J. Mod. Phys. C **8**, 1025–1036 (1997)

262. K. Fadhloun, H. Rakha, A. Loulizi, Transp. Lett. **6**, 185–196 (2014)
263. K. Fadhloun, H. Rakha, A. Loulizi, Transp. Res. Rec. **2422**, 61–70 (2014)
264. M. Fellendorf, in *Traffic Technology International'96* (UK & International, Dorking, England, 1996)
265. M. Fellendorf, P. Vortisch, in *Proc. of 80th TRB Meeting, Washington, 2001* (TRB, Washington, D.C., 2001)
266. A. Flannery, T. Datta, Transp. Res. Rec. **1572**, 68–75 (1997)
267. M. Florian, Transpn. Sci. **11**, 166–179 (1977)
268. G. Forbes, F.L. Hall, Transp. Res. A **24**, 338 (1990)
269. T.W. Forbes, J.J. Mullin, M.E. Simpson, in *Proc. 3rd International Symposium on the Theory of Traffic Flow*, ed. by L.C. Edie (Elsevier, New York, 1967), pp. 97–108
270. T.W. Forbes, M.E. Simpson, Transp. Sci. **2**, 77–104 (1968)
271. J. Foster, in *Procs. 1st Conference of the Australian Road Research Board*, (1962), pp. 229–257
272. B. Friedrich, in *Proc. of the VIIIth Int. Conf. of the European Working Group on Transportation*, (Rome, 2000)
273. B. Friedrich, *Schriftenreihe des Fachgebiets Verkehrsplanung und Verkehrstechnik der TU München* (TU München, 2000)
274. B. Friedrich, in *Autonomes Fahren*, ed. by M. Maurer, J.Ch. Gerdes, B. Lenz, H. Winner (Springer, Berlin, 2015), pp. 331–350
275. T.L. Friesz, Transp. Res. A **19**, 413–427 (1985)
276. T.L. Friesz, D. Bernstein, *Foundations of Network Optimization and Games*, Complex Networks and Dynamic Systems, vol. 3 (Springer, New York, Berlin, 2016)
277. T.L. Friesz, D. Bernstein, N.J. Mehta, R.L. Tobin, S. Ganjalizadeh, Oper. Res. **42**, 1120–1136 (1994)
278. T.L. Friesz, D. Bernstein, T.E. Smith, R.L. Tobin, B.W. Wie, Oper. Res. **41**, 179–191 (1993)
279. T.L. Friesz, D. Bernstein, T.E. Smith R.L. Tobin, B.W. Wie, Oper. Res. **41**, 179–191 (1993)
280. T.L. Friesz, H.-J. Cho, N.J. Mehta, R.L. Tobin, G. Anandalingam, Transp. Sci. **26**, 18–26 (1992)
281. T.L. Friesz, K. Han, P.A. Neto, A. Meimand, T. Yao, Transp. Res. B **47**, 102–126 (2013)
282. T.L. Friesz, J. Luque, R.L. Tobin, B.-W. Wie, Oper. Res. **37**, 893–901 (1989)
283. T.L. Friesz, R.L. Tobin, T.E. Smith, P.T. Harker, J. Regional Sci. **23**, 337–359 (1983)
284. H.-T. Fritzsche, Transp. Eng. Contrib. **5**, 317 (1994)
285. M. Fukui, Y. Ishibashi, J. Phys. Soc. Jpn. **62**, 3841–3844 (1993)
286. M. Fukui, Y. Ishibashi, J. Phys. Soc. Jpn. **63**, 2882 (1994)
287. M. Fukui, Y. Ishibashi, J. Phys. Soc. Jpn. **65**, 1868–1870 (1996)
288. M. Fukui, Y. Ishibashi, J. Phys. Soc. Jpn. **66**, 385–387 (1997)
289. M. Fukui, Y. Ishibashi, J. Phys. Soc. Jpn. **70**, 2793–2797 (2001)
290. M. Fukui, Y. Sugiyama, M. Schreckenberg, D.E. Wolf (eds.), *Traffic and Granular Flow'01* (Springer, Heidelberg, 2003)
291. K. Gao, R. Jiang, S.-X. Hu, B.-H. Wang, Q.-S. Wu, Phys. Rev. E **76**, 026105 (2007)
292. K. Gao, R. Jiang, B.-H. Wang, Q.-S. Wu, Physica A **388**, 3233–3243 (2009)
293. Z.-Y. Gao, K.-P. Li, X.-G. Li, H.-J. Huang, B.-H. Mao, J.-F. Zheng, Physica A **380**, 577–584 (2007)
294. N.H. Gartner, Highway Res. Rec. **445**, 12–23 (1973)
295. N.H. Gartner, Transp. Sci. **14**, 192–208 (1980)
296. N.H. Gartner, Transp. Sci. **14**, 174–191 (1980)
297. N.H. Gartner, Transp. Res. Rec. **906**, 75–81 (1983)
298. N.H. Gartner, Transp. Res. A **19**, 369–373 (1985)
299. N.H. Gartner, M. Al-Malik, Transp. Res. Rec. **1554**, 27–35 (1996)
300. N.H. Gartner, S.F. Assmann, F. Lasaga, D.L. Hou, Transp. Res. Rec. **1287**, 212–222 (1990)
301. N.H. Gartner, S.F. Assmann, F. Lasaga, D.L. Hou, Transp. Res. B **25**, 55–74 (1991)
302. N.H. Gartner, J.D.C. Little, Transp. Res. Rec. **531**, 58–69 (1975)
303. N.H. Gartner, J.D.C. Little, H. Gabbay, Transp. Sci. **9**, 321–343 (1975)

304. N.H. Gartner, J.D.C. Little, H. Gabbay, Transp. Res. Rec. **596**, 6–15 (1976)
305. N.H. Gartner, C.J. Messer, A.K. Rathi (eds.), *Traffic Flow Theory: A State-of-the-Art Report* (Transportation Research Board, Washington DC, 2001)
306. N.H. Gartner, F.J. Pooran, C.M. Andrews, in *2001 IEEE Intelligent transportation systems proceedings, Oakland* (California, 2001)
307. N.H. Gartner, F.J. Pooran, C.M. Andrews, Transp. Res. Rec. **1811**, 148–156 (2002)
308. N.H. Gartner, Ch. Stamatiadis, Transp. Res. Rec. **1644**, 150–156 (1998)
309. N.H. Gartner, Ch. Stamatiadis, Math. Comput. Model. **35**, 657–671 (2002)
310. N.H. Gartner, Ch. Stamatiadis, J. Intell. Transp. Sys. **8**, 77–86 (2004)
311. N.H. Gartner, Ch. Stamatiadis, in *Encyclopedia of Complexity and System Science*, ed. by R.A. Meyers (Springer, Berlin, 2009), pp. 9470–9500
312. N.H. Gartner, Ch. Stamatiadis, P.J. Tarnoff, Transp. Res. Rec. **1494**, 98–105 (1995)
313. N.H. Gartner, P.J. Tarnoff, C.M. Andrews, Transp. Res. Rec. **1324**, 105–114 (1991)
314. N.H. Gartner, N.H.M. Wilson (eds.), *Transportation and Traffic Theory* (Elsevier, New York, 1987)
315. A.V. Gasnikov, S.L. Klenov, E.A. Nurminski, Y.A. Kholodov, N.B. Shamray, *Introduction to Mathematical Simulations of Traffic Flow* (MCNMO, Moscow, 2013) (in Russian)
316. V.V. Gayah, C. Daganzo, Transp. Res. B **45**, 643–655 (2011)
317. D.C. Gazis, *Traffic Science* (Wiley, New York, 1974)
318. D.C. Gazis, *Traffic Theory* (Springer, Berlin, 2002)
319. D.C. Gazis, R.S. Foote, Transp. Sci. **3**, 255–275 (1969)
320. D.C. Gazis, R. Herman, Trans. Sci. **26**, 223 (1992)
321. D.C. Gazis, R. Herman, R.B. Potts, Oper. Res. **7**, 499–505 (1959)
322. D.C. Gazis, R. Herman, R.W. Rothery, Oper. Res. **9**, 545–567 (1961)
323. D.C. Gazis, R. Herman, G.H. Weiss, Oper. Res. **10**, 658–667 (1962)
324. H.-X. Ge, R.J. Cheng, Physica A **387**, 6952–6958 (2008)
325. H.-X. Ge, X.-p. Meng, R.-j. Cheng, S.-M. Lo, Physica A **390**, 3348–3353 (2011)
326. Y.E. Ge, B.R. Sun, H.M. Zhang, W.Y. Szeto, X. Zhou, Netw. Spat. Econ. **15**, 583–598 (2015)
327. J. Geistefeldt, W. Brilon, in *Transportation and Traffic Theory 2009*, ed. by W.H.K. Lam, S.C. Wong, H.K. Lo (Springer, Dordrecht, Heidelberg, London, New York, 2009), pp. 583–602
328. G. Gentile, Transp. Res. B **92**, 120–147 (2016)
329. D.L. Gerlough, M.J. Huber, *Traffic Flow Theory Special Report 165* (Transp. Res. Board, Washington D.C., 1975)
330. N. Geroliminis, C.F. Daganzo, Transp. Res. B **42**, 759–770 (2008)
331. N. Geroliminis, J. Haddad, M. Ramezani, IEEE Trans. ITS **14**, 348–359 (2013)
332. P.G. Gipps, Trans. Res. B **15**, 105–111 (1981)
333. P.G. Gipps, Trans. Res. B. **20**, 403–414 (1986)
334. C.D. van Goeverden, H. Botma, P.H.L. Bovy, Transp. Res. Rec. **1646**, 1–8 (1998)
335. G. Gomes, R. Horowitz, Transp. Res. C **14**, 244–262 (2006)
336. G. Gomes, A. May, R. Horowitz, Transp. Res. Rec. **1876**, 71–81 (2004)
337. E.J. Gonzales, C.F. Daganzo, Transp. Res. B. **46**, 1519–1534 (2012)
338. K.G. Goulias, in *Encyclopedia of Complexity and System Science*, ed. by R.A. Meyers (Springer, Berlin, 2009), pp. 9536–9565
339. R.B. Grafton, G.F. Newell, in *Vehicular Traffic Science* ed. by L C. Edie, R. Herman, R. Rothery (Elsevier, New York, 1967), pp. 239–257
340. H. Greenberg, A. Daou, Oper. Res. **8**, 524–532 (1960)
341. J. Greenberg, A. Klar, M. Rascle, SIAM J. Appl. Math. **63**, 818–833 (2003)
342. B.D. Greenshields, in *Highway Research Board Proceedings*, vol. 14, pp. 448–477 (1935)
343. B.D. Greenshields, D. Schapiro, E.L. Ericksen, *Technical Report No.1* (Yale Bureau of Highway Traffic, Yale University, 1947)
344. M. Guériau, R. Billot, N.-E. El Faouzi, J. Monteil, F. Armetta, S. Hassas, Transp. Res. C **67**, 266–279 (2016)
345. J.P.T. van der Gun, A.J. Pel, B. van Arem, Transp. Res. B **98**, 154–178 (2017)

346. H. Guo, Z. Cao, J. Zhang, D. Niyato, U. Fastenrath, in *17th Inter. IEEE Conference on ITS (ITSC)* (Qingdao, China, 2014), pp. 2180–2187
347. A.K. Gupta, V.K. Katiyar, Physica A **371**, 674–682 (2006)
348. A.K. Gupta, P. Redhu, Physica A **392**, 5622–5632 (2013)
349. A.K. Gupta, P. Redhu, Nonlinear Dyn. **76**, 1001–1011 (2014)
350. A.K. Gupta, S. Sharma, P. Redhu, Nonlinear Dyn. **80**, 1091–1108 (2015)
351. C. Habib-Mattar, A. Polus, M. Cohen, Transp. Res. Rec. **2124**, 58–66 (2009)
352. J. Haddad, M., Ramezani, N. Geroliminis, in *Proc. of the 91th TRB Annual Meeting* (TRB, Washington, D.C, USA, 2012)
353. M. Haghani, Z. Shahhoseini, M. Sarvi, J. Traffic Transp. Eng. (English Edition) **3**, 181–191 (2016)
354. F.A. Haight, Biometrika **46**, 420–424 (1959)
355. F.A. Haight, *Mathematical Theories of Traffic Flow* (Academic Press, New York, 1963)
356. F.L. Hall, Trans. Res. A **21**, 191–201 (1987)
357. F.L. Hall, in [305], pp. 2-1–2-36
358. F.L. Hall, K. Agyemang-Duah, Trans. Res. Rec. **1320**, 91–98 (1991)
359. F.L. Hall, B.L. Allen, M.A. Gunter, Trans. Res. A **20**, 197–210 (1986)
360. F.L. Hall, D. Barrow, Transp. Res. Rec. **1194**, 55–65 (1988)
361. F.L. Hall, W. Brilon, Transp. Res. Rec. **1457**, 35–42 (1994)
362. F.L. Hall, M.A. Gunter, Trans. Res. Rec. **1091**, 1–9 (1986)
363. F.L. Hall, L.M. Hall, Transp. Res. Rec. **1287**, 108–118 (1990)
364. F.L. Hall, V.F. Hurdle, J.H. Banks, Transp. Res. Rec. **1365**, 12–18 (1992)
365. F.L. Hall, B.N. Persaud, Transp. Res. Rec. **1232**, 9–16 (1989)
366. F.L. Hall, A. Pushkar, Y. Shi, Transp. Res. Rec. **1398**, 24–30 (1993)
367. S.H. Hamdar, H.S. Mahmassani, Transp. Res. Rec. **2088**, 45–56 (2008)
368. S.H. Hamdar, H.S. Mahmassani, Transp. Res. Rec. **2124**, 89–102 (2009)
369. S.H. Hamdar, H.S. Mahmassani, R.B. Chen, Accident Anal. Prevent. **40**, 315–326 (2008)
370. S.H. Hamdar, H.S. Mahmassani, M. Treiber, Transp. Res. B **78**, 32–53 (2015)
371. S.H. Hamdar, J.P. Schorr, Accident Anal. Prevent. **55**, 22–33 (2013)
372. S.H. Hamdar, A. Talebpour, J. Dong, IEEE Trans. ITS **16**, 264–273 (2014)
373. S.H. Hamdar, M. Treiber, H.S. Mahmassani, Transp. Res. Rec. **2088**, 208–217 (2008)
374. Y. Han, D. Chen, S. Ahn, Transp. Res. B **98**, 113–134 (2017)
375. K. Han, T.L. Friesz, T. Yao, Transp. Res. B **53**, 17–30 (2013)
376. J. Harders, *Schriftenreihe Straßenbau und Straßenverkehrstechnik*, Heft 76 (Bonn, Germany, 1968)
377. C. Hauck, Y. Sun, I. Timofeyev, Stoch. Dyn. **14**, 1350022 (2014)
378. K. Hausken, H. Rehborn, in *Game Theoretic Analysis of Congestion, Safety and Security*, Springer Series in Reliability Engineering (Springer, Berlin, 2015), pp. 113–141
379. J.J. Haynes, Highway Res. Rec. **99**, 59–80 (1965)
380. S. He, W. Guan, L. Song, Physica A **389**, 825–836 (2009)
381. A. Hegyi, T. Bellemans, B. De Schutter, in *Encyclopedia of Complexity and System Science*, ed. by R.A. Meyers (Springer, Berlin, 2009), pp. 3943–3964
382. A. Hegyi, B. De Schutter, J. Hellendoorn, Transp. Res. Rec. **1852**, 167–174 (2004)
383. A. Hegyi, B. De Schutter, J. Hellendoorn, IEEE Trans. Intell. Transp. Syst. **6**, 102–112 (2005)
384. A. Hegyi, B. De Schutter, H. Hellendoorn, Transp. Res. C **13**, 185–209 (2005)
385. D. Heidemann, Transp. Res. B **25**, 163–174 (1991)
386. D. Helbing, Phys. Rev. E **51**, 3164–3169 (1995)
387. D. Helbing, Phys. Rev. E **53**, 2366–2381 (1996)
388. D. Helbing, Physica A **233**, 253–282 (1996)
389. D. Helbing, *Verkehrsdynamik* (Springer, Berlin, 1997)
390. D. Helbing, Phys. Rev. E **55**, R25–R28 (1997)
391. D. Helbing, Phys. Rev. E **55**, 3735–3738 (1997)
392. D. Helbing, Physica A **242**, 175–194 (1997)

393. D. Helbing, Phys. Rev. E **57**, 6176–6179 (1998)
394. D. Helbing, Rev. Mod. Phys. **73**, 1067–1141 (2001)
395. D. Helbing, J. Phys. A Math. Gen. **36**, L593–L598 (2003)
396. D. Helbing, "Traffic flow", in *Encyclopedia of Nonlinear Science*, ed. by A. Scott (Routledge, New York, 2005)
397. D. Helbing, Eur. Phys. J. B **69**, 539–548 (2009)
398. D. Helbing, Eur. Phys. J. B **69**, 569–570 (2009)
399. D. Helbing, Eur. Phys. J. B **70**, 257–274 (2009)
400. D. Helbing, A. Greiner, Phys. Rev. E **55**, 5498–5507 (1997)
401. D. Helbing, A. Hennecke, M. Treiber, Phys. Rev. Lett. **82**, 4360–4363 (1999)
402. D. Helbing, A. Hennecke, V. Shvetsov, M. Treiber, Transp. Res. B **35**, 183–211 (2001)
403. D. Helbing, A. Hennecke, V. Shvetsov, M. Treiber, Math. Comput. Mod. **35**, 517–547 (2002)
404. D. Helbing, H.J. Herrmann, M. Schreckenberg, D.E. Wolf (eds.), *Traffic and Granular Flow'99* (Springer, Heidelberg, 2000)
405. D. Helbing, K. Nagel, Contemp. Phys. **45**, 405–426 (2004)
406. D. Helbing, M. Schreckenberg, Phys. Rev. E **59**, R2505 (1999)
407. D. Helbing B. Tilch, Phys. Rev. E **58**, 133–138 (1998)
408. D. Helbing, B. Tilch, Eur. Phys. J. B **68**, 577–586 (2009)
409. D. Helbing, M. Treiber, Phys. Rev. Lett. **81**, 3042–3045 (1998)
410. D. Helbing, M. Treiber, Science **282**, 2001–2003 (1998)
411. D. Helbing, M. Treiber, A. Kesting, Physica A **363**, 62–72 (2006)
412. D. Helbing, M. Treiber, A. Kesting, M. Schönhof, Eur. Phys. J. B. **69**, 583–598 (2009)
413. P. Hemmerle, M. Koller, H. Rehborn, B.S. Kerner, M. Schreckenberg, IET Intell. Transp. Syst. **10**, 122–129 (2016)
414. P. Hemmerle, M. Koller, G. Hermanns, H. Rehborn, B.S. Kerner, M. Schreckenberg, Collective Dyn. **1**, A7:1–27 (2016)
415. V. Henn, in [453], pp. 105–124
416. J. Henry, J. Farges, J. Tuffal, in *4th IFAC-IFIC–IFORS on Control in Transportation Systems* (Baden-Baden, 1983)
417. R. Herman, S. Ardekani, Transp. Sci. **18**, 101–140 (1984)
418. R. Herman, T. Lam, I. Prigogine, Transp. Sci. **6**, 440–451 (1972)
419. R. Herman, T. Lam, I. Prigogine, Science **179**, 918–920 (1973)
420. R. Herman, E.W. Montroll, R.B. Potts, R.W. Rothery, Oper. Res. **7**, 86–106 (1959)
421. M. Herrmann, B.S. Kerner, Physica A **255**, 163–188 (1998)
422. G. Hermanns, P. Hemmerle, H. Rehborn, M. Koller, B.S. Kerner, M. Schreckenberg, Transp. Res. Rec. **2490**, 47–55 (2015)
423. M. Herty, A. Klar, SIAM J. Sci. Comput. **25**, 1066–1087 (2003)
424. J.W. Hess, Highway Res. Rec. **27**, 69–115 (1963)
425. J.W. Hess, Highway Res. Rec. **99**, 81–116 (1965)
426. M.P. Heyes, R. Ashworth, Transp. Res. **6**, 287–291 (1972)
427. B.G. Heydecker, J.D. Addison, Transp. Sci. **39**, 39–57 (2005)
428. B.G. Heydecker, J.D. Addison, in *Transportation Networks: Recent Methodological Advances. Selected Proceedings of the 4th EURO Transportation Meeting*, ed. by M.G.H. Bell (Elsevier, Oxford, 1999), pp. 35–49
429. Highway Capacity Manual 2000 (National Research Council, Transportation Research Boad, Washington, D.C., 2000)
430. Highway Capacity Manual 2010 (National Research Council, Transportation Research Boad, Washington, D.C., 2010)
431. B.D. Hillegas, D.G. Houghton, P.J. Athol, Transp. Res. Rec. **495**, 53–63 (1974)
432. M. Hilliges, W. Weidlich, Transp. Res. B **29**, 407–431 (1995)
433. W. Himpe, R. Corthout, M.J.C. Tampére, Transp. Res. B **92**, 170–190 (2016)
434. Y. Hino, K. Tobita, T. Nagatani, Physica A **392**, 3223–3230 (2013)
435. S.P. Hoogendoorn, *Multiclass Continuum Modelling of Multilane Traffic Flow* (Delft University of Technology, Delft, 1999)

436. S.P. Hoogendoorn, Transp. Res. B. **39**, 709–727 (2005)
437. R. Hoogendoorn, B. van Arem, S.P. Hoogendoorn, Transp. Res. Rec. **2422**, 113–120 (2014)
438. S.P. Hoogendoorn, H. Botma, Transp. Res. Rec. **1591**, 14–22 (1997)
439. S.P. Hoogendoorn, P.H.L. Bovy, Transp. Res. Rec. **1646**, 18–28 (1998)
440. S.P. Hoogendoorn, P.H.L. Bovy, Transp. Res. Rec. **1678**, 150–159 (1999)
441. S.P. Hoogendoorn, P.H.L. Bovy, Transp. Res. B **34**, 123–146 (2000)
442. S.P. Hoogendoorn, P.H.L. Bovy, Netw. Spat. Econ. **1**, 137–166 (2001)
443. S.P. Hoogendoorn, P.H.L. Bovy, in *Proceedings of the Institution of Mechanical Engineers, Part I: Journal of Systems and Control Engineering*, vol. 216, pp. 283–303 (2001)
444. S.P. Hoogendoorn, P.H.L. Bovy, Transp. Res. B **35**, 317–336 (2001)
445. S.P. Hoogendoorn, R. Hoogendoorn, Phil. Trans. Roy. Soc. Lond. Math. Phys. Eng. Sci. **368**, 4497–4517 (2010)
446. S.P. Hoogendoorn, R. Hoogendoorn, K. Brookhuis, W. Daamen, Trans. Res. Rec. **2188**, 64–73 (2010)
447. S.P. Hoogendoorn, R. Hoogendoorn, W. Daamen, Trans. Res. Rec. **2260**, 152–162 (2011)
448. S.P. Hoogendoorn, R. Hoogendoorn, M. Wang, W. Daamen, Trans. Res. Rec. **2316**, 20–30 (2012)
449. S.P. Hoogendoorn, V.L. Knoop, in *Traffic Flow Theory and Modelling*, ed. by B. van Wee, J.A. Annema, D. Banister. (Edward Elgar Publishing, 2012), pp. 125–162
450. S.P. Hoogendoorn, V.L. Knoop, H. van Lint (eds.), *20th International Symposium on Transportation and Traffic Theory (ISTTT 2013)*. Procedia Soc. Behav. Sci. **80**, 1–996 (2013)
451. S.P. Hoogendoorn, V.L. Knoop, H.J. Van Zuylen, J. Adv. Transp. **42**, 357–377 (2008)
452. S. Hoogendoorn, H. van Lint, V.L. Knoop, Trans. Res. Rec. **2088**, 102–108 (2008)
453. S.P. Hoogendoorn, S. Luding, P.H.L. Bovy, M. Schreckenberg, D.E. Wolf (eds.), *Traffic and Granular Flow'03* (Springer, Heidelberg, 2005)
454. S.P. Hoogendoorn, S. Ossen, in [773] (2005), pp. 245–265
455. S.P. Hoogendoorn, S. Ossen, Eur. J Trans. Infrastructure Res. (EJTIR) **6**, 229–246 (2006)
456. S.P. Hoogendoorn, S. Ossen, M. Schreuder, Trans. Res. Rec. **1965**, 112–120 (2006)
457. S.P. Hoogendoorn, S. Ossen, M. Schreuder, in [16] (2007), pp. 583–606
458. S.P. Hoogendoorn, H. Van Zuylen, M. Schreuder, B. Gorte, G. Vosselman, Transp. Res. Rec. **1855**, 121–128 (2003)
459. Z. Hou, X. Xu, J. Yan, J.-X. Xu, G. Xiong, IEEE Trans. ITS **12**, 1305–1318 (2011)
460. Z. Hou, J. Yan, J.-X. Xu, Z. Li, IEEE Trans. ITS **13**, 606–618 (2012)
461. P. Hsu, J.H. Banks, Transp. Res. Rec. **1398**, 17–23 (1993)
462. T.-Y. Hu, C.-C. Tong, T.-Y. Liao, W.-M. Ho, IEEE Trans. ITS **13**, 1277–1286 (2012)
463. Hai-Jun Huang, W.H.K. Lam, Transp. Res. B **36**, 253–273 (2002)
464. H.-J. Huang, Z.-C. Li, Eur. J. Oper. Res. **176**, 1464–1477 (2007)
465. P.B. Hunt, D.I. Robertson, R.D. Bretherton, M.C. Royle, Traffic Eng. Contr. **23**, 190–192 (1982)
466. P.B. Hunt, D.I. Robertson, R.D. Bretherton, R.I. Winton, *TRRL Report No. LR 1014* (Transp. and Road Res. Lab., Crowthorne, UK, 1981)
467. V.F. Hurdle, P.K. Datta, Transp. Res. Rec. **905**, 127–137 (1983)
468. V.F. Hurdle, D.Y. Solomon, Transp. Sci. **20**, 153–163 (1986)
469. N. Huynh, H.S. Mahmassani, H. Tavana, Transp. Res. Rec. **1783**, 55–65 (2002)
470. R. Illner, A. Klar, T. Materne, Comm. Math. Sci. **1**, 1–12 (2003)
471. P.A. Ioannou, C.C. Chien, IEEE Trans. Veh. Technol. **42**, 657–672 (1993)
472. T. Iryo, Transp. Res. B **45**, 867–879 (2011)
473. T. Iryo, Netw. Spat. Econ. **15**, 599–616 (2015)
474. Y. Ishibashi, M. Fukui, J. Phys. Soc. Jpn. **65**, 2793–2795 (1996)
475. S. Ishak, C. Mamidala, Y. Qi, Transp. Res. Rec. **2178**, 79–89 (2010)
476. M. Iwasaki, Transp. Res. Rec. **1320**, 242–250 (1991)
477. B. Jäggi, S. Hohmann, K.W. Axhausen, J. Geistefeldt, Transp. Res. Rec. **2161**, 16–24 (2014)
478. B.N. Janson, Transp. Res. B **25**, 143–161 (1991)
479. B. Janson, C. Zozaya-Gorostiza, Transp. Res. B **21**, 299–310 (1987)

480. R. Jayakrishnan, H.S. Mahmassani, T.-Y. Hu, Transp. Res. C **2**, 129–147 (1994)
481. R. Jayakrishnan, W.K. Tsai, A. Chen, Transpn. Res. C **3**, 51–72 (1995)
482. K. Jerath, S.N. Brennan, IEEE Trans. ITS **13**, 1782–1791 (2012)
483. X. Ji, C. Shao, B. Wang, Procedia Eng. **137**, 620–629 (2016)
484. B. Jia, X.-G. Li, T. Chen, R. Jiang, Z.-Y. Gao, Transportmetrica **7**, 127 (2011)
485. A. Jia, B. Williams, N. Rouphail, Transp. Res. Rec. **2188**, 148–155 (2010)
486. R. Jiang, M.-B. Hu, B. Jia, R. Wang, Z.Y. Gao, Comput. Phys. Comm. **185**, 1439–1442 (2014)
487. R. Jiang, M.-B. Hu, B. Jia, R.L. Wang, Q.-S. Wu, Eur. Phys. J. B **54**, 267–273 (2006)
488. R. Jiang, M.-B. Hu, B. Jia, R. Wang, Q.-S. Wu, Eur. Phys. J. B **58**, 197–206 (2007)
489. R. Jiang, M.-B. Hu, B. Jia, R. Wang, Q.-S. Wu, Transportmetrica **4**, 51–61 (2008)
490. R. Jiang, M.-B. Hu, R. Wang, Q.-S. Wu, Phys. Lett. A **365**, 6–9 (2007)
491. R. Jiang, M.-B. Hu, Y.H. Wu, Q.-S. Wu, Phys. Rev. E **77**, 041128 (2008)
492. R. Jiang, M.-B. Hu, H.M. Zhang, Z.-Y. Gao, B. Jia, Q.-S. Wu, B. Wang, M. Yang, PLOS One **9**, e94351 (2014)
493. R. Jiang, M.B. Hu, H.M. Zhang, Z.Y. Gao, B. Jia, Q.S. Wu, Transp. Res. B **80**, 338–354 (2015)
494. R. Jiang, B. Jia, X.L. Wang, Q.S. Wu, J. Phys. A Math. Gen. **37**, 5777 (2004)
495. R. Jiang, B. Jia, Q.S. Wu, J. Phys. A Math. Gen. **36**, 11713 (2003)
496. R. Jiang, B. Jia, Q.S. Wu, Int. J. Modern Phys. C **15**, 619–628 (2004)
497. R. Jiang, C.-J. Jin, H.M. Zhang, Y.-X. Huang, J.-F. Tian, W. Wang, M.-B. Hu, H. Wang, B. Jia, in *Proc. Int. Symposium on Traffic and Transportation Theory ISTTT22* (2017) (accepted)
498. R. Jiang, X.L. Wang, Q.S. Wu, J. Phys. A Math. Gen. **36**, 4763 (2003)
499. Y. Jiang, S.C. Wong, H.W. Ho, P. Zhang, R. Liu, A. Sumalee, Transp. Res. B **4**, 343–363 (2011)
500. R. Jiang, Q.-S. Wu, J. Phys. A Math. Gen. **36**, 381–390 (2003)
501. R. Jiang, Q.-S. Wu, Phys. Rev. E **68**, 026135 (2003)
502. R. Jiang, Q.-S. Wu, Transp. Res. B **37**, 85–99 (2003)
503. R. Jiang, Q.-S. Wu, J. Phys. A Math. Gen. **37**, 8197–8213 (2004)
504. R. Jiang, Q.-S. Wu, Transp. Res. Rec. **1883**, 78–84 (2004)
505. R. Jiang, Q.-S. Wu, Int. J. Modern Phys. B **18**, 1991–2001 (2004)
506. R. Jiang, Q.-S. Wu, Phys. Rev. E **72**, 067103 (2005)
507. R. Jiang, Q.-S. Wu, Eur. Phys. J. B **46**, 581–584 (2005)
508. R. Jiang, Q.-S. Wu, Physica A **355**, 551–564 (2005)
509. R. Jiang, Q.-S. Wu, Physica A **364**, 493–496 (2006)
510. R. Jiang, Q.-S. Wu, Physica A **368**, 239–246 (2006)
511. R. Jiang, Q.-S. Wu, Phys. Lett. A **359**, 99–102 (2006)
512. R. Jiang, Q.-S. Wu, Physica A **366**, 523–529 (2006)
513. R. Jiang, Q.-S. Wu, Physica A **377**, 633–640 (2007)
514. R. Jiang, Q.-S. Wu, Physica A **375**, 297–306 (2007)
515. R. Jiang, Q.-S. Wu, B. Jia, Physica D **237**, 467–474 (2008)
516. R. Jiang, Q.-S. Wu, B.-H. Wang, Phys. Rev. E **66**, 036104 (2002)
517. R. Jiang, Q.S. Wu, Z.J. Zhu, Phys. Rev. E **64**, 017101 (2001)
518. R. Jiang, Q.-S. Wu, Z.J. Zhu, Chin. Sci. Bull. **46**, 345–348 (2001)
519. R. Jiang, Q.-S. Wu, Z.J. Zhu, Transp. Res. B **36**, 405–419 (2002)
520. W.-L. Jin, Transp. Res. B **46**, 1000–1022 (2012)
521. W.-L. Jin, Transp. Res. B **46**, 1360–1373 (2012)
522. W.-L. Jin, Netw. Spat. Econ. **15**, 617–634 (2015)
523. C.-J. Jin, W. Wang, Physica A **390**, 4184–4191 (2011)
524. S. Jin, D.-H. Wang, Z.-Y. Huang, P.-F. Tao, Physica A **390**, 1931–1940 (2011)
525. C.-J. Jin, W. Wang, R. Jiang, K. Gao, J. Stat. Mech. P03018 (2010)
526. C.-J. Jin, W. Wang, R. Jiang, H.M. Zhang, H. Wang, M.-B. Hu, Transp. Res. C **60**, 324–338 (2015)

527. E.R. Jones, M.E. Goolsby, Highway Res. Rec. **321**, 74–82 (1970)
528. T.R. Jones, R.B. Potts, Oper. Res. **10**, 745–763 (1962)
529. I. Juran, J.N. Prashker, S. Bekhor, I. Ishai, Transp. Res. C. **17**, 240–258 (2009)
530. P. Kachroo, S. Sastry, IEEE Trans. ITS **17**, 1438–1447 (2016)
531. G. Kalafatos, S. Peeta, in [694] (2009), pp. 541–558
532. Y. Kan, Y. Wang, M. Papageorgiou, I. Papamichail, Transp. Res. C **62**, 149–170 (2016)
533. M. Kanai, K. Nishinari, T. Tokihiro, Phys. Rev. E **72**, 035102 (2005)
534. A. Karoonsoontawong, D.-Y. Lin, Netw. Spat. Econ. **15**, 1011–1048 (2015)
535. S. Kaufmann, B.S. Kerner, M. Koller, H. Rehborn, S.L. Klenov, in *Proc. of 96th Transportation Research Board Annual Meeting*, TRB Paper 17-03078 (2017)
536. O. Kaumann, K. Froese, R. Chrobok, J. Wahle, L. Neubert, M. Schreckenberg, in [404] (2000), pp. 351–356
537. J. Kaupužs, R. Mahnke, R.J. Harris Phys. Rev. E **72**, 056125 (2005)
538. B.S. Kerner, in *Transportation Systems 1997*, ed. by M. Papageorgiou, A. Pouliezos (Elsevier Science, London, 1997), pp. 765–770
539. B.S. Kerner, Phys. Rev. Lett. **81**, 3797–3800 (1998)
540. B.S. Kerner, in [1003] (1998), pp. 621–642
541. B.S. Kerner, Transp. Res. Rec. **1678**, 160–167 (1999)
542. B.S. Kerner, in *Transportation and Traffic Theory*, ed. by A. Ceder (Elsevier Science, Amsterdam, 1999), pp. 147–171
543. B.S. Kerner, Phys. World **12**, 25–30 (August 1999)
544. B.S. Kerner, J. Phys. A Math. Gen. **33**, L221–L228 (2000)
545. B.S. Kerner, in *Traffic and Granular Flow '99: Social, Traffic and Granular Dynamics*, ed. by D. Helbing, H.J. Herrmann, M. Schreckenberg, D.E. Wolf (Springer, Heidelberg, Berlin, 2000), pp. 253–284
546. B.S. Kerner, Transp. Res. Rec. **1710**, 136–144 (2000)
547. B.S. Kerner, in *2001 IEEE Intelligent Transp. Systems Conference Proc.* (Oakland, 2001), pp. 88–93
548. B.S. Kerner, Netw. Spat. Econ. **1**, 35–76 (2001)
549. B.S. Kerner, Transp. Res. Rec. **1802**, 145–154 (2002)
550. B.S. Kerner, in *Traffic and Transportation Theory in the 21st Century*, ed. by M.A.P. Taylor (Elsevier Science, Amsterdam, 2002), pp. 417–439
551. B.S. Kerner, Phys. Rev. E **65**, 046138 (2002)
552. B.S. Kerner, Math. Comput. Modell. **35**, 481–508 (2002)
553. B.S. Kerner, in *Traffic and Granular Flow' 01*, ed. by M. Schreckenberg, Y. Sugiyama, D. Wolf (Springer, Berlin, 2003), pp. 13–50
554. B.S. Kerner, in *Proceedings of the 10th World Congress on Intelligent Transport Systems*, Paper No. 2043 T (Madrid, Spain 2003); cond-mat/0309017, e-print in http://arxiv.org/abs/cond-mat/0309017 (2003)
555. B.S. Kerner, Physica A **333**, 379–440 (2004)
556. B.S. Kerner, *The Physics of Traffic* (Springer, Berlin, Heidelberg, New York, 2004)
557. B.S. Kerner, Verfahren zur Ansteuerung eines in einem Fahrzeug befindlichen verkehrsadaptiven Assistenzsystems, German patent publication DE 10308256A1 (2004) (https://google.com/patents/DE10308256A1); Patent WO 2004076223A1 (2004) (https://google.com/patents/WO2004076223A1); EU Patent EP 1597106B1 (2006); German patent DE 502004001669D1 (2006)
558. B.S. Kerner, Physica A **355**, 565–601 (2005)
559. B.S. Kerner, in [773] (2005), pp. 181–203
560. B.S. Kerner, Straßenverkehrstechnik, Nr. **3**, 117–123 (2006)
561. B.S. Kerner, IEEE Trans. ITS **8**, 308–320 (2007)
562. B.S. Kerner, Transp. Res. Rec. **1999**, 30–39 (2007)
563. B.S. Kerner, Traffic Eng. Contr. **48**, 28–35 (2007)
564. B.S. Kerner, Traffic Eng. Contr. **48**, 68–75 (2007)
565. B.S. Kerner, Traffic Eng. Contr. **48**, 114–120 (2007)

566. B.S. Kerner, Method for actuating a traffic-adaptive assistance system which is located in a vehicle, USA patent US 20070150167A1 (2007) (https://google.com/patents/US20070150167A1); USA patent US 7451039B2 (2008)
567. B.S. Kerner, Betriebsverfahren für ein fahrzeugseitiges verkehrsadaptives Assistenzsystem, German patent publication DE 102007008253A1 (2007) (https://register.dpma.de/DPMAregister/pat/PatSchrifteneinsicht?docId=DE102007008253A1)
568. B.S. Kerner, Betriebsverfahren für ein fahrzeugseitiges verkehrsadaptives Assistenzsystem, German patent publication DE 102007008257A1 (2007) (https://register.dpma.de/DPMAregister/pat/PatSchrifteneinsicht?docId=DE102007008257A1)
569. B.S. Kerner, Transp. Res. Rec. **2088**, 80–89 (2008)
570. B.S. Kerner, J. Phys. A Math. Theor. **41**, 215101 (2008)
571. B.S. Kerner, in *Transportation Research Trends*, ed. by P.O. Inweldi (Nova Science Publishers, New York, 2008), pp. 1–92
572. B.S. Kerner, *Introduction to Modern Traffic Flow Theory and Control* (Springer, Heidelberg, Dordrecht, London, New York, 2009)
573. B.S. Kerner, in *Encyclopedia of Complexity and System Science*, ed. by R.A. Meyers (Springer, Berlin, 2009), pp. 9302–9355
574. B.S. Kerner, in *Encyclopedia of Complexity and System Science*, ed. by R.A. Meyers (Springer, Berlin, 2009), pp. 9355–9411
575. B.S. Kerner, J. Phys. A Math. Theor. **44**, 092001 (2011)
576. B.S. Kerner, Phys. Rev. E **84**, 045102(R) (2011)
577. B.S. Kerner, Phys. Rev. E **85**, 036110 (2012)
578. B.S. Kerner, Physica A **392**, 5261–5282 (2013)
579. B.S. Kerner, Europhys. Lett. **102**, 28010 (2013)
580. B.S. Kerner, Physica A **397**, 76–110 (2014)
581. B.S. Kerner, Traffic Eng. Contr. **55**, 139–141 (2014)
582. B.S. Kerner, in *Vehicular Communications and Networks*, ed. by W. Chen (Woodhead Publishings, Cambridge, 2015), pp. 223–254
583. B.S. Kerner, Phys. Rev. E **92**, 062827 (2015)
584. B.S. Kerner, Elektrotechnik Informationstechnik **132**, 417–433 2015)
585. B.S. Kerner, in *Traffic and Granular Flow 13*, ed. by M. Chraibi, M. Boltes, A. Schadschneider, A. Seyfried (Springer, Berlin, 2015), pp. 369–377
586. B.S. Kerner, Physica A **450**, 700–747 (2016)
587. B.S. Kerner, Eur. Phys. J. B **89**, 199 (2016)
588. B.S. Kerner, Physica A **466**, 626–662 (2017)
589. B.S. Kerner, M. Aleksić, H. Rehborn, A. Haug, in *Proc. of the 7th World Congress on Intelligent Transport Systems* (Turin, Italy, 2000), Paper No. 2035
590. B.S. Kerner, C. Demir, R.G. Herrtwich, S.L. Klenov, H. Rehborn, M. Aleksić, in *Proc. of 2005 IEEE Intelligent Transportation Systems*, pp. 44–49
591. B.S. Kerner, R.G. Herrtwich, Automatisierungstechnik **49**, 505–511 (2001)
592. B.S. Kerner, P. Hemmerle, M. Koller, G. Hermanns, S.L. Klenov, H. Rehborn, M. Schreckenberg, Phys. Rev. E **90**, 032810 (2014)
593. B.S. Kerner, S.L. Klenov, J. Phys. A Math. Gen. **35**, L31–L43 (2002)
594. B.S. Kerner, S.L. Klenov, Phys. Rev. E **68**, 036130 (2003)
595. B.S. Kerner, S.L. Klenov, J. Phys. A Math. Gen. **37**, 8753–8788 (2004)
596. B.S. Kerner, S.L. Klenov, J. Phys. A Math. Gen. **39**, 1775–1809 (2006)
597. B.S. Kerner, S.L. Klenov, Phys. Rev. E **80**, 056101 (2009)
598. B.S. Kerner, S.L. Klenov, in *Encyclopedia of Complexity and System Science*, ed. by R.A. Meyers (Springer, Berlin, 2009), pp. 9282–9302
599. B.S. Kerner, S.L. Klenov, J. Phys. A Math. Theor. **43**, 425101 (2010)
600. B.S. Kerner, S.L. Klenov, A. Brakemeier, in *Proc. of 2008 IEEE Intelligent Vehicles Symposium, 2008*, pp. 180–185
601. B.S. Kerner, S.L. Klenov, A. Brakemeier, in *Proc. of 4th Int. Workshop V2VCOM 2008*, pp. 57–63

602. B.S. Kerner, S.L. Klenov, A. Brakemeier, Traffic Eng. Contr. **6**, 217–222 (2010)
603. B.S. Kerner, S.L. Klenov, A. Brakemeier, in *Proc. of the Transportation Research Board 2010*, TRB Paper 10-0456. (Annual TRB Meeting, TRB, Washington DC, 2010)
604. B.S. Kerner, S.L. Klenov, G. Hermanns, P. Hemmerle, H. Rehborn, M. Schreckenberg, Phys. Rev. E **88**, 054801 (2013)
605. B.S. Kerner, S.L. Klenov, G. Hermanns, M. Schreckenberg, Physica A **392**, 4083–4105 (2013)
606. B.S. Kerner, S.L. Klenov, A. Hiller, J. Phys. A Math. Gen. **39**, 2001–2020 (2006)
607. B.S. Kerner, S.L. Klenov, A. Hiller, H. Rehborn, Phys. Rev. E **73**, 046107 (2006)
608. B.S. Kerner, S.L. Klenov, A. Hiller, Nonlinear Dyn. **49**, 525–553 (2007)
609. B.S. Kerner, S.L. Klenov, P. Konhäuser. Phys. Rev. E. **56**, 4200–4216 (1997)
610. B.S. Kerner, S.L. Klenov, M. Schreckenberg, Phys. Rev. E **84**, 046110 (2011)
611. B.S. Kerner, S.L. Klenov, M. Schreckenberg, Phys. Rev. E, **89**, 052807 (2014)
612. B.S. Kerner, S.L. Klenov, M. Schreckenberg, J. Stat. Mech. P03001 (2014)
613. B.S. Kerner, S.L. Klenov, D.E. Wolf, J. Phys. A Math. Gen. **35**, 9971–10013 (2002)
614. B.S. Kerner, M. Koller, S.L. Klenov, H. Rehborn, M. Leibel, Physica A **438**, 365–397 (2015)
615. B.S. Kerner, P. Konhäuser, Phys. Rev. E **48**, 2335–2338 (1993)
616. B.S. Kerner, P. Konhäuser, Phys. Rev. E **50**, 54–83 (1994)
617. B.S. Kerner, P. Konhäuser, bild der wissenschaft, Heft **11**, 86–89 (1994)
618. B.S. Kerner, P. Konhäuser, M. Schilke, Phys. Rev. E **51**, 6243–6246 (1995)
619. B.S. Kerner, P. Konhäuser, M. Schilke, in *Modeling Transport Systems*, vol. 2, ed. by D. Hensher, J. King, T. Oum (Elsevier Science, Oxford, 1996), pp. 167–182
620. B.S. Kerner, P. Konhäuser, M. Schilke, in *Transportation and Traffic Theory*, ed. by J.-B. Lesort (Elsevier Science, Oxford, 1996), pp. 119–145
621. B.S. Kerner, P. Konhäuser, M. Schilke, Phys. Lett. A. **215**, 45–56 (1996)
622. B.S. Kerner, H. Rehborn, Phys. Rev. E **53**, R1297–R1300 (1996)
623. B.S. Kerner, H. Rehborn, Phys. Rev. E **53**, R4275–R4278 (1996)
624. B.S. Kerner, H. Rehborn, Phys. Rev. Lett. **79**, 4030–4033 (1997)
625. B.S. Kerner, H. Rehborn, Int. Verkehrswesen **50**(Nr. 5), 196–203 (1998)
626. B.S. Kerner, H. Rehborn, Int. Verkehrswesen **50**(Nr. 7+8), 347–348 (1998)
627. B.S. Kerner, H. Rehborn, M. Aleksić, in *Traffic and Granular Flow'99*, ed. by D. Helbing, H.J. Herrmann, M. Schreckenberg, D.E. Wolf (Springer, Heidelberg, 2000), pp. 339–344
628. B.S. Kerner, H. Rehborn, M. Aleksić, in *Transportation Systems 2000*, ed. by E. Schieder, U. Becker (Elsevier Science, London, 2000), pp. 501–506.
629. B.S. Kerner, H. Rehborn, M. Aleksić, A. Haug, Traffic Eng. Contr. **42**, 282–287 (2001)
630. B.S. Kerner, H. Rehborn, M. Aleksić, A Haug, Transp. Res. C **12**, 369–400 (2004)
631. B.S. Kerner, H. Rehborn, M. Aleksić, A Haug, in *Human behaviour and traffic networks*, ed. by M. Schreckenberg, R. Selten (Springer, Berlin, 2004), pp. 251–284
632. B.S. Kerner, H. Rehborn, M. Aleksić, A. Haug, in *Proceedings of the 8th Int. IEEE conference on IST* (Vienna, Austria, 2005), pp. 251–256
633. B.S Kerner, H. Rehborn, M. Aleksić, A. Haug, R. Lange, Traffic Eng. Contr. **42**, 345–350 (2001)
634. B.S. Kerner, H. Rehborn, A. Haug, I. Maiwald-Hiller, Traffic Eng. Contr. **11**, 380–385 (2006)
635. B.S. Kerner, H. Rehborn, J. Palmer, S.L. Klenov, Traffic Eng. Contr. **52**, 141–148 (2011)
636. B.S. Kerner, H. Rehborn, R.-P. Schäfer, S.L. Klenov, J. Palmer, S. Lorkowski, N. Witte, Physica A **392**, 221–251 (2013)
637. A. Kesting, "Microscopic Modeling of Human and Automated Driving: Towards Traffic-Adaptive Cruise Control", *Doctoral Thesis* (Technical University of Dresden, Germany, 2008)
638. A. Kesting, M. Treiber, D. Helbing, Phil. Trans. R. Soc. A **368**, 4585–4605 (2010)
639. A. Kesting, M. Treiber, M. Schönhof, D. Helbing, Transp. Res. C **16**, 668–683 (2008)
640. A. Kesting, M. Treiber, M. Schönhof, D. Helbing, Transp. Res. Rec. **2000**, 16–24 (2007)
641. A. Kesting, M. Treiber, M. Schönhof, F. Kranke, D. Helbing, in [1012] (2007), pp. 633–643

642. M. Keyvan-Ekbatani, M. Papageorgiou, I. Papamichail, in *Proc. of TRB 2013 Annual Meeting* (TRB, Washington, D.C., 2013)
643. M. Keyvan-Ekbatani, A. Kouvelas, I. Papamichail, M. Papageorgiou, Transp. Res. B **46**, 1393–1403 (2012)
644. B. Khondaker, L. Kattan, Transp. Lett. **7**, 264–278 (2015)
645. B. Khondaker, L. Kattan, Transp. Res. C **58**, 146–159 (2015)
646. R. Kianfar, B. Augusto, A. Ebadighajari, U. Hakeem, J. Nilsson, A. Raza, R.S. Tabar, N.V. Irukulapati, C. Englund, P. Falcone, S. Papanastasiou, L. Svensson, H. Wymeersch, IEEE Trans. ITS **13**, 994–1007 (2012)
647. Y. Kim, H. Keller, Straßenverkehrstechnik **9**, 433–442 (2001)
648. R.M. Kimber, E.M. Hollis, Traffic Eng. Contr. **19**, 442–446 (1978)
649. R. Kimber, E. Hollis, "Traffic Queues and Delays at Road Junctions" (TRRL Laboratory Report, 909, UK, 1979)
650. A. Klar, R. Wegener, J. Stat. Phys. **87**, 91–114 (1997)
651. A. Klar, R. Wegener, SIAM J. Appl. Math. **59**, 983 (1999)
652. A. Klar, R. Wegener, SIAM J. Appl. Math. **59**, 1002 (1999)
653. S.L. Klenov, in *Proc. of Moscow Inst. of Phys. and Technology (State University)*, ed. by V.V. Kozlov, vol. 2 (2010) pp. 75–90 (in Russian)
654. K.M. Kockelman, Transp. Res. Rec. **1644**, 47–56 (1998)
655. U. Kohler, Transp. Sci. **13**, 146–162 (1979)
656. S. Kokubo, J. Tanimoto, A. Hagishima, Physica A **390**, 561–568 (2011)
657. M. Koller, PhD Thesis (University Tübingen, Germany, 2015)
658. E. Kometani, T. Sasaki, J. Oper. Res. Soc. Jpn. **2**, 11 (1958)
659. E. Kometani, T. Sasaki, Oper. Res. **7**, 704 (1959)
660. E. Kometani, T. Sasaki, in *Theory of Traffic Flow*, ed. by R. Herman (Elsevier, Amsterdam, 1961), pp. 105
661. A. Kondyli, L. Elefteriadou, W. Brilon, F.L. Hall, B. Persaud, S. Washburn, J. Transp. Eng. **139**, 931–940 (2013)
662. M. Koshi, M. Iwasaki, I. Ohkura, in *Proc. 8th International Symposium on Transportation and Traffic Theory*, ed. by V.F. Hurdle. (University of Toronto Press, Toronto, Ontario, 1983), pp. 403
663. M. Koshi, IATSS Res. **9**, 50–56 (1985)
664. M. Koshi, in [290] (2003), pp. 199–210
665. V.V. Kozlov, A.P. Buslaev, A.S. Bugaev, M.V. Yashina, A. Schadschneider, M. Schreckenberg (eds.), *Traffic and Granular Flow '11* (Springer, Heidelberg, 2013)
666. V.L. Knoop, H. van Lint, S.P. Hoogendoorn, Physica A **438**, 236–250 (2015)
667. F. Knorr, M. Schreckenberg, J. Stat. Mech. P07002 (2013)
668. W. Knospe, L. Santen, A. Schadschneider, M. Schreckenberg, Physica A **265**, 614 (1999)
669. W. Knospe, L. Santen, A. Schadschneider, M. Schreckenberg J. Phys. A **33**, L477 (2000)
670. W. Knospe, L. Santen, A. Schadschneider, M. Schreckenberg, Phys. Rev. E **70**, 016115 (2004)
671. S. Krauß, PhD thesis, (DRL-Forschungsbericht 98-08, 1998). http://www.zaik.de/~paper
672. S. Krauß, P. Wagner, C. Gawron, Phys. Rev. E **54**, 3707–3712 (1996)
673. S. Krauß, P. Wagner, C. Gawron, Phys. Rev. E **55**, 5597–5602 (1997)
674. M. Krbálek, P. Šeba, J. Phys. A **33**, L229 (2000)
675. M. Krbálek, P. Šeba, P. Wagner, Phys. Rev. E **64**, 066119 (2001)
676. I.A. Ktousakis, I.K. Nokolos, M. Papageorgiou, Transp. Res. Procedia **9**, 111–127 (2015)
677. T.S. Kuhn, *The Structure of Scientific Revolutions* 4th edn. (The University of Chicago Press, Chicago, London, 2012)
678. R.D. Kühne, in *Procs. of the 9th International Symposium on Transportation and Traffic Theory* ed. by J. Volmer, R. Hammerslag (VNU Scientific Press, Utrecht, The Netherlands, 1984), pp. 21–42
679. R.D. Kühne, in *Proc. 10th Int. Sympos. on Traffic and Trans. Theory* ed. by N.H. Gartner (Elsevier, New York, 1987), pp. 119–137

680. R. Kühne, in *Highway Capacity and Level of Service*, ed. by U. Brannolte (A.A. Balkema, Rotterdam, 1991), pp. 211
681. R.D. Kühne, A. Kroen, in *Proc. Int. Conf. Artificial Intelligence Appl. Transp. Engineering* (San Buenaventura, California, 1992), pp. 173–192
682. R. Kühne, R. Mahnke, I.A. Lubashevsky, J. Kaupužs, Phys. Rev. E **65**, 066125 (2002)
683. R.D. Kühne, M.B. Rödiger, in *Proc. Int. Conf. Artificial Intelligence Appl. Transp. Engineering* ed. by B.L. Nelson (Phoenix, Society for Comput. Simul. Int., 1991), pp. 762–770
684. S. Kukida, J. Tanimoto, A. Hagishima, Int. J. Mod. Phys. C **22**, 1 (2011)
685. A. Kumar, S. Peeta, Y. Nie, Transp. Res. Rec. **2283**, 131–142 (2012)
686. D. Kupsizewska, D. Van Vliet, in *Transport Planning Methods* (Proc. PTRC Planning and Transport Summer Ann. Meeting, Univ. Sussex, 1999)
687. D.A. Kurtze, D.C. Hong, Phys. Rev. E **52**, 218–221 (1995)
688. A.A. Kurzhanskiy, P. Varaiya, Phil. Trans. R. Soc. A **368**, 4607–4626 (2010)
689. M. Kuwahara, H. Kita, Y. Asakura (eds.), *21st International Symposium on Transportation and Traffic Theory*. Transp. Res. Procedia **7**, 1–704 (2015)
690. W.H.K. Lam, K.S. Chan, Z.-C. Li, M.G.H. Bell, J. Adv. Transp. **44**, 219–230 (2010)
691. W.H.K. Lam, H.-J. Huang, Transp. Res. B **26**, 275–287 (1992)
692. W.H.K. Lam, H.K. Lo, S.C. Wong, Transp. Res. B **66**, 1–3 (2014)
693. T.N. Lam, R.W. Rothery, Transp. Sci. **4**, 293–310 (1970)
694. W.H.K. Lam, S.C. Wong, H.K. Lo (eds.), *Transportation and Traffic Theory 2009* (Springer, Dordrecht, Heidelberg, London, New York, 2009)
695. M. Lampis, Transpn. Sci. **12**, 16–28 (1978)
696. R. Lapierre, G. Steierwald, *Verkehrsleittechnik für den Straßenverkehr, Band I, Grundlagen und Technologien der Verkehrsleittechnik* (Springer, Berlin, 1987)
697. L.J. LeBlanc, E.K. Morlok, W. Pierskalla, Transp. Res. B **9**, 309–318 (1975)
698. J.A. Laval, Trans. Res. Rec. **1988**, 86–91 (2006)
699. J.A. Laval, Trans. Res. B **40**, 937–944 (2006)
700. J.A. Laval, in *Traffic and Granular Flow'05*, ed. by A. Schadschneider, T. Pöschel, R. Kühne, M. Schreckenberg, D.E. Wolf (Springer, Berlin, 2007), pp. 521–526
701. J.A. Laval, Trans. Res. B **43**, 720–728 (2009)
702. J.A. Laval, F. Castrillón, Trans. Res. B **81**, 904–916 (2015)
703. J.A. Laval, B.R. Chilukuri, Trans. Sci. **48**, 217–224 (2013)
704. J.A. Laval, B.R. Chilukuri, Trans. Res. B **89**, 168–177 (2016)
705. J.A. Laval, C.F. Daganzo, Trans. Res. B **40**, 251–264 (2006)
706. J.A. Laval, L. Leclercq, Trans. Res. B **42**, 511–522 (2008)
707. J.A. Laval, L. Leclercq, Phil. Trans. Math. Phys. Eng. Sci. **368**, 4519–4541 (2010)
708. J.A. Laval, L. Leclercq, Trans. Res. B **52**, 17–30 (2013)
709. J.A. Laval, C.S. Toth, Y. Zhou, Trans. Res. B **70**, 228–238 (2014)
710. J.P. Lebacque, J.B. Lesort, F. Giorgi, Transp. Res. Rec. **1644**, 70–79 (1998)
711. L. Leclercq, Transp. Res. B **41**, 701–709 (2007)
712. L. Leclercq, S. Chanut, J.B. Lesort, Transp. Res. Rec. **1883**, 3–13 (2004)
713. L. Leclercq, N. Chiabaut, B. Trinquier, Transp. Res. B **62**, 1–12 (2014)
714. L. Leclercq, V.L. Knoop, F. Marczak, S.P. Hoogendoorn, Transp. Res. C **62**, 171–181 (2016)
715. L. Leclercq, J.A. Laval, in [22], pp. 151–160
716. H.K. Lee, R. Barlović, M. Schreckenberg, D. Kim, Phys. Rev. Lett. **92**, 238702 (2004)
717. S. Lee, B. Heydecker, Y.H. Kim, E.-Y. Shon, J. Adv. Trans. **4**, 143–158 (2011)
718. K. Lee, P. Hui, B. Wang, N. Johnson, J. Phys. Soc. Jpn. **70**, 3507–3510 (2001)
719. H.-K. Lee, B.-J. Kim, Physica A **390**, 4555–4561 (2011)
720. H.Y. Lee, H.-W. Lee, D. Kim, Phys. Rev. Lett. **81**, 1130 (1998)
721. H.Y. Lee, H.-W. Lee, D. Kim, Phys. Rev. E **59**, 5101–5111 (1999)
722. H.Y. Lee, H.-W. Lee, D. Kim, Physica A **281**, 78 (2000)
723. H.Y. Lee, H.-W. Lee, D. Kim, Phys. Rev. E **62**, 4737 (2000)
724. W.-H. Lee, S.-S. Tseng, J.-L. Shieh, H.-H. Chen, IEEE Trans. ITS **12**, 1047–1056 (2011)
725. J.-B. Lesort (ed.), *Transportation and Traffic Theory* (Elsevier Science, Oxford, 1996)

726. F. Leurent, H. Boujnah, Transp. Res. C **47**, 28–46 (2014)
727. W. Leutzbach, *Introduction to the Theory of Traffic Flow* (Springer, Berlin, 1988)
728. M.W. Levin, S.D. Boyles, Transp. Res. C **62**, 103–116 (2016)
729. L. Li, X. Chen, Transp. Res. C **76**, 170188 (2017)
730. J. Li, Q.-Y. Chen, H. Wang, D. Ni, Transportmetrica **8**, 387–405 (2012)
731. X.G. Li, Z.Y. Gao, K.P. Li, X.M. Zhao, Phys. Rev. E **76**, 016110 (2007)
732. M. Li, N.M. Rouphail, M. Mahmoudi, J. Liu, X. Zhou, Transp. Res. C **77**, 113–133 (2017)
733. P.Y. Li, A. Shrivastava, Transp. Res. C **10**, 275–301 (2002)
734. Q.-L. Li, B.-H. Wang, M.-R. Liu, Physica A **390**, 1356–1362 (2011)
735. M.J. Lighthill, G.B. Whitham, Proc. Roy. Soc. A **229**, 281–345 (1955)
736. W.H. Lin, C.F. Daganzo, Transp. Res. A **31**, 141–155 (1997)
737. T.-W. Lin, S.-L. Hwang, P. Green, Saf. Sci. **47**, 620–625 (2009)
738. R.V. Lindgren, R.L. Bertini, D. Helbing, M. Schönhof, Transp. Res. Rec. **1965**, 12–22 (2006)
739. J. van Lint, S.P. Hoogendoorn, Comput. Aid. Civil Infrast. Eng. **24**, 1–17 (2009)
740. E.M. Linzer, R.P. Roess, W.R. McShane, Transp. Res. Rec. **699**, 17–24 (1979)
741. J.D.C. Little, Oper. Res. **14**, 568–594 (1966)
742. J.D.C. Little, M.D. Kelson, N.H. Gartner, Transp. Res. Rec. **795**, 40–46 (1981)
743. S. Liu, H. Hellendoorn, B. De Schutter, IEEE Trans. ITS **18**, 306–320 (2017)
744. H. Liu, D.Z.W. Wang Transp. Res. B **72**, 20–39 (2015)
745. H. Liu, D.Z.W. Wang, J. Adv. Transp. **50**, 1295–1313 (2016)
746. H.K. Lo, E. Chang, Y.-Ch. Chan, Transp. Res. A **35**, 721–744 (2001)
747. H.K. Lo, W.Y. Szeto, Transp. Res. B **36**, 421–443 (2002)
748. J. Long, Z. Gao, P. Orenstein, H. Ren, IEEE Trans. ITS **13**, 469–481 (2012)
749. J. Long, Z. Gao, X. Zhao, A. Lian, P. Orenstein, Netw. Spat. Econ. **11**, 43–64 (2011)
750. J. Long, W.Y. Szeto, Z. Gao, H.-J. Huang, Q. Shi, Transp. Res. B **83**, 179–206 (2016)
751. J. Long, W.Y. Szeto, Q. Shi, Z. Gao, H.-J. Huang, Transportmetrica A **11**, 388–419 (2015)
752. M. Lorenz, L. Elefteriadou, Transp. Res. Circ. **E-C018**, 84–95 (2000)
753. C.-C. Lu, J. Liu, Y. Qu, S. Peeta, N.M. Rouphail, X. Zhou, Transp. Res. B **94**, 217–239 (2016)
754. C.-C. Lu, H.S. Mahmassani, X. Zhou, Transp. Res. C. **16**, 371–389 (2008)
755. L. Lu, Y. Xu, C. Antoniou, M. Ben-Akiva Transport. Res. C **51**, 149–166 (2015)
756. P. Luathep, A. Sumalee, W.H.K. Lam, Z-Ch. Li, H.K. Lo, Transp. Res. B **45**, 808–827 (2011)
757. I.A. Lubashevsky, S. Kalenkov, R. Mahnke, Phys. Rev. E **65**, 036140 (2002)
758. I.A. Lubashevsky, R. Mahnke, Phys. Rev. E **62**, 6082 (2000)
759. I.A. Lubashevsky, R. Mahnke, P. Wagner, S. Kalenkov, Phys. Rev. E **66**, 016117 (2002)
760. I.A. Lubashevsky, P. Wagner, R. Mahnke, Eur. Phys. J. B **32**, 243–247 (2003)
761. I.A. Lubashevsky, P. Wagner, R. Mahnke, Phys. Rev. E **68**, 056109 (2003)
762. S. Lübeck, M. Schreckenberg, K.D. Usadel, Phys. Rev. E **57**, 1171 (1998)
763. L. Ludovic, C. Estelle, L. Jorge, in *Hyperbolic Problems: Theory, Numerics, Applications*, ed. by S. Benzoni-Gavage, D. Serre (Springer, Berlin, Heidelberg, 2008), pp. 671–678
764. M. Lujak, S. Giordani, S. Ossowski, Neurocomputing **151**, 449–460 (2015)
765. R.T. Luttinen, Transp. Res. Rec. **1365**, 92–97 (1992)
766. W. Lv, W.-g. Song, Z.-m. Fang, Physica A **390**, 2303–2314 (2011)
767. M.J. Maher, P.C. Hughes, Transp. Res. B **31**, 341–355 (1997)
768. S. Maerivoet, B. De Moor, Phys. Rep. **419**, 1–64 (2005)
769. D. Mahalel, A.S. Hakkert, Transp. Sci. **17**, 71–86 (1983)
770. H.S. Mahmassani, Transp. Res. A **24**, 465–484 (1990)
771. H.S. Mahmassani, in *Operations Research and Decision Aid Methodologies in Traffic and Transportation Management*, Vol. 166 of the series NATO ASI Series (Springer, Berlin, 1998), pp. 104–135
772. H.S. Mahmassani, Netw. Spat. Econ. **1**, 267–292 (2001)
773. H.S. Mahmassani (ed.), *Traffic and Transportation Theory* (Elsevier Science, Amsterdam, 2005)
774. H.S. Mahmassani, G.-L. Chang, Transp. Res. B **20**, 297–320 (1986)

775. H.S. Mahmassani, G.-L. Chang, Transp. Sci. **21**, 89–99 (1987)
776. H.S. Mahmassani, G.-L. Chang, R. Herman, Transp. Sci. **20**, 258–271 (1986)
777. H.S. Mahmassani, P.S.-T. Chen, Transp. Res. Rec. **1306**, 69–81 (1991)
778. H.S. Mahmassani, R. Herman, Transp. Sci. **18**, 362–384 (1984)
779. H.S. Mahmassani, R. Jayakrishnan, Transp. Res. A **25**, 293–307 (1991)
780. H.S. Mahmassani, R. Jayakrishnan, R. Herman, Transp. Res. A **24**, 149–162 (1990)
781. H.S. Mahmassani, R.-C. Jou, Transp. Res. A **34**, 243–260 (2000)
782. H.S. Mahmassani, Y.-H. Liu, Transp. Res. C **7**, 91–107 (1999)
783. H.S. Mahmassani, P. Michalopoulos, Transp. Res. A **19**, 377–380 (1985)
784. H.S. Mahmassani, K.C. Mouskos, Transp. Res. B **22**, 275–290 (1988)
785. H.S. Mahmassani, S. Peeta, Transp. Res. Rec. **1408**, 83–93 (1993)
786. H.S. Mahmassani, M. Saberi, A. Zockaie, Transp. Res. C **36**, 480–497 (2013)
787. H.S. Mahmassani, Y. Sheffi, Transp. Res. B **15**, 143–148 (1981)
788. H.S. Mahmassani, D.G. Stephan, Transp. Res. Rec. **1203**, 69–84 (1988)
789. H.S. Mahmassani, C.C. Tong, Transp. Res. Rec. **1085**, 33–49 (1986)
790. H.S. Mahmassani, J.C. Williams, R. Herman, Transp. Res. Rec. **971**, 121–130 (1984)
791. H.S. Mahmassani, J.C. Williams, R. Herman, in *Proc. of the 10th Int. Symposium on Transp. and Traffic Theory*, ed. by N.H. Gartner (Elsevier, New York, 1987), pp. 1–20
792. H.S. Mahmassani, X. Zhou, in *Advances in Control, Communication Networks, and Transportation Systems*, (In Honor of Pravin Varaiya), ed. by E.H. Abed. (Birkhäuser, Boston, 2005), pp. 305–328
793. H.S. Mahmassani, X. Zhou, C.-C. Lu, Transp. Res. Rec. **1923**, 28–36 (2005)
794. R. Mahnke, J. Kaupužs, Phys. Rev. E **59**, 117 (1999)
795. R. Mahnke, J. Kaupužs, Netw. Spat. Econ. **1**, 103–136 (2001)
796. R. Mahnke, J. Kaupužs, I. Lubashevsky, Phys. Rep. **408**, 1–130 (2005)
797. R. Mahnke, J. Kaupužs, I.A. Lubashevsky, *Physics of Stochastic Processes: How Randomness Acts in Time* (Wiley-VCH, Weinheim, 2009)
798. R. Mahnke, N. Pieret, Phys. Rev. E **56**, 2666 (1997)
799. F.L. Mannering, W.P. Kilareski, *Principles of Highway Engineering and Traffic Analysis*, 2nd edn. (Wiley, New York, 1998)
800. D. Manstetten, W. Krautter, T. Schwab, in *Proc. of the 4th World Congress on Intelligent Transport Systems* (ITS Congress Association, Brussels 1997), Paper No. 00648
801. D.B. Martin, L. Newman, R.T. Johnson, Highway Res. Rec. **432**, 25–31 (1973)
802. J.-J. Martinez, C. Canudas-do-Wit, IEEE Trans. Contr. Syst. Technol. **15**, 246–258 (2007)
803. A.D. May, Highway Res. Rec. **59**, 9–38 (1964)
804. A.D. May, *Traffic Flow Fundamentals* (Prentice-Hall, New Jersey, 1990)
805. A.D. May, P. Athol, W. Parker, J.B. Rudden, Highway Res. Rec. **21**, 48–70 (1963)
806. A.D. May, H.E.M. Keller, Highway Res. Rec. **199**, 19–32 (1967)
807. A.D. May, H.E.M. Keller, Transp. Res. **1**, 117–128 (1967)
808. A.D. May, H.E.M. Keller, in *Proc. of the 3rd Inter. Symposium on Transp. and Traffic Theory* (Karlsruhe, 1968), pp. 37–47
809. J.M. McDermott, Transp. Eng. J. **106**, 333–348 (1980)
810. W.R. McShane, R.P. Roess, *Traffic Engineering* (Prentice Hall, Englewood Cliffs, 1990)
811. A.R. Mendez, R.M. Velasco, J. Phys. A Math. Theor. **46**, 462001 (2013)
812. D.K. Merchant, G.L. Nemhauser, Transp. Sci. **12**, 187–199 (1978)
813. D.K. Merchant, G.L. Nemhauser, Transp. Sci. **12**, 200–207 (1978)
814. P.G. Michalopoulos, G. Stephanopoulos, V.B. Pisharody. Transp. Sc. **14**, 9–41 (1980)
815. P.G. Michalopoulos, G. Stephanopoulos, G., Stephanopoulos. Transp. Res. B **15**, 35–51 (1981)
816. L. Mihaylova, R. Boel, A. Hegyi, Automatica **43**, 290–300 (2007)
817. H.S. Mika, J.B. Kreer, L.S. Yuan, Highway Res. Rec. **279**, 1–13 (1969)
818. A.J. Miller, Transp. Sci. **4**, 164–186 (1970)
819. J.A. Miller, R. Rothery, Transp. Sci. **1**, 81–94 (1967)
820. E.D. Miller-Hooks, H.S. Mahmassani, Transp. Sci. **34**, 198–215 (2000)

821. J.T. Morgan, J.D.C. Little, Oper. Res. **12**, 896–912 (1964)
822. N. Motamedidehkordi, T. Benz, M. Margreiter, in *Advanced Microsystems for Automotive Applications 2015*, Lecture Notes in Mobility 2016 (Springer, Berlin, 2016), pp 37–52
823. J. Mück, in *Tagungsband der Heureka 2002* (Forschungsgesellschaft für Straßen- und Verkehrswesen, Köln, 2002)
824. R. Müller, G. Nöcker, in *Proceedings of the Intelligent Vehicles '92 Symposium*, ed. by I. Masaki (IEEE, Detroit, USA, 1992), pp. 173–178
825. P.K. Munjal, Y.S. Hsu, Transp. Res. Rec. **509**, 29–41 (1974)
826. P.K. Munjal, Y.S. Hsu, R.L. Lawrence, Transp. Res. **5**, 257–266 (1971)
827. P. Munjal, L. Pipes, Transp. Res. **5**, 241–255 (1971)
828. P.K. Munjal, L.A. Pipes, Transp. Sci. **5**, 390–402 (1971)
829. J.C. Muñoz, C.F. Daganzo, in *Proc. of the 80th Annual Meeting of the Transportation Research Board* (TRB, Washington, D.C., 2001)
830. J.C. Muñoz, C.F. Daganzo, in *Traffic and Transportation Theory*, ed. by M.A.P. Taylor. (Pergamon, Oxford, 2002) pp 441–462
831. T. Musha, H. Higuchi, Jpn. J. Appl. Phys. **15**, 1271 (1976)
832. T. Musha, H. Higuchi, Jpn. J. Appl. Phys. **17**, 811 (1978)
833. T. Nagatani, Physica A **261**, 599–607 (1998)
834. T. Nagatani, Phys. Rev. E **59**, 4857–4864 (1999)
835. T. Nagatani, Physica A **271**, 200–221 (1999)
836. T. Nagatani, Physica A **272**, 592–611 (1999)
837. T. Nagatani, Phys. Rev. E **60**, 180–187 (1999)
838. T. Nagatani, Phys. Rev. E **60**, 6395–6401 (1999)
839. T. Nagatani, Physica A **280**, 602–613 (2000)
840. T. Nagatani, Phys. Rev. E **61**, 3564–3570 (2000)
841. T. Nagatani, Phys. Rev. E **61**, 3534–3540 (2000)
842. T. Nagatani, Phys. Rev. E **63**, 036115 (2001)
843. T. Nagatani, Phys. Rev. E **64**, 016106 (2001)
844. T. Nagatani, Physica A **290**, 501–511 (2001)
845. T. Nagatani, Rep. Prog. Phys. **65**, 1331–1386 (2002)
846. T. Nagatani, K. Nakanishi, Phys. Rev. E **57**, 6415–6421 (1998)
847. K. Nagel, Int. J. Modern Phys. C **5**, 567–580 (1994)
848. K. Nagel, Phys. Rev. E **53**, 4655 (1996)
849. K. Nagel, in *Int. Conference on Cellular Automata* (Springer, Berlin, Heidelberg, 2002), pp. 20–31
850. K. Nagel, C.L. Barrett, Int. J. Modern Phys. C **8**, 505–525 (1997)
851. K. Nagel, H.J. Herrmann, Physica A **199**, 254–269 (1993)
852. K. Nagel, M. Paczuski, Phys. Rev. E **51**, 2909 (1995)
853. K. Nagel, M. Rickert, Parallel Comput. **27**, 1611–1639 (2001)
854. K. Nagel, M. Rickert, C.L. Barrett, in *Int. Conference on Vector and Parallel Processing* (Springer, Berlin, Heidelberg, 1996), pp. 380–402
855. K. Nagel, M. Schreckenberg, J. Phys. (France) I **2**, 2221–2229 (1992)
856. K. Nagel, P. Wagner, R. Woesler, Oper. Res. **51**, 681–716 (2003)
857. K. Nagel, D.E. Wolf, P. Wagner, P. Simon, Phys. Rev. E **58**, 1425–1437 (1998)
858. A. Nagurney, J. Dong, Transp. Res. B **36**, 445–469 (2002)
859. Y. Naito, T. Nagatani, Physica A **391**, 1626–1635 (2012)
860. S. Nakayama, R.D. Connors, D. Walting, in [694] (2009), pp. 39–56
861. A. Nakayama, M. Fukui, M. Kikuchi, K. Hasebe, K. Nishinari, Y. Sugiyama, S-i. Tadaki, S. Yukawa, New J. Phys. **11**, 083025 (2009)
862. A. Nantes, D. Ngoduy, A. Bhaskar, M. Miska, E. Chung, Transp. Res. C **66**, 99–118 (2016)
863. P. Nelson, Transp. Theor. Stat. Phys. **24**, 383–409 (1995)
864. P. Nelson, Phys. Rev. E **61**, R6052 (2000)
865. Y. Nesterov, A. de Palma, Netw. Spat. Econ. **3**, 371–395 (2003)
866. J.P.L. Neto, M.L. Lyra, C.R. da Silva, Physica A **390**, 3558–3565 (2011)

867. L. Neubert, H.Y. Lee, M. Schreckenberg, J. Phys. A Math. Gen. **32**, 6517–6525 (1999)
868. L. Neubert, L. Santen, A. Schadschneider, M. Schreckenberg, Phys. Rev. E **60**, 6480–6490 (1999)
869. L. Neubert, L. Santen, A. Schadschneider, M. Schreckenberg, in [404] (2000), pp. 307–314
870. G.F. Newell, Oper. Res. **3**, 176–188 (1955)
871. G.F. Newell, Q. Appl. Math. **13**, 353–369 (1956)
872. G.F. Newell, Oper. Res. **7**, 589–598 (1959)
873. G.F. Newell, Q. Appl. Math. **17**, 67–76 (1959)
874. G.F. Newell, Ann. Math. Stat. **31**, 589–597 (1960)
875. G.F. Newell, Oper. Res. **8**, 390–405 (1960)
876. G.F. Newell, Oper. Res. **9**, 209–229 (1961)
877. G.F. Newell, Oper. Res. Soc. Jpn. **5**, 9–54 (1962)
878. G.F. Newell, "Instability in dense highway traffic, a review", in *Proc. Second Internat. Sympos. on Traffic Road Traffic Flow* (OECD, London, 1963), pp. 73–83
879. G.F. Newell, Q. Appl. Math. **21**, 315–324 (1964)
880. G.F. Newell, SIAM Rev. **575**, 223–240 (1965)
881. G.F. Newell, J. Appl. Probab. **3**, 247–260 (1966)
882. G.F. Newell, *Traffic Signal Synchonization for High Flows on a Two-Way Street* (Itte. Calif. Univ. Research Reports, 1967)
883. G.F. Newell, Transp. Sci. **3**, 30–52 (1969)
884. G.F. Newell, *Applications of Queuing Theory* (Chapman Hall, London, 1982)
885. G.F. Newell, Transp. Res. B **32**, 531–537 (1988)
886. G.F. Newell, Transp. Res. B **23**, 386–389 (1989)
887. G.F. Newell, *Theory of Highway Traffic Signals* (Institute of Transportation Studies, University of California, Davis CA 95616, USA, 1989)
888. G.F. Newell, in *Transportation and Traffic Theory*, ed. by M. Koshi (Elsevier Science, Amsterdam, 1990), pp. 589–598
889. G.F. Newell, *A Simplified Theory of Kinematic Waves* (Institute of Transportation Studies, University of California, Davis CA 95616, USA, 1991)
890. G.F. Newell, Transp. Res. B **27**, 281–287 (1993)
891. G.F. Newell, Transp. Res. B **27**, 289–303 (1993)
892. G.F. Newell, Transp. Res. B **27**, 305–313 (1993)
893. G.F. Newell, *Inst. of Transp. Studies Research Report UCB ITS-RR-93-3* (University of California, Berkley, CA, 1993)
894. G.F. Newell, *Theory of Highway Traffic Flow 1945–1965* (ARRB Group Limited 500 Burwood Highway Vermont South, Victoria, USA, 1995)
895. G.F. Newell, Trans. Res. B **32**, 531–537 (1998)
896. G.F. Newell, Trans. Res. A **32**, 39–44 (1998)
897. G.F. Newell, Trans. Res. B **33**, 337–350 (1999)
898. G.F. Newell, *A Simplified Car-Following Theory: A Lower Order Model*, Research report UCB-ITS-RR-99-1 (Inst. of Transp. Studies, University of California, Berkeley, 1999)
899. G.F. Newell, Trans. Res. B **36**, 195–205 (2002)
900. L. Newman, Highway Res. Rec. **27**, 14–43 (1963)
901. L. Newman, A.M. Dunnet, G.J. Meis, Highway Res. Rec. **303**, 44–55 (1970)
902. D. Ngoduy, Transpormetrica **8**, 43–60 (2012)
903. D. Ngoduy, Commun. Nonlinear Sci Numer. Simul. **18**, 2838–2851 (2013)
904. D. Ngoduy, N.H. Hoang, H.L. Vu, D. Watling, Transp. Res. B **92**, 148–169 (2016)
905. D. Ngoduy, S.P. Hoogendoorn, R. Liu, Physica A **388**, 2705–2716 (2009)
906. D. Ngoduy, S.P. Hoogendoorn, H. Van Zuylen, Trans. Res. Rec. **1965**, 91–102 (2006)
907. Y. Nie, Transp. Res. B **45**, 329–342 (2011)
908. Y. Nie, H.M. Zhang, Netw. Spat. Econ. **10**, 49–71 (2010)
909. K. Nishinari, J. Phys. A Math. Gen. **34**, 10727 (2001)
910. K. Nishinari, M. Hayashi (eds.), *Traffic Statistics in Tomei Express Way* (The Math. Soc. of Traffic Flow, Japan, 1999)

911. K. Nishinari, D. Takahashi, J. Phys. A Math. Gen. **32**, 93 (1999)
912. K. Nishinari, D. Takahashi, J. Phys. A **33**, 7709 (2000)
913. K. Nishinari, M. Treiber, D. Helbing, Phys. Rev. E **68**, 067101 (2003)
914. G. Nöcker, in *Band Elektronik im Kraftfahrzeugwesen*, ed. by G. Walliser, vol. 437 (Expert Verlag, 1994), pp. 299–322 (ISBN 3-8169-1024-6)
915. A. Ober-Sundermeier, H. Zackor, in *2001 IEEE Intelligent Transportation Systems Conference Proceedings* (Oakland (CA), USA, 2001), pp. 240–244
916. P.S. Olszewski, Transp. Res. Rec. **1287**, 42–53 (1990)
917. S. Ossen, S.P. Hoogendoorn, Transp. Res. C **19**, 182–195 (2011)
918. D. Owens, M.J. Schofield, Traffic Eng. Contr. **29**, 616–623 (1988)
919. G.M. Pacey, *Research Note No. Rn/2665/GMP* (Road Research Laboratory, London, 1956)
920. J. Pahl, T. Sands, Transp. Sci. **5**, 403–417 (1971)
921. J. Palmer, H. Rehborn, I. Gruttadauria, Int. J. Adv. Syst. Measur. **4**, 168–181 (2011)
922. J. Palmer, H. Rehborn, B.S. Kerner, Traffic Eng. Contr. **52**, 183–191 (2011)
923. M. Papageorgiou, *Application of Automatic Control Concepts in Traffic Flow Modeling and Control* (Springer, Berlin, New York, 1983)
924. M. Papageorgiou, J.-M. Blosseville, H. Hadj-Salem, Trans. Res. A. **24**, 361–370 (1990)
925. M. Papageorgiou, H. Hadj-Salem, J.-M. Blosseville, Transp. Res. Rec. **1320**, 58–64 (1991)
926. M. Papageorgiou, H. Hadj-Salem, F. Middleham, Transp. Res. Rec. **1603**, 99–98 (1997)
927. M. Papageorgiou, E. Kosmatopoulos, I. Papamichail, Y. Wang, IEEE Trans. ITS. **9**, 360–365 (2008)
928. M. Papageorgiou, A. Kotsialos, IEEE Trans. ITS. **3**, 271–281 (2002)
929. M. Papageorgiou, Y. Wang, E. Kosmatopoulos, I. Papamichail, Traffic Eng. Contr. **48**, 271–276 (2007)
930. M. Patriksson, *The Traffic Assignment Problem–Models and Methods* (VSP, Utrecht, 1994)
931. J.O. Paul, J. Transp. Eng. **142**, 04016025 (2016)
932. S.L. Paveri-Fontana, Transp. Res. **9**, 225–235 (1975)
933. H.J. Payne, in *Mathematical Models of Public Systems*, ed. by G.A. Bekey, vol. 1 (Simulation Council, La Jolla, 1971), pp. 51–60
934. H.J. Payne, in *Research Directions in Computer Control of Urban Traffic Systems*, ed. by W.S. Levine (Am. Soc. of Civil Engineers, New York, 1979), pp. 251–265
935. H.J. Payne, Tran. Res. Rec. **772**, 68 (1979)
936. S. Peeta, H.S. Mahmassani, Transp. Res. C **3**, 83–98 (1995)
937. S. Peeta, H.S. Mahmassani, Ann. Oper. Res. **60**, 81–113 (1995)
938. S. Peeta, A.K. Ziliaskopoulos, Netw. Spat. Econ. **1**, 233–265 (2001)
939. G.H. Peng, Physica A **391**, 5971–5977 (2013)
940. G.H. Peng, Nonlinear Dyn. **73**, 1035–1043 (2013)
941. G.H. Peng, X.H. Cai, B.F. Cao, C.Q. Liu, Physica A **391**, 656–663 (2012)
942. G.H. Peng, X.H. Cai, C.Q. Liu, B.F. Cao, M.X. Tuo, Phys. Lett. A **375**, 3973–3977 (2011)
943. G.H. Peng, X.H. Cai, C.Q. Liu, M.X. Tuo, Phys. Lett. A **375**, 2823–2827 (2011)
944. G.H. Peng, X.H. Cai, C.Q. Liu, M.X. Tuo, Phys. Lett. A **375**, 2153–2157 (2011)
945. G.H. Peng, X.H. Cai, C.Q. Liu, M.X. Tuo, Phys. Lett. A **376**, 447–451 (2012)
946. G.H. Peng, D.H. Sun, Phys. Lett. A **374**, 1694–1698 (2010)
947. B.N. Persaud, "Study of a Freeway Bottleneck to Explore Some Unresolved Traffic Flow Issues", *Ph.D. Dissertation* (University of Toronto, Toronto, Canada, 1986)
948. B.N. Persaud, F.L. Hall, Trans. Res. A **23**, 103–113 (1989)
949. B.N. Persaud, F.L. Hall, L.M. Hall, Transp. Res. Rec. **1287**, 167–175 (1990)
950. B.N. Persaud, V.F. Hurdle, Transp. Res. Rec. **1194**, 191–198 (1988)
951. B.N. Persaud, V.F. Hurdle, in *Proc. of International Symposium on Highway Capacity* (Karlsruhe, Germany, 1991)
952. B.N. Persaud, S. Yagar, R. Brownlee, Transp. Res. Rec. **1634**, 64–69 (1998)
953. B.N. Persaud, S. Yagar, D. Tsui, H. Look, Transp. Res. Rec. **1748**, 110–115 (2001)
954. W.F. Phillips, Transp. Plan. Technol. **5**, 131–138 (1979)

955. W.F. Phillips, in *Proc. 1978 IEEE Conf. Decision and Control* (IEEE, New York, 1979), pp. 1032–1036
956. B. Piccoli, A. Tosin, in *Encyclopedia of Complexity and System Science*, ed. by R.A. Meyers (Springer, Berlin, 2009), pp. 9727–9749
957. C. Pinnell, D.R. Drew, W.R. McCasland, J.A. Wattleworth, Highway Res. Rec. **157**, 22–70 (1967)
958. L.A. Pipes, J. Appl. Phys. **24**, 274–287 (1953)
959. A.W. Plank, E.A. Catchpole, Traffic Eng. Contr. **25**, 327–329 (1984)
960. R. Pothering, S. Gao, in *Proc. TRB 92nd Annual Meeting*, Paper # 13-3641 (TRB, Washington DC, 2013)
961. A. Pottmeier, C. Thiemann, A. Schadschneider, M. Schreckenberg, in *Traffic and Granular Flow'05*, ed. by A. Schadschneider, T. Pöschel, R. Kühne, M. Schreckenberg, D.E. Wolf (Springer, Berlin, 2007), pp. 503–508
962. I. Prigogine, in *Theory of Traffic Flow*, ed. by R. Herman (Elsevier, Amsterdam 1961), p. 158
963. I. Prigogine, F.C. Andrews, Oper. Res. **8**, 789–797 (1960)
964. I. Prigogine, R. Herman, *Kinetic Theory of Vehicular Traffic* (American Elsevier, New York, 1971)
965. Z. Qian, J. Li, X. Li, M. Zhang, H. Wang, Transp. Res. B **99**, 183–204 (2017)
966. Z. Qian, R. Rajagopal, Transp. Res. C **48**, 468–490 (2014)
967. Z. Qian, W. Shen, H.M. Zhang, Transp. Res. B **46**, 874–893 (2012)
968. Z. Qian, H.M. Zhang, Transp. Res. B **46**, 1489–1503 (2012)
969. Y. Qiu, J. Non-Newtonian Fluid Mech. **197**, 1–4 (2013)
970. Y. Qu, X. Zhou Transp. Res. C **75**, 1–16 (2017)
971. H. Rakha, Transp. Lett. **1**, 227–244 (2009)
972. H. Rakha, M. Van Aerde, Transp. Res. Rec. **1510**, 26–34 (1995)
973. H. Rakha, M. Arafeh, Transp. Sci. **44**, 151–168 (2010)
974. H. Rakha, B. Crowther, Transp. Res. Rec. **1802**, 248–262 (2002)
975. H. Rakha, B. Crowther, Transp. Res. A **37**, 1–27 (2003)
976. H. Rakha, P. Pasumarthy, S. Adjerid, Transp. Lett. **1**, 95–110 (2009)
977. H. Rakha, C. Pecker, H. Cybis, Transp. Res. Rec. **1999**, 115–127 (2007)
978. H. Rakha, A. Tawfik, in *Encyclopedia of Complexity and System Science*, ed. by R.A. Meyers (Springer, Berlin, 2009), pp. 9429–9470
979. H. Rakha, W. Wang, Transp. Res. Rec. **2124**, 113–124 (2009)
980. B. Ran, D. Boyce, *Modeling Dynamic Transportation Networks* (Springer, Berlin, 1996)
981. B. Ran, D.E. Boyce, L.J. LeBlanc, Oper. Res. **41**, 192–202 (1993)
982. A.K. Rathi, E.B. Lieberman, M. Yedlin, Transp. Res. Rec. **1112**, 61–71 (1987)
983. H. Rehborn, S.L. Klenov, in *Springer Encyclopedia of Complexity and System Science*, ed. by R.A. Meyers (Springer, Berlin, 2009), pp. 9500–9536
984. H. Rehborn, S.L. Klenov, J. Palmer, Physica A **390**, 4466–4485 (2011)
985. H. Rehborn, M. Koller, J. Adv. Transp. **48**, 1107–1120 (2014)
986. W.R. Reilly, C.C. Gardner, J.H. Kell, *FHWA Report No. RD-76-135/137* (Federal Highway Administration, Washington, D.C., 1976)
987. W.R. Reilly, D.W. Harwood, J.M. Schoen, R.0. Kuehl, K. Bauer, A.D.St. John, *Capacity and Level of Service Procedures for Multilane Rural and Suburban Highways*, Preliminary draft final report (NCHRP, TRB, National Research Council, Washington, D.C., 1988)
988. R.A. Reiss, N.H. Gartner, S.L. Cohen, Transp. Res. A **25**, 267–276 (1991)
989. F. Rempe, P. Franeck, U. Fastenrath, K. Bogenberger, Presented in 96th Transp. Res. Board Annual Meeting, TRB Paper 17-05542 (2017) (unpublished)
990. A. Reuschel, Österreichisches Ingenieur-Archiv **4**(3/4), 193–215 (1950)
991. A. Reuschel, Z. Österr. Ing.-Archit.-Ver. **95**, 59–62 (1950)
992. A. Reuschel, Z. Österr. Ing.-Archit.-Ver. **95**, 73–77 (1950)
993. C. Rhoades, X. Wang, Y. Ouyang, Transp. Res. C **69**, 328–342 (2016)
994. P.I. Richards, Oper. Res. **4**, 42–51 (1956)
995. M. Rickert, K. Nagel, M. Schreckenberg, A. Latour, Physica A **231**, 534–550 (1996)

996. J. Ringert, T. Urbanik II, Transp. Res. Rec. **1398**, 31–41 (1993)
997. D.I. Robertson, Traffic Eng. Contr. **10**, 276 (1969)
998. D.I. Robertson, *TRRL Report No. LR 2530* (Transportation and Road Research Laboratory, Crowthorne, UK, 1969)
999. D.I. Robertson, in *Proc. of the International Symposium on Traffic Control Systems, 1979*, pp. 262–288
1000. C. Roncoli, M. Papageorgiou, I. Papamichail, Transp. Res. C **57**, 241–259 (2015)
1001. C. Roncoli, M. Papageorgiou, I. Papamichail, Transport. Res. C **57**, 260–275 (2015)
1002. C. Roncoli,I. Papamichail, M. Papageorgiou, Transp. Res. C **62**, 117–132 (2016)
1003. R. Rysgaard (ed.), *Proceedings of the 3rd Symposium on Highway Capacity and Level of Service* (Road Directorate, Copenhagen, Ministry of Transport, Denmark, 1998)
1004. M. Saifuzzaman, Z. Zheng, Transp. Res. C **48**, 379–403 (2014)
1005. M. Sasvari, J. Kertész, Phys. Rev. E. **56**, 4104–4110 (1997)
1006. A. Schadschneider, Eur. Phys. J B **10**, 573–582 (1999)
1007. A. Schadschneider, Physica A **285**, 101–120 (2000)
1008. A. Schadschneider, Physica A **313**, 153–187 (2002)
1009. A. Schadschneider, Physica A **372**, 142–150 (2006)
1010. A. Schadschneider, D. Chowdhury, K. Nishinari, *Stochastic Transport in Complex Systems* (Elsevier Science, New York, 2011)
1011. A. Schadschneider, W. Knospe, L. Santen, M. Schreckenberg, Physica A **346**, 165–173 (2005)
1012. A. Schadschneider, T. Pöschel, R. Kühne, M. Schreckenberg, D.E. Wolf (eds.), *Traffic and Granular Flow'05* (Springer, Heidelberg, 2007)
1013. A. Schadschneider, M. Schreckenberg, J. Phys. A Math. Gen. **26**, L679–L684 (1993)
1014. A. Schadschneider, M. Schreckenberg, J. Phys. A Math. Gen. **30**, L69 (1997)
1015. A. Schadschneider, M. Schreckenberg, J. Phys. A Math. Gen. **31**, L225 (1998)
1016. J.-D. Schmöcker, M.G.H. Bell, F. Kurauchi, H. Shimamoto, in [694] (2009), pp. 1–18
1017. J. Schoen, A.D. May, W. Reilly, T. Urbanik, "Speed–Flow Relationships for Basic Freeway Segments", *Final Report, NCHRP Project 3–45*, (JHK & Associates and Texas Transportation Institute, 1995)
1018. M.J. Schofield, Trafic Eng. Contr. **27**, 509–513 (1986)
1019. M. Schönhof, D. Helbing, Transp. Sci. **41**, 135–166 (2007)
1020. M. Schönhof, D. Helbing, Transp. Res. B **43**, 784–797 (2009)
1021. M. Schönhof, M. Treiber, A. Kesting, D. Helbing, Transp. Res. Rec. **1999**, 3–12 (2007)
1022. J.P. Schorr, S.H. Hamdar, Accident Anal. Prevent. **71**, 93–105 (2014)
1023. M. Schreckenberg, R. Barlović, W. Knospe, H. Klüpfel, in *Computational Statistical Physics*, ed. by K.H. Hoffmann, M. Schreiber (Springer, Berlin, 2001), pp. 113–126
1024. M. Schreckenberg, A. Schadschneider, K. Nagel, N. Ito, Phys. Rev. E **51**, 2939–2949 (1995)
1025. M. Schreckenberg, D.E. Wolf (eds.), *Traffic and Granular Flow'97* (Springer, Singapore, 1998)
1026. H. Shao, W.H.K. Lam, M.L. Tam, Netw. Spat. Econ. **6**, 173–204 (2006)
1027. M. Shawky, H. Nakamura, Transp. Res. Rec. **2012**, 11–19 (2007)
1028. Y. Sheffi, *Urban Transportation Networks: Equilibrium Analysis with Mathematical Programming Methods* (Prentice-Hall, New Jersey, 1984)
1029. W. Shen, H.M. Zhang, Transp. Res. B **65**, 1–17 (2014)
1030. Y. Shiomi, T. Yoshii, R. Kitamura, Transp. Res. B **45**, 1314–1330 (2011)
1031. S.E. Shladover, D. Su, X.-T. Lu, Transp. Res. Rec. **2324**, 6370 (2012)
1032. V.I. Shvetsov, Automation Remote Contr. **64**, 1651–1689 (2003)
1033. V.I. Shvetsov, Automation Remote Contr. **70**, 1728–1736 (2009)
1034. V.I. Shvetsov, D. Helbing, Phys. Rev. E **59**, 6328–6339 (1999)
1035. W. Siegloch, *Schriftenreihe Straßenbau und Straßenverkehrstechnik*, Heft 154 (Bonn, Germany, 1973)
1036. R. Sipahi, S.-I. Niculescu, Phil. Trans. R. Soc. A **368**, 4563–4583 (2010)
1037. E. Smaragdis, M. Papageorgiou, Transp. Res. Rec. **1856**, 74–86 (2003)

1038. K.R. Smilowitz, C.F. Daganzo, M.J. Cassidy, R.L. Bertini, Transp. Res. Rec. **1678**, 225–233 (1999)
1039. M.J. Smith, Transp. Res. C **29**, 131–147 (2013)
1040. Ch. Stamatiadis, N.H. Gartner, Transp. Res. Rec. **1554**, 9–17 (1996)
1041. J.A. Stewart, H. Rakha, M. Van Aerde, Transp. Res. Rec. **1494**, 1–10 (1995)
1042. M. Strating, *Coordinated Signal Control for Urban Networks by Using MFD* (M.Sc. Thesis) (Delft University of Technology, Delft, The Netherlands, 2010)
1043. I. Strnad, M.K. Fijavž, M. Žura, J. Adv. Transp. **50**, 841–856 (2016)
1044. Y. Sugiyama, M. Fukui, M. Kikuchi, K. Hasebe, A. Nakayama, K. Nishinari, S.-I. Tadaki, S. Yukawa, New J. Phys. **10**, 033001 (2008)
1045. N. Sugiyama, T. Nagatani, Physica A **392**, 1848–1857 (2013)
1046. G. Stephanopoulos, P.G. Michalopoulos, Transp. Res. A **13**, 295–307 (1979)
1047. Y. Sugiyama, H. Yamada, Phys. Rev. E **55**, 7749–7752 (1997)
1048. D. Sun, A. Clinet, A.M. Bayen, Transp. Res. B **45**, 880–902 (2011)
1049. J. Sun, L. Zhao, H. Zhang, Transp. Res. Rec. **2421**, 64–73 (2014)
1050. D.-H. Sun, X.-Y. Liao, G.H. Peng, Physica A **390**, 631–635 (2011)
1051. Ch. Suwansirikul, T.L. Friesz, R.L. Tobin, Transp. Sci. **21**, 254–263 (1987)
1052. H. Suzuki, JSAE Rev. **24**, 403–410 (2003)
1053. W.Y. Szeto, L. O'Brien, M. O'Mahony, in [16] (2007), pp. 127–154
1054. W.Y. Szeto, X. Jaber, S.C. Wong, Transp. Rev. **32**, 491–518 (2012)
1055. S. Tadaki, M. Kikuchi, M. Fukui, A. Nakayama, K. Nishinari, A. Shibata, Y. Sugiyama, T. Yoshida, S. Yukawa, New J. Phys. **15**, 103034 (2013)
1056. S.I. Tadaki, K. Nishinari, M. Kikuchi, Y. Sugiyama, S. Yukawa, Physica A **315**, 156–162 (2002)
1057. M. Takayasu, H. Takayasu, Fractals **1**, 860–866 (1993)
1058. A. Talebpour, H.S. Mahmassani, Transp. Res. C **71**, 143–163 (2016)
1059. Z. Tan, H. Yang, R.-Y. Guo, Transp. Res. C **61**, 87–105 (2015)
1060. C.F. Tang, R. Jiang, Q.S. Wu, Physica A **377**, 641–650 (2007)
1061. T.Q. Tang, C.Y. Li, Y.H. Wu, H.J. Huang, Physica A **390**, 3362–3368 (2011)
1062. T.-Q. Tang, Y.-P. Wang, G.-Z. Yu, H.-J. Huang, Commun. Theor. Phys. **58**, 583 (2012)
1063. J.C. Tanner, Biometrica **54**, 657–658 (1967)
1064. M.A.P. Taylor (ed.), *Transportation and Traffic Theory in the 21st Century* (Elsevier Science, Amsterdam, 2002)
1065. L.-J. Tian, H.-J. Huang, Z.-Y. Gao, Netw. Spat. Econ. **12**, 589–608 (2012)
1066. J.-F. Tian, B. Jia, X.-G. Li, R. Jiang, X.-M. Zhao, Z.-Y. Gao, Physica A **388**, 4827–4837 (2009)
1067. J.-F. Tian, B. Jia, S.F. Ma, C.Q. Zhu, R. Jiang, Y.X. Ding, Transp. Sci. (2016). http://dx.doi.org/10.1287/trsc.2015.0667
1068. J.-F. Tian, R. Jiang, B. Jia, Z.-Y. Gao, S.F. Ma, Transp. Res. B **93**, 338–354 (2016)
1069. J.-F. Tian, R. Jiang, G. Li, M. Treiber, B. Jia, C.Q. Zhu, Transp. Rec. F **41**, 55–65 (2016)
1070. J.-F. Tian, G. Li, M. Treiber, R. Jiang, N. Jia, S.F. Ma, Transp. Rec. B **93**, 560–575 (2016)
1071. J.-F. Tian, M. Treiber, S.F. Ma, B. Jia, B. Jia, W.Y. Zhang, Transp. Res. B **71**, 138–157 (2015)
1072. J.-F. Tian, Z.-Z. Yuan, B. Jia, M.-h. Li, G.-j. Jiang, Physica A **391**, 4476–4482 (2012)
1073. J.-F. Tian, Z.-Z. Yuan, M. Treiber, B. Jia, W.-Y. Zhanga, Physica A **391**, 3129–3139 (2012)
1074. J.-F. Tian, Z.-Z. Yuan, B. Jia, H.-q. Fan, T. Wang, Phys. Lett. A **376**, 2781–2787 (2012)
1075. B. Tilch, D. Helbing, in *Traffic and Granular Flow' 99*, ed. by D. Helbing, H.J. Herrmann, M. Schreckenberg, D.E. Wolf (Springer, Berlin, 2000), pp. 333–338
1076. R.L. Tobin, T.L. Friesz, Transp. Sci. **22**, 242–250 (1988)
1077. K. Tobita, T. Nagatani, Physica A **391**, 6137–6145 (2012)
1078. J.R. Tolle, Transp. Res. **8**, 91–96 (1974)
1079. E. Tomer, L. Safonov, S. Havlin, Phys. Rev. Lett. **84**, 382 (2000)
1080. E. Tomer, L. Safonov, N. Madar, S. Havlin, Phys. Rev. E **65**, 065101 (R) (2002)
1081. http://en.wikipedia.org/wiki/Traffic_simulation

1082. M. Treiber, D. Helbing, J. Phys. A Math. Gen. **32**, L17–L23 (1999)
1083. M. Treiber, D. Helbing, Automatisierungstechnik **49**, 478–484 (2001)
1084. M. Treiber, D. Helbing, Phys. Rev. E **68**, 046119 (2003)
1085. M. Treiber, D. Helbing, Eur. Phys. J. B **68**, 607–618 (2009)
1086. M. Treiber, A. Hennecke, D. Helbing, Phys. Rev. E **59**, 239–253 (1999)
1087. M. Treiber, A. Hennecke, D. Helbing, Phys. Rev. E **62**, 1805–1824 (2000)
1088. M. Treiber, A. Kesting, *Verkehrsdynamik und -simulation: Daten, Modelle und Anwendungen der Verkehrsflussdynamik* (Springer, Heidelberg, Dordrecht, London, New York, 2010)
1089. M. Treiber, A. Kesting, *Traffic Flow Dynamics* (Springer, Berlin, 2013)
1090. M. Treiber, A. Kesting, D. Helbing, Phys. Rev. E **74**, 016123 (2006)
1091. M. Treiber, A. Kesting, D. Helbing, Transp. Res. B **44**, 983–1000 (2010)
1092. J. Treiterer, Transp. Res. **1**, 231–251 (1967)
1093. J. Treiterer, "Investigation of Traffic Dynamics by Aerial Photogrammetry Techniques", *Ohio State University Technical Report PB 246 094* (Columbus, Ohio, 1975)
1094. J. Treiterer, J.A. Myers, in *Procs. 6th International Symposium on Transportation and Traffic Theory*, ed. by D.J. Buckley (A.H. & AW Reed, London, 1974), pp. 13–38
1095. J. Treiterer, J.I. Taylor, Highway Res. Rec. **142**, 1–12 (1966)
1096. J. VanderWerf, S.E. Shladover, M.A. Miller, N. Kourjanskaia, Transp. Res. Rec. **1800**, 78–84 (2002)
1097. H.R. Varia, P.J. Gundaliya, S.L. Dhingra, Res. Transp. Econ. **38**, 35–44 (2013)
1098. J. Vasic, H.J. Ruskin, Physica A **391**, 2720–2729 (2012)
1099. Ö. Verbas, H.S. Mahmassani, M.F. Hyland, Transp. Res. B **93**, 1–16 (2016)
1100. http://de.wikipedia.org/wiki/VISSIM
1101. S.V. Ukkusuri, S.T. Waller, Netw. Spat. Econ. **8**, 383–406 (2008)
1102. T. Urbanik II, W. Hinshaw, K. Barnes, Transp. Res. Rec. **1320**, 110–118 (1991)
1103. C. Wagner, Physica A **245**, 124–138 (1997)
1104. P. Wagner, Transp. Res. B **46**, 1384–1392 (2012)
1105. P. Wagner, in *Autonomes Fahren*, ed. by M. Maurer, J.Ch. Gerdes, B. Lenz, H. Winner (Springer, Berlin, 2015), pp. 313–330
1106. C. Wagner, C. Hoffmann, R. Sollacher, J. Wagenhuber, B. Schürmann, Phys. Rev. E **54**, 5073–5085 (1996)
1107. P. Wagner, I.A. Lubashevsky, arXiv:cond-mat/0311192 [cond-mat.stat-mech] (2003)
1108. P. Wagner, K. Nagel, Eur. Phys. J. B **63**, 315–320 (2008)
1109. P. Wagner, K. Nagel, D. Wolf, Physica A **234**, 687–698 (1997)
1110. P. Wagner, R. Nippold, S. Gabloner, M. Margreiter, Chaos Solitons Fractals 1–9 (2016)
1111. J. Wahle, A.L.C. Bazzan, F. Klugl, M. Schreckenberg, Physica A **287**, 669–681 (2000)
1112. J. Wahle, A. Bazzan, F. Klugl, M. Schreckenberg, Transp. Res. C **10**, 399–417 (2002)
1113. J. Wahle, R. Chrobok, A. Pottmeier, M. Schreckenberg, Netw. Spat. Econ., **2**, 371–386 (2002)
1114. Y. Wang, Y.-Y. Chen, Physica A **463**, 12–24 (2016)
1115. M. Wang, W. Daamen, S.P. Hoogendoorn, B. van Arem, Transp. Res. C **40**, 271–289 (2014)
1116. M. Wang, W. Daamen, S.P. Hoogendoorn, B. van Arem, Transp. Res. C **40**, 290–311 (2014)
1117. M. Wang, W. Daamen, S.P. Hoogendoorn, B. van Arem, R. Happee, Transp. Res. C **58**, 73–92 (2015)
1118. J. Wang, W. Deng, in *Proc. TRB 92nd Annual Meeting*, Paper # 13-0836 (TRB, Washington DC, 2013)
1119. T. Wang, Z.Y. Gao, W.Y. Zhang, J. Zhang, S.B. Li, Nonlinear Dyn. **77**, 635–642 (2014)
1120. T. Wang, Z.Y. Gao, J. Zhang, Nonlinear Dyn. **73**, 2197–2205 (2013)
1121. Y. Wang, N. Geroliminis, L. Leclercq, Transp. Res. C **68**, 507–508 (2016)
1122. R. Wang, R. Jiang, Q.-S. Wu, M. Liu, Physica A **378**, 475–484 (2007)
1123. D.Z. Wang, H. Liu, in *Proc. TRB 92nd Annual Meeting*, Paper # 13-1443 (TRB, Washington DC, 2013)
1124. S. Wang, Q. Meng, H. Yang, Transp. Res. B **50**, 42–60 (2013)
1125. W. Wang, B. Wang, W. Zheng, C. Yin, T. Zhou, Phys. Rev. E **72**, 066702 (2005)

1126. Y. Wang, Y.I. Zhang, J. Hu, L. Li, Int. J. Mod. Phys. C **23**, 1250060 (2012)
1127. J.G. Wardrop, in *Proc. of Inst. of Civil Eng. II.* **1** 325–378 (1952)
1128. P. Wasielewski, Transp. Sci. **15**, 364–378 (1981)
1129. J.A. Wattleworth, Highway Res. Rec. **157**, 1–21 (1967)
1130. J.A. Wattleworth, D.S. Berry, Highway Res. Rec. **89**, 1–25 (1965)
1131. F.V. Webster, *Road Research Technical Paper No. 39* (Road Research Laboratory, London, UK, 1958)
1132. F.V. Webster, B.M. Cobbe, "UK Capacity Analysis Techniques and Delay Estimation Traffic Signals" (The Majesty's Stationary Office, London, England, 1966)
1133. C. Wei, Y. Asakura, T. Iryo, J. Adv. Transp. **46**, 222–235 (2012)
1134. E.A. Wemple, A.M. Morris, A.D. May, in *Procs. of the International Symposium on Highway Capacity*, ed. by U. Brannolte (Karlsruhe, Germany, 1991), pp. 439–455
1135. D. Westland, in *Proc. of the 3rd Sym. on Highway Capacity and Level of Service*, ed. by R. Rysgaard, vol. 2 (Road Directorate, Copenhagen, Ministry of Transport, Denmark, 1998), pp. 1095–1116
1136. G.B. Whitham, *Linear and Nonlinear Waves* (Wiley, New York, 1974)
1137. G.B. Whitham, Proc. R. Soc. Lond. A **428**, 49 (1990)
1138. B.-W. Wie, T.L. Friesz, R.L. Tobin, Transp. Res. B **24**, 431–442 (1990)
1139. R. Wiedemann, *Simulation des Verkehrsflusses* (University of Karlsruhe, Karlsruhe, 1974)
1140. R.E. Wilson, IMA J. Appl. Math. **66**, 509–537 (2001)
1141. R.E. Wilson, Phil. Trans. R. Soc. A **366**, 2017–2032 (2008)
1142. J.R. Windover, "Empirical Studies of the Dynamic Features of Freeway Traffic", *Ph.D. Thesis* (Dept. of CEE, ITS, UC, Berkeley, 1998)
1143. D.E. Wolf, Physica A **263**, 438–451 (1999)
1144. D.E. Wolf, M. Schreckenberg, A. Bachem (eds.), *Traffic and Granular Flow* (World Scientific, Singapore, 1995)
1145. G.C.K. Wong, S.C. Wong, Transp. Res. A **36**, 827–841 (2002)
1146. X. Wu, Y. Nie, in *Proc. TRB 92nd Annual Meeting*, Paper # 13-4100 (TRB, Washington DC, 2013)
1147. J.J. Wu, H.J. Sun, Z.Y. Gao, Phys. Rev. E **78**, 036103 (2008)
1148. C. Xie, S.T. Waller, Transp. Res. B **46**, 1023–1042 (2012)
1149. C. Xiong, X. Chen, X. He, X. Lin, L. Zhang, Transp. Res. C **64**, 148–163 (2016)
1150. Z.-T. Xiang, Y.-J. Li, Y.-F. Chen, L. Xiong, Physica A **392**, 5399–5413 (2013)
1151. X. Xu, A. Chen, L. Cheng, C. Yang, Transp. Res. B **95**, 53–75 (2017)
1152. X. Xu, A. Chen, S. Kitthamkesorn, H. Yang, H.K. Lo, Transp. Res. B **81**, 686–703 (2015)
1153. Ch. Xu, P. Liu, W. Wang, Zh. Li, Accident Anal. Prevent. **85**, 45–57 (2015)
1154. H. Yang, M.G.H. Bell, Transp. Res. B **31**, 303–314 (1997)
1155. H. Yang, M.G.H. Bell, Transp. Res. B **32**, 539–545 (1998)
1156. H. Yang, M.G.H. Bell, Transp. Rev. **18**, 257–278 (1998)
1157. H. Yang, M.G.H. Bell, Q. Menga, Transp. Res. B **34**, 255–275 (2000)
1158. J. Yang, G. Jiang, Transp. Res. C **47**, 168–178 (2014)
1159. H. Yang, W.-L. Jin, Transp. Res. C **53**, 19–34 (2015)
1160. D. Yang, P. Jin, Y. Pu, B. Ran, Eur. Phys. J. B **86**, 92 (2013)
1161. H. Yang, J. Lu, X. Hu, J. Jiang, Physica A **392**, 4009–4018 (2013)
1162. B. Yang, Ch. Monterola, Phys. Rev. E **92**, 042802 (2015)
1163. B. Yang, Ch. Monterola, Transp. Res. C **72**, 283–305 (2016)
1164. H. Yang, X. Wang, Y. Yin, Transp. Res. B **46**, 1295–1307 (2012)
1165. B. Yang, X. Xu, J.Z.F. Pang, Ch. Monterola, arXiv:1504.01256 (2015)
1166. B. Yang, J.W. Yoon, Ch. Monterola, arXiv:1603.04272 (2016)
1167. H. Yang, X. Zhang, Q. Meng, Transp. Res. B **38**, 191–213 (2004)
1168. M. Yildirimoglu, M. Ramezani, N. Geroliminis, Transp. Res. C **59**, 404–420 (2015)
1169. Q. Yu, D. Fang, W. Du, Eur. J. Oper. Res. **239**, 112–118 (2014)
1170. Y. Yufei, J.W.C. van Lint, R.E. Wilson, F. van Wageningen-Kessels, S.P. Hoogendoorn, IEEE Trans. ITS **13**, 59–70 (2012)

1171. S. Yukawa, M. Kikuchi, A. Nakayama, K. Nishinari, Y. Sugiyama, S. Tadaki, in [290] (2003), pp. 243–256
1172. H. Zackor, W. Balz, Th. Pischner, in *Forschung Straßenbau und Straßenverkehrstechnik*, Heft 434 (Bundesminister für Verkehr, Germany, 1985)
1173. H.M. Zhang, Transp. Res. B **32**, 485–498 (1998)
1174. H.M. Zhang, Transp. Res. B **33**, 387–398 (1999)
1175. H.M. Zhang, Transp. Res. B **34**, 583–603 (2000)
1176. H.M. Zhang, Transp. Res. B **35**, 337–365 (2001)
1177. H.M. Zhang, Transp. Res. B **36**, 275–290 (2002)
1178. H.M. Zhang, Transp. Res. B **37**, 27–41 (2003)
1179. H.M. Zhang, Transp. Res. B **37**, 101–105 (2003)
1180. C. Zhang, X. Chen, A. Sumalee, Transp. Res. B **45**, 534–552 (2011)
1181. H.M. Zhang, W.L. Jin, Transp. Res. Rec. **1802**, 197–204 (2002)
1182. H.M. Zhang, T. Kim, Transp. Res. B **39**, 385–399 (2005)
1183. K. Zhang, H.S. Mahmassani, C.-C. Lu, Transp. Res. Rec. **2085**, 86–94 (2008)
1184. K. Zhang, H.S. Mahmassani, C.-C. Lu, Transp. Res. C **27**, 189–204 (2013)
1185. H.M. Zhang, S.G. Ritchie, Transp. Res. C **5**, 273–286 (1997)
1186. M. Zhang, C.-W. Shu, G.C.K. Wong, S.C. Wong, J. Comput. Phys. **191**, 639–659 (2003)
1187. M. Zhang, D.H. Sun, W.N. Liu, Nonlinear Dyn. **81**, 1623–1633 (2015)
1188. M. Zhang, D.H. Sun, W.N. Liu, M. Zhao, S.L. Chen, Physica A **422**, 16–24 (2015)
1189. M. Zhang, D.H. Sun, C. Tian, Nonlinear Dyn. **77**, 839–847 (2014)
1190. P. Zhang, C.-X. Wu, S.C. Wong, Physica A **391**, 456–463 (2012)
1191. X. Zhang, H. Yang, Transp. Res. B **38**, 517–537 (2004)
1192. X. Zhang, H. Yang, H.-J. Huang, Eur. J. Oper. Res. **189**, 146–158 (2008)
1193. D. Zhao, X. Bai, F.-Y. Wang, J. Xu, W. Yu, IEEE Trans. ITS **12**, 990–999 (2011)
1194. C. Zhao, B. Fu, T. Wang, Transp. Res. E **61**, 135–141 (2014)
1195. C.-L. Zhao, H.-J. Huang, Transp. Res. C **68**, 22–37 (2016)
1196. J. Zhao, Y. Liu, Transp. Res. C **73**, 219–238 (2017)
1197. X.-m. Zhao, D.-f. Xie, Q. Li, Comput. Phys. Comm. **187**, 106–114 (2015)
1198. L. Zheng, Z. He, T. He, Transp. Res. C **75**, 136–167 (2017)
1199. R. Zhong, C. Chen, A.H.F. Chow, T. Pan, F. Yuan, Z. He, J. Adv. Transp. **50**, 363–385 (2016)
1200. R.X. Zhong, A. Sumalee, T.L. Friesz, W.H.K. Lam, Transp. Res. B **45**, 1035–1061 (2011)
1201. R. Zhong, A. Sumalee, T. Maruyama, J. Adv. Transp. **46**, 191–221 (2012)
1202. X. Zhou, H.S. Mahmassani, K. Zhang, Transp. Res. C **16**, 167–186 (2008)
1203. J. Zhou, Z.-K. Shi, Nonlinear Dyn. **83**, 1217–1236 (2015)
1204. F. Zhu, S.V. Ukkusuri, IEEE Trans. ITS **18**, 282–291 (2017)
1205. W.X. Zhu, L.D. Zhang, Int. J. Mod. Phys. C **23**, 1250025 (2012)
1206. B. Zielke, R. Bertini, M. Treiber, Transp. Res. Rec. **2088**, 57–67 (2008)
1207. A.K. Ziliaskopoulos, Transp. Sci. **34**, 37–49 (2000)
1208. A.K. Ziliaskopoulos, D. Kotzinos, H.S. Mahmassani, Transp. Res. C **5**, 95–107 (1997)

Chapter 2
Achievements of Empirical Studies of Traffic Breakdown at Highway Bottlenecks

2.1 Introduction

Traffic breakdown is almost daily observed in traffic networks of any industrial country of the world. As already emphasized in the book introduction (Sect. 1.1), traffic breakdown is a transition from free flow to congested traffic. Therefore, highway capacity of free flow is limited by traffic breakdown [23–25, 53–55, 73, 76, 77, 79, 83–85, 93, 94, 115, 129, 137, 167]. Traffic breakdown with resulting traffic congestion occurs usually at a road bottleneck (see, e.g., [1–14, 16–26, 28–38, 40–52, 55–75, 77–101, 103–106, 108, 109, 112–114, 116–160, 162–171, 174–181, 183, 185–200]). Thus, to understand the nature of highway capacity of real traffic, empirical features of traffic breakdown at a highway bottleneck should be known.

Traffic breakdown at a highway bottleneck is a transition from free flow to congested traffic in a neighborhood of the bottleneck [23–25, 53–55, 73, 79, 85, 105, 106, 115, 129, 137, 167].

Beginning from the classical work by Greenshields [76], a great effort has been made by many scientific groups to understand empirical features of traffic breakdown, in particular, by May et al. [136–138], by Hall et al. [77, 79, 81–86],

B.S. Kerner, *Breakdown in Traffic Networks*, DOI 10.1007/978-3-662-54473-0_2

by Persaud et al. [162–168], by Elefteriadou et al. [55, 129], by Brilon et al. [23–25, 73], by Banks et al. [5–14], by Cassidy et al. [33] and by many other scientific groups (see, e.g., [15, 27, 38, 39, 46, 98–100, 102, 161, 172, 173, 175, 178, 182, 184] as well as references in reviews and books [53, 93, 94, 105, 106, 137]).

As mentioned in Sect. 1.2, highway bottlenecks are caused, for example, by roadwork, on- and off-ramps, road gradients, reduction of lane number, slow moving vehicles [23–25, 53–55, 73, 79, 85, 105, 106, 115, 129, 137, 167]. In the letter case, a bottleneck caused by a slow moving vehicle is called a moving bottleneck [43, 61, 62, 71, 109, 121, 122, 147, 154, 155].

The objective of this chapter is a discussion of some of important achievements of empirical studies of traffic breakdown at highway bottlenecks in real measured traffic data.

2.2 Empirical Features of Traffic Breakdown

2.2.1 Traffic Breakdown—Transition from Free to Synchronized Flow at Highway Bottleneck

Empirical studies show that during traffic breakdown vehicle speed sharply decreases (Figs. 2.1a,b). In contrast, after traffic breakdown has occurred the flow rate can remain as large as in an initial free flow (traffic breakdown is labeled by "traffic breakdown" in Fig. 2.1c). For this reason, traffic breakdown is also called *speed drop* or *speed breakdown* [23–25, 53–55, 73, 79, 85, 105, 106, 115, 129, 137, 167].

Traffic breakdown at a highway bottleneck is a local phase transition from free flow (F) to congested traffic whose downstream front is usually fixed at the bottleneck location (Fig. 2.1a) [23–25, 53–55, 73, 77, 79, 83–85, 93, 94, 115, 129, 137, 167].

As mentioned in Sect. 1.3, in the three-phase theory such congested traffic is called synchronized flow (S) [105, 106]. In other words, using the terminology of the three-phase theory, traffic breakdown is a transition from free flow to synchronized flow (F→S transition) [105–107]. However, it should be emphasized that as long as nucleation features of synchronized flow are not discussed (this discussion will be done in Chap. 3), the term *synchronized flow* is nothing more as only the definition of congested traffic whose downstream front is fixed at the bottleneck.

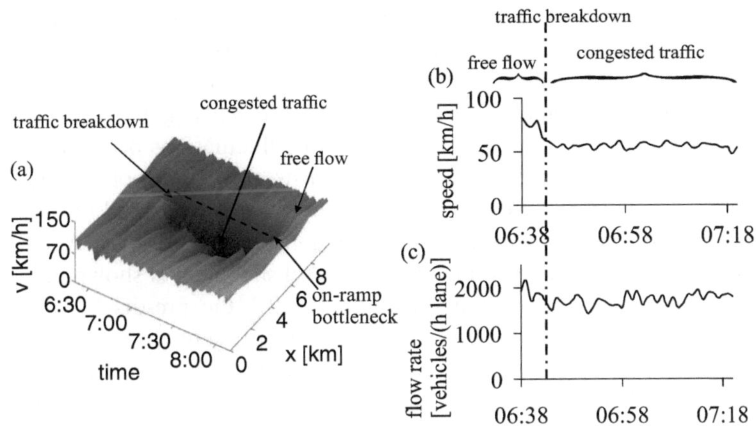

Fig. 2.1 Empirical example of traffic breakdown at on-ramp bottleneck: (**a**) Averaged vehicle speed in space and time (1-min averaging interval). (**b, c**) Vehicle speed (**b**) and averaged flow rate (per road lane) (**b**) measured on freeway in a neighborhood of the on-ramp bottleneck at detector location $x = 6.4$ km. Real measured traffic data of road detectors adapted from [105] that was measured on three-lane freeway A5-South in Germany on March 26, 1996; the data exhibits common features that are qualitatively the same as those measured on a variety of highways in different countries (see, e.g., [23–25, 53–55, 73, 79, 85, 115, 129, 137, 167])

2.2.2 Time-Dependence of Flow Rate During Empirical Traffic Breakdown at Highway Bottleneck

The flow rate in free flow downstream of a highway bottleneck measured just before traffic breakdown occurs is called the *pre-discharge flow rate*. The flow rate in free flow downstream of the bottleneck after traffic breakdown has occurred at this bottleneck, i.e., the flow rate in the congested pattern outflow is called the *discharge flow rate* [53, 79, 167]. We consider some empirical features of the pre-discharge flow rate and discharge flow rate measured in free flow downstream of on-ramp bottlenecks. An example of such on-ramp bottleneck is shown in Fig. 2.1. Other examples can be found in Chap. 10 of the book [105].

Banks [6] as well as Hall and Agyemang-Duah have found [79] that in the most cases, the discharge flow rate is smaller than the pre-discharge flow: "A flow-rate drop" is observed during the breakdown.[1] However, in some other cases that are statistically not significant one, the discharge flow rate is larger than the pre-discharge flow rate: "A flow-rate jump" is observed during traffic breakdown.

[1]Rather than the term "flow-rate drop" used here, the empirical flow-rate drop during the breakdown is usually called as "capacity drop" [6, 53–55, 77, 79, 85, 93, 115, 129, 167]. The reason why we do not use the well-known term "capacity drop" will be explained in Sect. 4.11.2.

An empirical example of "flow-rate jump" during the breakdown at an on-ramp bottleneck can be seen on Fig. 10.1 (e) of the book [105].

It must be stressed that empirical values of "flow-rate drop" and "flow-rate jump" observed during traffic breakdown are on average usually as large as fluctuations in the flow rate at the bottleneck. These flow-rate fluctuations occur, for example, due to a time-dependence of traffic demand upstream of the bottleneck. For this reason, it can be stated that the discharge flow rate that is measured *just* after traffic breakdown has occurred can be *on average* as large as the pre-discharge flow rate. A representative empirical example of this general statement is shown in Fig. 2.1c.

Indeed, during the breakdown shown in Fig. 2.1 the pre-discharge flow rate averaged over the road and given per highway lane is equal to 1880 vehicles/(h lane); the discharge flow rate just after the breakdown has occurred is equal to 1840 vehicles/(h lane). This means that the flow-rate drop during the breakdown is only 40 vehicles/(h lane), i.e., in this particular case the flow-rate drop is only 2% of the pre-discharge flow rate. For this particular empirical example, the average flow rate during the time interval 6:35–7:20 (Fig. 2.1c) is equal to 1740 vehicles/(h lane) and the standard deviation for the flow rate fluctuations is equal to 8.8% (153 vehicles/(h lane)). Our other empirical studies of flow-rate drops occurring during traffic breakdown that have been observed in 1-min averaged traffic data measured at on-ramp bottlenecks on German highways on about 200 days during 1996–2014 confirm the result of the work of Hall and Agyemang-Duah [79] who wrote "there appears to be roughly a 6% reduction in maximum flow rates after the onset of congestion".

In another empirical example of traffic breakdown at an on-ramp bottleneck shown on Fig. 10.1 (e) of the book [105], a flow-rate jump during traffic breakdown has been observed. The pre-discharge flow rate averaged over the road is equal to 1720 vehicles/(h lane); the discharge flow rate just after the breakdown has occurred is equal to 2100 vehicles/(h lane). This means that in this particular empirical example the flow-rate jump is about 380 vehicles/(h lane). However, as above-mentioned, our empirical results [105] have confirmed results of Banks [6] as well as Hall and Agyemang-Duah [79] that a flow-rate jump has considerably seldom been observed during the breakdown than a flow-rate drop.

Thus, empirical values of flow-rate drops and flow-rate jumps during the breakdown are nearly the same as flow rate fluctuations at the bottleneck. Therefore, we can conclude that in most empirical observations of traffic breakdown at highway bottlenecks the flow rate remains during the breakdown on average as large as in an initial free flow (Fig. 2.1c).[2]

[2]However, it should be emphasized that the above statement that the discharge flow rate can remain as large as the pre-discharge flow rate is related often *only* to the discharge flow rate measured during some limited time interval after the breakdown has occurred. Indeed, due to traffic breakdown a congested pattern emerges and further develops upstream of the bottleneck. Therefore, over time (usually 10–30 min after the breakdown has occurred at the bottleneck) the discharge flow rate can decrease considerably during the propagation of the congested pattern upstream of the bottleneck. In particular, this occurs often when due to the so-called pinch effect in

2.3 Stochastic Behavior and Probability of Traffic Breakdown at Highway Bottleneck

In 1995, Elefteriadou et al. found that traffic breakdown at a highway bottleneck has a stochastic (probabilistic) behavior [53, 55]. This means the following: At a given flow rate in free flow at the bottleneck traffic breakdown can occur but it should not necessarily occur. Thus on one day traffic breakdown occurs, however, on another day at the same flow rates traffic breakdown is not observed.

Studying the probability for the probabilistic breakdown phenomenon at a freeway bottleneck, Persaud et al. [167] have analyzed the flow rate q in free flow on different days at which traffic breakdown occurs at the bottleneck. All observed flow rates have been divided into groups in multiples of 100 vehicles/h. The frequency of flow rates, N_i, in each group i in free flow has been studied before traffic breakdown occurs. For each flow rate group i over a large number of days, the number of instances n_i is found in which the flow rate q_i of traffic breakdown at the bottleneck falls within the range of flow rate group i. Thus, for the calculation of the probability $P^{(B)}$ of traffic breakdown for the flow rate group i the following formula has been used [167]:

$$P^{(B)}(q_i) = \frac{n_i}{N_i}. \tag{2.1}$$

It can be seen from Fig. 2.2 that probability $P^{(B)}$ of traffic breakdown at the bottleneck is a sharp increasing function of the flow rate in free flow in the vicinity of the point, where the breakdown probability $P^{(B)} = 1$. This behavior of the breakdown probability $P^{(B)}$ is qualitatively independent of the averaging time interval T_{av} for the flow rate. However, the increase in this averaging time interval

synchronized flow upstream of the bottleneck [105], moving jams emerge in the synchronized flow. In this case, the congested pattern can exhibit a very complex spatiotemporal structure consisting of the synchronized flow and wide moving jams. The maximum flow rate in the outflow from a wide moving jam is considerably smaller than the maximum possible flow rate in synchronized flow. This is one of the reasons why the discharge flow rate in the outflow of a well-developed congested pattern at the bottleneck, as well-known from many empirical observations (see, e.g. [93, 94, 137] and references there), can become over time considerably smaller than the pre-discharge flow rate. A brief theoretical discussion of the development of congested patterns required for a comparison of two-phase and three-phase traffic flow models will be done in Sect. 8.4. However, a detailed consideration of the physics of the development of congested patterns and their empirical features are out of scope of this book. The physics of the development of congested patterns and their empirical features can be found in the book [105].

One of the exclusions of the above case of the decrease in the discharge flow rate over time is shown in Fig. 2.1. In this case, the synchronized flow is localized at the bottleneck (we call such a congested pattern as a localized synchronized flow pattern (LSP) [105]). No pinch effect occurs within the synchronized flow. For this reason, the statement that the discharge flow rate is on average as large as in an initial free flow is valid for this particular case for the whole time of the congested pattern existence (about one hour in Fig. 2.1a).

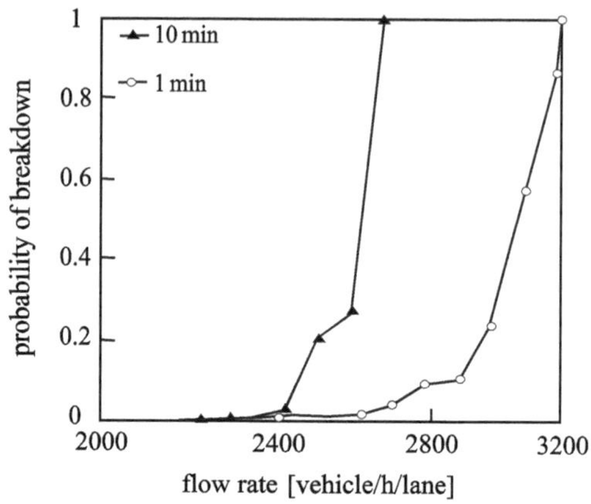

Fig. 2.2 Probability for traffic breakdown at on-ramp bottleneck for two different averaging time intervals for traffic variables $T_{av} = 1$ and 10 min. Adapted from Persaud et al. [167]

leads to a decrease in the flow rate at which $P^{(B)} = 1$ (curves $T_{av} = 1$ min and 10 min in Fig. 2.2). One of the reasons for this behavior of the flow-rate function of the breakdown probability $P^{(B)}$ is that in empirical data the flow rate in free flow usually changes considerably over time. As a result, in 10 minutes averaged data (Fig. 2.2) there can be points of 1 minute averaged data where the flow rate is considerably higher than the average flow rate in 10 minutes averaged data [167].

Growing flow-rate functions of the probability for traffic breakdown at on-ramp bottlenecks has also been found for freeways in the USA by Lorenz and Elefteriadou [129] as well as for 5 minutes averaged traffic data measured on German freeways by Brilon et al. [23–25, 73]. The results are qualitatively similar to the findings of Persaud et al. [167].

A growing character of the empirical flow-rate dependence of the probability of traffic breakdown at a highway bottleneck discovered by Persaud et al. [167] and further studied by Lorenz and Elefteriadou [129] as well as by Brilon et al. [23–25, 73] has also been confirmed in our later empirical studies of traffic breakdown at off- and on-ramp bottlenecks with the use of 1-min averaged data measured on German highways (black points in Fig. 2.3) [111]. The empirical probabilities of traffic breakdown are well fitted with the theoretical flow-rate dependence of the breakdown probability (dashed curves in Fig. 2.3) that was revealed in 2002 by Kerner et al. [110] in the framework of the three-phase theory (see explanations in Sect. 5.4).

The empirical probability of traffic breakdown at highway bottlenecks is a growing flow-rate function [23–25, 53, 54, 73, 115, 129, 167].

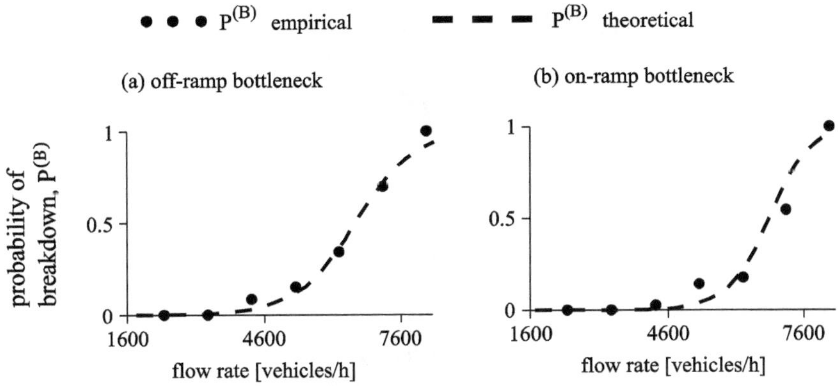

Fig. 2.3 Empirical flow-rate dependence of the probability of traffic breakdown $P^{(B)}$ measured at an off-ramp bottleneck (**a**) and an on-ramp bottleneck (**b**): (**a**) Empirical breakdown probability (*black points*) was found from a study of traffic data in which traffic breakdown at the off-ramp bottleneck was observed on 89 different days. (**b**) Empirical breakdown probability (*black points*) at the on-ramp bottleneck was found from a study of traffic data in which traffic breakdown was observed on 56 different days. Empirical breakdown probabilities (*black points*) are related to real field traffic data measured by road detectors installed along a section of three-lane freeway A5-South. Dashed curves are the theoretical flow-rate dependencies of the probability of traffic breakdown $P^{(B)}$ related to formula (5.12) of Sect. 5.4.1. Adapted from [111]

There is also a well-known hysteresis phenomenon associated with traffic breakdown and a return transition to free flow (e.g., [23–25, 53–55, 73, 79, 85, 115, 129, 137, 167]) (Fig. 2.4).

Traffic parameters, like weather, percentage of long vehicles in traffic flow, shares of aggressive and timid drivers are stochastic time-functions. Thus, it is generally assumed that the stochastic nature of real traffic breakdown might be explained by classical traffic flow theories, in which stochastic traffic parameters should be taken into account (see, e.g. [23–25, 53–55, 73, 79, 85, 115, 129, 167] and references there).

- In contrast with this general accepted assumption [23–25, 53–55, 73, 79, 85, 115, 129, 167], in Chap. 3 we will explain that the sole knowledge of the above-mentioned features of empirical traffic breakdown at highway bottlenecks revealed in [23–25, 53–55, 73, 76, 77, 79, 83–85, 93, 94, 115, 129, 137, 167] is *not* sufficient to disclose the physical nature of traffic breakdown and associated stochastic highway capacity.
- Indeed, in Sect. 4.10 we will show that the classical understanding of stochastic highway capacity (Sect. 1.7.1) that at any given time instant there is a particular value of highway capacity of free flow at a bottleneck [23–25, 53–55, 73, 115, 129, 167] is invalid for real traffic.

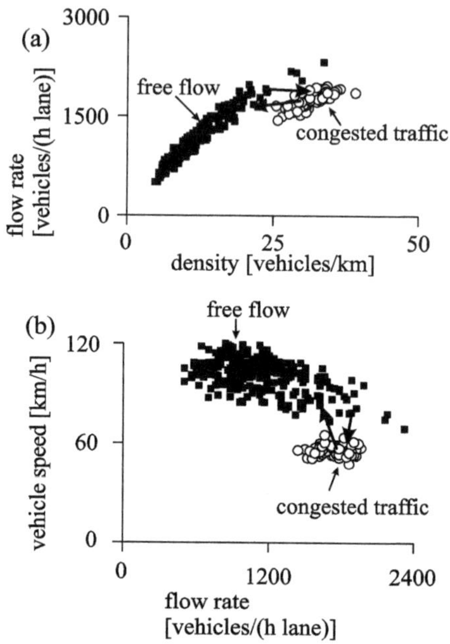

Fig. 2.4 Typical empirical example of hysteresis phenomenon [23–25, 53–55, 73, 79, 85, 115, 129, 137, 167] presented in the flow–density (**a**) and speed–flow planes (**b**). This empirical example of hysteresis phenomenon, which is caused by traffic breakdown (F→S transition) and return transition to from congested traffic (synchronized flow) to free flow (S→F transition), is related to traffic data measured in a neighborhood of the on-ramp bottleneck at detector location $x = 6.4$ km within traffic pattern shown in Fig. 2.1a. Black points—free flow, white circles—congested traffic (synchronized flow)

2.4 Conclusions

The achievements of empirical studies of traffic breakdown at road bottlenecks (see, e.g., [23–25, 53–55, 73, 79, 85, 115, 129, 137, 167]) can be summarized as follows:

1. Traffic breakdown at a highway bottleneck is a local phase transition from free flow (F) to congested traffic whose downstream front is usually fixed at the bottleneck location. In the three-phase theory, such congested traffic has been called "synchronized flow" (S) [105, 106]. However, as long as nucleation features of synchronized flow are not discussed (see Chap. 3), the term *synchronized flow* is nothing more as only the definition of congested traffic whose downstream front is fixed at the bottleneck.
2. During empirical traffic breakdown the vehicle speed decreases sharply. The flow rate in the emergent congested traffic, which is observed after the breakdown at the bottleneck location, can be as large as the flow rate has been in an initially free flow at the bottleneck before the breakdown has occurred (see, e.g., [23–25, 53–55, 73, 79, 85, 115, 129, 137, 167] and references there).

3. Empirical breakdown phenomenon exhibits the probabilistic nature: At the same highway bottleneck traffic breakdown occurs at very different flow rates on different days of data observations as discovered firstly by Elefteriadou et al. [55] and confirmed later in many empirical observations (see, e.g., [23–25, 54, 73, 115, 129, 167] and references in the book by Elefteriadou [53]).
4. The probability of empirical traffic breakdown is an increasing flow rate function (Fig. 2.2), as firstly discovered in 1998 by Persaud et al. [167]. Later this important empirical result has been confirmed in studies of real field traffic data measured in different countries [23–25, 54, 73, 115, 129].
5. There is a well-known hysteresis phenomenon associated with traffic breakdown and a return transition to free flow (e.g., [79, 85, 137]) (Fig. 2.4).

References

1. R.E. Allsop, M.G.H. Bell, B.G. Heydecker (eds.), *Transportation and Traffic Theory 2007* (Elsevier Science, Amsterdam, 2007)
2. C. Appert-Rolland, F. Chevoir, P. Gondret, S. Lassarre, J.-P. Lebacque, M. Schreckenberg (eds.), *Traffic and Granular Flow'07* (Springer, Heidelberg, 2009)
3. P. Athol, Highway Res. Rec. **72**, 58–87 (1965)
4. P. Athol, Highway Res. Rec. **72**, 137–155 (1965)
5. J.H. Banks, Transp. Res. Rec. **1225**, 53–60 (1989)
6. J.H. Banks, Transp. Res. Rec. **1287**, 20–28 (1990)
7. J.H. Banks, "Evaluation of the Two-Capacity Phenomenon as a Basis for Ramp Metering", *Final Report, Civil Engineering Report Series 9002* (San Diego State University, San Diego, CA, 1990)
8. J.H. Banks, Transp. Res. Rec. **1320**, 234–241 (1991)
9. J.H. Banks, Transp. Res. Rec. **1320**, 83–90 (1991)
10. J.H. Banks, Transp. Res. Rec. **1394**, 17–25 (1993)
11. J.H. Banks, Transp. Res. Rec. **1510**, 1–10 (1995)
12. J.H. Banks, Transp. Res. Rec. **1678**, 128–134 (1999)
13. J.H. Banks, Transp. Res. Rec. **1802**, 225–232 (2002)
14. J.H. Banks, Transp. Res. Rec. **2099**, 14–21 (2009)
15. S. Bassan, A. Faghri, A. Polus, Civil Eng. Environ. Syst. **24**, 261–274 (2007)
16. C.R. Bennett, C.M. Dunn, Transp. Res. Rec. **1510**, 70–75 (1995)
17. R. Bertini, S. Hansen, K. Bogenberger, Transp. Res. Rec. **1934**, 97–107 (2005)
18. P.H.L. Bovy (ed.), *Motorway Analysis: New Methodologies and Recent Empirical Findings* (Delft University Press, Delft, 1998)
19. D. Branston, Transp. Sci. **10**, 125–148 (1976)
20. L. Breiman, R.L. Lawrence, Transp. Res. **7**, 77–105 (1973)
21. L. Breiman, R.L. Lawrence, Transp. Res. **11**, 177–182 (1977)
22. L. Breiman, R.L. Lawrence, D. Goodwin, B. Bailey, Transp. Res. **11**, 221–228 (1977)
23. W. Brilon, J. Geistefeldt, M. Regler, in [133] (2005) pp. 125–144
24. W. Brilon, M. Regler, J. Geistefeld, Straßenverkehrstechnik, Heft **3**, 136 (2005)
25. W. Brilon, H. Zurlinden, Straßenverkehrstechnik, Heft **4**, 164 (2004)
26. D.J. Buckley, Transp. Sci. **2**, 107–133 (1968)
27. G. Carofiglio, V. Gehlen, D. Perino, in *2011 IEEE Intern. Conf. on Communications (ICC)*, pp. 1–6 (2011)

28. M.J. Cassidy, in *2001 IEEE Inter. Transp. Sys. Proc.* (IEEE, Oakland, USA, 2001), pp. 513–535
29. M.J. Cassidy, S. Ahn, Transp. Res. Rec. **1934**, 140–147 (2005)
30. M.J. Cassidy, R.L. Bertini, Transp. Res. B **33**, 25–42 (1998)
31. M.J. Cassidy, B. Coifman, Transp. Res. Rec. **1591**, 1–6 (1997)
32. M.J. Cassidy, A. Skabardonis (eds.), in *Papers Selected for the 19th Int. Sym. on Transp. and Traffic Theory*. Procedia Soc. Behav. Sci. **17**, 1–716 (2011)
33. M.J. Cassidy, A. Skabardonis, A.D. May, Transp. Res. Rec. **1225**, 61–72 (1989)
34. M.J. Cassidy, J.R. Windover, Transp. Res. Rec. **1484**, 73–79 (1995)
35. A. Ceder, *Investigation of Two-Regime Traffic Flow Models at the Micro- and Macroscopic Levels* (Ph.D. Thesis, UC, Berkeley University, Berkeley, CA, 1975)
36. A. Ceder (ed.), *Transportation and Traffic Theory* (Elsevier Science, Oxford, 1999)
37. A. Ceder, A.D. May, Transp. Res. Rec. **567**, 1–15 (1976)
38. H.C. Chin, A.D. May, Transp. Res. Rec. **1320**, 75–82 (1991)
39. A.H.F. Chow, X.-Y. Lu, T.Z. Qiu, in *IFAC Proceedings Volumes*, vol. 42, pp. 472–477 (2009)
40. D. Chowdhury, L. Santen, A. Schadschneider, Phys. Rep. **329**, 199 (2000)
41. K.G. Courage, Highway Res. Rec. **279**, 118–120 (1969)
42. C.F. Daganzo, M.J. Cassidy, R.L. Bertini, Transp. Res. A. **33**, 365–379 (1999)
43. C.F. Daganzo, J.A. Laval, *Report UCB-ITS-RR-93-7* (Institute of Transportation Studies, University of California, Berkeley, 2003)
44. A. Daou, Oper. Res. **12**, 360–361 (1964)
45. D.S. Dendrinos, Transp. Res. **12**, 191–194 (1978)
46. J. Drake, J. Shofer, A.D. May, Highway Res. Rec. **154**, 53–87 (1967)
47. D.R. Drew, Highway Res. Rec. **99**, 47–58 (1965)
48. D.R. Drew, C.J. Keese, Highway Res. Rec. **99**, 1–47 (1965)
49. R.E. Dudash, A.G.R. Bullen, Transp. Res. Rec. **905**, 115–117 (1983)
50. C.L. Dudek, C.J. Messer, Transp. Res. Rec. **469**, 1–15 (1973)
51. C.L. Dudek, S.H. Richards, Transp. Res. Rec. **869**, 14–18 (1982)
52. S.M. Easa, A.D. May, Transp. Res. Rec. **772**, 24–37 (1980)
53. L. Elefteriadou, in *An Introduction to Traffic Flow Theory*, Springer Optimization and Its Applications, Vol. 84 (Springer, Berlin, 2014)
54. L. Elefteriadou, A. Kondyli, W. Brilon, F.L. Hall, B. Persaud, S. Washburn, J. Transp. Eng. **140**, 04014003 (2014)
55. L. Elefteriadou, R.P. Roess, W.R. McShane, Transp. Res. Rec. **1484**, 80–89 (1995)
56. L.C. Edie, Oper. Res. **9**, 66–77 (1961)
57. L.C. Edie, E. Baverez, in *Third Int. Sym. on the Theory of Traffic Flow*, pp. 26–37 (1995)
58. L.C. Edie, R.S. Foote, in *Highway Research Board Proceedings* (HRB, National Research Council, Washington, D.C., 1958), vol. 37, pp. 334–344
59. L.C. Edie, R.S. Foote, in *Highway Research Board Proceedings* (HRB, National Research Council, Washington, D.C., 1960), vol. 39, pp. 492–505
60. L.C. Edie, P. Herman, T.N. Lam, Transp. Sci. **14**, 55–76 (1980)
61. K. Fadhloun, H. Rakha, A. Loulizi, Transp. Lett. **6**, 185–196 (2014)
62. K. Fadhloun, H. Rakha, A. Loulizi, Transp. Res. Rec. **2422**, 61–70 (2014)
63. G. Forbes, F.L. Hall, Transp. Res. A **24**, 338 (1990)
64. T.W. Forbes, J.J. Mullin, M.E. Simpson, in in *Proc. 3rd International Symposium on the Theory of Traffic Flow* ed. by L.C. Edie (Elsevier, New York, 1967), pp. 97–108
65. T.W. Forbes, M.E. Simpson, Transp. Sci. **2**, 77–104 (1968)
66. M. Fukui, Y. Sugiyama, M. Schreckenberg, D.E. Wolf (eds.), *Traffic and Granular Flow'01* (Springer, Heidelberg, 2003)
67. A.V. Gafarian, R.L. Lawrence, P.K. Munjal, J. Pahl, Highway Res. Rec. **349**, 13–30 (1971)
68. N.H. Gartner, C.J. Messer, A.K. Rathi (eds.), *Traffic Flow Theory: A State-of-the-Art Report* (Transportation Research Board, Washington DC, 2001)

69. N.H. Gartner, N.H.M. Wilson (eds.), *Transportation and Traffic Theory* (Elsevier, New York, 1987)
70. D.C. Gazis, R.S. Foote, Transp. Sci. **3**, 255–275 (1969)
71. D.C. Gazis, R. Herman, Trans. Sci. **26**, 223 (1992)
72. D.C. Gazis, R. Herman, G.H. Weiss, Oper. Res. **10**, 658–667 (1962)
73. J. Geistefeldt, W. Brilon, in [120], pp. 583–602
74. C.D. van Goeverden, H. Botma, P.H.L. Bovy, Transp. Res. Rec. **1646**, 1–8 (1998)
75. H. Greenberg, A. Daou, Oper. Res. **8**, 524–532 (1960)
76. B.D. Greenshields, in *Highway Research Board Proceedings*, vol. 14, pp. 448–477 (1935)
77. F.L. Hall, Trans. Res. A **21**, 191–201 (1987)
78. F.L. Hall, in [68], pp. 2-1–2-36
79. F.L. Hall, K. Agyemang-Duah, Trans. Res. Rec. **1320**, 91–98 (1991)
80. F.L. Hall, B.L. Allen, M.A. Gunter, Trans. Res. A **20**, 197–210 (1986)
81. F.L. Hall, D. Barrow, Transp. Res. Rec. **1194**, 55–65 (1988)
82. F.L. Hall, W. Brilon, Transp. Res. Rec. **1457**, 35–42 (1994)
83. F.L. Hall, M.A. Gunter, Trans. Res. Rec. **1091**, 1–9 (1986)
84. F.L. Hall, L.M. Hall, Transp. Res. Rec. **1287**, 108–118 (1990)
85. F.L. Hall, V.F. Hurdle, J.H. Banks, Transp. Res. Rec. **1365**, 12–18 (1992)
86. F.L. Hall, A. Pushkar, Y. Shi, Transp. Res. Rec. **1398**, 24–30 (1993)
87. J.J. Haynes, Highway Res. Rec. **99**, 59–80 (1965)
88. D. Helbing, Rev. Mod. Phys. **73**, 1067–1141 (2001)
89. D. Helbing, H.J. Herrmann, M. Schreckenberg, D.E. Wolf (eds.), *Traffic and Granular Flow'99* (Springer, Heidelberg, 2000)
90. J.W. Hess, Highway Res. Rec. **27**, 69–115 (1963)
91. J.W. Hess, Highway Res. Rec. **99**, 81–116 (1965)
92. M.P. Heyes, R. Ashworth, Transp. Res. **6**, 287–291 (1972)
93. Highway Capacity Manual 2000 (National Research Council, Transportation Research Boad, Washington, D.C., 2000)
94. Highway Capacity Manual 2010 (National Research Council, Transportation Research Boad, Washington, D.C., 2010)
95. B.D. Hillegas, D.G. Houghton, P.J. Athol, Transp. Res. Rec. **495**, 53–63 (1974)
96. S.P. Hoogendoorn, P.H.L. Bovy, Transp. Res. Rec. **1646**, 18–28 (1998)
97. S.P. Hoogendoorn, S. Luding, P.H.L. Bovy, M. Schreckenberg, D.E. Wolf (eds.), *Traffic and Granular Flow'03* (Springer, Heidelberg, 2005)
98. P. Hsu, J.H. Banks, Transp. Res. Rec. **1398**, 17–23 (1993)
99. V.F. Hurdle, P.K. Datta, Transp. Res. Rec. **905**, 127–137 (1983)
100. V.F. Hurdle, D.Y. Solomon, Transp. Sci. **20**, 153–163 (1986)
101. M. Iwasaki, Transp. Res. Rec. **1320**, 242–250 (1991)
102. A. Jia, B. Williams, N. Rouphail, Transp. Res. Rec. **2188**, 148–155 (2010)
103. E.R. Jones, M.E. Goolsby, Highway Res. Rec. **321**, 74–82 (1970)
104. T.R. Jones, R.B. Potts, Oper. Res. **10**, 745–763 (1962)
105. B.S. Kerner, *The Physics of Traffic* (Springer, Berlin, Heidelberg, New York, 2004)
106. B.S. Kerner, *Introduction to Modern Traffic Flow Theory and Control* (Springer, Heidelberg, Dordrecht, London, New York, 2009)
107. B.S. Kerner, Physica A **392**, 5261–5282 (2013)
108. B.S. Kerner Physica A **450**, 700–747 (2016)
109. B.S. Kerner, S.L. Klenov, J. Phys. A Math. Theor. **43**, 425101 (2010)
110. B.S. Kerner, S.L. Klenov, D.E. Wolf, J. Phys. A Math. Gen. **35**, 9971–10013 (2002)
111. B.S. Kerner, M. Koller, S.L. Klenov, H. Rehborn, M. Leibel, Physica A **438**, 365–397 (2015)
112. K.M. Kockelman, Transp. Res. Rec. **1644**, 47–56 (1998)
113. U. Kohler, Transp. Sci. **13**, 146–162 (1979)
114. M. Koller, PhD Thesis (University Tübingen, Germany) (2015)

115. A. Kondyli, L. Elefteriadou, W. Brilon, F.L. Hall, B. Persaud, S. Washburn, J. Transp. Eng. **139**, 931–940 (2013)
116. M. Koshi, M. Iwasaki, I. Ohkura, in *Proc. 8th International Symposium on Transportation and Traffic Theory*, ed. by V.F. Hurdle (University of Toronto Press, Toronto, Ontario, 1983), pp. 403
117. M. Koshi, Inter. Ass. Traffic Safety Sci. **10**(1), (1984)
118. M. Koshi, in [66] (2003), pp. 199–210
119. T.N. Lam, R.W. Rothery, Transp. Sci. **4**, 293–310 (1970)
120. W.H.K. Lam, S.C. Wong, H.K. Lo (eds.), *Transportation and Traffic Theory 2009* (Springer, Dordrecht, Heidelberg, London, New York, 2009)
121. J.P. Lebacque, J.B. Lesort, F. Giorgi, Transp. Res. Rec. **1644**, 70–79 (1998)
122. L. Leclercq, S. Chanut, J.B. Lesort, Transp. Res. Rec. **1883**, 3–13 (2004)
123. L. Leclercq, V.L. Knoop, F. Marczak, S.P. Hoogendoorn, Transp. Res. C **62**, 171–181 (2016)
124. J.-B. Lesort (ed.), *Transportation and Traffic Theory* (Elsevier Science, Oxford, 1996)
125. W. Leutzbach, *Introduction to the Theory of Traffic Flow* (Springer, Berlin, 1988)
126. R.V. Lindgren, R.L. Bertini, D. Helbing, M. Schönhof, Transp. Res. Rec. **1965**, 12–22 (2006)
127. J. van Lint, S.P. Hoogendoorn, Comput. Aid. Civil Infrast. Eng. **24**, 1–17 (2009)
128. E.M. Linzer, R.P. Roess, W.R. McShane, Transp. Res. Rec. **699**, 17–24 (1979)
129. M. Lorenz, L. Elefteriadou, Transp. Res. Circ. **E-C018**, 84–95 (2000)
130. R.T. Luttinen, Transp. Res. Rec. **1365**, 92–97 (1992)
131. S. Maerivoet, B. De Moor, Phys. Rep. **419**, 1–64 (2005)
132. D. Mahalel, A.S. Hakkert, Transp. Sci. **17**, 71–86 (1983)
133. H.S. Mahmassani (ed.), *Traffic and Transportation Theory* (Elsevier Science, Amsterdam, 2005)
134. R. Mahnke, J. Kaupužs, I. Lubashevsky, Phys. Rep. **408**, 1–130 (2005)
135. D.B. Martin, L. Newman, R.T. Johnson, Highway Res. Rec. **432**, 25–31 (1973)
136. A.D. May, Highway Res. Rec. **59**, 9–38 (1964)
137. A.D. May, *Traffic Flow Fundamentals* (Prentice-Hall, New Jersey, 1990)
138. A.D. May, P. Athol, W. Parker, J.B. Rudden, Highway Res. Rec. **21**, 48–70 (1963)
139. J.M. McDermott, Transp. Eng. J. **106**, 333–348 (1980)
140. H.S. Mika, J.B. Kreer, L.S. Yuan, Highway Res. Rec. **279**, 1–13 (1969)
141. A.J. Miller, Transp. Sci. **4**, 164–186 (1970)
142. P.K. Munjal, Y.S. Hsu, Transp. Res. Rec. **509**, 29–41 (1974)
143. P.K. Munjal, Y.S. Hsu, R.L. Lawrence, Transp. Res. **5**, 257–266 (1971)
144. P. Munjal, L. Pipes, Transp. Res. **5**, 241–255 (1971)
145. P.K. Munjal, L.A. Pipes, Transp. Sci. **5**, 390–402 (1971)
146. J.C. Muñoz, C.F. Daganzo, in *Proc. of the 80th Annual Meeting of the Transportation Research Board* (TRB, Washington, D.C., 2001)
147. J.C. Muñoz, C.F. Daganzo, in *Traffic and Transportation Theory*, ed. by M.A.P. Taylor (Pergamon, Oxford, 2002) pp 441–462
148. T. Nagatani, Rep. Prog. Phys. **65**, 1331–1386 (2002)
149. K. Nagel, P. Wagner, R. Woesler, Oper. Res. **51**, 681–716 (2003)
150. A. Nakayama, M. Fukui, M. Kikuchi, K. Hasebe, K. Nishinari, Y. Sugiyama, S-i. Tadaki, S. Yukawa, New J. Phys. **11**, 083025 (2009)
151. L. Neubert, H.Y. Lee, M. Schreckenberg, J. Phys. A Math. Gen. **32**, 6517–6525 (1999)
152. L. Neubert, L. Santen, A. Schadschneider, M. Schreckenberg, Phys. Rev. E **60**, 6480–6490 (1999)
153. L. Neubert, L. Santen, A. Schadschneider, M. Schreckenberg, in [89] (2000), pp. 307–314
154. G.F. Newell, Transp. Res. B **32**, 531 (1988)
155. G.F. Newell, *Inst. of Transp. Studies Research Report UCB ITS-RR-93-3* (University of California, Berkley, 1993)
156. L. Newman, Highway Res. Rec. **27**, 14–43 (1963)

157. K. Nishinari, M. Hayashi (eds.), *Traffic Statistics in Tomei Express Way* (The Math. Soc. of Traffic Flow, Japan, 1999)
158. K. Nishinari, M. Treiber, D. Helbing, Phys. Rev. E **68**, 067101 (2003)
159. D. Owens, M.J. Schofield, Traffic Eng. Cont. **29**, 616–623 (1988)
160. J. Pahl, T. Sands, Transp. Sci. **5**, 403–417 (1971)
161. J.O. Paul, J. Transp. Eng. **142**, 04016025 (2016)
162. B.N. Persaud, "Study of a Freeway Bottleneck to Explore Some Unresolved Traffic Flow Issues", *Ph.D. Dissertation*, (University of Toronto, Toronto, Canada, 1986)
163. B.N. Persaud, F.L. Hall, Trans. Res. A **23**, 103–113 (1989)
164. B.N. Persaud, F.L. Hall, L.M. Hall, Transp. Res. Rec. **1287**, 167–175 (1990)
165. B.N. Persaud, V.F. Hurdle, Transp. Res. Rec. **1194**, 191–198 (1988)
166. B.N. Persaud, V.F. Hurdle, in *Proc. of International Symposium on Highway Capacity* (Karlsruhe, Germany, 1991)
167. B.N. Persaud, S. Yagar, R. Brownlee, Trans. Res. Rec. **1634**, 64–69 (1998)
168. B.N. Persaud, S. Yagar, D. Tsui, H. Look, Transp. Res. Rec. **1748**, 110–115 (2001)
169. H. Rehborn, S.L. Klenov, in *Springer Encyclopedia of Complexity and System Science*, ed. by R.A. Meyers (Springer, Berlin, 2009), pp. 9500–9536
170. H. Rehborn, S.L. Klenov, J. Palmer, Physica A **390**, 4466–4485 (2011)
171. H. Rehborn, M. Koller, J. Adv. Transp. **48**, 1107–1120 (2014)
172. W.R. Reilly, C.C. Gardner, J.H. Kell, *FHWA Report No. RD-76-135/137* (Federal Highway Administration, Washington, D.C., 1976)
173. W.R. Reilly, D.W. Harwood, J.M. Schoen, R.0. Kuehl, K. Bauer, A.D.St. John, *Capacity and Level of Service Procedures for Multilane Rural and Suburban Highways*, Preliminary draft final report (NCHRP, TRB, National Research Council, Washington, D.C., 1988)
174. P.I. Richards, Oper. Res. **4**, 42–51 (1956)
175. J. Ringert, T. Urbanik II, Transp. Res. Rec. **1398**, 31–41 (1993)
176. A. Schadschneider, D. Chowdhury, K. Nishinari, *Stochastic Transport in Complex Systems* (Elsevier Science, New York, 2011)
177. A. Schadschneider, T. Pöschel, R. Kühne, M. Schreckenberg, D.E. Wolf (eds.), *Traffic and Granular Flow'05* (Springer, Heidelberg, 2007)
178. J. Schoen, A.D. May, W. Reilly, T. Urbanik, *Final Report, NCHRP Project 3–45*, (JHK & Associates and Texas Transportation Institute, 1995)
179. M. Schönhof, D. Helbing, Transp. Sci. **41**, 135–166 (2007)
180. M. Schönhof, D. Helbing, Transp. Res. B **43**, 784–797 (2009)
181. M. Schreckenberg, D.E. Wolf (eds.), *Traffic and Granular Flow'97* (Springer, Singapore, 1998)
182. M. Shawky, H. Nakamura, Transp. Res. Rec. **2012**, 11–19 (2007)
183. K.R. Smilowitz, C.F. Daganzo, M.J. Cassidy, R.L. Bertini, Transp. Res. Rec. **1678**, 225–233 (1999)
184. J. Sun, L. Zhao, H. Zhang, Transp. Res. Rec. **2421**, 64–73 (2014)
185. M.A.P. Taylor (ed.), *Transportation and Traffic Theory in the 21st Century* (Elsevier Science, Amsterdam, 2002)
186. J.R. Tolle, Transp. Res. **8**, 91–96 (1974)
187. M. Treiber, A. Kesting, *Traffic Flow Dynamics* (Springer, Berlin, 2013)
188. M. Treiber, A. Kesting, R.E. Wilson, Comput. Aid. Civil Infrast. Eng. **26**, 408–419 (2011)
189. J. Treiterer, Transp. Res. **1**, 231–251 (1967)
190. J. Treiterer, "Investigation of Traffic Dynamics by Aerial Photogrammetry Techniques", *Ohio State University Technical Report PB 246 094* (Columbus, Ohio, 1975)
191. J. Treiterer, J.A. Myers, in *Procs. 6th International Symposium on Transportation and Traffic Theory*, ed. by D.J. Buckley (A.H. & AW Reed, London, 1974), pp. 13–38
192. J. Treiterer, J.I. Taylor, Highway Res. Rec. **142**, 1–12 (1966)
193. T. Urbanik II, W. Hinshaw, K. Barnes, Transp. Res. Rec. **1320**, 110–118 (1991)
194. P. Wasielewski, Transp. Sci. **15**, 364–378 (1981)

195. E.A. Wemple, A.M. Morris, A.D. May, in *Procs. of the International Symposium on Highway Capacity*, ed. by U. Brannolte (Karlsruhe, Germany, 1991), pp. 439–455
196. D. Westland, in *Proc. of the 3rd Sym. on Highway Capacity and Level of Service*, ed. by R. Rysgaard, vol. 2, (Road Directorate, Copenhagen, Ministry of Transport, Denmark, 1998), pp. 1095–1116
197. J.R. Windover, "Empirical Studies of the Dynamic Features of Freeway Traffic", *Ph.D. Thesis* (Dept. of CEE, ITS, UC, Berkeley, 1998)
198. D.E. Wolf, M. Schreckenberg, A. Bachem (eds.), *Traffic and Granular Flow* (World Scientific, Singapore, 1995)
199. S. Yukawa, M. Kikuchi, A. Nakayama, K. Nishinari, Y. Sugiyama, S. Tadaki, in [66] (2003), pp. 243–256
200. B. Zielke, R. Bertini, M. Treiber, Transp. Res. Rec. **2088**, 57–67 (2008)

Chapter 3
Nucleation Nature of Traffic Breakdown—Empirical Fundamental of Transportation Science

3.1 Introduction

We have already mentioned that the solely knowledge of features of empirical traffic breakdown that have been discussed in Chap. 2 is not sufficient to disclose the nature of the stochastic (probabilistic) behavior of empirical traffic breakdown at highway bottlenecks.

In this chapter, based on an analysis of real field traffic data we will disclose the nature of probabilistic empirical traffic breakdown at highway bottlenecks. We will show that in accordance with hypotheses of the three-phase theory [4–26], free flow at a highway bottleneck is indeed metastable with respect to an F→S transition. We will find that the F→S transition occurs only when a nucleus for this transition appears in a metastable state of free flow. In other words, we will prove that empirical traffic breakdown (F→S transition) exhibits the nucleation nature.

> The main objective of this chapter is to show that the empirical evidence of the nucleation nature of a transition from free flow to synchronized flow (F→S transition) can be considered the empirical fundamental of transportation science.

However, before we consider the empirical proof of the metastability of free flow at highway bottlenecks with respect to an F→S transition, we should define the terms "empirical spontaneous traffic breakdown" and "empirical induced traffic breakdown" at highway bottlenecks (Sect. 3.2). Additionally, we should consider the term "nucleus" for traffic breakdown in more details (Sect. 3.3).

B.S. Kerner, *Breakdown in Traffic Networks*, DOI 10.1007/978-3-662-54473-0_3

3.2 Definitions of Empirical Spontaneous and Empirical Induced Traffic Breakdowns at Highway Bottlenecks

In observations of traffic breakdown (F→S transition) at a bottleneck in real field traffic data, we distinguish two different cases: (i) empirical *spontaneous* traffic breakdown (spontaneous F→S transition) and (ii) empirical induced traffic breakdown (induced F→S transition) [18, 20]:

- If before traffic breakdown occurs at the bottleneck, *free flow* exists at the bottleneck as well as upstream and downstream in a neighborhood of the bottleneck, then such an empirical traffic breakdown at the bottleneck is called empirical spontaneous traffic breakdown.
- An empirical induced traffic breakdown at the bottleneck is traffic breakdown induced by the propagation of a spatiotemporal *congested* traffic pattern. This congested pattern has occurred *earlier* than thc time instant of traffic breakdown at the bottleneck and at a *different* road location (for example at another bottleneck) than the bottleneck location. When this congested pattern reaches the bottleneck, the pattern induces traffic breakdown at the bottleneck.

An example of empirical spontaneous traffic breakdown has been presented in Fig. 2.1a of Chap. 2. Indeed, in this case before traffic breakdown occurs at a bottleneck, there is free flow at the bottleneck as well as upstream and downstream in a neighborhood of the on-ramp bottleneck. Other two examples of empirical spontaneous traffic breakdown, which will be considered below in more details, are shown in Fig. 3.1.

Two examples of empirical induced traffic breakdown are shown in Figs. 3.2 and 3.3. In the first example, a wide moving jam propagating through an on-ramp bottleneck *induces* the synchronized flow phase (F→S transition) at this bottleneck (induced traffic breakdown is labeled by "induced traffic breakdown" in Fig. 3.2).

In the second example, a moving synchronized flow pattern (MSP) that has initially emerged at the downstream off-ramp bottleneck propagates upstream. When the MSP reaches an upstream on-ramp bottleneck (labeled by "on-ramp bottleneck 1" in Fig. 3.3), the MSP induced the synchronized flow phase (F→S transition) at this on-ramp bottleneck (induced traffic breakdown is labeled by "ind" in Fig. 3.3b and by "induced traffic breakdown" in Fig. 3.3c,g).

In both examples of empirical induced traffic breakdown (induced F→S transition) (Figs. 3.2 and 3.3) the downstream front of resulting congested traffic is fixed at the bottleneck, i.e., in accordance with the definition [S] this congested traffic is associated with the synchronized flow phase (see Sect. 1.3). Synchronized flow is self-sustaining for a very long time (more than an hour) upstream of the bottleneck.

There are some differences between these two examples of empirical induced traffic breakdown (Figs. 3.2 and 3.3). In the case, when empirical induced traffic breakdown has been caused by the propagation of the wide moving jam through the bottleneck, the flow rate within the wide moving jam shown in Fig. 3.2d is very small. Contrarily, when empirical induced traffic breakdown has been caused by

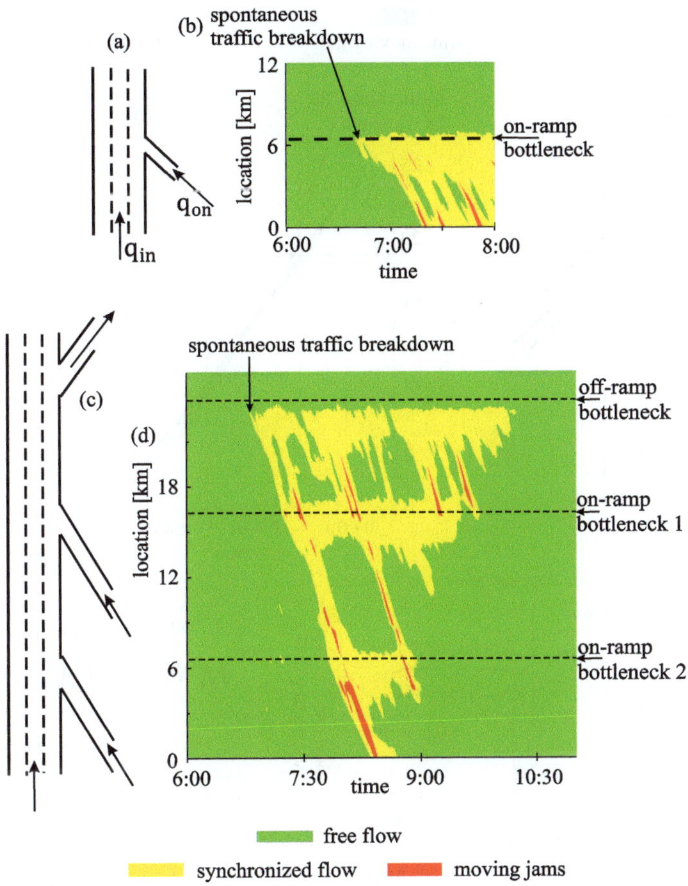

Fig. 3.1 Overview of features of empirical spontaneous traffic breakdown (F→S transition) at an on-ramp bottleneck (**a, b**) and an off-ramp bottleneck (**c, d**) (real field traffic data measured by road detectors on three-lane freeway A5-South in Germany on April 15, 1996 (**b**) and September 03, 1998 (**d**)): (**a, c**) Sketches of sections of three-lane highway in Germany with off- and on-ramps bottleneck. (**b, d**) Speed data measured with road detectors installed along road section in (**a, c**); data is presented in space and time with averaging method described in Sec. C.2 of [31]. (**b**) Empirical spontaneous traffic breakdown at on-ramp bottleneck. (**d**) Empirical spontaneous traffic breakdown at off-ramp bottleneck. The on-ramp bottleneck labeled by "on-ramp bottleneck" in (**a, b**) is the same as that labeled by "on-ramp bottleneck 2" in (**c, d**). Off-ramp bottleneck, on-ramp bottleneck 1 and on-ramp bottleneck 2 marked by dashed lines in (**c**) are, respectively, effectual bottlenecks explained on Fig. 2.1 of the book [18]

the propagation of the MSP to the bottleneck, the flow rate within the MSP shown in Fig. 3.3h is almost as large as in the surrounded free flow. Another difference between the wide moving jam shown in Fig. 3.2 and the MSP shown in Fig. 3.3 is as follows: The wide moving jam propagates through the bottleneck while maintaining

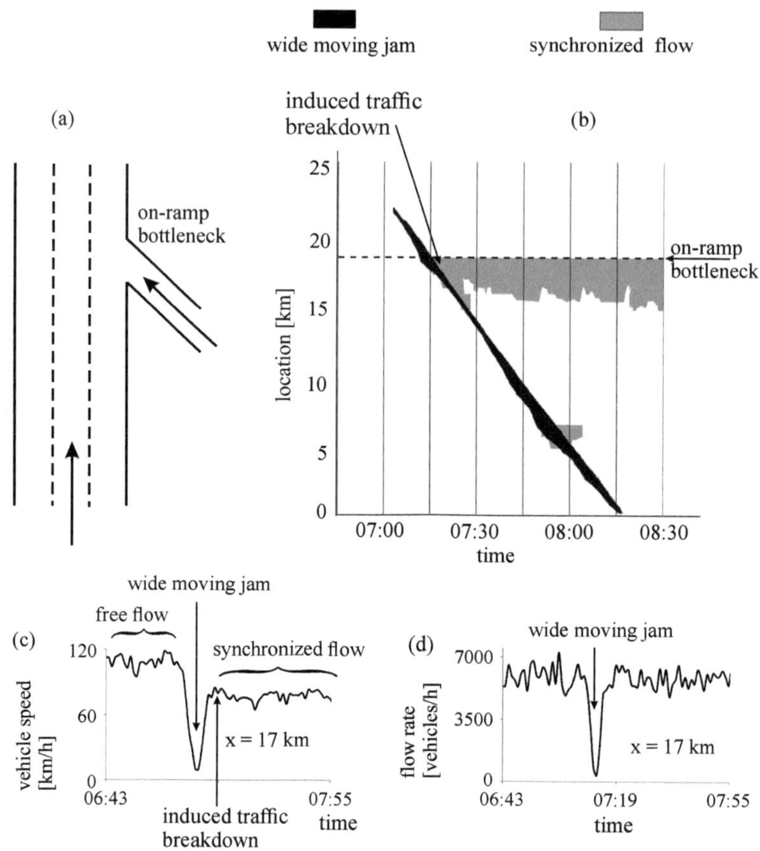

Fig. 3.2 Example of empirical induced traffic breakdown at on-ramp bottleneck that is induced by wide moving jam propagation: (**a**) Sketch of section of three-lane highway in Germany with an on-ramp bottleneck. (**b**) Speed data measured with road detectors installed along road section in (**a**); data is presented in space and time with averaging method described in Sec. C.2 of [31]. (**c, d**) Speed (**c**) and total flow rate (**d**) over time at the location of the bottleneck $x = 17$ km. Average data of vehicle speed in space and time (1-min averaging interval). Real field traffic data of road detectors measured on three-lane freeway A5-South in Germany on June 23, 1998. Adapted from [18]

the mean velocity of its downstream front (Fig. 3.2b). In contrast, the MSP is caught at the bottleneck (Fig. 3.3b). This effect has been called "catch effect".

However, as we will show in Sect. 3.6.1, in both examples of empirical induced traffic breakdown (Figs. 3.2 and 3.3) after induced traffic breakdown has occurred at the bottleneck, characteristics of synchronized flow that has emerged at the bottleneck due to the breakdown are qualitatively the same as those found in the case of empirical spontaneous traffic breakdown at the bottleneck.

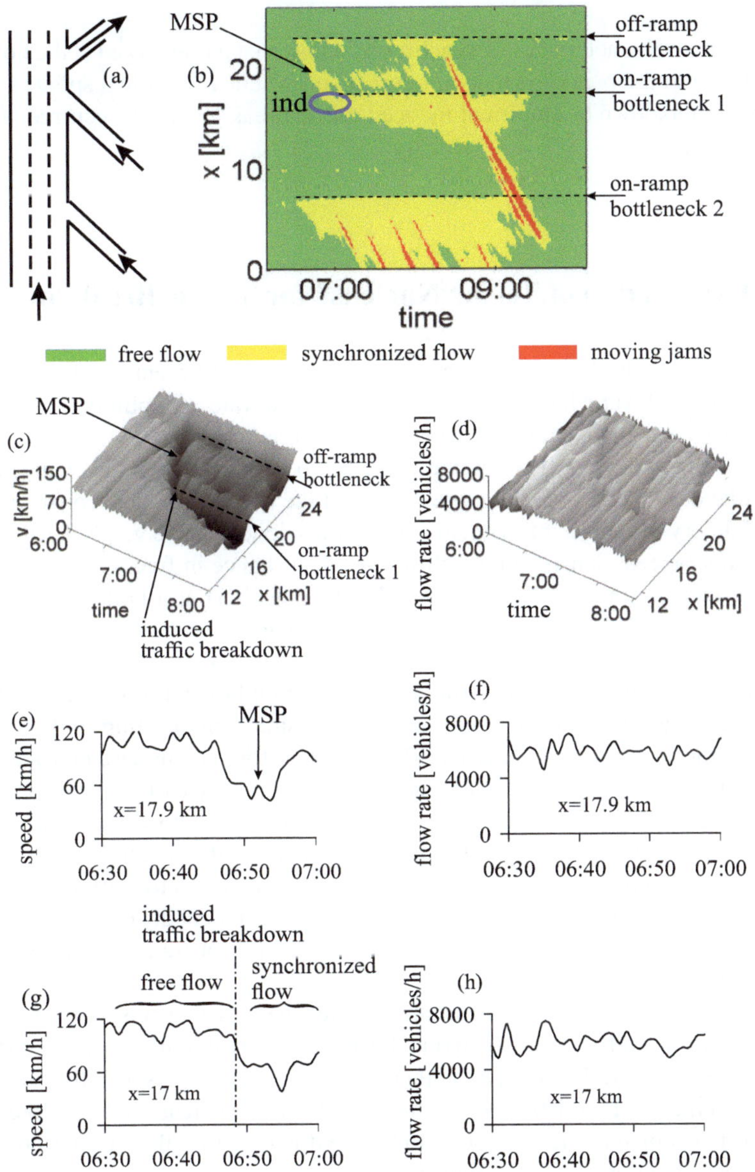

Fig. 3.3 Empirical example of traffic breakdown at on-ramp bottleneck induced by propagation of moving synchronized flow pattern (MSP): (**a**, **b**) Sketch of section of three-lane highway (**a**) and speed data (**b**) adapted from Fig. 1.5a,b; induced traffic breakdown is labeled by "ind" in (**b**). (**c**, **d**) Average speed (**c**) and total flow rate (**d**) on the main road in space and time related to a fragment of (**b**). (**e**–**h**) Time-functions of average speed (**e**, **g**) and total flow rate (**f**, **h**) at two road locations related to (**c**, **d**): 17.9 km (**e**, **f**) and 17 km (**g**, **h**). Location 17 km is about 100 m downstream of the end of the merging region of on-ramp bottleneck 1 (**c**). Real 1-min average data measured by detectors on three-lane freeway A5-South in Germany on April 20, 1998. Adapted from [18]

Qualitative characteristics of synchronized flow that has emerged at the bottleneck due to the breakdown do not depend on whether empirical spontaneous traffic breakdown or empirical induced traffic breakdown has occurred at the bottleneck.

3.3 Explanation of Term "Nucleus" for Traffic Breakdown

In free flow at a highway bottleneck, there are many different local disturbances (fluctuations) of vehicles speed, density, and flow rate. Amplitudes of different disturbances can change randomly over time in free flow.

In this chapter, we will show that there can be a local disturbances in free flow at a bottleneck that causes traffic breakdown at the bottleneck. In accordance with a general theory of metastable systems of natural science (see, e.g., [1–3, 28–30, 32–38]) such a (speed, density and/or flow rate) disturbance in free traffic flow can be called a *nucleus* for traffic breakdown (F→S transition) at a bottleneck. Because traffic breakdown is an F→S transition, the nucleus for traffic breakdown is a synonym of the term "the nucleus for the F→S transition".

Using this terminology, the term "metastable free flow with respect to an F→S transition" means that a local disturbance with a small enough amplitude occurring in metastable free flow at a bottleneck decays. Therefore, in this case metastable free flow persists at the bottleneck over time. Contrarily, if a local disturbance appears in metastable free flow in a neighborhood of the bottleneck whose amplitude is equal to or exceeds a critical amplitude of the disturbance, traffic breakdown is realized. This critical disturbance in free flow with respect to the F→S transition is a "critical nucleus" for traffic breakdown. Thus, the term "critical local disturbance" for traffic breakdown is equivalent to the term "critical nucleus" for traffic breakdown at the bottleneck.

In general, a local disturbance in metastable free flow that acts a nucleus for the F→S transition (traffic breakdown) can be associated with initial local disturbances of the speed, density, flow rate, and other traffic flow parameters, like the percentage of long vehicles. In other words, the source of a nucleus and spatiotemporal characteristics of the nucleus for traffic breakdown in free flow at the bottleneck can be very complex (see, e.g., Sect. 3.4).

A roughly simplified macroscopic explanation of the term *nucleus* for traffic breakdown (F→S transition) in free flow at the bottleneck that we use often in the book below is as follows. We consider a local speed disturbance in free flow that causes a *local decrease* in the speed at the bottleneck. The disturbance is localized at the bottleneck. We use the term "amplitude of local speed disturbance". The amplitude of the local speed disturbance is the speed difference between the speed in free flow outside the disturbance and the minimum speed within the disturbance.

In other words, the larger the disturbance amplitude, the smaller the free flow speed within the disturbance in comparison with the free flow speed outside the disturbance. If the amplitude of the disturbance is small enough, then no F→S transition occurs at the bottleneck.

However, when the amplitude of a local speed disturbance in the metastable free flow is equal to or exceeds a critical amplitude, the F→S transition does occur. The critical amplitude of the disturbance is the speed difference between the speed outside the disturbance and the speed within the disturbance required for traffic breakdown (F→S transition) at the bottleneck. A local speed disturbance occurring in the metastable free flow that leads to the F→S transition is called a nucleus for the F→S transition. Respectively, a local speed disturbance in the metastable free flow with a critical amplitude is called a critical nucleus.

This definition of a nucleus for traffic breakdown (F→S transition) explains why a local speed disturbance in the metastable free flow whose amplitude is smaller than the critical amplitude decays. Contrarily, when the amplitude of a local speed disturbance in the metastable free flow is equal to or exceeds the critical amplitude, this disturbance is a nucleus for the F→S transition. The term "the metastability of free flow with respect to the F→S transition" means also that traffic breakdown at the bottleneck exhibits the nucleation nature: If the nucleus for traffic breakdown occurs in free flow at the bottleneck, traffic breakdown does occur. In contrast, as long as no nucleus appears, no traffic breakdown occurs in a metastable state of free flow. As above mentioned, the term *traffic breakdown at a highway bottleneck* is a synonym of the term *F→S transition at the bottleneck*. Therefore, the term *a nucleus for F→S transition at the bottleneck* is a synonym of the term *a nucleus for traffic breakdown at the bottleneck*.

It must be noted that there are two ways for nucleus occurrence:

1. The nucleus for traffic breakdown (F→S transition) can occur spontaneously in free flow. The term *spontaneously* means that the nucleus results from random fluctuations of the free flow speed, the density, or/and the flow rate at the bottleneck. Empirical nuclei for spontaneous traffic breakdown will be revealed in next Sect. 3.4.
2. The nucleus for traffic breakdown can be a local congested pattern, when the pattern reaches the bottleneck location. In this case, as above-mentioned, traffic breakdown is called induced traffic breakdown.

One of possible scenarios for empirical induced traffic breakdown in real free flow at the bottleneck is as follows. Firstly, a local congested pattern occurs at a downstream bottleneck. Then the pattern propagates upstream to the location of the bottleneck under consideration. When this congested pattern reaches the bottleneck, the pattern induces traffic breakdown at the bottleneck. Empirical examples of this scenario are shown in Figs. 3.2 and 3.3. In accordance with above consideration of the nucleation nature of traffic breakdown, this local congested pattern can be considered the nucleus that induces traffic breakdown at the bottleneck. A more detailed consideration of empirical induced traffic breakdown will be made in Sect. 3.6.

3.4 Nucleation of Empirical Spontaneous Traffic Breakdown at Highway Bottlenecks

Results of this section are based on an empirical study made in [27]. In [27], an empirical analysis of more than 160 empirical spontaneous traffic breakdowns at on- and off-ramp bottlenecks on different highways in Germany measured during 1996–2014 has been made. This empirical analysis allows us to explain the nature and physics of empirical nuclei for spontaneous traffic breakdown at highway bottlenecks.

3.4.1 Waves in Empirical Free Flow

In each of the freeway lanes (Fig. 3.1a,c), road detectors measure the following 1-min averaged data: the flow rate of all vehicles q, the average speed v as well as the flow rate q_{slow} of long vehicles. Respectively, we can calculate the percentage of long vehicles $(q_{\text{slow}}/q)100\%$. Long vehicles can be considered *slow vehicles*. On the working days (Fig. 3.1b,d), the most of long vehicles are trucks. Trucks have a speed limit 80 km/h on German highways (in reality, trucks move usually at the speed within a range 80–90 km/h).

To find nuclei for traffic breakdown, we study possible waves in free flow. Additionally with possible waves of 1-min average traffic variables q, q_{slow}, and v, we investigate also waves of the following variables

$$\Delta q_{\text{wave}} = q - \bar{q}, \ \Delta v_{\text{wave}} = \bar{v} - v, \tag{3.1}$$

$$\Delta\psi_{\text{wave}} = \begin{cases} \Delta q^{(\text{slow})}_{\text{wave}}/\Delta q_{\text{wave}} & \text{at } \Delta q_{\text{wave}} > \delta_{\text{wave}}, \\ 0 & \text{at } \Delta q_{\text{wave}} \leq \delta_{\text{wave}}, \end{cases} \tag{3.2}$$

where $\Delta\psi_{\text{wave}}$ is a dimensionless characteristic of the share of long (slow) vehicles, δ_{wave} is constant ($\delta_{\text{wave}} > 0$); $\Delta q^{(\text{slow})}_{\text{wave}} = q_{\text{slow}} - \bar{q}_{\text{slow}}$; traffic variables $\bar{q}$, $\bar{q}_{\text{slow}}$, and $\bar{v}$ are related to 20-min average data with the use of the well-known procedure of "moving averaging". The dimensionless characteristic of the share of long (slow) vehicles $\Delta\psi_{\text{wave}}$ allows us to make a wave of the percentage of long (slow) vehicles visible in space x and time t.

To make possible wave propagation in free flow visible in space and time, we apply an approximate wave reconstruction procedure of [27] (Fig. 3.4). We have found that empirical waves can propagate through the whole road section. Because vehicles enter the main road from on-ramps and leave the main road to off-ramps, some of the waves can appear at on-ramps or disappear at off-ramps.[1]

[1]Although the number of trucks must be conserved between ramps, we observe sometimes that a wave can also reform or even disperse over time between road intersections. This might be

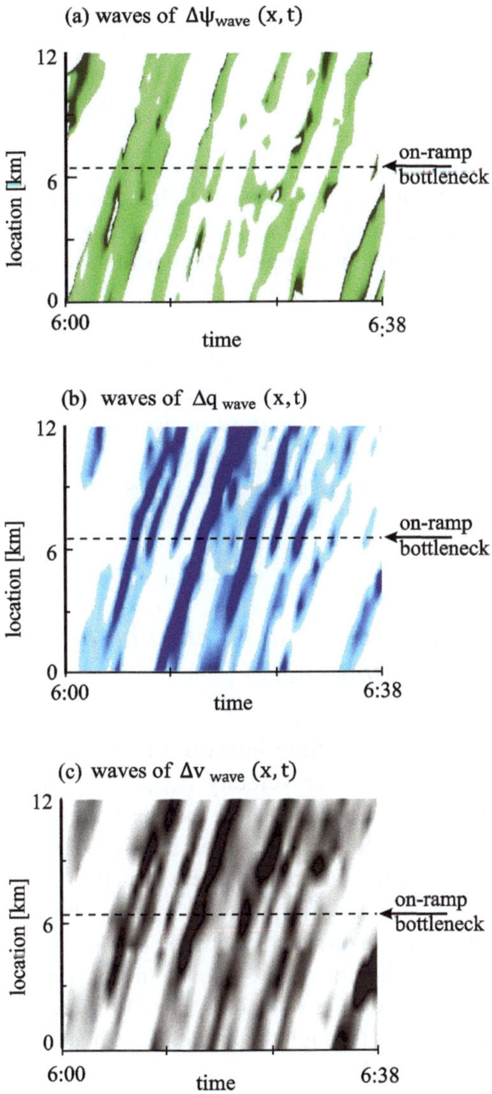

Fig. 3.4 Empirical waves of $\Delta\psi_{\rm wave}$ (**a**), $\Delta q_{\rm wave}$ (**b**), and $\Delta v_{\rm wave}$ (**c**) for data in Fig. 3.1b averaged across the road (real field traffic data measured by road detectors installed along three-lane freeway): (**a**) Waves of $\Delta\psi_{\rm wave}(x, t)$ are presented by regions with variable shades of gray (green in the on-line version) (in white regions $\Delta\psi_{\rm wave} \leq 0.1$, in black (*dark green*) regions $\Delta\psi_{\rm wave} \geq 1$). (**b**) Waves of $\Delta q_{\rm wave}(x, t)$ are presented by regions with variable shades of gray (blue in the on-line version) (in white regions $\Delta q_{\rm wave} \leq 600$ vehicles/h, in black (*dark blue*) regions $\Delta q_{\rm wave} \geq 1500$ vehicles/h). (**c**) Waves of $\Delta v_{\rm wave}(x, t)$ are presented by regions with variable shades of gray (in white regions $\Delta v_{\rm wave} \leq 1$ km/h, in black regions $\Delta v_{\rm wave} \geq 15$ km/h)

We find out that the waves of the flow rate $\Delta q_{\rm wave}$ and the speed $\Delta v_{\rm wave}$ almost coincide with the waves of the share of slow vehicles in free flow $\Delta\psi_{\rm wave}$ (Fig. 3.4). We see that each of the waves propagates downstream with the mean wave velocity $v_{\rm wave}$ that is approximately equal to the mean speed of slow vehicles

explained by a complex spatiotemporal dynamics of fast cars that overtake the trucks. However, this microscopic spatiotemporal dynamics of the waves cannot be resolved in 1 minute averaged data. Empirical vehicle trajectories of (almost) all vehicles in free flow, which are required for an analysis of this wave dynamics in free flow, are not currently available.

$v_{\mathrm{slow}}^{(\mathrm{mean})}$ (Fig. 3.4):

$$v_{\mathrm{wave}} = v_{\mathrm{slow}}^{(\mathrm{mean})}. \tag{3.3}$$

In all empirical data measured on working days, $v_{\mathrm{slow}}^{(\mathrm{mean})}$ is given by the average speed of long vehicles that changes within range 85–88 km/h. Within any of the waves propagating with the velocity v_{wave} (3.3) the percentage of long vehicles and the flow rate are larger, whereas the average speed is lower than outside the wave.

In free flow, there are waves of the flow rate and the speed that coincide approximately with waves of the share of slow vehicles. The waves propagate in the flow direction on average with the average speed of slow vehicles. Within any of the waves, the share of slow vehicles and the flow rate are larger, whereas the average vehicle speed is lower than outside the wave.

3.4.2 Empirical Nucleation of Traffic Breakdown at On-Ramp Bottleneck

During a long time interval, the waves of traffic variables in free flow propagate with the positive velocity v_{wave} (3.3) through the on-ramp bottleneck. No traffic breakdown occurs at the bottleneck (Fig. 3.4). This changes crucially when we consider a longer time interval (Fig. 3.5).

Indeed, when one of the waves propagates through the on-ramp bottleneck, the wave initiates traffic breakdown at the bottleneck. During the subsequent wave propagation downstream of the bottleneck, the structure of the wave and its features do not change. Thus, one of the waves in free flow studied above leads to the occurrence of a nucleus for traffic breakdown at the bottleneck, when the wave propagates through the bottleneck (labeled by "nucleus initiating traffic breakdown" in Fig. 3.5c).

3.4.3 Empirical Nucleation of Traffic Breakdown at Off-Ramp Bottleneck

The empirical result shown in Fig. 3.5 remains qualitatively the same for the case of wave propagation through an off-ramp bottleneck. In Fig. 3.1d, there are three bottlenecks: an off-ramp bottleneck and two upstream on-ramp bottlenecks. In this case, traffic breakdown occurs at the off-ramp bottleneck. This traffic breakdown

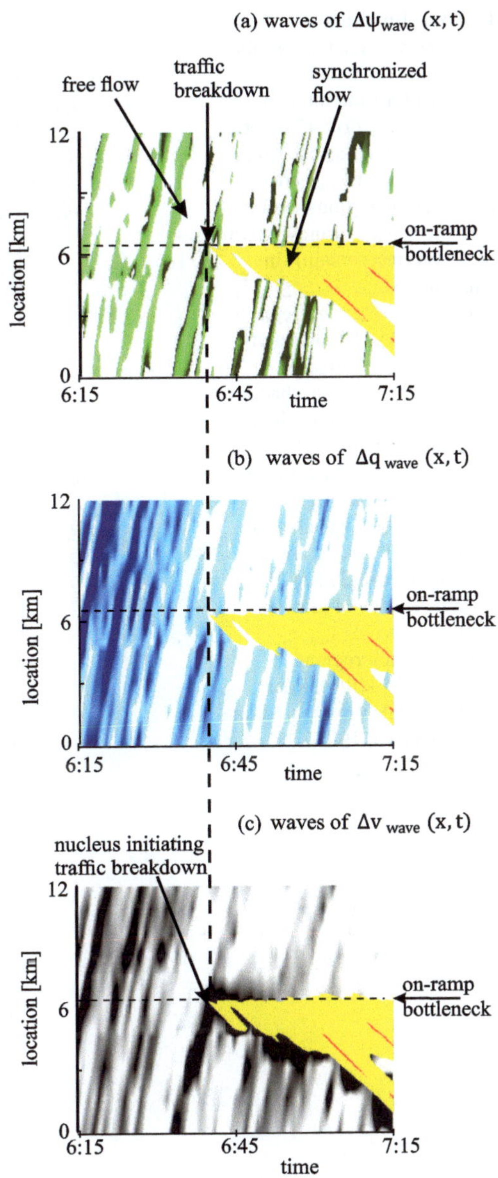

Fig. 3.5 Empirical nucleus in free flow for data in Fig. 3.1b. Empirical waves of $\Delta\psi_{\text{wave}}$ (**a**), Δq_{wave} (**b**), and Δv_{wave} (**c**) in free flow for a longer time interval as that in Fig. 3.4. In (**a–c**), regions labeled by "synchronized flow" show symbolically synchronized flow. Parameters of the presentation of empirical waves in (**a–c**) are the same as those in Fig. 3.4a–c, respectively

leads to the emergence of a complex spatiotemporal congested pattern upstream of the off-ramp bottleneck (Fig. 3.1d).

Before the breakdown has occurred, there is also a complex sequence of waves of traffic variables $\Delta\psi_{\text{wave}}$, Δq_{wave}, and Δv_{wave} in free flow. Some of the waves propagate through the whole 25 km long highway section (Fig. 3.6).

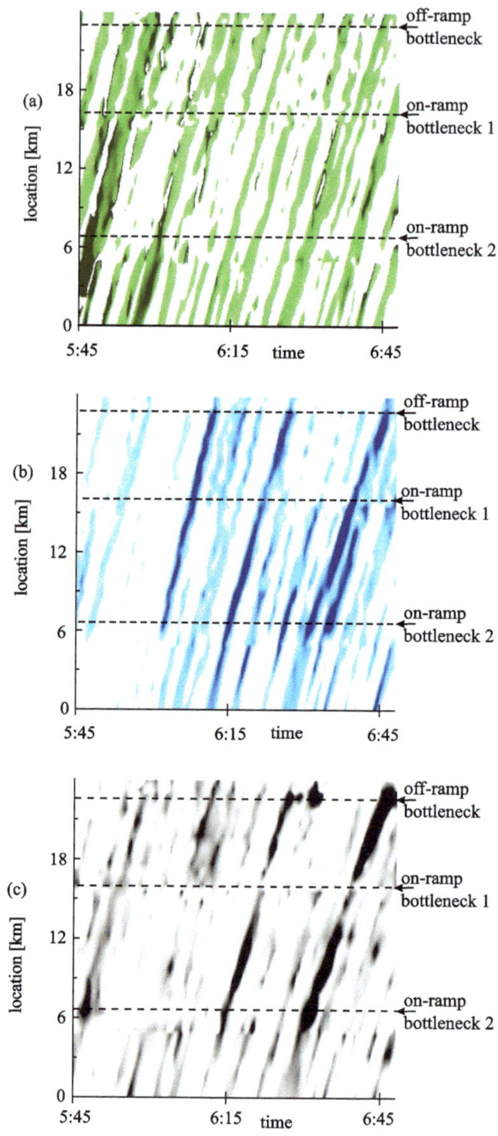

Fig. 3.6 Empirical waves in free flow averaged across the road related to data in Fig. 3.1d for time interval $05:45 \leq t \leq 06:45$ before the breakdown has occurred: (**a**) Waves of $\Delta\psi_{\rm wave}$ are presented by regions with variable shades of gray (green in the on-line version) (in white regions $\Delta\psi_{\rm wave} \leq 0.1$, in black (*dark green*) regions $\Delta\psi_{\rm wave} \geq 1$). (**b**) Waves of $\Delta q_{\rm wave}$ are presented by regions with variable shades of gray (blue in the on-line version) (in white regions $\Delta q_{\rm wave} \leq 700$ vehicles/h, in black (*dark blue*) regions $\Delta q_{\rm wave} \geq 2000$ vehicles/h). (**c**) Waves of $\Delta v_{\rm wave}$ are presented by regions with variable shades of gray (in white regions $\Delta v_{\rm wave} \leq 2$ km/h, in black regions $\Delta v_{\rm wave} \geq 15$ km/h). Real field traffic data measured on September 03, 1998

When we consider a longer time interval as that shown in Fig. 3.6, we find that while one of the waves approaches the off-ramp bottleneck, the wave initiates traffic breakdown at the bottleneck (Fig. 3.7). The structure of the wave and its features do not change after the wave is downstream of the off-ramp bottleneck. Thus, as in the case of the on-ramp bottleneck (Fig. 3.5c), the wave in free flow mentioned above leads to the occurrence of a nucleus for traffic breakdown at the off-ramp bottleneck, when the wave propagates through this bottleneck (Fig. 3.7c).

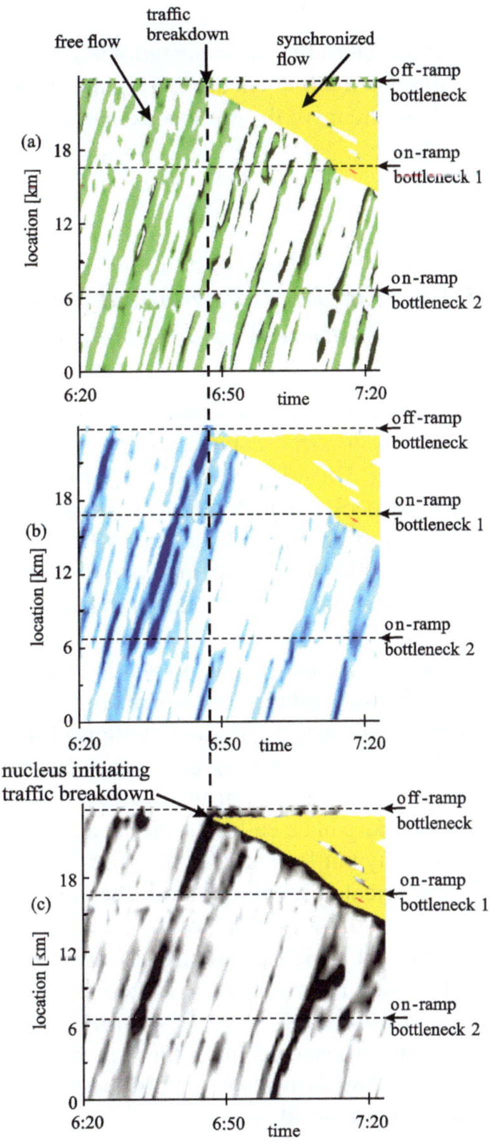

Fig. 3.7 Empirical nucleus in free flow related to data in Fig. 3.1d. Empirical waves of $\Delta\psi_{\text{wave}}$ (**a**), Δq_{wave} (**b**), and Δv_{wave} (**c**) in free flow averaged across the road for a longer time interval as that shown in Fig. 3.6. In (**a–c**), region labeled by "synchronized flow" shows symbolically synchronized flow. Parameters of presentation of empirical waves in (**a–c**) are the same as those in Fig. 3.6a–c, respectively. Real field traffic data measured by road detectors on September 03, 1998

During a long time interval, the waves of traffic variables in free flow do not lead to traffic breakdown while the wave propagate with the positive velocity through the bottleneck. However, randomly one of the waves initiates traffic breakdown at the bottleneck. In other words, the wave leads to the occurrence

(continued)

of a nucleus for traffic breakdown at the bottleneck, when the wave propagates through the bottleneck. During the subsequent wave propagation in free flow downstream of the bottleneck, the structure of the wave and its features do not change.

3.4.4 Empirical Permanent Speed Disturbance at Highway Bottleneck and Nucleation of Traffic Breakdown

In empirical data sets, a wave moving in free flow at the velocity (3.3) acts as a nucleus for traffic breakdown *only* at some effective location of a highway bottleneck. The effective bottleneck location is a road location in a neighborhood of the bottleneck at which traffic breakdown is observed with the largest probability[2]: In the most empirical data, no traffic breakdown has been observed between the bottleneck locations. To understand this empirical result, rather than waves of Δv_{wave} (Fig. 3.6), we consider empirical waves of the speed $v(x, t)$ averaged across the road (Fig. 3.8).

We see that additionally to waves of the speed propagating downstream, there are three narrow road regions, which are localized in neighborhoods of the locations of off-ramp bottleneck, on-ramp bottleneck 1, and on-ramp bottleneck 2, respectively. Within these narrow regions, the speed is smaller than outside them (Fig. 3.8). These narrow regions of the decrease in the speed at the effective locations of the bottlenecks can be called empirical permanent local speed disturbances in free flow at highway bottlenecks (see also Sect. 5.3.1).

Empirical observations show that a wave acts as a nucleus for traffic breakdown only when the wave reaches the location of a permanent local speed disturbance in free flow at a highway bottleneck. For this reason, the location of the permanent disturbance determines the effective location of the bottleneck at which traffic breakdown occurs. A decrease in the free flow speed within the permanent local speed disturbance becomes larger, when the wave reaches the effective bottleneck location. This is because within the wave the flow rate is larger and the speed is smaller than outside the wave.

The physics of the occurrence of empirical nuclei for traffic breakdown at highway bottlenecks can be explained by an interaction of a wave in free flow

(continued)

[2]Empirical features of effective bottleneck location have been explained in Sec. 9.2 of the book [18].

with a permanent speed disturbance localized at the effective location of the bottleneck.

Empirical results presented in Figs. 3.8b and 3.9 for the off-ramp bottleneck confirm the above conclusion that a wave becomes a nucleus for traffic breakdown

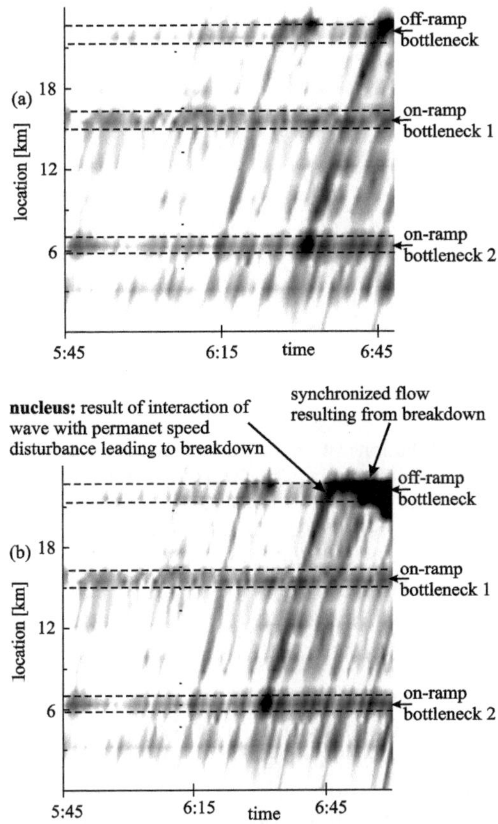

Fig. 3.8 Explanation of physics of nuclei for traffic breakdown with empirical data: (**a**) Empirical permanent local speed disturbances in free flow at highway bottlenecks for the data set for which traffic breakdown is shown in Fig. 3.7 (real field traffic data measured on September 03, 1998). (**b**) The same data as in (**a**), however, for a longer time interval showing that the nucleus for the breakdown at the off-ramp bottleneck appears due to some interaction of the wave with a permanent speed disturbance at the bottleneck. In (**a, b**), empirical data for the speed $v(x, t)$ presented by regions with variable shades of gray; in white regions $v \geq 115$ km/h, in black regions $v \leq 80$ km/h. Narrow road regions of a smaller speed (permanent local speed disturbances), which are localized in neighborhoods of the effective locations of the bottlenecks, are marked by double dashed lines. Off-ramp bottleneck, on-ramp bottleneck 1, and on-ramp bottleneck 2 are the same as those in Fig. 3.6c,d

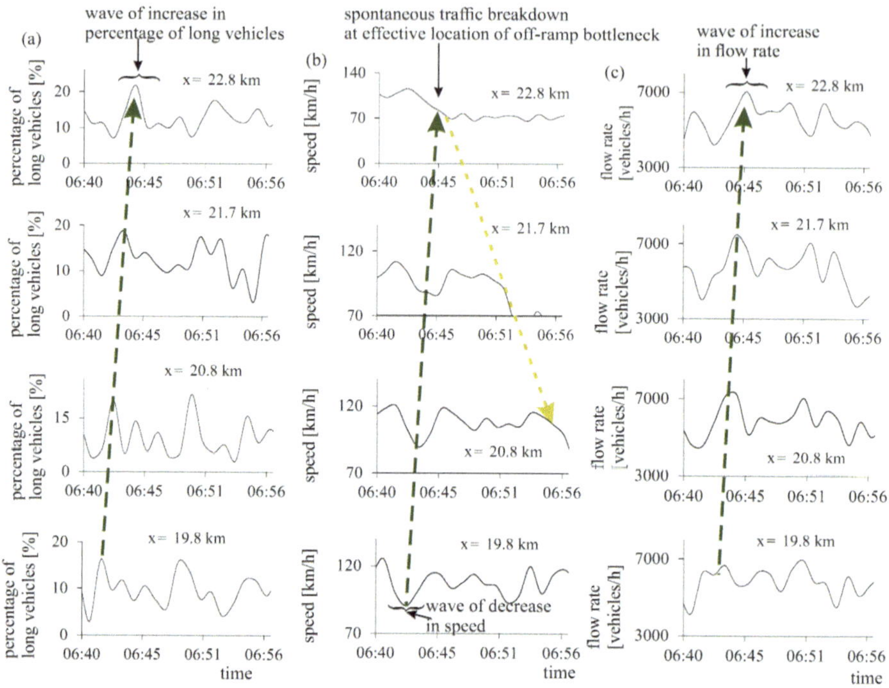

Fig. 3.9 Empirical time-distributions of traffic variables averaged across the road at different detector locations within a wave that initiates the breakdown at the location of a permanent speed disturbance localized at effective location of the off-ramp bottleneck; data is related to Figs. 3.1d and 3.7: (**a**) The percentage of long vehicles. (**b**) The speed. (**c**) The total flow rate. Arrows in downstream direction show the downstream propagation of the wave. The arrow in upstream direction in (**b**) shows the propagation of synchronized flow that has occurred due to the breakdown at the off-ramp bottleneck

only at the effective location of the bottleneck at which the permanent speed disturbance is localized. The same conclusion is valid for traffic breakdown at the on-ramp bottleneck related to data in Figs. 3.1b and 3.5.

For both off- and on-ramp bottlenecks, within a wave propagating in free flow the percentage of long vehicles (Figs. 3.9a and 3.10a) and the flow rate (Figs. 3.9c and 3.10c) are larger, whereas the average speed (Figs. 3.9b and 3.10b) is lower than outside the wave. The wave becomes a nucleus for traffic breakdown only at the effective location of the bottleneck at which a permanent speed disturbance is localized (Fig. 3.8b).

Thus we have found the empirical evidence of the effect of a permanent local speed disturbance in free flow at the effective location of a bottleneck on the breakdown: The interaction of the permanent local speed disturbance at a bottleneck with a wave in free flow propagating through this bottleneck results in the nucleation of the breakdown at the bottleneck. This confirms the theoretical explanation of an F→S transition by the nucleation effect as stated in the three-phase theory. In the

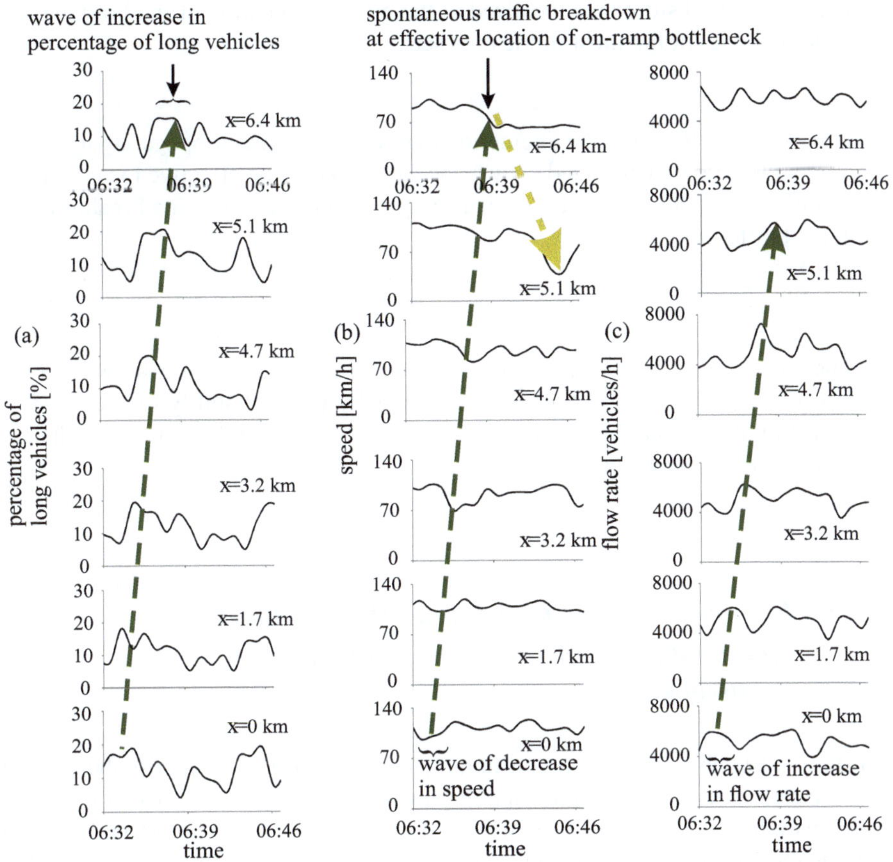

Fig. 3.10 Empirical time-distributions of traffic variables at different road detector locations within a wave that initiates the breakdown at the on-ramp bottleneck; data is related to Figs. 3.1b and 3.5: (**a**) The percentage of long vehicles. (**b**) The speed averaged across the road. (**c**) The total flow rate. Arrows in downstream direction show regions of the downstream propagation of the wave. The arrow in upstream direction in (**b**) shows the propagation of synchronized flow that has occurred due to the breakdown at the bottleneck. The effective location of the on-ramp bottleneck is approximately equal to $x = 6.4$ km

three-phase theory, the assumption about the existence of a permanent local speed disturbance in free flow at the effective location of a bottleneck should explain why the probability of the F→S transition in metastable free flow is considerably larger at the bottleneck than outside the bottleneck [18].

More than 160 traffic breakdowns at on- and off-ramp bottlenecks on different highways in Germany measured during 1996–2014 have been studied. It turns out that the empirical result of the nucleation of traffic breakdown (F→S transition) at a highway bottleneck due to the interaction of one of the waves in free flow with a permanent speed disturbance localized at the effective location of the bottleneck is the common one for the most of the data sets.

In real field traffic data, the interaction of one of the waves propagating in free flow with a permanent speed disturbance localized at the effective location of a bottleneck is the origin of the occurrence of a nucleus for traffic breakdown at the bottleneck.

The existence of a permanent local speed disturbance in free flow at the effective location of a bottleneck explains why the probability of the F→S transition in metastable free flow is considerably larger at the bottleneck than that outside the bottleneck.

3.4.5 Empirical Two-Dimensional (2D) Asymmetric Spatiotemporal Structure of Nuclei for Traffic Breakdown

To study a possible effect of a non-homogeneity of traffic flow across the road on the nucleation of traffic breakdown at highway bottlenecks, we consider empirical traffic variables in different freeway lanes (Figs. 3.11 and 3.12). We should mention that the most of long (slow) vehicles move in the right lane (sometimes traffic flow in the right lane consists of almost 100% (slow) long vehicles) (Fig. 3.11a). The percentage of long vehicles in the middle lane is considerably smaller than in the right lane; almost no long vehicles move in the left lane (Fig. 3.11a).

The following empirical result has been found: An empirical wave in free flow exhibits a two-dimensional (2D) structure: Wave's characteristics are different in different highway lanes (Fig. 3.12). This wave structure is asymmetric for different traffic variables in the perpendicularly direction to the flow direction. The most waves of the share of slow vehicles $\Delta\psi_{\text{wave}}$ are observed in the right lane, while in the left lane almost no waves of the share of slow vehicles $\Delta\psi_{\text{wave}}$ exist. On contrary, the most waves of the flow rate Δq_{wave} and vehicle speed Δv_{wave} are observed in the left lane, while in the right lane almost no waves of Δq_{wave} and Δv_{wave} exist.

An empirical wave in free flow exhibits a 2D-structure in the perpendicularly direction to the traffic flow direction. The most waves of the share of slow

(continued)

vehicles are observed in the right lane, while in the left lane almost no waves of the share of slow vehicles exist. On contrary, the most waves of the flow rate and vehicle speed are observed in the left lane.

However, to understand the effect of 2D asymmetric structure of nuclei on features of traffic breakdown, a study of *microscopic* empirical data is required, in which lane changing and vehicle acceleration (deceleration) in a neighborhood of a bottleneck can be resolved. Unfortunately, such microscopic (single-vehicle) empirical data for free flow at bottlenecks is currently not available. Therefore, in Sect. 5.14.1 with the use of a three-phase stochastic microscopic traffic flow model, we study theoretical predictions about the effect of 2D asymmetric structure of nuclei on microscopic features of traffic breakdown at an on-ramp bottleneck.

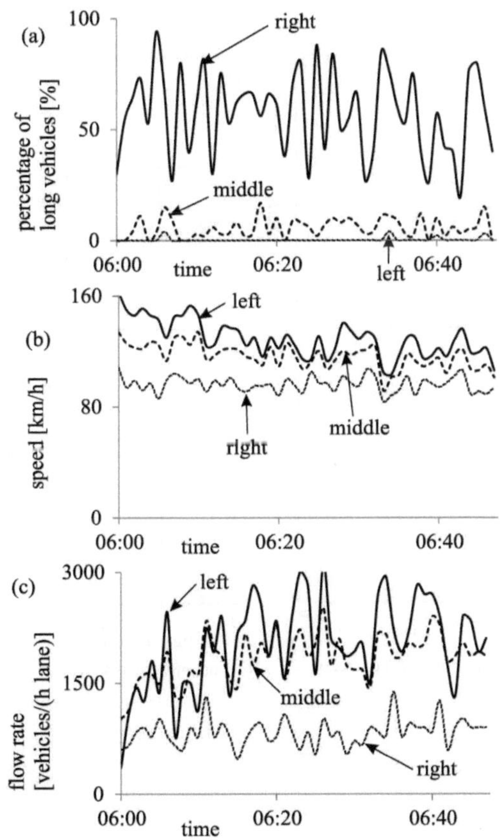

Fig. 3.11 Empirical time-dependencies of percentage of long vehicles (**a**), speed (**b**), and flow rate (**c**) in different highway lanes at location $x = 0$ km for real field traffic data shown in Fig. 3.1b

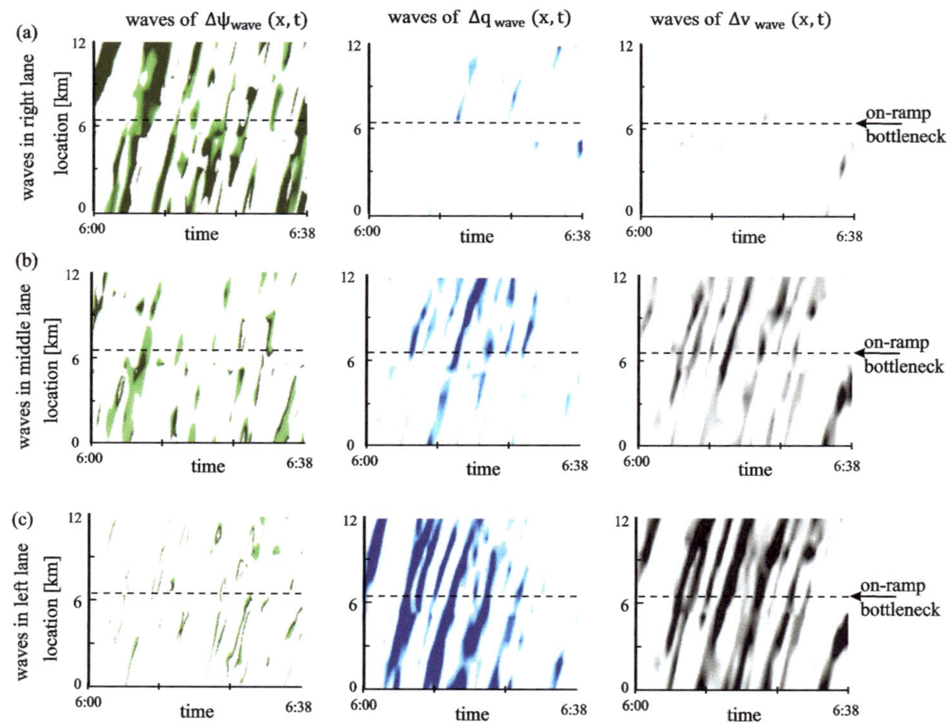

Fig. 3.12 Empirical waves of the percentage of slow (long) vehicles $\Delta\psi_{\text{wave}}$ (*left column*), the flow rate Δq_{wave} (*middle column*), and the vehicle speed Δv_{wave} (*right column*) in free flow for different road lanes for real field traffic data shown in Fig. 3.1b during the same time interval as that in Fig. 3.4: (**a**) Right lane. (**b**) Middle lane. (**c**) Left lane. Waves of $\Delta\psi_{\text{wave}}(x, t)$ are presented by regions with variable shades of gray (green in the on-line version) (in white regions $\Delta\psi_{\text{wave}} \leq 0.3$, in black (*dark green*) regions $\Delta\psi_{\text{wave}} \geq 1$). Waves of $\Delta q_{\text{wave}}(x, t)$ are presented by regions with variable shades of gray (blue in the on-line version) (in white regions $\Delta q_{\text{wave}} \leq 480$ vehicles/h, in black (*dark blue*) regions $\Delta q_{\text{wave}} \geq 800$ vehicles/h). Waves of $\Delta v_{\text{wave}}(x, t)$ are presented by regions with variable shades of gray (in white regions $\Delta v_{\text{wave}} \leq 7$ km/h, in black regions $\Delta v_{\text{wave}} \geq 20$ km/h)

3.5 Waves in Free Flow and Empirical Spontaneous Traffic Breakdown in Flow Without Trucks

Real field traffic data obtained on highways shows that waves of slowly moving long vehicles (trucks) cause the breakdown of highway traffic at on- and off-ramp bottlenecks. However, the following question can arise:

- If trucks were forbidden on highways, would no traffic breakdowns at on- and off-ramp bottlenecks occur?

Real field traffic data shows that empirical traffic breakdown at on- and off-ramp bottlenecks does occur even when truck traffic is forbidden on highways.

On Sundays, truck traffic is generally forbidden on German highways.[3] Nevertheless, traffic breakdown (F→S transition) at highway bottlenecks is also observed (Fig. 3.13a). We find also a lot of slow moving vehicles in traffic on Sundays that move mostly in the right lane (Fig. 3.13b). To explain this, we should mention that in real field traffic data studied in this chapter a road detector identifies a vehicle passing the detector as a "long vehicle", when the vehicle length is about 5.5 m or longer. Therefore, most of motor-home vehicles (caravans) in Sunday traffic are measured as long vehicles ("long vehicles" in Fig. 3.13c). In accordance with Fig. 3.13c, the average speed of these slow vehicles $v_{\text{slow}}^{(\text{mean})} \approx 98$ km/h is higher than trucks whose maximum speed is limited on German highways (Sect. 3.4.1).

Due to the existence of slow vehicles in empirical Sunday traffic, we find waves in free flow (Fig. 3.14). We find also that traffic breakdown occurs when one of the waves reaches a permanent speed disturbance at the bottleneck location (Fig. 3.14c). As in free flow with trucks (Fig. 3.12), there is a 2D-structure of the waves in free flow without trucks (Fig. 3.15) that exhibits qualitatively the same features as those explained in Sect. 3.4.5.

3.6 Induced Traffic Breakdown—Empirical Proof of Nucleation Nature of Empirical Traffic Breakdown

The author has often confronted with the following question of colleagues:

- Is there some general empirical proof of the nucleation nature of empirical traffic breakdown (F→S transition)?

To perform a clear empirical proof of the nucleation nature of traffic breakdown that is independent of differences in vehicle and driver characteristics in free flow, we compare *empirical induced* traffic breakdown (Fig. 3.16a) with *empirical spontaneous* traffic breakdown caused by wave propagation (Figs. 3.5, 3.7, and 3.14).

---

[3]Only a few trucks with a special permit from authorities may be allowed.

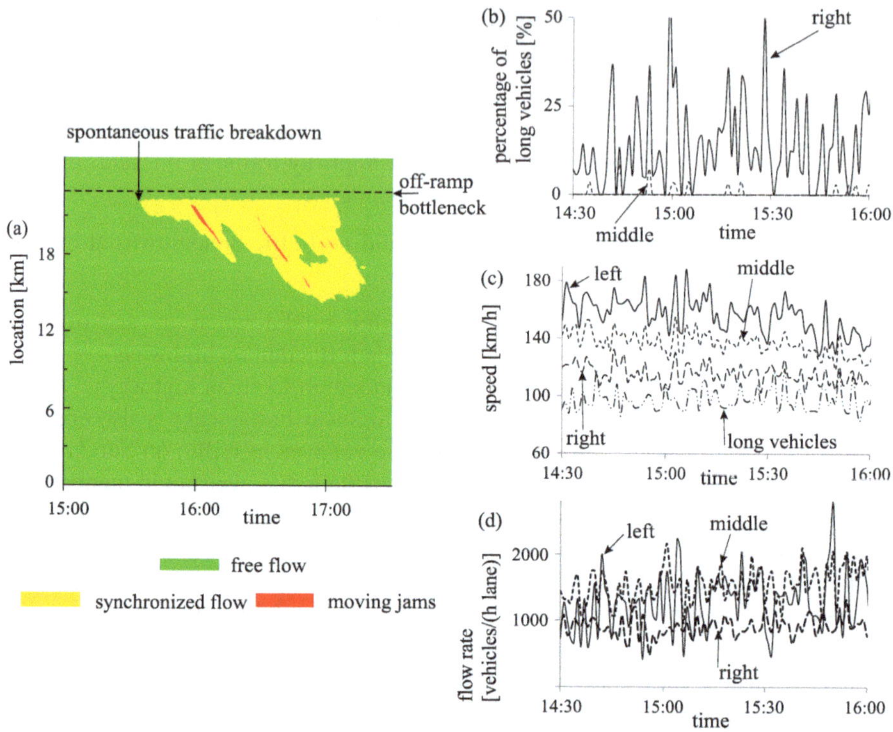

Fig. 3.13 Empirical real field traffic data measured by road detectors on Sunday, July 10, 2011: (**a**) Speed data measured with road detectors installed along road section; data is presented in space and time with averaging method described in Sec. C.2 of [31]; location of off-ramp bottleneck is the same as that in Fig. 3.1c,d. (**b**–**d**) Time-dependencies of percentage of slow vehicles (**b**), speed (**c**), and flow rate (**d**) in different highway lanes at location $x = 0$ km

3.6.1 *Sources of Nucleus for Empirical Traffic Breakdown*

Both an empirical wave in free flow (Fig. 3.5) and a localized congested pattern (wide moving jam in Fig. 3.16a) become nuclei for traffic breakdown in metastable free flow, when they reach the effective location of a highway bottleneck. The propagation of a single congested pattern to the effective bottleneck location is usually sufficient for the inducing of the breakdown at the bottleneck (Fig. 3.16a). In contrast, many waves in free flow can propagate through the bottleneck while initiating no breakdown at the bottleneck (Figs. 3.4 and 3.6).

The latter empirical result allows us to assume that at a given flow rate in metastable free flow at a highway bottleneck there is a *critical wave* related to a *critical nucleus* for traffic breakdown. Therefore, if the wave amplitude is smaller than that of the critical wave for the given flow rate in the metastable free flow at the bottleneck, then no breakdown occurs while the wave propagates through the bottleneck. However, amplitudes of the waves that become nuclei for the breakdown

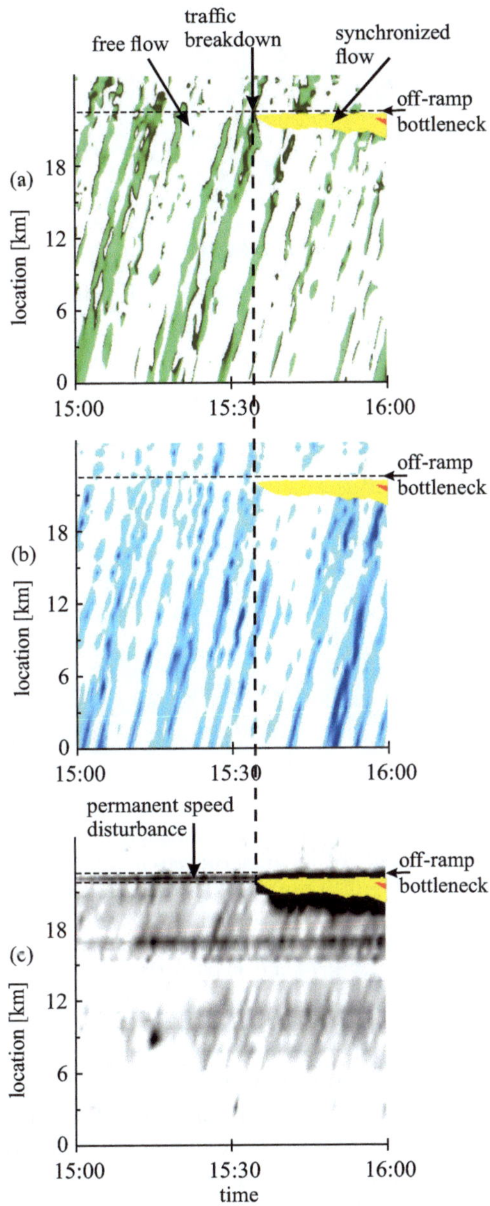

Fig. 3.14 Empirical nucleus for traffic breakdown (F→S transition) in free flow associated with data in Fig. 3.13a. Empirical waves of $\Delta\psi_{\text{wave}}$ (**a**), Δq_{wave} (**b**), and v (**c**). Waves of $\Delta\psi_{\text{wave}}(x, t)$ are presented by regions with variable shades of gray (green in the on-line version) (in white regions $\Delta\psi_{\text{wave}} \leq 0.1$, in black (*dark green*) regions $\Delta\psi_{\text{wave}} \geq 1$). Waves of $\Delta q_{\text{wave}}(x, t)$ are presented by regions with variable shades of gray (blue in the on-line version) (in white regions $\Delta q_{\text{wave}} \leq 300$ vehicles/h, in black (*dark blue*) regions $\Delta q_{\text{wave}} \geq 1500$ vehicles/h). Empirical data for the speed $v(x, t)$ in (**c**) are presented by regions with variable shades of gray (in white regions $v \geq 120$ km/h, in black regions $v \leq 100$ km/h). In (**a–c**), region labeled by "synchronized flow" shows symbolically synchronized flow

at the effective locations of the bottlenecks in Figs. 3.5 and 3.7 should be equal to or larger than amplitudes of critical waves required for the breakdown in the metastable free flow at the related bottlenecks, respectively.

In contrast with waves in free flow, within a congested pattern the speed is usually smaller than the speed within a critical nucleus required for the breakdown

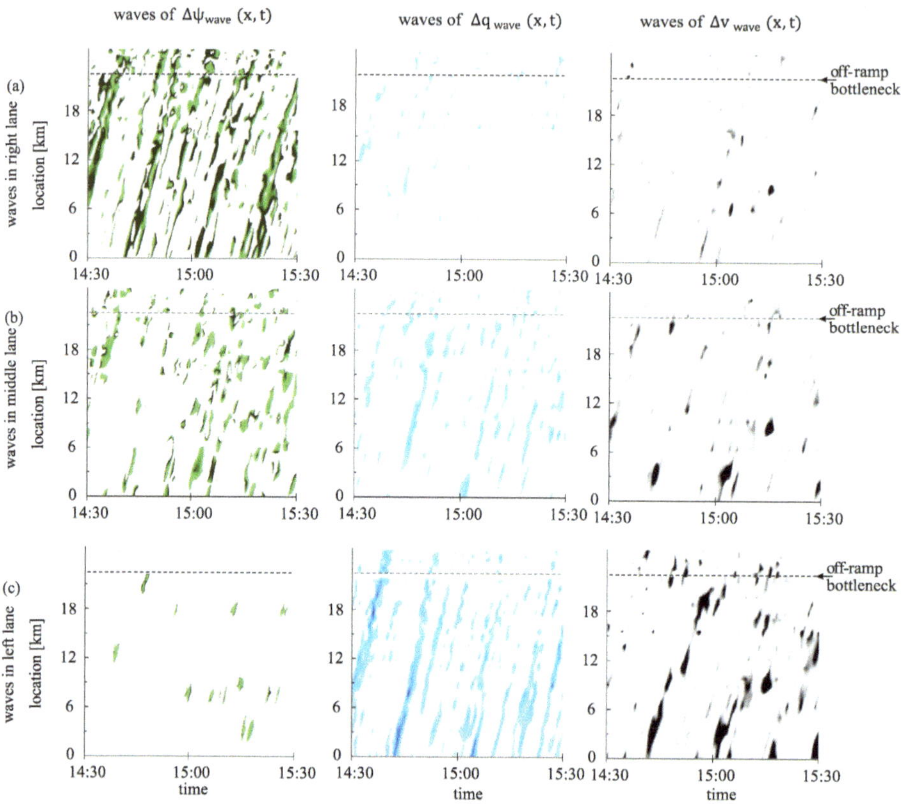

Fig. 3.15 Empirical waves of $\Delta\psi_{\text{wave}}$ (*left column*), the flow rate Δq_{wave} (*middle column*), and the vehicle speed Δv_{wave} (*right column*) in free flow for different road lanes for real field traffic data shown in Fig. 3.13a: (**a**) Right lane. (**b**) Middle lane. (**c**) Left lane. Parameters of the presentation of empirical waves in (**a**, **b**) are the same as those in Fig. 3.14 for $\Delta\psi_{\text{wave}}(x,t)$ and $\Delta q_{\text{wave}}(x,t)$, respectively. Waves of $\Delta v_{\text{wave}}(x,t)$ are presented by regions with variable shades of gray (in white regions $\Delta v_{\text{wave}} \leq 7$ km/h, in black regions $\Delta v_{\text{wave}} \geq 15$ km/h). Other parameters are the same as those in Fig. 3.14

in metastable free flow at a bottleneck. For this reason, a localized congested pattern becomes a nucleus for traffic breakdown in the metastable free flow, when the pattern reaches the effective bottleneck location.

Thus, a basic difference between *empirical spontaneous* breakdown (Figs. 3.1b,d, 3.5, 3.7, and 3.14) and *empirical induced* breakdown (Fig. 3.16a) is as follows: To initiate spontaneous traffic breakdown at the bottleneck in metastable free flow, i.e., to be a nucleus for the breakdown, a wave should be equal to or a larger one than a critical wave. In contrast, a localized congested pattern is usually a nucleus for the breakdown in the metastable free flow at the bottleneck.

However, after the breakdown has occurred, characteristics of synchronized flow that has been formed at the bottleneck do not depend on whether synchronized flow

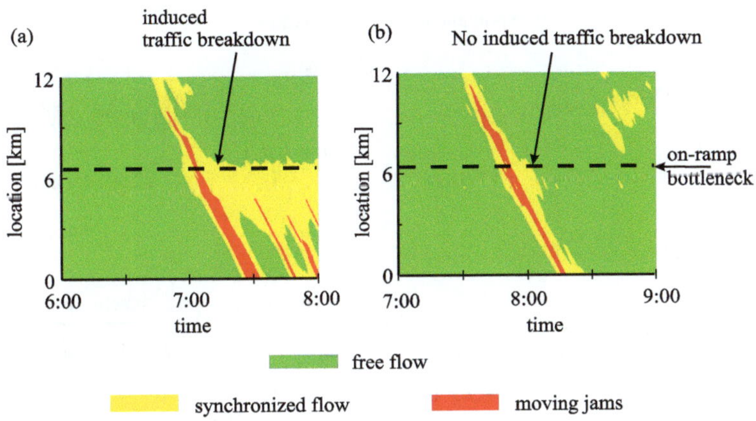

Fig. 3.16 Empirical jam propagation through on-ramp bottleneck: (**a**) Empirical induced traffic breakdown. (**b**) Moving jam propagation through the bottleneck without induced traffic breakdown. Real field traffic data measured by road detectors on three-lane freeway A5-South in Germany on March 22, 2001 (**a**) and June 23, 1998 (**b**). On-ramp bottleneck marked by dashed lines in (**a, b**) is the same one as that in Fig. 3.1a,b. Speed data measured with road detectors installed along road section; data is presented in space and time with averaging method described in Sec. C.2 of [31]

has occurred due to empirical spontaneous breakdown or due to empirical induced breakdown. This statement is illustrated by empirical data presented in Figs. 3.17 and 3.18.

In Fig. 3.17, synchronized flow has occurred at the on-ramp bottleneck (Fig. 3.1b) due to empirical spontaneous breakdown. The spontaneous traffic breakdown is realized when a wave reaches the effective bottleneck location (Fig. 3.5). In contrast, in Fig. 3.18 synchronized flow has been induced at the on-ramp bottleneck due to the propagation of a wide moving jam through the bottleneck (Fig. 3.16a).

In both cases (Figs. 3.17 and 3.18), synchronized flow resulting from the breakdown at the bottleneck is self-maintained under free flow conditions downstream of the bottleneck. Empirical features of synchronized flow resulting from the induced breakdown (at $t > 7{:}07$ in Fig. 3.16a) are qualitatively identical to those found in synchronized flow resulting from empirical spontaneous traffic breakdown (Fig. 3.1b). In particular, as in the case of empirical spontaneous breakdown, in the case of empirical induced traffic breakdown the flow rate in synchronized flow resulting from the breakdown can be as high as the flow rate in free flow just before the breakdown has occurred (location $x = 6.4$ km in Figs. 3.17 and 3.18); this is in contrast with a moving jam within which the flow rate is very small (Fig. 3.18, right column).

A study of a huge number of other empirical data has confirmed this conclusion: Qualitative the same features of synchronized flow resulting from the induced breakdown (at $t > 7{:}07$ in Fig. 3.16a) and those found in synchronized flow resulting from empirical spontaneous traffic breakdown (Fig. 3.1b) have also been found in all other real field traffic data. An example is shown in Fig. 3.19 for empirical

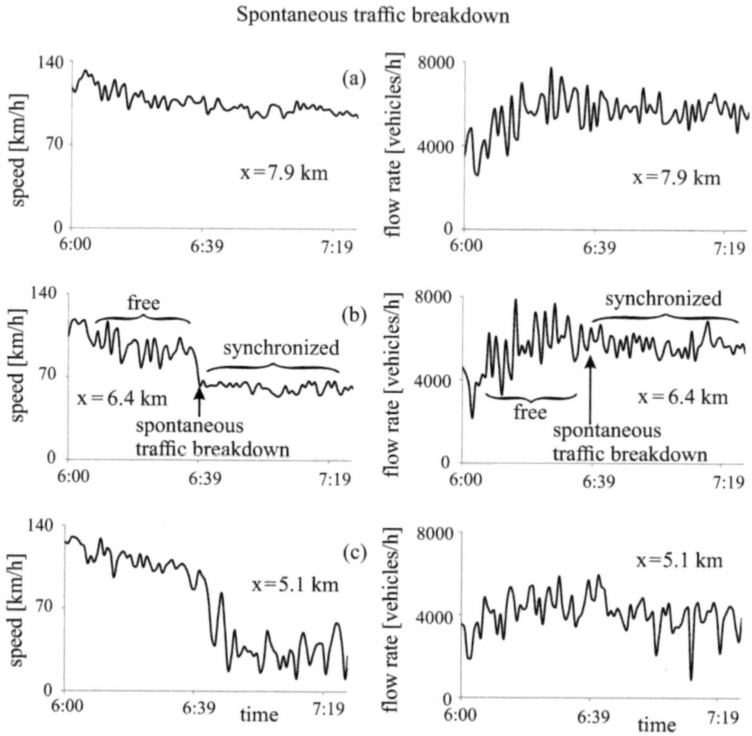

Fig. 3.17 Empirical spontaneous traffic breakdown related to Fig. 3.1b: (**a–c**) Empirical 1-min average speed (*left column*) and flow rate (*right column*) as time-functions measured by detectors installed on freeway A5-South downstream of the bottleneck ($x = 7.9$ km) (**a**), at the bottleneck location ($x = 6.4$ km) (**b**), and upstream of the bottleneck ($x = 5.1$ km) (**c**). Data was measured on April 15, 1996. Free flow and synchronized flow are labeled by, respectively, "free" and "synchronized"

induced traffic breakdown caused by the propagation of a moving synchronized flow pattern (MSP) to the location of the upstream on-ramp bottleneck (Fig. 3.3b–d). In particular, in Fig. 3.19 as in Figs. 3.17 and 3.18, synchronized flow resulting from the breakdown at the bottleneck is self-maintained under free flow conditions downstream of the bottleneck (location $x = 17.9$ km in Fig. 3.19a). Empirical features of synchronized flow resulting from empirical induced traffic breakdown (at $t > 7{:}00$ in Fig. 3.3b–d) are qualitatively identical to those found in synchronized flow resulting from empirical spontaneous traffic breakdown (Fig. 3.1b) and empirical induced traffic breakdown caused by the propagation of the wide moving jam (Fig. 3.16a). In particular, as in the case of empirical spontaneous breakdown, in the case of empirical induced traffic breakdown caused by the MSP propagation the flow rate in synchronized flow resulting from the breakdown can be as high as the flow rate in free flow just before the breakdown has occurred (compare time-dependence of the

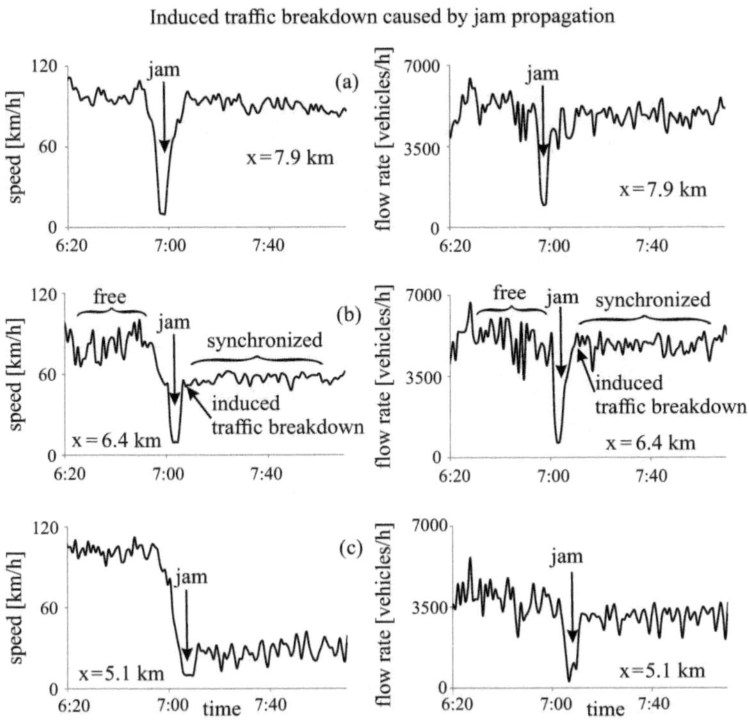

Fig. 3.18 Empirical induced traffic breakdown caused by the propagation of a wide moving jam related to Fig. 3.16a: (**a–c**) Empirical 1-min average speed (*left column*) and flow rate (*right column*) as time-functions measured by detectors installed on freeway A5-South downstream of the bottleneck ($x = 7.9$ km) (**a**), at the bottleneck location ($x = 6.4$ km) (**b**), and upstream of the bottleneck ($x = 5.1$ km) (**c**). Data was measured on March 22, 2001. Free flow, synchronized flow, and the wide moving jam are labeled by, respectively, "free", "synchronized", and "jam"

flow rates at location $x = 17$ km in Fig. 3.19 and at location $x = 6.4$ km in Figs. 3.17 and 3.18).

This shows that rather than the nature of traffic breakdown, the terms *empirical spontaneous* and *empirical induced* traffic breakdowns at a bottleneck distinguish different *sources* of a nucleus that occurrence leads to traffic breakdown: In Fig. 3.1b, the source of empirical spontaneous traffic breakdown is one of the waves in free flow shown in Fig. 3.5. In Fig. 3.16a, the source of empirical induced traffic breakdown is the wide moving jam. In Fig. 3.3, the source of empirical induced traffic breakdown is the MSP.

In contrast with the wide moving jam shown in Fig. 3.16a, a wide moving jam shown in Fig. 3.16b does not induce traffic breakdown at the bottleneck: After the wide moving jam shown Fig. 3.16b is far away upstream of the bottleneck, free flow returns both at the effective bottleneck location as well as downstream and upstream of the bottleneck (Fig. 3.20).

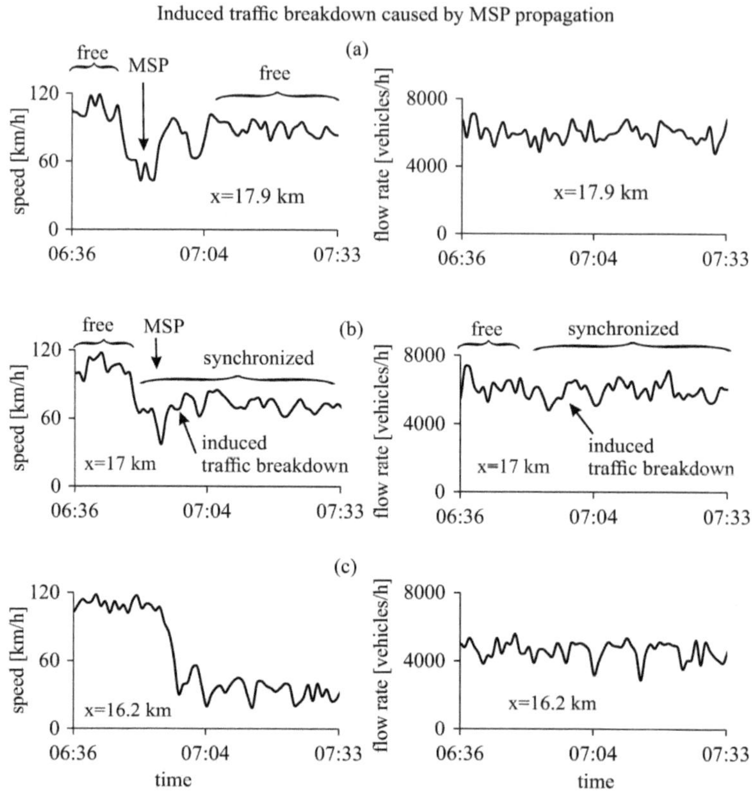

Fig. 3.19 Empirical induced traffic breakdown caused by the propagation of a moving synchronized flow pattern (MSP) related to Fig. 3.3b–d: (**a–c**) Empirical 1-min average speed (*left column*) and flow rate (*right column*) as time-functions measured by detectors installed on freeway A5-South downstream of the bottleneck ($x = 17.9$ km) (**a**), at the bottleneck location ($x = 17$ km) (**b**), and upstream of the bottleneck ($x = 16.2$ km) (**c**). Data was measured on April 20, 1998. Free flow, synchronized flow, and the MSP are labeled by, respectively, "free", "synchronized", and "MSP"

However, if free flow is in a metastable state at the bottleneck, the wide moving jam should be a nucleus for traffic breakdown at the bottleneck. Therefore, the case shown in Fig. 3.16b, in which no traffic breakdown has been induced, should be related to the opposite condition: Rather than free flow is metastable, this free flow is in a stable state with respect to traffic breakdown.

We can conclude that there should be some critical flow rate that separates metastable and stable states of free flow with respect to traffic breakdown (F→S transition) at the bottleneck. When the flow rate at the bottleneck is smaller than this critical flow rate, there can be no nuclei for traffic breakdown at the bottleneck. This means that regardless how large a time-limited disturbance in free flow at the bottleneck is, no traffic breakdown occurs at the bottleneck: After the disturbance has disappeared at the bottleneck, free flow recovers at the bottleneck. This critical

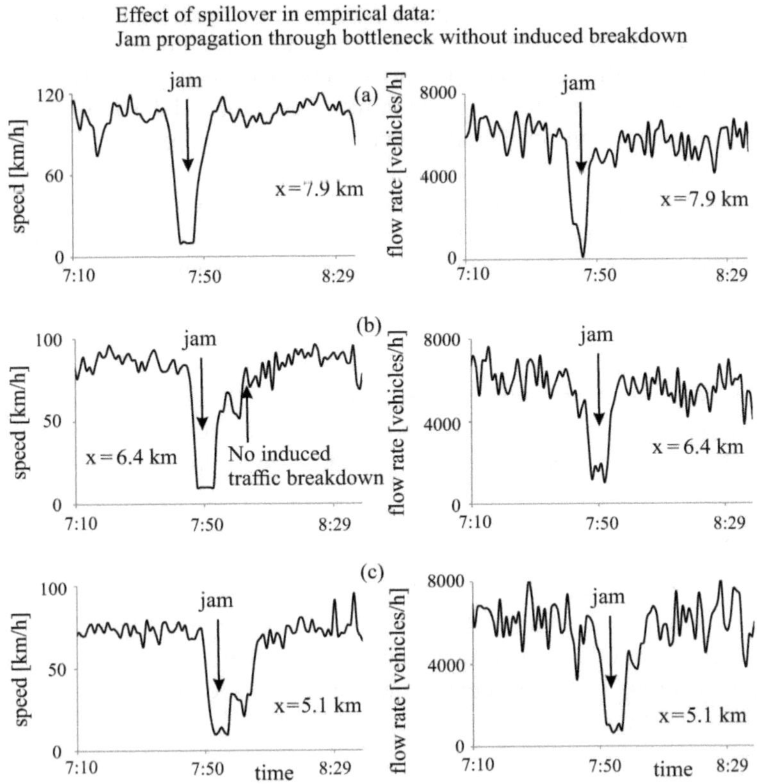

Fig. 3.20 Empirical example of spillover without induced traffic breakdown (real field traffic data measured by road detectors on three-lane freeway A5-South in Germany): (**a–c**) Wide moving jam propagation through on-ramp bottleneck without induced traffic breakdown related to Fig. 3.16b. Data was measured on June 23, 1998. Empirical 1-min average speed (*left column*) and flow rate (*right column*) as time-functions measured by detectors installed on freeway A5-South at the bottleneck location ($x = 6.4$ km) (**a**), downstream ($x = 7.9$ km) (**b**) and upstream of the bottleneck ($x = 5.1$ km) (**c**) are shown; the wide moving jam is labeled by "jam"

flow rate determines the minimum highway capacity $C_{\min}$ briefly discussed in Sect. 1.10. This emphasizes another difference between empirical spontaneous and empirical induced traffic breakdowns at a highway bottleneck that is as follows.

As we have seen above, there are many waves in free flow that propagate through the bottleneck without initiating of empirical spontaneous traffic breakdown. From this empirical result, we *cannot* state whether free flow is in a metastable state or free flow is in a stable state with respect to traffic breakdown. Indeed, if all these waves are smaller than a critical wave, then no traffic breakdown occurs in the metastable free flow at the bottleneck. No traffic breakdown occurs also, when free flow is in a stable state with respect to traffic breakdown, i.e., if the flow rate in free flow at the bottleneck is smaller than the minimum highway capacity $C_{\min}$.

In contrast, when a wide moving jam propagates through the bottleneck without inducing of traffic breakdown, we can assume that free flow is in a stable state with respect to traffic breakdown. This means that when the wide moving jam propagates through a bottleneck without inducing the breakdown, we can assume that the flow rate is smaller than the minimum highway capacity $C_{\rm min}$ of free flow at the bottleneck.

The empirical induced traffic breakdown at a highway bottleneck discussed above does not depend on the degree of the heterogeneity of real vehicular traffic. For this reason, empirical induced traffic breakdown is a general empirical proof of the nucleation nature of empirical traffic breakdown.

Rather than the nature of traffic breakdown, the terms *empirical spontaneous* and *empirical induced* traffic breakdowns at a bottleneck distinguish different *sources* of a nucleus that occurrence leads to traffic breakdown at the bottleneck.

The empirical induced traffic breakdown at a highway bottleneck is a general empirical proof of the nucleation nature of empirical traffic breakdown: This empirical proof is independent of the degree of the heterogeneity of real vehicular traffic.

3.6.2 *Induced Traffic Breakdown as One of Different Consequences of Spillover in Real Traffic*

Upstream propagation of traffic congestion occurring at a downstream bottleneck is often called *spillback*. If this traffic congestion propagation forces congested traffic at an upstream bottleneck, it is called *spillover*. When the wide moving jams shown in Fig. 3.16 reach the bottleneck, any of the wide moving jams can be considered *spillover*: The wide moving jam forces congested traffic at the bottleneck. Due to the upstream jam propagation, the jams can be considered as spillover *only* during a short time interval: When the jams are far away upstream of the bottleneck, they do not force congested traffic at the bottleneck.

However, we do not use the term *spillover*. This is because there can be at least the following *qualitatively different* empirical effects of spillover that should be considered separately each other:

(i) An empirical induced traffic breakdown occurs due to the propagation of a wide moving jam through a bottleneck (Figs. 3.16a and 3.18b).

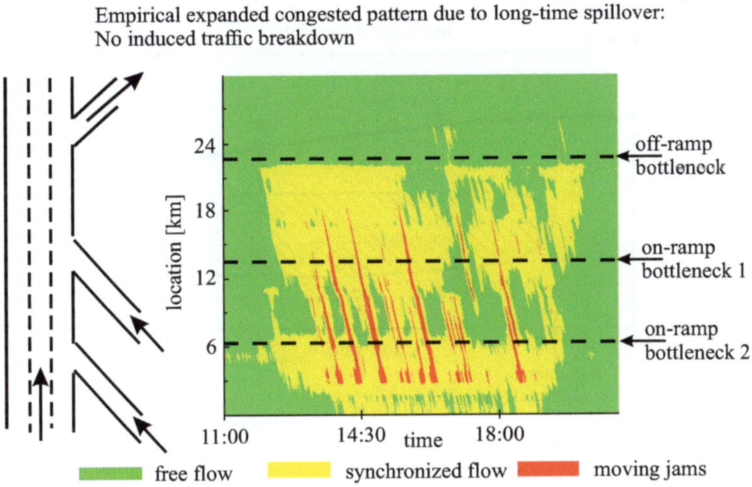

Fig. 3.21 Empirical example of spillover without induced traffic breakdown (real field traffic data measured by road detectors on three-lane freeway A5-North in Germany): Expanded congested pattern (EP) (*right*) (data was measured on March 23, 2001) and scheme of a section of freeway A5-North with three bottlenecks (*left*). Bottlenecks have been explained in Sec. 9.2.2 of [18]. The term "expanded congested pattern (EP)" has been explained in the book [18]

(ii) The propagation of a wide moving jam through a bottleneck does not lead to induced traffic breakdown (Figs. 3.16b and 3.20).

(iii) An expanded congested pattern (EP) occurs due to spillover at on-ramp bottleneck 1 (Fig. 3.21) [14]. This spillover *cannot* be considered as induced traffic breakdown. This is because during the whole time of the existence of traffic congestion at on-ramp bottleneck 1 this traffic congestion is forced by downstream traffic congestion occurring initially at the off-ramp bottleneck (Fig. 3.21).

3.7 Empirical Nucleation Nature of Traffic Breakdown as Origin of the Infinity of Highway Capacities

In Sect. 3.3, we have explained that the empirical nucleation nature of traffic breakdown (F→S transition) observed above in real field traffic data means that free flow is in a metastable state with respect to the F→S transition at the bottleneck. In its turn, the term "the metastability of free flow" means that if a nucleus for traffic breakdown occurs in a metastable state of free flow at the bottleneck, traffic breakdown does occur. In contrast, as long as no nucleus appears, no breakdown occurs in the metastable state of free flow.

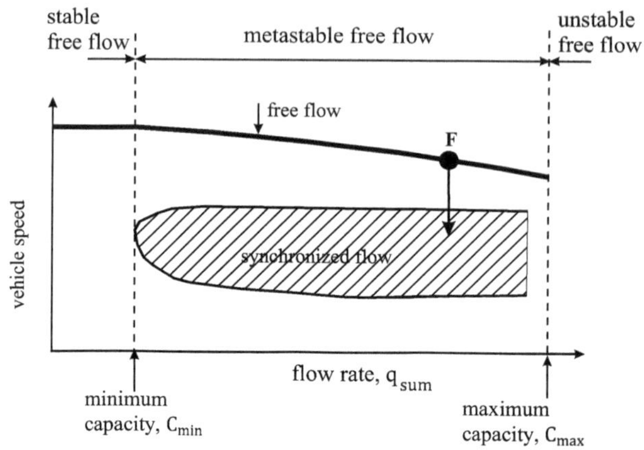

Fig. 3.22 Qualitative explanation of the empirical metastability of free flow with respect to traffic breakdown (F→S transition) at bottleneck [4–26]. Qualitative Z-speed–flow-rate characteristic for traffic breakdown; F—free flow, S—synchronized flow (two-dimensional (2D) hatched region) (results of simulations of the Z-speed–flow-rate characteristic for traffic breakdown will be discussed in Sect. 5.4.2; see Fig. 5.8). Arrow F→S illustrates symbolically one of possible F→S transitions occurring in a metastable state of free flow when a nucleus appears in the free flow at the bottleneck

Clearly, there should a *limited range* of the flow rate q_{sum} in free flow at the bottleneck within which empirical free flow is in a metastable state (Fig. 3.22). Indeed, in real field traffic data at small enough values of the flow rate q_{sum} no traffic breakdown (F→S transition) is observed at the bottleneck. This means that no nuclei can occur in free flow that can lead to traffic breakdown (F→S transition) at the bottleneck. This free flow is stable with respect to traffic breakdown (F→S transition) at the bottleneck. As explained above, the minimum flow rate of the flow rate that separates the stable and metastable free flow is equal to a minimum highway capacity C_{min} of free flow at the bottleneck (Fig. 3.22). Therefore, at

$$q_{sum} < C_{min}, \tag{3.4}$$

free flow is stable at the bottleneck. In other words, under condition (3.4) no nuclei can appear in the free flow. Consequently, no traffic breakdown can occur at the bottleneck.

Contrarily, at

$$q_{sum} \geq C_{min} \tag{3.5}$$

free flow is in a metastable state with respect to traffic breakdown at the bottleneck (Fig. 3.22). Therefore, under condition (3.5) traffic breakdown can occur at the bottleneck.

There should be a large enough flow rate at which any small disturbance in free flow is a nucleus for traffic breakdown (F→S transition) at the bottleneck. At this flow rate that we denote by a maximum highway capacity $C_{\max}$ of free flow at the bottleneck free flow can be considered as unstable free flow (Fig. 3.22). This means that the flow rate

$$q_{\rm sum} = C_{\max} \tag{3.6}$$

separates the metastable free flow and unstable free flow with respect to traffic breakdown (F→S transition) at the bottleneck (Fig. 3.22): At

$$q_{\rm sum} \geq C_{\max} \tag{3.7}$$

free flow is unstable with respect to traffic breakdown (F→S transition) at the bottleneck. Correspondingly, at

$$q_{\rm sum} < C_{\max} \tag{3.8}$$

free flow is metastable with respect to traffic breakdown (F→S transition) at the bottleneck.

Thus, from formulas (3.5) and (3.8) it follows that free flow is in a metastable state with respect to the F→S transition at the bottleneck within the flow rate range (Fig. 3.22)

$$C_{\min} \leq q_{\rm sum} < C_{\max}. \tag{3.9}$$

Any flow rate $q_{\rm sum}$ in free flow at a highway bottleneck that satisfies (3.9) is highway capacity. This is because at each of the flow rates (3.9) traffic breakdown can occur at the bottleneck. At any time instant, there are the infinite number of the flow rates (3.9) at which traffic breakdown can occur at the bottleneck. Therefore, there are the infinite number of highway capacities at any time instant. The existence of an infinite number of highway capacities at any time instant means that highway capacity is stochastic. This conclusion is one of the important consequences of the empirical nucleation nature of traffic breakdown discussed in this chapter.

Formulas (3.4)–(3.9) result from the empirical nucleation nature of traffic breakdown (F→S transition) at the bottleneck discussed above. A more detailed consideration of the physical sense of formulas (3.4)–(3.9) [4–26] as well as the highway capacity definitions $C_{\min}$ and $C_{\max}$ will be considered in Chap. 5.

3.8 Conclusions

1. Empirical traffic breakdown (F→S transition) at a highway bottleneck can occur in a metastable state of free flow with respect to the F→S transition. In the metastable state of free flow, traffic breakdown occurs only if a nucleus required for the breakdown appears. Empirical traffic breakdown (F→S transition) at the bottleneck can be either spontaneous or induced.
2. In real field traffic data, a nucleus for empirical spontaneous traffic breakdown at a highway bottleneck occurs through an interaction of one of the waves in free flow with a permanent speed disturbance localized at the bottleneck. When the wave reaches the location of the permanent speed disturbance at the bottleneck (effective bottleneck location), spontaneous traffic breakdown, i.e., spontaneous phase transition from free flow to synchronized flow (F→S transition) occurs.
3. Waves in free flow, which can be nuclei for empirical spontaneous traffic breakdown at highway bottlenecks, appear often due to slow vehicles moving in traffic flow. These waves propagate with the average speed of slow vehicles in free flow. Within a wave, the total flow rate is larger and the speed averaged across the highway is smaller than outside the wave.
4. The waves in free flow exhibit a two-dimensional (2D) asymmetric spatiotemporal structure whose characteristics are different in different highway lanes.
5. A wave that amplitude is large enough can become a nucleus for the breakdown while the wave propagates through the bottleneck; small enough waves in free flow do not become nuclei for traffic breakdown.
6. In contrast with empirical *spontaneous* traffic breakdown, any localized congested pattern becomes usually a nucleus for empirical *induced* traffic breakdown in the metastable free flow at a highway bottleneck: When the congested pattern reaches the location of the bottleneck, the pattern induces the breakdown at the bottleneck.
7. Rather than the nature of traffic breakdown, the terms *empirical spontaneous* and *empirical induced* traffic breakdowns at a bottleneck distinguish different *sources* of a nucleus that occurrence leads to traffic breakdown at the bottleneck. Therefore, the source of a nucleus for traffic breakdown (F→S transition) is the *solely* difference between empirical spontaneous and induced traffic breakdowns at a highway bottleneck.
8. Qualitative features of empirical synchronized flow at a highway bottleneck resulting from either empirical induced traffic breakdown or empirical spontaneous traffic breakdown are identical ones.
9. There is a limited range of the flow rate in free flow at which free flow is in a metastable state with respect to traffic breakdown (F→S transition) at the bottleneck. All flow rates within this flow-rate range are highway capacities. This means that at any time instant there are the infinity of highway capacities. The existence of the infinite number of highway capacities at any time instant means that highway capacity is stochastic highway capacity. This conclusion is

one of the important consequences of the empirical nucleation nature of traffic breakdown.

10. The nature and features of nuclei for empirical traffic breakdown at highway bottlenecks studied in this chapter confirm the basic assumption of three-phase traffic theory about the nucleation nature of traffic breakdown (F→S transition) at highway bottlenecks.

The empirical induced F→S transition at the bottleneck proves the empirical nucleation nature of traffic breakdown at the bottleneck. The empirical nucleation nature of traffic breakdown at a highway bottleneck can be considered the empirical fundamental of transportation science.

References

1. S. Chandrasekhar, *Hydrodynamic and Hydromagnetic Stability* (Oxford University Press, Oxford 1961)
2. C.W. Gardiner, *Handbook of Stochastic Methods*, 2nd edn. (Springer, Berlin, 1990)
3. H. Haken, *Synergetics* (Springer, Berlin, 1977)
4. B.S. Kerner, in *Proceedings of the 3rd Symposium on Highway Capacity and Level of Service*, ed. by R. Rysgaard, vol. 2, (Road Directorate, Ministry of Transport, Denmark, 1998), pp. 621–642
5. B.S. Kerner, Trans. Res. Rec. **1678**, 160–167 (1999)
6. B.S. Kerner, in *Transportation and Traffic Theory*, ed. by A. Ceder. (Elsevier Science, Amsterdam, 1999), pp. 147–171
7. B.S. Kerner, Physics World **12**, 25–30 (August 1999)
8. B.S. Kerner. J. Phys. A Math. Gen. **33**, L221–L228 (2000)
9. B.S. Kerner, in *Traffic and Granular Flow '99: Social, Traffic and Granular Dynamics*, ed. by D. Helbing, H.J. Herrmann, M. Schreckenberg, D.E. Wolf (Springer, Heidelberg, Berlin, 2000), pp. 253–284
10. B.S. Kerner, Transp. Res. Rec. **1710**. 136–144 (2000)
11. B.S. Kerner, Netw. Spat. Econ. **1**, 35–76 (2001)
12. B.S. Kerner, Transp. Res. Rec. **1802**, 145–154 (2002)
13. B.S. Kerner, in *Traffic and Transportation Theory in the 21st Century*, ed. by M.A.P. Taylor (Elsevier Science, Amsterdam, 2002), pp. 417–439
14. B.S. Kerner, Phys. Rev. E. **65**, 046138 (2002)
15. B.S. Kerner, Math. Comput. Modell. **35**, 481–508 (2002)
16. B.S. Kerner, in *Traffic and Granular Flow' 01*, ed. by M. Schreckenberg, Y. Sugiyama, D. Wolf (Springer, Berlin, 2003), pp. 13–50
17. B.S. Kerner, Physica A **333**, 379–440 (2004)
18. B.S. Kerner, *The Physics of Traffic* (Springer, Berlin, New York, 2004)
19. B.S. Kerner, in *Transportation Research Trends*, ed. by P.O. Inweldi. (Nova Science Publishers, New York, 2008), pp. 1–92
20. B.S. Kerner, *Introduction to Modern Traffic Flow Theory and Control* (Springer, Heidelberg, Dordrecht, London, New York, 2009)
21. B.S. Kerner, in *Encyclopedia of Complexity and System Science*, ed. by R.A. Meyers (Springer, Berlin, 2009), pp. 9302–9355

22. B.S. Kerner, in *Encyclopedia of Complexity and System Science*, ed. by R.A. Meyers (Springer, Berlin, 2009), pp. 9355–9411
23. B.S. Kerner, Physica A **392**, 5261–5282 (2013)
24. B.S. Kerner, Phys. Rev. E **92**, 062827 (2015)
25. B.S. Kerner, Elektrotechnik und Informationstechnik **132**, 417–433 (2015)
26. B.S. Kerner Physica A **450**, 700–747 (2016)
27. B.S. Kerner, M. Koller, S.L. Klenov, H. Rehborn, M. Leibel, Physica A **438**, 365–397 (2015)
28. B.S. Kerner, V.V. Osipov, Sov. Phys. Usp. **32**, 101–138 (1989)
29. B.S. Kerner, V.V. Osipov, Sov. Phys. Usp. **33**, 679–719 (1990)
30. B.S. Kerner, V.V. Osipov, *Autosolitons: A New Approach to Problems of Self-Organization and Turbulence* (Kluwer, Dordrecht, Boston, London, 1994)
31. B.S. Kerner, H. Rehborn, R.-P. Schäfer, S.L. Klenov, J. Palmer, S. Lorkowski, N. Witte, Physica A **392**, 221–251 (2013)
32. A.S. Mikhailov, *Foundations of Synergetics I*, 2nd edn. (Springer, Berlin, 1994)
33. A.S. Mikhailov, A.Yu. Loskutov, *Foundation of Synergetics II. Complex Patterns* (Springer, Berlin, 1991)
34. G. Nicolic, I. Prigogine, *Self-Organization in Non-equilibrium Systems* (Wiley, New York, 1977)
35. F.-J. Niedernostheide (ed.) *Nonlinear Dynamics and Pattern Formation in Semiconductors and Devices* (Springer, Berlin, 1995)
36. G.M. Pound, V.K. La Mer, J. Am. Chem. Soc. **74**, 2323 (1952)
37. E. Sanz, C. Vega, J.R. Espinosa, R. Cabellero-Bernal, J.L.F. Abascal, C. Valeriani, J. Am. Chem. Soc. **135**, 15008–15017 (2013)
38. V.A. Vasil'ev, Yu.M. Romanovskii, D.S. Chernavskii, V.G. Yakhno, *Autowave Processes in Kinetic Systems* (Springer, Berlin, 1990)

Chapter 4
Failure of Generally Accepted Classical Traffic Flow Theories

4.1 Introduction

Traffic researchers have developed a huge number of traffic theories for optimization and control of traffic and transportation networks. In particular, to generally accepted fundamentals and methodologies of traffic and transportation theory belong the following theories and associated methodologies:

(i) The Lighthill-Whitham-Richards (LWR) model [336, 448].

(ii) The General Motors (GM) model class.

In traffic flow models within the framework of the GM model class traffic breakdown is explained by a free flow instability that causes a growing wave of vehicle speed reduction propagating upstream in traffic flow. This classical traffic flow instability has been introduced in the GM car-following model by Herman, Gazis et al. [170, 171, 233] (see, e.g., [24, 28–30, 167, 300, 301, 347, 378, 379, 386–391, 394, 396–398, 414, 415, 443, 535] as well as references in reviews, books, and conference proceedings [13, 16, 57, 83, 88, 103, 109, 141, 161, 163, 165, 168, 188, 213, 223, 236, 237, 245, 311, 328, 351, 353, 361, 362, 365, 377, 380, 392, 407, 434, 439, 444, 446, 459, 469, 495, 508, 509, 536, 539, 540]).

(iii) The understanding of highway capacity of free flow at a bottleneck as a stochastic value (see, e.g., [69–71, 147, 174, 189, 191, 195, 197, 236, 237, 343, 365, 432]).

(iv) Principles for traffic and transportation network optimization and control in which travel cost should be minimized (examples are Wardrop's user equilibrium (UE) and system optimum (SO) principles [529]).

These classical theories and methodologies are the basis for dynamic traffic assignment, traffic control and optimization in traffic and transportation networks (see, e.g., [2, 4, 9, 11, 12, 21–23, 49–54, 56, 64, 66, 90, 91, 98, 101, 102, 104, 105, 110, 122, 128, 129, 136, 137, 153–159, 181, 187, 201, 235, 244, 251, 253, 261, 288, 289, 306, 308, 309, 346, 352, 354, 355, 357, 367, 368, 370, 383, 384,

B.S. Kerner, *Breakdown in Traffic Networks*, DOI 10.1007/978-3-662-54473-0_4

399, 403, 408–411, 413, 416, 417, 435, 436, 440, 441, 461, 463, 470, 477, 478, 482, 484, 488, 489, 496, 498, 513, 519, 524, 527, 528, 530–533, 541, 542, 545–548, 558–561, 568–570, 576] and references in reviews and books [55, 82, 164, 202, 407, 412, 418, 445, 447]) as well as for development and simulations of intelligent transportation systems (ITS) (e.g., [14, 44, 95, 96, 176, 246, 247, 249, 257, 285–287, 292, 341, 503, 557, 565]), traffic simulation tools (see, e.g., [5, 6, 40, 48, 62, 151, 180, 250, 450, 451, 501, 515]), and for many other traffic engineering applications (e.g., [143, 256, 356, 552, 577]).

The fundamentals and associated methodologies of these theories have made a great impact on the understanding of many traffic phenomena. However, approaches for control, dynamic traffic assignment, and network optimization based on these fundamentals and methodologies have failed by their applications in the real world. Even several decades of a very intensive effort to improve and validate network optimization models have no success. Indeed, there can be found no examples where on-line implementations of the network optimization models based on these fundamentals and methodologies could reduce congestion in real traffic and transportation networks.

The objective of this chapter is to explain the failure of applications of the generally accepted classical traffic flow theories for the control and optimization of traffic and transportation networks in the real world. We show that there is *a fundamental problem* for these classical theories that is as follows:

1. Generally accepted fundamentals and methodologies of classical traffic and transportation theories are not consistent with the empirical nucleation nature of traffic breakdown (F→S transition) at a highway bottleneck.
2. For this reason, the fundamentals and methodologies of traffic and transportation theory cannot be applied for reliable control, dynamic traffic assignment, and optimization in traffic and transportation networks.

It should be noted that there are a number of reviews and books, in which the classical traffic flow models have been presented and their mathematical features have been discussed (see, e.g., [5, 6, 13, 16, 40, 48, 57, 62, 67, 83, 88, 103, 106, 109, 141, 151, 161, 163, 165, 168, 180, 188, 213, 223, 236, 237, 245, 250, 311, 328, 329, 351, 353, 361, 362, 365, 377, 380, 392, 407, 434, 439, 444, 446, 450, 451, 454, 457, 459, 469, 495, 501, 508, 509, 515, 536, 539, 540]). In this chapter, we study traffic flow models in relation to their description of traffic breakdown at highway bottlenecks. It has been found that very different mathematical traffic flow models exhibit qualitatively the same features of traffic breakdown at a highway bottleneck [268, 269]. For this reason, rather than a consideration of the mathematical formulations of the models, we focus on a discussion of the reason for the failure of the models in the explanation of the empirical fundamental of transportation science—the nucleation nature of traffic breakdown (F→S transition) at the bottleneck (Chap. 3).

4.2 Fundamental Diagram of Traffic Flow

4.2.1 Empirical Features of Fundamental Diagram of Traffic Flow

Beginning from the classic work of Greenshields [96], the fundamental diagram of traffic flow is the basis for traffic flow theories and models as well as the basic methodology for empirical studies of measured traffic data (see, e.g., reviews, books, and conference proceedings [13, 16, 57, 83, 88, 103, 109, 141, 161, 163, 165, 168, 188, 213, 223, 236, 237, 245, 311, 328, 351, 353, 361, 362, 365, 377, 380, 392, 407, 434, 439, 444, 446, 459, 469, 495, 508, 509, 536, 539, 540]).

The fundamental diagram is a flow-rate–density relationship, i.e., a correspondence between a given vehicle density and the flow rate in traffic flow (Fig. 4.1a). The fundamental diagram reflects the result of empirical observations that the greater the density, the lower the averaged speed in vehicular traffic (Fig. 4.1b).

Thus in the flow–density plane, the fundamental diagram should pass through the origin (when the density is zero so is the flow rate) and should have at least one maximum (Fig. 4.1a). The fundamental diagram gives also connections between the average space gap between vehicles and the average speed (Fig. 4.1c) as well as between link travel time (travel time on a link of a traffic network) and flow rate

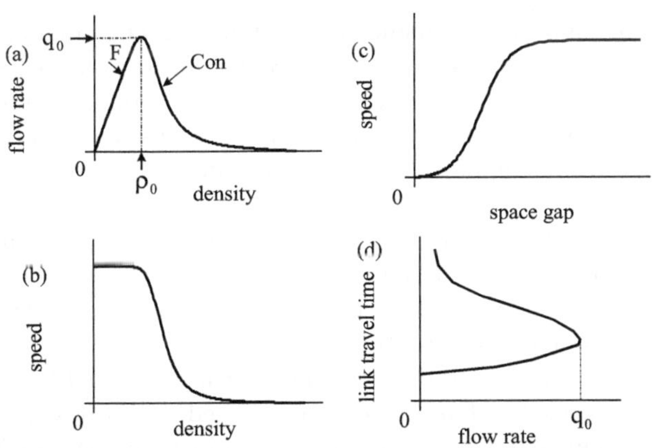

Fig. 4.1 Qualitative example of fundamental diagram: (**a**) Flow–density relationship (fundamental diagram). (**b–d**) The speed–density (**b**), speed–space-gap (**c**) and link-travel-time–flow (**d**) relationships associated with (**a**)

(Fig. 4.1d).[1] The fundamental diagram is up to now the most often subject of traffic flow measurements and it is the teaching basis in transportation engineering.

The fundamental diagram of traffic flow reflects the result of empirical observations that the greater the density, the lower the averaged speed in vehicular traffic.

Empirical observations show that the shape of empirical fundamental diagrams of congested traffic depends qualitatively on the location at which data measurements have been performed within a congested pattern at a road bottleneck (see, e.g., a review by Hall in [163]). An example of different fundamental diagram types associated with a congested pattern at an on-ramp bottleneck is shown in Fig. 4.2 (see explanations of the empirical fundamental diagram in Chap. 15 of the book [268]). Depending on a road location various empirical fundamental diagram types exist. Moreover, if data measured at many different locations within congested traffic is used for averaging to one empirical fundamental diagram, then there is almost no possibility to find a relation of such a fundamental diagram to spatiotemporal features of real traffic patterns. Therefore, in a general case the validation of a theoretical fundamental diagram of a traffic flow model cannot be performed.

We can make the following conclusions:

1. The shape of empirical fundamental diagrams depends qualitatively on the location at which data has been measured within a traffic pattern. However, the spatiotemporal correspondence between traffic variables within the pattern is lost in the empirical fundamental diagram. For this reason, it is not possible to reconstruct spatiotemporal characteristics of the pattern from the fundamental diagram.
2. Spatiotemporal characteristics of waves in free flow, in particular, the wave interaction with a speed disturbance localized at a highway bottleneck that leads to the formation of the nucleus for traffic breakdown at the bottleneck (Chap. 3) are lost in the empirical fundamental diagram.
3. The nucleation nature of traffic breakdown at highway bottlenecks cannot be revealed from a study of traffic breakdown and highway capacity in which empirical fundamental diagrams or other empirical relationships between average characteristics of traffic variables (for example, the speed and the flow rate) are analyzed.

[1]There is a huge number of publications devoted to measurements of flow–density (fundamental diagram), speed-density, speed–space-gap, and link-travel-time–flow relationships (see, e.g., a review by Hall in [163]).

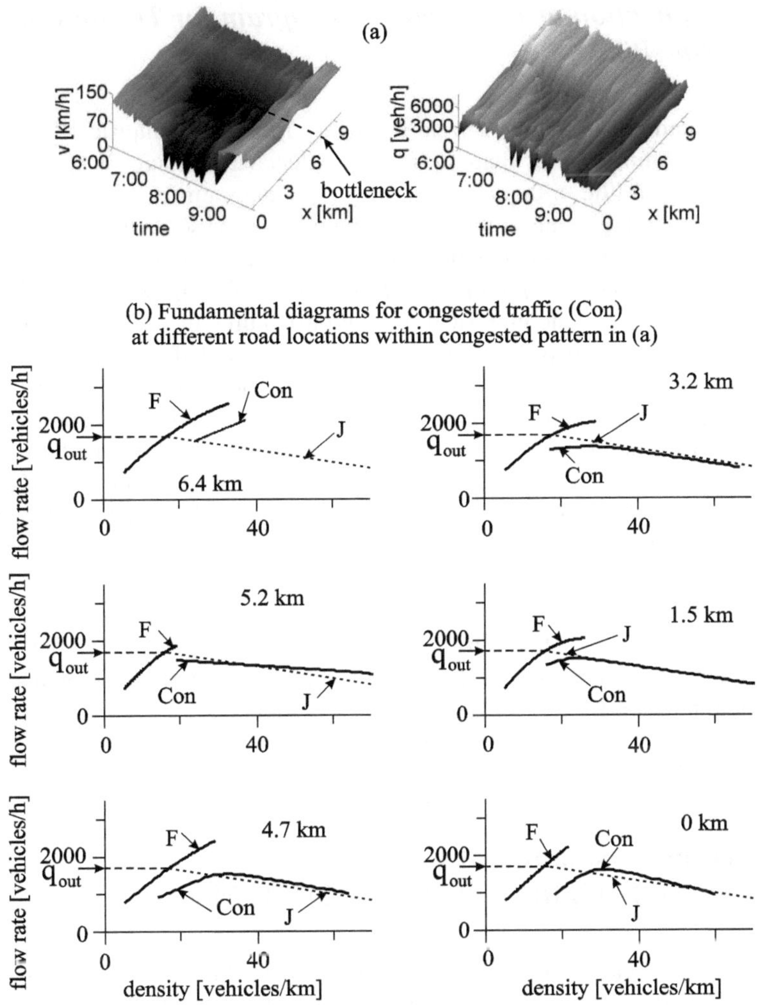

Fig. 4.2 Dependence of empirical fundamental diagram on measurements of congested traffic at different road locations: (**a**) Average speed (*left*) and total flow rate (*right*) in space and time for a congested pattern occurring on three-lane freeway A5-South in Germany (real field traffic data measured by road detectors on April 15, 1996). (**b**) Empirical fundamental diagrams at different locations within the congested traffic pattern shown in (**a**). F—free flow. Con—congested traffic. J—line *J* (*dashed line*) (see explanations of line *J* in Sect. 4.5.1 below). On-ramp bottleneck in (**a**) is at the location about 6.4 km. In (**b**), data has been averaged across the highway and presented per highway lane. Adapted from [268]

4.2.2 Application of Fundamental Diagram for Traffic Flow Modelling

Thus, spatiotempotal features of real traffic are lost in the empirical fundamental diagram. Therefore, rather than some empirical characteristic of spatiotempotal features of real traffic, a theoretical fundamental diagram for traffic flow used in a traffic flow model is only a theoretical assumption for some hypothetical model solutions. The empirical nucleation features of traffic breakdown at the bottleneck can be explained only, if *spatiotemporal features* of free flow are considered. As explained in Sect. 1.6, the empirical nucleation features of traffic breakdown at a highway bottleneck (Chap. 3) is the empirical fundamental of transportation science.

For this reason, to understand whether a traffic flow model is consistent with the empirical fundamental of transportation science, simulations of spatiotemporal dynamics of traffic breakdown at a bottleneck with a traffic flow model should be analyzed. Therefore, in this chapter a discussion of the description of traffic breakdown will be made through the analysis of spatiotemporal traffic flow dynamics of classical traffic flow models.

> A traffic flow model should be consistent with the empirical fundamental of transportation science—the empirical nucleation nature of traffic breakdown studied in Chap. 3.

4.3 Traffic Breakdown at Bottleneck in Lighthill-Whitham-Richards (LWR) Model

One of the first classical traffic flow models is the famous Lighthill-Whitham-Richards (LWR) traffic flow model introduced in 1955–1956 [336, 448]. In 1955 Lighthill and Whitham [336] wrote in their classic work (see p. 319 in [336]): "...The fundamental hypothesis of the theory is that at any point of the road the flow (vehicles per hour) is a function of the concentration (vehicles per mile)..." (Fig. 4.1a).

4.3.1 Basic Assumption of LWR Model

The basic idea of the classic LWR model [336, 448] is as follows: The maximum flow rate q_0 associated with the maximum point (ρ_0, q_0) at the fundamental diagram (Fig. 4.1a) determines highway capacity of free flow at a highway bottleneck denoted by C:

$$C = q_0. \tag{4.1}$$

If the flow rate in free flow upstream of the bottleneck reaches the maximum flow rate q_0 at the fundamental diagram (Fig. 4.1a), i.e, it reaches the bottleneck capacity (4.1), then a further increase in the upstream flow rates must lead to congestion formation and upstream congestion propagation.

This hypothesis of the LWR theory means that congested traffic occurs only, when the flow rate in free flow upstream of the bottleneck exceeds highway capacity C determined by (4.1). This assumption of the LWR traffic flow theory about the reason for traffic breakdown (the onset of congestion) determines the highway capacity and this is very often used as the definition of congested traffic.

- In accordance with the LWR traffic flow theory, highway capacity is determined by the maximum flow rate q_0 at the fundamental diagram of traffic flow.

4.3.2 *Achievements of LWR Theory in Description of Traffic Breakdown*

The LWR model consists of the law of conservation of the number of vehicles on the road

$$\frac{\partial \rho}{\partial t} + \frac{\partial Q(\rho)}{\partial x} = 0, \tag{4.2}$$

in which there is a relationship between the flow rate Q and density ρ

$$Q = Q(\rho) \tag{4.3}$$

associated with the fundamental diagram of traffic flow. Here, x is a spatial coordinate in the direction of traffic flow, and t is time. The LWR model (4.2), (4.3) has discontinuous solutions in the form of shock waves [336, 392, 448, 534]. For this reason, the application of the LWR model for numerical simulations of large traffic and transportation networks has been limited before Daganzo has introduced a cell-transmission model that is consistent with the LWR theory [107, 108].

With the use of numerical simulations of Daganzo's cell-transmission model [107], one can derive features of traffic breakdown and resulting traffic congestion that the LWR theory exhibits (Fig. 4.3). Numerical simulations of Daganzo's cell-transmission model presented in Fig. 4.3 are made on a single-lane road with an on-ramp bottleneck at a given flow rate in free flow on the main road upstream of the bottleneck q_{in} and different on-ramp inflow rates q_{on}. We choose the flow rate q_{in} in free flow upstream of the bottleneck that is smaller than

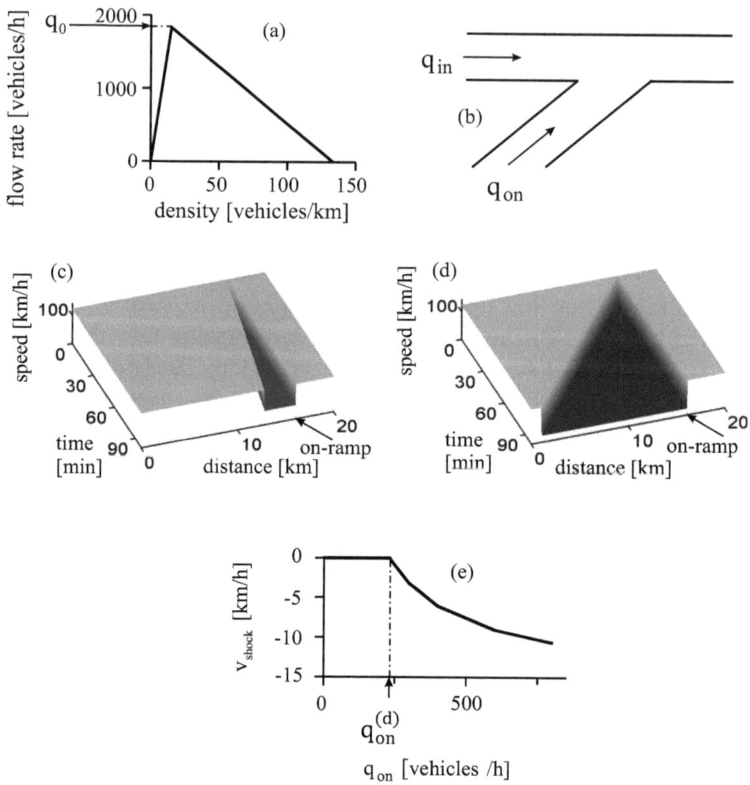

Fig. 4.3 Simulations of traffic breakdown at on-ramp bottleneck on single-lane road in the LWR theory [270]: (**a**) The fundamental diagram used in simulations of Daganzo's cell-transmission model. (**b**) Qualitative schema of the on-ramp bottleneck. (**c, d**) Congested traffic at the bottleneck under condition (4.7) for two different values of the on-ramp inflow rate q_{on} at a given flow rate q_{in} upstream of the bottleneck: Vehicle speed distributions in space and time that show shock wave emergence and propagation upstream of the on-ramp bottleneck. (**e**) Dependence of the velocity of the shock wave v_{shock} on the flow rate q_{on}. $q_0 = 1830$ vehicles/h. $q_{in} = 1600$ vehicles/h. $q_{on} =$ 270 (**c**), 600 vehicles/h (**d**)

q_0 (Fig. 4.3), i.e., we assume that condition

$$q_{in} < q_0 \tag{4.4}$$

is satisfied.

If the on-ramp inflow rate $q_{on} = 0$, the speed and density are spatially homogeneous. When the flow rate q_{on} increases beginning from zero, the sum of the flow rates $q_{on} + q_{in}$ increases too as long as condition

$$q_{on} + q_{in} < q_0 \tag{4.5}$$

is satisfied. This is because there is a critical on-ramp inflow rate

$$q_{\rm on} = q_{\rm on}^{\rm (d)} = q_0 - q_{\rm in}. \tag{4.6}$$

When $q_{\rm on} < q_{\rm on}^{\rm (d)}$, i.e., condition (4.5) is satisfied, then no congestion occurs at the bottleneck.

However, when $q_{\rm on} > q_{\rm on}^{\rm (d)}$, i.e., condition

$$q_{\rm on} + q_{\rm in} > q_0 \tag{4.7}$$

is satisfied, which is opposite one to condition (4.5), then a shock wave of lower speed and greater density appears (Fig. 4.3c,d). The upstream propagation of the shock wave leads to the widening of congested traffic pattern at the bottleneck (Fig. 4.3c,d).

The more the flow rate $q_{\rm on}$ exceeds the value $q_{\rm on}^{\rm (d)}$ (4.6), i.e., the larger the value

$$\Delta q = q_{\rm in} + q_{\rm on} - q_0 > 0, \tag{4.8}$$

the larger the absolute value of the shock velocity $| v_{\rm shock} |$ (Fig. 4.3c–e).

Thus, numerical simulations of Daganzo's cell-transmission model show that when traffic breakdown has occurred, i.e., under condition (4.7), a shock wave propagates upstream of the bottleneck resulting in congested traffic at the bottleneck (Fig. 4.3c–e). The downstream front of this congested traffic is fixed at the bottleneck (Fig. 4.3c,d). Therefore, in accordance with the definition of the synchronized flow phase [S] made in the three-phase theory, this congested traffic belongs to the synchronized flow phase.

4.3.3 *Failure of LWR Theory in Explanation of Empirical Nucleation Nature of Traffic Breakdown*

Because of above-mentioned achievements of the LWR theory and Daganzo's model in the description of traffic breakdown at a highway bottleneck, a question arises:

- Why does the author make a statement about the *failure* of the LWR theory in the explanation of the empirical fundamental of transportation science?

As explained in Sect. 1.6, the empirical nucleation features of traffic breakdown at a highway bottleneck is the empirical fundamental of transportation science.

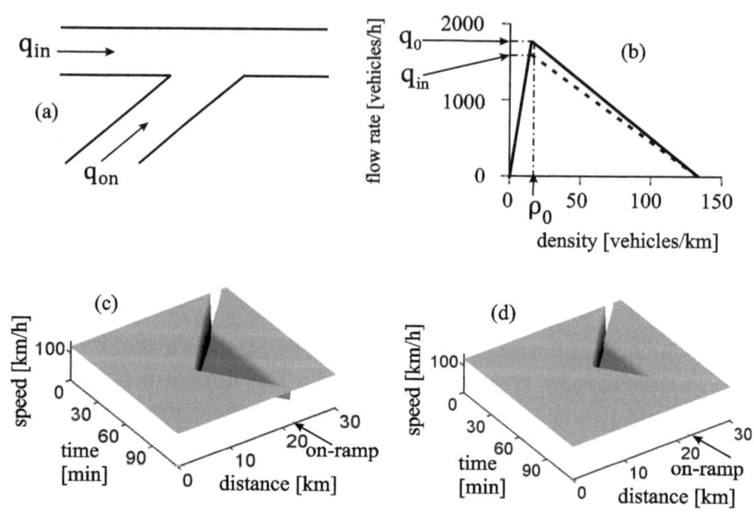

Fig. 4.4 Continuation of simulations of traffic breakdown at on-ramp bottleneck on single-lane road in the LWR theory [270]: (**a**) Qualitative schema of the on-ramp bottleneck. (**b**) The fundamental diagram (*solid curves*) used in simulations of Daganzo's cell-transmission model that is the same as that in Fig. 4.3a. (**c**, **d**) Congested traffic at the bottleneck caused by jam propagation through the bottleneck under condition (4.5): Vehicle speed in space and time; the jam width (in the longitudinal direction) at $t = 0$ is equal to $W_J = 2$ (**c**) and 1.5 km (**d**). A dashed line in (**b**) is related to the upstream front of wide moving jams in (**c**, **d**), while the jams propagate upstream of the bottleneck. $q_0 = 1830$ vehicles/h, $q_{on} = 120$ vehicles/h, $q_{in} = 1600$ vehicles/h

The main disadvantages of the LWR theory in the explanation of traffic breakdown at a bottleneck are as follows:

(i) The LWR theory cannot show the empirical metastability of free flow with respect to traffic breakdown at the bottleneck.
(ii) The LWR theory cannot show the empirical induced traffic breakdown at the bottleneck.

In other words, the LWR theory cannot show and explain the evidence of empirical nucleation nature of traffic breakdown at highway bottlenecks (Chap. 3).

This conclusion is explained in Fig. 4.4 in which at the same given flow rate q_{in} as that given in Fig. 4.3 we have chosen smaller q_{on} satisfying condition (4.5). As follows from Figs. 4.4c,d, no induced traffic breakdown occurs in free flow at the on-ramp bottleneck.

In simulations, a wide moving jam has been induced at $t = 0$ downstream of the on-ramp bottleneck. As in empirical data (see Fig. 3.2), while propagating upstream of the on-ramp the jam causes congested traffic at the on-ramp (Fig. 4.4c,d). However, in contrast with empirical data, in which congested traffic persists at the bottleneck independent of whether the jam propagates further or the jam dissolves, in the LWR-theory congested traffic always dissolves after the jam dissolution. In the LWR model, congested traffic dissolves independent of the initial jam width W_J

(in the longitudinal direction) (Figs. 4.4c,d) and on the value $q_{on} + q_{in}$ as long as condition (4.5) is satisfied.

To explain the dissolution of congested traffic, firstly we consider the reason for the dissolution of moving jams that can be seen in Figs. 4.4c,d. Note that the jam dissolution results from condition $q_{in} < q_0$ (Fig. 4.4b). Indeed, the downstream jam front moving with the velocity v_g is associated with the line for congested traffic on the fundamental diagram (solid line with negative slope in Fig. 4.4b). In contrast, the upstream jam front moving with the velocity v_{up} is associated with a line with negative slope in the flow–density plane that, due to condition $q_{in} < q_0$, is below the line for congested traffic: When the wide moving jams are upstream of the bottleneck, the line in the flow–density plane related to the upstream jam front is a dashed line with negative slope shown in Fig. 4.4b. Therefore, condition $| v_g | > | v_{up} |$ is satisfied. As a result, the jam width (in the longitudinal direction) decreases over time, i.e., the wide moving jam dissolves upstream of the bottleneck as shown in Figs. 4.4c,d.

Before the wide moving jam reaches the bottleneck, i.e., when the wide moving jam is still downstream of the bottleneck, the flow rate in the jam outflow $q^{(J)}_{out}$ is equal to q_0. When the jam is upstream of the bottleneck, then within the road region between the bottleneck and the downstream jam front congested traffic occurs. The flow rate within this congestion is equal to the flow rate in the jam outflow, which decreases to $q^{(J)}_{out} = q_0 - q_{on}$. This is because in the LWR-model the flow rate downstream of the bottleneck cannot exceed q_0 (Fig. 4.4b). However, after the wide moving jam has dissolved, the flow rate upstream of the bottleneck decreases to q_{in} resulting in the dissolution of congestion due to condition (4.5).

The LWR theory cannot show and explain the empirical evidence of the nucleation nature of empirical traffic breakdown at highway bottlenecks that is the empirical fundamental of transportation science.

Thus, the LWR model cannot show the nucleation nature of traffic breakdown at highway bottlenecks. For this reason, the LWR theory as well as further theoretical approaches based on the LWR theory, like Daganzo's cell transmission model [107, 108], so-called N-curves [85, 337, 372], a macroscopic fundamental diagram (MFD) [116, 175], as well as ITS-applications based on the LWR theory (see, e.g., [25, 63, 111–115, 117, 118, 142, 179, 299, 312–320, 323, 342, 543, 556] and references there) are inconsistent with the nucleation nature of real traffic breakdown at road bottlenecks. Applications of these approaches for an analysis of the effect of ITS on traffic flow, which are widely used by many researchers, can lead to invalid (and sometimes incorrect) conclusions about the ITS performance in real traffic.

Because the LWR theory is inconsistent with the empirical fundamental of transportation science, theoretical approaches based on the LWR theory cannot be used for a reliable analysis of ITS-applications.

4.4 Description of Traffic Breakdown with General Motors (GM) Model Class

4.4.1 Classical Traffic Flow Instability: Growing Wave of Local Speed Reduction in Traffic Flow Due to Over-Deceleration Effect

As long ago as 1958–1961, Herman, Gazis, Montroll, Potts, Rothery, and Chandler [93, 170, 171, 233] assumed that traffic breakdown is associated with an instability of free flow. This instability is related to a finite reaction time of drivers. The driver reaction time is responsible for an over-deceleration effect.

The over-deceleration effect is as follows. If a vehicle begins to decelerate unexpectedly, then due to the finite driver reaction time the following vehicle starts deceleration with a delay. When the driver reaction time is long enough, the driver of the following vehicle decelerates stronger than it is needed to avoid collisions. As a result, the speed of the following vehicle becomes lower than the speed of the preceding vehicle. If this over-deceleration effect is realized for the following drivers, a wave of vehicle speed reduction should appear and grow over time. In the book, we call the traffic flow instability due to the over-deceleration effect (driver reaction time) that can lead to a growing wave of a local decrease in the speed in traffic flow as the classical traffic flow instability of Herman, Gazis, Montroll, Potts, Rothery, and Chandler [93, 170, 171, 233].

In the car-following GM model of Herman, Gazis, Montroll, Potts, Rothery, and Chandler [93, 170, 171, 233], a driver reaction time denoted by τ_0 is explicitly used in vehicle deceleration (acceleration) $a(t+\tau_0)$: The vehicle reacts with the time delay τ_0 on any changes in the space gap to the preceding vehicle $g(t)$ and the speed difference $\Delta v(t) = v_\ell(t) - v(t)$ between the speed of the preceding vehicle $v_\ell(t)$ and the vehicle speed $v(t)$. As a result, the GM model reads as follows [171]:

$$a(t+\tau_0) = \frac{\Delta v(t)[v(t+\tau_0)]^{\mathrm{m}_1}}{T_0[g(t)+d]^{\mathrm{m}_2}}, \tag{4.9}$$

$$a(t) = \frac{dv(t)}{dt}, \tag{4.10}$$

where d is the vehicle length; T_0, m_1, m_2 are constants. By integrating Eqs. (4.9), (4.10) at $m_1 < 1$, $m_2 > 1$, one gets model solutions for steady states related to the fundamental diagram [171]

$$V_0(\rho) = v_0[1 - (\rho/\rho_{\max})^{(m_2-1)}]^{(1-m_1)^{-1}}, \tag{4.11}$$

where v_0 is constant, $\rho_{\max} = 1/d$.

The idea of the GM model with a driver reaction time has been incorporated either explicitly or implicitly in many other traffic flow models (see, e.g., [24, 28–30, 42, 58, 60, 72, 73, 75, 92, 144, 148, 160, 162, 167, 172, 173, 184–186, 199, 206–212, 214–222, 224–232, 234, 239, 240, 242, 243, 260, 263, 281, 293, 300, 301, 307, 324–327, 333, 347–349, 375, 376, 378, 379, 382, 385–391, 394, 396–398, 402, 405, 414, 415, 419–426, 443, 449, 458, 460, 464–468, 474, 475, 479–481, 483, 490, 493, 494, 497, 499, 500, 502, 504–506, 511, 514, 516, 518, 520, 525, 526, 535, 537, 538, 551, 562–564, 572, 573, 575] as well as references in reviews, books, and conference proceedings [13, 16, 57, 67, 83, 88, 103, 106, 109, 141, 161, 163, 165, 168, 188, 213, 223, 236, 237, 245, 311, 328, 329, 351, 353, 361, 362, 365, 377, 380, 392, 407, 434, 439, 444, 446, 454, 457, 459, 469, 495, 508, 509, 536, 539, 540]). These traffic flow models can be considered belonging to the same GM model class. The reasons for this conclusion are as follows.

1. All these models exhibit qualitatively the same classical traffic flow instability. The classical traffic flow instability occurs when the vehicle density in free flow is large enough. Due to this instability, a growing wave of a local *speed decrease* in traffic flow occurs. This growing wave of a local speed decrease can be considered "growing deceleration wave" in traffic flow. Although a mathematical description of the classical traffic flow instability can be very different in different traffic flow models, the physical reason of this instability is associated with a time delay in driver reaction (driver over-deceleration effect) as introduced by Herman, Gazis, Montroll, Potts, Rothery, and Chandler [93, 170, 171, 233] in their GM model. Therefore, this classical traffic flow instability can be considered the traffic flow instability of the GM model class.
2. In traffic flow models of the GM model class, this classical traffic flow instability causes a phase transition from free flow to moving jam(s) as revealed firstly in 1993–1994 by Kerner and Konhäuser from numerical simulations of a version of the Payne's macroscopic model [275, 276]. As we explain below in Sects. 4.4.2 and 4.4.3, in the models belonging to the GM model class the development of this classical traffic flow instability leads to the formation of wide moving jam(s) in an initially free flow (F→J transition).

A numerical simulation of the classical free flow instability of Herman, Gazis, Montroll, Potts, Rothery, and Chandler [93, 170, 171, 233] presented in Fig. 4.5 has been made with the use of a version of the optimal velocity (OV) model by Bando et al. [28–30]. In this simulation example, vehicle 1 (dashed vehicle trajectory in Fig. 4.5) moves initially in homogeneous (in space and time) free flow

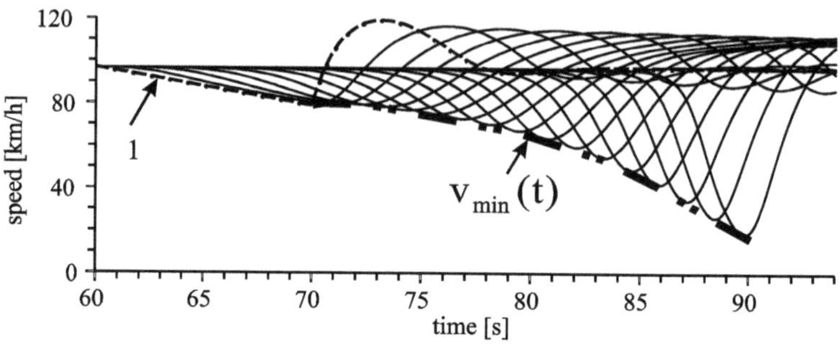

Fig. 4.5 Simulations of growing wave of local speed reduction caused by the classical traffic flow instability of Herman, Gazis, Montroll, Potts, Rothery, and Chandler [93, 170, 171, 233]. Microscopic vehicle speeds along trajectories of vehicles moving on a single-lane road. Simulations are made with Bando et al. OV model

of a relatively large density. At time instant $t = 60$ s the vehicle begins to decelerate with constant deceleration $a = 0.5\ \mathrm{m/s^2}$ during time interval 10 s. This vehicle deceleration results in a speed decrease of vehicle 1 (speed decrease on a dashed curve for the time-function of the speed of vehicle 1 in Fig. 4.5). At time $t > 70$ s, vehicle 1 begins to accelerate approaching finally the initial free flow speed. However, the initial deceleration of vehicle 1 causes a local speed disturbance within which the speed is smaller than the speed in the initial free flow. Due to a finite driver reaction time, while approaching the local speed disturbance the speed of each of the following vehicles becomes lower than the speed of the preceding vehicle. In the simulation example, this over-deceleration effect is realized for all following drivers resulting in a growing wave of local speed reduction (labeled by dotted-dashed curve $v_{\mathrm{min}}(t)$ in Fig. 4.5).

4.4.2 "Boomerang" Effect

In 1993–1994 from a numerical study of a version of Payne's macroscopic model Kerner and Konhäuser found that the classical traffic flow instability of Herman, Gazis, Montroll, Potts, Rothery, and Chandler [93, 170, 171, 233] leads to the emergence of a wide moving jam in free flow (F→J transition) [275, 276].

The F→J transition is associated with a "boomerang" behavior of a local disturbance in free flow (Figs. 4.6 and 4.7) [276]. The boomerang effect in traffic flow models of the GM model class revealed in 1994 [276, 277] was simulated in an initially homogeneous free flow on a homogeneous road without bottlenecks (Fig. 4.6): A small initial localized disturbance of the speed reduction and increase in the density in free flow has been initiated at some road location. The disturbance propagates downstream in free flow while only slightly growing over time. Later, the disturbance comes to a stop while beginning to grow strongly. Due to this disturbance growth, the disturbance transforms into a "narrow" moving jam: Within

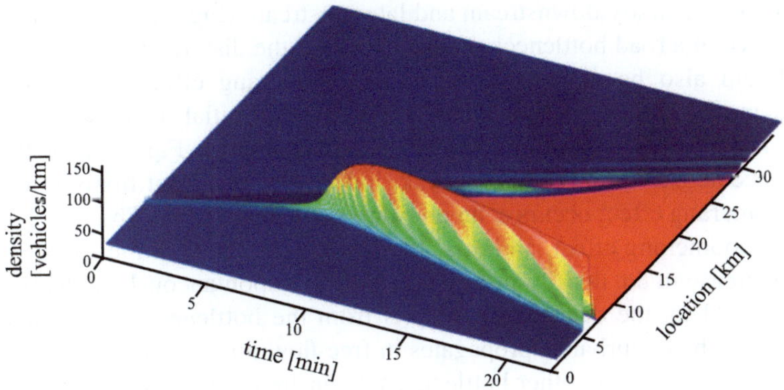

Fig. 4.6 "Boomerang" behavior of the growing disturbance in metastable free flow with respect to wide moving jam formation [276, 277]. Vehicle density development in space and time. Simulations of a version of the Payne model (see Sec. 10.3.3 of the book [269]). Homogeneous road without bottlenecks

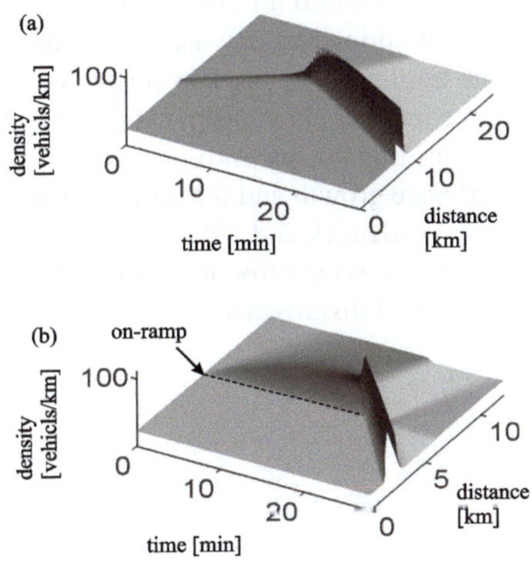

Fig. 4.7 Comparison of the boomerang effect on homogeneous road (**a**) (qualitatively the same simulations as those in Fig. 4.6) and a road with on-ramp bottleneck (**b**) [276–278]. Vehicle density development in space and time. Simulations of a version of the Payne model (see Sec. 10.3.3 of the book [269])

a small local road region the speed and the flow rate become very low whereas the density is large. The moving jam begins to propagate upstream. Finally, the narrow moving jam transforms into a wide moving jam. While propagating upstream, the wide moving jam passes through the location at which the initial disturbance has occurred. For this reason, this effect resembles the boomerang effect (Figs. 4.6 and 4.7a).

It must be noted that there are *no* bottlenecks on the road in simulations presented in Figs. 4.6 and 4.7a. Therefore, while the localized disturbance in free flow propagates firstly downstream and later upstream (Figs. 4.6 and 4.7a), there is no influence of a road bottleneck on the growth of the disturbance.

It should also be emphasized that the boomerang effect in the models of the GM model class is independent of whether the initial disturbance occurs on homogeneous road without any bottlenecks as shown in Fig. 4.7a, or the initial disturbance appears at a road bottleneck. Indeed, as found out firstly in [278], the same boomerang effect occurs at an on-ramp bottleneck (Fig. 4.7b).

In the boomerang effect, an initial localized disturbance of the speed reduction and an increase in the density in free flow emerges spontaneously at the bottleneck (Fig. 4.7b). Then, the disturbance departs from the bottleneck in the direction of traffic flow. The disturbance propagates in free flow downstream of the bottleneck (Fig. 4.7b). There are no other bottlenecks downstream of the on-ramp bottleneck. Nevertheless, we observe qualitative the same spatiotemporal behavior of the disturbance growth as on homogeneous road (Fig. 4.7a): The initial disturbance propagates downstream in free flow while only slightly growing in the amplitude over time. Then, the disturbance comes to a stop while beginning to grow strongly. Finally, the disturbance transforms into a moving jam. This jam emergence occurs on homogeneous road several kilometers downstream of the bottleneck. Therefore, there is no influence of the bottleneck of the jam formation. When the moving jam becomes a large enough, the jam begins to propagate upstream.

Finally, the moving jam transforms into a wide moving jam. As abovementioned, both the disturbance growth and the wide moving jam emergence occur downstream of the on-ramp bottleneck and without any influence of the bottleneck. While propagating upstream, the wide moving jam passes through the location of the bottleneck at which the initial disturbance has occurred.[2]

[2]While propagating though the bottleneck, the wide moving jam can cause a variety of diverse effects of congested pattern formation at the bottleneck found out in 1999–2001 in the models of the GM model class by Helbing, Treiber et al. [222, 507]. This is because the free flow at the bottleneck is in a metastable state with respect to an F→J transition. These diverse effects of congested pattern formation, which are considered in details in papers of Helbing and Treiber et al. [464, 465, 511], in the review by Helbing [213] and the book by Treiber and Kesting [508, 509], are out of scope of the book under consideration. The reason of this is as follows. All the effects of the congested pattern formation at road bottlenecks found with the simulations of the models of the GM model class are associated with the metastability of free flow at the bottleneck with respect to the F→J transition. Contrarily, in real traffic as proven in Chap. 3, rather than the F→J transition, the metastability of free flow at the bottleneck with respect to the F→S transition is responsible for traffic breakdown at a highway bottleneck in real traffic (see Sect. 4.5.1). For this reason, all theoretical results of the congested pattern formation at road bottlenecks found by Helbing, Treiber, Kesting et al. as well as by other authors in many different traffic flow models of the GM model class (see e.g. [206–222, 224–231, 464, 465, 504, 505, 507–509, 511]) are invalid for real traffic. This critical conclusion has been considered in more detailed in Sec. 10.3.10 and 10.3.11 of the book [269] (see also Sects. 4.7, 4.9.1, and 4.11.4 below).

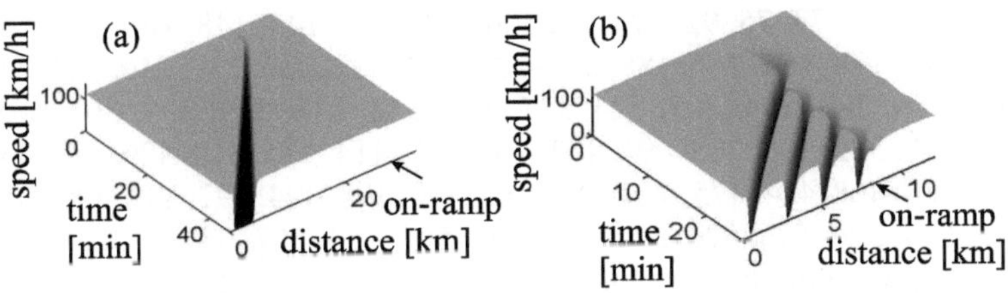

Fig. 4.8 Simulations of traffic breakdown at on-ramp bottleneck with the use of the same traffic flow model as that used in Fig. 4.7: Speed in space and time during spontaneous wide moving jam emergence at the bottleneck. The flow rate on the main road upstream of the bottleneck q_{in} is larger in (**a**) than that in (**b**), whereas on-ramp inflow q_{on} is smaller in (**a**) than that in (**b**) at the same other model parameters. Adapted from [269]

4.4.3 Moving Jam Emergence at Bottleneck

It should be noted that at the same flow rate on the main road upstream of the bottleneck q_{in} as that used for the simulation of the boomerang effect shown in Fig. 4.7b and a greater flow rate to the on-ramp a wide moving jam can emerge at the bottleneck without the boomerang effect: A local disturbance occurring in free flow at the bottleneck propagates upstream of the bottleneck while growing in the amplitude. Finally, the initial disturbance transforms into a wide moving jam (F→J transition) (Fig. 4.8a). When the flow rate to the on-ramp increases, rather than a single wide moving jam (Fig. 4.8a), a sequence of wide moving jams emerges spontaneously at the bottleneck (Fig. 4.8b) [213, 278].

As above-mentioned, microscopic, macroscopic as well as all other traffic flow models, in which the classical traffic flow instability leads to an F→J transition (Fig. 4.8), are related to the GM model class.[3] It must be emphasized that an F→J transition occurs in a metastable free flow with respect to the classical traffic flow instability of Herman, Gazis, Montroll, Potts, Rothery, and Chandler [93, 170, 171, 233]. In other words, the F→J transition exhibits the nucleation nature. However, rather than the F→J transition of the GM model class, in real data traffic breakdown is a phase transition from free flow to synchronized flow (F→S transition) (see empirical examples of spontaneous traffic breakdown at highway bottlenecks in Chap. 3). Therefore, traffic flow models of the GM model class [7, 28–30, 42, 43,

[3]Some of the models belonging to the GM model class like the OV model [28–30] and IDM (intelligent driving model) [213, 507–509] have a unique speed–density relationship, whereas the initial GM model of Herman, Montroll, Potts, and Rothery does not have [233]. Rather than a relation between steady speed and density in a model, the common feature of any traffic flow model belonging to the GM model is a free flow instability that causes a growing wave of vehicle speed reduction propagating upstream in traffic flow. Note that to describe traffic beyond of instabilities, Gazis, Herman and Rothery made a further development of the GM model in which steady-state solutions lie on a one-dimensional curve in the flow-density plane (fundamental diagram) [171] (see the review by Nagel et al. [380] for more details).

57, 58, 60, 67, 72, 73, 75, 92, 97, 103, 106, 123, 124, 144, 148, 160, 162, 163, 168, 172, 173, 177, 178, 184–186, 188, 206–222, 224–232, 236, 239, 240, 242, 243, 258, 260, 263, 276, 293, 304, 307, 324–327, 329, 333, 348, 349, 351, 361, 365, 373–377, 379, 380, 382, 385, 388, 392, 402, 405, 407, 414, 415, 419–426, 434, 439, 449, 458, 460, 464–468, 474, 475, 479–481, 483, 490, 493, 494, 497, 499, 500, 502, 504–509, 511, 514, 520, 525, 526, 534–539, 551, 562–564, 567, 572, 573, 575] failed to show the empirical fundamental of transportation science—the nucleation nature of traffic breakdown at a bottleneck with respect to an F→S transition in metastable free flow at the bottleneck.

The classical traffic flow instability in traffic flow models of the GM model class leads to the emergence of wide moving jam(s) in free flow at highway bottlenecks (F→J transition). In other words, traffic flow models of the GM model class have failed to show the empirical fundamental of transportation science—the nucleation nature of traffic breakdown at a bottleneck with respect to an F→S transition in metastable free flow at the bottleneck.

Nevertheless, the traffic flow models of the GM model class have *great achievements* in the understanding of some other real traffic flow phenomena like the characteristic parameters of wide moving jam propagation (Sect. 4.5.1) and a study of driver behavioral assumptions (Sect. 4.5.2).

4.5 Achievements of Generally Accepted Classical Traffic Models

In the LWR-model as well as many other classical traffic flow models, in particular, the models of the GM model class incorporate diverse driver behavioral characteristics related to real traffic [13, 16, 24, 28–30, 57, 83, 88, 103, 107–109, 141, 161, 163, 165, 167, 168, 170, 171, 188, 213, 223, 233, 236, 237, 245, 294–296, 300, 301, 311, 328, 336, 347, 351, 353, 361, 362, 365, 377–380, 386–392, 394, 396–398, 407, 414, 415, 434, 437–439, 443, 444, 446, 448, 459, 469, 495, 508, 509, 535, 536, 539, 540]. Because achievements of the LWR theory [336, 448] and the associated Daganzo's cell transmission model [107, 108] have already been discussed in Sect. 4.3.2, we consider here achievements of traffic flow models belonging to the GM model class.

4.5.1 Metastability of Free Flow with Respect to Moving Jam Emergence and Line J

In the Kerner and Konhäuser theory of wide moving jams developed in 1994 [276], it was found that under conditions (Fig. 4.9)

$$q_{\text{out}} \leq q < q_{\text{cr}}^{(\text{J})} \; (\rho_{\min} \leq \rho < \rho_{\text{cr}}^{(\text{J})}), \tag{4.12}$$

free flow is *metastable* with respect to an F→J transition. In (4.12), q_{out} and $\rho_{\min}$ are the flow rate and the vehicle density in free flow formed in the jam outflow; $q_{\text{cr}}^{(\text{J})}$ and $\rho_{\text{cr}}^{(\text{J})}$ are, respectively, the critical flow rate and the vehicle density in free flow at which free flow becomes unstable with respect to the classical traffic flow instability of the GM model class. The metastability of free flow with respect to the F→J transition means that under condition (4.12) small enough disturbances in an initial homogeneous free flow decay. However, if a nucleus required for the F→J transition, i.e., a great enough local disturbance appears in the free flow, this disturbance grows leading to the occurrence of a wide moving jam (Figs. 4.6, 4.7, and 4.8).

The metastability of free flow with respect to the F→J transition is associated with characteristic parameters of wide moving jam propagation. This metastability of free flow and characteristic parameters of wide moving jam propagation can be explained with the use of so-called line *J* in the flow–density plane (Fig. 4.9). The line *J* represents the propagation of the downstream front of a wide moving jam in the flow–density plane (Fig. 4.9) [276].

To understand features of the metastability of free flow with respect to the F→J transition under condition (4.12) found out firstly in the Kerner and Konhäuser

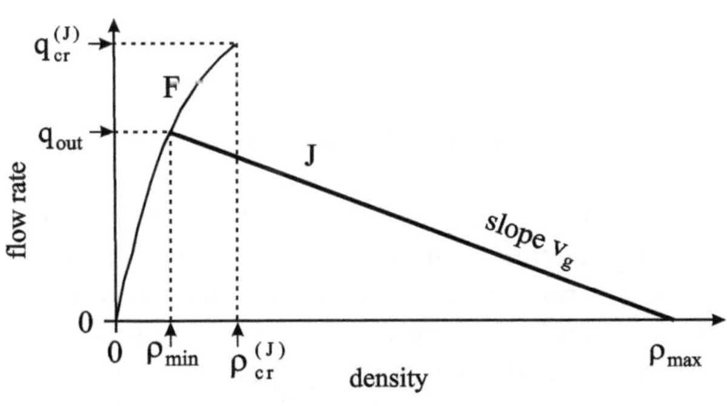

Fig. 4.9 Qualitative explanation of metastability of free flow with respect to moving jam emergence and line *J*: Qualitative representation of the part of the fundamental diagram for free flow (F) together with the line *J* (line J) whose slope is equal to the mean downstream jam front velocity v_{g} [276]. q_{out} and $\rho_{\min}$ are, respectively, the flow rate and the vehicle density in free flow formed in the jam outflow (see Fig. 4.10). $q_{\text{cr}}^{(\text{J})}$ and $\rho_{\text{cr}}^{(\text{J})}$ are, respectively, the critical flow rate and the vehicle density in free flow at which free flow becomes unstable with respect to the classical traffic flow instability of the GM model class. $\rho_{\max}$ is the jam density that is assumed to be constant

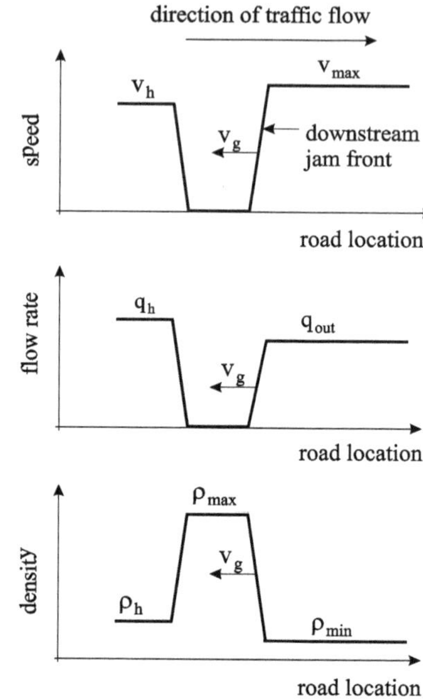

Fig. 4.10 Qualitative illustration of characteristic parameters of wide moving jam propagation [276]. Schematic representation of a wide moving jam at a fixed time instant. Spatial distributions of vehicle speed v, flow rate q, and vehicle density ρ in the wide moving jam, which propagates through an initially homogeneous state of free flow with speed $v_{\rm h}$, flow rate $q_{\rm h}$, and density $\rho_{\rm h}$

theory of wide moving jams (Fig. 4.9) [276], we discuss briefly characteristic parameters of the propagation of wide moving jams on highways (Figs. 4.10 and 4.11). In particular, one of the characteristic parameters of wide moving propagation is as follows: A wide moving jam propagates through a road bottleneck while maintaining the mean velocity $v_{\rm g}$ of the downstream jam front. In other words, the jam satisfied the characteristic feature [J] of the wide moving jam of the three-phase theory (Sect. 1.3).

4.5.1.1 Characteristic Parameters of Wide Moving Jam

The main common feature of the characteristic parameters of wide moving jam propagation is that at given traffic parameters (weather, percentage of long vehicles, etc.) the characteristic parameters of wide moving jam propagation do not depend on traffic variables in traffic flow upstream of a wide moving jam and they do not change while the wide moving jam propagates on a road [276, 282]. These characteristic parameters are the same for different wide moving jams. The jam characteristic parameters are as follows [276, 282]:

(i) The mean velocity of the downstream front of a wide moving jam denoted by $v_{\rm g}$. The constancy of $v_{\rm g}$ while the jam propagates on the road is consistent with the characteristic jam feature [J] of Sect. 1.3.

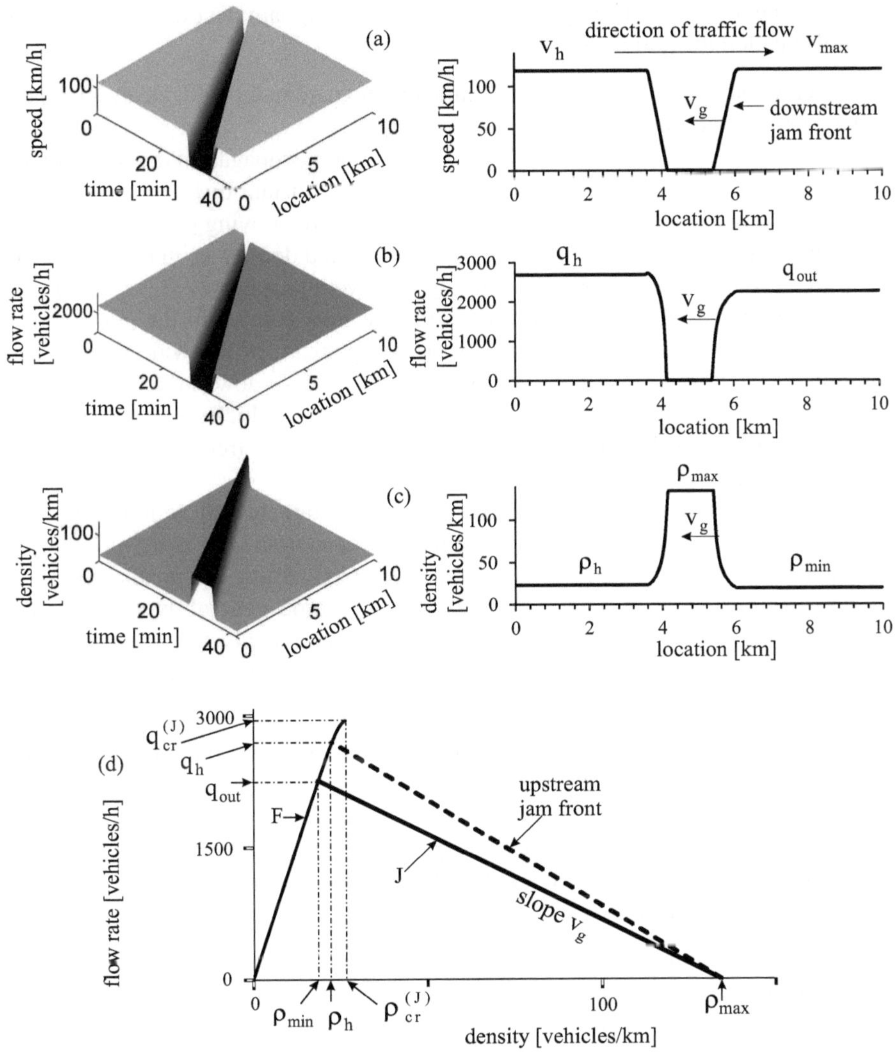

Fig. 4.11 Numerical simulations that summarize qualitative illustrations of Kerner-Konhäuser theory of wide moving jams shown in Figs. 4.9 and 4.10 [276]. (**a–c**) Propagation of a wide moving jam on a single-lane road: speed (**a**), flow rate (**b**) and density (**c**) in space and time (*left column*) and the same traffic variables as a function of road location (*right column*). (**d**) Representation of the part of the fundamental diagram for free flow (F) together with the line *J* (line J) whose slope is equal to the mean downstream jam front velocity v_g related to the wide moving jam shown in (**a–c**). Simulations of Bando et al. OV model

(ii) The flow rate q_{out}, density ρ_{min}, and average vehicle speed v_{max} in the outflow from the wide moving jam. These traffic variables are the characteristic parameters only under condition that free flow is formed in the jam outflow.

(iii) The mean vehicle density within the wide moving jam denoted by ρ_{max} that is also called the jam density.

However, we should mention that the characteristic parameters of wide moving jams can depend considerably on traffic parameters.

The characteristic parameters of wide moving jam propagation are qualitatively illustrated in Fig. 4.10. In this figure, at a given time instant distributions of traffic variables along a road are shown associated with a wide moving jam propagating in an initial homogeneous free flow. The flow rate q_h and density ρ_h in this initial free flow are chosen to be greater, respectively, the average speed v_h to be lower than the related characteristic traffic variables of free flow formed in the jam outflow: $q_h > q_{out}$, $\rho_h > \rho_{min}$, $v_h < v_{max}$. Obviously, upstream of the wide moving jam the initial free flow occurs. However, this is not the case downstream of the wide moving jam. This is because during the jam propagation vehicles escaping from the wide moving jam at the downstream jam front form a new free flow with the flow rate q_{out}, density ρ_{min}, and average speed v_{max}.

As above-mentioned, one of the characteristic parameters of wide moving jam propagation is the mean velocity of the downstream front of a wide moving jam v_g (called also as a characteristic jam velocity). Concerning the mean velocity of the downstream front of a wide moving jam v_g, some general assumptions can be made [264]. Each driver within the wide moving jam can start to accelerate to either free flow or synchronized flow downstream after two conditions have been satisfied:

(i) The preceding vehicle has already begun to move away from the wide moving jam.
(ii) Due to the preceding vehicle motion, after some time the space gap between the two drivers has exceeded a space gap at which a safety condition for driver acceleration is satisfied. This means that there is some time delay in vehicle acceleration at the downstream front of the wide moving jam.

The mean time delay in vehicle acceleration at the downstream front of a wide moving jam will be denoted by $\tau_{del}^{(acc)}(0)$. In empirical data [268],

$$\tau_{del}^{(acc)}(0) \approx 1.5 - 2 \text{ sec}. \tag{4.13}$$

The motion of the downstream front of a wide moving jam results from acceleration of drivers from the standstill within the wide moving jam to flow downstream of the wide moving jam. Because the average distance between vehicles inside the wide moving jam, including average length of each vehicle, equals $1/\rho_{max}$, the velocity of the downstream front of the wide moving jam is

$$v_g = -\frac{1}{\rho_{max}\tau_{del}^{(acc)}(0)}. \tag{4.14}$$

In empirical observations [268, 282],

$$v_g \sim -15\ \mathrm{km/h}. \tag{4.15}$$

4.5.1.2 Line *J*

The characteristic parameters of wide moving jam propagation and the jam feature [J], which defines the wide moving jam phase in congested traffic (Sect. 1.1), can be presented by the line J in the flow–density plane. The slope of the line J is equal to the characteristic velocity v_g. If in the wide moving jam outflow free flow is formed, then the flow rate q_{out} in this jam outflow and the related density ρ_{min} (Fig. 4.10) give the left coordinates $(\rho_{min},\ q_{out})$ of the line J (Fig. 4.9); q_{out}, ρ_{min}, and the related average vehicle speed v_{max} are characteristic parameters of wide moving jam propagation, which do not depend on initial conditions and on a state of free flow upstream of the wide moving jam (speed v_h, flow rate q_h, and density ρ_h in Fig. 4.10). The right coordinates $(\rho_{max},\ 0)$ of the line J are related to the traffic variables within the wide moving jam, the density ρ_{max} and flow rate $q_{min} = \rho_{max} v_{min}$, where the average vehicle speed within the wide moving jam v_{min} is here assumed to be zero that results in $q_{min} = 0$ within the wide moving jam (Fig. 4.9).

Because the slope of the line J is equal to the characteristic jam velocity v_g (4.14), on the line J the derivative

$$\frac{dq}{d\rho} = v_g, \tag{4.16}$$

i.e., in accordance with the line J definition it satisfies the equation

$$q(\rho) = \mid v_g \mid (\rho_{max} - \rho) \tag{4.17}$$

that using (4.14) can be rewritten in the equivalent form as

$$q(\rho) = \frac{1}{\tau_{del}^{(acc)}(0)} \left(1 - \frac{\rho}{\rho_{max}}\right). \tag{4.18}$$

From Eq. (4.17), for the flow rate in free flow formed in the jam outflow we get the formula

$$q_{out} = \mid v_g \mid (\rho_{max} - \rho_{min}), \tag{4.19}$$

which, as follows from (4.18), can also be written as

$$q_{out} = \frac{1}{\tau_{del}^{(acc)}(0)} \left(1 - \frac{\rho_{min}}{\rho_{max}}\right). \tag{4.20}$$

The empirical evidence of the line J and associated characteristic parameters of wide moving jam propagation was found in 1996 by Kerner and Rehborn [282].[4] A result of empirical observations of wide moving jam propagation is that the maximum flow rate in free flow denoted by $q_{\max}^{(\text{free, emp})}$ is considerably greater than the flow rate in free flow in the wide moving jam outflow q_{out} [282] (see Fig. 8.3e in Sect. 8.3.1):

$$q_{\max}^{(\text{free, emp})} > q_{\text{out}}. \tag{4.21}$$

In particular, for 1-min average measured traffic data has been found that [282]

$$\frac{q_{\max}^{(\text{free, emp})}}{q_{\text{out}}} \approx 1.5. \tag{4.22}$$

The theoretical result about the free flow metastability with respect to an F→J transition within the flow rate range (4.12) is confirmed by empirical condition (4.21) [282].

When the flow rate in free flow q exceeds the jam outflow rate q_{out} (Fig. 4.9), specifically, conditions (4.12) are satisfied, then free flow is metastable one with respect to the formation of a moving jam. In other words, the flow rate in the jam outflow q_{out} (Fig. 4.9) is a threshold flow rate for the formation of the moving jam [276].

The conclusion of [274, 276, 278–280] that the classical traffic flow instability of Herman, Gazis, Montroll, Potts, Rothery, and Chandler [93, 170, 171, 233] leads to an F→J transition in a metastable free flow is the *general* one for all models of the GM model class. For this reason, the line J, the metastability of free flow with respect to the F→J transition under condition (4.12) (Fig. 4.9), and the characteristic parameters of wide moving propagation of the Kerner-Konhäuser theory of wide moving jams [276] have also been found and studied in many other models of GM model class [42, 58, 60, 72, 73, 75, 92, 144, 148, 160, 162, 172, 173, 184–186, 206–212, 214–222, 224–232, 239, 240, 242, 243, 260, 263, 293, 307, 324–327, 333, 348, 349, 375, 376, 382, 385, 402, 405, 419–426, 449, 458, 460, 464–468, 474, 475, 479–481, 483, 490, 493, 494, 497, 499, 500, 502, 504–506, 511, 514, 520, 525, 526, 537, 538, 551, 562–564, 572, 573, 575] and reviews [57, 103, 213, 351, 361, 377, 380, 434]. There models include a huge number of traffic flow models

[4] Empirical characteristic parameters of wide moving jam propagation will be considered in more details in Sect. 8.3.1.

like Gipps's model [177, 178], Payne's model [414, 415], Newell's optimal velocity (OV) model [388], Wiedemann's model [536], Whitham's model [535], the Nagel-Schreckenberg (NaSch) cellular automaton (CA) model [43, 379], Bando et al. OV model [28–30], a lattice model by Nagatani [373, 374], Treiber's intelligent driving model (IDM) [213, 507–509], Krauß's model [300, 301], the Aw-Rascle model [24], a full velocity difference OV model by Jiang et al. [258]. The results of the Kerner-Konhäuser theory of moving traffic jams [276] discussed above with the use of qualitative illustrations (Figs. 4.9 and 4.10) are summarized in Fig. 4.11 with the use of numerical simulations of Bando et al. OV model.

Many other different traffic flow models of the GM model class as well as results of their analysis can be found, for example, in [20, 27–30, 41–43, 58, 60, 61, 68, 72, 73, 75, 92, 123, 124, 144, 148, 160, 162, 172, 173, 177, 178, 182, 184–186, 206–212, 214–222, 224–232, 239, 240, 242, 243, 258, 260, 263, 276, 293, 304, 307, 324–327, 333, 348, 349, 373–376, 379, 382, 385, 388, 402, 405, 414, 415, 419–426, 449, 458, 460, 464–468, 474, 475, 479–481, 483, 490, 493, 494, 497, 499, 500, 502, 504–509, 511, 514, 520–523, 525, 526, 535, 537, 538, 551, 553, 554, 562–564, 572, 573, 575] and reviews [57, 67, 103, 106, 163, 168, 188, 213, 236, 329, 351, 361, 377, 380, 392, 407, 434, 439, 534, 536, 539].

4.5.2 Driver Behavioral Assumptions

Other great achievements of traffic flow model of the GM model class are associated with the understanding of some important common driver behavioral assumptions. Additionally to the driver reaction time that explains the classical traffic flow instability of Herman, Gazis, Montroll, Potts, Rothery, and Chandler [93, 170, 171, 233], the following driver behavioral assumptions and their mathematical descriptions allows us to simulate the characteristic parameters of wide moving jams observed in real traffic flow.

To explain the characteristic parameters of wide moving jam propagation, the mean time delay in driver acceleration $\tau_{\text{del}}^{(\text{acc})}(0)$ at the downstream front of a wide moving jam should be longer than the time delay in vehicle acceleration $\tau_{\text{del}}^{(\text{acc})}$ at higher speeds in traffic flow. This is known as a slow-to-start rule in traffic flow modeling [43, 491].

In the theory of wide moving jam propagation derived in [276] from a numerical study of a version of the macroscopic Payne's model, the slow-to-start rule has been simulated through the use of a special form of the fundamental diagram. At the fundamental diagram, there is a wide density range in a neighborhood of the jam density within which the steady speed is nearly zero. As a result, vehicles could not almost accelerate from the initial speed $v = 0$ within the wide moving jam before the density decreases considerably, i.e., the mean space gap between vehicles increases. Thus, only after a relatively long mean time delay in vehicle acceleration $\tau_{\text{del}}^{(\text{acc})}(0)$ vehicles could escape from the wide moving jam.

Rather than the implicit simulation of the slow-to-start rule through the use of a special form of the fundamental diagram discussed above, another mathematical idea of simulation of slow-to-start rule firstly introduced by Takayasu and Takayasu [491] has been used in a further development of the Nagel-Schreckenberg (NaSch) CA model [43]. In the NaSch CA model, there is probability of fluctuations p that simulates the time delay in driver acceleration. Thus to simulate the slow-to-start rule, this probability has been taken considerably greater at $v = 0$ than at $v > 0$ [43]:

$$p(v) = \begin{cases} p_1 \text{ for } v = 0, \\ p_2 \text{ for } v > 0, \end{cases} \tag{4.23}$$

where p_1, p_2 are constants, $p_1 > p_2$.

Most of the achievements of these classical approaches are also used in three-phase traffic flow models [268, 269]. In particular, for simulations of moving jam emergence in synchronized flow (S→J transitions) three-phase traffic flow models incorporate over-deceleration effect (driver's over-reaction) introduced by Herman, Gazis, Montroll, Potts, Rothery, and Chandler [93, 170, 171, 233].

A stochastic approach to traffic flow modeling introduced by Nagel and Schreckenberg in the NaSch CA model [43, 379] is the basic approach for the mathematical description of diverse driver's delays in stochastic three-phase traffic flow models (see Appendices A and B). A detailed review of the Nagel-Schreckenberg approach to simulations of driver's delays has recently been made in the book by Schadschneider et al. [103].

Additionally, the stochastic three-phase traffic flow models (see Appendices A and B) incorporate the mathematical formulation for the description of a safe speed in traffic flow firstly introduced by Gipps [177] and further developed by Krauß et al. [301] as well as lane changing rules developed by Nagel et al. [381].

4.6 Summary of Achievements of Classical Traffic Flow Models

We can summarize the main achievements of the classical traffic flow models as follows.

1. The LWR-theory [107–109, 336, 365, 392, 448, 534] can show a transition to traffic congestion whose downstream front is fixed at a bottleneck.
2. Traffic flow models in the framework of the GM model class can explain at least the following empirical features of traffic congestion:

- Characteristic parameters of wide moving jams and line J [276] (see also, e.g., [29, 30, 43, 213, 301, 324–326, 464–466, 505, 506, 508, 509, 511, 553, 554]).
- A broad spread of traffic data in the flow–density plane associated with congested traffic (e.g., [220, 221, 502, 510, 553]).[5]

3. Most of the achievements of these classical approaches are also used in three-phase traffic flow models. In particular, three-phase traffic flow models incorporate over-deceleration effect (driver's over-reaction) of the GM model, a stochastic approach by Nagel and Schreckenberg for simulations of driver delay times, the Gipps [177] and Krauß et al. [301] formulations for the description of a safe speed as well as lane changing rules of Nagel et al. [381].

4.7 Why Are Generally Accepted Classical Two-Phase Traffic Flow Models Inconsistent with Features of Real Traffic?

Because of great achievements of generally accepted classical traffic flow models discussed in Sects. 4.3.2, 4.5, and 4.6, a question arises:

- Why does the author state that the generally accepted classical traffic flow theories and models are not applicable for a reliable description of traffic breakdown, highway capacity, traffic control, dynamic traffic assignment, and optimization of real traffic and transportation networks?

The failure of the LWR-theory, the Daganzo's cell transmission model, traffic flow models of the GM model class as well as other two-phase traffic flow models[6] is explained as follows[7]:

[5]In simulations, this broad spread of traffic data is usually associated with different driver behavioral characteristics (heterogeneous flow of different drivers and vehicles) used in a traffic flow model as well as with complex dynamics of moving jams and other dynamic spatiotemporal effects in congested traffic, rather than with steady states of a traffic flow model.

[6]Recall that in Sect. 1.12 to distinguish classical traffic flow models from three-phase traffic flow models, we have called traffic flow models in the frameworks of the LWR model, Daganzo's cell-transmission model, traffic flow models of the GM model class as well as any other traffic flow models that can be considered a combination of these classical approaches to traffic flow modeling as "two-phase traffic flow models". In other words, in the book the term *a two-phase traffic flow model* is a synonym of the term *a classical traffic flow model* and the term *a classical traffic flow theory*.

[7]In the books [268, 269], for two-phase traffic flow models we have used the term *traffic flow models in the framework of the "fundamental diagram approach"*. We do not use the term *the fundamental diagram approach* any more. It occurs that the use of the term *the fundamental diagram approach* leads to confusions for many authors that have begun to work in the framework of the three-phase theory. One of the reasons for this confusion is as follows. As shown in Sec. 12.1

1. The LWR-model and the Daganzo's cell transmission model failed because these models cannot show the metastability of free flow with respect to traffic breakdown as well as induced traffic breakdown at a highway bottleneck observed in real traffic.
2. Traffic flow models of the GM model class failed because the models cannot show the metastability of free flow with respect to an F→S transition at a highway bottleneck as observed in all real field traffic data (Chap. 3). Recall that the metastability of free flow with respect to the F→S transition at the bottleneck explains the nucleation nature of traffic breakdown at highway bottlenecks that is the empirical fundamental of transportation science.

The classical traffic flow models failed because they are inconsistent with the empirical fundamental of transportation science—the nucleation nature of traffic breakdown at highway bottlenecks.

These critical conclusions are independent of the choice of model parameters. For example, due to an appropriate choice of model parameters in a traffic flow model of the GM model class, the classical traffic flow instability of the GM model class can be realized in the model only at a vehicle density that is a larger one than the density associated with the maximum on the fundamental diagram of the model. As shown in [273], in this case the traffic flow model exhibits features of both the LWR model and a model of the GM model class. However, such a "combined" traffic flow model cannot also show the metastability of free flow with respect to an F→S transition (traffic breakdown) at the bottleneck (for more detailed explanations, see Sect. 4.8).

The classical traffic flow models cannot show the metastability of free flow with respect to the F→S transition at the bottleneck. For this reason, the classical traffic flow models are not consistent with the empirical fundamental of transportation science.

Contrarily to the classical traffic flow models, the three-phase theory is consistent with the empirical fundamental of transportation science. Three-phase traffic flow models incorporate the metastability of free flow with respect to the F→S transition at the bottleneck.

---

of the book [269], there can be derived three-phase traffic flow models that exhibit a fundamental diagram for steady state model solutions. Therefore, instead of the term *the fundamental diagram approach*, as mentioned in footnote 6, in this book for classical traffic flow models and theories we use the term *two-phase traffic flow models*.

Because the classical traffic flow models cannot explain the empirical fundamental of transportation science, in the subsequent chapters of the book we will consider simulations of traffic breakdown at highway bottlenecks with the use of three-phase traffic flow models in the framework of the three-phase theory *only*.[8] The above critical comment is also related to the effect of a moving bottleneck on traffic breakdown: *None* of the traffic flow models and theories that have been used, in particular, in [120, 149, 150, 169, 321, 322, 372, 393, 395] can describe the metastability of free flow with respect to an F→S transition at a highway bottleneck. Therefore, in the book simulations and explanations of the effects of road bottlenecks and moving bottlenecks on traffic breakdown are made in the framework of the three-phase theory.

4.8 Model Validation with Empirical Data

The main point of the criticism of the classical traffic flow models explained in Sect. 4.7 was known since years [268, 269]. Nevertheless, this criticism has not persuaded many researches working in the field. One of the disagreements of the researches can be explained by the fact that many of the classical traffic flow models can show some of empirical features of congested traffic patterns, by the appropriate choice of model parameters (e.g., [199, 216, 464, 465, 511]).

To illustrate this, we consider a traffic flow model of the GM model class in which the classical traffic flow instability with respect to moving jam formation is realized. As mentioned in Sect. 4.4.3, the traffic flow instability occurs in an initially homogeneous traffic flow when the vehicle density in this flow reaches a critical density $\rho_{cr}^{(J)}$; the critical flow rate related to the critical density $\rho_{cr}^{(J)}$ has been denoted by $q_{cr}^{(J)}$. There are two possibilities for the arrangement of the point of the classical traffic flow instability $(\rho_{cr}^{(J)}, q_{cr}^{(J)})$ on the fundamental diagram [273]:

(i) The point $(\rho_{cr}^{(J)}, q_{cr}^{(J)})$ lies left of the maximum point of the fundamental diagram (ρ_0, q_0), i.e., on the branch of the diagram with a positive slope (Fig. 4.12a):

$$\rho_{cr}^{(J)} < \rho_0. \tag{4.24}$$

(ii) The point $(\rho_{cr}^{(J)}, q_{cr}^{(J)})$ lies right of the maximum point of the fundamental diagram (ρ_0, q_0), i.e., on the branch of the diagram with a negative slope (Fig. 4.12b):

$$\rho_{cr}^{(J)} > \rho_0. \tag{4.25}$$

[8] The exclusion is Chap. 8, in which we will explain why the three-phase theory is incommensurable with classical traffic flow theories.

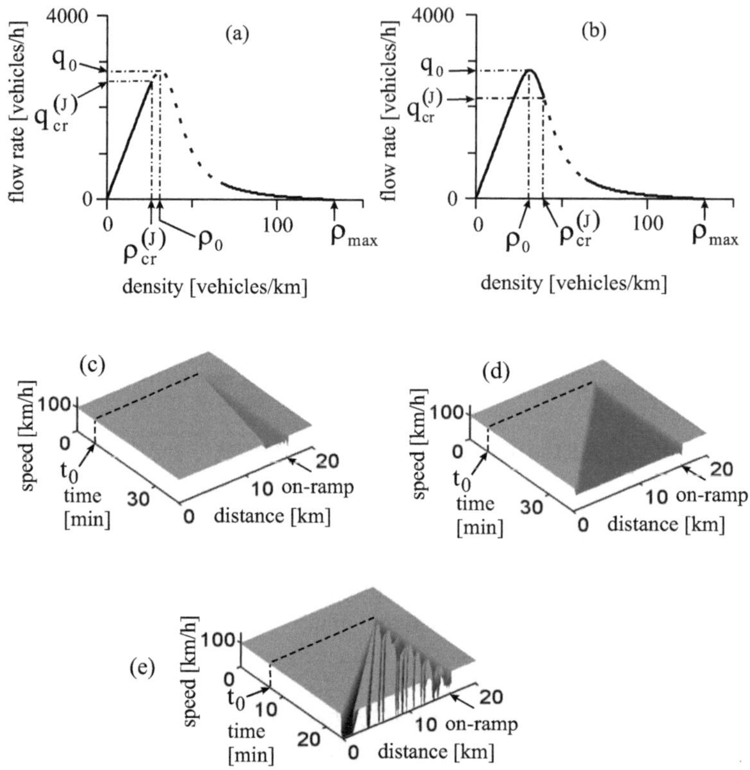

Fig. 4.12 Traffic breakdown and resulting congested patterns in some traffic flow models with the classical traffic flow instability of the GM model class at on-ramp bottleneck: (**a, b**)—Fundamental diagrams when the condition (4.24) (**a**) and (4.25) (**b**) are satisfied, respectively; dashed parts of the fundamental diagrams are related to unstable states with respect to the classical traffic flow instability of GM model class. (**c–e**) Speed in space and time in congested patterns under condition (4.25). (**c, d**)—Widening synchronized flow patterns (WSP) upstream of the bottleneck at two flow rates q_{on} =160 (**c**) and 400 (**d**) vehicles/h. (**e**)—Formation of wide moving jams within the WSP upstream of the bottleneck at q_{on} =700 vehicles/h. In (**c–e**), the flow rate q_{in} =2676 vehicles/h; the on-ramp inflow has been switched on at time instant $t = t_0$. Figures (**c–e**) are related to simulations of Bando et al. OV-model. Adapted from [273]

Note that in both cases (i) and (ii) all states on the fundamental diagram, in which the density satisfies conditions

$$\rho_{\min} \leq \rho < \rho_{\text{cr}}^{(\text{J})}, \tag{4.26}$$

are metastable states with respect to moving jam emergence; in (4.26), $\rho_{\min}$ is the vehicle density in free flow formed in the outflow of a wide moving jam.

The case (i) has already been considered in Sect. 4.4.3: Due to the classical traffic flow instability of Herman, Gazis, Montroll, Potts, Rothery, and Chandler [93, 170, 171, 233], wide moving jams emerge in free flow (F→J transition).

In the case (ii) (Fig. 4.12b), we find a qualitative different dynamics of the model. As in the LWR-model and the Daganzo's cell transmission model (see Figs. 4.3c,d of Sect. 4.3), traffic breakdown results in a widening synchronized flow pattern (WSP) at the bottleneck (Fig. 4.12c,d). However, when the flow rate $q_{\rm on}$ increases subsequently, moving jams emerge spontaneously within the WSP (Fig. 4.12e). The moving jams emerge when the average density within synchronized flow of the WSP approaches the critical density $\rho_{\rm cr}^{\rm (J)}$ for the classical traffic flow instability of Herman, Gazis, Montroll, Potts, Rothery, and Chandler [93, 170, 171, 233]. Thus, already a change of the model parameters, at which the critical density $\rho_{\rm cr}^{\rm (J)}$ becomes larger than the density ρ_0 at the maximum point of the fundamental diagram, has a great effect on the traffic flow dynamics that the model exhibits. Therefore, traffic flow models under condition (4.25) exhibits features of the LWR model (in a neighborhood of the density ρ_0 at the maximum of the fundamental diagram) *and* features of a traffic flow model of the GM model class (at some larger vehicle densities than the density ρ_0) (Figs. 4.12c,d). We can call such a traffic flow model that exhibits both features of the LWR model and of a traffic flow model of the GM model class as a "combined" two-phase traffic flow model.[9]

With the use of such a complex model, it is indeed possible to find a set of model parameters that permits to reproduce satisfactory a given time-dependence of traffic variables within an empirical congested pattern (e.g., [216, 464, 465, 511]). Thus, by the appropriate choice of model parameters, a traffic flow model, in which congested traffic whose downstream front is fixed at the bottleneck (synchronized flow) *and* the classical traffic flow instability of the GM model class can occur, can be validated to show model solutions, which at the first glance look similar to many empirical measured spatiotemporal congested traffic patterns (Figs. 4.12c,d). We should mention that as explained in Secs. 10.3.10–10.3.12 of the book [269], features of many model solutions for congested traffic patterns found and discussed in [213, 216, 464, 465, 506–509, 511] have no sense for real traffic. Nevertheless, this is *not* the main point of the criticism of the classical traffic flow theories.

The main point of the criticism of the classical traffic flow models including "combined" two-phase traffic flow models is as follows. In real traffic, the synchronized flow occurs due to the metastability of free flow with respect to an F→S transition (traffic breakdown) at the bottleneck. Contrarily, regardless of the choice in model parameters, any two-phase traffic flow model fails to show the metastability of free flow at a bottleneck with respect F→S transition (traffic breakdown) at the bottleneck.

[9]See the explanation of the term *two-phase traffic flow model* in Sect. 1.12.

To explain this, it must be emphasized that the occurrence of synchronized flow under conditions (4.25) in the model with the classical traffic flow instability of the GM model class is qualitatively the same as that in the LWR model and the Daganzo's cell transmission model (Sect. 4.3). In particular, there is *no* nucleation effect by traffic breakdown (F→S transition) when synchronized flow (WSP) occurs at the bottleneck (Figs. 4.12c,d). In other words, synchronized flow, which can be simulated with the "combined" two-phase traffic flow model (Figs. 4.12c,d), exhibits qualitatively different features in comparison with the synchronized flow found in real field traffic data.

The classical traffic flow models including "combined" two-phase traffic flow model are inconsistent with the empirical fundamental of transportation science: Traffic breakdown (F→S transition) at the bottleneck does not exhibit the nucleation nature observed in all empirical data for traffic breakdown (Chap. 3). The critical conclusion is regardless of the choice of model parameters.

Regardless of the choice of model parameters, *none* of the classical traffic flow models shows the metastability of free flow at a highway bottleneck with respect to an F→S transition. This metastability explains the empirical nucleation nature of real traffic breakdown at the bottleneck.

4.9 Applications of Classical Traffic Flow Theories for Development of Intelligent Transportation systems (ITS)

4.9.1 *Simulations of ITS Performance*

Because none of the classical traffic flow models can show the metastability of free flow with respect to an F→S transition (traffic breakdown) at the bottleneck, the application of these classical models for an analysis of the effect of ITS on traffic flow, which is generally accepted by traffic and transportation researchers, do lead to invalid (and sometimes incorrect) conclusions about the ITS performance in real traffic.

This criticism is related to all ITS that affect traffic flow, for example, on-ramp metering (see, e.g., [21, 79, 262, 408–411, 413]) (see also Sect. 4.9.2), variable speed limit control (see, e.g., [1, 10, 77–81, 86, 94, 204, 290, 291]) and many other ITS-applications, like the effect of cooperative driving with the use of vehicle-two-vehicle (V2V) communication on traffic flow characteristics as well as simulations of the performance of dynamic traffic assignment and optimization in traffic and transportation networks (e.g., [2, 4, 8, 9, 15, 26, 47, 49–56, 59, 76, 90, 91, 98, 101, 102, 104, 105, 110, 137, 138, 153–159, 181, 182, 187, 200, 201, 235, 251, 253, 259, 261, 288, 289, 306, 308–310, 339, 340, 344–346, 352, 354, 355, 357, 367, 368, 370, 383, 384, 403, 416, 417, 440, 441, 445, 447, 461, 470, 471, 477, 482, 484, 488, 489, 498, 513, 528, 532, 533, 542, 544–550, 555, 556, 559, 560, 566, 568–570, 574, 576]). These ITS applications and results of their analyzes are inconsistent with the empirical fundamental of transportation science.

Unfortunately, this critical conclusion is also related to most studies of the effect of adaptive cruise control (ACC) and other vehicle systems on traffic flow, in particular, considered and/or reviewed in [133, 152, 283, 284, 302, 330, 332, 338, 363, 369, 400, 401, 452, 453, 473, 485, 508, 509, 512, 517, 571]. In other words, because the classical traffic flow models cannot show the empirical features of the metastability of free flow with respect to an F→S transition (traffic breakdown) at a highway bottleneck, the application of these generally accepted models and associated simulation tools for a study of the effect of automatic driving vehicles on traffic flow can lead to incorrect conclusions. For this reason, such simulations (see, for example, [27, 61, 68, 133, 283, 284, 302, 330, 332, 369, 400, 401, 452, 453, 473, 508, 509, 521–523, 571]) cannot also be used for the development of reliable systems for automatic driving vehicles (see also Sect. 4.9.3). This criticism is also related to the use of well-known simulation tools based on the classical traffic flow theories like simulation tools VISSIM (Wiedemann model [536]) and SUMO (Krauß model [301]) (see, e.g., [152, 369, 517]).

To study the performance of ITS with a traffic flow model, the model should simulate conditions at which the application of ITS can prevent traffic breakdown in a traffic network. However, none of the classical traffic flow models can simulate the empirical nucleation nature of real traffic breakdown at the bottleneck.

The three-phase theory is devoted to the explanation of the empirical nucleation nature of real traffic breakdown at the bottleneck. Therefore, only three-phase traffic flow models can be used for a reliable study of ITS applications.

4.9.2 On-Ramp Metering

One of the well-known traffic control methods is on-ramp metering at an on-ramp bottleneck [21, 202, 365, 407, 476]. Various models and methods based on the "free flow control approach" to on-ramp metering have been developed. The basic idea of this approach is to maintain free flow on the main road at the bottleneck (e.g., [21, 408–412, 476]). Free flow at the bottleneck should be maintained at the maximum possible throughput in free flow downstream of the bottleneck. In other words, traffic breakdown and resulting congested patterns at the bottleneck should be prevented.

An example of this approach is the ALINEA method of Papageorgiou et al. [408–412, 476] used in many real installations. The theoretical basis of ALINEA method is associated with the LWR theory: It is assumed that the capacity of an on-ramp bottleneck is related to the maximum point on the fundamental diagram. Therefore, in ALINEA method traffic breakdown should not occur as long as a well-known traffic variable *occupancy* (see, for example, the definition of the occupancy in Sec. 2.1 of the book [269]) downstream of the bottleneck does not exceed some critical occupancy. However, this understanding of *critical occupancy* that follows from the LWR theory contradicts fundamentally the empirical nucleation nature of traffic breakdown at the bottleneck. Indeed, at *any occupancy* in free flow related to the flow rate range between the minimum and maximum highway capacities traffic breakdown can occur with some finite probability at the bottleneck (Chap. 3). For this reason, under application of ALINEA traffic breakdown does randomly occur at the bottleneck within this flow rate range. This critical conclusion is also related to other free flow control methods.

This is because independent of a control method and on a mathematical formulation of a controller it is usually assumed that in free flow there is either critical flow rate, or critical density, or else critical occupancy that determines a choice of optimal flow rate, or optimal density, or else optimal occupancy: In the control method, either the optimal flow rate, or the optimal density, or else the optimal occupancy is a target value for the controller, which should be chosen smaller than the associated critical flow rate, or the critical density, or else the critical occupancy, respectively (see, e.g., [3, 21, 202, 365, 407, 476]).

However, as follows from the empirical nucleation nature of traffic breakdown (Chap. 3), when the flow rate at a highway bottleneck is larger than the minimum highway capacity, there are the infinite number of the flow rates, the densities, and the occupancies in free flow at which traffic breakdown (F→S transition) can occur at the bottleneck. These flow rates, the densities, and the occupancies are related to different flow rates q_{sum} (3.9) in metastable free flow at the bottleneck (see Sect. 3.7). This means that there can be chosen neither particular "optimal" flow rate, nor "optimal" density, nor else "optimal" occupancy that can be chosen as a target value for the controller. For this reason, the methods of traffic control (see, e.g., [3, 21, 202, 365, 407, 476]) are inconsistent with the empirical nucleation nature of traffic breakdown. Therefore, these control methods cannot reduce traffic congestion at network bottlenecks.

Due to the metastability of free flow with respect to an F→S transition at a highway bottleneck, there is neither a particular optimal flow rate, nor an optimal density, nor else an optimal occupancy in free flow that can be chosen as a target value for a controller of a traffic control method. For this reason, standard methods of traffic control (see, e.g., [3, 21, 202, 365, 407, 476] and references there) are inconsistent with the empirical nucleation nature of traffic breakdown. These generally accepted control methods cannot reduce traffic congestion in traffic networks.

A detailed consideration of this criticism can be found in Sec. 10.6.5 of the book [269]. Moreover, in Sec. 10.7 of the book [269], we have made a comparison of the performance of ALINEA and on-ramp metering ANCONA based on so-called "congested pattern control approach" (see Sect. 12.6).

In ANCONA, no on-ramp metering and no other traffic control is realized as long as free flow conditions are measured at a bottleneck. Only after traffic breakdown has occurred at the bottleneck, on-ramp inflow control starts [268, 269]. The congested pattern control approach can also be called as "control of traffic breakdown at a bottleneck". Under this traffic control, traffic breakdown (F→S transition) is allowed to occur at the bottleneck. However, after traffic breakdown has occurred, traffic control begins that should cause a return transition from synchronized flow to free flow (S→F transition) at the bottleneck. In some cases, when the return S→F transition is not possible to achieve at the bottleneck, traffic control should lead to a spatial limitation of congested pattern at the bottleneck. A detailed consideration of ANCONA approach can be found in Sec. 23.3 of the book [268]. We will use "control of traffic breakdown at a bottleneck" in Sect. 12.6.

4.9.3 Effect of Automatic Driving on Traffic Flow

In a simple case of a single-lane road, a vehicle system called an adaptive cruise control (ACC) governs the motion of an automatic driving vehicle (called also as a self-driving vehicle) in traffic flow on the road.

There can be many different ACC dynamics rules of motion behind the preceding vehicle (e.g., [17–19, 45, 46, 65, 99, 125–127, 132, 134, 135, 139, 238, 241, 252, 254, 284, 305, 334, 335, 350, 371, 404, 406, 442, 455, 456, 472, 486, 487, 503, 508, 509]). We limit the consideration by a classical model of ACC vehicle. In the classical ACC model, acceleration (deceleration) $a^{(ACC)}(t)$ of the ACC vehicle is determined by the space gap to a preceding vehicle $g(t)$ and the relative speed $\Delta v(t) = v_\ell(t) - v(t)$

measured by the ACC vehicle as well as by a desired space gap $g^{(\mathrm{ACC})} = v(t)\tau_{\mathrm{d}}^{(\mathrm{ACC})}$, where $v(t)$ is the speed of the ACC-vehicle, $v_{\ell}(t)$ is the speed of the preceding vehicle, and $\tau_{\mathrm{d}}^{(\mathrm{ACC})}$ is a desired net time gap (desired time headway) of the ACC-vehicle to the preceding vehicle (e.g., [17–19, 45, 46, 65, 99, 125–127, 132–135, 139, 241, 283, 284, 302, 305, 331, 332, 334, 335, 338, 350, 363, 400, 401, 442, 452, 455, 456, 472, 473, 485–487, 503, 508, 509, 512, 571]):

$$a^{(\mathrm{ACC})}(t) = K_1(g(t) - v(t)\tau_{\mathrm{d}}^{(\mathrm{ACC})}) + K_2(v_{\ell}(t) - v(t)), \tag{4.27}$$

where K_1 and K_2 are coefficients of ACC adaptation.

To find the effect of automatic driving vehicles (self-driving vehicles) on traffic flow, a mixed traffic flow consisting of automatic driving vehicles randomly distributed between human driving vehicles should be considered. It should be noted that coefficients K_1 and K_2 of ACC adaptation in (4.27) can be chosen large enough at which free flow consisting of 100% of ACC vehicles is stable with respect to traffic breakdown at a highway bottleneck. Therefore, we can expect that there should be a critical percentage of ACC-vehicles in mixed traffic flow, when the ACC-vehicles suppress any instabilities in traffic flow caused by manual driving vehicles.

In other words, this assumption should be valid regardless of features of a traffic flow model used for simulations of human driving vehicles. It turns out that this critical percentage of ACC-vehicles is about 30% of ACC vehicles (see, e.g., [133, 152, 283, 284, 302, 332, 338, 363, 369, 400, 401, 452, 473, 485, 508, 509, 512, 517, 571]). The work by Davis [125] was one of the first to obtain this result with another human driver model that had some of the features of the three-phase theory. However, the case of 30% (or more) of ACC vehicles in mixed traffic flow is currently a non-realistic one. This is because there are still no self-driving vehicles on the market. Thus, the analysis of mixed traffic flow with a considerable smaller percentage of self-driving vehicles is of the interest.

To find the effect of a considerably smaller percentage of automatic driving vehicles (self-driving vehicles) on traffic flow, a traffic flow model of human driving vehicles has to describe the metastability of an F→S transition in free flow at the bottleneck found in real field traffic data (Chap. 3). However, as we have proven above in this chapter, none of the classical traffic flow models can show the metastability of the F→S transition in free flow at the bottleneck.

Unfortunately, this critical conclusion is related to most studies of the effect of ACC-vehicles on traffic flow (see, e.g., [133, 152, 283, 284, 302, 332, 338, 363, 369, 400, 401, 452, 473, 485, 508, 509, 512, 517, 571]). For this reason, such simulations (see, for example, [133, 283, 284, 302, 332, 369, 400, 401, 452, 473, 492, 508, 509, 571]) cannot also be used for the development of reliable systems for automatic driving vehicles. This criticism is also related to the use of well-known simulation tools based on the classical traffic flow theories like simulation tools VISSIM (Wiedemann model [536]) and SUMO (Krauß model [301]) (see, e.g., [152, 369, 517]).

The generally accepted classical traffic and transportation theories cannot show the metastability of the F→S transition in free flow at the bottleneck. Therefore, the classical traffic flow models cannot show real traffic dynamics associated with the interaction of human driving vehicles in traffic flow. For this reason, the classical traffic flow models cannot be used for a reliable study of the effect of automatic driving vehicles on traffic flow.

To perform a study the effect of ACC-vehicles on traffic flow that is in the agreement with empirical features of real traffic flow, simulations of mixed traffic flow should be made with a three-phase traffic flow model. Such an analysis of the effect of ACC-vehicles on traffic flow will be made in Chap. 6.

4.10 Classical Understanding of Stochastic Highway Capacity

The classical highway capacity is defined through the occurrence of traffic breakdown at a highway bottleneck: The highway capacity is equal to the flow rate in an initially free flow at the bottleneck at which traffic breakdown is observed at the bottleneck (see, e.g., [69–71, 145–147, 174, 189, 191, 195, 197, 236, 237, 297, 343, 365, 432]).

As already emphasized in Sect. 2.3, in 1995 Elefteriadou et al. found out that empirical traffic breakdown at a highway bottleneck has a stochastic (probabilistic) behavior [145–147, 297]. Respectively, as firstly found by Persaud et al. [432] the probability of empirical spontaneous traffic breakdown is an increasing flow rate function (Fig. 2.2) [432]. To explain the stochastic (probabilistic) behavior of traffic breakdown at highway bottleneck, Brilon has introduced the following concept for a stochastic highway capacity [69–71, 174].

In accordance with the classical capacity definition, Brilon's stochastic highway capacity is equal to the flow rate in an initially free flow at the bottleneck at which traffic breakdown is observed at the bottleneck. At any time instant, there is a particular value of stochastic capacity of free flow at the bottleneck. However, as long as free flow is observed at the bottleneck, this particular value of stochastic capacity cannot be measured. Therefore, stochastic capacity is defined through a capacity distribution function $F_{\rm C}^{\rm (B)}(q_{\rm sum})$, where $q_{\rm sum}$ is the flow rate at a highway bottleneck [69–71, 174]:

$$F_{\rm C}^{\rm (B)}(q_{\rm sum}) = p(C \leq q_{\rm sum}), \tag{4.28}$$

where $p(C \leq q_{\rm sum})$ is the probability that stochastic highway capacity C is equal to or smaller than the flow rate $q_{\rm sum}$.

Thus, the basic theoretical assumption of the classical understanding of stochastic highway capacity is that traffic breakdown is observed at a time instant t at which the flow rate $q_{\rm sum}$ reaches the capacity C. This means that the flow rate function of the probability of traffic breakdown should be determined by the capacity distribution function [69–71, 174]:

$$P^{\rm (B)}(q_{\rm sum}) = F_{\rm C}^{\rm (B)}(q_{\rm sum}). \tag{4.29}$$

It must be noted that as found in empirical observations [69–71, 174, 343, 432] the breakdown probability $P^{\rm (B)}(q_{\rm sum})$ is an increasing function of the flow rate (left in Fig. 4.13). Thus this flow rate function $P^{\rm (B)}(q_{\rm sum})$ is the empirical evidence. However, Eq. (4.29) is a theoretical hypothesis only. This is because in contrast with the breakdown probability function $P^{\rm (B)}(q_{\rm sum})$, the capacity distribution function $F_{\rm C}^{\rm (B)}(q_{\rm sum})$ cannot be measured. This understanding of stochastic capacity of free flow at a bottleneck, which is currently well accepted in the community of traffic and transportation engineers [145, 146, 297], is illustrated in Fig. 4.13.

In Fig. 4.13, we show a qualitative hypothetical fragment of the time-dependence of stochastic capacity C over time t (Fig. 4.13, right). Left in Fig. 4.13, we show a qualitative flow rate dependence of the probability of spontaneous traffic breakdown $P^{\rm (B)}(q_{\rm sum})$.

It is often assumed that a stochastic behavior of highway capacity is associated with a stochastic change in traffic parameters over time [69–71, 145, 146, 174, 297]. Examples of the traffic parameters, which can indeed be stochastic time-functions in real traffic, are weather, mean driver's characteristics (e.g., mean driver reaction time), share of long vehicles, etc.

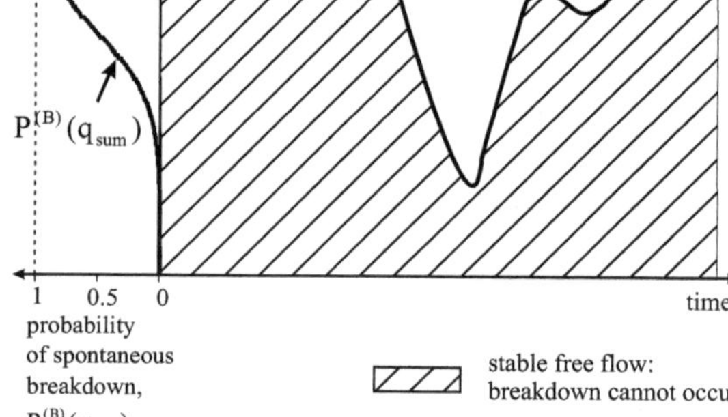

Fig. 4.13 Qualitative explanation of Brilon's stochastic highway capacity of free flow at a highway bottleneck (*right figure*). Left figure—a qualitative flow-rate function of the probability of the spontaneous breakdown $P^{\rm (B)}(q_{\rm sum})$ (see explanations of the function $P^{\rm (B)}(q_{\rm sum})$ in Chap. 5)

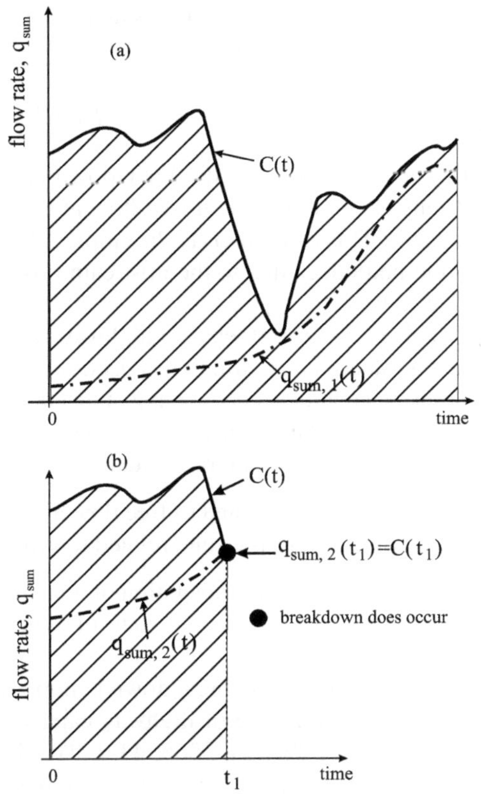

Fig. 4.14 Qualitative explanation of traffic breakdown with the use of Brilon's stochastic highway capacity of free flow at a highway bottleneck. The fragment of the hypothetical time-function of Brilon's stochastic highway capacity $C(t)$ is adapted from Fig. 4.13. Adapted from [271, 272]

In accordance with the definition of stochastic capacity (4.28), (4.29), no traffic breakdown can occur, when the time dependence of the flow rate is given by a hypothetical time dependence $q_{\text{sum}}(t) = q_{\text{sum},\,1}(t)$. This is because at all-time instants $q_{\text{sum},\,1}(t) < C(t)$ (Fig. 4.14a).

In contrast, for another hypothetical time dependence $q_{\text{sum}}(t) = q_{\text{sum},\,2}(t)$ traffic breakdown should occur at time instant $t = t_1$ at which this flow rate is equal to the capacity value: $q_{\text{sum},\,2}(t_1) = C(t_1)$ (Fig. 4.14b).

In other words, the classical understanding of stochastic capacity can be explained as follows: At a given time instant no traffic breakdown can occur at a highway bottleneck if the flow rate in free flow at the bottleneck at the time instant is smaller than the value of the capacity at this time instant.

The basic importance of the words "at a given time instant" in the capacity definition is as follows: Because Brilon's stochastic capacity $C(t)$ changes stochastically over time (Fig. 4.13, right), clearly that at a given time instant traffic breakdown can occur at the flow rate that is smaller than the value of the stochastic capacity was at another time instant (Fig. 4.14b).

Thus, in the classical understanding of stochastic highway capacity, free flow is stable under condition

$$q_{\text{sum}}(t) < C(t). \tag{4.30}$$

This means that no traffic breakdown can occur or be induced at the bottleneck at long as the flow rate in free flow at the bottleneck is smaller than the stochastic capacity. This contradicts to the empirical fact that traffic breakdown can be induced at the bottleneck due to the upstream propagation of a localized congested pattern (Sect. 3.6).

This is because stochastic highway capacity cannot depend on whether there is a congested pattern, which has occurred outside the bottleneck and independent of the bottleneck existence, or not. Indeed, the empirical evidence of induced traffic breakdown is the empirical proof that at a given flow rate at a bottleneck there can be one of two different traffic states at the bottleneck:

(i) A traffic state related to free flow and
(ii) a congested traffic state related to synchronized flow (Sect. 3.6).

Due to the upstream propagation of a localized congested pattern, a transition from the state of free flow to the state of synchronized flow, i.e., traffic breakdown is induced.

The induced traffic breakdown is impossible to occur under the classical understanding of the nature of highway capacity [69–71, 145–147, 174, 189, 191, 195, 197, 236, 237, 297, 343, 365, 432]. This is because in the classical understanding of highway capacity, free flow is stable under condition (4.30), i.e., no traffic breakdown can occur (Fig. 4.14a).

In contrast with the classical understanding of the nature of highway capacity, the evidence of the empirical induced breakdown means that free flow is in a metastable state with respect to the breakdown. The metastability of free flow at the bottleneck should exist for all flow rates at which traffic breakdown can be induced at the bottleneck. This empirical evidence of the metastability of free flow at the bottleneck contradicts fundamentally the concept of Brilon's stochastic capacity, in which free flow is stable under condition (4.30).

Thus, the currently accepted understanding of stochastic highway capacity [69–71, 74, 145–147, 174, 189, 191, 195, 197, 236, 237, 255, 297, 343, 365, 432] failed because this understanding about the nature of highway capacity contradicts the empirical evidence that traffic breakdown can be induced at a highway bottleneck as observed in real traffic (Sect. 3.6).

4.11 Strict Belief in Classical Theories as Reason for Defective Analysis of Empirical Traffic Phenomena

It is obvious that real traffic flow phenomena observed in real field traffic data should be the basis for any traffic flow theory. In other words, the analysis of empirical traffic data should determine conclusions about possibilities and limitations of a traffic flow model in the explanation of the real world. Almost each of the traffic models can explain some real traffic phenomena and each of the models exhibits a limited area of the applicability for the explanation of real traffic and transportation phenomena. This is also related to the classical traffic flow theories that are able to explain some of empirical traffic flow phenomena. As we have shown above, the classical traffic flow theories failed for ITS-applications because these theories and models are not consistent with the empirical nucleation nature of traffic breakdown (F→S transition) at a highway bottleneck.

In the most fields of science, new experimental and/or empirical phenomena initiate the development of theories and models of the phenomena. Unfortunately, it is different in the field of traffic flow theories: Usually, results of classical traffic flow theories and solutions of associated traffic flow models determine the methodologies of analyzing of empirical traffic data. This prevents the understanding of those new real traffic flow phenomena that cannot be explained by the classical theories. In this section, this statement of the author will be illustrated based on a critical analysis of some well-known terms of the classical theories like "capacity drop" and "macroscopic fundamental diagram (MFD)".

4.11.1 A Possible Origin of Failure of Classical Traffic Flow Models

Traffic occurs in space and time. Thus, empirical spatiotemporal features of traffic phenomena should be the empirical basis for a traffic flow model. The basic spatiotemporal phenomenon in traffic network is empirical traffic breakdown at a network bottleneck (Chap. 3). Therefore, the nucleation spatiotemporal features of real traffic breakdown should determine the development of traffic theories and models.

However, all latest developments in the traffic and transportation theory ignore the nucleation spatiotemporal features of real traffic breakdown in traffic and transportation networks. A possible reason for this invalid development of traffic and transportation science is as follows.

A strict belief in classical traffic theories in traffic and transportation community leads to a defective (and sometimes invalid) analysis of empirical traffic phenomena.

In particular, classical traffic flow theories and models do not show the nucleation nature (F→S transition) at a highway bottleneck. This is probably the reason why the empirical evidence of the metastability of free flow with respect to the F→S transition at the bottleneck is simply ignored in the state-of-the-art of traffic and transportation research.

In general, we can distinguished two different standard invalid methodologies of a study of empirical traffic flow phenomena:

(i) All empirical traffic flow phenomena that contradict solutions of a traffic model appear to be ignored.
(ii) Real empirical spatiotemporal traffic flow phenomena are averaged either in space or/and in time. As a result of this averaging, important empirical spatiotemporal traffic flow phenomena cannot not be found in the averaged data any more. Often this averaging of real measured data makes the resulting "empirical traffic flow phenomenon" in the agreement with theoretical conclusions made from a traffic model. In other words, all empirical features of real traffic phenomena, which cannot be explained by the model, are eliminated through the data averaging.

From the point of view of the author, the invalid methodologies of such a study of traffic flow phenomena can be explained by a very long time interval between the development of the classical traffic theories made in 1950s-1960s and the understanding of the empirical nucleation nature of traffic breakdown at road bottlenecks made at the end of 1990s [265–267] (see other references in the book [268]). During this time interval, several generations of traffic researches developed a huge number of traffic flow models based on the classical approaches. Thus, in accordance with historical analysis of scientific revolutions in other fields of science made by Kuhn [303], we may assume that it is very difficult for traffic researchers to see and accept that there is empirical traffic phenomenon that contradicts basically all fundamentals of the classical traffic flow theories and models. Below we consider a few examples of these invalid methodologies of empirical analyzes of traffic flow phenomena.

4.11.2 Capacity Drop

A classical example of a methodology related to item (i) of Sect. 4.11.1 is the theoretical concept about stochastic highway capacity discussed in Sect. 4.10 as well as the terms "two-capacity phenomenon" and "capacity drop" (see, e.g., [31–39, 69–71, 84, 87, 89, 100, 140, 145–147, 174, 189–198, 236, 237, 248, 297, 343, 364–366, 427–433, 462]). There are many achievements of empirical data analysis of traffic breakdown at highway bottlenecks mentioned in Chap. 2. However, in this analysis of empirical traffic breakdown and highway capacity the phenomenon of empirical induced traffic breakdown at the bottlenecks has been ignored (see, e.g., [33, 69–71, 74, 145–147, 174, 189, 191, 195, 197, 236, 237, 297, 343, 365, 432]). From the point of view of the author of this book, the reason for this is as follows. The empirical phenomenon of induced traffic breakdown contradicts the classical understanding of highway capacity as well as hypotheses about "two-capacity phenomenon" and "capacity drop" that are related to the state-of-the-art in traffic and transportation research (see, e.g., [33, 69–71, 145–147, 174, 189, 191, 195, 197, 236, 237, 297, 343, 365, 432]).

One of the consequences of this invalid empirical analysis of real traffic data is the concept of the existence of "capacity drop" associated with the classical understanding of highway capacity. The term "capacity drop" that is very often used in the literature (see, e.g., [33, 145–147, 191, 197, 236, 297, 343, 432]) should explain a flow-rate drop that is often observed during traffic breakdown at a highway bottleneck (Sect. 2.2.2).

The capacity drop means that the pre-discharge flow rate measured at the bottleneck in free flow just before traffic breakdown occurs at this bottleneck is larger than the discharge flow rate from synchronized flow (congested traffic) that emerges at the bottleneck just after the breakdown has occurred at the bottleneck (see, e.g., [33, 191]).

The term "capacity drop" is related to the assumption that capacity of free flow at the bottleneck at the time instant of the breakdown is related to this pre-discharge flow rate. Respectively, it is assumed that capacity of this bottleneck after the breakdown has occurred is related to the discharge flow rate that is smaller than the pre-discharge flow rate. In other words, the term "capacity drop" reflects the basic assumption of the classical theory of highway capacity of free flow at the bottleneck that at any time instant there is a particular value of (stochastic) highway capacity.

It must be emphasized that a flow-rate drop that can occur during traffic breakdown at a highway bottleneck [191] is the empirical evidence. However, the assumption that this flow-rate drop can be explained within the framework of the classical understanding of highway capacity is a theoretical hypothesis only. As we have explained in Sect. 4.10, the classical understanding of highway capacity is invalid for real traffic. This is because the classical understanding of highway capacity contradicts basically the empirical fundamental of transportation science—the empirical nucleation nature of traffic breakdown at highway bottlenecks (Chap. 3). The terms "capacity drop" and "two-capacity phenomenon" result from the classical understanding of stochastic highway capacity. For this reason, in the book we do not use the terms "capacity drop" and "two-capacity phenomenon".

The concept "capacity drop" contradicts basically the empirical fundamental of transportation science—the nucleation nature of traffic breakdown at a network bottleneck.

Indeed, as explained in Chap. 3, the empirical nucleation nature of traffic breakdown at a network bottleneck is associated with the metastability of free flow at the bottleneck with respect to an F→S transition (traffic breakdown). This metastability of free flow at the bottleneck should exist for all flow rates at which traffic breakdown can be induced at the bottleneck. This flow rate is between some minimum highway capacity C_{min} and some maximum highway capacity C_{max} (3.9). This empirical evidence of the metastability of free flow at the bottleneck contradicts fundamentally the concept of the classical understanding of stochastic highway capacity (Sect. 4.10), in which free flow is stable under condition (4.30). Therefore, the concept "capacity drop" resulting from the classical understanding of stochastic highway capacity is also inconsistent with the evidence of the empirical metastability of free flow with respect to an F→S transitions at the bottleneck.[10]

[10]There are many theoretical works in which rather than a flow-rate drop observed during traffic breakdown at a highway bottleneck (Sect. 2.2.2) as "capacity drop" another flow-rate drop $q_{\mathrm{cr}}^{(\mathrm{J})} - q_{\mathrm{out}}$ shown in Fig. 4.9 is considered. This application of the term "capacity drop" is incorrect for another reason. To explain this critical statement, we should recall that traffic breakdown occurs at the effective location of a highway bottleneck (see explanations of the effective location of the bottleneck in Sect. 3.4.4). In Fig. 4.9, free flow states together with the line J that characterizes some features of wide moving jams have been presented. As explained in details in Sec. 15.4 of the book [268], rather than traffic breakdown at the bottleneck, the free flow states together with the line J shown in Fig. 4.9 explain the Koshi's reverse-λ fundamental diagram found by Koshi et al. in empirical data [298]. In Sec. 15.4 of the book [268], it has also been shown that the Koshi's reverse-λ fundamental diagram can be measured at road locations at which wide moving jams have already been formed within a congested pattern localized at a highway bottleneck. The road locations are several kilometers upstream of the effective bottleneck location at which traffic breakdown has initially occurred. In the empirical example of a congested pattern shown in Fig. 4.2a, these road locations at which the Koshi's reverse-λ fundamental diagram can be measured (locations $x = 0$ and $x = 1.5$ km in Fig. 4.2b) are about 5 km upstream of the effective location of the bottleneck at which traffic breakdown has been observed (location $x = 6.4$ km in Fig. 4.2b). Thus, the flow-rate drop between the maximum flow rate that is reachable in free flow and the outflow rate in the wide moving jam outflow (Figs. 4.2b ($x = 0$ and $x = 1.5$ km) and 4.9) has in general *no* relation to a flow-rate drop observed [33, 191] during traffic breakdown at the effective location of traffic breakdown at a highway bottleneck (Sect. 2.2.2).

4.11.3 Macroscopic Fundamental Diagram

Another classical example of a methodology related to items (i) and (ii) of Sect. 4.11.1 is a theoretical analysis of empirical traffic flow phenomena in traffic and transportation networks with a macroscopic fundamental diagram (MFD) [116, 119, 121, 142, 166, 175, 299, 323, 543, 556].

To illustrate this critical conclusion, we recall some features of the fundamental diagram of traffic flow (Fig. 4.1a). The fundamental diagram of traffic flow is related to a macroscopic presentation of measurements of traffic data at a road location. The fundamental diagram reflects the well-known empirical macroscopic features of traffic flow that the larger the vehicle density, the lower is the average vehicle speed on the road (Fig. 4.1b) (see, e.g., reviews, books, and conference proceedings [13, 16, 57, 83, 88, 103, 109, 141, 161, 163, 165, 168, 188, 213, 223, 236, 237, 245, 311, 328, 351, 353, 361, 362, 365, 377, 380, 392, 407, 434, 439, 444, 446, 459, 469, 495, 508, 509, 536, 539, 540]).

Differently to the fundamental diagram of traffic flow (Fig. 4.1a), the macroscopic fundamental diagram (MFD) [116, 175] that is also called as a network fundamental diagram (NFD) [358–360] presents a relation between macroscopic traffic variables related to an urban network. In other words, to find an MFD, some average vehicle density in the whole urban network and the total flow (either inflow or outflow) rate in the network should be measured. Empirical studies have shown that the MFD can indeed show a similar relationship between the average vehicle density in the whole network and the total inflow rate in the network [116, 175] as the classical fundamental diagram (Fig. 4.1a).

Thus, studies of the MFD permit to make the conclusion that the larger the total average vehicle density in the network, the lower is the average speed in the network (see, e.g., [116, 142, 175, 299, 323, 358–360, 543, 556] and references there). However, this conclusion as well as other consequences from the MFD cannot be used for the reliable analysis of ITS-applications. Indeed, in the MFD all individual empirical features of traffic breakdown at different network bottlenecks in the urban network have been averaged. Therefore, all complex spatiotemporal traffic phenomena occurring at the network bottlenecks have been totally lost in the MFD.

The complexity of some of empirical spatiotemporal traffic phenomena occurring at network bottlenecks have been shown and proven in Chap. 3. Indeed, from the empirical analysis of traffic breakdown at highway bottlenecks we have seen that to understand traffic flow phenomena in a network, a detailed analysis of spatiotemporal traffic phenomena at each of the network bottlenecks should be made. Only in this case, a valid conclusion about features of the traffic network can be made. As mentioned, in the MFD, all these complex spatiotemporal traffic phenomena are averaged. This means that basic characteristics of the empirical traffic phenomena occurring in an urban network are lost in the MFD. Therefore, no valid conclusions about real traffic phenomena in the network can be found from an analysis of the MFD.

Spatiotemporal traffic phenomena occurring at network bottlenecks are lost in the MFD. However, spatiotemporal traffic phenomena occurring at network bottlenecks determine the basic empirical features of a traffic network, like traffic breakdown at network bottlenecks and the resulting spatiotemporal development of traffic congestion in the network. Therefore, the MFD cannot be used for the reliable analysis of ITS-applications.

A study of the MFD might have a sense about 20–30 years ago, when spatiotemporal features of traffic breakdown at network bottlenecks have not still been understood. However, we do know now a lot of empirical features of spatiotemporal traffic phenomena at network bottlenecks. Because spatiotemporal traffic features that are needed for dynamic traffic assignment, control, and optimization of urban networks are lost in the MFD, the MFD cannot be used for reliable ITS-applications.

Rather than the MFD, empirical features of spatiotemporal traffic phenomena at network bottlenecks should be used for the reliable analysis of ITS-applications. In particular, the nucleation nature of traffic breakdown at a network bottleneck that is the fundamental spatiotemporal traffic phenomenon in traffic networks should be the empirical basis for the development of reliable methods of traffic control and dynamic traffic assignment in traffic networks.

4.11.4 Boomerang Effect, Homogeneous Congested Traffic, and Diagram of Congested Traffic States

An additional example of the invalid methodology related to items (i) and (ii) of Sect. 4.11.1 is an empirical analysis of real field measured data made in [464, 465]. In these works, based on traffic data measured on German highways the authors of [464, 465] have tried to make an empirical proof of the existence of a set of theoretical congested traffic states at a highway bottleneck. This set of congested traffic states results from theoretical diagrams of congested traffic patterns of Helbing, Treiber et al.[213, 216, 464, 465, 506–509, 511]. The diagrams have been derived based on simulations of two-phase traffic flow models [213, 216, 464, 465, 506–509, 511].

For the analysis of the theory of traffic flow resulting from numerical simulations of the two-phase traffic flow models [213, 216, 276–281, 464, 465, 506–509, 511], at least two conclusions of this theory are important to study:

1. The theoretical boomerang effect revealed by Kerner and Konhäuser from a study of a traffic flow model of the GM model class (Fig. 4.7b) [276–278].
2. Theoretical solutions of the traffic flow models of the GM model class called HCT (homogeneous congested traffic) revealed by Helbing, Treiber et al. [213, 506–509] (see Fig. 10.9 of the book [269]).

A proof of the boomerang effect (Fig. 4.7b) in empirical data should show that a direct spontaneous phase transition from free flow (F) to a wide moving jam (J) (F→J transition) might be possible in real field traffic data. An observation of HCT solutions might prove another result of this theory [213, 216, 464, 465, 506–509, 511]: When the density of traffic flow increases continuously, then there should be the transformation of so-called oscillating congested traffic (OCT) into homogeneous congested traffic (HCT). In its turn, a desired proof of the boomerang effect and HCT might prove the diagrams of congested traffic states of Helbing, Treiber et al. [213, 216, 464, 465, 506–509, 511].

It must be stressed that the term "boomerang effect" (Fig. 4.7) has a sense only when traffic breakdown occurs without any influence of a bottleneck. Otherwise, when the bottleneck is the reason for traffic breakdown and the resulting upstream propagation of traffic congestion, then this well-known effect of the occurrence of traffic congestion has nothing to do with the *boomerang* effect. Indeed, in the boomerang effect the transformation of a local disturbance propagating initially downstream in free flow into a wide moving jam propagating upstream should occur on a homogeneous road section without any bottlenecks (Fig. 4.7).

By a study of the "boomerang effect" in empirical data, the authors of [464, 465] have not taken into account that traffic congestion occurs only when a disturbance in free flow reaches an off-ramp bottleneck (this case is shown in Fig. 3.8b). Thus, in real field traffic data studied in [464, 465] traffic breakdown occurs at the off-ramp bottleneck.[11] As explained in Chap. 3, the study of this real field traffic data shows that in the reality traffic congestion has occurred through an interaction of a wave (disturbance) propagating downstream in free flow with the permanent speed disturbance localized at the bottleneck. This real feature of traffic breakdown has nothing to do with the boomerang effect resulting from simulations of traffic flow models of the GM model class (Fig. 4.7b). Thus, the empirical study made in [464, 465] is related to our statement of item (i) of Sect. 4.11.1: Empirical traffic flow phenomena that contradict solutions of a traffic model, which should be "empirically proven", have been ignored in [464, 465].

[11]This can be seen on Fig. 7.7 (b, c) of the book [269] in which the same empirical data as that studied in [464, 465] have been presented.

The empirical proof of the boomerang effect made in [464, 465] is invalid.

An example of our another statement (item (ii)) of Sect. 4.11.1 is the "empirical proof" of theoretical solutions HCT claimed in [464, 465]. The real field traffic data used in [464, 465] has been presented on Fig. 10.8 of the book [269]. This empirical data is strongly non-homogeneous in space and time. The authors of [464, 465] have averaged this real data in space and time. Due to this data averaging, the real data that is strongly non-homogeneous in space and time transforms into "homogeneous in space and time" data. This averaged data should prove the existence of theoretical HCT (for more details, see Sec. 10.3.10 of the book [269]). A similar invalid methodology of the empirical data analysis related to item (ii) of Sect. 4.11.1 has also been applied in [464, 465, 508, 509, 511] to "prove" Helbing's theoretical diagrams of congested traffic states at highway bottlenecks.

The empirical proof of homogeneous congested traffic (HCT) and diagrams of congested states made in [464, 465] is invalid.

4.11.5 Driver Behavioral Assumptions

There is a number of publications and reviews devoted to an analysis of the role of different driver behaviors for traffic flow modeling. In particular, in a recent review by Saifuzzaman and Zheng [454] many traffic flow models are compared with this objective. However, almost all of these well-known models are belong to the same GM model class (Sect. 4.4). In other words, the models that seem to be very different from driver behavioral assumptions exhibits the same classical traffic flow instability of Herman, Gazis, Montroll, Potts, Rothery, and Chandler [93, 170, 171, 233] (Sect. 4.4.1).

Moreover, under appropriate choice of model parameters the different models show the same characteristic features of wide moving jam propagation as well as many other common features of congested traffic patterns. Therefore, the following questions arise to the analysis of the role of different driver behaviors for traffic flow modeling made, for example, in [13, 16, 28–30, 42, 43, 57, 58, 60, 67, 72, 73, 75, 83, 88, 92, 103, 106, 123, 124, 141, 144, 148, 160–163, 165, 168, 172, 173, 177, 178, 184–186, 188, 199, 206–232, 236, 239, 240, 242, 243, 245, 258, 260, 263, 276, 293, 304, 307, 311, 324–329, 333, 348, 349, 351, 353, 361, 362, 365, 373–377, 379, 380, 382, 385, 388, 392, 402, 405, 407, 414, 415, 419–426, 434, 439, 444, 446, 449, 454, 457–460, 464–469, 474, 475, 479–481, 483, 490, 493–495, 497, 499, 500, 502, 504–509, 511, 514, 520, 525, 526, 534–539, 539, 540, 551, 562–564, 572, 573, 575]:

1. If these traffic flow models are really very different, then why are the models exhibit qualitatively the same behavior of traffic flow instability and resulting congested traffic patterns?
2. If all these traffic flow models really incorporate *basic* driver behaviors that determine fundamental traffic flow characteristics, then why can none of these traffic flow models show the empirical fundamental of transportation science—the empirical nucleation nature of traffic breakdown at highway bottleneck? Recall that the nucleation nature of traffic breakdown results from the metastability of free flow with respect to an F→S transition at a highway bottleneck (Chap. 3).

The only one possible answer on these questions is that the driver behaviors incorporated in classical traffic flow theories and models (e.g., [13, 16, 28–30, 42, 43, 57, 58, 60, 67, 72, 73, 75, 83, 88, 92, 103, 106, 123, 124, 130, 131, 141, 144, 148, 160–163, 165, 168, 172, 173, 177, 178, 183–186, 188, 188, 199, 206–232, 236, 239, 240, 242, 243, 245, 258, 260, 263, 276, 293, 304, 307, 311, 324–329, 333, 348, 349, 351, 353, 361, 362, 365, 373–377, 379, 380, 382, 385, 388, 392, 402, 405, 407, 414, 415, 419–426, 434, 439, 444, 446, 449, 454, 457–460, 464–469, 474, 475, 479–481, 483, 490, 493–495, 497, 499, 500, 502, 504–509, 511, 514, 520, 525, 526, 534–539, 539, 540, 551, 562–564, 572, 573, 575]) do *not* include basic empirical driver behaviors that are responsible for the empirical fundamental of transportation science.

Most of the traffic researchers assume that the classical traffic flow models incorporate *basic driver behaviors* in traffic flow. This generally accepted assumption is invalid: None of these traffic flow models can show the empirical fundamental of transportation science—the empirical nucleation nature of traffic breakdown at a highway bottleneck, which results from *real basic driver behaviors* in traffic flow.

The generally accepted methodology to the choice of driver behavioral assumptions for the development of traffic flow models is invalid.

This classical methodology cannot distinguish between real basic driver behaviors, which are responsible for traffic breakdown at network bottlenecks, and ancillary driver behaviors. This can explain the failure of the classical traffic flow models in the description of the real world.

4.12 Conclusions

1. Although there are great achievements of classical traffic flow theories and models in the understanding of many important empirical traffic phenomena, the classical theories and models cannot explain and show the fundamental of transportation science—the nucleation nature of traffic breakdown at highway bottleneck. This explains the failure of applications of the classical theories and models in the real world.
2. In particular, the failure of the LWR-theory, the Daganzo's cell transmission model as well as traffic flow models of the GM model class is explained as follows:

   (i) The LWR-model and the Daganzo's cell transmission model failed because these models cannot show induced traffic breakdown at a highway bottleneck observed in real traffic.

   (ii) Traffic flow models of the GM model class failed because the models cannot show the metastability of free flow with respect to an F→S transition at a highway bottleneck.

3. Therefore, the classical traffic flow models cannot be used for a reliable analysis of the effect of ITS (intelligent transportation systems) on traffic flow as well as for a reliable study of ITS performance in real traffic.
4. This criticism is related to all ITS that affect traffic flow, for example, on-ramp metering (see, e.g., [21, 79, 203, 205, 262, 408–411, 413]), variable speed limit control (see, e.g., [1, 10, 77–81, 86, 94, 203–205, 290, 291]) and many other ITS-applications, like the effect of cooperative driving with the use of vehicle-two-vehicle (V2V) communication on traffic flow characteristics, simulations of the performance of dynamic traffic assignment and optimization in traffic and transportation networks (e.g., [2, 4, 8, 9, 15, 49–56, 76, 90, 91, 98, 101, 102, 104, 105, 110, 137, 138, 153–159, 181, 187, 201, 235, 251, 253, 259, 261, 288, 289, 306, 308–310, 340, 345, 346, 352, 354, 355, 357, 367, 368, 370, 383, 384, 403, 416, 417, 440, 441, 445, 447, 461, 470, 471, 477, 482, 484, 488, 489, 498, 513, 528, 532, 533, 542, 544–550, 555, 556, 559, 560, 568–570, 576]) as well as the analysis of the effect of adaptive cruise control (ACC) and automatic driving vehicles on traffic flow (e.g., [133, 152, 283, 284, 302, 330, 332, 338, 363, 369, 400, 401, 452, 453, 473, 485, 508, 509, 512, 517, 571]).

References

1. M. Abdel-Aty, J. Dilmore, A. Dhindsa, Accid. Anal. Prevent. **38**, 335–345 (2006)
2. K. Abdelghany, D. Valdes, A. Abdelfatah, H.S. Mahmassani, Transp. Res. Rec. **1667**, 67–76 (1999)
3. H. Abouaïssa, M. Fliess, C. Join, Int. J. Contr., doi: 10.1080/00207179.2016.1193223 (2016)
4. L. D'Acierno, M. Gallo, B. Montella, Europ. J. Oper. Res. **217**, 459–469 (2012)
5. M. van Aerde, H. Rakha, INTEGRATION ©Release 2.30 for Windows: User's Guide - Volume I: Fundamental Model Features (M. Van Aerde & Assoc., Blacksburg, 2007)
6. M. van Aerde, H. Rakha, INTEGRATION ©Release 2.30 for Windows: User's Guide - Volume II: Advanced Model Features (M. Van Aerde & Assoc., Blacksburg, 2007)
7. K. Aghabayk, M. Sarvi, W. Young, J. Adv. Transp. **50**, 1432–1446 (2016)
8. S.D. Ahipas, R. Meskarian, T.L. Magnanti, K. Natarajan, Transp. Res. B **81**, 333–354 (2015)
9. T. Akamatzu, T. Nagae, in [13] (2007), pp. 87–110
10. P. Allaby, B. Hellinga, M. Bullock, IEEE Trans. Intell. Transp. Syst. **8**, 671–680 (2006)
11. B.L. Allen, G.F. Newell, Transp. Sci. **10**, 227–242 (1976)
12. B.L. Allen, G.F. Newell, Transp. Sci. **10**, 243–268 (1976)
13. R.E. Allsop, M.G.H. Bell, B.G. Heydecker (eds.), *Transportation and Traffic Theory 2007* (Elsevier Science, Amsterdam, 2007)
14. J. Al-Obaedi, S. Yousif, IEEE Trans. ITS **13**, 296–306 (2012)
15. M. Amirgholy, E.J. Gonzales, Transp. Res. B (2016), doi.org/10.1016/j.trb.2015.11.006
16. C. Appert-Rolland, F. Chevoir, P. Gondret, S. Lassarre, J.-P. Lebacque, M. Schreckenberg (eds.), *Traffic and Granular Flow'07* (Springer, Heidelberg, 2009)
17. B. van Arem, C. van Driel, R. Visser, IEEE Trans. Intell. Transp. Syst. **7**, 429–436 (2006)
18. B. van Arem, J.H. Hogema, S.A. Smulders, in *Proceedings of 3rd World Congress on Intelligent Transport Systems 1996*
19. B. van Arem, A.P. de Vos, M.J. Vanderschuren, "The microscopic traffic simulation model MIXIC 1.3", *TNO-Report, INRO-VVG 1997–02b* (Delft, 1997)
20. M. Ardelt, C. Coester, N. Kaempchen, IEEE Trans. ITS **13**, 1576–1585 (2012)
21. E.D. Arnold, "Ramp Metering: A Review of the Literature" (Virginia Transportation Research Council, 1998)
22. P.J. Athol, A.G.R. Bullen, Highway Res. Rec. **456**, 50–54 (1973)
23. A. Aungsuyanon, D. Boyce, B. Ran, in *Proc. TRB 92nd Annual Meeting*, Paper # 13-1423 (TRB, Washington DC, 2013)
24. A. Aw, M. Rascle, SIAM J. Appl. Math. **60**, 916–938 (2000)
25. H.M.A. Aziz, S.V. Ukkusuri, Transp. Res. Rec. **2311**, 73–84 (2012)
26. C. Backfrieder, G. Ostermayer, C.F. Mecklenbräuker, IEEE Trans. ITS **18**, 349–363 (2017)
27. M. Bahram, Z. Ghandeharioun, P. Zahn, M. Baur, W. Huber, F. Busch, *17th IEEE Intern. Conference on ITS (ITSC) 2014*, pp. 1752–1757
28. M. Bando, K. Hasebe, A. Nakayama, A. Shibata, Y. Sugiyama, Jpn. J. Appl. Math. **11**, 203–223 (1994)
29. M. Bando, K. Hasebe, A. Nakayama, A. Shibata, Y. Sugiyama, Phys. Rev. E **51**, 1035–1042 (1995)
30. M. Bando, K. Hasebe, A. Nakayama, A. Shibata, Y. Sugiyama, J. Phys. I France **5**, 1389–1399 (1995)
31. J.H. Banks, Transp. Res. Rec. **1510**, 1–10 (1995)
32. J.H. Banks, Transp. Res. Rec. **1225**, 53–60 (1989)
33. J.H. Banks, Transp. Res. Rec. **1287**, 20–28 (1990)
34. J.H. Banks, "Evaluation of the Two-Capacity Phenomenon as a Basis for Ramp Metering", *Final Report, Civil Engineering Report Series 9002* (San Diego State University, San Diego, CA, 1990)
35. J.H. Banks, Transp. Res. Rec. **1320**, 234–241 (1991)
36. J.H. Banks, Transp. Res. Rec. **1320**, 83–90 (1991)

37. J.H. Banks, Transp. Res. Rec. **1394**, 17–25 (1993)
38. J.H. Banks, Transp. Res. Rec. **1678**, 128–134 (1999)
39. J.H. Banks, Transp. Res. Rec. **1802**, 225–232 (2002)
40. J. Barceló, J. Casas, Oper. Res. Comp. Sci. Interf. Ser. **31**, 57–98 (2005)
41. J. Barceló, E. Codina, J. Casas, J.L. Ferrer, D. Garcia, J. Intel. Robotic Sys. **41**, 173–203 (2005)
42. R. Barlović, T. Huisinga, A. Schadschneider, M. Schreckenberg, Phys. Rev. E **66**, 046113 (2002)
43. R. Barlović, L. Santen, A. Schadschneider, M. Schreckenberg, Eur. Phys. J. B **5**, 793–800 (1998)
44. L.D. Baskar, B. De Schutter, H. Hellendoorn, IEEE Trans. ITS **13**, 838–847 (2012)
45. S. Becker, M. Bork, H.T. Dorissen, G. Geduld, O. Hofmann, K. Naab, G. Nöcker, P. Rieth, J. Sonntag, in *Proceedings of the 1st World Congress on Applications of Transport Telematics and Intelligent Vehicle-Highway Systems* (1994), pp. 1828–1835
46. S. Becker, M. Bork, H.T. Dorissen, G. Geduld, O. Hofmann, K. Naab, G. Nöcker, P. Rieth, J. Sonntag, in *Proceedings of the 1st World Congress on Applications of Transport Telematics and Intelligent Vehicle-Highway Systems* (1994), pp. 1836–1843
47. N. Bekiaris-Liberis, C. Roncoli, M. Papageorgiou, IEEE Trans. ITS **17**, 3484–3497 (2016)
48. M. Behrisch, L. Bieker, J. Erdmann, D. Krajzewicz, "SUMO - Simulation of Urban MObility: An Overview", in *SIMUL 2011, The Third Inter. Conf. on Adv. in Sys. Simulation* (2011)
49. M.G.H. Bell, Transp. Res. B **29**, 125–137 (1995)
50. M.G.H. Bell, Transp. Res. B **29**, 287–295 (1995)
51. M.G.H. Bell, Transp. Res. B **34**, 533–545 (2000)
52. M.G.H. Bell, Ch. Cassir, in *2nd Int. Conf. on Traffic and Transportation Studies (ICTTS), Beijing, China*, ed. by K.C. Wang, G. Xiao, J. Ji. (American Society of Civil Engineers, USA, 2000), pp. 9–16
53. M.G.H. Bell, Ch. Cassir, *Reliability of Transport Networks* (Research Studies Press, Baldock, Hertfordshire, 2000)
54. M.G.H. Bell, Ch. Cassir, Transp. Res. B **36**, 671–681 (2002)
55. M.G.H. Bell, Y. Iida, *Transportation Network Analysis* (Wiley, Hoboken, 1997)
56. M.G.H. Bell, C.M. Shielda, F. Busch, G. Kruse, Transp. Res. C **5**, 197–210 (1997)
57. N. Bellomo, V. Coscia, M. Delitala, Math. Mod. Meth. App. Sc. **12**, 1801–1843 (2002)
58. P. Berg, A. Mason, A.W. Woods, Phys. Rev. E **61**, 1056–1066 (2000)
59. V.A.C. van den Berg, E.T. Verhoef, Transp. Res. B **94**, 43–60 (2016)
60. P. Berg, A.W. Woods, Phys. Rev. E **64**, 035602(R) (2001)
61. G.N. Bifulco, L. Pariota, F. Simonelli, R.D. Pace, Transp. Res. C **29**, 156–170 (2013)
62. L. Bloomberg, J. Dale, Transp. Res. Rec. **1727**, 52–60 (2000)
63. R. Boel, L. Mihaylova, Transp. Res. B **40**, 319–334 (2006)
64. K. Bogenberger, A.D. May, *Working Paper UCB-ITS-PWP-99-19, California Partners for Advanced Transit and Highways (PATH)* (Institute of Transportation Studies (UCB), UC Berkeley, 1999)
65. A. Bose, P. Ioannou, Transp. Res. C **11**, 439 (2003)
66. S.D. Boyles, in *Proc. TRB 92nd Annual Meeting*, Paper # 13-4519 (TRB, Washington DC, 2013)
67. M. Brackstone, M. McDonald, Transp. Res. F **2**, 181 (1998)
68. M. Brackstone, P. Sykes, P. Vortisch, Traffic Eng. Contr. **55**, 165–171 (2013)
69. W. Brilon, J. Geistefeld, H. Zurlinden, Transp. Res. Rec. **2027**, 1–8 (2007)
70. W. Brilon, M. Regler, J. Geistefeld, Straßenverkehrstechnik, Heft **3**, 136 (2005)
71. W. Brilon, H. Zurlinden, Straßenverkehrstechnik, Heft **4**, 164 (2004)
72. E. Brockfeld, R.D. Kühne, A. Skabardonis, P. Wagner, Trans. Res. Rec. **1852**, 124–129 (2003)
73. E. Brockfeld, R.D. Kühne, P. Wagner, in *Proceeding of the Transportation Research Board 84th Annual Meeting*, TRB Paper No. 05-2152 (TRB, Washington, DC, 2005)
74. S.C. Calvert, H. Taale, S.P. Hoogendoorn, J. Adv. Transp. **50**, 570–588 (2016)
75. J.L. Cao, Z.K. Shi, Int. J. Mod. Phys. C **26**, 1550121 (2015)

76. M. Carey, P. Humphreys, M. McHugh, R. McIvor, Transp. Res. B **65**, 90–104 (2014)
77. R.C. Carlson, I. Papamichail, M. Papageorgiou, IEEE Trans. Intell. Transp. Syst. **12**, 1261–1276 (2011)
78. R.C. Carlson, I. Papamichail, M. Papageorgiou, J. Intell. Transport. Syst. **17**, 268–281 (2013)
79. R.C. Carlson, I. Papamichail, M. Papageorgiou, Transp. Res. C **46**, 209–221 (2014)
80. R.C. Carlson, I. Papamichail, M. Papageorgiou, A. Messmer, Transp. Sci. **44**, 238–253 (2010)
81. R.C. Carlson, I. Papamichail, M. Papageorgiou, A. Messmer, Transp. Res. C **18**, 193–212 (2010)
82. E. Cascetta, *Transportation Systems Engineering: Theory and Methods* (Kluwer Academic, Dordrecht, 2001)
83. M.J. Cassidy, A. Skabardonis (eds.), *Papers selected for the 19th Int. Sym. on Transp. and Traffic Theory*. Procedia Soc. Behav. Sci. **17**, 1–716 (2011)
84. M.J. Cassidy, A. Skabardonis, A.D. May, Transp. Res. Rec. **1225**, 61–72 (1989)
85. M.J. Cassidy, J.R. Windover, Transp. Res. Rec. **1484**, 73–79 (1995)
86. A.G. Castro, A. Monzon, in *Proc. of the 13th International Conference on Reliability and Statistics in Transportation and Communication (RelStat'13)*, Riga, Latvia, October 2013, pp. 117–127
87. A. Ceder, *Investigation of Two-Regime Traffic Flow Models at the Micro- and Macroscopic Levels*, (Ph.D. Thesis, UC, Berkeley University, Berkeley, CA, 1975)
88. A. Ceder (ed.), *Transportation and Traffic Theory. Proceedings of the 14th International Symposium on Transportation and Traffic Theory* (Elsevier Science, Oxford, 1999)
89. A. Ceder, A.D. May, Transp. Res. Rec. **567**, 1–15 (1976)
90. H. Ceylan, M.G.H. Bell, J. Adv. Transp. **38**, 291–321 (2004)
91. H. Ceylan, M.G.H. Bell, Transp. Res. B **39**, 169–185 (2005)
92. E. Chamberlayne, H. Rakha, D. Bish, Transp. Lett. **4**, 227–242 (2012)
93. R.E. Chandler, R. Herman, E.W. Montroll, Oper. Res. **6**, 165–184 (1958)
94. D. Chen, S. Ahn, A. Hegyi, Transp. Res. B **70**, 340–358 (2014)
95. L.-W. Chen, T.-Y. Hu, IEEE Trans. ITS **13**, 1266–1276 (2012)
96. X. Chen, L. Li, Z. Li, IEEE Trans. ITS **13**, 1705–1716 (2012)
97. Q. Chen, Y. Wang, J. Adv. Transp. **50**, 949–966 (2016)
98. A. Chen, Z. Zhou, W.H.K. Lam, Transp. Res. B **45**, 1619–1640 (2011)
99. C.C. Chien, Y. Zhang, P.A. Ioannou, Automatica **33**, 1273–1285 (1997)
100. H.C. Chin, A.D. May, Transp. Res. Rec. **1320**, 75–82 (1991)
101. Y.-C. Chiu, H.S. Mahmassani, Transp. Res. Rec. **1783**, 89–97 (2002)
102. A.H.F. Chow, in [13] (2007), pp. 301–326
103. D. Chowdhury, L. Santen, A. Schadschneider, Phys. Rep. **329**, 199 (2000)
104. B.D. Chung, T. Yao, B. Zhang, Netw. Spat. Econ. **12**, 167–181(2012)
105. E. Cipriani, S. Gori, M. Petrelli, Transp. Res. C **20**, 3–14 (2012)
106. M. Cremer, *Der Verkehrsfluss auf Schnellstrassen* (Springer, Berlin, 1979)
107. C.F. Daganzo, Transp. Res. B **28**, 269–287 (1994)
108. C.F. Daganzo, Transp. Res. B **29**, 79–93 (1995)
109. C.F. Daganzo, *Fundamentals of Transportation and Traffic Operations* (Elsevier Science, New York, 1997)
110. C.F. Daganzo, Transp. Sci. **32**, 3–11 (1998)
111. C.F. Daganzo, Transp. Res. B **36**, 131–158 (2002)
112. C.F. Daganzo, Transp. Res. B **36**, 159–169 (2002)
113. C.F. Daganzo, Transp. Res. B **39**, 934–950 (2005)
114. C.F. Daganzo, Transp. Res. B **39**, 187–196 (2005)
115. C.F. Daganzo, Transp. Res. B **40**, 396–403 (2006)
116. C.F. Daganzo, Transp. Res. B **41**, 49–62 (2007)
117. C.F. Daganzo, Transp. Res. B **45**, 782–788 (2011)
118. C.F. Daganzo, Transp. Res. B **69**, 50–59 (2014)
119. C.F. Daganzo, N. Geroliminis, Transp. Res. B **42**, 771–781 (2008)

120. C.F. Daganzo, J.A. Laval, "On the Numerical Treatment of Moving Bottlenecks". *Report UCB-ITS-RR-93-7* (Institute of Transportation Studies, University of California, Berkeley, 2003)
121. C.F. Daganzo, L.J. Lehe, Transp. Res. B **90**, 56–69 (2016)
122. C.F. Daganzo, Y. Sheffi, Transp. Sc. **11**, 253–274 (1977)
123. L.C. Davis, Phys. Rev. E **66**, 038101 (2002)
124. L.C. Davis, Physica A **319**, 557 (2003)
125. L.C. Davis, Phys. Rev. E. **69**, 066110 (2004)
126. L.C. Davis, Physica A **368**, 541–550 (2006)
127. L.C. Davis, Physica A **379**, 274–290 (2007)
128. L.C. Davis, Physica A **388**, 4459 (2009)
129. L.C. Davis, Physica A **389**, 3588 (2010)
130. L.C. Davis, Physica A **390**, 943–950 (2011)
131. L.C. Davis, Physica A **390**, 3624 (2011)
132. L.C. Davis, Physica A **405**, 128–139 (2014)
133. A.I. Delis, I.K. Nikolos, M. Papageorgiou, Comp. Math. Appl. **70**, 1921–1947 (2015)
134. C. Demir, in [161], pp. 305–317
135. C. Demir, M. Cremer, S. Donikian, S. Espie, in *Proceedings of 5th World Congress on Intelligent Transport Systems* (Seoul, 1998)
136. V. Dixit, L.M. Gardner, S.T. Waller, in *Proc. TRB 92nd Annual Meeting*, Paper # 13-5103 (TRB, Washington DC, 2013)
137. K. Doan, S.V. Ukkusuri, Transp. Res. B **46**, 1218–1238 (2012)
138. K. Doan, S.V. Ukkusuri, Transp. Res. C **51**, 41–65 (2015)
139. S. Donikian, S. Espie, M. Parent, G. Rousseau, in *Proceedings of 5th World Congress on Intelligent Transport Systems* (Seoul, 1998)
140. J. Drake, J. Shofer, A.D. May, Highway Res. Rec. **154**, 53–87 (1967)
141. D. Drew, *Traffic Flow Theory and Control* (McGraw Hill, New York, 1968)
142. J. Du, H. Rakha, V.V. Gayah, Transp. Res. C **66**, 136–149 (2016)
143. A. Ehlert, M.G.H. Bell, S. Grosso, Transp. Res. B **40**, 460–479 (2006)
144. B. Eisenblätter, L. Santen, A. Schadschneider, M. Schreckenberg, Phys. Rev. E **57**, 1309 (1998)
145. L. Elefteriadou, *An Introduction to Traffic Flow Theory*, Springer Optimization and Its Applications, vol. 84 (Springer, Berlin, 2014)
146. L. Elefteriadou, A. Kondyli, W. Brilon, F.L. Hall, B. Persaud, S. Washburn, J. Transp. Eng. **140**, 04014003 (2014)
147. L. Elefteriadou, R.P. Roess, W.R. McShane, Transp. Res. Rec. **1484**, 80–89 (1995)
148. J. Esser, M. Schreckenberg, Int. J. Mod. Phys. C **8**, 1025 (1997)
149. K. Fadhloun, H. Rakha, A. Loulizi, Transp. Lett. **6**, 185–196 (2014)
150. K. Fadhloun, H. Rakha, A. Loulizi, Transp. Res. Rec. **2422**, 61–70 (2014)
151. M. Fellendorf, in *Traffic Technology International'96* (UK & International, Dorking, England, 1996)
152. B. Friedrich, in *Autonomes Fahren*, ed. by M. Maurer, J.Ch. Gerdes, B. Lenz, H. Winner (Springer, Berlin, 2015), pp. 331–350
153. T.L. Friesz, Transp. Res. A **190**, 413–427 (1985)
154. T.L. Friesz, D. Bernstein, N.J. Mehta, R.L. Tobin, S. Ganjalizadeh, Oper. Res. **42**, 1120–1136 (1994)
155. T.L. Friesz, D. Bernstein, T.E. Smith, R.L. Tobin, B.W. Wie, Oper. Res. **41**, 179–191 (1993)
156. T.L. Friesz, H.-J. Cho, N.J. Mehta, R.L. Tobin, G. Anandalingam, Transp. Sci. **26**, 18–26 (1992)
157. T.L. Friesz, K. Han, P.A. Neto, A. Meimand, T. Yao, Transp. Res. B **47**, 102–126 (2013)
158. T.L. Friesz, J. Luque, R.L. Tobin, B.-W. Wie, Oper. Res. **37**, 893–901 (1989)
159. T.L. Friesz, R.L. Tobin, T.E. Smith, P.T. Harker, J. of Regional Sci. **23**, 337–359 (1983)
160. H.-T. Fritzsche, Transp. Eng. Contrib. **5**, 317 (1994)

161. M. Fukui, Y. Sugiyama, M. Schreckenberg, D.E. Wolf (eds.), *Traffic and Granular Flow' 01* (Springer, Heidelberg, 2003)
162. Z.-Y. Gao, K.-P. Li, X.-G. Li, H.-J. Huang, B.-H. Mao, J.-F. Zheng, Physica A **380**, 577–584 (2007)
163. N.H. Gartner, C.J. Messer, A.K. Rathi (eds.), *Traffic Flow Theory: A State-of-the-Art Report* (Transportation Research Board, Washington DC, 2001)
164. N.H. Gartner, Ch. Stamatiadis, in *Encyclopedia of Complexity and System Science*, ed. by R.A. Meyers (Springer, Berlin, 2009), pp. 9470–9500
165. N.H. Gartner, N.H.M. Wilson (eds.) *Transportation and Traffic Theory* (Elsevier, New York, 1987)
166. V.V. Gayah, C. Daganzo, Transp. Res. B **45**, 643–655 (2011)
167. D.C. Gazis, *Traffic Science* (Wiley, New York, 1974)
168. D.C. Gazis, *Traffic Theory* (Springer, Berlin, 2002)
169. D.C. Gazis, R. Herman, Trans. Sci. **26**, 223 (1992)
170. D.C. Gazis, R. Herman, R.B. Potts, Oper. Res. **7**, 499–505 (1959)
171. D.C. Gazis, R. Herman, R.W. Rothery, Oper. Res. **9**, 545–567 (1961)
172. H.X. Ge, R.J. Cheng, Physica A **387**, 6952–6958 (2008)
173. H.-X. Ge, X.-P. Meng, R.-J. Cheng, S.-M. Lo, Physica A **390**, 3348–3353 (2011)
174. J. Geistefeldt, W. Brilon, in [311] (2009), pp. 583–602
175. N. Geroliminis, C.F. Daganzo, Transp. Res. B **42**, 759–770 (2008)
176. N. Geroliminis, J. Haddad, M. Ramezani, IEEE Trans. ITS **14**, 348–359 (2013)
177. P.G. Gipps, Trans. Res. B **15**, 105–111 (1981)
178. P.G. Gipps, Trans. Res. B. **20**, 403–414 (1986)
179. G. Gomes, R. Horowitz, Transp. Res. C **14**, 244–262 (2006)
180. G. Gomes, A. May, R. Horowitz, Transp. Res. Rec. **1876**, 71–81 (2004)
181. E.J. Gonzales, C.F. Daganzo, Trans. Res. B. **46**, 1519–1534 (2012)
182. M. Guériau, R. Billot, N.-E. El Faouzi, J. Monteil, F. Armetta, S. Hassas, Transp. Res. C **67**, 266–279 (2016)
183. A.K. Gupta, V.K. Katiyar, Physica A **371**, 674–682 (2006)
184. A.K. Gupta, P. Redhu, Physica A **392**, 5622–5632 (2013)
185. A.K. Gupta, P. Redhu, Nonlinear Dyn. **76**, 1001–1011 (2014)
186. A.K. Gupta, S. Sharma, P. Redhu, Nonlinear Dyn. **80**, 1091–1108 (2015)
187. J. Haddad, M. Ramezani, N. Geroliminis, in *Proc. of the 91th TRB Annual Meeting* (TRB, Washington, D.C, USA, 2012)
188. F.A. Haight, *Mathematical Theories of Traffic Flow* (Academic Press, New York, 1963)
189. F.L. Hall, Transp. Res. A **21**, 191–201 (1987)
190. F.L. Hall, in [163] (2001), pp. 2-1–2-36
191. F.L. Hall, K. Agyemang-Duah, Transp. Res. Rec. **1320**, 91–98 (1991)
192. F.L. Hall, B.L. Allen, M.A. Gunter, Trans. Res. A **20**, 197–210 (1986)
193. F.L. Hall, D. Barrow, Transp. Res. Rec. **1194**, 55–65 (1988)
194. F.L. Hall, W. Brilon, Transp. Res. Rec. **1457**, 35–42 (1994)
195. F.L. Hall, M.A. Gunter, Trans. Res. Rec. **1091**, 1–9 (1986)
196. F.L. Hall, L.M. Hall, Transp. Res. Rec. **1287**, 108–118 (1990)
197. F.L. Hall, V.F. Hurdle, J.H. Banks, Transp. Res. Rec. **1365**, 12–18 (1992)
198. F.L. Hall, A. Pushkar, Y. Shi, Transp. Res. Rec. **1398**, 24–30 (1993)
199. S.H. Hamdar, H.S. Mahmassani, M. Treiber, Transp. Res. B. **78**, 32–53 (2015)
200. Y. Han, D. Chen, S. Ahn, Transp. Res. B **98**, 113–134 (2017)
201. K. Han, T.L. Friesz, T. Yao, Transp. Res. B **53**, 17–30 (2013)
202. A. Hegyi, T. Bellemans, B. De Schutter, in *Encyclopedia of Complexity and System Science*, ed. by R.A. Meyers (Springer, Berlin, 2009), pp. 3943–3964
203. A. Hegyi, B. De Schutter, J. Hellendoorn, Transp. Res. Rec. **1852**, 167–174 (2004)
204. A. Hegyi, B. De Schutter, J. Hellendoorn, IEEE Trans. Intell. Transp. Syst. **6**, 102–112 (2005)
205. A. Hegyi, B. De Schutter, H. Hellendoorn, Transp. Res. C **13**, 185–209 (2005)
206. D. Helbing, Phys. Rev. E **51**, 3164–3169 (1995)

207. D. Helbing, Phys. Rev. E **53**, 2366–2381 (1996)
208. D. Helbing, Physica A **233**, 253–282 (1996)
209. D. Helbing, Phys. Rev. E **55**, R25–R28 (1997)
210. D. Helbing, Phys. Rev. E **55**, 3735–3738 (1997)
211. D. Helbing, Physica A **242**, 175–194 (1997)
212. D. Helbing, Phys. Rev. E **57**, 6176–6179 (1998)
213. D. Helbing, Rev. Mod. Phys. **73**, 1067–1141 (2001)
214. D. Helbing, J. Phys. A Math. Gen. **36**, L593–L598 (2003)
215. D. Helbing, "Traffic flow", in *Encyclopedia of Nonlinear Science*, ed. by A. Scott (Routledge, New York, 2005)
216. D. Helbing, Eur. Phys. J. B **69**, 539–548 (2009)
217. D. Helbing, Eur. Phys. J. B **69**, 569–570 (2009)
218. D. Helbing, Eur. Phys. J. B **70**, 257–274 (2009)
219. D. Helbing, A. Greiner, Phys. Rev. E **55**, 5498–5507 (1997)
220. D. Helbing, A. Hennecke, V. Shvetsov, M. Treiber, Transp. Res. B **35**, 183–211 (2001)
221. D. Helbing, A. Hennecke, V. Shvetsov, M. Treiber, Math. Comp. Mod. **35**, 517–547 (2002)
222. D. Helbing, A. Hennecke, M. Treiber, Phys. Rev. Lett. **82**, 4360–4363 (1999)
223. D. Helbing, H.J. Herrmann, M. Schreckenberg, D.E. Wolf (eds.), *Traffic and Granular Flow' 99* (Springer, Heidelberg, 2000)
224. D. Helbing, K. Nagel, Contemp. Phys. **45**, 405–426 (2004)
225. D. Helbing, M. Schreckenberg, Phys. Rev. E **59**, R2505 (1999)
226. D. Helbing B. Tilch, Phys. Rev. E **58**, 133–138 (1998)
227. D. Helbing, B. Tilch, Eur. Phys. J. B **68**, 577–586 (2009)
228. D. Helbing, M. Treiber, Phys. Rev. Lett. **81**, 3042–3045 (1998)
229. D. Helbing, M. Treiber, Science **282**, 2001–2003 (1998)
230. D. Helbing, M. Treiber, A. Kesting, Physica A **363**, 62–72 (2006)
231. D. Helbing, M. Treiber, A. Kesting, M. Schönhof, Eur. Phys. J. B **69**, 583–598 (2009)
232. R. Herman, S. Ardekani, Transp. Sci. **18**, 101–140 (1984)
233. R. Herman, E.W. Montroll, R.B. Potts, R.W. Rothery, Oper. Res. **7**, 86–106 (1959)
234. M. Herrmann, B.S. Kerner, Physica A **255**, 163–188 (1998)
235. B.G. Heydecker, J.D. Addison, Transp. Sci. **39**, 39–57 (2005)
236. Highway Capacity Manual 2000, (National Research Council, Transportation Research Board, Washington, D.C., 2000)
237. Highway Capacity Manual 2010, (National Research Council, Transportation Research Board, Washington, D.C., 2010)
238. A. Hiller, in *Tagungsband des Workshops Fahrerassistenzsysteme FAS2003*, Hrsg.: C. Stiller und M. Maurer (Universität Karlsruhe, Germany, 2003), S. 31–34 (ISBN 3-9809121-0-8)
239. M. Hilliges, W. Weidlich, Trans. Res. B **29**, 407 (1995)
240. Y. Hino, K. Tobita, T. Nagatani, Physica A **392**, 3223–3230 (2013)
241. J.H. Hogema, W.H. Janssen, in *Proceedings of 3rd World Congress on Intelligent Transport Systems* (1996)
242. S.P. Hoogendoorn, *Multiclass Continuum Modelling of Multilane Traffic Flow* (Delft University of Technology, Delft, 1999)
243. S.P. Hoogendoorn, R. Hoogendoorn, Phil. Trans. R. Soc. A **368**, 4497–4517 (2010)
244. S.P. Hoogendoorn, V.L. Knoop, H.J. Van Zuylen, J. Adv. Transp. **42**, 357–377 (2008)
245. S.P. Hoogendoorn, S. Luding, P.H.L. Bovy, M. Schreckenberg, D.E. Wolf (eds.), *Traffic and Granular Flow' 03* (Springer, Heidelberg, 2005)
246. Z. Hou, X. Xu, J. Yan, J.-X. Xu, G. Xiong, IEEE Trans. ITS **12**, 1305–1318 (2011)
247. Z. Hou, J. Yan, J.-X. Xu, Z. Li, IEEE Trans. ITS **13**, 606–618 (2012)
248. P. Hsu, J.H. Banks, Transp. Res. Rec. **1398**, 17–23 (1993)
249. T.-Y. Hu, C.-C. Tong, T.-Y. Liao, W.-M. Ho, IEEE Trans. ITS **13**, 1277–1286 (2012)
250. P.B. Hunt, D.I. Robertson, R.D. Bretherton, M.C. Royle, Traffic Eng. Contr. **23**, 190–192 (1982)
251. N. Huynh, H.S. Mahmassani, H. Tavana, Transp. Res. Rec. **1783**, 55–65 (2002)

252. P.A. Ioannou, C.C. Chien, IEEE Trans. Veh. Technol. **42**, 657–672 (1993)
253. T. Iryo, Transp. Res. B **45**, 867–879 (2011)
254. ISO 15622, 2010. Intelligent transport systems - Adaptive Cruise Control systems - Performance requirements and test procedures
255. B. Jäggi, S. Hohmann, K.W. Axhausen, J. Geistefeldt, Transp. Res. Rec. **2161**, 16–24 (2014)
256. R. Jayakrishnan, H.S. Mahmassani, T.-Y. Hu, Transp. Res. C **2**, 129–147 (1994)
257. K. Jerath, S.N. Brennan, IEEE Trans. ITS **13**, 1782–1791 (2012)
258. R. Jiang, Q.S. Wu, Z.J. Zhu, Phys. Rev. E **64**, 017101 (2001)
259. W.-L. Jin, Transp. Res. B **46**, 1360–1373 (2012)
260. S. Jin, D.-H. Wang, Z.-Y. Huang, P.-F. Tao, Physica A **390**, 1931–1940 (2011)
261. G. Kalafatos, S. Peeta, in [311] (2009), pp. 541–558
262. Y. Kan, Y. Wang, M. Papageorgiou, I. Papamichail, Transp. Res. C **62**, 149–170 (2016)
263. O. Kaumann, K. Froese, R. Chrobok, J. Wahle, L. Neubert, M. Schreckenberg, in [223] (2000), pp. 351–356
264. B.S. Kerner, in *Transportation Systems 1997*, ed. by M. Papageorgiou, A. Pouliezos. (Elsevier Science, London, 1997), pp. 765–770
265. B.S. Kerner, Phys. Rev. Lett. **81**, 3797–3800 (1998)
266. B.S. Kerner, in *Proceedings of the 3rd Symposium on Highway Capacity and Level of Service*, ed. by R. Rysgaard, vol. 2 (Road Directorate, Ministry of Transport, Denmark, 1998), pp. 621–642
267. B.S. Kerner, Trans. Res. Rec. **1678**, 160–167 (1999)
268. B.S. Kerner, *The Physics of Traffic* (Springer, Berlin, New York, 2004)
269. B.S. Kerner, *Introduction to Modern Traffic Flow Theory and Control* (Springer, Heidelberg, Dordrecht, London, New York, 2009)
270. B.S. Kerner, Physica A **392**, 5261–5282 (2013)
271. B.S. Kerner, in *Vehicular Communications and Networks*, ed. by W. Chen (Woodhead Publishings, Cambridge, 2015), pp. 223–254
272. B.S. Kerner, Elektrotechnik Informationstechnik **132**, 417–433 (2015)
273. B.S. Kerner, S.L. Klenov, J. Phys. A Math. Gen. **39**, 1775–1809 (2006)
274. B.S. Kerner, S.L. Klenov, P. Konhäuser, Phys. Rev. E. **56**, 4200–4216 (1997)
275. B.S. Kerner, P. Konhäuser, Phys. Rev. E **48**, 2335–2338 (1993)
276. B.S. Kerner, P. Konhäuser, Phys. Rev. E **50**, 54–83 (1994)
277. B.S. Kerner, P. Konhäuser, bild der wissenschaft, Heft **11**, 86–89 (1994)
278. B.S. Kerner, P. Konhäuser, M. Schilke, Phys. Rev. E **51**, 6243–6246 (1995)
279. B.S. Kerner, P. Konhäuser, M. Schilke, in *Modeling Transport Systems*, vol.2, ed. by D. Hensher, J. King, T. Oum (Elsevier Science, Oxford, 1996), pp. 167–182
280. B.S. Kerner, P. Konhäuser, M. Schilke, in *Transportation and Traffic Theory*, ed. by J.-B. Lesort (Elsevier Science, Oxford, 1996), pp. 119–145
281. B.S. Kerner, P. Konhäuser, M. Schilke, Phys. Lett. A. **215**, 45–56 (1996)
282. B.S. Kerner, H. Rehborn, Phys. Rev. E **53**, R1297–R1300 (1996)
283. A. Kesting, "Microscopic Modeling of Human and Automated Driving: Towards Traffic-Adaptive Cruise Control", *Doctoral Thesis* (Technical University of Dresden, Germany, 2008)
284. A. Kesting, M. Treiber, D. Helbing, Phil. Trans. R. Soc. A **368**, 4585–4605 (2010)
285. A. Kesting, M. Treiber, M. Schönhof, D. Helbing, Transp. Res. C **16**, 668–683 (2008)
286. A. Kesting, M. Treiber, M. Schönhof, D. Helbing, Transp. Res. Rec. **2000**, 16–24 (2007)
287. A. Kesting, M. Treiber, M. Schönhof, F. Kranke, D. Helbing, in [459] (2007), pp. 633–643
288. M. Keyvan-Ekbatani, A. Kouvelas, I. Papamichail, M. Papageorgiou, Transp. Res. B **46**, 1393–1403 (2012)
289. M. Keyvan-Ekbatani, M. Papageorgiou, I. Papamichail, in *Proc. of TRB 2013 Annual Meeting* (TRB, Washington, D.C., 2013)
290. B. Khondaker, L. Kattan, Transp. Lett. **7**, 264–278 (2015)

291. B. Khondaker, L. Kattan, Transp. Res. C **58**, 146–159 (2015)
292. R. Kianfar, B. Augusto, A. Ebadighajari, U. Hakeem, J. Nilsson, A. Raza, R.S. Tabar, N.V. Irukulapati, C. Englund, P. Falcone, S. Papanastasiou, L. Svensson, H. Wymeersch, IEEE Trans. ITS **13**, 994–1007 (2012)
293. W. Knospe, L. Santen, A. Schadschneider, M. Schreckenberg, Phys. Rev. E **70**, 016115 (2004)
294. E. Kometani, T. Sasaki, J. Oper. Res. Soc. Jpn. **2**, 11 (1958)
295. E. Kometani, T. Sasaki, Oper. Res. **7**, 704–720 (1959)
296. E. Kometani, T. Sasaki, in *Theory of Traffic Flow*, ed. by R. Herman (Elsevier, Amsterdam, 1961), pp. 105
297. A. Kondyli, L. Elefteriadou, W. Brilon, F.L. Hall, B. Persaud, S. Washburn, J. Transp. Eng. **139**, 931–940 (2013)
298. M. Koshi, M. Iwasaki, I. Ohkura, in *Proc. 8th International Symposium on Transportation and Traffic Theory*, ed. by V.F. Hurdle. (University of Toronto Press, Toronto, Ontario, 1983), pp. 403
299. V.L. Knoop, H. van Lint, S.P. Hoogendoorn, Physica A **438**, 236–250 (2015)
300. S. Krauß, PhD thesis, DRL-Forschungsbericht 98-08 (1998). http://www.zaik.de/~paper
301. S. Krauß, P. Wagner, C. Gawron, Phys. Rev. E **55**, 5597–5602 (1997)
302. I.A. Ktousakis, I.K. Nokolos, M. Papageorgiou, Transp. Res. Procedia **9**, 111–127 (2015)
303. T.S. Kuhn, *The Structure of Scientific Revolutions*, 4th edn. (The University of Chicago Press, Chicago, London, 2012)
304. R.D. Kühne, in *Procs. of the 9th International Symposium on Transportation and Traffic Theory*, ed. by J. Volmer, R. Hammerslag (VNU Scientific Press, Utrecht, The Netherlands, 1984), pp. 21–42
305. S. Kukuchi, N. Uno, M. Tanaka, J. Transp. Eng. **129**, 146 (2003)
306. A. Kumar, S. Peeta, Y. Nie, Transp. Res. Rec. **2283**, 131–142 (2012)
307. D.A. Kurtze, D.C. Hong, Phys. Rev. E **52**, 218–221 (1995)
308. A.A. Kurzhanskiy, P. Varaiya, Phil. Trans. R. Soc. A **368**, 4607–4626 (2010)
309. W.H.K. Lam,K.S. Chan, Z.-C. Li, M.G.H. Bell, J. Adv. Transp. **44**, 219–230 (2010)
310. W.H.K. Lam, H.K. Lo, S.C. Wong, Transp. Res. B **66**, 1–3 (2014)
311. W.H.K. Lam, S.C. Wong, H.K. Lo (eds.), *Transportation and Traffic Theory 2009* (Springer, Dordrecht, Heidelberg, London, New York, 2009)
312. J.A. Laval, Transp. Res. B **40**, 937–944 (2006)
313. J.A. Laval, Transp. Res. Rec. **1988**, 86–91 (2006)
314. J.A. Laval, Transp. Res. B **43**, 720–728 (2009)
315. J.A. Laval, F. Castrillón, Transp. Res. B **81**, 904–916 (2015)
316. J.A. Laval, B.R. Chilukuri, Transp. Sci. **48**, 217–224 (2013)
317. J.A. Laval, B.R. Chilukuri, Transp. Res. B **89**, 168–177 (2016)
318. J.A. Laval, C.F. Daganzo, Transp. Res. B **40**, 251–264 (2006)
319. J.A. Laval, L. Leclercq, Transp. Res. B **42**, 511–522 (2008)
320. J.A. Laval, L. Leclercq, Transp. Res. B **52**, 17–30 (2013)
321. J.P. Lebacque, J.B. Lesort, F. Giorgi, Transp. Res. Rec. **1644**, 70–79 (1998)
322. L. Leclercq, S. Chanut, J.B. Lesort, Transp. Res. Rec. **1883**, 3–13 (2004)
323. L. Leclercq, N. Chiabaut, B. Trinquier, Transp. Res. B **62**, 1–12 (2014)
324. H.Y. Lee, H.-W. Lee, D. Kim, Phys. Rev. Lett. **81**, 1130 (1998)
325. H.Y. Lee, H.-W. Lee, D. Kim, Phys. Rev. E **59**, 5101–5111 (1999)
326. H.Y. Lee, H.-W. Lee, D. Kim, Physica A **281**, 78 (2000)
327. H.Y. Lee, H.-W. Lee, D. Kim, Phys. Rev. E **62**, 4737 (2000)
328. J.-B. Lesort (ed.), *Transportation and Traffic Theory. Proceedings of the 13th International Symposium on Transportation and Traffic Theory* (Elsevier Science, Oxford, 1996)
329. W. Leutzbach, *Introduction to the Theory of Traffic Flow* (Springer, Berlin, 1988)
330. M.W. Levin, S.D. Boyles, Transp. Res. C **62**, 103–116 (2016)
331. W.Levine, M. Athans, IEEE Trans. Automat. Contr. **11**, 355–361 (1966)
332. P.Y. Li, A. Shrivastava, Transp. Res. C **10**, 275–301 (2002)
333. Q.-L. Li, B.-H. Wang, M.-R. Liu, Physica A **390**, 1356–1362 (2011)

334. C.-Y. Liang, H. Peng, Veh. Syst. Dyn. **32**, 313–330 (1999)
335. C.-Y. Liang, H. Peng, JSME Int. J. Ser. C **43**, 671–677 (2000)
336. M.J. Lighthill, G.B. Whitham, Proc. Roy. Soc. A **229**, 281–345 (1955)
337. W.H. Lin, C.F. Daganzo, Transp. Res. A **31**, 141–155 (1997)
338. T.-W. Lin, S.-L. Hwang, P. Green, Saf. Sci. **47**, 620–625 (2009)
339. S. Liu, H. Hellendoorn, B. De Schutter, IEEE Trans. ITS **18**, 306–320 (2017)
340. H. Liu, D.Z.W. Wang Transp. Res. B **72**, 20–39 (2015)
341. J. Long, Z. Gao, P. Orenstein, H. Ren, IEEE Trans. ITS **13**, 469–481 (2012)
342. J. Long, Z. Gao, X. Zhao, A. Lian, P. Orenstein, Netw. Spat. Econ. **11**, 43–64 (2011)
343. M. Lorenz, L. Elefteriadou, Transp. Res. Circ. **E-C018**, 84–95 (2000)
344. C.-C. Lu, J. Liu, Y. Qu, S. Peeta, N.M. Rouphail, X. Zhou, Transp. Res. B **94**, 217–239 (2016)
345. L. Lu, Y. Xu, C. Antoniou, M. Ben-Akiva, Transp. Res. C **51**, 149–166 (2015)
346. P. Luathep, A. Sumalee, W.H.K. Lam, Z-Ch. Li, H.K. Lo, Transp. Res. B **45**, 808–827 (2011)
347. I. Lubashevsky, P. Wagner, R. Mahnke, Eur. Phys. J. B **32**, 243–247 (2003)
348. S. Lübeck, M. Schreckenberg, K.D. Usadel, Phys. Rev. E **57**, 1171 (1998)
349. W. Lv, W.-g. Song, Z.-m. Fang, Physica A **390**, 2303–2314 (2011)
350. M. McDonald, J. Wu, in *Proceedings of the ISC 97 Conference* (Boston, USA, 1997)
351. S. Maerivoet, B. De Moor, Phys. Rep. **419**, 1–64 (2005)
352. H.S. Mahmassani, Netw. Spat. Econ. **1**, 267–292 (2001)
353. H.S. Mahmassani (ed.), *Traffic and Transportation Theory* (Elsevier Science, Amsterdam, 2005)
354. H.S. Mahmassani, G.L. Chang. Transp. Sci. **21**, 89–99 (1987)
355. H.S. Mahmassani, R. Herman, Transp. Sci. **18**, 362–384 (1984)
356. H.S. Mahmassani, R. Jayakrishnan, Transp. Res. A **25**, 293–307 (1991)
357. H.S. Mahmassani, S. Peeta, Transp. Res. Rec. **1408**, 83–93 (1993)
358. H.S. Mahmassani, M. Saberi, A. Zockaie, Transp. Res. C **36**, 480–497 (2013)
359. H.S. Mahmassani, J.C. Williams, R. Herman, Transp. Res. Rec. **971**, 121–130 (1984)
360. H.S. Mahmassani, J.C. Williams, R. Herman, in *Proc. of the 10th Int. Symposium on Transp. and Traffic Theory*, ed. by N.H. Gartner (Elsevier, New York, 1987), pp. 1–20
361. R. Mahnke, J. Kaupužs, I. Lubashevsky, Phys. Rep. **408**, 1–130 (2005)
362. F.L. Mannering, W.P. Kilareski, *Principles of Highway Engineering and Traffic Analysis*, 2nd edn. (Wiley, New York, 1998)
363. J.-J. Martinez, C. Canudas-do-Wit, IEEE Trans. Contr. Syst. Technol. **15**, 246–258 (2007)
364. A.D. May, Highway Res. Rec. **59**, 9–38 (1964)
365. A.D. May, *Traffic Flow Fundamentals* (Prentice-Hall, New Jersey, 1990)
366. A.D. May, P. Athol, W. Parker, J.B. Rudden, Highway Res. Rec. **21**, 48–70 (1963)
367. D.K. Merchant, G.L. Nemhauser. Transp. Sci. **12**, 183–199 (1978)
368. D.K. Merchant, G.L. Nemhauser, Transp. Sci. **12**, 200–207 (1978)
369. N. Motamedidehkordi, T. Benz, M. Margreiter, in *Advanced Microsystems for Automotive Applications 2015, Lecture Notes in Mobility 2016* (Springer, Berlin, 2016), pp 37–52
370. R. Mounce, in [311] (2009), pp. 327–344
371. R. Müller, G. Nöcker, in *Proceedings of the Intelligent Vehicles '92 Symposium*, ed. by I. Masaki (IEEE, Detroit, USA, 1992), pp. 173–178
372. J.C. Muñoz, C.F. Daganzo, in *Traffic and Transportation Theory*, ed. by M.A.P. Taylor (Pergamon, Oxford, 2002) pp. 441–462
373. T. Nagatani, Physica A **261**, 599–607 (1998)
374. T. Nagatani, Phys. Rev. E **59**, 4857–4864 (1999)
375. T. Nagatani, Physica A **271**, 200–221 (1999)
376. T. Nagatani, Physica A **272**, 592–611 (1999)
377. T. Nagatani, Rep. Prog. Phys. **65**, 1331–1386 (2002)
378. T. Nagatani, K. Nakanishi, Phys. Rev. E **57**, 6415–6421 (1998)
379. K. Nagel, M. Schreckenberg, J. Phys. (France) I **2**, 2221–2229 (1992)
380. K. Nagel, P. Wagner, R. Woesler, Oper. Res. **51**, 681–716 (2003)
381. K. Nagel, D.E. Wolf, P. Wagner, P. Simon, Phys. Rev. E **58**, 1425–1437 (1998)

382. Y. Naito, T. Nagatani, Physica A **391**, 1626–1635 (2012)
383. S. Nakayama, R.D. Connors, D. Walting, in [311] (2009), pp. 39–56
384. Y. Nesterov, A. de Palma, Netw. Spat. Econ. **3**, 371–395 (2003)
385. J.P.L. Neto, M.L. Lyra, C.R. da Silva, Physica A **390**, 3558–3565 (2011)
386. G.F. Newell, Oper. Res. **3**, 176–188 (1955)
387. G.F. Newell, Ann. Math. Stat. **31**, 589–597 (1960)
388. G.F. Newell, Oper. Res. **9**, 209–229 (1961)
389. G.F. Newell, Oper. Res. Soc. Jpn. **5**, 9–54 (1962)
390. G.F. Newell, SIAM Rev. **575**, 223–240 (1965)
391. G.F. Newell, in *Proc. Second Internat. Sympos. on Traffic Road Traffic Flow* (OECD, London, 1965), pp. 73–83
392. G.F. Newell, *Applications of Queuing Theory* (Chapman Hall, London, 1982)
393. G.F. Newell, Transp. Res. B **32**, 531 (1988)
394. G.F. Newell, Transp. Res. B **27**, 289–303 (1993)
395. G.F. Newell, *Inst. of Transp. Studies Research Report UCB ITS-RR-93-3* (University of California, Berkley, CA, 1993)
396. G.F. Newell, "Theory of highway traffic flow 1945 to 1965". *Course Notes UCB-ITS-CN-95-1* (Inst. of Transp. Studies, University of California, Berkeley, 1995)
397. G.F. Newell, "A simplified car-following theory: A lower order model". *Research Report UCB-ITS-RR-99-1* (Inst. of Transp. Studies, University of California, Berkeley, 1999)
398. G.F. Newell, Trans. Res. B **36**, 195–205 (2002)
399. L. Newman, A.M. Dunnet, G.J. Meis, Highway Res. Rec. **303**, 44–55 (1970)
400. D. Ngoduy, Transpormetrica **8**, 43–60 (2012)
401. D. Ngoduy, Commun. Nonlinear Sci. Numer. Simul. **18**, 2838–2851 (2013)
402. D. Ngoduy, S.P. Hoogendoorn, R. Liu, Physica A **388**, 2705–2716 (2009)
403. Y. Nie, Transp. Res. B **45**, 329–342 (2011)
404. G. Nöcker, in *Band Elektronik im Kraftfahrzeugwesen*, ed. by G. Walliser, vol. 437 (Expert Verlag, 1994), pp. 299–322 (ISBN 3-8169-1024-6)
405. S. Ossen, S.P. Hoogendoorn, Transp. Res. C **19**, 182–195 (2011)
406. R. Ourulingesh, *Adaptive Cruise Control*, M.Sc. Thesis (Kamval Rekhi School of Information Technology, Indian Institute of Technology Bombay, 2004)
407. M. Papageorgiou, *Application of Automatic Control Concepts in Traffic Flow Modeling and Control* (Springer, Berlin, New York, 1983)
408. M. Papageorgiou, J.-M. Blosseville, H. Hadj-Salem, Trans. Res. A. **24**, 361–370 (1990)
409. M. Papageorgiou, H. Hadj-Salem, J.-M. Blosseville, Transp. Res. Rec. **1320**, 58–64 (1991)
410. M. Papageorgiou, H. Hadj-Salem, F. Middleham, Transp. Res. Rec. **1603**, 99–98 (1997)
411. M. Papageorgiou, E. Kosmatopoulos, I. Papamichail, Y. Wang, IEEE Trans. ITS. **9**, 360–365 (2008)
412. M. Papageorgiou, A. Kotsialos, IEEE Trans. ITS. **3**, 271–281 (2002)
413. M. Papageorgiou, Y. Wang, E. Kosmatopoulos, I. Papamichail, Traf. Eng. Cont. **48**, 271–276 (2007)
414. H.J. Payne, in *Mathematical Models of Public Systems*, ed. by G.A. Bekey, vol. 1 (Simulation Council, La Jolla, 1971), pp. 51–60
415. H.J. Payne, Transp. Res. Rec. **772**, 68 (1979)
416. S. Peeta, H.S. Mahmassani, Transp. Res. C **3**, 83–98 (1995)
417. S. Peeta, H.S. Mahmassani, Ann. Oper. Res. **60**, 81–113 (1995)
418. S. Peeta, A.K. Ziliaskopoulos, Netw. Spat. Econ. **1**, 233–265 (2001)
419. G.H. Peng, Physica A **391**, 5971–5977 (2013)
420. G.H. Peng, Nonlinear Dyn. **73**, 1035–1043 (2013)
421. G.H. Peng, X.H. Cai, B.F. Cao, C.Q. Liu, Physica A **391**, 656–663 (2012)
422. G.H. Peng, X.H. Cai, C.Q. Liu, B.F. Cao, M.X. Tuo, Phys. Lett. A **375**, 3973–3977 (2011)
423. G.H. Peng, X.H. Cai, C.Q. Liu, M.X. Tuo, Phys. Lett. A **375**, 2823–2827 (2011)
424. G.H. Peng, X.H. Cai, C.Q. Liu, M.X. Tuo, Phys. Lett. A **375**, 2153–2157 (2011)
425. G.H. Peng, X.H. Cai, C.Q. Liu, M.X. Tuo, Phys. Lett. A **376**, 447–451 (2012)

426. G.H. Peng, D.H. Sun, Phys. Lett. A **374**, 1694–1698 (2010)
427. B.N. Persaud, "Study of a Freeway Bottleneck to Explore Some Unresolved Traffic Flow Issues". *Ph.D. Dissertation* (University of Toronto, Toronto, Canada, 1986)
428. B.N. Persaud, F.L. Hall, Trans. Res. A **23**, 103–113 (1989)
429. B.N. Persaud, F.L. Hall, L.M. Hall, Transp. Res. Rec. **1287**, 167–175 (1990)
430. B.N. Persaud, V.F. Hurdle, Transp. Res. Rec. **1194**, 191–198 (1988)
431. B.N. Persaud, V.F. Hurdle, in *Proceedings of International Symposium on Highway Capacity* (Karlsruhe, Germany, 1991)
432. B.N. Persaud, S. Yagar, R. Brownlee, Transp. Res. Rec. **1634**, 64–69 (1998)
433. B.N. Persaud, S. Yagar, D. Tsui, H. Look, Transp. Res. Rec. **1748**, 110–115 (2001)
434. B. Piccoli, A. Tosin, in *Encyclopedia of Complexity and System Science*, ed. by R.A. Meyers (Springer, Berlin, 2009), pp. 9727–9749
435. C. Pinnell, D.R. Drew, W.R. McCasland, J.A. Wattleworth, Highway Res. Rec. **157**, 22–70 (1967)
436. R. Pothering, S. Gao, in *Proc. TRB 92nd Annual Meeting*, Paper # 13-3641 (TRB, Washington DC, 2013)
437. I. Prigogine, in *Theory of Traffic Flow*, ed. by R. Herman (Elsevier, Amsterdam, 1961), p. 158
438. I. Prigogine, F.C. Andrews, Oper. Res. **8**, 789 (1960)
439. I. Prigogine, R. Herman, *Kinetic Theory of Vehicular Traffic* (American Elsevier, New York, 1971)
440. Z. Qian, W. Shen, H.M. Zhang, Transp. Res. B **46**, 874–893 (2012)
441. Z. Qian, H.M. Zhang, Transp. Res. B **46**, 1489–1503 (2012)
442. R. Rajamani, *Vehicle Dynamics and Control, Mechanical Engineering Series* (Springer US, Boston, 2012)
443. H. Rakha, M. Van Aerde, Transp. Res. Rec. **1510**, 26–34 (1995)
444. H. Rakha, P. Pasumarthy, S. Adjerid, Transp. Lett. **1**, 95–110 (2009)
445. H. Rakha, A. Tawfik, in *Encyclopedia of Complexity and System Science*, ed. by R.A. Meyers (Springer, Berlin, 2009), pp. 9429–9470
446. H. Rakha, W. Wang, Transp. Res. Rec. **2124**, 113–124 (2009)
447. B. Ran, D. Boyce, *Modeling Dynamic Transportation Networks* (Springer, Berlin, 1996)
448. P.I. Richards, Oper. Res. **4**, 42–51 (1956)
449. M. Rickert, K. Nagel, M. Schreckenberg, A. Latour, Physica A **231**, 534–550 (1996)
450. D.I. Robertson, Traffic Eng. Contr. **10**, 276 (1969)
451. D.I. Robertson, *TRRL Report No. LR 2530* (Transportation and Road Research Laboratory, Crowthorne, UK, 1969)
452. C. Roncoli, M. Papageorgiou, I. Papamichail, Transp. Res. C **57**, 241–259 (2015)
453. C. Roncoli,M. Papageorgiou, I. Papamichail, Transp. Res. C **57**, 260–275 (2015)
454. M. Saifuzzaman, Z. Zheng, Transp. Res. C **48**, 379–403 (2014)
455. G. Sala, P. Fabio, in *Proceedings of 3rd World Congress on Intelligent Transport Systems* (1996)
456. S.P. Sathiyan, S.S. Kumar, A.I. Selvakumar, Int. J. Innovative Technol. Exploring Eng. (IJITEE) **2**, 89–96 (2013)
457. A. Schadschneider, D. Chowdhury, K. Nishinari, *Stochastic Transport in Complex Systems* (Elsevier Science, New York, 2011)
458. A. Schadschneider, W. Knospe, L. Santen, M. Schreckenberg, Physica A **346**, 165–173 (2005)
459. A. Schadschneider, T. Pöschel, R. Kühne, M. Schreckenberg, D.E. Wolf (eds.), *Traffic and Granular Flow'05* (Springer, Heidelberg, 2007)
460. A. Schadschneider, M. Schreckenberg, J. Phys. A **26**, L679–L684 (1993)
461. J.-D. Schmöcker, M.G.H. Bell, F. Kurauchi, H. Shimamoto, in [311] (2009), pp. 1–18
462. J. Schoen, A.D. May, W. Reilly, T. Urbanik, "Speed–Flow Relationships for Basic Freeway Segments". *Final Report, NCHRP Project 3–45*, (JHK & Associates and Texas Transportation Institute, 1995)
463. M.J. Schofield, Traffic Eng. Contr. **27**, 509–513 (1986)
464. M. Schönhof, D. Helbing, Transp. Sci. **41**, 135–166 (2007)

465. M. Schönhof, D. Helbing, Transp. Res. B **43**, 784–797 (2009)
466. M. Schönhof, M. Treiber, A. Kesting, D. Helbing, Transp. Res. Rec. **1999**, 3–12 (2007)
467. M. Schreckenberg, R. Barlović, W. Knospe, H. Klüpfel, in *Computational Statistical Physics*, ed. by K.H. Hoffmann, M. Schreiber (Springer, Berlin, 2001), pp. 113–126
468. M. Schreckenberg, A. Schadschneider, K. Nagel, N. Ito, Phys. Rev. E **51**, 2939–2949 (1995)
469. M. Schreckenberg, D.E. Wolf (eds.), *Traffic and Granular Flow' 97. Proceedings of the International Workshop on Traffic and Granular Flow* (Springer, Singapore, 1998)
470. Y. Sheffi, *Urban Transportation Networks: Equilibrium Analysis with Mathematical Programming Methods* (Prentice-Hall, Englewood Cliffs, 1985)
471. W. Shen, H.M. Zhang, Transp. Res. B **65**, 1–17 (2014)
472. S.E. Shladover, Veh. Syst. Dyn. **24**, 551–595 (1995)
473. S.E. Shladover, D. Su, X.-T. Lu, Transp. Res. Rec. **2324**, 6370 (2012)
474. V. Shvetsov, D. Helbing, Phys. Rev. E **59**, 6328–6339 (1999)
475. R. Sipahi, S.-I. Niculescu, Phil. Trans. R. Soc. A **368**, 4563–4583 (2010)
476. E. Smaragdis, M. Papageorgiou, Transp. Res. Rec. **1856**, 74–86 (2003)
477. M.J. Smith, Transp. Res. C **29**, 131–147 (2013)
478. M. Strating, *Coordinated Signal Control for Urban Networks by Using MFD* (M.Sc. Thesis) (Delft University of Technology, Delft, The Netherlands, 2010)
479. Y. Sugiyama, M. Fukui, M. Kikuchi, K. Hasebe, A. Nakayama, K. Nishinari, S.-I. Tadaki, S. Yukawa, New J. Phys. **10**, 033001 (2008)
480. N. Sugiyama, T. Nagatani, Physica A **392**, 1848–1857 (2013)
481. Y. Sugiyama, H. Yamada, Phys. Rev. E **55**, 7749–7752 (1997)
482. D. Sun, A. Clinet, A.M. Bayen, Transp. Res. B **45**, 880–902 (2011)
483. D.-H. Sun, X.-Y. Liao, G.H. Peng, Physica A **390**, 631–635 (2011)
484. Ch. Suwansirikul, T.L. Friesz, R.L. Tobin, Transp. Sci. **21**, 254–263 (1987)
485. H. Suzuki, JSAE Rev. **24**, 403–410 (2003)
486. D. Swaroop, J.K. Hedrick, IEEE Trans. Automat. Contr. **41**, 349–357 (1996)
487. D. Swaroop, J.K. Hedrick, S.B. Choi, IEEE Trans. Veh. Technol. **50**, 150–161 (2001)
488. W.Y. Szeto, L. O'Brien, M. O'Mahony, in [13] (2007), pp. 127–154
489. W.Y. Szeto, X. Jaber, S.C. Wong, Transp. Rev. **32**, 491–518 (2012)
490. S.I. Tadaki, K. Nishinari, M. Kikuchi, Y. Sugiyama, S. Yukawa, Physica A, **315**, 156–162 (2002)
491. M. Takayasu, H. Takayasu, Fractals **1**, 860–866 (1993)
492. A. Talebpour, H.S. Mahmassani, Transp. Res. C **71**, 143–163 (2016)
493. C.F. Tang, R. Jiang, Q.S. Wu, Physica A **377**, 641–650 (2007)
494. T.Q. Tang, C.Y. Li, Y.H. Wu, H.J. Huang, Physica A **390**, 3362–3368 (2011)
495. M.A.P. Taylor (ed.), *Transportation and Traffic Theory in the 21st Century. Proceedings of the 15th International Symposium on Transportation and Traffic Theory* (Elsevier Science, Amsterdam, 2002)
496. L.-J. Tian, H.-J. Huang, Z.-Y. Gao, Netw. Spat. Econ. **12**, 589–608 (2012)
497. J.-f. Tian, Z.-z. Yuan, B. Jia, M.-h. Li, G.-j. Jiang, Physica A **391**, 4476–4482 (2012)
498. R.L. Tobin, T.L. Friesz, Transp. Sci. **22**, 242–250 (1988)
499. E. Tomer, L. Safonov, S. Havlin, Phys. Rev. Lett. **84**, 382 (2000)
500. E. Tomer, L. Safonov, N. Madar, S. Havlin, Phys. Rev. E **65**, 065101 (R) (2002)
501. http://en.wikipedia.org/wiki/Traffic$_$simulation
502. M. Treiber, D. Helbing, J. Phys. A Math. Gen. **32**, L17–L23 (1999)
503. M. Treiber, D. Helbing, Automatisierungstechnik **49**, 478–484 (2001)
504. M. Treiber, D. Helbing, Phys. Rev. E **68**, 046119 (2003)
505. M. Treiber, D. Helbing, Eur. Phys. J. B **68**, 607–618 (2009)
506. M. Treiber, A. Hennecke, D. Helbing, Phys. Rev. E **59**, 239–253 (1999)
507. M. Treiber, A. Hennecke, D. Helbing, Phys. Rev. E **62**, 1805–1824 (2000)
508. M. Treiber, A. Kesting, *Verkehrsdynamik und -simulation: Daten, Modelle und Anwendungen der Verkehrsflussdynamik* (Springer, Heidelberg, Dordrecht, London, New York, 2010)

509. M. Treiber, A. Kesting, *Traffic Flow Dynamics* (Springer, Heidelberg, New York, Dordrecht, London, 2013)
510. M. Treiber, A. Kesting, D. Helbing, Phys. Rev. E **74**, 016123 (2006)
511. M. Treiber, A. Kesting, D. Helbing, Transp. Res. B **44**, 983–1000 (2010)
512. J. VanderWerf, S.E. Shladover, M.A. Miller, N. Kourjanskaia, Transp. Res. Rec. **1800**, 78–84 (2002)
513. H.R. Varia, P.J. Gundaliya, S.L. Dhingra, Res. Transp. Econ. **38**, 35–44 (2013)
514. J. Vasic, H.J. Ruskin, Physica A **391**, 2720–2729 (2012)
515. http://de.wikipedia.org/wiki/VISSIM
516. P. Wagner, Transp. Res. B **46**, 1384–1392 (2012)
517. P. Wagner, in *Autonomes Fahren*, ed. by M. Maurer, J.Ch. Gerdes, B. Lenz, H. Winner (Springer, Berlin 2015), pp. 313–330
518. P. Wagner, R. Nippold, S. Gabloner, M. Margreiter, Chaos Solitons Fractals, 1–9 (2016)
519. J. Wahle, A.L.C. Bazzan, F. Klugl, M. Schreckenberg, Physica A **287**, 669–681 (2000)
520. J. Wahle, R. Chrobok, A. Pottmeier, M. Schreckenberg, Netw. Spat. Econ. **2**, 371–386 (2002)
521. M. Wang, W. Daamen, S.P. Hoogendoorn, B. van Arem, Transp. Res. C **40**, 271–289 (2014)
522. M. Wang, W. Daamen, S.P. Hoogendoorn, B. van Arem, Transp. Res. C **40**, 290–311 (2014)
523. M. Wang, W. Daamen, S.P. Hoogendoorn, B. van Arem, R. Happee, Transp. Res. C **58**, 73–92 (2015)
524. J. Wang, W. Deng, in *Proc. TRB 92nd Annual Meeting*, Paper # 13-0836 (TRB, Washington DC, 2013)
525. T. Wang, Z.Y. Gao, W.Y. Zhang, J. Zhang, S.B. Li, Nonlinear Dyn. **77**, 635–642 (2014)
526. T. Wang, Z.Y. Gao, J. Zhang, Nonlinear Dyn. **73**, 2197–2205 (2013)
527. D.Z. Wang, H. Liu, in *Proc. TRB 92nd Annual Meeting*, Paper # 13-1443 (TRB, Washington DC, 2013)
528. S. Wang, Q. Meng, H. Yang, Transp. Res. B **50**, 42–60 (2013)
529. J.G.Wardrop, in *Proc. of Inst. of Civil Eng. II*, vol. 1, pp. 325–378 (1952)
530. J.A. Wattleworth, Highway Res. Rec. **157**, 1–21 (1967)
531. J.A. Wattleworth, D.S. Berry, Highway Res. Rec. **89**, 1–25 (1965)
532. C. Wei, Y. Asakura, T. Iryo, J. Adv. Transp. **46**, 222–235 (2012)
533. B.-W. Wie, T.L. Friesz, R.L. Tobin, Transp. Res. B **24**, 431–442 (1990)
534. G.B. Whitham, *Linear and Nonlinear Waves* (Wiley, New York, 1974)
535. G.B. Whitham, Proc. R. Soc. Lond. A **428**, 49 (1990)
536. R. Wiedemann, *Simulation des Verkehrsflusses* (University of Karlsruhe, Karlsruhe, 1974)
537. R.E. Wilson, IMA J. Appl. Math. **66**, 509–537 (2001)
538. R.E. Wilson, Phil. Trans. R. Soc. A **366**, 2017–2032 (2008)
539. D.E. Wolf, Physica A **263**, 438–451 (1999)
540. D.E. Wolf, M. Schreckenberg, A. Bachem (eds.), *Traffic and Granular Flow. Proceedings of the International Workshop on Traffic and Granular Flow* (World Scientific, Singapore, 1995)
541. X. Wu, Y. Nie, in *Proc. TRB 92nd Annual Meeting*, Paper # 13-4100 (TRB, Washington DC, 2013)
542. C. Xie, S.T. Waller, Transp. Res. B **46**, 1023–1042 (2012)
543. C. Xiong, X. Chen, X. He, X. Lin, L. Zhang, Transp. Res. C **64**, 148–163 (2016)
544. X. Xu, A. Chen, S. Kitthamkesorn, H. Yang, H.K. Lo, Transp. Res. B **81**, 686–703 (2015)
545. H. Yang, M.G.H. Bell, Transp. Res. B **31**, 303–314 (1997)
546. H. Yang, M.G.H. Bell, Transp. Rev. **18**, 257–278 (1998)
547. H. Yang, M.G.H. Bell, Transp. Res. B **32**, 539–545 (1998)
548. H. Yang, M.G.H. Bell, Q. Menga, Transp. Res. B **34**, 255–275 (2000)
549. J. Yang, G. Jiang, Transp. Res. C **47**, 168–178 (2014)
550. H. Yang, W.-L. Jin Transp. Res. C **53**, 19–34 (2015)
551. D. Yang, P. Jin, Y. Pu, B. Ran, Eur. Phys. J. B **86**, 92 (2013)
552. H. Yang, Q. Menga, M.G.H. Bell, Transp. Sci. **35**, 107–123 (2001)
553. B. Yang, Ch. Monterola, Phys. Rev. E **92**, 042802 (2015)
554. B. Yang, Ch. Monterola, Transp. Res. C **72**, 283–305 (2016)

555. H. Yang, X. Wang, Y. Yin, Transp. Res. B **46**, 1295–1307 (2012)
556. M. Yildirimoglu, M. Ramezani, N. Geroliminis, Transp. Res. C **59**, 404–420 (2015)
557. Y. Yufei, J.W.C. van Lint, R.E. Wilson, F. van Wageningen-Kessels, S.P. Hoogendoorn, IEEE Trans. ITS **13**, 59–70 (2012)
558. C. Zhang, X. Chen, A. Sumalee, Transp. Res. B **45**, 534–552 (2011)
559. K. Zhang, H.S. Mahmassani, C.-C. Lu, Transp. Res. Rec. **2085**, 86–94 (2008)
560. K. Zhang, H.S. Mahmassani, C.-C. Lu, Transp. Res. C **27**, 189–204 (2013)
561. H.M. Zhang, S.G. Ritchie, Transp. Res. C **5**, 273–286 (1997)
562. M. Zhang, D.H. Sun, W.N. Liu, Nonlinear Dyn. **81**, 1623–1633 (2015)
563. M. Zhang, D.H. Sun, W.N. Liu, M. Zhao, S.L. Chen, Physica A **422**, 16–24 (2015)
564. M. Zhang, D.H. Sun, C. Tian, Nonlinear Dyn. **77**, 839–847 (2014)
565. D. Zhao, X. Bai, F.-Y. Wang, J. Xu, W. Yu, IEEE Trans. ITS **12**, 990–999 (2011)
566. J. Zhao, Y. Liu, Transp. Res. C **73**, 219–238 (2017)
567. L. Zheng, Z. He, T. He, Transp. Res. C **75** 136–167 (2017)
568. R.X. Zhong, A. Sumalee, T.L. Friesz, W.H.K. Lam, Transp. Res. B **45**, 1035–1061 (2011)
569. R. Zhong, A. Sumalee, T. Maruyama, J. Adv. Transp. **46**, 191–221 (2012)
570. X. Zhou, H.S. Mahmassani, K. Zhang, Transp. Res. C **16**, 167–186 (2008)
571. J. Zhou, H. Peng, IEEE Trans. Intell. Transp. Syst. **6**, 229–237 (2005)
572. J. Zhou, Z.-K. Shi, Nonlinear Dyn. **83**, 1217–1236 (2015)
573. J. Zhou, Z.-K. Shi, C.-P. Wang, Nonlinear Dyn. 1–20 (2016)
574. F. Zhu, S.V. Ukkusuri, IEEE Trans. ITS **18**, 282–291 (2017)
575. W.X. Zhu, L.D. Zhang, Int. J. Mod. Phys. C **23**, 1250025 (2012)
576. A.K. Ziliaskopoulos, D. Kotzinos, H.S. Mahmassani, Transp. Res. C **5**, 95–107 (1997)
577. A.K. Ziliaskopoulos, H.S. Mahmassani, Transp. Res. Rec. **1408**, 94–100 (1993)

Chapter 5
Theoretical Fundamental of Transportation Science—The Three-Phase Theory

5.1 Introduction—Definition of Stochastic Highway Capacity

In this chapter, following [4–17, 20–25, 31, 35, 39–44, 46, 47] we present a theory of traffic breakdown at highway bottlenecks in the framework of the three-phase theory.

Among empirical features of traffic breakdown at highway bottlenecks discussed in Chaps. 2 and 3, the following important empirical features of empirical traffic breakdown at highway bottlenecks should be distinguished:

(i) The downstream front of congested traffic resulting from traffic breakdown is usually fixed at the bottleneck location (Chap. 2). This congested traffic is called synchronized flow (S): The term *traffic breakdown at a highway bottleneck* is a synonym of the term *F→S transition at the bottleneck.*
(ii) At the same highway bottleneck, there can be either empirical spontaneous traffic breakdown or empirical induced traffic breakdown. Empirical traffic breakdown can occur within a broad range of the flow rate at a highway bottleneck (Chap. 3).
(iii) The empirical evidence of induced traffic breakdown at a bottleneck proves the nucleation nature of traffic breakdown (Chap. 3).

Nucleation features of empirical traffic breakdown (empirical F→S transition) discussed in Chap. 3 are as follows [4–17, 20–28]:

1. Empirical traffic breakdown occurs, when a nucleus appears in free flow in a neighborhood of a highway bottleneck.
2. The nucleus for traffic breakdown is a local disturbance in the initial free flow in a neighborhood of the bottleneck, whose development results in an F→S transition at the bottleneck. The term *a nucleus for traffic breakdown at a highway bottleneck* is a synonym of the term *a nucleus for F→S transition at the bottleneck.*

B.S. Kerner, *Breakdown in Traffic Networks*, DOI 10.1007/978-3-662-54473-0_5

One of the most important consequences of the empirical features of traffic breakdown is the empirical proof of the nucleation nature of empirical traffic breakdown at a highway bottleneck: Traffic breakdown is an F→S transition occurring in metastable free flow at the bottleneck. This metastability of free flow with respect to the F→S transition is realized within a flow rate range (3.9) that we rewrite here

$$C_{\min} \le q_{\text{sum}} < C_{\max}, \tag{5.1}$$

where q_{sum} is the flow rate in free flow at a highway bottleneck, $C_{\min}$ is a minimum highway capacity, $C_{\max}$ is a maximum highway capacity. Under conditions (5.1), traffic breakdown can be induced at the bottleneck.

At any time instant, there are the infinite number of the flow rates q_{sum} (5.1) in free flow at a highway bottleneck at which traffic breakdown can be induced at the bottleneck. These flow rates are highway capacities of free flow at the bottleneck.

This conclusion of empirical observations of traffic breakdown (F→S transition) at the bottleneck leads to the following definition of stochastic highway capacity of free flow at a highway bottleneck made in the three-phase theory [16, 17, 20–28]:

At any time instant, there are the infinite number of highway capacities C of free flow at the bottleneck. The range of the infinite number of highway capacities is limited by the minimum highway capacity $C_{\min}$ and the maximum highway capacity $C_{\max}$ (Fig. 3.22):

$$C_{\min} \le C \le C_{\max}, \tag{5.2}$$

where $C_{\min} < C_{\max}$. The existence of an infinite number of highway capacities at any time instant means that highway capacity is stochastic.

The physical sense of this capacity definition is as follows (Fig. 3.22). Highway capacity is limited by traffic breakdown in an initial free flow at a highway bottleneck. In other words, any flow rate q_{sum} in free flow at the bottleneck at which traffic breakdown can occur is highway capacity. At any time instant, there are the infinite number of such highway capacities $C = q_{\text{sum}}$ at which traffic breakdown *can* occur. These capacities satisfy conditions (5.2). Thus, in the three-phase theory any flow rate q_{sum} in free flow at a highway bottleneck that satisfies conditions

$$C_{min} \le q_{sum} \le C_{max} \tag{5.3}$$

is equal to one of the stochastic highway capacities of free flow at the bottleneck (Fig. 3.22).

Under condition (5.1), free flow is in a *metastable* state with respect to an to F→S transition (traffic breakdown). When condition

$$q_{sum} = C_{max} \tag{5.4}$$

is satisfied, we assume that the probability of traffic breakdown during a given time interval is equal to 1 (see Sect. 5.4.8). For this reason, free flow can be considered as an *unstable* one with the respect to traffic breakdown at the bottleneck. This explains the difference between conditions (5.1) that include only metastable states of free flow and conditions (5.3) that include both metastable states of free flow (5.1) and the unstable state of free flow related to condition (5.4).

Figure 3.22 together with conditions (5.1) explain the term *metastable free flow with respect to an F→S transition (traffic breakdown) at the bottleneck* as follows (labeled by "metastable free flow" in Fig. 3.22). For each flow rate q_{sum} in free flow at the bottleneck that satisfies conditions (5.1) there can be a transition from a state of the free flow traffic phase (F) to a state of the synchronized flow traffic phase (S) at the bottleneck. This F→S transition (traffic breakdown) at the bottleneck occurs only, when a nucleus required for the breakdown appears at the bottleneck.

In accordance with Sect. 3.3, the sense of the term *nucleus required for traffic breakdown (F→S transition) at the bottleneck* is as follows. We consider a local speed disturbance in free flow that causes a local decrease in the speed at the bottleneck. The disturbance is localized at the bottleneck. The larger the disturbance amplitude, the smaller the free flow speed within the disturbance in comparison with the free flow speed outside the disturbance. If the amplitude of the disturbance is small enough, then no F→S transition occurs at the bottleneck. However, when the amplitude of a local speed disturbance in the metastable free flow is equal to or exceeds a critical amplitude, the F→S transition does occur. A local speed disturbance occurring in the metastable free flow that leads to the F→S transition is called a nucleus for the F→S transition. In more details, such a macroscopic consideration of a nucleus required for F→S transition will be done in Sect. 5.3. A microscopic theory of a nucleus required for F→S transition will be presented in Sect. 5.12 and 5.13.

When condition

$$q_{\text{sum}} < C_{\text{min}} \tag{5.5}$$

is satisfied, no traffic breakdown can occur. This is independent of the amplitude of a time-limited local disturbance in free flow at the bottleneck. Indeed, we assume that under condition (5.5) free flow is *stable* with respect to traffic breakdown (F→S transition) at the bottleneck (labeled by "stable free flow" in Fig. 3.22).

Respectively, when condition

$$q_{\text{sum}} \geq C_{\text{max}} \tag{5.6}$$

is satisfied, traffic breakdown does occur at the bottleneck during a given time interval (see Sect. 5.4.8). This is because we assume that under condition (5.6) free flow is *unstable* with respect to traffic breakdown (F→S transition) at the bottleneck (labeled by "unstable free flow" in Fig. 3.22).

In accordance with features of empirical induced traffic breakdown formulated in Sect. 3.6, we can make the following statements about the empirical proof of the definition of the stochastic highway capacities in the three-phase theory.

Any flow rate q_{sum} in free flow at a highway bottleneck at which traffic breakdown can be induced at a highway bottleneck is equal to one of the stochastic highway capacities of free flow at the bottleneck. There are the infinite number of such flow rates that satisfy conditions (5.1).

The empirical induced traffic breakdown is the empirical proof of the statement of the three-phase theory that *at any time instant* there are the infinite number of highway capacities C of free flow at a highway bottleneck that are between the minimum highway capacity C_{min} and the maximum highway capacity C_{max} (5.2) (Fig. 3.22). The existence of an infinite number of highway capacities at any time instant means that highway capacity is stochastic.

The assumption of the three-phase theory [4–17, 20–28] about the nucleation nature of traffic breakdown (F→S transition) as well as the consequence of this assumption that *at any time instant* there are the infinite number of stochastic highway capacities in free flow at a road bottleneck (5.2) are consistent with the empirical nucleation nature of traffic breakdown at a highway bottleneck.

The empirical nucleation nature of traffic breakdown at a highway bottleneck is an empirical fundamental of transportation science (Chap. 3). Therefore, the three-phase theory can be considered a *theoretical fundamental* of transportation science.

The main aim of this chapter is to present results of the three-phase theory that explains the empirical nucleation nature of traffic breakdown at highway bottlenecks. The chapter is organized as follows:

(i) In Sects. 5.2 and 5.3 we present *qualitative three-phase theory* of the nucleation nature of traffic breakdown at highway bottlenecks.
(ii) A *probabilistic three-phase theory* of traffic breakdown at highway bottlenecks is presented in Sects. 5.4–5.7.
(iii) *Driver behavioral characteristics* that are responsible for the nucleation nature of traffic breakdown at highway bottlenecks are considered in Sects. 5.8–5.10.
(iv) A discussion of results of a *microscopic stochastic three-phase theory* of traffic breakdown is the subject of Sects. 5.11–5.13.

The microscopic stochastic three-phase theory describes spatiotemporal traffic phenomena through microscopic interactions of single vehicles in traffic flow. One of the basic features of this theory is the description of the effect of *fluctuations in traffic flow* on traffic breakdown and on other critical phenomena in traffic flow [23, 28, 31, 35–37, 41, 44, 45, 47]. With the use of the microscopic stochastic three-phase theory, an analysis and spatiotemporal description of nuclei required for traffic breakdown at highway bottlenecks will be made in Sects. 5.11–5.13.
(v) Results of the three-phase theory are used in Sect. 5.14 for the explanation of empirical features of traffic breakdown in *heterogeneous traffic flow*.
(vi) A summary of driver behavioral assumptions made in the three-phase theory for the explanation of the empirical nucleation nature of real traffic breakdown at highway bottlenecks is presented in Sect. 5.15.

5.2 The Basic Assumption of Three-Phase Traffic Theory

To explain features of empirical spontaneous and empirical induced traffic breakdowns at highway bottlenecks, in the three-phase theory the following basic assumption has been made [4–17, 20–28]

$$P^{(\mathrm{B})}(q_{\mathrm{sum}}) = P^{(\mathrm{B})}_{\mathrm{nucleus}}(q_{\mathrm{sum}}). \tag{5.7}$$

In (5.7), $P^{(B)}(q_{sum})$ is the flow-rate dependence of the probability that during a given time interval T_{ob} for observing traffic flow[1] spontaneous traffic breakdown occurs at a highway bottleneck, $P^{(B)}_{nucleus}(q_{sum})$ is the flow-rate dependence of the probability that during the time interval T_{ob} a nucleus for traffic breakdown occurs spontaneously in free flow at the bottleneck.[2]

5.3 Qualitative Theory of Critical Nucleus for Traffic Breakdown at Bottleneck

For simplicity, we restrict a qualitative analysis of critical nucleus for traffic breakdown at highway bottlenecks by a consideration of a local *speed* disturbance in free flow at the bottleneck. Naturally, a local disturbance of other traffic variables like the density, time headway, or the flow rate can be the origin of the occurrence of a nucleus for traffic breakdown at the bottleneck.

5.3.1 Permanent Speed Disturbance at Bottleneck

For qualitative explanations of the basic assumption of the three-phase theory (5.7), we recall that in empirical observations a nucleus for traffic breakdown appears at a location of a permanent speed disturbance at a highway bottleneck (Sect. 3.4.4). Therefore, in a qualitative theory of a critical nucleus for traffic breakdown at a highway bottleneck [9], there is a permanent speed disturbance ("permanent disturbance" for short) at the bottleneck (Figs. 5.1 and 5.2).

The existence of the permanent speed disturbance explains why a nucleus for traffic breakdown has a considerably greater probability to occur spontaneously in a neighborhood of the location of the permanent disturbance. The occurrence of the permanent speed disturbance at highway bottlenecks has been proven both in empirical studies (Sect. 3.4.4) and in simulations of free flow at highway bottlenecks

[1] In the theoretical definition of the breakdown probability $P^{(B)}(q_{sum})$ (5.7), we assume that the flow rate q_{sum} does not depend on time during the time interval T_{ob}. In contrast, in real field traffic data the flow rate is a very complex time-function. For this reason, in the empirical definition of the flow-rate dependence of the probability of spontaneous traffic breakdown at the bottleneck $P^{(B)}(q_i)$ (2.1) (see Figs. 2.2 and 2.3 of Sect. 2.3), rather than the time interval T_{ob}, we should use an averaging time interval for traffic variables T_{av}. A comparison of the theoretical and empirical definitions of the flow-rate dependence of the probability of spontaneous traffic breakdown at a bottleneck has been considered in Sec. 10.3.1 of the book [17].

[2] A mathematical nucleation theory of traffic breakdown at a highway bottleneck can be found in [39–41].

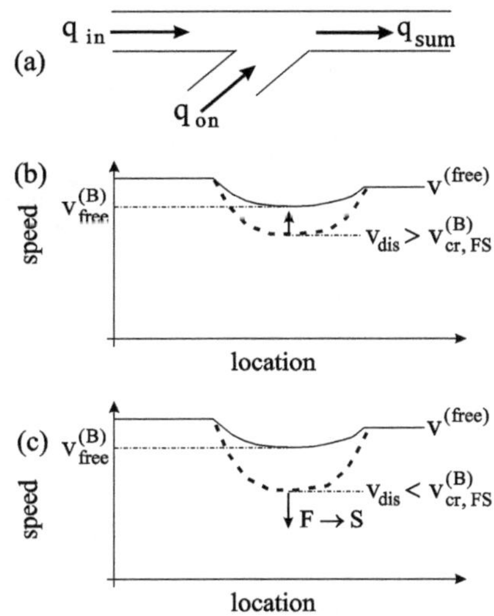

Fig. 5.1 A model of permanent speed disturbance at on-ramp bottleneck: (**a**) Sketch of bottleneck. (**b**, **c**) Qualitative speed distributions in space within the permanent speed disturbance with the minimum speed $v_{\rm dis}$ at a given time instant (*dashed curves*) and within the deterministic component of the permanent disturbance with the minimum speed $v^{\rm (B)}_{\rm free}$ (*solid curves*). For on-ramp bottleneck $q_{\rm sum} = q_{\rm in} + q_{\rm on}$

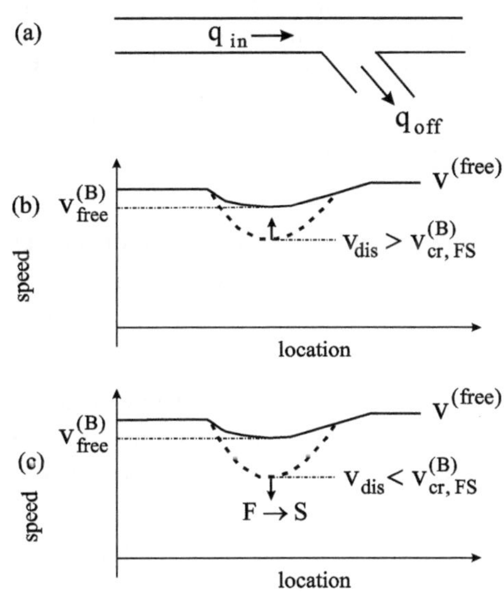

Fig. 5.2 A model of permanent speed disturbance at off-ramp bottleneck: (**a**) Sketch of bottleneck. (**b**, **c**) Qualitative speed distributions in space within the permanent speed disturbance with the minimum speed $v_{\rm dis}$ at a given time instant (*dashed curves*) and within the deterministic component of the permanent disturbance with the minimum speed $v^{\rm (B)}_{\rm free}$ (*solid curves*). For off-ramp bottleneck $q_{\rm sum} = q_{\rm in}$

with a stochastic microscopic traffic flow model in the framework of the three-phase theory (Figs. 5.3 and 5.4).

In general, the permanent local speed disturbance in free flow exhibits a complex spatiotemporal dynamics at the bottleneck (Sect. 5.13). For simplicity of the qualitative discussion presented in this section we assume that at a given time instant the permanent disturbance can be considered consisting of two components (dashed curves in Figs. 5.1b,c and 5.2b,c):

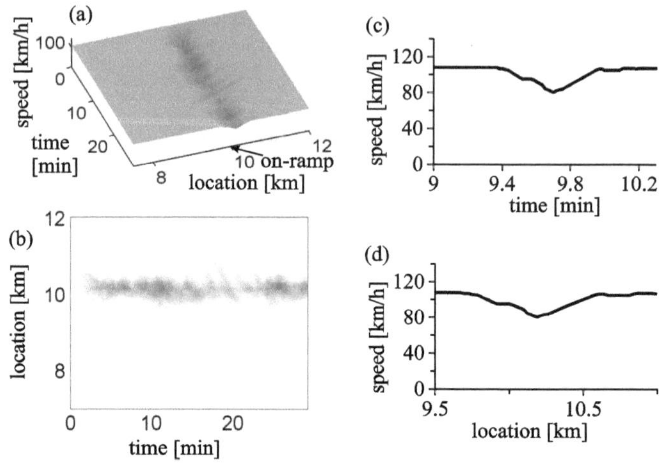

Fig. 5.3 Simulations of a permanent speed disturbance on single-lane road with on-ramp bottleneck: (**a, b**) Vehicle speed in space and time (**a**) and the same speed data presented by regions with variable shades of gray (**b**) (shades of gray vary from white to black when the speed decreases from 107 km/h (*white*) to 40 km/h (*black*)). (**c, d**) Microscopic speed of one of the vehicles in time (**c**) and in space (**d**) that moves through the permanent speed disturbance. The merging region of the on-ramp begins at $x = 10$ km. Simulations of the Kerner-Klenov stochastic microscopic three-phase model (see Appendix A)

(a) A deterministic component of the permanent local speed disturbance ("deterministic local speed disturbance" for short) with the minimum speed $v_{\text{free}}^{(\text{B})}$ within the deterministic local speed disturbance.
(b) A random component of the permanent local speed disturbance.

We consider random components of the permanent local speed disturbance, which lead to a *speed decrease* in comparison with the speed $v_{\text{free}}^{(\text{B})}$. Then the amplitude of the random component of the permanent local speed disturbance is equal to

$$\Delta v_{\text{free}}^{(\text{B})} = v_{\text{free}}^{(\text{B})} - v_{\text{dis}}, \tag{5.8}$$

where v_{dis} is the minimum speed within the permanent speed disturbance at the given time instant (Figs. 5.1b,c and 5.2b,c).

5.3.2 Critical Nucleus at Location of Permanent Speed Disturbance

As already explained in Sect. 3.3, a critical nucleus required for traffic breakdown at a highway bottleneck is a local speed disturbance in free flow at the bottleneck that amplitude is equal to a critical amplitude: If the amplitude of a local speed

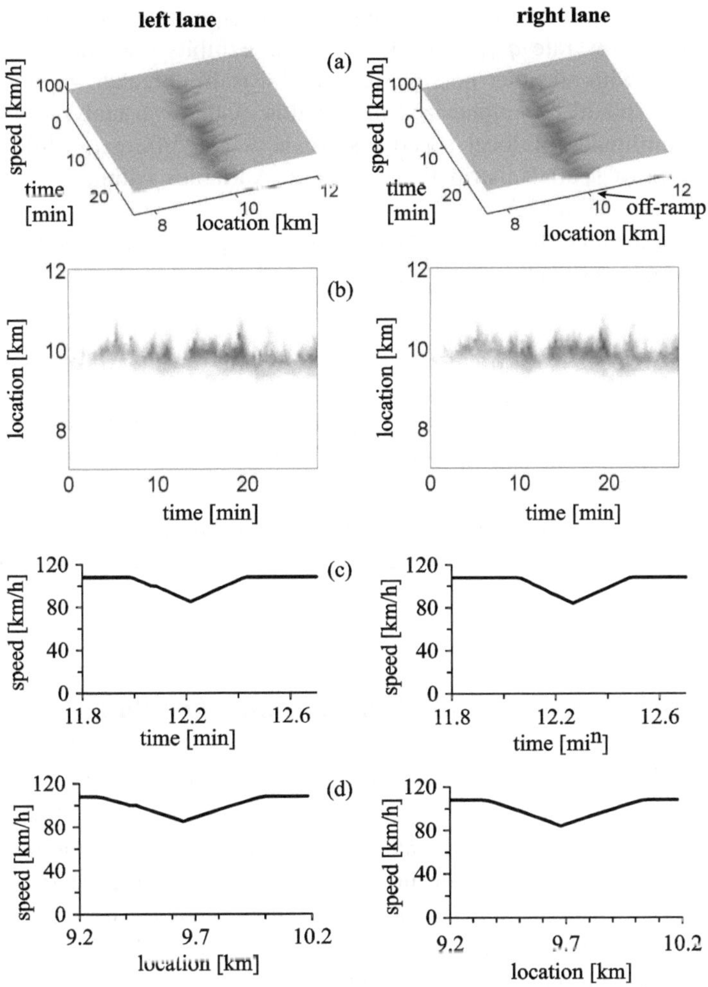

Fig. 5.4 Simulations of a permanent speed disturbance on two-lane road with off-ramp bottleneck: (**a, b**) Vehicle speed in space and time (**a**) and the same speed data presented by regions with variable shades of gray (**b**) (shades of gray vary from white to black when the speed decreases from 107 km/h (*white*) to 40 km/h (*black*)). (**c, d**) Microscopic speeds of two of the vehicles moving through the permanent speed disturbance, respectively, in the left and the right lanes in time (**c**) and in space (**d**). Figures in left column are related to the left road lane, in right column—the right lane. The off-ramp lane begins at $x = 10$ km. Simulations of the Kerner-Klenov stochastic microscopic three-phase model (see Appendix A)

disturbance in free flow is smaller than the critical amplitude, the disturbance decays, i.e., no traffic breakdown occurs. In contrast, if the amplitude of a local speed disturbance in free flow is equal to or larger than the critical amplitude, traffic breakdown does occur.

A nucleus for traffic breakdown is equal to or larger than the critical nucleus: At each given flow rate q_{sum}, a critical nucleus exhibits the smallest amplitude in comparison with all other possible nucleus for traffic breakdown at a highway bottleneck. It should be emphasized that in this qualitative analysis, we assume that only amplitudes of local speed disturbances are different, while all other characteristics of spatial distributions of traffic variables within the local speed disturbances are the same.[3]

We denote the amplitude of a critical random component of the permanent local speed disturbance by $\Delta v^{(B)}_{\text{free}} = \Delta v^{(B)}_{\text{cr, FS}}$ (Fig. 5.5a):

$$\Delta v^{(B)}_{\text{cr, FS}} = v^{(B)}_{\text{free}} - v^{(B)}_{\text{cr, FS}}, \tag{5.9}$$

where $v_{\text{dis}} = v^{(B)}_{\text{cr, FS}}$ is the average speed within the critical permanent local speed disturbance (Fig. 5.5a). This critical permanent local speed disturbance can be considered a critical nucleus with the amplitude $\Delta v^{(B)}_{\text{cr, FS}}$ (5.9). Respectively, the average speed within the critical permanent local speed disturbance $v_{\text{dis}} = v^{(B)}_{\text{cr, FS}}$ can be considered the speed within the critical nucleus for traffic breakdown (F→S transition).

5.3.3 *Dependence of Critical Nucleus on Flow Rate*

For a qualitative discussion of a critical nucleus for traffic breakdown, we assume that traffic parameters (weather, mean driver's characteristics, share of long vehicles, etc.) remain the same for all flow rates in free flow. Under this condition, we can also assume that the larger is the flow rate q_{sum} in free flow, the smaller is the amplitude $\Delta v^{(B)}_{\text{cr, FS}}(q_{\text{sum}})$ of the critical nucleus required to initiate spontaneous traffic breakdown in metastable free flow at a bottleneck. This explains the decreasing flow-rate function of the amplitude $\Delta v^{(B)}_{\text{cr, FS}}(q_{\text{sum}})$ of a critical nucleus shown in Fig. 5.5a.

Indeed, the amplitude $\Delta v^{(B)}_{\text{cr, FS}}(q_{\text{sum}})$ of the critical nucleus reaches its maximum value at the flow rate that is equal to the minimum highway capacity $q_{\text{sum}} = C_{\text{min}}$. When the flow rate q_{sum} increases, the amplitude $\Delta v^{(B)}_{\text{cr, FS}}(q_{\text{sum}})$ of the critical nucleus (5.9) decreases continuously. The amplitude $\Delta v^{(B)}_{\text{cr, FS}}(q_{\text{sum}})$ of the critical nucleus (5.9) tends to zero at the flow rate that is equal to the maximum highway capacity $q_{\text{sum}} = C_{\text{max}}$ (5.4). This explains why under condition (5.4) we can

[3]For simplicity of this qualitative consideration, we neglect here a complex spatiotemporal dynamics of the permanent local speed disturbance, which will be discussed in Sect. 5.13.

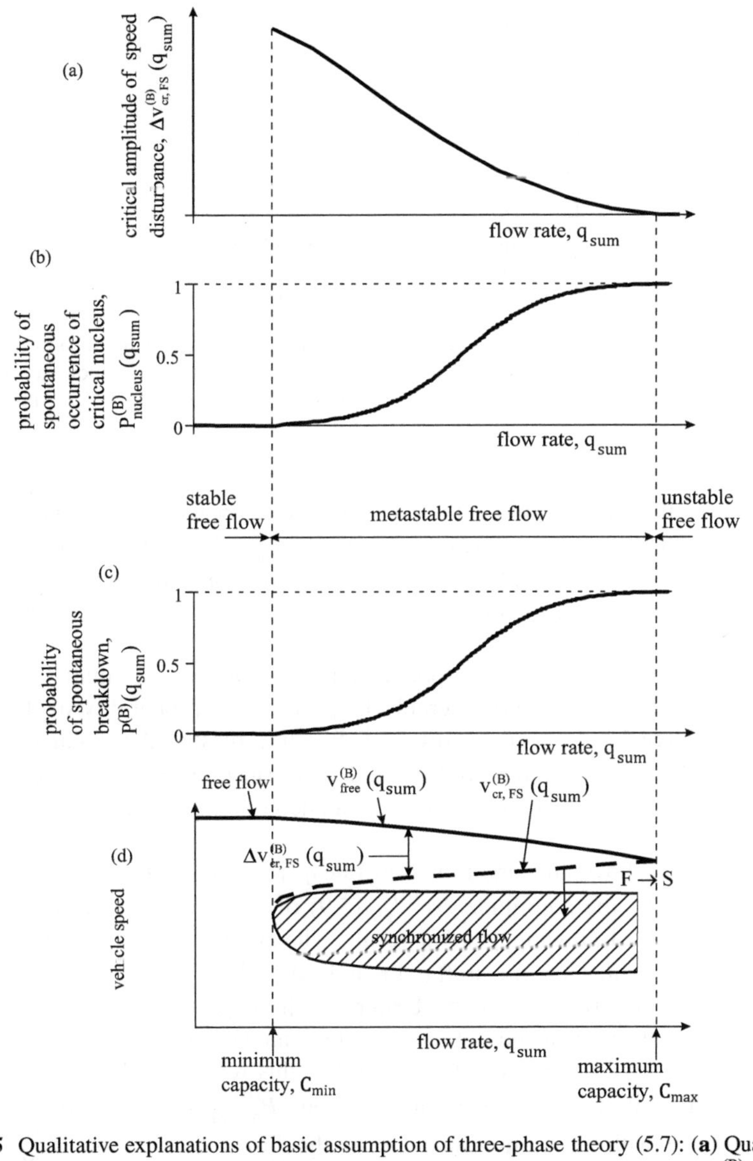

Fig. 5.5 Qualitative explanations of basic assumption of three-phase theory (5.7): (**a**) Qualitative flow-rate dependence of the amplitude of critical nucleus for traffic breakdown $\Delta v^{(B)}_{cr, FS}(q_{sum})$. (**b**) Qualitative flow-rate dependence of the probability of nucleus occurrence $P^{(B)}_{nucleus}(q_{sum})$ in (5.7). (**c**) Qualitative flow-rate function of the breakdown probability $P^{(B)}(q_{sum})$. (**d**) Z-Characteristic of traffic breakdown with states for free flow and synchronized flow adapted from Fig. 3.22 together with a branch for critical amplitude of local speed disturbance for an F→S transition (*dashed curve*)

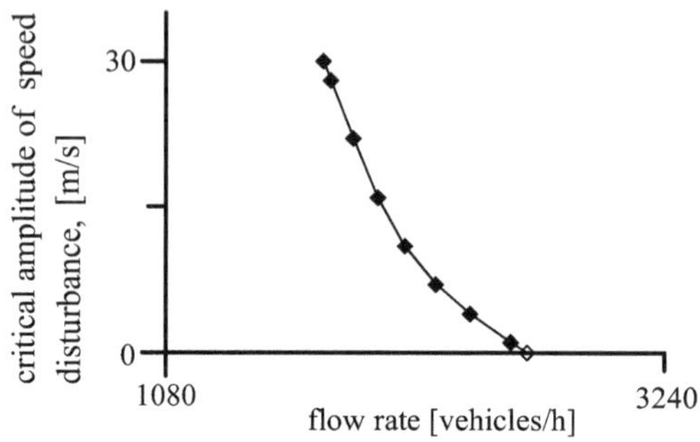

Fig. 5.6 Simulations of the critical amplitude of speed disturbance (the amplitude of the critical nucleus) required for traffic breakdown (F→S transition) in free flow as a flow-rate function. Simulations of the Kerner-Klenov stochastic microscopic three-phase model (see Appendix A). Adapted from [35]

consider free flow as unstable with respect to traffic breakdown (for more details, see Sect. 5.4). The decreasing flow-rate function $\Delta v^{(B)}_{\text{cr, FS}}(q_{\text{sum}})$ (Fig. 5.5a) following from this qualitative theory of the nucleation of traffic breakdown has indeed been confirmed in simulations with the Kerner-Klenov stochastic microscopic three-phase traffic flow model (Fig. 5.6).

Obviously, the probability of spontaneous occurrence of a small local speed disturbance in free flow is considerably larger than the probability of the occurrence of a large local speed disturbance. For this reason, the smaller the critical nucleus $\Delta v^{(B)}_{\text{cr, FS}}$ is required for the F→S transition at the bottleneck, the more probable this nucleus can occur spontaneously in free flow at the bottleneck.

As explained above, the larger the flow rate q_{sum} in free flow at a bottleneck, the smaller should be the amplitude $\Delta v^{(B)}_{\text{cr, FS}}$ (5.9) of the critical nucleus (Fig. 5.5a). This leads to the following general result: The probability of the spontaneous occurrence of a nucleus for traffic breakdown $P^{(B)}_{\text{nucleus}}(q_{\text{sum}})$ is an increasing function of the flow rate (Fig. 5.5b). Under condition $q_{\text{sum}} < C_{\text{min}}$ (5.5), the F→S transition cannot occur at the bottleneck. Thus, under condition (5.5) there is no nucleus for the F→S transition (Fig. 5.5a). For this reason, under condition (5.5) the probability of the occurrence of a nucleus $P^{(B)}_{\text{nucleus}}(q_{\text{sum}}) = 0$ (Fig. 5.5b). In accordance with (5.7), this analysis of the flow-rate function $P^{(B)}_{\text{nucleus}}(q_{\text{sum}})$ (Fig. 5.5b) explains the increasing flow-rate function of the breakdown probability $P^{(B)}(q_{\text{sum}})$ (Fig. 5.5c).

The probability of the spontaneous occurrence of a nucleus $P^{(B)}_{\text{nucleus}}(q_{\text{sum}})$ is an increasing function of the flow rate q_{sum} in free flow at a highway bottleneck (Fig. 5.5b).

In accordance with the basic assumption of three-phase theory (5.7), the probability of spontaneous traffic breakdown $P^{(B)}(q_{\text{sum}})$ is an increasing function of the flow rate q_{sum} in free flow at a highway bottleneck (Fig. 5.5c).

5.3.4 Z-Characteristic for Traffic Breakdown

In accordance with formula (5.9) and the flow-rate dependence of the amplitude of the critical nucleus $\Delta v^{(B)}_{\text{cr, FS}}(q_{\text{sum}})$ for traffic breakdown at a highway bottleneck (Fig. 5.5a) explained in Sect. 5.3.3, the speed within the critical nucleus $v^{(B)}_{\text{cr, FS}}(q_{\text{sum}})$ is an increasing function of the flow rate (dashed curve in Fig. 5.5d).

In the speed–flow plane (Fig. 5.5d), flow rate dependencies of the speed within the deterministic local speed disturbance in free flow at the bottleneck $v^{(B)}_{\text{free}}(q_{\text{sum}})$, the speed within the critical nucleus $v^{(B)}_{\text{cr, FS}}(q_{\text{sum}})$, and a 2D-region of states of synchronized flow resulting from traffic breakdown at the bottleneck represent together a *Z-characteristic for traffic breakdown* at the bottleneck (Fig. 5.5d) that in a more simplified case has already been shown in Fig. 3.22. The Z-characteristic for traffic breakdown is a *macroscopic characteristic* of traffic breakdown at the bottleneck.

With the use of the Z-characteristic for traffic breakdown at the bottleneck (Fig. 5.5d) traffic breakdown is explained as follows. The nucleus for traffic breakdown appears randomly when the speed v_{dis} within a local speed disturbance at the bottleneck becomes equal to or lower than the speed $v^{(B)}_{\text{cr, FS}}$ within a critical nucleus for traffic breakdown (dashed curves in Figs. 5.1c and 5.2c), i.e.,

$$v_{\text{dis}} \leq v^{(B)}_{\text{cr, FS}}. \tag{5.10}$$

The development of the nucleus leads to traffic breakdown at the bottleneck (labeled by down-arrows F→S in Figs. 5.1c, 5.2c and 5.5d).

When the opposite condition

$$v_{\text{dis}} > v^{(B)}_{\text{cr, FS}} \tag{5.11}$$

is satisfied (dashed curves in Figs. 5.1b and 5.2b), no nucleus appears. Therefore, no traffic breakdown occurs at the bottleneck. In this case, the random component of the permanent local speed disturbance decays over time (labeled by up-arrows in Figs. 5.1b and 5.2b).

An Z-characteristic of traffic breakdown (Fig. 5.5d) is a macroscopic traffic flow characteristic that explains the nucleation nature of traffic breakdown at a highway bottleneck.

5.4 Probabilistic Characteristics of Spontaneous Traffic Breakdown at Bottleneck

In Sect. 5.3, we have considered some *qualitative* results of the three-phase theory about the nucleation nature of traffic breakdown at highway bottlenecks [4–17]. These qualitative results have been confirmed in a microscopic stochastic three-phase theory [18, 19, 23, 28, 29, 31, 35–37, 41, 44, 45, 47].

In this section, we consider *theoretical* macroscopic probabilistic characteristics of traffic breakdown resulting from the microscopic stochastic three-phase theory like a theoretical probability of traffic breakdown at a highway bottleneck (Sect. 5.4.1) and a theoretical Z-characteristic for traffic breakdown at the bottleneck (Sect. 5.4.2).

Microscopic characteristics of nuclei for traffic breakdown at highway bottlenecks resulting from the microscopic stochastic three-phase theory [23, 28, 31, 35–37, 41, 44, 45, 47] will be considered in Sect. 5.12 and 5.13.

5.4.1 Theoretical Probability of Spontaneous Traffic Breakdown

A theoretical probability of spontaneous traffic breakdown in the framework of the microscopic stochastic three-phase theory was firstly found in 2002 (Fig. 5.7) [47]. The flow-rate function of the breakdown probability (black points in Fig. 5.7) is

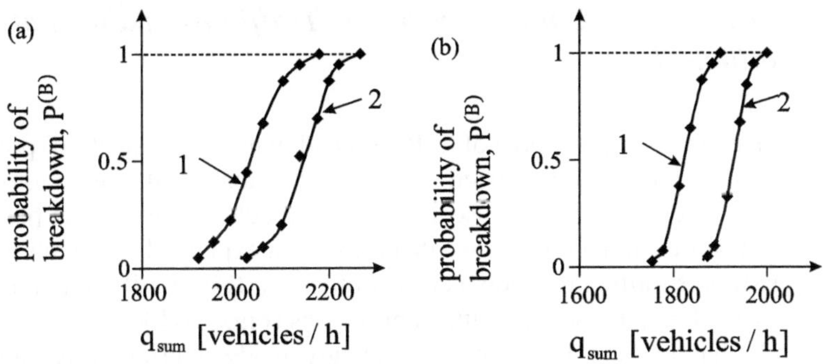

Fig. 5.7 Simulated flow-rate dependencies of the probability of traffic breakdown at an on-ramp bottleneck on single-lane road: (**a**) On-ramp inflow rate $q_{on} = 60$ vehicles/h. (**b**) On-ramp inflow rate $q_{on} = 200$ vehicles/h. Curves 1 and 2 in (**a, b**) are related, respectively, to two different values of time interval for observing traffic flow $T_{ob} = 30$ (curves 1) and 15 min (curves 2). $q_{sum} = q_{in} + q_{on}$. Adapted from [47]

well fitted by a function [47]:

$$P^{(B)}(q_{sum}) = \frac{1}{1 + \exp[\beta(q_P - q_{sum})]}, \tag{5.12}$$

where q_{sum} is the flow rate at the bottleneck,[4] β and q_P are parameters.[5]

A growing character of the flow-rate dependence of the theoretical probability of traffic breakdown (5.12) corresponds qualitatively to the empirical flow-rate dependence of the probability of traffic breakdown discussed in Sect. 2.3 (see Fig. 2.3).

---

[4]Obviously, formula (5.12) can be rewritten as follows (this equivalent form for formula (5.12) has been used in [47]; see caption to Fig. 18 of [47]):

$$P^{(B)}(q_{sum}) = (1 + \tanh[\chi(q_{sum} - q_P)])/2,$$

where $\chi = \beta/2$.

[5] For an on-ramp bottleneck (Fig. 5.1a), in (5.12) parameters q_P and β depend on the on-ramp inflow rate q_{on}, the flow rate $q_{sum} = q_{on} + q_{in}$ is the flow rate in free flow downstream of the bottleneck, where q_{in} is the flow rate in free flow on the main road upstream of the bottleneck. For the off-ramp bottleneck (Fig. 5.2a), in (5.12) parameters q_P and β depend on the percentage of vehicles leaving the main road to off-ramp at an off-ramp bottleneck $\eta_{off} = (q_{off}/q_{in})100\%$, the flow rate $q_{sum} = q_{in}$, q_{off} is the flow rate of vehicles leaving the main road to off-ramp at the off-ramp bottleneck [17].

5.4.2 *Theoretical Z-Characteristic for Traffic Breakdown at Bottleneck*

The qualitative macroscopic model of traffic breakdown (F→S transition) presented in Sect. 5.3 [3, 4] has in details been confirmed by numerical simulations made with different microscopic three-phase traffic flow models [35, 36, 47]. In particular, macroscopic features of the breakdown that have been explained with the use of the Z-characteristic for traffic breakdown at the bottleneck (Fig. 5.5d) are qualitatively the same as found in microscopic traffic simulations (Fig. 5.8) [36].

We recall that in Sect. 1.12 we have already made a brief discussion of a Z-characteristic for traffic breakdown at an on-ramp bottleneck derived with simulations of the Kerner-Klenov stochastic microscopic three-phase traffic flow model [36]: For the simplification of the consideration of the Z-characteristic, in Fig. 1.8b we have averaged all 2D-states of different speeds in synchronized flow for a given on-ramp inflow q_{on} to some average synchronized flow speed. The Z-characteristic for traffic breakdown at the on-ramp bottleneck on a single lane road shown in Fig. 5.8 is derived without the averaging of synchronized flow states.

As a qualitative Z-characteristic for traffic breakdown shown in Fig. 5.5d, the theoretical Z-characteristic for traffic breakdown (Fig. 5.8) is built by metastable

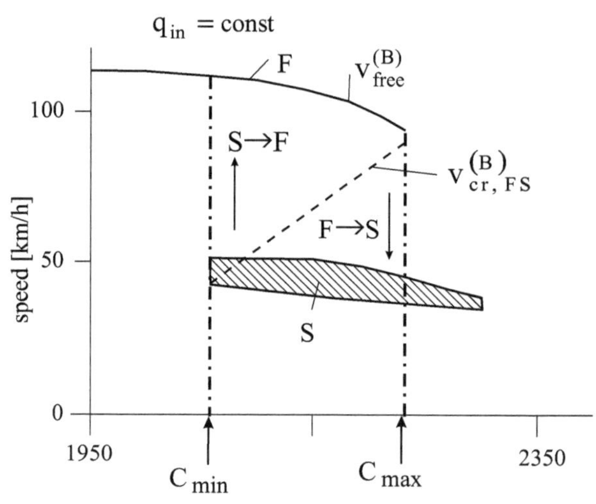

Fig. 5.8 Simulations of Z-characteristic for traffic breakdown at on-ramp bottleneck with the Kerner-Klenov stochastic microscopic three-phase model: Z-characteristic as function of the flow rate at the on-ramp bottleneck $q_{in} + q_{on}$ when the on-ramp inflow rate q_{on} is a variable and the flow rate upstream of the on-ramp bottleneck q_{in} is a given value (q_{in} = 1756 vehicles/h). F—free flow phase, S—synchronized flow phase. Arrows F→S and S→F illustrate qualitatively the F→S and S→F transitions, respectively. C_{min} is the minimum highway capacity, C_{max} is the maximum highway capacity. Adapted from [36]

states of the traffic phases F and S (solid curves F and S, respectively) together with a branch for the speed within critical nuclei $v^{(B)}_{cr, FS}(q_{sum})$ for the F→S transition (traffic breakdown) (dashed curve $v^{(B)}_{cr, FS}$ in Fig. 5.8 between the metastable states of the phases F and S).

The theoretical Z-characteristic for traffic breakdown (Fig. 5.8) derived through simulations of the Kerner-Klenov microscopic stochastic three-phase model confirms the qualitative theory of the nucleation of traffic breakdown at a highway bottleneck (Sect. 5.3) as follows.

(i) When the initial state of traffic flow at the bottleneck is free flow, simulations show that if the speed within a local speed disturbance in a free flow state at the bottleneck is equal to or smaller than the speed $v^{(B)}_{cr, FS}$ within a critical nucleus for traffic breakdown, this disturbance is a nucleus for an F→S transition: The F→S transition (traffic breakdown) does occur at the bottleneck.

(ii) We have found that the flow-rate dependence of the amplitude of the critical nucleus $\Delta v^{(B)}_{cr, FS}(q_{sum})$ (5.9) for the F→S transition is qualitatively the same as shown in Fig. 5.5a:

- At the flow rate $q_{sum} < C_{min}$, there is no nucleus that can lead to an F→S transition at the bottleneck.
- The speed $v^{(B)}_{cr, FS}$ within the critical nucleus for traffic breakdown is the smallest one at the flow rate $q_{sum} = C_{min}$.
- Within the flow rate range $C_{min} \leq q_{sum} < C_{max}$ (5.1), when the flow rate q_{sum} increases, the amplitude of the critical nucleus $\Delta v^{(B)}_{cr, FS}(q_{sum})$ (5.9) required for the F→S transition at the bottleneck decreases.

(iii) This decrease in the amplitude of the critical nucleus for traffic breakdown with the increase in the flow rate $\Delta v^{(B)}_{cr, FS}(q_{sum})$ (5.9) found in simulations explains another result of simulations that the probability of traffic breakdown $P^{(B)}(q_{sum})$ during a given time interval T_{ob} is a decreasing function of the flow rate q_{sum} (Fig. 5.7). This theoretical result is the same as that of the qualitative three-phase theory (Fig. 5.5a,c).

It should be emphasize that as the qualitative Z-characteristic shown in Fig. 5.5d, the theoretical Z-characteristic for traffic breakdown (Fig. 5.8) is a macroscopic characteristic of traffic breakdown. This means that it has been derived based on averaging of many simulation realizations (runs) in which traffic breakdown has been observed in simulations (Sect. 5.4.4). These different simulation realizations (runs) have been made for each of the flow rates q_{sum}. For this reason, the branch for the speed within the critical nuclei required for the breakdown (dashed curve in Fig. 5.8) represents averaged values of the speed within the critical nuclei. In the reality, a nucleus for traffic breakdown exhibits a very complex spatiotemporal microscopic structure. The spatiotemporal microscopic structure of the nucleus for traffic breakdown at the bottleneck as well as the microscopic traffic flow features that determine the occurrence of the nucleus are the subject of Sects. 5.12 and 5.13.

5.4.3 *Flow-Rate Dependence of Characteristics of Spontaneous Traffic Breakdown*

Under conditions (5.1), free flow is in a metastable state with respect to an F→S transition at a bottleneck. Numerical simulations of traffic flow models in the framework of the three-phase theory show that in this metastable free flow there can be two different flow rate regions in which traffic breakdown at a highway bottleneck exhibits different features (Fig. 5.9). In accordance with the results of numerical simulations, we can distinguish four ranges I–IV of the flow rate q_{sum} at the bottleneck in which traffic breakdown can exhibit qualitatively different probabilistic characteristics (Fig. 5.10). The objective of this Sect. 5.4.3 is an explanation of the meaning and consequences of these different ranges I–IV of the flow rate q_{sum} at the bottleneck.

5.4.3.1 Flow-Rate Region I

Under condition $q_{sum} < C_{min}$ (5.5), there is no traffic state associated with the synchronized flow phase. Therefore, no traffic breakdown (F→S transition) can occur at the bottleneck. This is independent of the amplitude of an initial time-limited speed disturbance at the bottleneck. This means that there exist no nuclei for traffic breakdown. Thus, as mentioned above, under condition (5.5) free flow is *stable* with respect to an F→S transition.

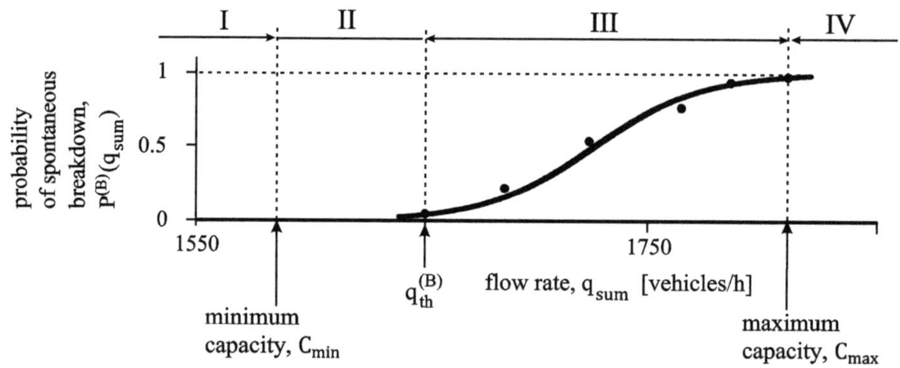

Fig. 5.9 Simulations of probability of traffic breakdown $P^{(B)}(q_{sum})$ made with the KKSW CA model (Appendix B). Different values of the flow rate q_{sum} are related to the change in the flow rate q_{in} at a given on-ramp inflow rate $q_{on} = 400$ vehicles/h. $T_{ob} = 30$ min. Number of simulation realizations used for the calculation of the breakdown probability for each of the flow rates is equal to $N_r = 40$ (see explanations in Sect. 5.4.5 below). Function $P^{(B)}(q_{sum})$ is well fitted through formula (5.12). Adapted from [44, 46]

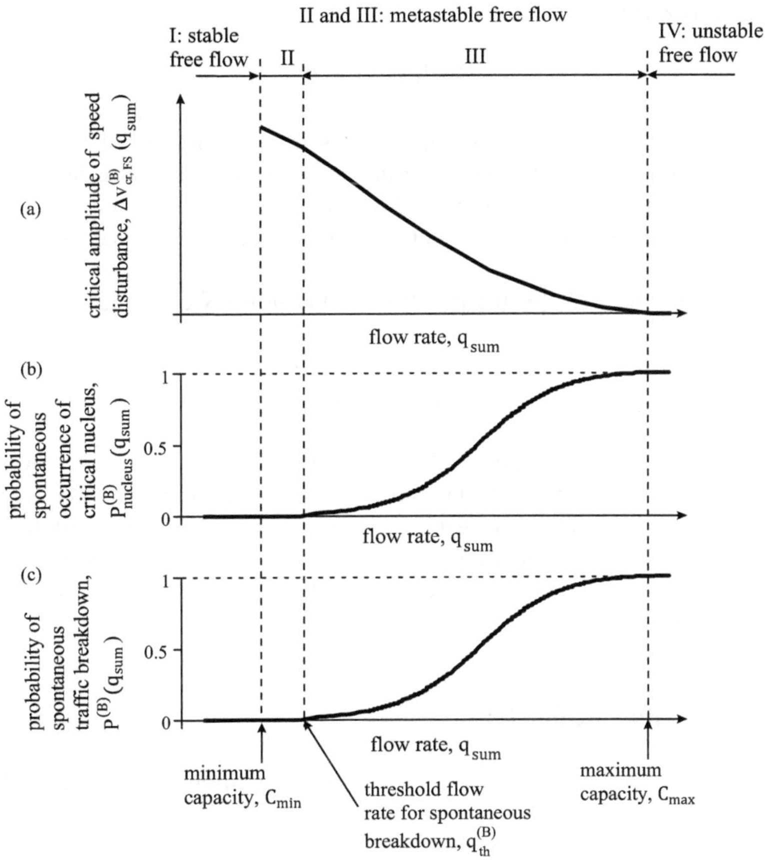

Fig. 5.10 Qualitative explanations of two different flow rate regions II and III within the flow rate range of a metastable free flow with respect to traffic breakdown (F→S transition) at the bottleneck associated with results of numerical simulations (Fig. 5.9): (**a**, **b**) Qualitative flow-rate dependencies of the critical nucleus $\Delta v^{(B)}_{\text{cr, FS}}(q_{\text{sum}})$ (**a**), of function $P^{(B)}_{\text{nucleus}}(q_{\text{sum}})$ (**b**), and qualitative breakdown probability $P^{(B)}(q_{\text{sum}})$ (**c**) have been adapted from Fig. 5.5, respectively

5.4.3.2 Flow-Rate Region II

When the flow rate q_{sum} exceeds the minimum capacity $C_{\min}$, however, the difference $q_{\text{sum}} - C_{\min}$ is small enough, conditions

$$C_{\min} \leq q_{\text{sum}} < q^{(B)}_{\text{th}} \tag{5.13}$$

can be satisfied. In (5.13), $q_{\text{sum}} = q^{(B)}_{\text{th}}$ is a threshold value of the flow rate q_{sum} for spontaneous traffic breakdown at a bottleneck. Under conditions (5.13) the probability of spontaneous traffic breakdown at the bottleneck during the time interval T_{ob} is zero:

$$P^{(B)}(q_{sum}) = 0. \tag{5.14}$$

However, under condition (5.13) free flow is in a metastable sate with respect to traffic breakdown (F→S transition) at the bottleneck. Because in accordance with (5.14) no spontaneous traffic breakdown occurs at the bottleneck during the time interval T_{ob}, traffic breakdown can only be induced in the metastable free flow at the bottleneck.

From condition (5.13) for the free flow metastability we can state that the flow rate

$$q_{sum} = C_{min} \tag{5.15}$$

can also be considered a threshold flow rate for an *induced* F→S transition at the bottleneck: The minimum highway capacity C_{min} is the *minimum flow* rate q_{sum} downstream of the bottleneck at which an F→S transition (traffic breakdown) can still be induced. This is because under condition $q_{sum} < C_{min}$(5.5) free flow is stable with respect to an F→S transition at the bottleneck: At $q_{sum} < C_{min}$ no traffic breakdown can be induced.

5.4.3.3 Flow-Rate Region III

When the flow rate q_{sum} increases, rather than conditions (5.13), the following conditions

$$q_{th}^{(B)} \leq q_{sum} < C_{max} \tag{5.16}$$

can be satisfied. As explained in Sects. 5.3 and 5.4.2, the increase in the flow rate q_{sum} in metastable free flow leads to a decrease in the amplitude of the critical nucleus $\Delta v^{(B)}_{cr,\,FS}(q_{sum})$ for traffic breakdown (Fig. 5.5a). Therefore, we can assume that under conditions (5.16), the probability of the spontaneous traffic breakdown at the bottleneck during the time interval T_{ob} can satisfy conditions

$$0 < P^{(B)}(q_{sum}) < 1. \tag{5.17}$$

This means that traffic breakdown can occur spontaneously during the time interval T_{ob}. Because under conditions (5.17) free flow is in a metastable state with respect to traffic breakdown at the bottleneck, the breakdown can also be induced at the bottleneck. Thus, in range III (5.17) of the flow rate, either spontaneous or induced traffic breakdown is possible.

Simulations of traffic breakdown at a highway bottleneck with microscopic three-phase traffic flow models show that when the time interval T_{ob} for observing traffic flow increases, the difference $q_{th}^{(B)} - C_{min}$ decreases. In this case, the width of range III of the flow rate increases, while the width of range II of the flow rate decreases, correspondingly. The physical meaning of this result is as follows. The longer the time interval T_{ob}, the more probable that a nucleus for traffic breakdown appears randomly at the bottleneck. In other words, at a given flow rate q_{sum} the longer the time interval T_{ob}, the larger the probability $P_{nucleus}^{(B)}$ for the nucleus occurrence. In its turn, correspondingly to the basic assumption of three-phase theory (5.7), the breakdown probability $P^{(B)}$ increases, when the time interval T_{ob} becomes longer. Thus, the increase in the time interval T_{ob} causes a decrease in the threshold flow rate $q_{th}^{(B)}$ with a related shift of the boundary between ranges III and II of the flow rate in Fig. 5.10 to the left in the flow-rate axis.

5.4.3.4 Flow-Rate Region IV

Under condition

$$q_{sum} \geq C_{max}, \tag{5.18}$$

the probability of the spontaneous occurrence of a nucleus during the time interval T_{ob} satisfies condition $P_{nucleus}^{(B)}(q_{sum}) = 1$. This means that the probability of the spontaneous traffic breakdown at the bottleneck occurring during the given time interval T_{ob} is equal to

$$P^{(B)}(q_{sum}) = 1. \tag{5.19}$$

5.4.4 *Time-Delayed Traffic Breakdown and Calculation of Breakdown Probability at Bottleneck*

In accordance with the basic assumption of the three-phase theory (5.7), when a nucleus appears at a highway bottleneck, spontaneous traffic breakdown does occur in free flow at the bottleneck. However, the appearance of the nucleus in free flow is a random event. This means that there should be a random time delay in the appearance of the nucleus. We denote this time delay by $T^{(B)}$. In other words, spontaneous traffic breakdown is a time-delayed phase transition from free flow to synchronized flow with a random value of the time delay. Thus, spontaneous traffic breakdown (F→S transition) can be called as a random time-delayed traffic breakdown (time-delayed traffic breakdown for short).

To understand a random time delay of spontaneous traffic breakdown, we consider numerical simulations of traffic breakdown in traffic flow of identical drivers and identical vehicles moving on a single-lane road with an on-ramp bottleneck. We assume that flow rates q_{on} and q_{in} in free flow at the bottleneck

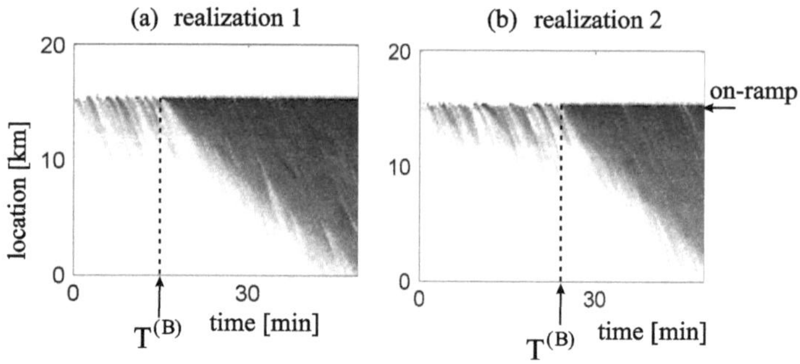

Fig. 5.11 Simulations of random time delay of traffic breakdown on single-lane road with on-ramp bottleneck with the KKSW CA model (see Appendix B): (**a, b**) Speed data presented by regions with variable shades of gray (shades of gray vary from white to black when the speed decreases from 120 km/h (*white*) to zero (*black*)). Two different simulation realizations 1 (**a**) and 2 (**b**) are related to the same set of the flow rates $q_{\rm on}$, $q_{\rm in}$. $T_{\rm ob} = 45$ min. Adapted from [44, 46]

(Fig. 5.1a) are time-independent. We assume also that the flow rate $q_{\rm sum} = q_{\rm on} + q_{\rm in}$ satisfies conditions (5.16). Under these conditions, during a given long enough time interval $T_{\rm ob}$ for observing traffic flow, traffic breakdown can occur spontaneously with the probability satisfying conditions (5.17) or the breakdown can be induced at the bottleneck.

A time delay $T^{(\rm B)}$ for traffic breakdown is equal to a time interval between the begin of simulations with time-independent flow rates $q_{\rm on}$ and $q_{\rm in}$ at time instant $t = 0$ and the occurrence of spontaneous traffic breakdown at the bottleneck (time delay $T^{(\rm B)}$ in Fig. 5.11a).

We repeat simulations for the same flow rates $q_{\rm on}$ and $q_{\rm in}$ and the same other model parameters. However, we change an initial condition for a random function in the model.[6] We have found that the change in the initial conditions, which is made *only* at time instant $t = 0$, leads to another time delay $T^{(\rm B)}$ for traffic breakdown (time delay $T^{(\rm B)}$ in Fig. 5.11b).

Simulations made at the same flow rates and the same other model parameters, however, at different initial conditions can be called different *simulation realizations* or different *simulation runs*. Making many different simulation realizations, we find that values of time delays $T^{(\rm B)}$ for traffic breakdown at the bottleneck are indeed random values.

Different simulations realizations have been used for the simulation of the probability of traffic breakdown at the bottleneck as the flow-rate function $P^{(\rm B)}(q_{\rm sum})$ (Fig. 5.9). For a given flow rate $q_{\rm sum}$, the breakdown probability $P^{(\rm B)}$ has been

[6] Initial values for random function *rand*() in the model (see Appendix B) are different for different simulation realizations.

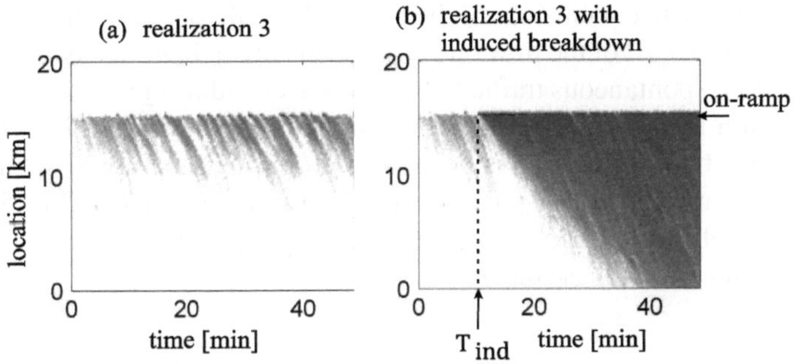

Fig. 5.12 Continuation of simulations of traffic breakdown at on-ramp bottleneck presented in Fig. 5.11. Simulations of induced traffic breakdown: (**a**, **b**) Speed data presented by regions with variable shades of gray (shades of gray vary from white to black when the speed decreases from 120 km/h (*white*) to zero (*black*)). In (**a**), one of the simulation realizations (realization 3) is shown that has been calculated at the same model parameters as those in Fig. 5.11. However, in contrast with simulation realizations 1 and 2 shown in Fig. 5.11, in simulation realization 3 no spontaneous traffic breakdown occurs during the time interval $T_{\mathrm{ob}} = 45$ min. In (**b**), traffic breakdown has been induced for the same initial conditions as that in realization 3 (**a**). Adapted from [44, 46]

calculated through formula

$$P^{(\mathrm{B})} = \frac{n_{\mathrm{r}}}{N_{\mathrm{r}}}. \tag{5.20}$$

In (5.20), N_{r} (where $N_{\mathrm{r}} \gg 1$) is the number of all different simulation realizations, n_{r} is the number of simulation realizations in which traffic breakdowns have occurred during a chosen time interval T_{ob}. Based on formula (5.20), simulations of the breakdown probability $P^{(\mathrm{B})}$ has been repeated for each flow rate q_{sum}. This simulation procedure allows us to find the flow-rate dependence of the breakdown probability $P^{(\mathrm{B})}(q_{\mathrm{sum}})$ (Fig. 5.9). For the calculation of the breakdown probability $P^{(\mathrm{B})}(q_{\mathrm{sum}})$ at the on-ramp bottleneck, either the flow rate q_{in} or the flow rate q_{on} has been chosen as a constant value.[7]

As explained in Sect. 5.4.3, there are four regions I–IV of the flow rate $q_{\mathrm{sum}} = q_{\mathrm{on}} + q_{\mathrm{in}}$ in which different conditions for traffic breakdown are satisfied. In Fig. 5.11, we have chosen the flow rate q_{sum} that satisfies conditions (5.16) of range III of the flow rate q_{sum}. For flow-rate range III, the breakdown probability $P^{(\mathrm{B})}$ is less than 1 (Fig. 5.9). In accordance with formula (5.20), we have indeed found that in some simulation realizations traffic breakdown has occurred (realizations 1 and 2 in Figs. 5.11a,b), however, in some other simulation realizations *no* traffic breakdown occurs at the bottleneck during the time interval T_{ob} (realization 3 shown in Fig. 5.12a).

[7]Simulations of the function $P^{(\mathrm{B})}(q_{\mathrm{sum}}, q_{\mathrm{on}})$ for the case when both flow rates q_{in} and q_{on} are changed can be found in Sect. 12.4 (see Fig. 12.3e).

In accordance with conclusions of the probabilistic theory of traffic breakdown of Sect. 5.4.3, for range III of the flow rate q_{sum}, traffic breakdown can be induced at the bottleneck, if no spontaneous traffic breakdown occurs during the time interval T_{ob}. This conclusion is confirmed in numerical simulations shown in Fig. 5.12b. In the simulations, traffic breakdown has been induced through the use of a time-limited increase in the on-ramp inflow rate q_{on} applied at time instant T_{ind} (Fig. 5.12b). We can make the following conclusion of simulations with traffic flow models in the framework of the three-phase theory.

Spontaneous traffic breakdown at a highway bottleneck is a random time-delayed phase transition from free flow (F) to synchronized flow (S) (random time-delayed F→S transition).

5.4.5 *Effect of Number of Simulation Realizations on Threshold Flow Rate and Maximum Highway Capacity*

As in a study of the flow-rate dependence of the empirical breakdown probability $P^{(B)}$ (Figs. 2.2 and 2.3),[8] in numerical calculations of the breakdown probability $P^{(B)}(q_{sum})$ [47] only a finite number N_r of simulation realizations (runs) can be made for the calculation of the value $P^{(B)}$ (5.20) for each given flow rate q_{sum}. In accordance with the definition of the threshold flow rate $q_{th}^{(B)}$, the smallest value of the breakdown probability $P^{(B)}$, which is still larger than zero, is related to the flow rate $q_{sum} = q_{th}^{(B)}$. Thus, the smallest value of the breakdown probability satisfying condition (5.17) is given by formula

$$P^{(B)} \Big|_{q_{sum}=q_{th}^{(B)}} = \frac{1}{N_r}. \tag{5.21}$$

In other words, the larger the number N_r of simulation realizations (runs), the more exactly the threshold flow rate $q_{sum} = q_{th}^{(B)}$ can be calculated. However, because $N_r \neq \infty$, the exact value of the threshold flow rate $q_{sum} = q_{th}^{(B)}$ cannot be found.

[8]The probability of the spontaneous occurrence of traffic breakdown at a highway bottleneck during a given time interval is calculated from a *finite number* of different days of the observation of free flow at a highway bottleneck. This is because a calculation of the breakdown probability for an *infinite number* of different days of the observation of free flow at the bottleneck has no sense for real traffic flow.

For the same reason, the exact value of the maximum capacity $C_{\max}$ cannot also be found: An approximate value of $C_{\max}$ is found from condition

$$P^{(\mathrm{B})}\,|_{q_{\mathrm{sum}}=C_{\max}} = \frac{N_{\mathrm{r}}-1}{N_{\mathrm{r}}}. \tag{5.22}$$

In real field traffic data, the flow rate q_{sum} in free flow at a highway bottleneck cannot usually be considered time-independent. Therefore, a random time delay in traffic breakdown is difficult to prove by a study of real field traffic data.

However, as we have shown above the existence of a random time delay in traffic breakdown can be proven in numerical simulations of spontaneous traffic breakdown at a highway bottleneck with traffic flow models in the framework of the three-phase theory. The reliability of this proof is as follows. These traffic flow models are able to reproduce all empirical features of spontaneous traffic breakdown studied in Chap. 3.

5.4.6 Mean Time Delay for Occurrence of Traffic Breakdown

As explained in Sect. 5.3, the smaller the amplitude of the critical nucleus for traffic breakdown $\Delta v^{(\mathrm{B})}_{\mathrm{cr,\ FS}}$, the larger the probability $P^{(\mathrm{B})}_{\mathrm{nucleus}}(q_{\mathrm{sum}})$ of the occurrence of the critical nucleus (Figs. 5.5a,b). Therefore, the larger the probability of the occurrence of the critical nucleus $P^{(\mathrm{B})}_{\mathrm{nucleus}}$ during a time interval T_{ob}, the shorter should be the mean time delay of the occurrence of the nucleus that we denote by $T^{(\mathrm{B,\ mean})}$. This means that the mean time delay $T^{(\mathrm{B,\ mean})}(q_{\mathrm{sum}})$ of traffic breakdown is a decreasing flow rate function. This qualitative conclusion is confirmed with simulations of traffic flow models in the framework of the three-phase theory (Fig. 5.13).

However, we should mention that a calculation of the mean time delay $T^{(\mathrm{B,\ mean})}(q_{\mathrm{sum}})$ of traffic breakdown has the physical sense only as long as it is

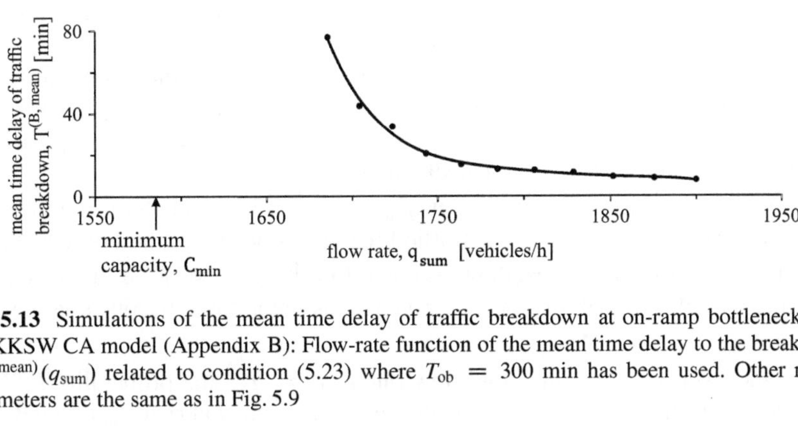

Fig. 5.13 Simulations of the mean time delay of traffic breakdown at on-ramp bottleneck with the KKSW CA model (Appendix B): Flow-rate function of the mean time delay to the breakdown $T^{(\mathrm{B,\ mean})}(q_{\mathrm{sum}})$ related to condition (5.23) where $T_{\mathrm{ob}} = 300$ min has been used. Other model parameters are the same as in Fig. 5.9

considerably shorter than the time interval T_{ob} for observing traffic flow, i.e., when

$$T^{(B,\, mean)}(q_{sum}) \ll T_{ob}. \tag{5.23}$$

Formula (5.23) has been taken into account by the calculation of the flow-rate function $T^{(B,\, mean)}(q_{sum})$ in Fig. 5.13. To determine function $T^{(B,\, mean)}(q_{sum})$ (Fig. 5.13), we have chosen a very long time interval for observing free flow $T_{ob} = 300$ min at which the probability of traffic breakdown $P^{(B)}(q_{sum}) = 1$.

As shown in [39–41], the mean time for the random occurrence of the nucleus $T^{(B,\, mean)}$ in free flow can be described by a classical formula of metastable systems of natural science [1].

5.4.7 *Definition and Physical Meaning of Threshold Flow Rate for Spontaneous Traffic Breakdown*

The threshold flow rate $q^{(B)}_{th}$ is defined through conditions

$$\begin{aligned} P^{(B)}(q_{sum}) &= 0 \quad \text{at} \quad q_{sum} < q^{(B)}_{th}, \\ P^{(B)}(q_{sum}) &> 0 \quad \text{at} \quad q_{sum} \geq q^{(B)}_{th}. \end{aligned} \tag{5.24}$$

Conditions (5.24) mean that the probability of spontaneous traffic breakdown during a time interval T_{ob} as a flow-rate function $P^{(B)}(q_{sum})$ becomes larger than zero at the flow rate $q_{sum} = q^{(B)}_{th}$. The physics of this behavior of the function $P^{(B)}(q_{sum})$ is as follows. Under condition

$$q_{sum} = C_{min}, \tag{5.25}$$

the amplitude of a critical nucleus $\Delta v^{(B)}_{cr,\, FS}$ required for the spontaneous breakdown can be assumed to be a very large one (Fig. 5.5a). Indeed, simulations show that a *spontaneous occurrence* of such a large nucleus during a given final time of the observing traffic flow T_{ob} within a final number of different simulation realizations N_r is not possible. This conclusion leading to condition $P^{(B)} = 0$ (5.14) is valid within some flow-rate range (5.13). Within this flow-rate range such a large nucleus can be induced at a highway bottleneck only.

However, simulations show that at $q_{sum} = q^{(B)}_{th}$ in one of the different simulation realizations N_r spontaneous traffic breakdown occurs at the bottleneck. The probability of this traffic breakdown is equal to $P^{(B)}(q^{(B)}_{th}) = \frac{1}{N_r} \ll 1$ (5.21). This explains why the probability of the spontaneous breakdown becomes larger than zero at $q_{sum} = q^{(B)}_{th}$ (5.24).

In other words, at the threshold flow rate $q_{sum} = q^{(B)}_{th}$ the probability $P^{(B)}$ that spontaneous traffic breakdown occurs during a given time interval T_{ob} is small but it is still larger than zero. However, when the flow rate $q_{\rm sum}$ becomes smaller than the threshold flow rate $q^{(\rm B)}_{\rm th}$, then the probability of spontaneous occurrence of a nucleus for traffic breakdown in free flow during the time interval $T_{\rm ob}$ is zero. Therefore, the probability of spontaneous traffic breakdown during this time interval $T_{\rm ob}$ is also zero.

For this reason, the threshold flow rate $q^{(\rm B)}_{\rm th}$ separates two regions of the flow rates $q_{\rm sum}$ satisfying conditions (5.1) for the free flow metastability with respect to traffic breakdown (F→S transition) at the bottleneck:

(i) When the flow rate $q_{\rm sum}$ is larger than the threshold value $q_{\rm sum} = q^{(\rm B)}_{\rm th}$, then spontaneous traffic breakdown can occur at the bottleneck with some probability during the given time interval $T_{\rm ob}$.
(ii) On contrary, under conditions (5.13), when the flow rate $q_{\rm sum}$ is smaller than the threshold value $q_{\rm sum} = q^{(\rm B)}_{\rm th}$, no spontaneous traffic breakdown can occur in metastable free flow at the bottleneck during the given time interval $T_{\rm ob}$. In the latter case, traffic breakdown can only be induced at the bottleneck.

5.4.8 *Definition and Physical Meaning of Maximum Highway Capacity of Free Flow at Bottleneck*

Within the flow-rate range IV (5.18), we consider the difference

$$\Delta q_{\rm sum} = q_{\rm sum} - C_{\rm max}. \tag{5.26}$$

Correspondingly to condition (5.18), $\Delta q_{\rm sum} \geq 0$. Clearly, the larger the difference $\Delta q_{\rm sum}$, the smaller the amplitude of the critical nucleus required for traffic breakdown at the bottleneck. Therefore, the larger the difference $\Delta q_{\rm sum}$, the shorter the mean time delay $T^{(\rm B,\ mean)}(q_{\rm sum})$ of traffic breakdown at the bottleneck (Fig. 5.13).

Indeed, when $\Delta q_{\rm sum}$ (5.26) is a large enough value, any small speed disturbance is a nucleus for traffic breakdown. In this case, the initial free flow can be considered unstable with respect to traffic breakdown (F→S transition). This explains why under condition (5.6) we consider free flow as unstable with respect to traffic breakdown (F→S transition) at the bottleneck (labeled by "unstable free flow" in Figs. 5.5 and 5.10).

Condition (5.18) defines the maximum capacity $C_{\rm max}$ of free flow at a highway bottleneck as follows.

The maximum capacity $C_{\rm max}$ is equal to the minimum flow flow in free flow at the bottleneck $q_{\rm sum}$ at which traffic breakdown at the bottleneck occurs spontaneously with probability $P^{(\rm B)}(q_{\rm sum}) = 1$ during a given time interval $T_{\rm ob}$.

The larger the difference Δq_{sum} (5.26), the shorter the mean time delay $T^{(B,\ mean)}(q_{sum})$ of traffic breakdown at the bottleneck (Fig. 5.13).

5.4.9 Summary of Probabilistic Characteristics of Traffic Breakdown in Three-Phase Theory

To summarize results of the analysis of traffic breakdown (F→S transition) in metastable free flow at a bottleneck, we can conclude that in the three-phase theory we distinguish the following basic characteristics of traffic breakdown at the bottleneck [16, 17, 20–28]:

- The minimum highway capacity C_{min}.
- The threshold flow rate for spontaneous traffic breakdown $q_{th}^{(B)}$.
- The maximum highway capacity C_{max}.
- A random time delay of traffic breakdown $T^{(B)}$ at a bottleneck.
- The mean time delay $T^{(B,\ mean)}$ of traffic breakdown at the bottleneck.
- The probability $P^{(B)}$ that spontaneous traffic breakdown occurs at the bottleneck during a given observation time interval T_{ob}.
- The Z-characteristic for traffic breakdown at the bottleneck.

Due to the existence of the random time delay of traffic breakdown $T^{(B)}$, spontaneous traffic breakdown at the bottleneck can be called a *time-delayed traffic breakdown* or a *random time-delayed traffic breakdown* at the bottleneck.

5.5 Induced Traffic Breakdown at Bottleneck in Empirical Traffic Data and Numerical Simulations

The nucleation nature of traffic breakdown means that traffic breakdown does occur when and only when a nucleus for the breakdown appears in free flow at the bottleneck. However, there can be many different sources of the occurrence of the nucleus that initiates empirical traffic breakdown observed in real field traffic data. In particular, as already emphasized in Sect. 3.6.1, the difference in the source of a nucleus for empirical traffic breakdown is the solely difference between empirical spontaneous and induced traffic breakdowns at a highway bottleneck.

In particular, to be a nucleus for empirical spontaneous traffic breakdown at a highway bottleneck, a local disturbance in metastable free flow, in which the flow rate q_{sum} satisfied conditions $C_{min} \leq q_{sum} < C_{max}$ (5.1), should be equal to or a larger one than a critical local disturbance (critical nucleus). Only large enough local disturbances in free flow can be equal to or a larger one than the critical nucleus. Small enough local disturbances in free flow cannot be nuclei for traffic breakdown. This feature of the source of nucleus occurrence distinguishes empirical spontaneous traffic breakdown from empirical induced traffic breakdown at highway bottlenecks. Indeed, in contrast with empirical spontaneous traffic breakdown, under conditions (5.1) almost any localized congested pattern becomes usually a nucleus for induced empirical traffic breakdown, when the pattern reaches the location of the bottleneck.

It must be stressed that in real free flow we cannot artificially induce a nucleus for traffic breakdown. Contrarily to real free flow, in theoretical and/or numerical studies of free flow with traffic flow models we can induce a local speed disturbance in a neighborhood of the bottleneck. This disturbance can artificially be chosen a large enough to be a nucleus for traffic breakdown at the bottleneck. Therefore, such traffic breakdown can be considered "induced" traffic breakdown. This is true, even if free flow is realized both upstream and downstream of the bottleneck before in simulations the nucleus has been induced in a neighborhood of the bottleneck.

To explain this point, we consider conditions at which traffic breakdown has been induced at the on-ramp bottleneck in simulations shown in Fig. 5.12. In the simulation realization shown in Fig. 5.12a no spontaneous traffic breakdown occurs at the on-ramp bottleneck during a chosen time interval T_{ob}. Then, at time instant T_{ind} we have increased the on-ramp inflow during a limited time interval. This on-ramp inflow impulse leads to a decrease in the speed in the neighborhood of the bottleneck below the speed $v^{(\text{B})}_{\text{cr, FS}}$ within the critical nucleus (Sect. 5.3.2), i.e., to the occurrence of a nucleus for traffic breakdown. As a result, induced traffic breakdown is realized at the bottleneck (Fig. 5.12b).

However, we should emphasize that the use of time-independent flow rates at the bottleneck and the organization of a special short-time increasing in the on-ramp inflow, which have been made in numerical simulations (Fig. 5.12b), are not realistic for real traffic.

Contrarily to numerical simulations, in empirical free flow there can be many picks of the on-ramp inflow rate. One of these picks can lead to traffic breakdown at the bottleneck. Because before the breakdown occurs free flow has been both upstream and downstream of the bottleneck, we should call this empirical traffic breakdown as empirical spontaneous traffic breakdown. This emphasize once more that the definitions of induced and spontaneous traffic breakdowns depend on the source of a nucleus that occurrence leads to the breakdown: If this nucleus appears in free flow at a bottleneck due to some random fluctuations in the flow rate, the vehicle speed, or the vehicle density at the bottleneck, we consider the breakdown as empirical spontaneous traffic breakdown. If there is an external nucleation of the breakdown at the bottleneck, we consider the breakdown as empirical induced traffic breakdown. For example, the breakdown at a bottleneck caused by the propagation of a congested pattern to the bottleneck location is induced traffic breakdown both in real field data and in numerical simulations. In contrast, as above-mentioned, traffic breakdown caused by an impulse of the on-ramp inflow can be considered as induced traffic breakdown only in numerical simulations.

5.6 Large Fluctuations in Free Flow: Minimum Highway Capacity as Threshold Flow Rate for Spontaneous Traffic Breakdown at Bottleneck

It should be noted that the definition of the threshold flow rate $q_{\text{sum}} = q_{\text{th}}^{(\text{B})}$ (Sect. 5.4.7) follows from the assumption that within the flow-rate region II (5.13) very large nuclei are required for the breakdown (Fig. 5.5a). Therefore, we have assumed that when conditions (5.13) are satisfied, a nucleus for traffic breakdown cannot occur *spontaneously* at the bottleneck. In this case, under conditions (5.13) *only* induced traffic breakdown can occur at the bottleneck. This fact has been proven in all numerical simulations in which some large but finite number of simulation realizations made during a finite time interval T_{ob} have been performed.

Nevertheless, in simulations of different three-phase traffic flow models we have found that the larger model fluctuations, the smaller the difference $q_{\text{th}}^{(\text{B})} - C_{\text{min}}$ between the threshold flow rate $q_{\text{th}}^{(\text{B})}$ and the minimum capacity C_{min}. The physics of this results is as follows. The larger the fluctuations in traffic flow, the larger the probability that a large nucleus for traffic breakdown occurs spontaneously during the observation time T_{ob}. The larger nuclei can occur spontaneously in traffic flow, the smaller the difference $q_{\text{th}}^{(\text{B})} - C_{\text{min}}$.

However, the fact that in numerical simulations there cannot be found large nuclei required for the breakdown, when conditions (5.13) are satisfied (Fig. 5.5a), does not prove that this result exhibits a general character. Indeed, under conditions (5.13) free flow is in a metastable state with respect to traffic breakdown. Therefore, for traffic flow with very large fluctuations we cannot exclude *the spontaneous occurrence* of large nuclei (Fig. 5.5a) required for the breakdown at the bottleneck during a finite observation time T_{ob}.

In a particular limit case of very large traffic flow fluctuations, we should assume that

$$C_{\text{min}} = q_{\text{th}}^{(\text{B})}. \tag{5.27}$$

In this case, the minimum capacity becomes both the flow-rate boundary of free flow metastability and the threshold flow rate for spontaneous traffic breakdown at the bottleneck during the observation time T_{ob}. Therefore, under condition (5.27), the threshold flow rate as a different characteristic flow rate in comparison with the minimum capacity has lost the physical sense. The same conclusion can be made about conditions (5.13): In this particular traffic flow with large fluctuations, the flow-rate ranges II (5.13) and III (5.16) in Fig. 5.10 merge into one flow rate range $C_{\text{min}} \leq q_{\text{sum}} < C_{\text{max}}$ (5.1).

In the range of the flow rate (5.1), free flow is in a metastable state with respect to traffic breakdown at the bottleneck. Under condition (5.27), either spontaneous or induced traffic breakdown is possible within the flow rate range (5.1). In other words, within the flow rate range (5.1) there is a finite probability that traffic breakdown occurs spontaneously during the time interval T_{ob} at the bottleneck.

Thus, we cannot exclude that in some future models in the framework of the three-phase theory we will find that the case (5.27) could be realized in model simulations. We can assume that condition (5.27) should be realized in many observations of traffic breakdown found in real field traffic data. This is because large flow rate fluctuations is often the case in real traffic. However, whether and when condition (5.27) or condition

$$C_{min} < q_{th}^{(B)} \tag{5.28}$$

is realized in real traffic is not known. An empirical study that could answer this question is an interesting subject of further investigations of real traffic.

Under condition (5.27), the minimum capacity C_{min} can be defined through conditions

$$\begin{aligned} P^{(B)}(q_{sum}) &= 0 \quad \text{at} \quad q_{sum} < C_{min}, \\ P^{(B)}(q_{sum}) &> 0 \quad \text{at} \quad q_{sum} \geq C_{min}. \end{aligned} \tag{5.29}$$

Conditions (5.29) mean that the probability that during the time interval T_{ob} spontaneous traffic breakdown occurs in free flow at the bottleneck $P^{(B)}(q_{sum})$ becomes larger than zero at the flow rate $q_{sum} = C_{min}$.

However, in all three-phase traffic flow models whose simulation results known and used in this book there are separated flow rate ranges II and III as discussed in Sect. 5.4. Therefore, in the book we distinguish usually traffic phenomena occurring in flow rage range II, in which only induced traffic breakdown is possible, and traffic phenomena occurring in flow rage range III, in which either induced or spontaneous traffic breakdown is possible at the bottleneck. This distinguishing does not limit a general character of the further analysis. Indeed, condition (5.27) is related to traffic flow with very large fluctuations. In this case, the threshold flow rate and the minimum capacity become the same characteristic of traffic breakdown at the bottleneck.

5.7 Stochastic Minimum and Maximum Highway Capacities

In the three-phase theory, the term "stochastic" in the definition of stochastic highway capacity of free flow at a highway bottleneck (Sect. 5.1) is associated with the result of the theory that *at any time instant* there are the infinite number of highway capacities at the bottleneck that are between the minimum highway capacity and the maximum highway capacity.

It must be noted that the maximum capacity C_{min}, the minimum capacity C_{max}, and the value $q_{th}^{(B)}$ considered in Sect. 5.4 [32–34] depend on traffic parameters.

These traffic parameters are weather, mean driver's characteristics (e.g., mean driver reaction time), share of long vehicles, etc. In real traffic flow, these traffic parameters change over time. For this reason, the values of the minimum highway capacity C_{min}, the maximum highway capacity C_{max}, and the value $q_{\text{th}}^{(\text{B})}$ change also over time. Moreover, in real traffic flow, the traffic parameters are stochastic time functions. Therefore, in real traffic flow we should consider some stochastic maximum capacity $C_{\text{max}}^{(\text{stoch})}(t)$, stochastic minimum capacity $C_{\text{min}}^{(\text{stoch})}(t)$, and a stochastic threshold flow rate $q_{\text{th}}^{(\text{B, stoch})}(t)$ whose time dependencies are determined by stochastic characteristics of traffic parameters. Qualitative hypothetical fragment of these time-functions within a time interval is shown in Fig. 5.14.

Thus, we should generalize the above definition of stochastic highway capacity (Sect. 5.1) as follows:

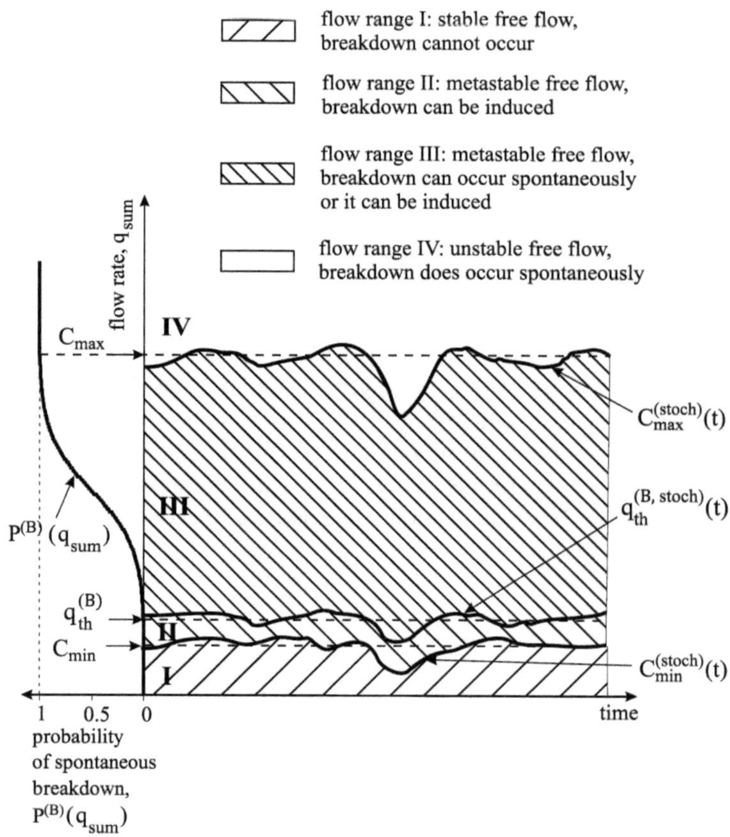

Fig. 5.14 Qualitative explanations of a connection between the probability of traffic breakdown at bottleneck and stochastic highway capacity. Breakdown probability $P^{(\text{B})}(q_{\text{sum}})$ (*left hand of figure*) is adapted from Fig. 5.5b. Flow rate ranges I, II, III, and IV (*right hand of figure*) have the same sense as those shown in Fig. 5.10 and discussed in Sect. 5.4

At any time instant t, there are the infinite number of stochastic highway capacities C of free flow at a highway bottleneck. The range of these capacities is limited by a minimum stochastic highway capacity $C_{min}^{(stoch)}(t)$ and a maximum stochastic highway capacity $C_{max}^{(stoch)}(t)$ (Fig. 5.14):

$$C_{min}^{(stoch)}(t) \le C \le C_{max}^{(stoch)}(t), \tag{5.30}$$

where $C_{min}^{(stoch)}(t) < C_{max}^{(stoch)}(t)$.

It should be noted that stochastic functions $C_{max}^{(stoch)}(t)$, $C_{min}^{(stoch)}(t)$, and $q_{th}^{(B,\ stoch)}(t)$ (Fig. 5.14) cannot be measured in empirical observations. Only their mean values, respectively, C_{max}, C_{min}, and $q_{th}^{(B)}$ can be found in empirical studies of measured traffic data. In particular, the mean values C_{max} and $q_{th}^{(B)}$ can be found from an empirical study of the flow-rate function of the breakdown probability $P^{(B)}(q_{sum})$.

Qualitative features of ranges I–IV of the flow rate q_{sum} in free flow at a bottlenecks considered in Sect. 5.4 can be generalized for stochastic functions $C_{max}^{(stoch)}(t)$, $C_{min}^{(stoch)}(t)$, and $q_{th}^{(B,\ stoch)}(t)$ as follows.

If at a given time instant t the flow rate $q_{sum}(t)$ is related to range I of flow rate in Fig. 5.14, i.e., this flow rate satisfies condition

$$q_{sum}(t) < C_{min}^{(stoch)}(t), \tag{5.31}$$

no traffic breakdown can occur at the bottleneck. Thus, this flow rate is smaller than highway capacity.

If at a given time instant t the flow rate $q_{sum}(t)$ is related to range II of flow rate in Fig. 5.14, i.e., this flow rate satisfies conditions

$$C_{min}^{(stoch)}(t) \le q_{sum}(t) < q_{th}^{(B,\ stoch)}(t), \tag{5.32}$$

then no spontaneous traffic breakdown can occur during the time interval T_{ob}. This is because in this case the probability of traffic breakdown $P^{(B)}(q_{sum}) = 0$ (5.14). However, under conditions (5.32), free flow is in a metastable state with respect to traffic breakdown at the bottleneck. In other words, this means that in this case traffic breakdown can *only* be induced at the bottleneck.

If at a given time instant t the flow rate $q_{sum}(t)$ is related to range III of flow rate in Fig. 5.14, i.e., this flow rate satisfies conditions

$$q_{th}^{(B,\ stoch)}(t) \le q_{sum}(t) < C_{max}^{(stoch)}(t), \tag{5.33}$$

then spontaneous traffic breakdown can occur during the time interval T_{ob} with the probability that is larger than zero but smaller than 1. The spontaneous traffic breakdown does occur, if at the given time instant a nucleus for traffic breakdown appears *randomly* in free flow. This random occurrence of the nucleus explains the term "stochastic" in the definition of stochastic highway capacity.

The term "stochastic" in the definition of stochastic highway capacity made in the framework of the three-phase theory is explained by the possibility of *random occurrence* of a nucleus for traffic breakdown. The nucleus occurs with the probability $0 < P^{(B)}(q_{sum}) < 1$ at any flow rate $q_{sum}(t)$ in free flow at the bottleneck that satisfies conditions (5.33).

If at a given time instant t the flow rate $q_{sum}(t)$ is related to range IV of flow rate in Fig. 5.14, i.e., this flow rate satisfies condition

$$q_{sum}(t) \geq C_{max}^{(stoch)}(t), \tag{5.34}$$

traffic breakdown does occur spontaneously during the time interval T_{ob}. This is because in this case the probability of the breakdown reaches the value 1.

5.8 Competition of Driver Over-Acceleration and Driver Speed Adaptation: A Qualitative Model

Up to now, while discussing the nucleation model of traffic breakdown introduced in the three-phase theory, we have not considered driver behavioral characteristics that are responsible for the nucleation nature of traffic breakdown at the bottleneck. In this and next sections of this chapter we analysis these driver behavioral characteristics as well as their effects on nucleation features of free flow and synchronized flow at a highway bottleneck. For simplicity, we start from a qualitative consideration of these effects.

The metastability of free flow with respect to an F→S transition has been explained in the three-phase traffic theory through the following hypotheses [4–17, 20–28, 31]:

- The metastability of free flow with respect to an F→S transition at a highway bottleneck is caused by a competition between two opposing tendencies—the tendency towards free flow due to the *over-acceleration effect* (driver over-acceleration) and the tendency towards synchronized flow due to the *speed adaptation effect* (driver speed adaptation).
- The tendency towards free flow due to the over-acceleration effect is associated with a traffic flow instability in synchronized flow caused by a discontinuous character of a density (flow-rate) function of the probability of driver over-acceleration. The discontinuous character of driver over-acceleration is associated with the existence of a driver time delay in over-acceleration.

- This instability in synchronized flow causes a growing speed wave of a local increase in the speed in synchronized flow. The growth of the speed wave leads to a phase transition from synchronized flow (S) to free flow (F) (S→F transition). For this reason, we call this instability as an *S→F instability*.

In other words, it is assumed in the three-phase theory [4–17, 20–28, 31] that due to a driver time delay in over-acceleration, the probability of driver over-acceleration in car following exhibits a discontinuous character. The discontinuous character of the over-acceleration leads to the S→F instability, i.e., to the emergence of a growing wave of a local increase in the speed in synchronized flow. This traffic flow instability (S→F instability) can be considered the result of the over-acceleration effect that determines the tendency towards free flow within a local speed increase in synchronized flow.

When we apply these hypotheses of the three-phase theory to the qualitative model of the nucleation of traffic breakdown discussed in Sect. 5.3, we can state that a local speed disturbance in free flow is a nucleus required for traffic breakdown, if the over-acceleration effect within this speed disturbance is weaker than the speed adaptation effect. Otherwise, if within a local speed disturbance the over-acceleration effect is stronger than the speed adaptation effect, the local speed disturbance decays over time, i.e., this disturbance is not a nucleus required for traffic breakdown.

5.9 Driver Speed Adaptation

5.9.1 Two-Dimensional (2D) Synchronized Flow States

For a qualitative consideration of the speed adaptation effect, we introduce the term a *steady state of synchronized flow*:

- A steady state of synchronized flow is a *hypothetical* state of synchronized flow of identical vehicles and drivers in which all vehicles move with the same time-independent speed and have the same space gaps, i.e., this synchronized flow is homogeneous in time and space. Obviously that vehicle acceleration (deceleration) is equal to zero in steady states of synchronized flow.

In the three-phase theory, the following hypothesis (hypothesis 11 of Sect. 1.9) about two-dimensional (2D) steady states of synchronized flow has been introduced [2–7][9]:

[9]It should be noted that for 2D steady states of synchronized flow (Fig. 5.15a) we have often used the term "the fundamental hypothesis of three-phase traffic theory" [17, 25]. The term should emphasize the qualitative difference between the fundamental hypothesis of earlier traffic flow theories about a theoretical fundamental diagram of traffic flow, i.e., 1D-curve(s) in the flow–density plane (see a discussion of classical traffic flow models with a theoretical fundamental

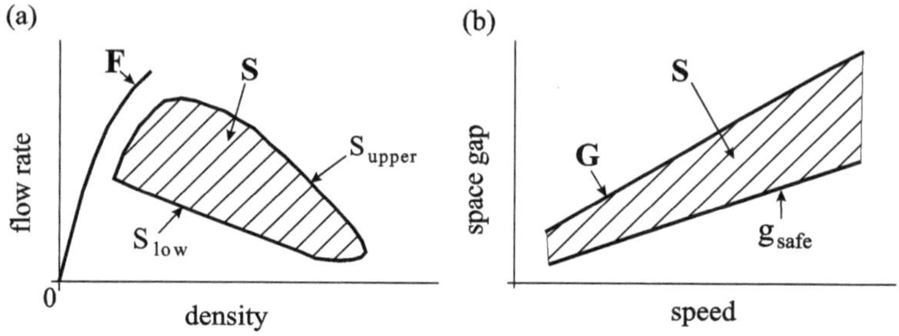

Fig. 5.15 Hypothesis about steady states of synchronized flow in the three-phase theory [2–7]: (**a**) Qualitative representation of free flow states (*F*) and 2D steady states of synchronized flow (dashed region *S*) on a multi-lane road in the flow–density plane. (**b**) A part of the 2D steady states of synchronized flow shown in (**a**) in the space-gap–speed plane (dashed region *S*)

Steady states of synchronized flow cover a two-dimensional (2D) region in the flow–density plane (Fig. 5.15a).

---

diagram (1D steady states) of traffic flow in Chap. 4). Thus, the term "fundamental hypothesis of three-phase traffic theory" has a sense *only*, if a comparison between 2D steady states of synchronized flow of the three-phase theory and 1D steady states of traffic flow (theoretical fundamental diagram) of earlier traffic flow theories is made. To emphasize this comparison, we have also used the term "fundamental diagram approach to traffic flow modeling" in the application to two-phase traffic flow models with 1D steady states of traffic flow.

Unfortunately, over last years some of traffic researchers have "extended" the use of the terms "fundamental hypothesis of three-phase traffic theory" and "fundamental diagram approach to traffic flow modelling" *behind* the above-mentioned comparison. In particular, some of the researchers equalize the three-phase theory and the fundamental hypothesis of three-phase traffic theory. This confusion leads to invalid conclusions and results. For example, while making an assumption that the 2D-states of traffic flow is the main difference between three-phase and two-phase traffic flow models, some of the researchers make an incorrect conclusion that "two-phase traffic models are approximations of the three-phase traffic models".

For these reasons, in this book we do not use the terms "the fundamental hypothesis of three-phase traffic theory" and "fundamental diagram approach to traffic flow modelling". Instead of the term "the fundamental hypothesis of three-phase traffic theory", we use the term "hypothesis of three-phase theory about 2D-states of synchronized flow". Instead of the term "fundamental diagram approach to traffic flow modelling", we use the term "two-phase traffic flow models".

As emphasized in Sect. 1.9, the main reason for the three-phase theory is the explanation of the nucleation nature of traffic breakdown at highway bottlenecks. In other words, the basic assumption of the three-phase theory is the nucleation nature of traffic breakdown at highway bottlenecks, i.e., the explanation of the metastability of free flow with respect to an F→S transition at a highway bottleneck [17, 25]. Moreover, it turns out [38] that (see also Sec. 12.1 of the book [25]), the metastability of free flow with respect to the F→S transition can be explained even if for each vehicle density all synchronized flow states are averaged to one average state. Indeed, even when *no* 2D-region of steady states of traffic flow it is assumed in a traffic flow model, the metastability of free flow with respect to the F→S transition can be mathematically described in a three-phase model with 1D-region of steady states [38] through a competition between driver speed adaptation to such averaged states of synchronized flow and driver over-acceleration.

This hypothesis means that alone interactions between identical vehicles and drivers are responsible for 2D-steady states in synchronized flow in accordance with the following driver behavioral assumptions (see Sects. 4.3 and 8.6 of the book [17] for detailed explanations): Even if the speed difference between vehicles is negligible, a driver recognizes whether the space gap to the preceding vehicle becomes larger or smaller. This is true for each synchronized flow speed within a finite space-gap range. Therefore, we can assume that a given steady speed can be related to the infinite number of space gaps; respectively, a given space gap between vehicles can be related to the infinite number of steady speeds (Fig. 5.15b).

The upper boundary of the 2D region of the steady states of synchronized flow (labeled by $S_{\rm upper}$ in Fig. 5.15a) is determined by a *safe space gap* denoted by $g_{\rm safe}$ (Fig. 5.15b). The safe space gap is associated with a safe time headway denoted by $\tau_{\rm safe}$ that is determined from the equation:

$$g_{\rm safe}(v) = v\tau_{\rm safe}(v). \tag{5.35}$$

The lower boundary of the 2D-region for steady states of synchronized flow in the flow–density plane (labeled by $S_{\rm low}$) is determined by a *synchronization space gap* between vehicles denoted by G (Fig. 5.15b). The synchronization space gap is associated with a synchronization time headway denoted by $\tau_{\rm G}$ that is determined from the equation:

$$G(v) = v\tau_{\rm G}(v). \tag{5.36}$$

In accordance with this hypothesis of the three-phase theory [2–7], at a given steady speed v in synchronized flow there are the *infinite number* of space gaps g within the range[10]

$$g_{\rm safe}(v) \le g \le G(v) \tag{5.37}$$

at which a driver can move with this steady, i.e., time-independent speed v. Conditions (5.37) are equivalent to

$$\tau_{\rm safe}(v) \le \tau^{\rm (net)} \le \tau_{\rm G}(v), \tag{5.38}$$

where we have taken into account that in a general case the safe time headway $\tau_{\rm safe}$ and/or the synchronization time headway $\tau_{\rm G}$ can be speed functions.

Respectively, at a given space gap g between vehicles in synchronized flow there are the *infinite number* of vehicle steady speeds v within the range

$$v_{\rm G}(g) \le v \le v_{\rm safe}(g), \tag{5.39}$$

[10] A more detailed consideration of features of the safe gap $g_{\rm safe}$, the synchronization space gap G in 2D-region of synchronized flow, related time headway between vehicles ($\tau_{\rm safe}$ and $\tau_{\rm G}$), and driver behavioral assumptions explaining this hypothesis can be found in the book [25].

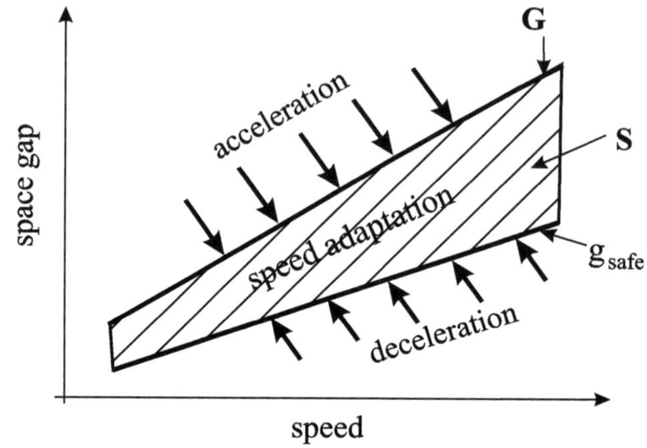

Fig. 5.16 Qualitative explanation of car-following in the three-phase theory [2–7]: A vehicle accelerates at $g > G$ (labeled by "acceleration") and decelerates at $g < g_{\rm safe}$ (labeled by "deceleration"), whereas under condition (5.37) the vehicle adapts its speed to the speed of the preceding vehicle without caring what the precise space gap is (labeled by "speed adaptation"). A dashed region of synchronized flow is adapted from Fig. 5.15b

where the speed $v_{\rm G}(g)$ is a solution of the equation

$$g = G(v_{\rm G}) \tag{5.40}$$

and the safe speed $v_{\rm safe}(g)$ is a solution of the equation

$$g = g_{\rm safe}(v_{\rm safe}). \tag{5.41}$$

A more detailed consideration of Fig. 5.16 together with conditions (5.37) and (5.39) shown in Fig. 5.17 allows us to understand some of the consequences of the hypothesis of the three-phase theory about 2D-steady states of synchronized flow.

In Fig. 5.17b, we assume that all vehicles move at a given space gap (dashed line $g = \text{const}$ in Fig. 5.17a) and the speed difference $\Delta v = v_\ell - v$ between the speed of the preceding vehicle v_ℓ and the vehicle speed v is zero: $\Delta v = 0$. Now under conditions

$$g = \text{const}, \ \Delta v = 0 \tag{5.42}$$

we consider steady states of synchronized flow at different vehicle speeds. As long as conditions (5.39) and (5.42) are satisfied, in accordance with the hypothesis of the three-phase theory about 2D-steady states of synchronized flow, the vehicle acceleration (deceleration) is equal to zero (Fig. 5.17b). When the steady speed v is smaller then the steady speed at the synchronization gap

$$v < v_{\rm G}(g), \tag{5.43}$$

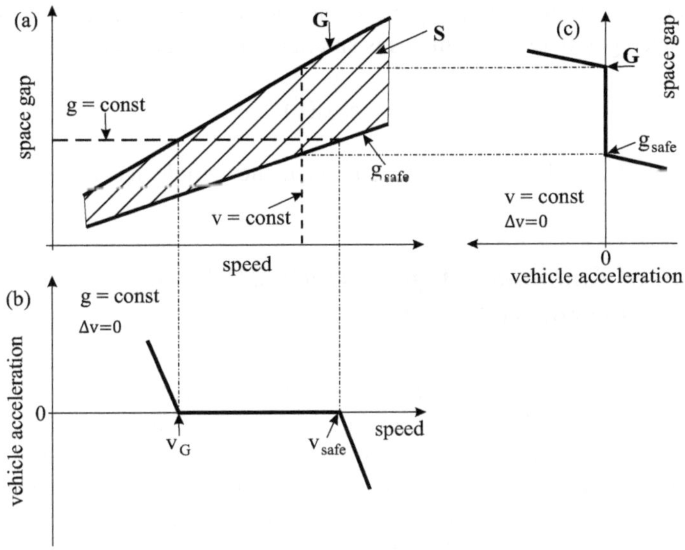

Fig. 5.17 Qualitative explanations of formulas (5.37) and (5.39): (**a**) 2D-region of synchronized flow adapted from Fig. 5.16 with dashed lines $g = \text{const}$ and $v = \text{const}$. (**b**) Qualitative dependence of vehicle acceleration at $g = \text{const}$ and $\Delta v = 0$ (5.42). (**c**) Qualitative dependence of vehicle acceleration at $v = \text{const}$ and $\Delta v = 0$ (5.45)

conditions (5.39) are not valid any more and the vehicle accelerates. Contrarily, when the steady speed v is larger then the safe speed

$$v > v_{\text{safe}}(g), \tag{5.44}$$

conditions (5.39) are also not valid any more and the vehicle decelerates (Fig. 5.17b).

In Fig. 5.17c, we assume that all vehicles move at a given steady speed (dashed line $v = \text{const}$ in Fig. 5.17a). Now under conditions

$$v = \text{const}, \ \Delta v = 0 \tag{5.45}$$

we consider steady states of synchronized flow at different space gaps. As long as conditions (5.37) and (5.45) are satisfied, in accordance with the hypothesis of the three-phase theory about 2D-steady states of synchronized flow, the vehicle acceleration (deceleration) is equal to zero (Fig. 5.17c). However, when the space gap between vehicles is larger then the synchronization gap

$$g > G(v), \tag{5.46}$$

conditions (5.37) are not valid any more and the vehicle accelerates. Respectively, when the space gap between vehicles is smaller that the safe gap

$$g < g_{\text{safe}}(v), \tag{5.47}$$

conditions (5.37) are also not valid any more and the vehicle decelerates (Fig. 5.17c).

5.9.2 Speed Adaptation Effect Within 2D-States of Synchronized Flow

When a driver approaches a slower moving preceding vehicle and the driver cannot pass it, the driver decelerates within the synchronization space gap G adapting the speed to the speed of the preceding vehicle (Fig. 5.16). This speed adaptation within the synchronization gap is associated with the 2D-region of steady states of synchronized flow given by conditions (5.37) (Fig. 5.15b).

In general, the speed adaptation effect is defined as follows:

The speed adaptation effect is the adaptation of the vehicle speed to the speed of the preceding vehicle at any space gap within the space gap range between the synchronization space gap and the safe space gap. At each vehicle speed within synchronized flow, while adapting the vehicle speed to the speed of the preceding vehicle a driver can make an arbitrary choice in the space gap that satisfies conditions (5.37): the driver accepts different spaces gaps at different times and does not control a fixed space gap to the preceding vehicle. In other words, within the space gap range (5.37) the driver speed adaptation occurs without caring what the precise space gap to the preceding vehicle is.

If the preceding vehicle moves at a time-independent synchronized flow speed, and if we neglect fluctuations, the following driver moves with the speed of the preceding vehicle (the vehicle acceleration is equal to zero, Figs. 5.17b,c) at any space gap from the space gap range (5.37) i.e., at any time headway from the time headway range (5.38). This is due to the driver behavioral assumption associated with driver interactions in synchronized flow that at a given speed in steady states of synchronized flow, the driver can make an arbitrary choice in the space gap to the preceding vehicle within the gap range (5.37). Thus, under condition (5.37) drivers do not try to reach a particular space gap (particular time headway) to the preceding vehicle, but adapt the speed while keeping the space gap in a range.

It must be emphasized [4–6, 17] that 2D-steady states of synchronized flow are only some *hypothetical flow states*, which cannot be exactly found and, therefore, measured in real synchronized flow. This is at least because of fluctuations, which are always present in real traffic flow. Moreover, we should mention that moving during a long enough time interval the driver can arbitrarily change the space gap within the range (5.37): she/he moves usually at a particular space gap only during a finite time interval changing to the moving at another space gap within the range (5.37) during another time interval, and so on.

In other words, even at a given speed in synchronized flow there are always dynamic transitions over time between different 2D-states of synchronized flow [4–6, 17]. The dynamic 2D-states of synchronized flow are non-homogeneous in space and time. However, these dynamic synchronized flow states can approximately correspond to the hypothetical 2D-steady states of synchronized flow.

In the three-phase theory [4–6, 17] it is assumed that features of instabilities and resulting phase transitions that can occur in the dynamic 2D-states of synchronized flow should be qualitatively the same as those in 2D-steady states of synchronized flow. This assumption of the three-phase theory is also related to an S→F instability. This emphasizes the importance of a consideration of 2D-steady states of hypothetical synchronized flow for the understanding of the nature of traffic breakdown as well as other traffic flow phenomena in dynamic non-homogeneous 2D-states of real synchronized flow.

This explains the hypothesis of the three-phase theory about 2D-steady states of synchronized flow [4–6, 17] as follows.

At each vehicle speed within synchronized flow, while adapting the vehicle speed to the speed of the preceding vehicle a driver can make an arbitrary choice in the space gap that satisfies conditions (5.37): The driver accepts different spaces gaps at different times and does not control a fixed space gap to the preceding vehicle. This driver speed adaptation effect within 2D-states of synchronized flow is independent of whether steady states of hypothetical synchronized flow or dynamic non-homogeneous states of real synchronized flow are considered.

5.9.3 About Mathematical Modeling of 2D-States of Synchronized Flow

Rather than a mathematical model of traffic flow, the three-phase theory is a qualitative theory that consists of several hypotheses (Sect. 1.12). The basic assumption of the three-phase theory is that traffic breakdown at a highway bottleneck is an F→S transition that exhibits the nucleation nature (Sect. 5.2).

As it has been explained in Sec. 6.1 of the book [25], the hypothesis about 2D-steady states of synchronized flow discussed above can be considered the result of empirical definition [S] of the synchronized flow phase (Sect. 1.3). For this reason, as we have already mentioned in many our publications (see, e.g., footnote 1 in Sec. 4.3.4 of the book [17]), to simulate the nucleation nature of the F→S transition (traffic breakdown) at a highway bottleneck, a three-phase traffic flow model should not necessarily satisfy the hypothesis about 2D-steady states of synchronized flow of the three-phase theory.

In particular, steady states of the three-phase traffic flow model can lie on 1D-states of traffic flow (three-phase models with a theoretical fundamental diagram). Examples of the three-phase traffic flow models with a fundamental diagram are both deterministic model and stochastic speed adaptation (SA) models developed in [38] (see Sects. 3 and 5.3 of [38] as well as Sec. 12.1 of the book [25]). Although these deterministic and stochastic three phase models do not satisfy the hypothesis about 2D-steady states of synchronized flow of the three-phase theory, the models satisfy the hypothesis of the three-phase theory about the nucleation nature of the F→S transition (traffic breakdown) at the bottleneck. These models can show an 2Z-characteristic for phase transition. For this reason, these traffic flow models with a fundamental diagram are three-phase traffic flow models (Sect. 1.12).

Three-phase traffic flow models that satisfy the hypothesis of the three-phase theory about 2D-steady states of synchronized flow and three-phase traffic flow models with the fundamental diagram can be considered two "limit" cases in the relation to the use of the hypothesis about the 2D-steady states of synchronized flow by the model development. As above mentioned, the Kerner-Klenov stochastic microscopic model (Appendix A) incorporates the 2D-steady states of synchronized flow. Contrarily, the deterministic and stochastic SA three-phase models of [38] do not incorporate the hypothesis of the three-phase theory about 2D-steady states of synchronized flow.

As emphasized in Sect. 5.9.2 (see also [2] as well as Sec. 4.3.4 of the book [17]), steady states of synchronized flow do not exactly exist in real traffic flow. In real traffic flow, 2D-states of synchronized flow are always some *dynamic* synchronized flow states. In the three-phase theory it is assumed [2, 17] that real dynamic 2D-states of synchronized flow exhibit qualitatively the same non-linear features with respect to the phase transitions in traffic flow as those of the 2D-steady states of the qualitative three-phase theory. Some of the *qualitatively the same non-linear features* of real dynamic 2D-states and the hypothetical 2D-steady states of synchronized flow are as follows.

(i) Synchronized flow exhibits two qualitatively different kinds of traffic flow instabilities (see also Sect. 8.2):

  - The synchronized flow instability introduced in the three-phase theory associated with a growing wave of the speed *increase* in synchronized flow. The wave growth leads to an S→F transition.
  - The classical traffic flow instability of Herman, Gazis, Montroll, Potts, Rothery, and Chandler (Sect. 4.4.1) that leads to a growing wave of the speed *decrease* in synchronized flow. The wave growth results in an S→J transition (see Chap. 8).

(ii) Hypothesis 7 of the three-phase theory about a sequence of F→S→J transitions (Sect. 1.9).

(iii) Hypothesis 8 of the three-phase theory about a double Z-characteristic for phase transitions (Sect. 1.9).

As above-mentioned, to simulate these features of synchronized flow of the three-phase theory, a three-phase traffic flow model should not necessarily incorporate 2D-steady states of synchronized flow: There can be developed a diverse variety of mathematical approaches for the description of the dynamic 2D-states of synchronized flow in a three-phase traffic flow model that exhibits the above mentioned features.

For illustrations of different spatiotemporal features of the phenomenon *traffic breakdown in traffic networks* presented in this book, we use simulations of the Kerner-Klenov stochastic microscopic model (Appendix A) and the KKSW CA model (Appendix B) in the framework of the three-phase theory. This is because these models can show and simulated all features of traffic phenomena found up to now in real field traffic data.

In these three-phase models, the following rules of vehicle motion have been chosen to describe the speed adaptation effect:

(i) In the three-phase models under consideration (Appendices A and B), we have used a discrete time $t_{\mathrm{n}} = \tau n$, $n = 0, 1, \ldots$, where τ is time step (we use $\tau = 1$ s). When a time dependence of traffic variable is considered, conditions (5.37) should be replaced by

$$q_{\mathrm{safe},n} \leq g_n \leq G_n. \tag{5.48}$$

The speed adaptation effect takes place within the space gap range (5.48).

(ii) In the Kerner-Klenov stochastic microscopic model, to incorporate driver behaviors associated with the speed adaptation effect within 2D-states of synchronized flow, under conditions (5.48), when model fluctuations are not taken into account and the speed is lower than the speed v_{free} in free flow, the vehicle speed v_{n+1} at time step $n+1$ is equal to

$$v_{n+1} = v_n + \max(-b_n\tau, \min(a_n\tau,\ v_{\ell,n} - v_n)), \tag{5.49}$$

where b_n is a vehicle deceleration, a_n is a vehicle acceleration, $v_{\ell,n}$ is the speed of the preceding vehicle.

(iii) In accordance with (5.49), under conditions (5.48), the vehicle tends to adjust its speed v_n to the speed of the preceding vehicle $v_{\ell,n}$ without caring, what the precise space gap is, as long as it is safe. The vehicle accelerates, if it is slower than the preceding one and the vehicle decelerates, if its speed is higher than the speed of the preceding vehicle.[11]

[11]Thus, when a vehicle moves initially at a higher speed than the speed of the preceding vehicle and the vehicle cannot overtake the preceding vehicle, then under conditions (5.48) the vehicle decelerates to the speed of the preceding vehicle. Depending on the average value of this vehicle

(iv) We can see that when the speed difference $\Delta v_n = v_{\ell,n} - v_n$ at time step n is not zero, but it is small enough, from formula (5.49) we get

$$v_{n+1} = v_{\ell,n}. \tag{5.50}$$

This means that at the next time step the vehicle speed is equal to the speed of the preceding vehicle. This is the consequence of the driver speed adaptation to the speed of the preceding vehicle that is regardless of the space gap g_n between these vehicles within the space gap range (5.48).

(v) Another consequence of the driver speed adaptation is as follows. When within the space gap range (5.48) the speed difference

$$\Delta v_n = v_{\ell,n} - v_n = 0, \tag{5.51}$$

in accordance with (5.49), the driver does not necessarily accelerate or decelerate. This is because in accordance with the hypothesis about 2D-states of synchronized flow vehicle deceleration (acceleration) is not a function of a space gap between vehicles within the space gap range (5.48). Therefore, we can expect that under conditions (5.48) and (5.51) for dynamic 2D-states of traffic flow the vehicle acceleration or deceleration should be approximately equal to zero. This assumption follows from the hypothesis about 2D-states of traffic flow. Indeed, as shown in Figs. 5.17b,c and explained in Sect. 5.9.1, for hypothetical 2D-steady states of synchronized flow under conditions (5.48) and (5.51) the vehicle acceleration or deceleration is exactly equal to zero.

(vi) However, there are always fluctuations of driver acceleration and deceleration in real traffic. For this reason, due to model fluctuations incorporated in the three-phase models under consideration the condition $\Delta v_n = 0$ (5.51) cannot be satisfied during a long time interval. In other words, even when the preceding vehicle moves at a time independent speed, model fluctuations force the vehicle either to accelerate or to decelerate while adapting the vehicle speed to the speed of the preceding vehicle. Therefore, rather than 2D-steady states, dynamic 2D-states of synchronized flow are realized in both the Kerner-Klenov stochastic model (Appendix A) and the KKSW CA

---

deceleration, we distinguish "strong" speed adaptation and "weak" speed adaptation (see Sect. A.7 of Appendix A) [28]. When the vehicle decelerates quickly to the speed of the preceding vehicle, the average space gap between vehicles remains a large one. This case is called "strong" speed adaptation. Contrarily, when the vehicle adapts its speed slowly to the speed of the preceding vehicle, the average space gap between vehicles becomes a small one. This case is called "weak" speed adaptation. Although in both cases conditions (5.48) are satisfied, features of synchronized flow can be completely different. As shown in [28], under "weak" speed adaptation moving jams emerge easier spontaneously in synchronized flow. In contrast, under "strong" speed adaptation moving jams dissolve usually in synchronized flow (this effect of jam dissolution can also be called "jam absorption"). Because we do not make a comprehensive discussion of spatiotemporal features of congested patterns resulting from the breakdown, a more detailed consideration of "strong" speed adaptation and "weak" speed adaptation [28] is out of the scope of this book.

model (Appendix B). Simulations of the models show that the dynamic 2D-states of synchronized flow exhibit qualitatively the same features of phase transitions as those postulated in the three-phase theory for 2D-steady states of synchronized flow.

(vii) When the space gap to the preceding vehicle g_n is larger than the synchronization space gap G_n and the speed difference $\Delta v_n < 0$, the vehicle catches up the preceding vehicle. The vehicle decelerates adapting its speed to the speed of the preceding vehicle only when the space gap becomes to be equal to or smaller than the synchronization gap G_n.

(viii) Because vehicle deceleration is limited, the synchronization space gap G_n is chosen as an increasing function of the vehicle speed v_n and a decreasing function of the speed difference Δv_n. In particular, in the Kerner-Klenov model we have used the following function $G_n = G(v_n, v_{\ell,n})$ for the synchronization space gap:

$$G(v_n, v_{\ell,n}) = \max(0, k\tau v_n + a^{-1} v_n(v_n - v_{\ell,n})), \tag{5.52}$$

where a is a maximum acceleration, a and k are constants.

5.10 Driver Over-Acceleration

5.10.1 Hypothesis About Discontinuous Character of Over-Acceleration

When condition (5.37) is satisfied, a vehicle accelerates usually when it is slower than the preceding vehicle, and decelerates when it is faster than the preceding vehicle. However, under condition (5.37) the vehicle can also accelerate even if the vehicle speed $v(t)$ is *not* smaller than the speed of the preceding vehicle $v_\ell(t)$, i.e., when condition

$$v(t) \geq v_\ell(t) \tag{5.53}$$

is satisfied. Such vehicle acceleration can be considered *vehicle over-acceleration* [17]. In general, the over-acceleration effect is defined as follows[12]:

[12]It must be noted that in real traffic flow the speed adaptation and over-acceleration effects appear usually in their dynamic competition within a local disturbance in free flow at a bottleneck. For this reason, a separate consideration of the over-acceleration effect made here is a simplification of the reality (see Sects. 5.12 and 5.13 below).

The over-acceleration effect is driver maneuver leading to a higher speed from initial car-following at a lower speed occurring under condition (5.37).

Accordingly to the hypothesis of the three-phase theory about 2D-steady states of synchronized flow, at each given density there is a speed gap between states of free flow and synchronized flow. At a given vehicle density, probability of over-acceleration P_{OA} is greater in free flow than in synchronized flow. This assumption leads to the following hypothesis of the three-phase theory about a discontinuous character of probability of driver over-acceleration [4–6, 17]:

Probability of over-acceleration exhibits a *discontinuous* character. In particular, the probability is a Z-shaped density function (Fig. 5.18): At a given density, there is a *drop* in probability of over-acceleration P_{OA}, when free flow transforms into synchronized flow.

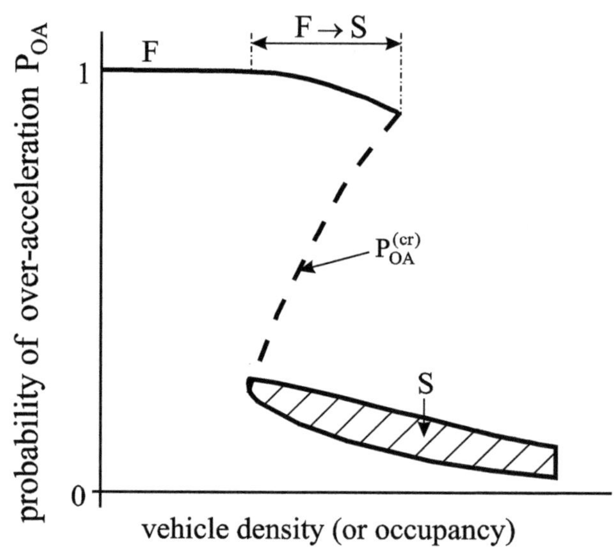

Fig. 5.18 Qualitative Z-shaped density function of probability of over-acceleration [4–6]. *F*—free flow, *S*—synchronized flow, dashed curve between states *F* and *S* is related to probability of over-acceleration $P_{OA}^{(cr)}$ within critical nuclei for traffic breakdown (F→S transition). Traffic breakdown is possible at any density in initial free flow that is within a density range labeled by "F→S". 2D-region for synchronized flow (*dashed region*) is associated with the 2D-region for steady states of synchronized flow shown in Fig. 5.15a

To understand this hypothesis of the three-phase theory, two assumptions about features of the over-acceleration should be mentioned:

(i) There should be a driver time delay in over-acceleration. Indeed, while moving in synchronized flow, a driver has to wait while searching for the opportunity to accelerate to a higher speed.
(ii) At a given vehicle density at which either a state of free flow or a state of synchronized flow is possible, probability of over-acceleration in free flow is greater than in synchronized flow.

These two features of over-acceleration lead to a conclusion that when a driver moves initially in synchronized flow, an S→F instability caused by over-acceleration should occur. This S→F instability should lead to a growing wave of the speed increase in synchronized flow leading to a transition from synchronized flow to free flow (S→F transition). In its turn, due to this traffic flow instability, some of states on a density function of the probability of over-acceleration cannot exist. Therefore, there should be a drop in the density function of the probability of over-acceleration (Fig. 5.18). This explains the above mentioned hypothesis about a discontinuous character of over-acceleration.

For a qualitative explanation of the S→F instability, we assume that a driver moves initially in synchronized flow. If the driver increases the speed due to over-acceleration, the probability of over-acceleration for the following drivers increases. Therefore, the subsequent over-acceleration of the following drivers becomes more probable. Due to the time delay in driver over-acceleration, it takes some time for the development of this over-acceleration effect. In other words, the S→F instability caused by over-acceleration should be associated with a growing wave of a local increase in the vehicle speed in synchronized flow. In Sects. 5.12 and 5.13, we will show that this assumption about the occurrence of a growing wave of the increase in the vehicle speed that results from the S→F instability in synchronized flow is indeed supported by microscopic numerical simulations of a three-phase traffic flow model.

It should be noted that in the above qualitative explanation of the S→F instability due to the over-acceleration effect we have not considered the speed adaptation effect. The speed adaptation effect can prevent the development of the S→F instability. Therefore, to describe the nucleation nature of traffic breakdown, we should consider a competition between vehicle deceleration associated with the speed adaptation effect and vehicle acceleration associated with the over-acceleration effect.

In the three-phase theory, the nucleation nature of traffic breakdown is explained by a competition between two opposing tendencies occurring within a random local disturbance in which the speed is lower than in an initial free flow:

(continued)

(i) A tendency towards synchronized flow due to vehicle deceleration associated with the speed adaptation effect.
(ii) A tendency towards the initial free flow due to vehicle acceleration associated with the over-acceleration effect.

As above mentioned, the over-acceleration effect leads to an S→F instability that causes a growing wave of a local increase of the speed in synchronized flow to the speed in free flow. In other words, the S→F instability can be considered a synonym of the over-acceleration effect that determines the tendency towards free flow within a local speed disturbance in free flow.

The S→F instability at a highway bottleneck governs the nucleation nature of an F→S transition. In other words, the S→F instability leads to the metastability of free flow with respect to an F→S transition (traffic breakdown) at the bottleneck.

The discontinuous character of over-acceleration probability leads to the increasing flow-rate function of breakdown probability (Fig. 5.5c) as well as to a Z-characteristic for traffic breakdown at a highway bottleneck (Fig. 5.5d) [35, 36, 47]. Already in first microscopic three-phase traffic flow models two mechanisms for over-acceleration have been introduced and studied [35, 36, 47]:

(i) Over-acceleration due to vehicle acceleration in car-following occurring without lane changing.
(ii) Over-acceleration due to lane changing to a faster lane that is possible on a multi-lane road.

On single-lane roads, the over-acceleration mechanism occurring without lane changing does lead to the Z-characteristic for traffic breakdown (F→S transition) as well as to the increasing flow-rate function of the breakdown probability, as this has firstly been found out in numerical simulations with stochastic three-phase traffic flow models [35, 47].

For multi-lane roads, the importance of the over-acceleration mechanism due to lane changing to a faster lane has been emphasized in [41]. Obviously the resulting discontinuous character of driver over-acceleration in traffic flow is determined by a combination of all possible different mechanisms of over-acceleration.

There can be developed a number of different mathematical models of driver over-acceleration leading to the discontinuously behavior of the probability of driver over-acceleration. In the remaining of this section, we discuss briefly some of them.

5.10.2 Mathematical Models of Over-Acceleration Effect on Single-Lane Road

In the KKSW CA model (Appendix B), a formulation for model fluctuations that simulates over-acceleration on a single-lane road is as follows. Each vehicle, which moves in synchronized flow with a space gap that satisfies conditions (5.48), accelerates randomly with some probability:

$$\text{if } v_n \geq v_{\ell,n}, \text{ then with probability } p_{\rm a}$$
$$v_{n+1} = \min(v_n + 1,\ v_{\rm free}), \tag{5.54}$$

where $p_{\rm a}$ is the probability of driver over-acceleration ($0 < p_{\rm a} < 1$), $v_{\rm free}$ is the vehicle speed in free flow assumed to be constant. This random vehicle acceleration occurs only under conditions (5.48). Thus, the vehicle accelerates with probability $p_{\rm a}$, even if the preceding vehicle does not accelerate and the vehicle speed is not lower than the speed of the preceding vehicle. Therefore, in accordance with the definition of over-acceleration made above, this vehicle acceleration is an example of driver over-acceleration.

Because the probability of over-acceleration $p_{\rm a} < 1$, there is *on average* a time delay in over-acceleration. The mean time delay in the over-acceleration is longer than time step of the KKSW CA model ($\tau = 1$ s) (see Appendix B). The over-acceleration effect results in the discontinuous character of the probability of over-acceleration as a density (and flow rate) function as required by the hypothesis of the three-phase theory (Fig. 5.18).

The probability of over-acceleration $p_{\rm a}$ can be chosen as an increasing function of the vehicle speed:

$$p_{\rm a}(v_n) = p_{\rm a,1} + p_{\rm a,2} \max(0, \min(1,\ (v_n - v^{(\rm syn)})/\Delta v^{(\rm syn)})), \tag{5.55}$$

where $p_{\rm a,1}$, $p_{\rm a,2}$, $v^{(\rm syn)}$ and $\Delta v^{(\rm syn)}$ are constants. This supports the over-acceleration within a local speed disturbance of the increase in the speed in synchronized flow. This model feature is related to a prediction of the three-phase theory that the stronger the driver over-acceleration, the more probable should be the occurrence of the S→F instability.

In this approach to a mathematical modeling of the discontinuous behavior of driver over-acceleration, a *dynamic* speed gap appears (gray region labeled by "dynamic speed gap" in Fig. 5.19) between free flow (line *F*) and steady states of synchronized flow of lower speeds. Within this dynamic speed gap, initially steady states of synchronized flow of higher speeds are destroyed and no long-time living synchronized flow states occur.

Thus, in the KKSW CA model a *discontinuous character* of driver over-acceleration is modeled (Appendix B) through the occurrence of the dynamic speed gap caused by stochastic model dynamics that destroys initially steady states

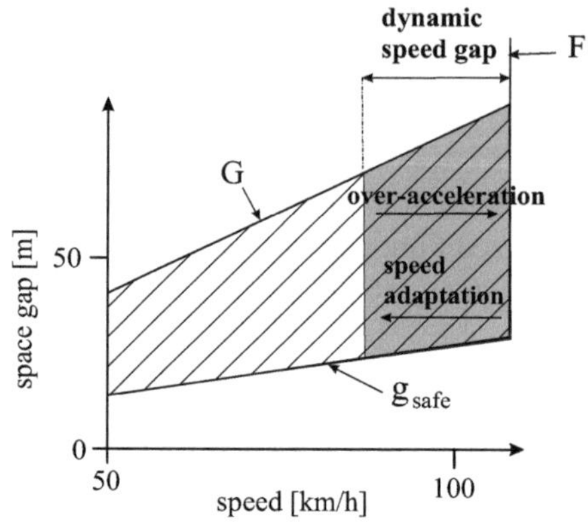

Fig. 5.19 Qualitative explanation of competition of over-acceleration and speed adaptation effects in stochastic three-phase traffic flow models considered in Appendices A and B. Parts of steady states of synchronized flow at higher speeds (*dashed region*) and free flow speed (F). *Gray region* shows qualitatively a dynamic speed gap in which steady states of synchronized flow are destroyed and no long-time living synchronized flow states occur. Adapted from [25]

of synchronized flow of higher speeds (gray region in Fig. 5.19). A competition between the speed adaptation and over-acceleration effects (labeled by "speed adaptation" and "over-acceleration" in Fig. 5.19), which occurs within a local disturbance in free flow, is associated with this dynamic speed gap. This simulates the nucleation nature of traffic breakdown.

Such a dynamic speed gap between states of free flow and synchronized flow (Fig. 5.19) resulting in a discontinuous character of driver over-acceleration on a single-lane road is also simulated in the Kerner-Klenov microscopic stochastic model (Appendix A). Due to a random fluctuation in driver acceleration given by formula

$$\xi_{\mathrm{a}} = a^{(\mathrm{a})}\tau\Theta(p_{\mathrm{a}} - r) \tag{5.56}$$

the driver accelerates with probability p_{a}. This random acceleration occurs additionally to a "deterministic component" of driver acceleration associated with dynamic rules of vehicle motion on a single-lane road. In (5.56), $r = \mathrm{rand}(0, 1)$ is a value randomly distributed between 0 and 1; $a^{(\mathrm{a})}$ is constant; $\Theta(z) = 0$ at $z < 0$ and $\Theta(z) = 1$ at $z \geq 0$. Because the probability of random acceleration $p_{\mathrm{a}} < 1$, there is a time delay in driver over-acceleration.

It should be noted that a dynamic speed gap between states of free flow and synchronized flow (Fig. 5.19) together with a discontinuous character of driver over-acceleration on a single-lane road can also be simulated in deterministic three-phase traffic flow models. In particular, probably the first deterministic three-phase traffic flow model that can simulate the metastability of free flow with respect to an F→S transition at the bottleneck has been the Kerner-Klenov microscopic deterministic model in the framework of the three-phase theory (called as the ATD model) [38]. In this model, there are a speed gap between steady states of synchronized flow and free flow states as well as a time delay in driver over-acceleration. These two model features allows us to simulate the nucleation nature of the F→S transition [38].

In comparison with the Kerner-Klenov stochastic microscopic model, the achievement of the deterministic ATD model is that the time delay in driver over-acceleration is simulated individually for each of the drivers. In contrast, in the Kerner-Klenov stochastic microscopic model the mean time delay in driver over-acceleration is simulated as collective effect due to stochastic model features.

However, there is an important achievement of the Kerner-Klenov stochastic microscopic model in comparison with the deterministic ATD model: The stochastic model allows us to simulate the probability of traffic breakdown as a function of traffic variables. Because the probability of traffic breakdown is of a great importance for the understanding of the most breakdown phenomena in traffic and transportation networks, in the book we consider results of simulations of the stochastic models only.

5.10.3 Mathematical Simulation of Over-Acceleration Effect Due to Lane Changing

On multi-lane roads, even if driver over-acceleration in a lane discussed in Sect. 5.10.2 is ignored, the discontinuous character of driver over-acceleration (Fig. 5.18) can be simulated through driver behavior by lane changing. Such simulations of driver over-acceleration have been firstly made in the Kerner-Klenov stochastic model (Appendix A) and then in the KKSW CA model (Appendix B). For example, we consider briefly simulations of the discontinuous character of driver over-acceleration (Fig. 5.18) with the KKSW CA model.

In the KKSW CA model of traffic flow on a two-lane road (Appendix B), driver over-acceleration through lane changing to faster lane is simulated as follows [45]: Lane changing (for passing) occurs with probability p_c when the following intention (5.57), (5.58) and safety conditions (5.59) are satisfied:

$$R \rightarrow L: v_n^+ \geq v_{\ell,n} + \delta \text{ and } v_n \geq v_{\ell,n}, \tag{5.57}$$

$$L \rightarrow R: v_n^+ \geq v_{\ell,n} + \delta \text{ or } v_n^+ \geq v_n + \delta, \tag{5.58}$$

$$g_n^+ \geq \min(v_n,\ g_c) \text{ and } g_n^- \geq \min(v_n^-,\ g_c), \tag{5.59}$$

where $R \to L$ and $L \to R$ denote, respectively, lane changing from the right lane to left lane and from the left lane to right lane; in lane changing rules (5.57)–(5.59), superscripts "+" and "−" denote variables and functions related, respectively, to the preceding vehicle and the trailing vehicle in the "target" lane between which the vehicle appears if the vehicle changes to this target lane. In particular, g_n^+ is the space gap between the vehicle and the preceding vehicle in the target lane, g_n^- is the space gap between the vehicle and the trailing vehicle in the target lane; the vehicle speed v_n^+ or the speed of the preceding vehicle $v_{\ell,n}$ in (5.57) and (5.58) is set to ∞ if the space gap g_n^+ or space gap g_n exceeds a given look-ahead distance L_a, respectively; L_a, δ, and g_c are constants.

For a qualitative explanation of the physics of the mathematical modeling of the discontinuous character of driver over-acceleration (Fig. 5.18) due to lane changing on multi-lane roads, we consider a scenario in which a vehicle that moves in free flow on a multi-lane road approaches a slower moving preceding vehicle. For simplicity, we assume in this qualitative consideration that there is only a mechanism of over-acceleration associated with lane changing to a faster lane. If firstly the vehicle cannot pass the preceding vehicle, then the vehicle decelerates within the synchronization space gap to the speed of the preceding vehicle, i.e., the speed adaptation effect is realized leading to car-following of the slow preceding vehicle (Fig. 5.20a).

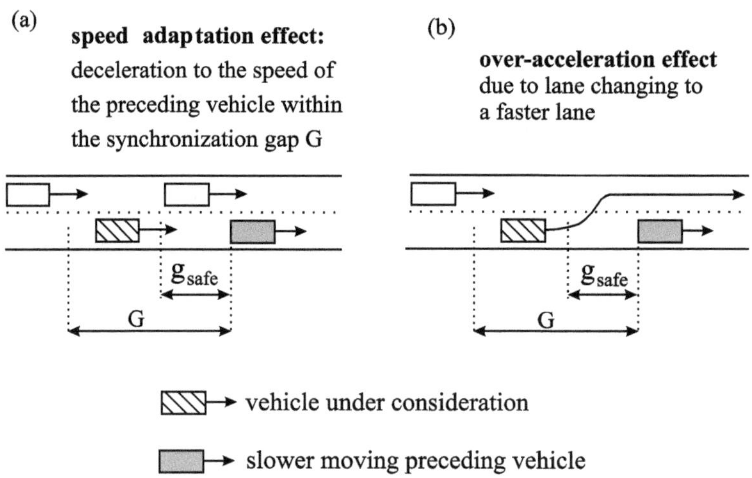

Fig. 5.20 Qualitative explanation of vehicle deceleration due to speed adaptation (**a**) and vehicle acceleration due to over-acceleration for a special case, when over-acceleration is associated with lane changing *only* (**b**) [4–6]

We assume in this scenario that later the vehicle can pass this slow moving preceding vehicle. To pass the preceding vehicle, the vehicle should change lane and accelerate. The vehicle acceleration takes place, even if the vehicle is not currently slower than the preceding vehicle and the preceding vehicle does not accelerate. For this reason, this vehicle acceleration is an example of over-acceleration.

It should be noted that the vehicle passing can be responsible also for over-acceleration of the following vehicle(s) that continue to move in the lane. There can be at least two examples of such over-acceleration. (i) When a vehicle changes the lane for passing, the space gap between the following and preceding vehicles increases. For this reason, under condition (5.37) the following vehicle can accelerate, even when the preceding vehicle is not slower than the preceding vehicle and the preceding vehicle does not accelerate. (ii) As over-acceleration can also be considered a case, when under condition (5.37) the following vehicle continues to move at a higher speed than the the preceding vehicle; the vehicle begins to decelerate adapting its speed to the slower preceding vehicle only when the space gap becomes close to the safe one g_{safe}.

In the cases under consideration, the over-acceleration effect includes vehicle acceleration for passing from car-following, i.e., lane changing to a faster lane (Fig. 5.20b). For this reason, the probability of the over-acceleration effect P_{OA} is equal to the passing probability from car-following .

However, in [44] it has been shown that if the discontinuous character of driver over-acceleration (Fig. 5.18) is simulated through lane changing *only* (driver over-acceleration in a lane discussed in Sect. 5.10.2 is ignored), then some non-realistic effects in heterogeneous traffic flow consisting of passenger vehicles and trucks are realized.

Additionally to the mathematical models for simulations of driver over-acceleration discussed above, there are also other mechanisms of driver over-acceleration that will be discussed in Sect. A.8 of Appendix A.

The discontinuous character of driver over-acceleration (Fig. 5.18) can be shown by different mathematical models of driver over-acceleration. In the Kerner-Klenov stochastic microscopic model (Appendix A) and the KKSW CA model (Appendix B) that simulation results are used in this book, driver over-acceleration is simulated through a random driver acceleration in car-following. There is a time delay in the driver over-acceleration occurring in a road lane. This leads to the discontinuous character of driver over-acceleration, even when vehicles move on a single lane road. On a multi-lane road, a lane changing to the faster lane facilitates driver over-acceleration.

To perform realistic simulations of heterogeneous traffic flow on a multi-lane road, additionally with driver over-acceleration through lane changing, driver over-acceleration in a road lane should be taken into account.

5.11 Microscopic Stochastic Features of S→F Instability Away of Bottlenecks

A qualitative discussion of the spatiotemporal competition between the over-acceleration and speed adaptation effects made above have been confirmed with the use of microscopic simulations with stochastic three-phase traffic flow models [41, 44, 45]. To illustrate a possibility of an S→F instability occurring in synchronized flow away of a road bottleneck on a single lane road discussed in Sect. 5.10.2, we consider briefly results of simulations of the KKSW CA model (Figs. 5.21, 5.22, 5.23).

To simulate an S→F instability away of a road bottleneck on a single lane road, firstly, we should simulate synchronized flow. With this objective we have simulated a widening synchronized flow pattern (WSP) at an on-ramp bottleneck (Fig. 5.21a,b). Synchronized flow of the WSP propagates upstream of the bottleneck. At 2–3 km upstream of the bottleneck, the following three scenarios of a competition between over-acceleration and speed adaptation within synchronized flow can be distinguished [44]:

Scenario 1. No wave of the increase in the speed occurs in synchronized flow (Fig. 5.21c,d).

Scenario 2. A wave of the increase in the speed occurs in synchronized flow. However, such an acceleration wave decays over time. As result, no S→F transition is realized (Fig. 5.22).

Scenario 3. A growing wave of the increase in the speed occurs in synchronized flow. This S→F instability leads to an S→F transition (Fig. 5.23).

In Scenario 1 (Fig. 5.21c,d), the competition between the over-acceleration and speed adaptation causes a random increase in the vehicle speed at one of the time steps and a decay of this speed increase at next time steps in synchronized flow. An example of this competition is as follows. At some time step, vehicle 3 accelerates (Fig. 5.21d, $t \approx 22.62$ min). Vehicle 3 accelerates although the preceding vehicle 2 decelerates at this time step (speed of vehicle 2 in Fig. 5.21d): Conditions (5.37) and (5.53) are satisfied at which the acceleration of vehicle 3 can be considered an example of over-acceleration (labeled by "over-acceleration" in Fig. 5.21d). At next time step, vehicle 3 decelerates due to the speed adaptation to the speed of vehicle 2.

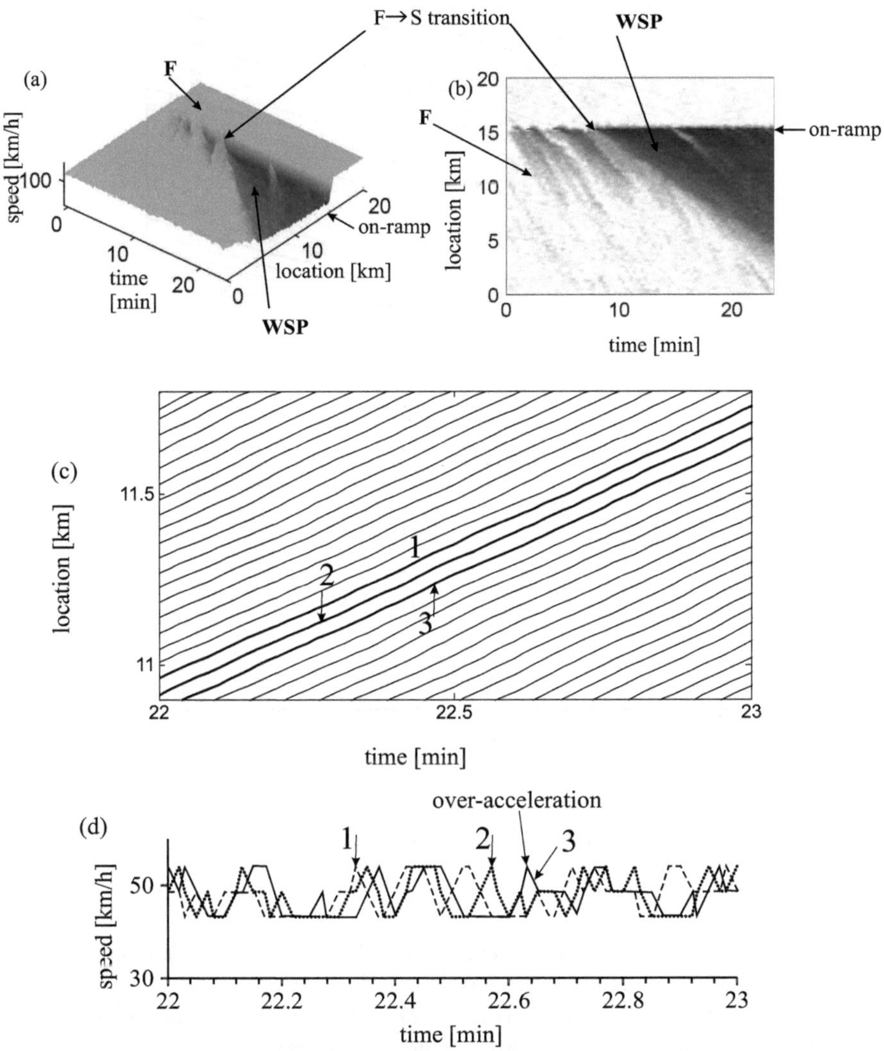

Fig. 5.21 Simulations of synchronized flow away of a on-ramp bottleneck on a single lane road made with the KKSW CA model (Appendix B): (**a**, **b**) Speed in space and time within widening synchronized flow pattern (WSP) at on-ramp bottleneck (**a**) and the same speed data presented by regions with variable shades of gray (shades of gray vary from white to black when the speed decreases from 135 km/h (*white*) to zero (*black*)) (**b**). (**c**) Fragment of vehicle trajectories in space and time within synchronized flow of the WSP shown in (**a**, **b**). (**d**) Single vehicle speed along vehicle trajectories 1–3 marked by bold in (**c**). F—free flow, S—synchronized flow. $q_{\text{on}} = 400$ vehicles/h, $q_{\text{in}} = 1452$ vehicles/h. On-ramp location $x_{\text{on}} = 15$ km. Adapted from [44]

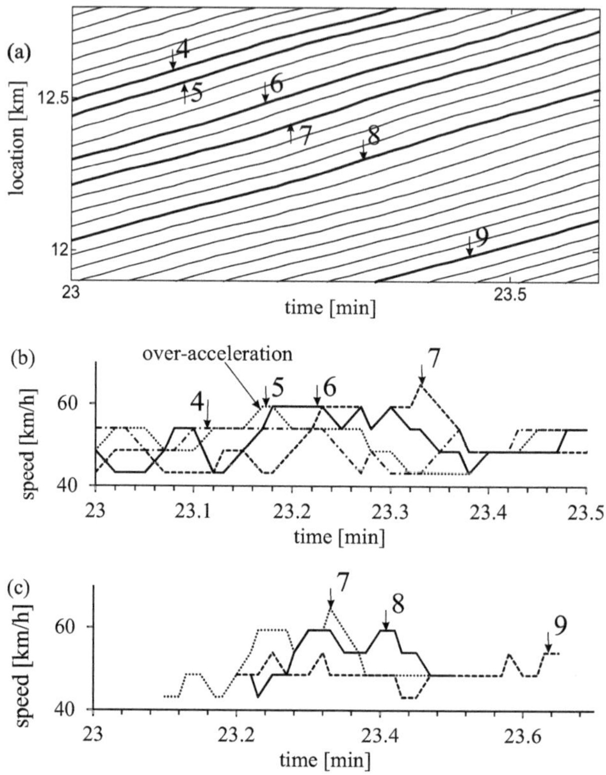

Fig. 5.22 Simulations of the emergence and decay of acceleration wave within synchronized flow of WSP shown in Fig. 5.21a,b: (**a**) Fragment of vehicle trajectories in space and time. (**b**, **c**) Single vehicle speed along vehicle trajectories 4–9 marked by bold in (**a**). Adapted from [44]

Due to the speed adaptation effect, synchronized flow can be self-maintained during a long time interval upstream of the bottleneck.[13]

In an example of Scenario 2 (Fig. 5.22), vehicle 5 accelerates (Fig. 5.22a,b, $t \approx 23.16$ min). Vehicle 5 accelerates although the preceding vehicle 4 does not accelerate (speed of vehicle 4 in Fig. 5.22b). In other words, conditions (5.37) and (5.53) are satisfied at which the acceleration of vehicle 5 can be considered an example of over-acceleration (labeled by "over-acceleration" in Fig. 5.22b).

Vehicle 6 catches up the preceding vehicle 5 while increasing the speed to the speed of vehicle 5 (trajectory 6 in Fig. 5.22a,b, $t \approx 23.2$ min). Although vehicle 6 begins to decelerate at $t \approx 23.8$ min, nevertheless, the following vehicle 7 increases

[13]Discrete values of the increase and the decrease in the speed that can be seen in Fig. 5.21d are associated with the discretization of the vehicle acceleration and the speed used in the KKSW CA model (Appendix B).

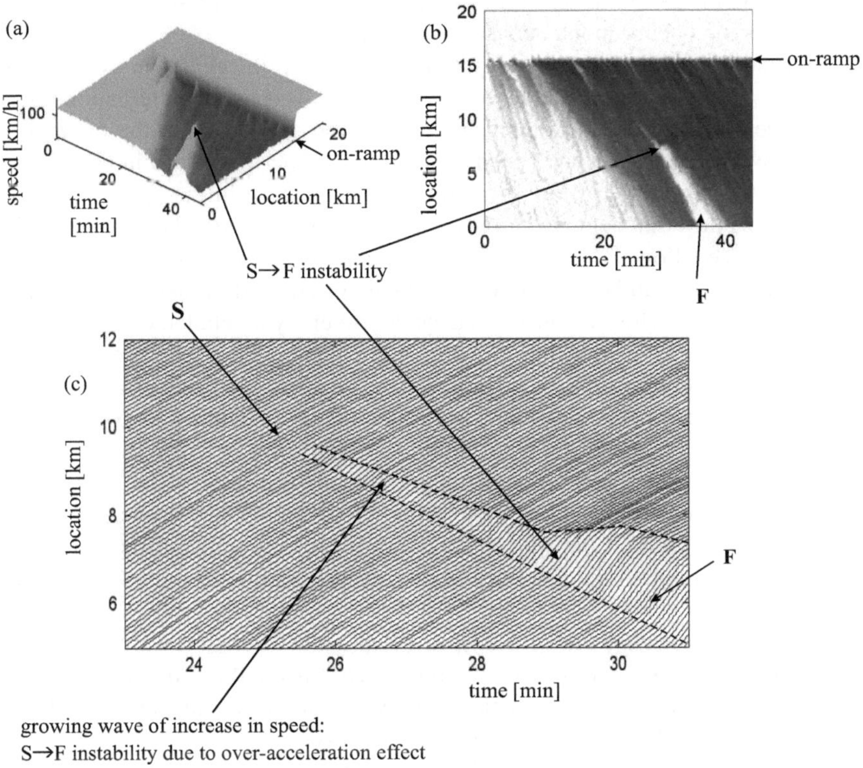

Fig. 5.23 Simulations of S→F instability away of a road bottleneck on a single lane road made with the KKSW CA model (Appendix B): (**a**, **b**) Speed in space and time within the WSP at on-ramp bottleneck presented in Fig. 5.21a,b for a longer time interval (**a**) and the same speed data presented by regions with variable shades of gray (shades of gray vary from white to black when the speed decreases from 135 km/h (*white*) to zero (*black*)) (**b**). (**c**) Fragment of vehicle trajectories in space and time related to (**a**, **b**). Adapted from [41]

the speed due to over-acceleration (trajectory 7 in Fig. 5.22a,b, $t \approx 23.35$ min). Therefore, a speed wave of the increase in the speed appears in synchronized flow.

However, because vehicle 6 continues deceleration (trajectory 6 in Fig. 5.22a,b, $23.3 < t < 23.38$ min), the following vehicle 7 must also decelerate strongly (trajectory 7 in Fig. 5.22a,b, $23.32 < t < 23.38$ min). This can explain why the speed wave of the increase in the speed decays: Vehicle 8 that follows vehicle 7 must also decelerate (trajectory 8 in Fig. 5.22a,c, $23.41 < t < 23.44$ min). This decay of the acceleration wave leads to the maintaining of synchronized flow (trajectory 9 in Fig. 5.22a,c).

However, if we continue simulations of the WSP shown in Fig. 5.21a,b for a longer time interval, we find scenario 3 mentioned above: A growing wave of the increase in the speed (growing acceleration wave) occurs within synchronized flow of the WSP. The growth of this acceleration wave within synchronized flow of the

WSP leads to an S→F transition (Fig. 5.23a,b). Indeed, we can see that over time a free flow region occurs spontaneously within the WSP.

An usual behavior of the development of the S→F instability in synchronized flow away of the bottleneck is as follows. Firstly, we observe alternations of the occurrence and decay of a random increase in the speed due to the competition between over-acceleration and speed adaptation as those discussed in Figs. 5.21c,d and 5.22. At some time instant, a random local increase in the speed appears in synchronized flow that begins to grow continuously over time—the S→F instability in synchronized flow is realized. This random local increase in the speed in synchronized flow can be considered a nucleus for an S→F transition in the initial synchronized flow. This conclusion has been proven by a detailed consideration of vehicular trajectories within the growing speed wave of the local increase in the speed within synchronized flow (Fig. 5.23d) that readers can find in [44].

This behavior of the development of the S→F instability in synchronized flow proves that synchronized flow within the WSP (Fig. 5.23) is in a metastable state with respect to the S→F transition: A small increase in the speed within synchronized flow decays over time (Figs. 5.21d and 5.22). However, when a random local increase in the speed is equal to or exceeds some critical value, i.e., a nucleus appears, then the S→F instability occurs leading to the S→F transition (Fig. 5.23).

Contrarily to scenario 2 (Fig. 5.22), in which an acceleration wave decays over time while propagating upstream within synchronized flow, in the case under consideration (Fig. 5.23), over-acceleration overcomes speed adaptation within the acceleration wave. This leads to the S→F instability.

It should be noted that during the upstream propagation of the speed wave of the local increase in the speed, the microscopic speed of the following each other vehicles should not necessary increases continuously to reach the occurrence of the nucleus. This means that we cannot exact state at which time instant a nucleus appears within synchronized flow. This incertitude in the determination of the nucleus for the development of the S→F instability can be explained as follows: The critical nucleus for the S→F instability is characterized by some critical spatiotemporal distributions of traffic variables.

5.12 Microscopic Stochastic Features of S→F Instability at Bottleneck

The S→F instability away of a highway bottleneck discussed in Sect. 5.11 does not explain the hypothesis of the three-phase theory that the S→F instability at the bottleneck should govern the metastability of free flow with respect to an F→S transition (traffic breakdown) at the bottleneck [4–7, 17, 25]. Indeed, the above consideration of the competition of the over-acceleration and speed adaptation effects leading to the S→F instability away of the bottleneck does not answer the following questions:

(i) What is a local disturbance in synchronized flow that can spontaneously initiate the S→F instability at the *bottleneck*?
(ii) Can be proven that the S→F instability at the bottleneck exhibits the nucleation nature?
(iii) How does the S→F instability occurring in *synchronized flow* governs the metastability of *free flow* with respect to the F→S transition at the bottleneck?
Indeed, in accordance with in the three-phase theory [17], the speed adaptation effect, which describes the tendency from free flow to synchronized flow, cannot lead to some traffic flow instability. Therefore, the speed adaptation effect cannot be the origin of the nucleation nature of the F→S transition at the bottleneck observed in real traffic.
(iv) What is the physics of a random time delay to the F→S transition at the bottleneck found in simulations with stochastic three-phase traffic flow models [17, 25, 35, 36, 38, 41, 47]?

In Sects. 5.12 and 5.13, we present results of a microscopic stochastic three-phase theory developed in [31] that answers the above questions (i)–(iv).

5.12.1 "Speed Peak"—Local Speed Disturbance in Synchronized Flow at Bottleneck Initiating S→F Instability

In simulations of traffic flow on a single-lane road with an on-ramp bottleneck, we find a sequence of F→S and S→F transitions at the bottleneck (labeled respectively by "F→S transitions" and "S→F transitions" in Fig. 5.24a–d). At chosen flow rates q_{on} and q_{in} (Fig. 5.24), each of the F→S transitions leads to the formation of a widening synchronized flow pattern (WSP) at the bottleneck (labeled by "WSP_1", "WSP_2", and "WSP_3" in Fig. 5.24). To understand *microscopic features* of the S→F instability, we consider of an S→F transition shown in Fig. 5.24c,d.

We have found that there is a local disturbance of the increase in the speed in synchronized flow at the bottleneck. The disturbance emerges at the downstream front of synchronized flow at the on-ramp bottleneck. This disturbance of the local increase in the speed in synchronized flow can initiate an S→F instability (Fig. 5.25). We call the local disturbance as "speed peak" (labeled by "speed peak" on trajectory 2 in Figs. 5.25b):

The origin of the speed peak is as follows. At time instant $t = t_1^{(1)}$ vehicle 1 begins to accelerate at the downstream front of synchronized flow (Fig. 5.25b). Within the downstream front of synchronized flow, vehicle 1 accelerates continuously from a synchronized flow speed to free flow downstream of the bottleneck. Vehicle 1 reaches a free flow speed at time instant $t_2^{(1)}$ (trajectory 1 in

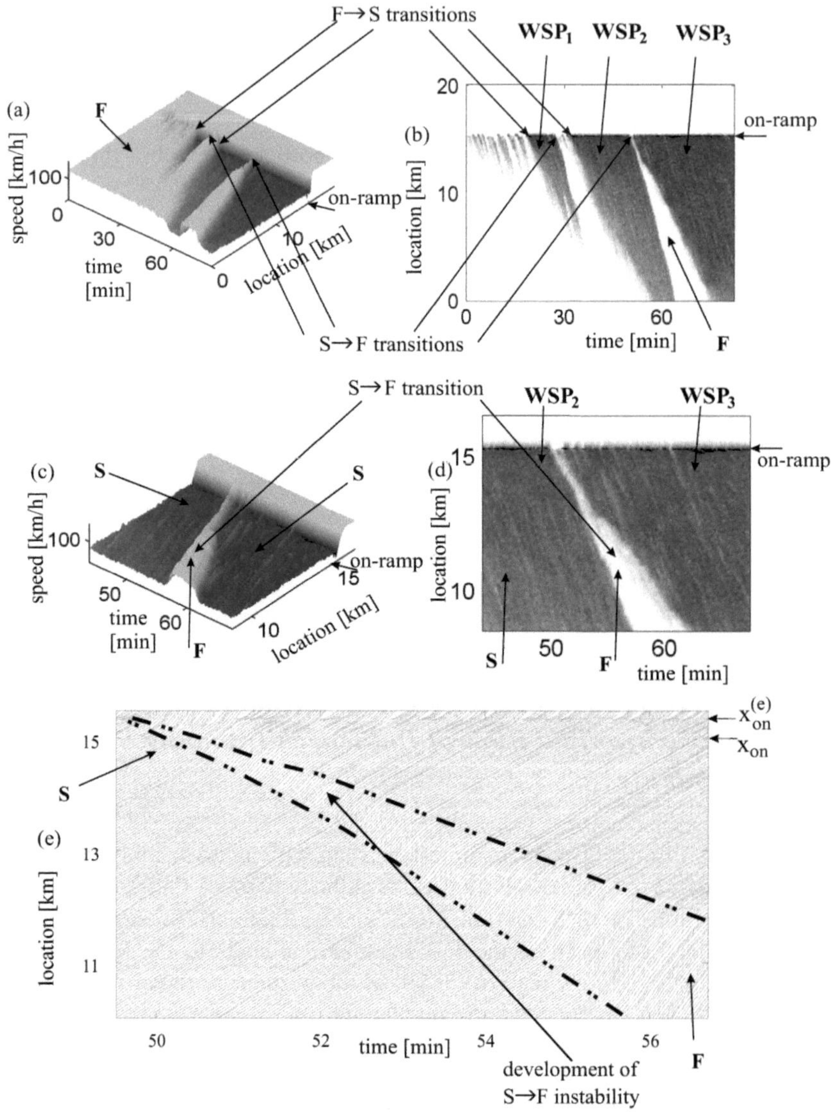

Fig. 5.24 Simulations of S→F instability in synchronized flow leading to S→F transition at on-ramp bottleneck on single-lane road with the KKSW CA model (Appendix B): (**a–d**) Speed in space and time (**a**, **c**) and the same speed data presented by regions with variable shades of gray (shades of gray vary from white to black when the speed decreases from 110 km/h (*white*) to zero (*black*)) (**b**, **d**); figures (**c**, **d**) are, respectively, fragments of (**a**, **b**) in larger scales in space and time. (**e**) Fragment of vehicle trajectories in space and time related to (**c**, **d**); *bold dashed-dotted curves* in (**e**) mark the development of S→F instability in synchronized flow leading to S→F transition. F—free flow, S—synchronized flow, WSP—widening synchronized flow pattern. q_{on} = 360 vehicles/h, q_{in} = 1406 vehicles/h. x_{on} = 15 km and $x_{on}^{(e)}$ = 15.3 km are, respectively, the beginning and the end of the merging region of the on-ramp which which vehicles can merge from the on-ramp onto the main road, i.e., merging region of the on-ramp is located within 15 km $\leq x \leq$ 15.3 km. Adapted from [31]

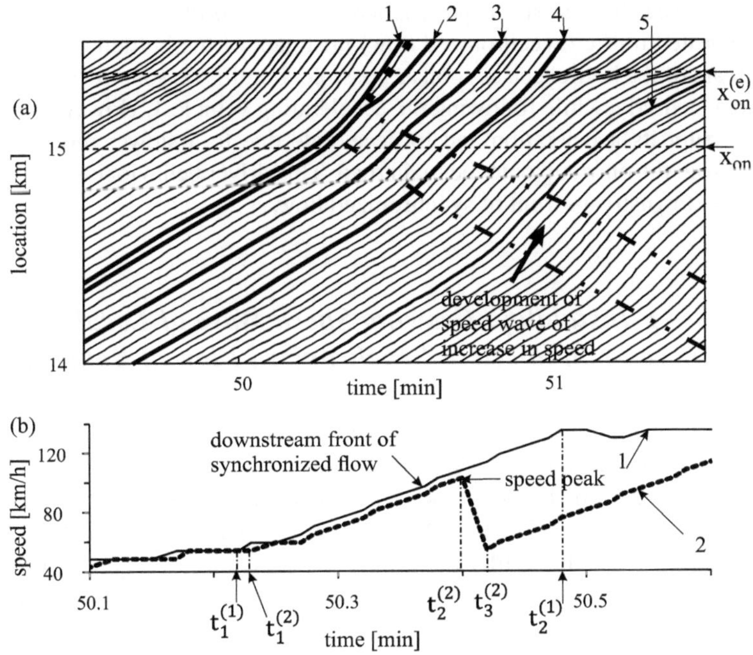

Fig. 5.25 Speed peak at the downstream front of synchronized flow at bottleneck that initiates S→F instability shown in Fig. 5.24c–e: (**a**) Fragment of vehicle trajectories related to Fig. 5.24e; *bold dashed-dotted curves* in (**a**) mark the development of the speed wave of increase in speed within synchronized flow. (**b**) Microscopic vehicle speed along trajectories as time-functions. In (**b**), vehicle trajectories are labeled by the same numbers as those in (**a**). Adapted from [31]

Fig. 5.25b). A different situation is realized for vehicle 2 that follows vehicle 1 on the main road. After vehicle 1 has begun to accelerate, vehicle 2 begins also to accelerate at the downstream front of synchronized flow at time instant $t_1^{(2)}$ (trajectory 2 in Fig. 5.25b). However, a slower moving vehicle merges from the on-ramp lane onto the main road between vehicles 1 and 2 (bold dotted vehicle trajectory between vehicle trajectories 1 and 2 in Fig. 5.25a). Because vehicles 1 and 2 move on single-lane road, vehicle 2 cannot overtake the vehicle merging from the on-ramp. As a result, vehicle 2 must decelerate at time $t_2^{(2)}$ (trajectory 2 in Fig. 5.25b). This effect leads to the occurrence of a speed peak at the downstream front of synchronized flow at the bottleneck (Fig. 5.25b). After the vehicle merging from the on-ramp increases its speed, vehicle 2 can continue acceleration to the free flow speed at time instant $t_3^{(2)}$ (trajectory 2 in Fig. 5.25b).

Thus, we see that the speed peak is associated with driver behavior to escape from congested traffic (synchronized flow) upstream of the bottleneck to free flow downstream of the bottleneck as soon as possible. For this reason, a driver of vehicle 2 accelerates firstly, while following the accelerating vehicle 1. However, the driver must interrupt its acceleration and to decelerate due to a slower moving vehicle merging from the on-ramp onto the main road. This driver behavior causes the speed peak at the downstream front of synchronized flow.

5.12.2 S→F Instability: Growing Speed Wave of Local Increase in Speed in Synchronized Flow at Bottleneck

The speed peak initiates a speed wave of a local increase in the speed within synchronized flow. This speed wave propagates upstream. This effect can be seen in Figs. 5.25a and 5.26. The speed wave begins to grow both in the amplitude and in the space (Figs. 5.26, 5.27, 5.28, 5.29). Finally, the growth of the wave leads to an S→F transition at the bottleneck. The S→F instability, i.e., the growth of the speed wave of a local increase in the speed within synchronized flow is caused by the over-acceleration effect. The growing speed wave of the local increase in the speed in synchronized flow can also be considered "growing acceleration wave" in synchronized flow. To show the effect of over-acceleration on the S→F instability, we consider vehicle trajectories 5–13 within the growing speed wave of the increase in the speed (Fig. 5.27).

The over-acceleration effect can be seen, if we compare the motion of vehicles 5, 6 with vehicle 7 that follow each other (Fig. 5.27a) within the speed wave of the local increase in the speed (Fig. 5.27b). Whereas vehicle 6 follows vehicle 5 without over-acceleration, vehicle 7 accelerates while reaching the speed that exceeds the speed of preceding vehicle 6 appreciably (trajectories 6 and 7 in Fig. 5.27b). Although vehicle 6 begins to decelerate, nevertheless vehicle 7 accelerates. This acceleration of vehicle 7 occurs under conditions (5.37) and (5.53). For this reason, the acceleration of vehicle 7 is an example of the over-acceleration effect (labeled by "over-acceleration" in Fig. 5.27b).

The effect of over-acceleration exhibits also vehicle 9 that follows vehicle 8, vehicle 10 that follows vehicle 11 as well as vehicle 13 that follows vehicle 12 (trajectories 9–13 in Fig. 5.27c,d). The subsequent effects of over-acceleration of different vehicles lead to the S→F instability, i.e., to a growing wave of the local increase in the speed within synchronized flow. The speed wave grows both in the amplitude and in the space extension during upstream propagation of the wave within synchronized flow. The subsequent development of the S→F instability caused by the over-acceleration effect can be seen in Figs. 5.28 and 5.29.

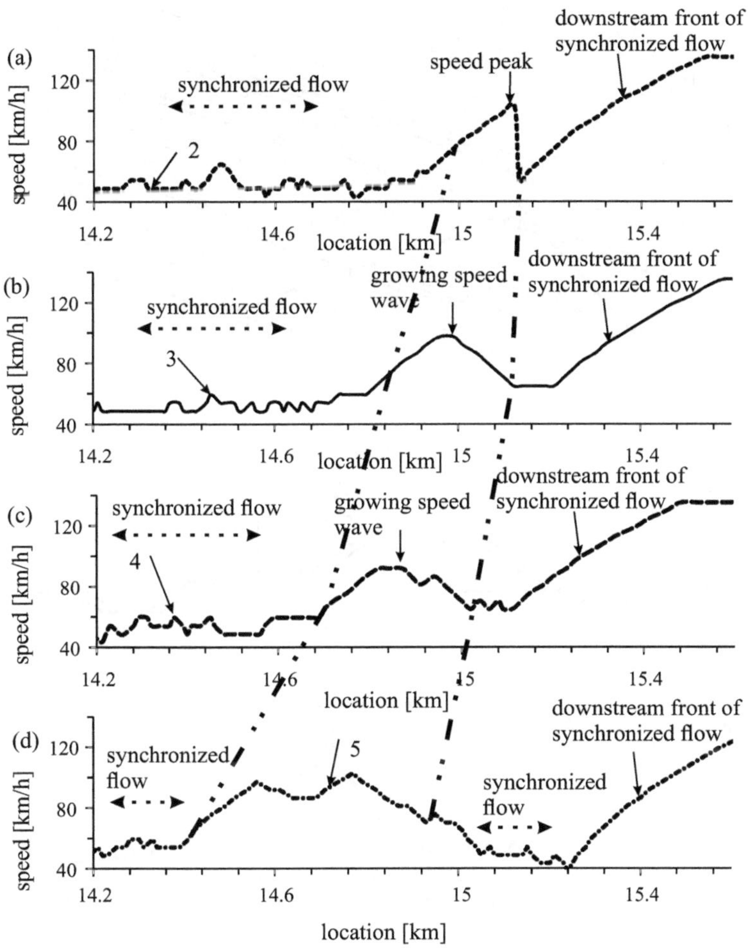

Fig. 5.26 Transformation of a speed peak shown in Fig. 5.25b into a growing wave of the local increase in the speed (labeled by "growing speed wave") propagating upstream within synchronized flow: (**a–d**) Microscopic vehicle speeds along trajectories as road location-functions labeled by the same numbers as those in Fig. 5.25a. In (**a–d**), *bold dashed-dotted curves* mark a growing wave of the local increase in the speed within synchronized flow as a function of road location. Adapted from [31]

The S→F instability leads to a growing wave of the local increase in the speed within synchronized flow. The growth of the speed wave is realized due to over-acceleration effect.

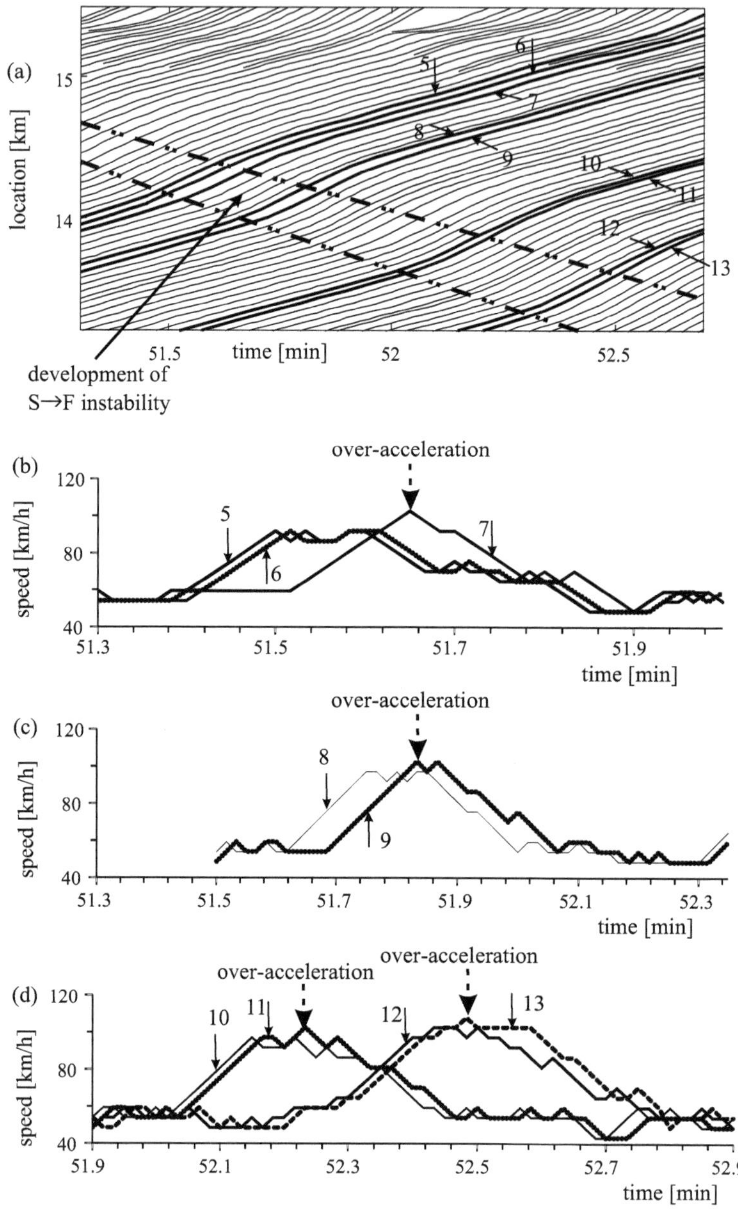

Fig. 5.27 Effect of over-acceleration on S→F instability shown in Fig. 5.24c–e: (**a**) Fragment of vehicle trajectories in space and time; *bold dashed-dotted curves* mark upstream propagation of the growing wave of the local increase in the speed within synchronized flow. (**b**–**d**) Microscopic vehicle speed along trajectories as time functions labeled by the same numbers as those in (**a**). Adapted from [31]

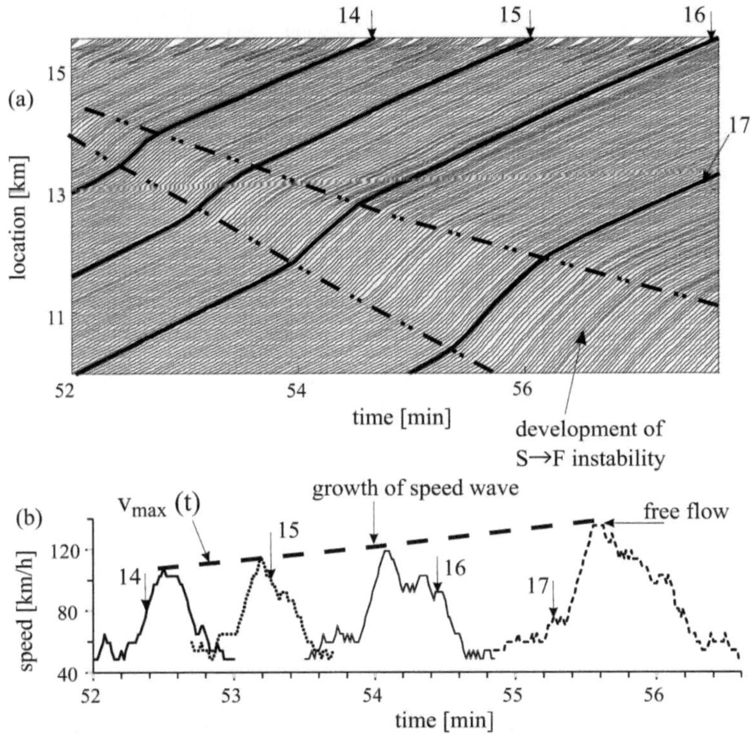

Fig. 5.28 Subsequent development of S→F instability shown in Fig. 5.27a: (**a**) Fragment of vehicle trajectories in space and time; *bold dashed-dotted curves* in (**a**) mark the development of a speed wave of the local increase in the speed within synchronized flow. (**b**) Microscopic vehicle speeds along trajectories as time functions labeled by the same numbers as those in (**a**); *bold dashed curve* marks the increase of the wave amplitude over time $v_{max}(t)$. Adapted from [31]

5.12.3 Dissolving Speed Wave of Local Increase in Speed Within Synchronized Flow at Bottleneck

There can be many speed peaks that occur randomly at the downstream front of synchronized flow at the on-ramp bottleneck (Fig. 5.30a). The physics of all speed peaks shown in Fig. 5.30 is the same as discussed above (Sect. 5.12.1). As an example, we consider a speed peak labeled by "speed peak A" in Fig. 5.30a. Speed peak A appears on vehicle trajectory 19 due to a slow vehicle merging from the on-ramp onto the main road (bold dotted vehicle trajectory between vehicle trajectories 18 and 19 in Fig. 5.30b).

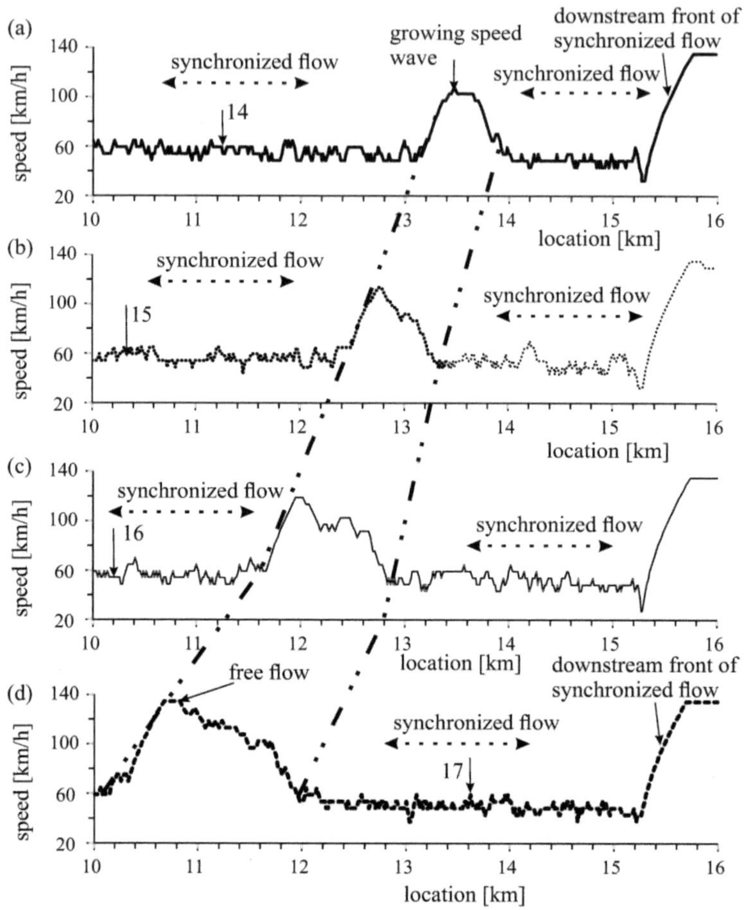

Fig. 5.29 Continuation of Fig. 5.28: (**a–d**) Microscopic vehicle speeds along trajectories as road location-functions labeled by the same numbers as those in Fig. 5.28a; *bold dashed-dotted curves* mark the development of growing speed wave of the increase in the speed within synchronized flow as a function of road location. Adapted from [31]

Speed peak "A" initiates a speed wave of the increase in the speed within synchronized flow that propagates upstream. However, rather than an S→F instability occurs as discussed in Sect. 5.12.2, the wave is fully dissolved about 0.3 km upstream of the beginning of the on-ramp merging region at $x = 15$ km. We call this wave as "dissolving speed wave" of the increase in the speed in synchronized flow (Figs. 5.30b and 5.31).

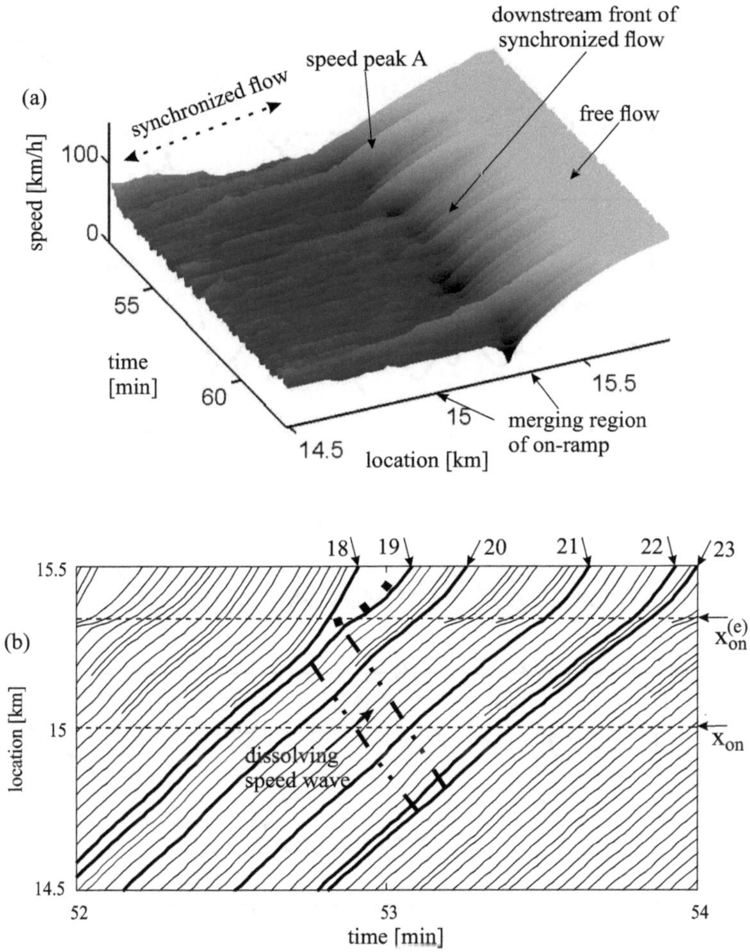

Fig. 5.30 Speed peaks at downstream front of synchronized flow at the on-ramp bottleneck: (**a**) Speed in space and time; fragment of Fig. 5.24a for a time interval that begins after time instant at which the S→F instability at bottleneck has occurred; one of the speed peaks in (**a**) is marked by "speed peak A". (**b**) Fragment of vehicle trajectories with a dissolving speed wave initiated by speed peak A in (**a**) (dissolving speed wave is marked by *dotted-dashed curves*). Adapted from [31]

The speed peak shown in Fig. 5.25, which initiates the S→F instability (Sects. 5.12.1 and 5.12.2), and speed peak A that does not initiate an S→F instability differ in their amplitudes: The speed within the peak shown in Fig. 5.25 is about 98 km/h; the speed within peak A is considerably smaller (about 70 km/h). All other speed peaks that emerge at the downstream front of synchronized flow (Fig. 5.30a) exhibit also considerably smaller amplitudes than that of the speed peak shown in Fig. 5.25. As a result of small amplitudes of speed peaks, all waves of the increase in the speed within synchronized flow shown in Fig. 5.30a are dissolving speed

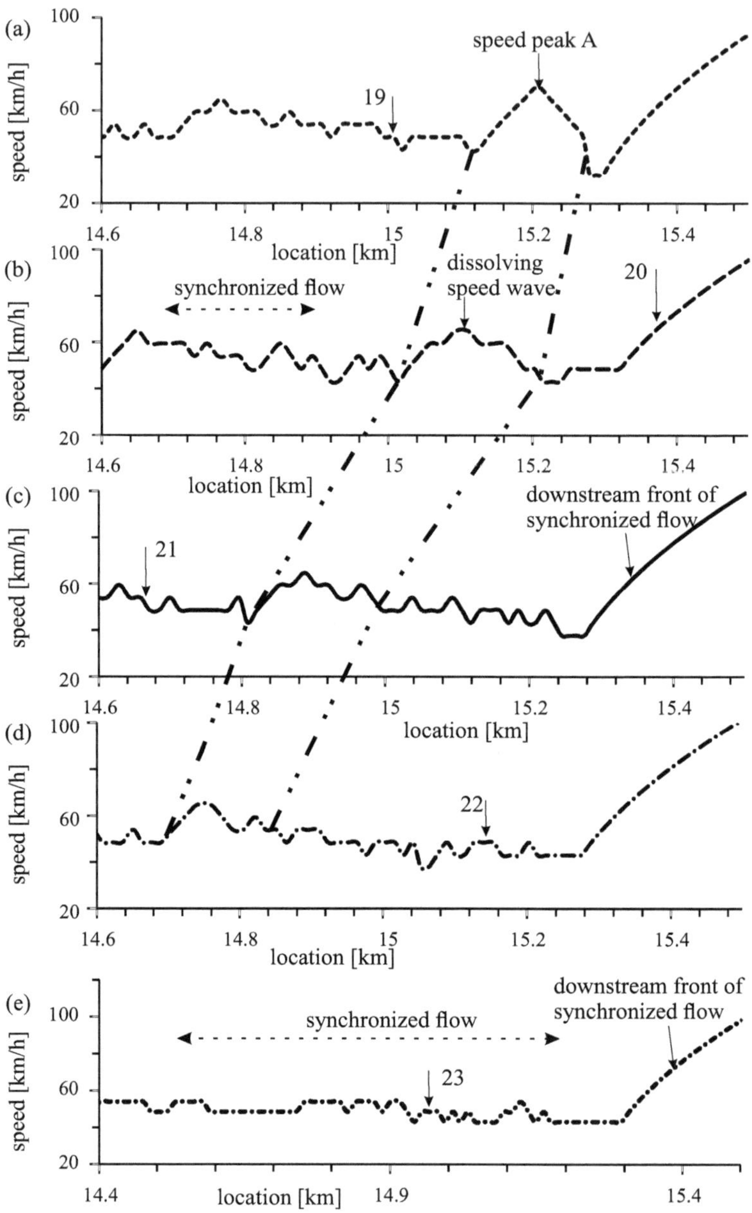

Fig. 5.31 Dissolving speed wave of the local increase in the speed within synchronized flow at the on-ramp bottleneck related to Fig. 5.30: (**a–e**) Microscopic speed along vehicle trajectories as road location functions. Vehicle trajectories are labeled by the same numbers as those in Fig. 5.30b. *Bold dashed-dotted curves* mark the propagation of dissolving speed wave in space. Adapted from [31]

waves. A dissolving speed wave of the increase in the speed in synchronized flow can also be considered "dissolving acceleration wave" in synchronized flow.

5.12.4 Nucleation Nature of S→F Instability

We have found that if the amplitude of a speed peak is equal to or larger than some critical amplitude of the speed peak, the speed peak is a nucleus for an S→F instability (Sects. 5.12.1 and 5.12.2). Contrarily, if the peak amplitude is smaller than the critical one (as this is the case for all speed peaks in Fig. 5.30a), the speed peak is smaller than a nucleus for an S→F instability: Instead of the S→F instability, the speed peak initiates a dissolving speed wave of the local increase in the speed within synchronized flow (Figs. 5.30b and 5.31).

The physics of the nucleation nature of an S→F instability is as follows. The over-acceleration effect is able to overcome speed adaptation between following each other vehicles (speed adaptation effect) only if the speed within the speed wave is large enough: When the over-acceleration effect is stronger than the speed adaptation effect within the speed wave, as that occurs in Fig. 5.27, the S→F instability is realized. Otherwise, when during the speed wave propagation the speed adaptation effect suppresses the over-acceleration within synchronized flow, the speed wave dissolves over time, i.e., no S→F instability is realized (Fig. 5.31b–e). Thus, in accordance with a hypothesis of the three-phase theory discussed in Sects. 5.8–5.10, a competition between the speed adaptation and over-acceleration effects determines the nucleation nature of the S→F instability.

The S→F instability exhibits the nucleation nature. Small speed peaks (small initial disturbances of the local increase in the speed) occurring at the downstream front of synchronized flow decay over time. When a large enough speed peak appears at the downstream front of synchronized flow, the S→F instability occurs. This large enough speed peak can be considered a nucleus for the S→F instability.

The S→F instability causes a growing wave of the local increase in the speed within synchronized flow. The growth of the speed wave is realized due to over-acceleration effect.

5.13 S→F Instability as Origin of Nucleation Nature of Traffic Breakdown at Bottleneck

In this section, we present results of a stochastic theory of the S→F instability at highway bottlenecks of [31]. We show that and how the S→F instability in synchronized flow at the bottleneck determines the random time delay of traffic breakdown. It turns out that the S→F instability governs traffic breakdown (F→S transition) at the bottleneck.

As already discussed in Sect. 5.4.4, there is a random time delay $T^{(B)}$ between the beginning of a simulation realization and the time instant at which traffic breakdown (F→S transition) occurs resulting in the emergence of a congested pattern at the bottleneck. At chosen flow rates q_{on} and q_{in}, the congested pattern is an WSP (Fig. 5.32).

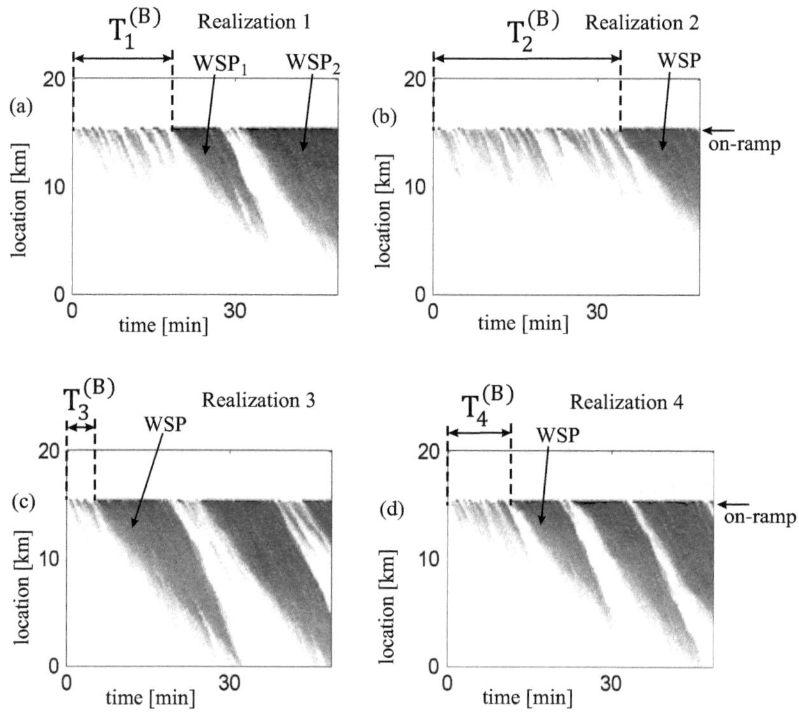

Fig. 5.32 Random time delay of traffic breakdown (F→S transition) at on-ramp bottleneck: (**a–d**) Speed in space and time for four different simulation realizations (runs) presented by regions with variable shades of gray (shades of gray vary from white to black when the speed decreases from 110 km/h (*white*) to zero (*black*)). Realization 1 in (**a**) is a fragment of Fig. 5.24b, i.e., realization 1 is the simulation realization studied in Figs. 5.24–5.31. Time delays of traffic breakdown $T^{(B)}$ in different simulation realizations 1–4 are $T_1^{(B)} = 19$ min (**a**), $T_2^{(B)} = 35$ min (**b**), $T_3^{(B)} = 7$ min (**c**), and $T_4^{(B)} = 13$ min (**d**). $q_{on} = 360$ vehicles/h, $q_{in} = 1406$ vehicles/h. Adapted from [31]

5.13.1 *Microscopic Nature of Permanent Local Speed Disturbance in Free Flow at Bottleneck*

We have found that during the time delay of traffic breakdown $0 < t < T_1^{(B)}$ (Figs. 5.32a and 5.33a) there is a permanent spatiotemporal competition between the speed adaptation effect supporting an F→S transition and the over-acceleration effect supporting an S→F instability that counteracts the emergence of synchronized flow. This competition results in the occurrence of a permanent local speed decrease in a neighborhood of the bottleneck that we have called in Sect. 5.3.1 as "permanent local speed disturbance" at the bottleneck. There can be distinguished two cases of this competition:

(i) There is a noticeable time lag between the beginning of an F→S transition due to the speed adaptation and the beginning of an S→F instability due to over-acceleration that prevents the formation of a congested pattern at the bottleneck; this case we call "a sequence of F→S→F transitions" at the bottleneck.
(ii) There is a spatiotemporal "overlapping" of the speed adaptation and over-acceleration effects.

A permanent local speed disturbance at a highway bottleneck results from a spatiotemporal competition of the speed adaptation and over-acceleration effects.

5.13.2 *Sequence of F→S→F Transitions at Bottleneck*

One of the sequences of F→S→F transitions within the permanent local speed disturbance at the on-ramp bottleneck is marked by dashed-dotted curves in Fig. 5.33a. An F→S transition and a return S→F transition that build the sequence of F→S→F transitions are explained as follows (Figs. 5.33b,c).

After several slow moving vehicles have merged from the on-ramp onto the main road, the following vehicles on the main road have to decelerate strongly due to the speed adaptation effect. This results in the upstream propagation of synchronized flow upstream of the bottleneck, i.e., an F→S transition occurs. The upstream front of the emergent synchronized flow begins to propagate upstream (the upstream front of this synchronized flow is labeled by "upstream front" in Fig. 5.33b).

For the same reason as explained in Sect. 5.12.1, within the downstream front of the synchronized flow localized at the bottleneck (labeled by "downstream front 1" in Fig. 5.33b) speed peaks appear. The speed peaks lead to the emergence of dissolving speed waves in the synchronized flow; the dissolving waves have also qualitatively the same microscopic features as shown in Fig. 5.31.

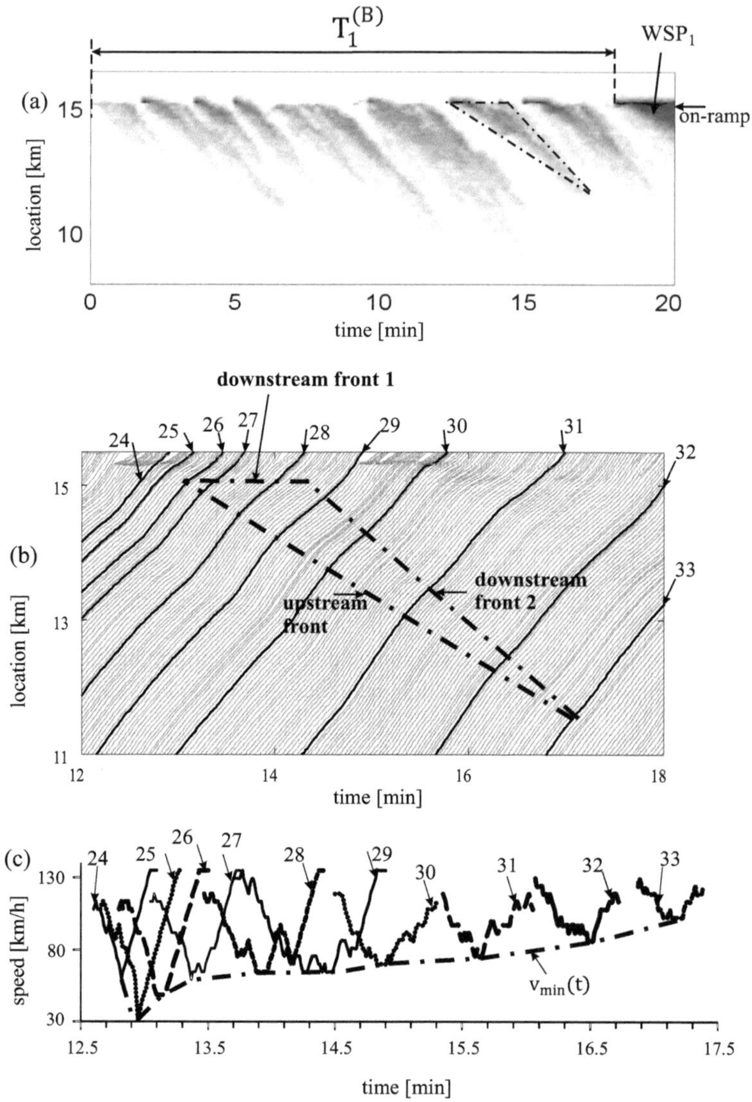

Fig. 5.33 Simulations of F→S→F transitions within a permanent local speed disturbance at on-ramp bottleneck related to Fig. 5.32a: (**a**) Speed in space and time presented by regions with variable shades of gray (shades of gray vary from white to black when the speed decreases from 110 km/h (*white*) to 20 km/h (*black*)) within time delay of traffic breakdown. (**b**) Fragment of vehicle trajectories in space and time related to a short time interval in (**a**). (**c**) Microscopic vehicle speeds along trajectories as time functions labeled by the same numbers as those in (**b**). In (**b**), "downstream front 1" and "downstream front 2" label the downstream front of synchronized flow for two cases, when the front is localized at the bottleneck and propagate upstream, respectively. Adapted from [31]

A crucial difference of the case under consideration (Fig. 5.33b) with traffic breakdown leading to the emergence of WSP_1 shown in Figs. 5.32a and 5.33a is as follows. We find that in Fig. 5.33b synchronized flow exists for a few minutes only: A speed peak occurs at the downstream front of this synchronized flow (labeled by "downstream front 1" in Fig. 5.33b) that initiates an S→F instability at the bottleneck. The S→F instability interrupts the formation of a congested pattern at the bottleneck.

Indeed, due to the S→F instability, rather than an WSP occurs, a localized region of synchronized flow *departs from the bottleneck*: The downstream front of this synchronized flow (labeled by "downstream front 2" in Fig. 5.33b) propagates upstream from the bottleneck. While propagating upstream from the bottleneck, synchronized flow dissolves over time.

Thus, rather than the WSP occurs at the bottleneck, dissolving synchronized flow appears at the bottleneck (region of this synchronized flow is limited by "upstream front" as well as "downstream front 1" and "downstream front 2" shown in Fig. 5.33b). As a result of such a dissolving synchronized flow, the minimum speed $v_{min}(t)$ within the permanent local speed disturbance firstly decreases and then increases over time (vehicle trajectories 24–33 in Fig. 5.33c).

The physics of the S→F instability is the same as disclosed in Sect. 5.12.2. In particular, the S→F instability leads to a growing speed wave of the local increase in the speed within synchronized flow. The growth of the speed wave is realized due to over-acceleration effect.

5.13.3 Nature of Random Time Delay of Traffic Breakdown at Bottleneck

There can be many sequences of F→S→F transitions that have been marked by dashed-dotted curves in Fig. 5.34. Each of them interrupts the formation of a congested pattern at the bottleneck. This explains the existence of a time delay of traffic breakdown: Rather than the congested pattern appears at the bottleneck, the sequences of F→S→F transitions result in a narrow region of the local decrease in the speed in free flow localized at the bottleneck (called as a "permanent local speed disturbance" in free flow at the bottleneck).

> An F→S transition at a highway bottleneck with a subsequent S→F transition is called a sequence F→S→F transitions. The sequence F→S→F transitions leads to the occurrence of a dissolving synchronized flow region at the bottleneck.

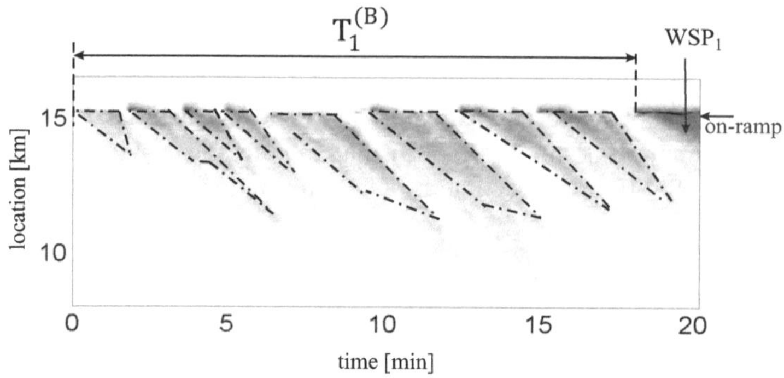

Fig. 5.34 Continuation of Fig. 5.33a: Many sequences of F→S→F transitions at on-ramp bottleneck. Speed in space and time has been adapted from Fig. 5.33a. Different sequences of F→S→F transitions with associated regions of dissolving synchronized flow at the bottleneck are marked by *dashed-dotted curves*. Adapted from [31]

There can be many sequences of F→S→F transitions following each other at a highway bottleneck. The sequences of F→S→F transitions cause many regions of dissolving synchronized flow at the bottleneck (regions of dissolving synchronized flow at the bottleneck are marked by dashed-dotted curves in Fig. 5.34).

The sequences of F→S→F transitions result in a random time delay of traffic breakdown at the bottleneck.

During time delay $0 < t < T_1^{(B)}$ of the breakdown (Fig. 5.32a), there are also time intervals within which there is no noticeable time lag between the beginning of the F→S transition and the S→F instability due to over-acceleration. In this case, rather than to distinguish a sequence of F→S→F transitions within the permanent local speed disturbance at the bottleneck, we find a spatiotemporal "overlapping" of the speed adaptation and over-acceleration effects (see illustrations of this case in [31]).

In this case, there is an upstream front of the permanent local speed disturbance within which vehicles moving on the main road at a free flow speed decelerate to a smaller speed due to slower moving vehicles that merge from the on-ramp onto the main road. There is also a downstream front of the permanent local speed disturbance within which vehicles accelerate to a free flow speed that is a higher one than the average speed within the permanent local speed disturbance. We have found that the distribution of the speed within the permanent local speed disturbance exhibits a complex spatiotemporal dynamics.

This complex dynamics of the permanent local speed disturbance at the bottleneck is as follows. As in the fully developed synchronized flow (Fig. 5.30a), within the permanent local speed disturbance there is a sequence of speed peaks that occur randomly at the downstream front of the permanent local speed disturbance. Due to the speed peaks, regions of the increase in the speed appear that propagate upstream within the permanent local speed disturbance. Within the regions of the speed increase, the over-acceleration effect occurs that prevents the upstream propagation of the upstream front of synchronized flow due to the speed adaptation. These over-acceleration effects can be considered short time S→F instabilities that increase the speed within the permanent local speed disturbance. These short time S→F instabilities prevent a continuous propagation of the upstream front of the permanent local speed disturbance, i.e., they prevent the development of traffic breakdown at the bottleneck. Therefore, rather than traffic breakdown resulting in the formation of WSP_1 (Fig. 5.32a), the permanent local speed disturbance persists at the bottleneck. Thus, the competition between speed adaptation and over-acceleration determines a random time delay of traffic breakdown at the bottleneck independent of whether sequences of F→S→F transitions can be distinguished or not within the permanent local speed disturbance at the bottleneck.

It should be noted that recently sequences of F→S→F transitions occurring before traffic breakdown is realized at a highway bottleneck have been found in empirical microscopic (single-vehicle) data measured on German highway A8 near Stuttgart [49].

As long as during a short time interval the S→F instability can destroy the development of the F→S transition, we consider that no "completed" F→S transition has occurred (time interval $0 < t < T_1^{(B)}$ in Fig. 5.34). This is because due to this interruption of the development of the F→S transition, only regions of dissolving synchronized flow appear upstream of the bottleneck (Fig. 5.34).

However, there can be a time instant $T^{(B)}$ at which there is no S→F instability that can prevent the development of the F→S transition: In this case, no interruption of an F→S transition is realized. Such a "completed" F→S transition[14] leads to the formation of a congested pattern at the bottleneck (WSP_1 in Fig. 5.34). In its turn, the S→F instability does not occur, if no nucleus required for the S→F instability appears in synchronized flow at the bottleneck (Sect. 5.12.4).

The occurrence of the nucleus required for the S→F instability in synchronized flow at the bottleneck is a random event. This explained why a time delay of traffic breakdown (F→S transition) $T^{(B)}$ is a random value. Thus, when no nucleus for the S→F instability appears during a long enough time interval after the beginning of the development of an F→S transition at the bottleneck, a congested pattern is formed at the bottleneck. Through this nucleation feature of the S→F instability, traffic breakdown (F→S transition) at a highway bottleneck exhibits also the

[14]With the exception of this Sect. 5.13, in the book the term *F→S transition* determines the "completed" F→S transition only. For this reason, we do not used the term *"completed" F→S transition* any more.

nucleation nature: The S→F instability governs traffic breakdown (F→S transition) at a highway bottleneck.

In the microscopic stochastic three-phase theory presented in Sects. 5.12 and 5.13, the occurrence of a nucleus for traffic breakdown at a highway bottleneck is characterized by some spatiotemporal distributions of traffic variables within the permanent local speed disturbance at the bottleneck at which the development of the F→S transition cannot be interrupted by the S→F instability. Thus, the microscopic stochastic three-phase theory shows that the nucleus for traffic breakdown (F→S transition) exhibits complex spatiotemporal structure. The spatiotemporal structure of the nucleus is associated with a spatiotemporal competition between the speed adaptation and over-acceleration effects within the permanent local speed disturbance at the bottleneck.

Therefore, the critical amplitude of a critical nucleus considered in the nucleation model of Sect. 5.3 of a qualitative three-phase theory as a solely parameter of the critical nucleus is a very rough simplification of the reality.

The S→F instability governs traffic breakdown (F→S transition) at a highway bottleneck.

The nucleation nature of the S→F instability is the origin of the nucleation nature of traffic breakdown (F→S transition) at a highway bottleneck.

A nucleus for traffic breakdown (F→S transition) at a highway bottleneck exhibits complex spatiotemporal structure. The spatiotemporal structure of the nucleus is associated with a spatiotemporal competition between the speed adaptation and over-acceleration effects within the permanent local speed disturbance at the bottleneck.

5.14 Explanation of Empirical Features of Traffic Breakdown at Bottleneck with Three-Phase Theory

Up to now in this chapter we have considered the theory of traffic breakdown at highway bottlenecks for a hypothetical case of traffic flow consisting of identical driver and identical vehicles. However, real traffic flow is a very heterogeneous one: There are different types of vehicles (for example, slow and fast vehicles, passenger vehicles and trucks) and there are different types of drivers (for example, timid and aggressive drivers) in real traffic. As explained in Chap. 3, in real field traffic data the nucleation nature of traffic breakdown at highway bottlenecks is independent of the degree of heterogeneity of real traffic. Therefore, the following question arises:

- Does the theoretical result about the nucleation nature of traffic breakdown at a road bottleneck discussed above for traffic flow consisting of identical driver and identical vehicles remain for real heterogeneous traffic flow?

Theoretical investigations made in [24–28, 30, 37, 44] have proven this statement: Traffic breakdown at a highway bottleneck in heterogeneous traffic flow exhibits the same nucleation nature as that discussed above for traffic flow consisting of identical driver and identical vehicles.

In this section, based on results of [43, 48], we illustrate this general conclusion based on a comparison of features of traffic breakdown in traffic flow consisting of identical driver and identical vehicles with probabilistic features of traffic breakdown at highway bottlenecks in heterogeneous traffic flow with a moving bottleneck.

5.14.1 Nucleation of Traffic Breakdown at Road Bottleneck in Traffic Flow with Moving Bottleneck

We consider results of simulations of traffic flow on two-lane road with an on-ramp bottleneck (Fig. 5.35a). The on-ramp bottleneck is at location $x = 10$ km. We assume that traffic flow consists of identical passenger vehicles in which there is only one slow vehicle moving in the right road lane. The moving bottleneck is caused by this slow vehicle that moves at a maximum speed $v_{\mathrm{M}} = 82.8$ km/h. This speed is lower than the free flow speed of other (identical) vehicles that is equal to 108 km/h.

One of the main theoretical results of this section is that in accordance with empirical data discussed in Chap. 3, a nucleus required for traffic breakdown in heterogeneous traffic flow appears due to the interaction of one of the waves in free flow with a permanent local speed disturbance localized at a highway bottleneck.

As expected from a discussion of a well-known effect of a moving bottleneck on free flow made in Sect. 1.2, simulations show that to pass the moving bottleneck, passenger vehicles moving initially in the right lane change to the left lane in a neighborhood of the moving bottleneck.[15]

[15] To avoid confusions, we do not consider results of simulations of the effect of a moving bottleneck on free flow studied with two-phase traffic flow models. This is because none of the two-phase traffic flow models can show the nucleation nature of traffic breakdown (F→S transition) (Chap. 4). In other words, the two-phase traffic flow models used earlier for a study of the effect of a moving bottleneck on traffic flow (see references in Sect. 1.2) cannot show the theoretical result of this section explaining the nucleation of the breakdown (F→S transition) due to the interaction of one of the waves in free flow with a permanent local speed disturbance localized at a

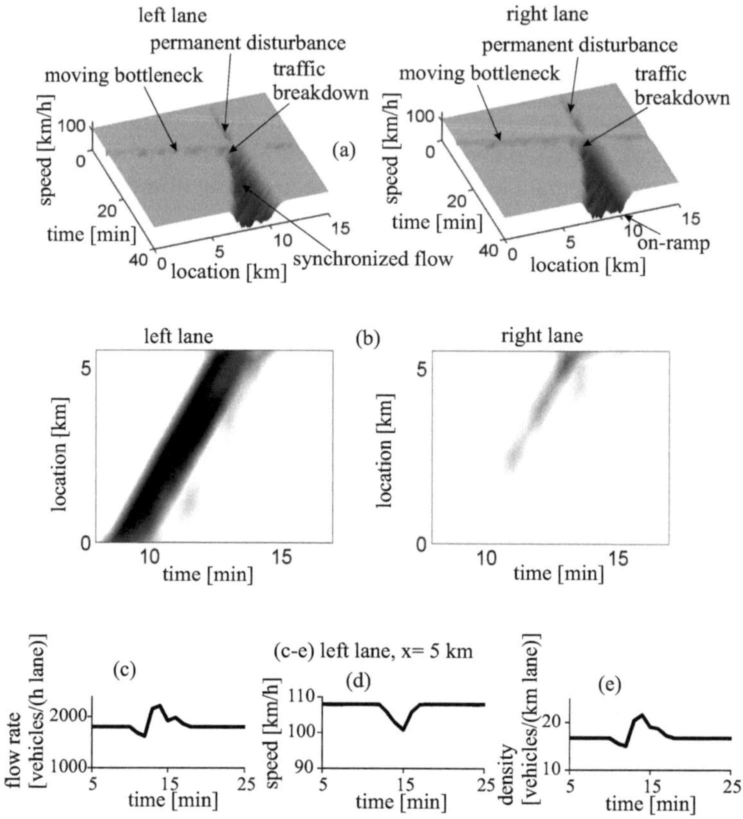

Fig. 5.35 Simulations of the nucleation of traffic breakdown on two-lane road with on-ramp and moving bottlenecks [48]: (**a**) Nucleation of traffic breakdown at on-ramp bottleneck through wave propagation caused by moving bottleneck (left—left lane, right—right lane). (**b**) Wave of the flow rate $q(x, t)$ in different road lanes (left—left lane, right—right lane) presented by regions with variable shades of gray (shades of gray vary from white to black when the flow rate increases from $q \leq 2000$ vehicles/(h lane) (*white*) to $q \geq 2150$ vehicles/(h lane) (*black*)). (**c**–**e**) Time-functions of the flow rate (**c**), the speed (**d**) and the vehicle density (**e**) within the wave in the left lane at location $x = 5$ km. $v_{\rm M} = 82.8$ km/h, $v_{\rm free} = 108$ km/h, $q_{\rm in} = 1800$ vehicles/(h lane), $q_{\rm on} = 750$ vehicles/h. The beginning and the end of the merging region of the on-ramp are, respectively, $x_{\rm on} = 10$ km and $x_{\rm on}^{(e)} = 10.3$ km

This lane changing increases the flow rate in the left lane in the neighborhood of the moving bottleneck: A wave of the increase in the flow rate occurs that moves at the velocity $v_{\rm M}$ (Fig. 5.35b). Due to lane changing from the right lane, the speed in the left lane within the wave decreases (this speed decrease is labeled as "moving

highway bottleneck as found in empirical data (Chap. 3). For this reason, we use the Kerner-Klenov stochastic microscopic model that can explain empirical nucleation features of traffic breakdown in heterogeneous traffic flow [48].

bottleneck" in Fig. 5.35a). Time-distributions of the flow rate, speed, and density within the wave in the left lane (Fig. 5.35c–e) confirm these conclusions about the wave features. Thus, the wave in free flow exhibits a 2D asymmetric structure whose characteristics are different in different highway lanes (Fig. 5.35b).

In accordance with results of Sect. 5.3.1, we find that before the moving bottleneck have not reached the effective location of the on-ramp bottleneck, there is a permanent local speed disturbance at the bottleneck. This disturbance exists both in the right and left road lanes (labeled as "permanent disturbance" in Fig. 5.35a). When the wave in free flow caused by the moving bottleneck reaches the effective location of the on-ramp bottleneck, the speed decreases at the bottleneck additionally to the speed within the permanent local speed disturbance localized at the bottleneck ($x \approx 10$ km in Fig. 5.35a). As a result, in the simulation realization presented in Fig. 5.35, when the wave reaches the bottleneck, the wave becomes a nucleus for spontaneous traffic breakdown at the bottleneck. Indeed, the interaction of the permanent local speed disturbance with the wave leads to traffic breakdown (F$\rightarrow$S transition). In particular, we have found [48] that traffic breakdown begins to develop in the left lane in which the flow rate is larger and the speed is smaller than outside the wave.

In contrast to simulations with a single slow vehicle (Fig. 5.35a), in real field data there are many slow vehicles (Sect. 3.4.5). However, in the empirical data there are large time-oscillations of the percentage of slow vehicles (Fig. 3.11a). For this reason, a sequence of waves of slow vehicles occurs in free flow (left column in Fig. 3.12). Based on a simple model with the single slow vehicle, we have simulated one of such waves of real traffic (Fig. 5.35). As in empirical observations (Fig. 3.12), in simulations the wave exhibits a 2D-structure of the flow rate and speed (Fig. 5.35b–e).[16] Moreover, as in empirical observations (Sect. 3.4), in simulations we have found that a nucleus for traffic breakdown (F$\rightarrow$S transition) at the bottleneck (Fig. 5.35b) occurs due to the interaction of this wave with the permanent local speed disturbance at the bottleneck. Thus the above simulations explain the physics of nuclei for traffic breakdown found in empirical data (Sect. 3.4.5).

In the simulation realization presented in Fig. 5.35, no spontaneous traffic breakdown occurs at the on-ramp before the moving bottleneck reaches the on-ramp bottleneck. This is because at chosen flow rates q_{in} and q_{on} the probability of spontaneous traffic breakdown in free flow due to model fluctuations only, i.e., without the moving bottleneck is small (although this probability is larger than zero; see Sect. 5.14.2).

[16]A more detailed comparison of simulations of the wave in free flow caused by the moving bottleneck (Fig. 5.35b,c) with empirical features of waves in free flow discussed in Sect. 3.4.5 can be found in [48].

A nucleus for traffic breakdown (F→S transition) at the bottleneck occurs due to an *interaction* of the wave in free flow caused by the moving bottleneck with the permanent local speed disturbance localized at the bottleneck.

5.14.2 Features of Flow-Rate Dependence of Probability of Traffic Breakdown at Bottleneck

In Sect. 5.14.1 we have shown that the statement of the three-phase theory about the nucleation nature of traffic breakdown at the bottleneck remains for heterogeneous traffic flow with moving bottleneck. Nevertheless, quantitative characteristics of traffic breakdown at a highway bottleneck can depend considerably on drivers and vehicles parameters. Therefore, the following question arises:

- Is there a general characteristic of the nucleation nature of traffic breakdown at highway bottlenecks that is qualitatively independent of the degree of heterogeneity of real traffic?

The three-phase theory shows that such a general characteristic of the nucleation nature of traffic breakdown at highway bottlenecks is the flow-rate dependence of the probability of traffic breakdown (F→S transition). The heterogeneity of real traffic can change *quantitative characteristics* of this probability function—the maximum and minimum capacities as well as the threshold flow rate [24–28, 30, 37, 44].

Qualitative features of the flow-rate dependence of the probability of traffic breakdown are independent of the degree of heterogeneity of real traffic.

We illustrate this general conclusion based on a comparison of the flow-rate dependencies of the probability of traffic breakdown for two traffic flows:

(i) Traffic flow consisting of identical driver and identical vehicles (traffic flow without moving bottlenecks, for short).
(ii) Traffic flow with a moving bottleneck considered in Sect. 5.14.1. One of the reasons for such a comparison is as follows.

It has been found that even at the same flow rates (q_{in}, q_{on}), as those used in simulations of traffic breakdown in traffic flow with the moving bottleneck (Fig. 5.35), in traffic flow without moving bottlenecks spontaneous traffic breakdown can nevertheless occur (Fig. 5.36).

The main result of this comparison is that independent of whether traffic flow without moving bottleneck or traffic flow with a moving bottleneck is considered,

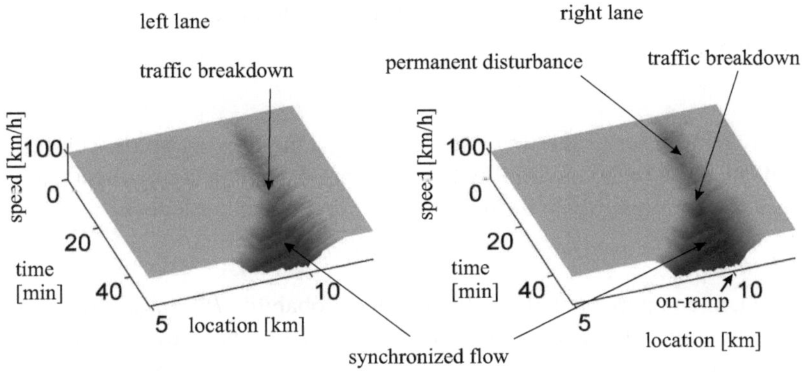

Fig. 5.36 Traffic breakdown in traffic flow without moving bottlenecks at the same flow rates as those in Fig. 5.35: Speed in space and time in the case when there are no moving bottlenecks in traffic flow of identical drivers and vehicles (left figure—left lane, right—right lane); $q_{in} = 1800$ vehicles/(h lane), $q_{on} = 750$ vehicles/h. Adapted from [48].

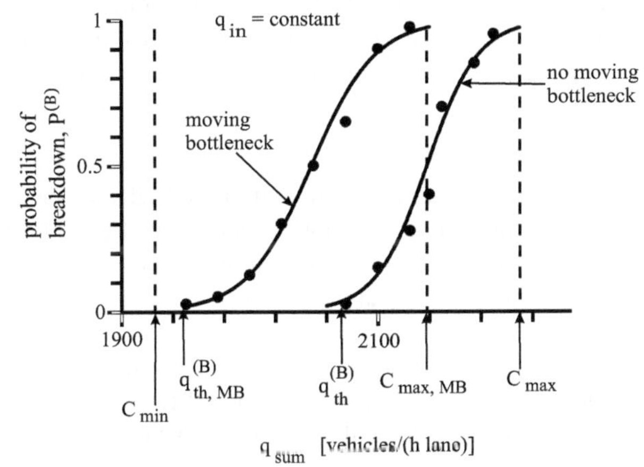

Fig. 5.37 Simulated probabilities of traffic breakdown $P^{(B)}(q_{sum})$ in traffic flow of identical drivers and vehicles with moving bottleneck (left curve labeled by "moving bottleneck") and without moving bottlenecks (right curve labeled by "no moving bottleneck") as functions of the flow rate downstream of the bottleneck $q_{sum} = q_{in} + 0.5q_{on}$. In simulations, the flow rate q_{sum} is varied through the change in the on-ramp inflow rate q_{on} at constant $q_{in} = 1800$ vehicles/h. Calculated minimum capacity $C_{min} \approx 1925$ vehicles/(h lane). Adapted from [48]

qualitative features of flow-rate functions of the probability of spontaneous traffic breakdown $P^{(B)}(q_{sum})$ (Fig. 5.37) for both traffic flows are identical and they are described by the same formula (5.12).

There is, however, an important quantitative difference between these different traffic flows:

At any chosen set of the flow rates (q_{in}, q_{on}), the moving bottleneck results in the increase in the probability for traffic breakdown in comparison with the breakdown probability in traffic flow without moving bottlenecks.

Indeed, in comparison with the breakdown probability $P^{(B)}(q_{sum})$ for traffic flow without moving bottlenecks (right curve labeled by "no moving bottleneck" in Fig. 5.37), the flow rate dependence of the breakdown probability $P^{(B)}(q_{sum})$ for traffic flow with the moving bottleneck is shifted to left in the flow rate axis (left curve labeled by "moving bottleneck" in Fig. 5.37). As above-mentioned, each of the functions $P^{(B)}(q_{sum})$ is well fitted by formula (5.12).[17]

We have also found that the moving bottleneck results in the decrease in both the maximum highway capacity C_{max} and the threshold flow rate for spontaneous traffic breakdown $q_{th}^{(B)}$. To distinguish the cases of traffic flows with the moving bottleneck and without moving bottlenecks, we denote the maximum highway capacity C_{max} and the threshold flow rate $q_{th}^{(B)}$ for traffic flow with the moving bottleneck by $C_{max,\ MB}$ and $q_{th,\ MB}^{(B)}$, respectively (Fig. 5.37).

In contrast with values C_{max} and $q_{th}^{(B)}$, the minimum highway capacity C_{min} does not depend on whether there is a moving bottleneck in traffic flow or not. This is because at the given flow rate, the minimum capacity C_{min} shown in Fig. 5.37 determines the smallest flow rate q_{on} at which traffic breakdown can still be *induced* at the bottleneck.

We can make the following conclusions about explanations of empirical probabilistic features of traffic breakdown at highway bottlenecks presented in Chap. 3 with the three-phase theory presented above in this chapter.

1. The basis characteristic of *spontaneous* traffic breakdown (spontaneous F→S transition) is the breakdown probability $P^{(B)}(q_{sum})$. Within the flow rate range $q_{th}^{(B)} \leq q_{sum} < C_{max}$, traffic breakdown occurs spontaneously during a given time interval T_{ob} with some probability, i.e., $0 < P^{(B)}(q_{sum}) < 1$: In some simulation realizations the breakdown occurs, however, in some other simulation realizations related to the same traffic flow parameters no breakdown occurs. This is valid for both traffic flows with the moving bottleneck and without moving bottlenecks (Fig. 5.37).
2. As in empirical data (Fig. 3.16a), in simulations a wide moving jam does induce traffic breakdown (F→S transition) in metastable free flow (5.3) at an on-ramp bottleneck (Fig. 5.38). This distinguishes induced traffic breakdown from spontaneous traffic breakdown.

[17] In the cases under consideration (Fig. 5.37), in (5.12) the sets of the parameters β and q_P are different for these two functions $P^{(B)}(q_{sum})$ (left and right curves in Fig. 5.37).

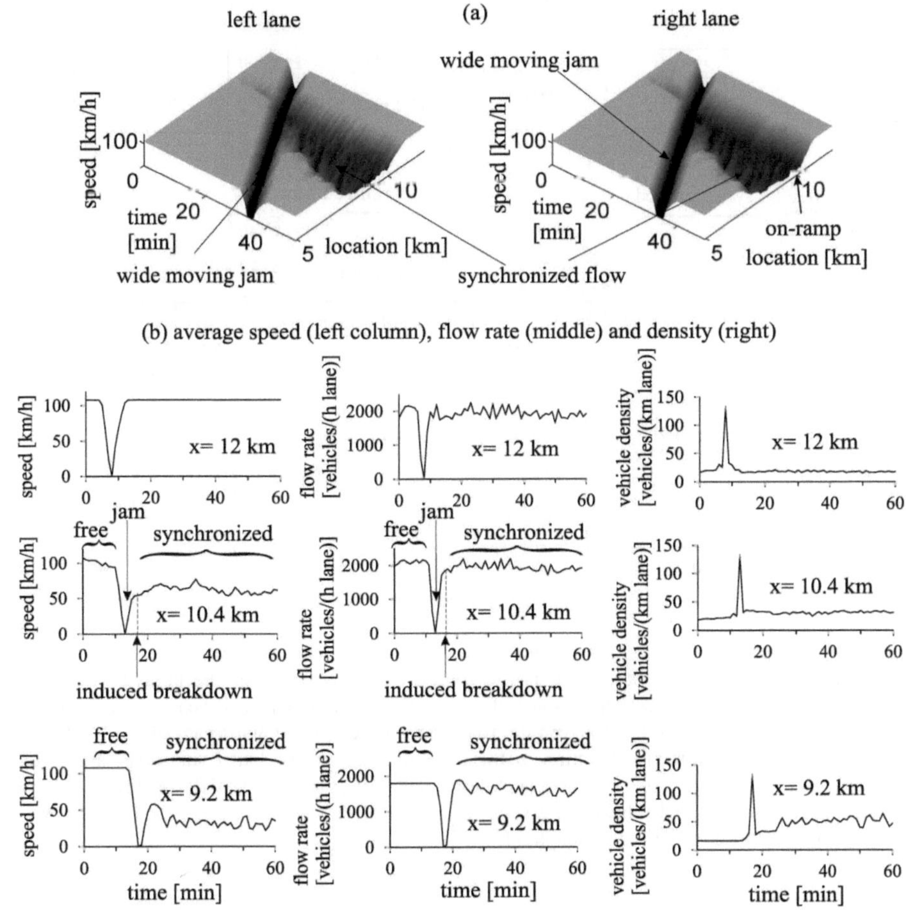

Fig. 5.38 Simulations of synchronized flow resulting from induced traffic breakdown: (**a**) Speed in space and time for the case of traffic breakdown induced by a wide moving jam propagating through the bottleneck (left figure—left lane, right—right lane). (**b**) Time-functions of speed (*figures in left column*), flow rate (*middle column*), and vehicle density (*right column*) related to (**a**) that have been averaged across the road at road location downstream of the bottleneck ($x = 12$ km), at the effective location of the bottleneck ($x = 10.4$ km), and at a road location upstream of the bottleneck ($x = 9.2$ km). $q_{\text{in}} = 1800$ vehicles/(h lane); $q_{\text{on}} = 700$ vehicles/h. Free flow, synchronized flow, and the wide moving jam are labeled by, respectively, "free", "synchronized", and "jam". Other model parameters are the same as those in Fig. 5.35. Adapted from [48]

3. In accordance with empirical results (Sect. 3.6.1), the spontaneous and induced traffic breakdowns are distinguished through different sources of a nucleus for the breakdown. However, the nature of the breakdown in both cases is the metastability of free flow at the bottleneck with respect to an F→S transition.

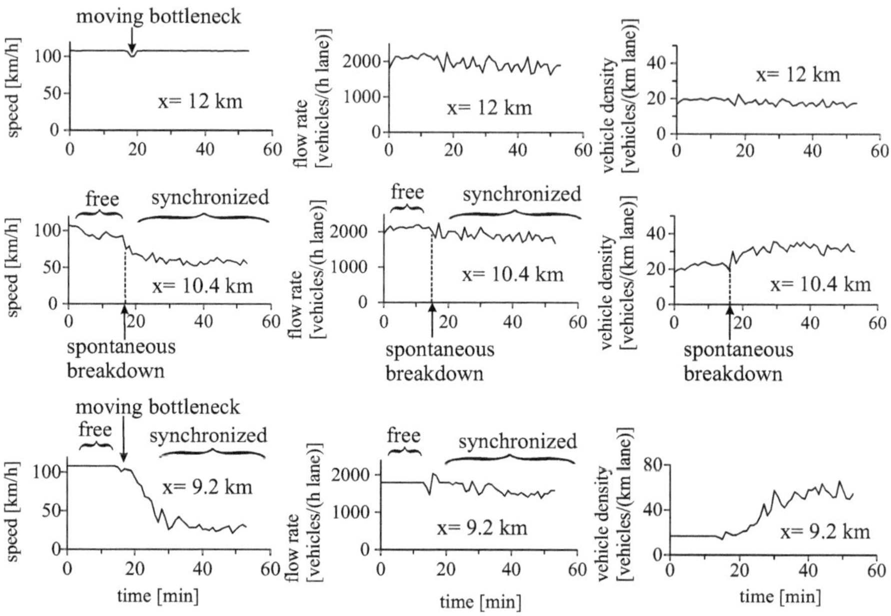

Fig. 5.39 Simulations of synchronized flow resulting from spontaneous traffic breakdown shown in Fig. 5.35, when moving bottleneck is the source of the nucleus for the breakdown. Time-functions of speed (*figures in left column*), flow rate (*middle column*), and vehicle density (*right column*) that have been averaged across the road at road locations downstream of the bottleneck ($x = 12$ km), at the effective location of the bottleneck ($x = 10.4$ km), and upstream of the bottleneck ($x = 9.2$ km). $q_{in} = 1800$ vehicles/(h lane); $q_{on} = 700$ vehicles/h. Free flow and synchronized flow are labeled by, respectively, "free" and "synchronized". Other model parameters are the same as those in Fig. 5.35. Adapted from [48]

4. After the breakdown has occurred, qualitative features of the emergent synchronized flow at the bottleneck do not depend either on the source of a nucleus for spontaneous traffic breakdown (whether there is the moving bottleneck (Fig. 5.39) or there is no moving bottlenecks in free flow (Fig. 5.40)) or on features of the localized congested pattern that has induced traffic breakdown (Fig. 5.38).

   In particular, as in empirical data (Figs. 3.17 and 3.18), the flow rate in synchronized flow that appears just after the breakdown has occurred is almost the same as the flow rate in the initial free flow ($x = 10.4$ in Figs. 5.38b, 5.39, and 5.40). This is an important feature of synchronized flow in comparison with a moving jam. Indeed, within the moving jam the flow rate is usually considerably smaller than the flow rate in free flows both upstream and downstream of the jam.

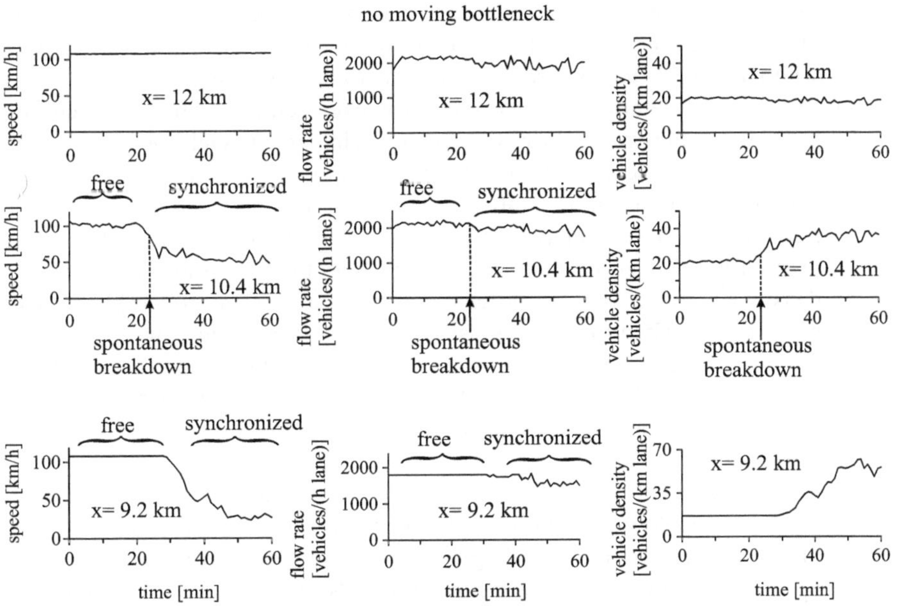

Fig. 5.40 Simulations of synchronized flow resulting from spontaneous traffic breakdown shown in Fig. 5.36, when random local disturbances in traffic flow in a neighborhood of the bottleneck determine the source of a nucleus in traffic flow without moving bottlenecks. Time-functions of speed (*figures in left column*), flow rate (*middle column*), and vehicle density (*right column*) that have been averaged across the road at road locations downstream of the bottleneck ($x = 12$ km), at the effective location of the bottleneck ($x = 10.4$ km), and upstream of the bottleneck ($x = 9.2$ km). $q_{in} = 1800$ vehicles/(h lane); $q_{on} = 700$ vehicles/h. Free flow and synchronized flow are labeled by, respectively, "free" and "synchronized". Other model parameters are the same as those in Fig. 5.35. Adapted from [48]

5.15 Conclusions: Driver Behaviors Explaining Nucleation Nature of Real Traffic Breakdown at Highway Bottlenecks

Driver behavioral characteristics used in the three-phase theory to explain the nucleation nature of traffic breakdown at highway bottlenecks are as follows.

(i) The nucleation nature of traffic breakdown is explained by a competition between two opposing tendencies occurring within a random local disturbance at a bottleneck in which the speed is lower than that in an initial free flow:

- A tendency towards synchronized flow due to vehicle deceleration associated with the speed adaptation effect.
- A tendency towards the initial free flow due to vehicle acceleration associated with the over-acceleration effect.

(ii) The speed adaptation effect is the adaptation of the vehicle speed to the speed of the preceding vehicle that is realized within the space gap range (5.37) without caring what the precise space gap to the preceding vehicle is.

(iii) The over-acceleration effect is driver maneuver leading to a higher speed from initial car-following at a lower speed occurring under condition (5.37).

(iv) There is a time delay in driver over-acceleration. At the same density, driver over-acceleration is more probable to occur in free flow than in synchronized flow. These driver behaviors result in a discontinuous character of the probability of over-acceleration. In particular, the probability is a Z-shaped density function: At a given density, there is a *drop* in probability of over-acceleration, when free flow transforms into synchronized flow.

(v) The initial local speed disturbance of the increase in the speed that initiates the S→F instability at the bottleneck occurs at the downstream front of synchronized flow at the bottleneck. We call the local speed disturbance as "speed peak".

(vi) The initial local speed disturbance of the increase in the speed within synchronized flow at the bottleneck can transform into a growing speed wave of the increase in the speed (growing acceleration wave) that propagates upstream within synchronized flow and leads to free flow at the bottleneck. This S→F instability is caused by driver over-acceleration. Due to the S→F instability, the downstream front of the initial synchronized flow begins to move upstream from the bottleneck, while free flow appears at the bottleneck. The S→F instability occurs due to a finite time delay in driver over-acceleration.

(vii) There can be many speed peaks with random amplitudes that occur randomly over time at the downstream front of synchronized flow at the bottleneck. Only when a large enough speed peak appears, the S→F instability occurs. Speed peaks of smaller amplitude cause dissolving speed waves of the local increase in the speed (dissolving acceleration waves) in synchronized flow: All these waves dissolve over time while propagating upstream within synchronized flow. As a result, the synchronized flow persists at the bottleneck. Thus, the S→F instability exhibits the nucleation nature.

(viii) When an F→S transition begins to develop, i.e., the upstream front of synchronized flow begins to propagate upstream from the bottleneck, an S→F instability can randomly occur. The S→F instability is realized when within the emergent synchronized flow at the bottleneck driver over-acceleration becomes on averaged stronger than driver speed adaptation. Due to the S→F instability, free flow appears at the bottleneck. As a result, the downstream front of the synchronized flow departs upstream from the bottleneck. In its turn, this results in the dissolution of the synchronized flow, i.e., in the interruption of the formation of a congested pattern due to the F→S transition. We call this effect as a sequence of F→S→F transitions. The sequence of F→S→F transitions interrupts the formation of a congested pattern at the bottleneck.

(ix) Because the S→F instability exhibits the nucleation nature (item (vii)), the S→F instability does not occur, if no nucleus required for the S→F instability appears in synchronized flow at the bottleneck. The occurrence of the nucleus required for the S→F instability in synchronized flow at the bottleneck is a random event. This explains why a time delay of traffic breakdown (F→S transition) $T^{(B)}$ is a random value. Thus, when no nucleus for the S→F instability appears during a long enough time interval after the beginning of the development of an F→S transition at the bottleneck, a congested pattern is formed at the bottleneck. Through the nucleation feature of the S→F instability, traffic breakdown (F→S transition) at a highway bottleneck exhibits also the nucleation nature: The S→F instability governs traffic breakdown (F→S transition) at a highway bottleneck.

The empirical fundamental of transportation science is the empirical nucleation nature of traffic breakdown at network bottlenecks. The empirical nucleation nature of traffic breakdown at network bottlenecks is explained in the three-phase theory through the basic assumption of this theory (5.7) about the metastability of free flow at a highway bottleneck with respect to an F→S transition (traffic breakdown) at the bottleneck.

The three-phase theory is the theoretical fundamental of transportation science that explains the empirical nucleation nature of traffic breakdown.

References

1. C.W. Gardiner, *Handbook of Stochastic Methods*, 2nd edn. (Springer, Berlin, 1990)
2. B.S. Kerner, Phys. Rev. Lett. **81**, 3797–3400 (1998)
3. B.S. Kerner, in *Proceedings of the 3rd Symposium on Highway Capacity and Level of Service*, ed. by R. Rysgaard, vol. 2 (Road Directorate, Ministry of Transport, Denmark, 1998), pp. 621–642
4. B.S. Kerner, Trans. Res. Rec. **1678**, 160–167 (1999)
5. B.S. Kerner, in *Transportation and Traffic Theory*, ed. by A. Ceder (Elsevier Science, Amsterdam, 1999), pp. 147–171
6. B.S. Kerner, Physics World **12**, 25–30 (August 1999)
7. B.S. Kerner. J. Phys. A Math. Gen. **33**, L221–L228 (2000)
8. B.S. Kerner, in *Traffic and Granular Flow '99: Social, Traffic and Granular Dynamics*, ed. by D. Helbing, H.J. Herrmann, M. Schreckenberg, D.E. Wolf (Springer, Heidelberg, Berlin, 2000), pp. 253–284
9. B.S. Kerner, Transp. Res. Rec. **1710**, 136–144 (2000)
10. B.S. Kerner, Netw. Spat. Econ. **1**, 35–76 (2001)
11. B. S. Kerner, Transp. Res. Rec. **1802**, 145–154 (2002)

12. B. S. Kerner, in *Traffic and Transportation Theory in the 21st Century*, ed. by M.A.P. Taylor (Elsevier Science, Amsterdam, 2002), pp. 417–439
13. B.S. Kerner, Phys. Rev. E **65**, 046138 (2002)
14. B. S. Kerner, Math. Comput. Modell. **35**, 481–508 (2002)
15. B.S. Kerner, in *Traffic and Granular Flow' 01*, ed. by M. Schreckenberg, Y. Sugiyama, D. Wolf (Springer, Berlin, 2003), pp. 13–50
16. B.S. Kerner, Physica A **333**, 379–440 (2004)
17. B.S. Kerner, *The Physics of Traffic* (Springer, Berlin, New York, 2004)
18. B.S. Kerner, Physica A **355**, 565–601 (2005)
19. B.S. Kerner, in *Traffic and Transportation Theory*, ed. by H.S. Mahmassani (Elsevier Science, Amsterdam, 2005), pp. 181–203
20. B.S. Kerner, IEEE Trans. ITS **8**, 308–320 (2007)
21. B.S. Kerner, Transp. Res. Rec. **1999**, 30–39 (2007)
22. B.S. Kerner, Transp. Res. Rec. **2088**, 80–89 (2008)
23. B.S. Kerner, J. Phys. A Math. Theor. **41**, 215101 (2008)
24. B.S. Kerner, in *Transportation Research Trends*, ed. by P.O. Inweldi. (Nova Science Publishers, New York, 2008), pp. 1–92
25. B.S. Kerner, *Introduction to Modern Traffic Flow Theory and Control* (Springer, Heidelberg, Dordrecht, London, New York, 2009)
26. B.S. Kerner, in *Encyclopedia of Complexity and System Science*, ed. by R.A. Meyers (Springer, Berlin, 2009), pp. 9302–9355
27. B.S. Kerner, in *Encyclopedia of Complexity and System Science*, ed. by R.A. Meyers (Springer, Berlin, 2009), pp. 9355–9411
28. B.S. Kerner, Phys. Rev. E. **85**, 036110 (2012)
29. B.S. Kerner, Traffic Eng. Contr. **54**, 47 (2013)
30. B.S. Kerner, Physica A **392**, 5261–5282 (2013)
31. B.S. Kerner, Phys. Rev. E **92**, 062827 (2015)
32. B.S. Kerner, in *Vehicular Communications and Networks*, ed. by W. Chen (Woodhead Publishings, Cambridge, 2015), pp. 223–254
33. B.S. Kerner, Elektrotechnik Informationstechnik **132**, 417–433 (2015)
34. B.S. Kerner, Physica A **450**, 700–747 (2016)
35. B.S. Kerner, S.L. Klenov, J. Phys. A Math. Gen. **35**, L31–L43 (2002)
36. B.S. Kerner, S.L. Klenov, Phys. Rev. E **68**, 036130 (2003)
37. B.S. Kerner, S.L. Klenov, J. Phys. A Math. Gen. **37**, 8753–8788 (2004)
38. B.S. Kerner, S.L. Klenov, J. Phys. A Math. Gen. **39**, 1775–1809 (2006)
39. B.S. Kerner, S.L. Klenov, Transp. Res. Rec. **1965**, 70–78 (2006)
40. B.S. Kerner, S.L. Klenov, Physica A **364**, 473–492 (2006)
41. B.S. Kerner, S.L. Klenov, Phys. Rev. E **80**, 056101 (2009)
42. B.S Kerner, S.L. Klenov, in *Encyclopedia of Complexity and System Science*, ed. by R.A. Meyers (Springer, Berlin, 2009), pp. 9282–9302
43. B.S. Kerner, S.L. Klenov, J. Phys. A Math. Theor. **43**, 425101 (2010)
44. B.S. Kerner, S.L. Klenov, G. Hermanns, M. Schreckenberg, Physica A **392**, 4083–4105 (2013)
45. B.S. Kerner, S.L. Klenov, M. Schreckenberg, Phys. Rev. E **84**, 046110 (2011)
46. B.S. Kerner, S.L. Klenov, M. Schreckenberg, Phys. Rev. E **89**, 052807 (2014)
47. B.S. Kerner, S.L. Klenov, D.E. Wolf, J. Phys. A Math. Gen. **35**, 9971–10013 (2002)
48. B.S. Kerner, M. Koller, S.L. Klenov, H. Rehborn, M. Leibel, Physica A **438**, 365–397 (2015)
49. S.-E. Molzahn, B.S. Kerner, H. Rehborn, S.L. Klenov, M. Koller (unpublished)

Chapter 6
Effect of Automatic Driving on Probability of Breakdown in Traffic Networks

6.1 Introduction

A current effort of many car-developing companies is devoted to the development of automatic driving vehicles.[1] It is assumed that future vehicular traffic in traffic and transportation networks is a mixed traffic flow consisting of human driving and automatic driving vehicles. Automatic driving vehicles should considerably enhance capacity of a traffic network. Capacity of the network is limited by traffic breakdown at network bottlenecks.

For an analysis of the effect of automatic driving vehicles on traffic flow, we consider a simple case of vehicular traffic on a single-lane road with an on-ramp bottleneck. On the single-lane road, no vehicles can pass. For this reason, automatic driving can be achieved through the use of an adaptive cruise control (ACC) in a vehicle: An ACC-vehicle follows a preceding vehicle automatically based on some ACC dynamics rules of motion. Depending on the dynamic behavior of the preceding vehicle, these ACC-rules determine either automatic acceleration or automatic deceleration of the ACC-vehicle or else the maintaining of a time-independent speed. The preceding vehicle can be either a human driving vehicle or an automatic driving vehicle through an ACC-system in the vehicle.

As explained in Sect. 4.9.3, simulations of different traffic flow models show that if a platoon of the ACC-vehicles is stable, then the ACC-vehicles prevent traffic breakdown in mixed traffic flow, when the percentage of the ACC-vehicles in mixed traffic flow exceeds a critical value, which is about 30% of the ACC-vehicles.[2] However, it is not very probable that such large percentage of ACC-vehicle in mixed

[1] Automatic driving is also called self-driving, autonomous driving, or automated driving. Respectively, automatic driving vehicles are also called self-driving vehicles, autonomous driving vehicles, or automated driving vehicles.

[2] The work by Davis [1] was one of the first to obtain this result with a human driver model that had some of the features of the three-phase theory.

B.S. Kerner, *Breakdown in Traffic Networks*, DOI 10.1007/978-3-662-54473-0_6

traffic flow can be reached in the next future. Therefore, we consider considerably smaller values of the percentage of ACC-vehicles than 30%. We would like to find the answer on the following question:

- Can automatic driving vehicles in mixed traffic flow reduce the probability of traffic breakdown at network bottlenecks?

The importance of this subject can be explained as follows. In previous Chap. 5 it has been shown that the probability of spontaneous traffic breakdown at a network bottleneck is one of the most important characteristics of stochastic highway capacity. Therefore, to find the effect of automatic driving vehicles on traffic flow, we should study the influence of automatic driving vehicles on the probability of traffic breakdown at a bottleneck. This is the main objective of this chapter.

As explained in Chap. 4, the classical theories failed in the understanding of stochastic highway capacity. It has also been proven in Chap. 4 that it is not possible to perform a reliable study of the effect of automatic driving vehicles and other ITS on traffic flow with the use of the classical traffic-flow theories. Therefore, for the analysis of the effect of automatic driving vehicles on traffic flow we use simulations of the Kerner-Klenov microscopic stochastic traffic flow model that is in the framework of the three-phase theory (Appendix A). The model can explain empirical features of stochastic highway capacity.

In this chapter, we consider automatic driving vehicles whose platoon is always stable: No traffic breakdown occurs at chosen parameters of the automatic driving vehicles and the flow rate at a road bottleneck, when traffic flow consists of 100% of the automatic driving vehicles. Nevertheless, following [10], we will find that depending on the dynamic characteristics of these automatic driving vehicles they can either improve or deteriorate the performance of mixed traffic flow consisting of both manual driving and automatic driving vehicles. We will show that automatic driving vehicles can considerably effect on the probability of traffic breakdown at a network bottleneck in mixed traffic flow. In particular, we will find that the effect of automatic driving vehicles on the probability of traffic breakdown can already be crucial one, when the percentage of automatic driving vehicles is only about 5%. This emphasizes once more the importance of studies of the probabilistic characteristics of traffic breakdown at network bottlenecks (Chap. 5).

6.2 Operating Points and String Stability of Adaptive Cruise Control (ACC)

In simulations of mixed traffic flow on a single lane road with an on-ramp bottleneck discussed below, there are vehicles that have no ACC system (human driving vehicles) and ACC-vehicles (automatic driving vehicles). The ACC vehicles are randomly distributed on the road between human driving vehicles. The percentage of automatic driving vehicles denoted by $\gamma^{(\mathrm{ACC})}$ is the same value in traffic flow upstream of the bottleneck and in the on-ramp inflow onto the main road.

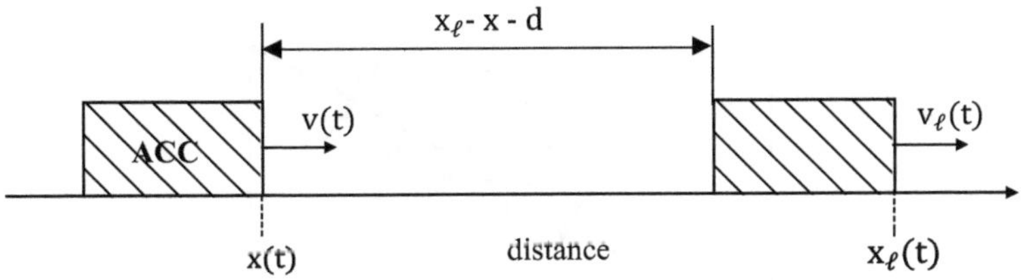

Fig. 6.1 Model of the ACC vehicle (see, e.g. [2–5, 16–20, 22, 23])

We use the classical ACC-model (4.27) that (for the simplicity of the further consideration) we rewrite here:

$$a^{(\mathrm{ACC})}(t) = K_1(g(t) - v(t)\tau_{\mathrm{d}}^{(\mathrm{ACC})}) + K_2(v_\ell(t) - v(t)), \tag{6.1}$$

where K_1 and K_2 are coefficients of ACC adaptation,[3] $v(t)$ is the speed of the ACC-vehicle, $v_\ell(t)$ is the speed of the preceding vehicle (Fig. 6.1), $\tau_{\mathrm{d}}^{(\mathrm{ACC})}$ is a desired net time gap (desired time headway) of the ACC-vehicle to the preceding vehicle, $a^{(\mathrm{ACC})}(t)$ is acceleration (deceleration) of the ACC vehicle that is determined by the space gap to a preceding vehicle $g(t) = x_\ell(t) - x(t) - d$ and the relative speed $\Delta v(t) = v_\ell(t) - v(t)$ measured by the ACC vehicle as well as by a desired space gap $g^{(\mathrm{ACC})} = v(t)\tau_{\mathrm{d}}^{(\mathrm{ACC})}$.

In Eq. (6.1), coefficients of ACC adaptation K_1 and K_2 have the following physical meaning (see, e.g., works [2–5, 16–20, 22, 23] and references there): The coefficients K_1 and K_2 describe the dynamic adaptation of the ACC vehicle when either the space gap g is different from the desired one $g^{(\mathrm{ACC})} = v\tau_{\mathrm{d}}^{(\mathrm{ACC})}$ (Fig. 6.2), i.e.,

$$g \neq v\tau_{\mathrm{d}}^{(\mathrm{ACC})} \tag{6.2}$$

or the vehicle speed is different from the speed of the preceding vehicle, i.e.,

$$v \neq v_\ell. \tag{6.3}$$

In contrast, if the time headway of the ACC-vehicle $\tau^{(\mathrm{net})}$ is equal to the desired time headway

$$\tau^{(\mathrm{net})} = \tau_{\mathrm{d}}^{(\mathrm{ACC})} \tag{6.4}$$

[3]In real ACC, the coefficients of ACC adaptation can be functions of the speed and the space gap. Moreover, the coefficients of ACC adaptation can depend on a driving situation. However, the objective of this chapter is a study of *qualitative* features of the effect of ACC-vehicles in mixed traffic flow on the probability of traffic breakdown at network bottlenecks. To reach this goal, we can assume that the coefficients of ACC adaptation do not depend on the speed and the space gap in mixed traffic flow.

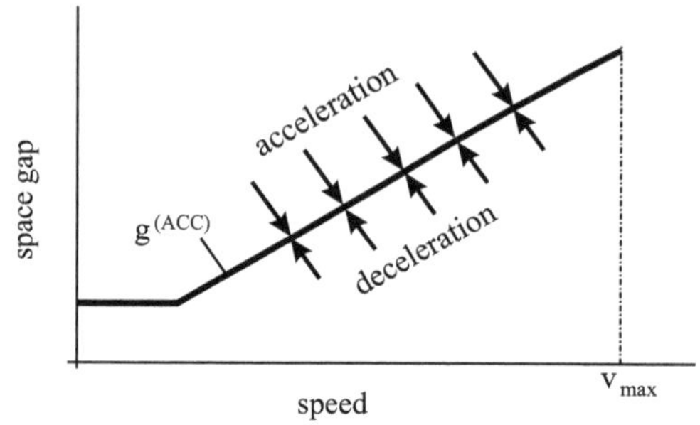

Fig. 6.2 Operating points of the classical model of ACC: Qualitative speed dependence of desired space gap $g^{(\mathrm{ACC})} = v\tau_{\mathrm{d}}^{(\mathrm{ACC})}$ (see, e.g. [2–5, 16–20, 22, 23]). When the speed of the ACC-vehicle is smaller than a speed associated with a given minimum space gap between vehicles, then the desired space gap does not depend on the speed as shown in figure for small enough speeds; we do not consider this speed range in the main text

and the relative speed $\Delta v = v_{\ell} - v$ of the ACC-vehicle v to the speed of the preceding vehicle v_{ℓ} is zero

$$\Delta v = 0, \tag{6.5}$$

then from (6.1) we obtain that

$$a^{(\mathrm{ACC})} = 0. \tag{6.6}$$

This means that the ACC-vehicle moves with a time-independent speed.

Conditions (6.4) and (6.5) determine so-called operating points of the ACC-vehicle (see, e.g., works [16–20, 22, 23] and references there). An operating point of the ACC-vehicle is a hypothetical steady state of the ACC-vehicle that follows the preceding vehicle moving at a time-independent speed. In the operating point, both the acceleration (deceleration) of the ACC-vehicle and the relative speed of the ACC-vehicle to the speed of the preceding vehicle are equal to zero, i.e., the ACC-vehicle moves at the time-independent speed of the preceding vehicle. Accordingly (6.4), in the classical model of the ACC-vehicle at each given speed of the ACC-vehicle that is larger than zero the operating point of the ACC-vehicle corresponds to a time headway that is equal to the desired time headway $\tau_{\mathrm{d}}^{(\mathrm{ACC})}$. For each of the operating points of the ACC-vehicle, formula (6.4) is equivalent to formula

$$g^{(\mathrm{ACC})} = v\tau_{\mathrm{d}}^{(\mathrm{ACC})}. \tag{6.7}$$

Using formula (6.7), the operating points of the classical ACC (6.4), (6.5) can be presented by a curve in the gap–speed flow plane (Fig. 6.2). For each given speed $v > 0$ of the ACC-vehicle, there is only one operating point of the classical ACC that is determined by formulas (6.4), (6.5) (see, e.g., works [16–20, 22, 23] and references there).

For each given speed of the ACC-vehicle, there is only one operating point for the classical ACC.

The physics of the dynamic equation for the ACC vehicle (6.1) is as follows (see, e.g., works [2, 4, 16–20, 22, 23] and references there). It can be seen that in Eq. (6.1) the current time headway $\tau^{(\mathrm{net})} = g/v$ (6.15) is compared with the desired time headway $\tau_{\mathrm{d}}^{(\mathrm{ACC})}$. If

$$\tau^{(\mathrm{net})} > \tau_{\mathrm{d}}^{(\mathrm{ACC})}, \tag{6.8}$$

then the ACC vehicle automatically accelerates to reduce the time headway $\tau^{(\mathrm{net})}$ to the desired value $\tau_{\mathrm{d}}^{(\mathrm{ACC})}$ (labeled by "acceleration" in Fig. 6.2). If

$$\tau^{(\mathrm{net})} < \tau_{\mathrm{d}}^{(\mathrm{ACC})}, \tag{6.9}$$

then the ACC vehicle decelerates automatically to increase the time headway (labeled by "deceleration" in Fig. 6.2). Moreover, the acceleration and deceleration of the ACC vehicle depend on the current difference between the speed of the ACC vehicle and the preceding vehicle. If the preceding vehicle has a higher speed than the ACC vehicle, i.e., when

$$v_{\ell} > v, \tag{6.10}$$

the ACC vehicle accelerates. Otherwise, if

$$v_{\ell} < v, \tag{6.11}$$

the ACC vehicle decelerates.

An important characteristic of the ACC-vehicles is a stability of a platoon of the ACC-vehicles called *string stability*. Liang and Peng [16] have found that for a string stability of the ACC vehicles coefficients of ACC adaptation K_1 and K_2 in (6.1) and the desired time headway of the ACC vehicles $\tau_{\mathrm{d}}^{(\mathrm{ACC})}$ should satisfy condition [16]

$$K_2 > \frac{2 - K_1(\tau_{\mathrm{d}}^{(\mathrm{ACC})})^2}{2\tau_{\mathrm{d}}^{(\mathrm{ACC})}}. \tag{6.12}$$

Below to limit the analysis, we consider the effect of the ACC vehicles on traffic flow *only* for a relatively short desired time headway of the ACC vehicles $\tau_{\mathrm{d}}^{(\mathrm{ACC})} = 1.1$ s. However, we will use different sets of coefficients K_1 and K_2 of ACC adaptation.

It should be noted that even when condition for string stability (6.12) is satisfied, nevertheless, traffic congestion occurs in traffic flow consisting of 100% ACC-vehicles, if the flow rate $q_{\rm sum} = q_{\rm on} + q_{\rm in}$ at the on-ramp bottleneck exceeds the value

$$q_0^{\rm (ACC)} = \frac{3600}{\tau_{\rm d}^{\rm (ACC)} + d/v_{\rm free}}. \tag{6.13}$$

For this reason, in all simulations of the effect of ACC-vehicles on traffic flow presented below, we have chosen model parameters at which traffic breakdown occurs in mixed traffic flow only, when the flow rate at the bottleneck $q_{\rm sum}$ is considerably smaller than $q_0^{\rm (ACC)}$ (6.13):

$$q_{\rm sum} < q_0^{\rm (ACC)} \approx 2667 \text{ [vehicles/h]}, \tag{6.14}$$

where in formula (6.13) we have taken into account that we have considered only ACC-vehicles with $\tau_{\rm d}^{\rm (ACC)} = 1.1$ s, $v_{\rm free} = 30$ m/s, and $d = 7.5$ m.

Under condition (6.14), it is often expected that if there is string stability of a platoon of the ACC-vehicles, then the ACC-vehicles should improve traffic flow.

- The main objective of our analysis of the effect of the ACC-vehicles on traffic flow presented below is to prove that this assumption should not necessarily be valid, even when condition of string stability for the ACC-vehicles (6.12) *and* condition (6.14) are satisfied.
- We will find that depending on the coefficients of ACC adaptation K_1 and K_2 in (6.1) (which all satisfy condition (6.12) for string stability) the ACC-vehicles can either improve or deteriorate the traffic system.

6.3 Decrease in Probability of Traffic Breakdown Through Automatic Driving Vehicles

All simulations of human driving vehicles presented below in this chapter are made with the Kerner-Klenov microscopic stochastic three-phase traffic flow model [11–15]. Because in the Kerner-Klenov model discrete time step is used (Appendix A), we use in the classical ACC-model (6.1) the discrete time $t = n\tau$, where $n = 0, 1, 2, ..$; $\tau = 1$ s is time step. Therefore, the space gap to the preceding vehicle is equal to $g_n = x_{\ell,n} - x_n - d$ and the relative speed is given by $\Delta v_n = v_{\ell,n} - v_n$, where x_n and v_n are coordinate and speed of the ACC-vehicle, $x_{\ell,n}$ and $v_{\ell,n}$ are coordinate and speed of the preceding vehicle, d is the vehicle length that is assumed the same one for automatic driving and human driving vehicles. Respectively, the current net time gap (time headway) between ACC-vehicle and the preceding vehicle calculated by ACC-vehicle is equal to

$$\tau_n^{\rm (net)} = g_n/v_n. \tag{6.15}$$

Correspondingly, the classical model of the dynamics of ACC-vehicle (6.1) can be rewritten as follows:

$$a_n^{(\mathrm{ACC})} = K_1(g_n - v_n\tau_\mathrm{d}^{(\mathrm{ACC})}) + K_2(v_{\ell,n} - v_n). \tag{6.16}$$

The ACC vehicles move in accordance with Eq. (6.16) where, in addition, the following formulas are used:

$$v_{\mathrm{c},n}^{(\mathrm{ACC})} = v_n + \tau \max(-b_\mathrm{ACC}, \min(\lfloor a_n^{(\mathrm{ACC})} \rfloor, a_\mathrm{ACC})), \tag{6.17}$$

$$v_{n+1} = \max(0, \min(v_\mathrm{free}, v_{\mathrm{c},n}^{(\mathrm{ACC})}, v_{\mathrm{s},n})), \tag{6.18}$$

$\lfloor z \rfloor$ denotes the integer part[4] of z. Through the use of formula (6.17), acceleration and deceleration of the ACC vehicles are limited by some maximum acceleration a_ACC and maximum deceleration b_ACC, respectively. Owing to the formula (6.18), the speed of the ACC vehicle v_{n+1} at time step $n+1$ is limited by the maximum speed in free flow v_free and by the safe speed $v_{\mathrm{s},n}$ to avoid collisions between vehicles.[5] The maximum speed in free flow v_free and the safe speed $v_{\mathrm{s},n}$ are chosen, respectively, the same as those in the microscopic model of human driving vehicles (Appendix A). It should be noted that the model of ACC-vehicle merging from the on-ramp onto the main road is similar to that for human driving vehicles (see Sect. A.6.4 of Appendix A).

We consider traffic flow of manual driving vehicles, in which traffic breakdown occurs with some probability within the observation time $T_\mathrm{ob} = 30$ min (Fig. 6.3a,b). At the same flow rates and parameters of the on-ramp bottleneck no congestion occurs in traffic flow consisting of 100% stable ACC vehicles (Fig. 6.3c,d), when conditions (6.12) and (6.14) are satisfied. Therefore, it should be expected that in mixed traffic flow, in which manual driving vehicles exhibit the same characteristics as those in Fig. 6.3a,b and ACC vehicles are the same as those in Fig. 6.3c,d, there should be a critical percentage of ACC-vehicles when the ACC-vehicles suppress any instabilities in traffic flow caused by manual driving

[4]In the discrete model version (Appendix A) used for all simulations in this chapter, the discretization cell $\delta x = 0.01$ m is used. Respectively, vehicle acceleration is measured in units $\delta a = \delta x/\tau^2$.

[5]Simulations show that the use of the safe speed in formula (6.18) does not influence on the dynamics of the ACC vehicles (6.1) in *free flow* outside the bottleneck. However, due to vehicle merging from the on-ramp onto the main road, time headway of the vehicle to the preceding vehicle can be considerably smaller than $\tau_\mathrm{d}^{(\mathrm{ACC})}$. Therefore, formula (6.18) allows us to avoid collisions of the ACC vehicle with the preceding vehicle in such dangerous situations. Moreover, very small values of time headway can occur in congested conditions; formula (6.18) prevents vehicle collisions in these cases also. While working at the Daimler Company, the author was lucky to take part in the development of real ACC vehicles, which are on the market; to avoid collisions in dangerous simulations, dynamics rules of all real ACC vehicles include some safety dynamic rules that can be similar to (6.17), (6.18).

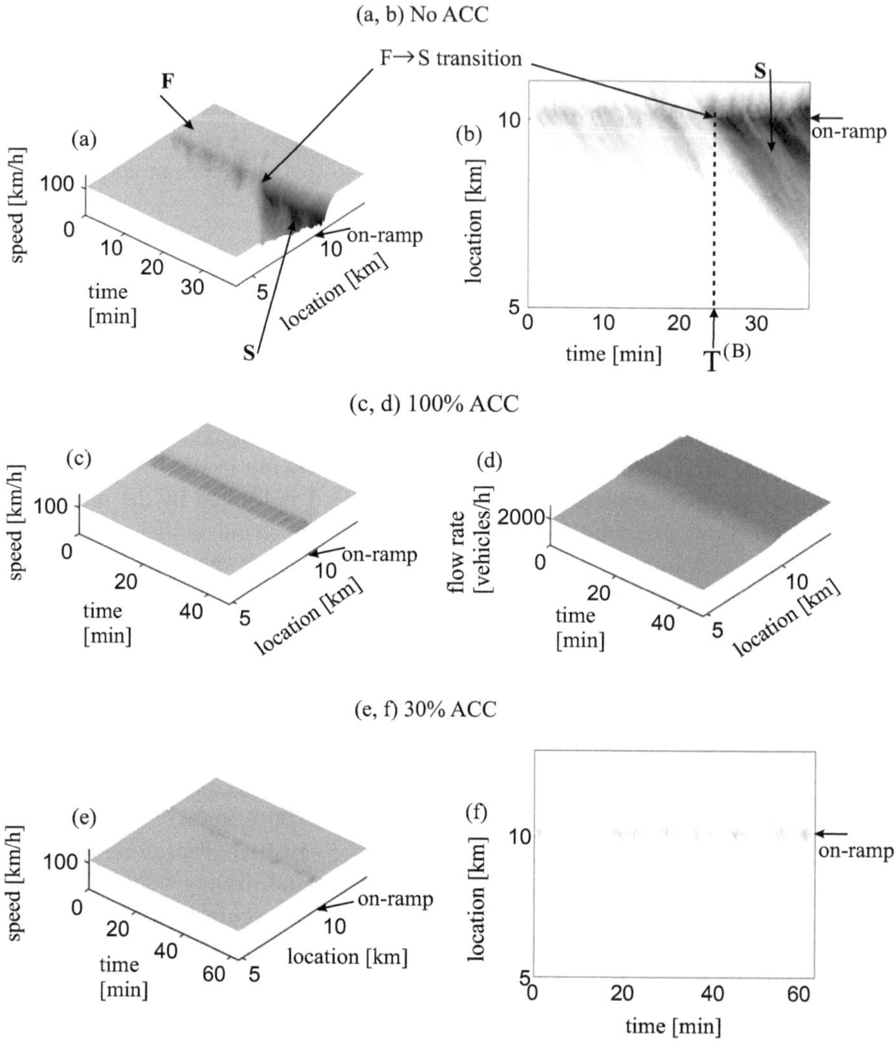

Fig. 6.3 Suppression of F→S transition in traffic flow on single-lane road with an on-ramp bottleneck through ACC-vehicles: Vehicle speed (**a**, **c**, **e**) and the flow rate (averaging over 20 vehicles) (**d**) in space and time. (**b**, **f**) The same speed data as those in (**a**, **e**) but presented by regions with variable shades of gray (shades of gray vary from white to black when the speed decreases from 105 km/h (*white*) to zero (*black*)). In (**a**, **b**), there are no ACC vehicles, i.e., traffic flow consists only of human driving vehicles. In (**c**, **d**), traffic flow with $\gamma^{(\mathrm{ACC})} = 100\%$ of ACC vehicles with $\tau_{\mathrm{d}}^{(\mathrm{ACC})} = 1.1$ s and coefficients $(K_1, K_2) = (0.14\ s^{-2}, 0.9\ s^{-1})$ satisfying (6.12); values $a_{\mathrm{ACC}} = b_{\mathrm{ACC}} = 3\ \mathrm{m/s^2}$. In (**e**, **f**) $\gamma^{(\mathrm{ACC})} = 30\%$ of ACC vehicles with the same parameters as those (**c**, **d**). Flow rates in all figures are $q_{\mathrm{on}} = 320$ vehicles/h, $q_{\mathrm{in}} = 2000$ vehicles/h ($q_{\mathrm{sum}} = q_{\mathrm{on}} + q_{\mathrm{in}} = 2320$ vehicles/h that satisfies (6.14)). F—free flow, S—synchronized flow. On-ramp location $x_{\mathrm{on}} = 10$ km. Adapted from [10]

vehicles. As found in earlier simulations with different traffic flow models and explained in Sect. 4.9.3, this critical percentage of the ACC-vehicles is $\gamma^{(\mathrm{ACC})} \approx 30\%$ (Fig. 6.3e,f).

We have found that as long as the percentage of ACC-vehicles is appreciably smaller than $\gamma^{(\mathrm{ACC})} = 10\%$, no considerable change in the probabilistic features of traffic breakdown at the bottleneck is observed. Even when the percentage of ACC-vehicles increases to $\gamma^{(\mathrm{ACC})} = 10\%$, features of traffic breakdown remain almost the same as those in traffic flow of manual driving vehicles (Fig. 6.4a,b) and only a relatively small decrease in the probability of the breakdown is observed (compare curves labeled by "10% ACC" and "No ACC" in Fig. 6.5a).

Now, in comparison with Fig. 6.4a,b, we increase the percentage $\gamma^{(\mathrm{ACC})}$ of automatic driving vehicles to $\gamma^{(\mathrm{ACC})} = 20\%$ without any other changes in simulations. We have found that at $\gamma^{(\mathrm{ACC})} = 20\%$ automatic driving vehicles no traffic breakdown occurs during the observation time $T_{\mathrm{ob}} = 30$ min (Fig. 6.4c,d). However, if we continue simulations shown in Fig. 6.4c,d during a longer time interval, we have found that after a long time delay $T^{(\mathrm{B})} \approx 47$ min traffic breakdown has nevertheless occurred at the bottleneck (Fig. 6.4e,f). We have found that the mean time delay of traffic breakdown for $\gamma^{(\mathrm{ACC})} = 20\%$ of ACC vehicles is considerably longer than for traffic flow consisting of only human driving vehicles.

For $\gamma^{(\mathrm{ACC})} = 20\%$ of automatic driving vehicles the probability that traffic breakdown in mixed traffic flow occurs during the observation time $T_{\mathrm{ob}} = 30$ min is equal to $P^{(\mathrm{B})} = 0.1$ for model parameters used in Fig. 6.4c,d. Thus, we can expect that in comparison with the simulation realization shown in Fig. 6.4c,d, in which no breakdown is observed, there should be other simulation realizations made at the same parameters of mixed traffic flow, in which traffic breakdown is observed during the time interval $T_{\mathrm{ob}} = 30$ min. Such simulation realizations with different random values of time delays $T^{(\mathrm{B})}$ to the breakdown indeed exist (Fig. 6.6).

The physics of these results can be explained as follows [10]. In traffic flow consisting of human driving vehicles only, traffic breakdown (F$\rightarrow$S transition) occurs (Fig. 6.3a,b), when a local speed disturbance that causes a large enough local speed decrease (nucleus for the breakdown) appears randomly in metastable free flow at the bottleneck. Free flow that consists of 100% automatic driving vehicles is stable (Fig. 6.3c,d). For this reason, we can assume that a long enough platoon of ACC-vehicles that propagates through a local disturbance at the bottleneck can prevent the growth of the local disturbance. The larger the percentage $\gamma^{(\mathrm{ACC})}$ of automatic driving vehicles in mixed traffic flow, the larger the probability of the appearance of the long platoon of ACC-vehicles, therefore, the larger the probability of the prevention of the disturbance growth and the smaller the breakdown probability $P^{(\mathrm{B})}$ are.

These qualitative explanations are confirmed by numerical simulations of the flow-rate dependence of the probability of traffic breakdown $P^{(\mathrm{B})}(q_{\mathrm{sum}})$ presented in Fig. 6.5. Indeed, we have found that the flow-rate dependence of the probability of traffic breakdown for mixed traffic flow with 20% automatic driving vehicles (right curve $P^{(\mathrm{B})}(q_{\mathrm{sum}})$ labeled by "20% ACC" in Fig. 6.5a) is shifted to larger flow rates in

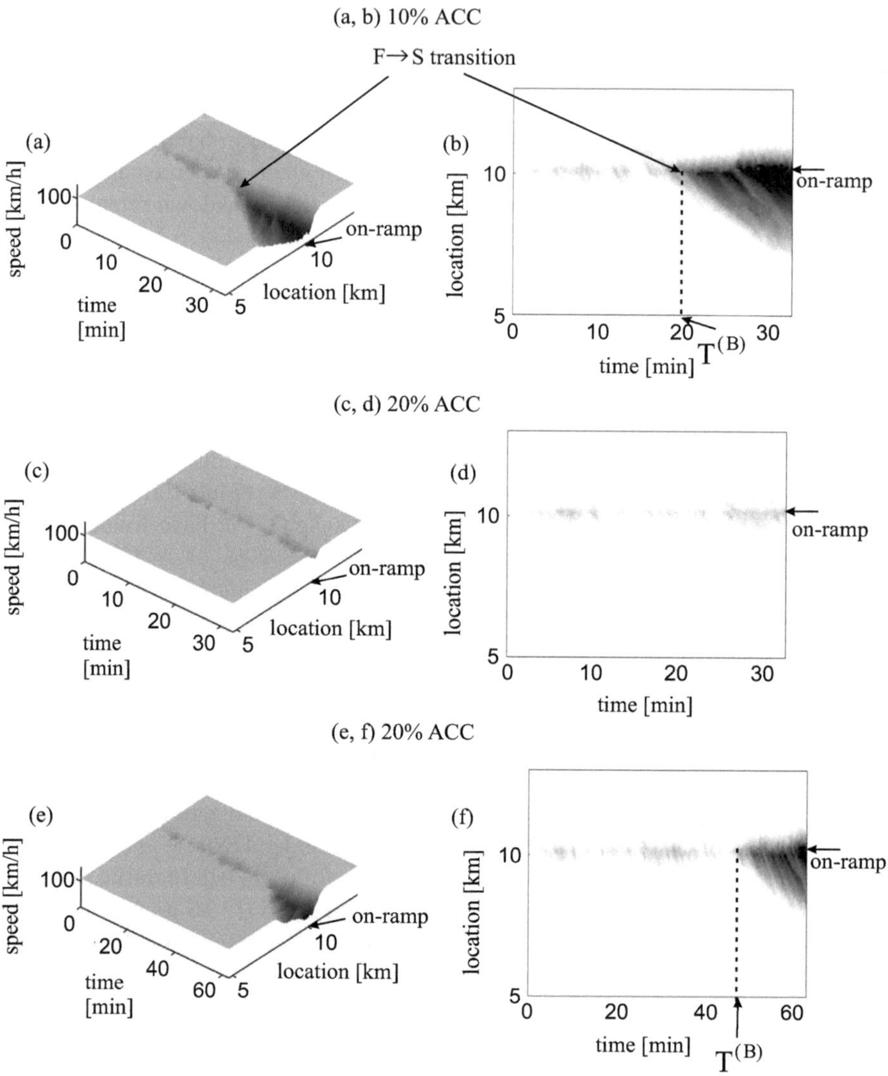

Fig. 6.4 Simulations of the effect of ACC-vehicles on traffic breakdown: Vehicle speed in space and time (**a**, **c**, **e**) and the same speed data presented by regions with variable shades of gray (**b**, **d**, **f**) (shades of gray vary from white to black when the speed decreases from 105 km/h (*white*) to zero (*black*)). (**a**, **b**) $\gamma^{(ACC)} = 10\%$. (c–f) $\gamma^{(ACC)} = 20\%$. Other parameters are the same as those in Fig. 6.3. Arrows F→S in (**a**, **b**) mark the F→S transition (traffic breakdown) at the location of on-ramp bottleneck. F—free flow, S—synchronized flow. Adapted from [10]

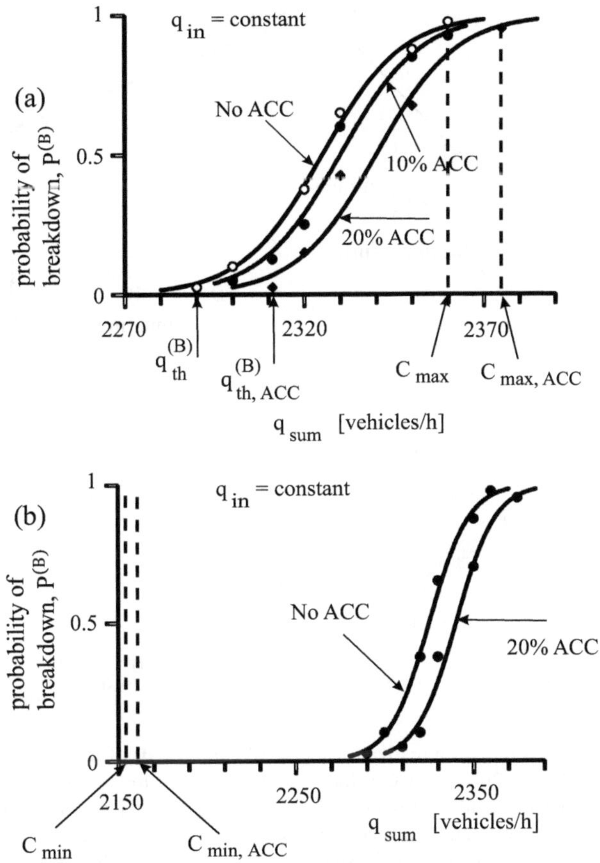

Fig. 6.5 Probabilities of traffic breakdown $P^{(B)}(q_{sum})$ in traffic flows without ACC-vehicles (left curves in (**a, b**) labeled by "No ACC") as well as with 10% and 20% of ACC-vehicles (right curves labeled by "10% ACC" and "20% ACC") as functions of the flow rate downstream of the bottleneck q_{sum}; curves "No ACC" and "20% ACC" are shown in different flow-rate scales in (**a**) and (**b**). The flow rate $q_{sum} = q_{in} + q_{on}$ is varied through the change in the on-ramp inflow rate q_{on} at constant $q_{in} = 2000$ vehicles/h. To distinguish the cases of traffic flows with the ACC-vehicles and without ACC-vehicles, we denote the maximum highway capacity C_{max}, the minimum highway capacity C_{min}, and the threshold flow rate $q_{th}^{(B)}$ for traffic flow with 20% of ACC-vehicles by $C_{max, ACC}$, $C_{min, ACC}$, and $q_{th, ACC}^{(B)}$, respectively. In (**a, b**), circles and points $P^{(B)}(q_{sum})$ are results of simulations, solid curves are related to formula (5.12). Other model parameters are the same as those in Fig. 6.3. Adapted from [10]

comparison with the function $P^{(B)}(q_{sum})$ for traffic flow consisting of human driving vehicles only (left curve labeled by "No ACC" in Fig. 6.5a).

Correspondingly, we have found that the maximum highway capacity C_{max} of free flow at the bottleneck and the threshold flow rate $q_{th}^{(B)}$ for spontaneous traffic breakdown at the bottleneck increase for mixed traffic flow (Fig. 6.5a). Traffic flow

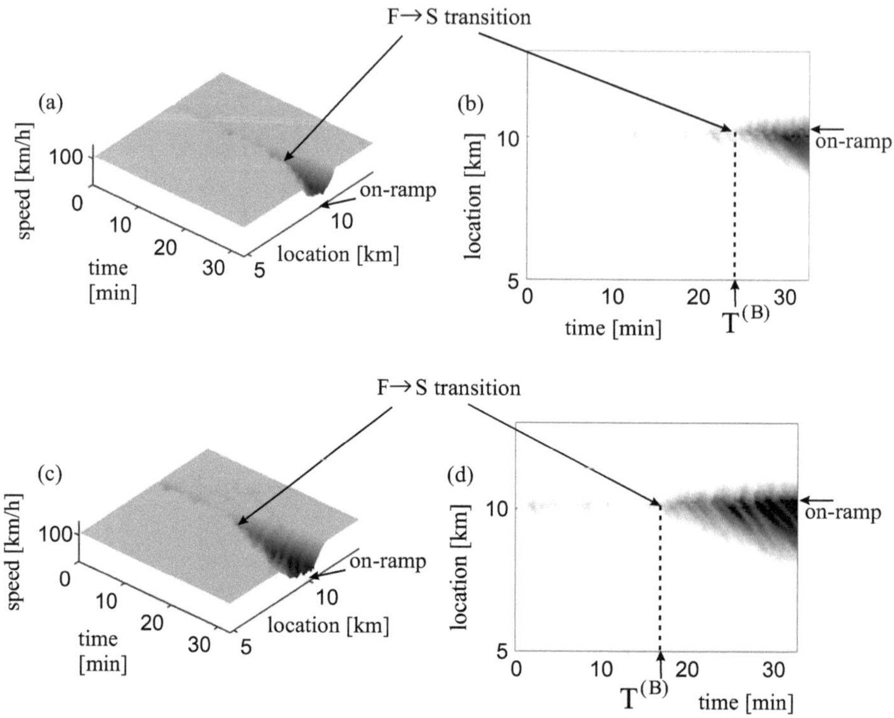

Fig. 6.6 Two different simulation realizations 2 (**a**, **b**) and 3 (**c**, **d**) of traffic breakdown in mixed traffic flow with $\gamma^{(ACC)} = 20\%$ of ACC vehicles with the same time of observation of traffic flow $T_{ob} = 30$ min as that in simulation realization 1 in which no breakdown occurs during $T_{ob} =$ 30 min (Fig. 6.4c,d): Vehicle speed in space and time (**a**, **c**) and the same speed data presented by regions with variable shades of gray (**b**, **d**) (shades of gray vary from white to black when the speed decreases from 105 km/h (*white*) to zero (*black*)). Other model parameters are the same as those in Fig. 6.4c,d. Adapted from [10]

consisting of human driving vehicles only and mixed traffic flow are different traffic flows. For this reason, we can expect that the minimum highway capacities C_{min} of free flow at the bottleneck can also be different values in these two different traffic flows. Indeed, we have found that in mixed traffic flow the minimum capacity of free flow at the bottleneck denoted by $C_{min,\, ACC}$ is slightly larger than the minimum highway capacity C_{min} of free flow at the bottleneck in traffic flow consisting of human driving vehicles only (Fig. 6.5b).

Automatic driving vehicles can decrease the probability of traffic breakdown in mixed free flow.

Automatic driving vehicles can increase the threshold flow rate for spontaneous traffic breakdown as well as the maximum and minimum capacities of free flow at the bottleneck.

However, a large enough percentage of automatic driving vehicles (about 20% or more) in mixed free flow is required to improve traffic flow characteristics appreciably.

6.4 Deterioration of Performance of Traffic System Through Automatic Driving Vehicles

Rather than the enhancement of traffic flow characteristics (Sect. 6.3), automatic driving vehicles can also result in the deterioration of the performance of the traffic system [10]. To show this, we consider mixed traffic flow under condition (6.14). In this mixed traffic flow, ACC-vehicles exhibit the same short desired time headway $\tau_{\rm d}^{(\rm ACC)} = 1.1$ s as used above in Sect. 6.3. Moreover, all sets of dynamics coefficients K_1 and K_2 of ACC-vehicles used below (see Figs. 6.7, 6.8, 6.9) satisfy condition of string stability (6.12). Nevertheless, we will find that the ACC-vehicles can lead to a considerable increase in the probability of traffic breakdown at road bottlenecks (Fig. 6.7).[6]

Indeed, we have found that even a relatively small percentage $\gamma^{(\rm ACC)} = 5\%$ of the ACC-vehicles in mixed traffic flow can *increase* considerably the probability of traffic breakdown (Fig. 6.7a, curves 1–4). The importance of this result is as follows: In Sect. 6.3, we have mentioned that the positive effect of the ACC-vehicles on traffic flow, in particular, the decrease in the probability of traffic breakdown is considerable only at large percentages of ACC-vehicles $\gamma^{(\rm ACC)} \approx 20\%$. In the next future, we could expect only much smaller percentages of automatic driving vehicles in mixed traffic flow, like $\gamma^{(\rm ACC)} \approx 5\%$. Therefore, the deterioration of the performance of mixed traffic flow shown Fig. 6.7a (curves 1–4) can be a subject of the development of automatic driving vehicles in car-development companies already during next years.

To understand the deterioration of the performance of the traffic system through automatic driving vehicles (Fig. 6.7), firstly we should note that for each set of dynamic coefficients $(K_1,\ K_2)$ of the ACC-vehicles used in simulations shown

[6] In this book, we do not discuss another possible case of the deterioration of the performance of traffic system that is often assumed to occur when automatic driving vehicles follow strictly all traffic regulation rules, like a given speed limit.

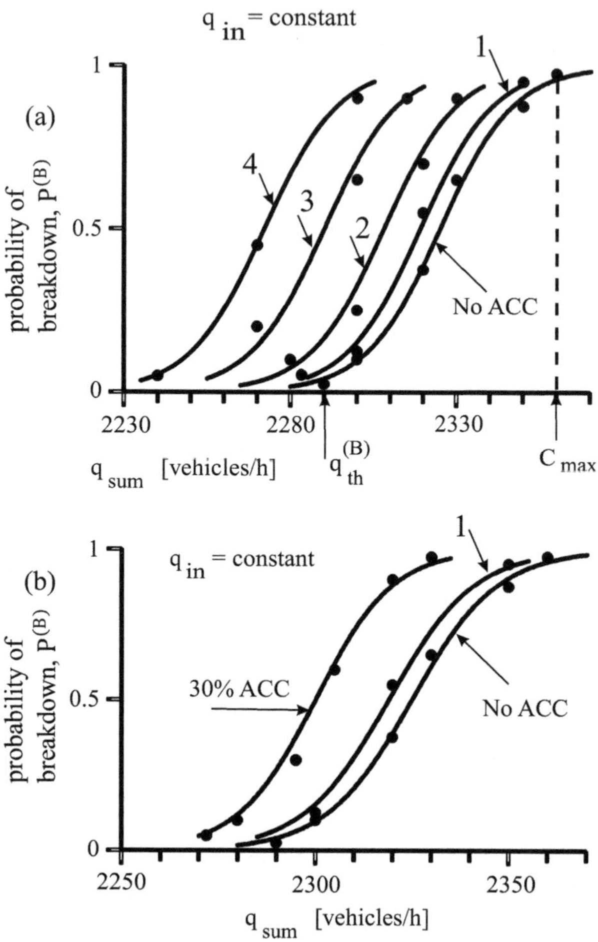

Fig. 6.7 Increase in the probability of traffic breakdown $P^{(B)}$ in mixed free traffic flow through ACC-vehicles with $\tau_d^{(ACC)} = 1.1$ s: In (**a, b**), the flow-rate dependence of the breakdown probability $P^{(B)}(q_{sum})$ labeled by "No ACC" is the same as that in Fig. 6.5 for free flow of manual driving vehicles without ACC-vehicles. In (**a**), curves $P^{(B)}(q_{sum})$ labeled by numbers 1–4 are related to mixed free flows with $\gamma^{(ACC)} = 5\%$ of ACC-vehicles with different sets of coefficients $(K_1, K_2) = (0.2\ s^{-2}, 0.82\ s^{-1})$ for curve 1, $(0.3\ s^{-2}, 0.77\ s^{-1})$ for curve 2, $(0.5\ s^{-2}, 0.65\ s^{-1})$ for curve 3, $(0.7\ s^{-2}, 0.55\ s^{-1})$ for curve 4, which satisfy (6.12). In (**b**), curve $P^{(B)}(q_{sum})$ labeled by "30% ACC" is related to mixed free flow with $\gamma^{(ACC)} = 30\%$ of ACC-vehicles with $(K_1, K_2) = (0.2\ s^{-2}, 0.82\ s^{-1})$; curve 1 is the same as that in (**a**). The flow rate $q_{sum} = q_{in} + q_{on}$ is varied through the change in the on-ramp inflow rate q_{on} at constant $q_{in} = 2000$ vehicles/h. Functions $P^{(B)}(q_{sum})$ are described by formula (5.12). Adapted from [10]

in Fig. 6.7a (curves 1–4), free flow consisting of $\gamma^{(ACC)} = 100\%$ of the ACC-vehicles is stable: We have found the same results for free flow stability as those presented in Fig. 6.3c,d. This means that when in mixed free flow the percentage

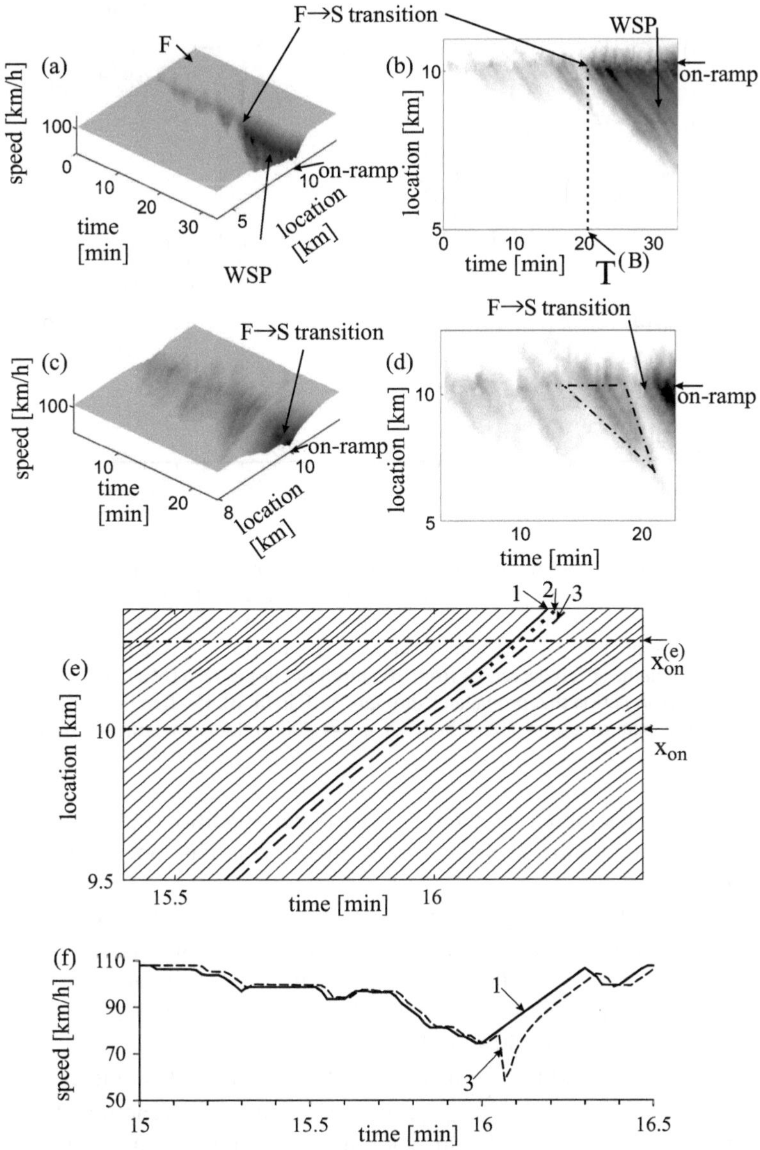

Fig. 6.8 Simulations of dynamics of permanent local speed disturbance at on-ramp bottleneck on single-lane road for a mixed traffic flow with $\gamma^{(ACC)} = 5\%$ of ACC-vehicles with coefficients $(K_1, K_2) = (0.5\ s^{-2}, 0.65\ s^{-1})$: (**a–d**) Speed in space and time (**a**, **c**) and the same data presented by regions with variable shades of gray (**b**, **d**) (shades of gray vary from white to black when the speed decreases from 105 km/h (*white*) to zero (*black*) in (**b**) and from 105 km/h (*white*) to 20 km/h (*black*) in (**d**)); in (**c**, **d**), we show the same data as those in (**a**, **b**), however, for smaller space and time intervals. (**e**) Fragment of vehicle trajectories in space and time related to (**c**, **d**). (**f**) Microscopic vehicle speeds along trajectories as time functions labeled by the same numbers as those in (**e**). In (**d**), *dashed-dotted lines* denote F→S→F transitions. F—free flow, WSP—widening synchronized flow pattern. $q_{on} = 320$ vehicles/h, $q_{in} = 2000$ vehicles/h. Adapted from [10]

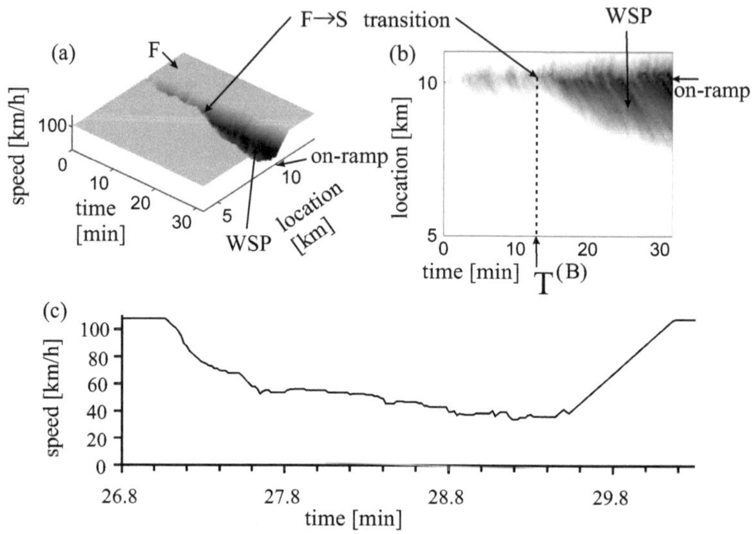

Fig. 6.9 Simulations of F→S transition (traffic breakdown) at on-ramp bottleneck on single-lane road for mixed traffic flow with $\gamma^{(ACC)} = 30\%$ of ACC-vehicles with $(K_1, K_2) = (0.2\ s^{-2}, 0.82\ s^{-1})$ related to curve labeled by "30% ACC" in Fig. 6.7b: (**a, b**) Speed in space and time (**a**) and the same data presented by regions with variable shades of gray (**b**) (shades of gray vary from white to black when the speed decreases from 105 km/h (*white*) to zero (*black*)). (**c**) One of vehicle trajectories as time-function that propagates through WSP in (**a, b**). F—free flow, WSP—widening synchronized flow pattern. $q_{on} = 305$ vehicles/h, $q_{in} = 2000$ vehicles/h. Adapted from [10]

of the ACC-vehicles increases, then, at least under condition $\gamma^{(ACC)} \to 100\%$, no traffic breakdown should be observed any more. This has indeed been found in simulations.

We have found that for any of the sets of dynamic coefficients (K_1, K_2) used in Fig. 6.7, when the percentage of the ACC-vehicles increases from $\gamma^{(ACC)} = 5\%$ to larger values, firstly, the flow rate dependence of the breakdown probability $P^{(B)}(q_{sum})$ is subsequently shifted to the left in the flow rate axis as shown in Fig. 6.7b (curve labeled by "30% ACC").

However, there should be a critical percentage of the ACC-vehicles denoted by $\gamma^{(ACC)}_{cr,\ increase}$. When $\gamma^{(ACC)} = \gamma^{(ACC)}_{cr,\ increase}$, then the shift of the function $P^{(B)}(q_{sum})$ to the left in the flow rate axis should reach its maximum. When the percentage of ACC-vehicles increases subsequently, i.e, $\gamma^{(ACC)} > \gamma^{(ACC)}_{cr,\ increase}$, then the function $P^{(B)}(q_{sum})$ should begin to be shifted to the right in the flow rate axis in comparison with the case $\gamma^{(ACC)} = \gamma^{(ACC)}_{cr,\ increase}$. Finally, as above-mentioned, at $\gamma^{(ACC)} \to 100\%$ no traffic breakdown occur at the bottleneck as long as condition (6.14) is satisfied.

The above assumption about the behavior of the function $P^{(B)}(q_{sum})$ under increase in the percentage of the ACC-vehicles is indeed confirmed by simulation results, which have been made for each of the sets of dynamic coefficients (K_1, K_2) of the ACC-vehicles used in Fig. 6.7. Moreover, it turns out that already for dynamic coefficients $(K_1, K_2) = (0.2\ s^{-2}, 0.82\ s^{-1})$, which do not considerably differ from $(K_1, K_2) = (0.14\ s^{-2}, 0.9\ s^{-1})$ used in Figs. 6.3–6.5, we have found that $\gamma^{(ACC)}_{cr,\ increase} \approx 30\%$ (curve $P^{(B)}(q_{sum})$ labeled by "30% ACC" in Fig. 6.7b). This leads to the following result:

The deterioration of the performance of the traffic system through automatic driving vehicles can occur within broad ranges of the percentage of ACC-vehicles and the set of coefficients (K_1, K_2), which satisfy condition (6.12) of string stability of ACC-vehicles.

To understand this negative effect of the ACC-vehicles on traffic flow, firstly note that as long as the percentage of the ACC-vehicles in mixed traffic flow is not very large, traffic breakdown at the bottleneck is qualitatively the same time-delayed F→S transition (Figs. 6.8a,b and 6.9a,b) as that in traffic flow consisting of manual driving vehicles only (Fig. 6.3a,b). In both cases, WSPs result from the breakdown (Figs. 6.3a,b, 6.8a,b, and 6.9a,b). Moreover, in both cases, before traffic breakdown occurs (time intervals $0 < t < T^{(B)}$ in Figs. 6.3a,b, 6.8a,b, and 6.9a,b), there are many F→S→F transitions at the bottlenecks (dashed-dotted lines shown in Figs. 6.8d and 6.10b denote some of the regions of dissolving synchronized flow occurring due to a sequence of the F→S→F transitions). As explained in [9] (see Sect. 5.13), the F→S→F transitions determine the dynamics of a permanent local speed disturbance at the bottleneck.

A crucial difference between mixed traffic flow and traffic flow without ACC-vehicles becomes clear, when we consider the dynamics of a permanent local speed disturbance at the bottleneck: We have found that the amplitude of the permanent local speed disturbance at the bottleneck occurring in mixed traffic flow can increase considerably (Figs. 6.8c–f) in comparison with that occurring in free flow consisting only of manual driving vehicles (Fig. 6.10).

We consider a vehicle moving at a low speed in the on-ramp lane that merges from the on-ramp onto the main road (bold dotted vehicle trajectory 2 in Fig. 6.10c). After merging, this vehicle 2 is between two vehicles 1 and 3 moving on the main road (trajectories 1 and 3 in Fig. 6.10c). The motion of vehicle 1 on the main road is not influenced by vehicle 2 that is upstream of vehicle 1. Contrarily, vehicle 3 (trajectory 3 in Fig. 6.10c) should decelerate to the speed of vehicle 2. As a result, due to the vehicle merging the speed decreases within the permanent speed disturbance localized at the bottleneck (compare speeds of vehicles 1 and 3 in Fig. 6.10d).

It turns out that the effect of the speed reduction caused by the merging vehicle can increase considerably, when vehicle 3 is an ACC-vehicle (dashed trajectory 3 in Fig. 6.8e). Indeed, the deceleration of the ACC-vehicle (vehicle 3 in Fig. 6.8f) due to the merging vehicle (dotted trajectory 2 in Fig. 6.8e) becomes considerably stronger than in the case of traffic flow without ACC-vehicles (compare speeds of vehicles

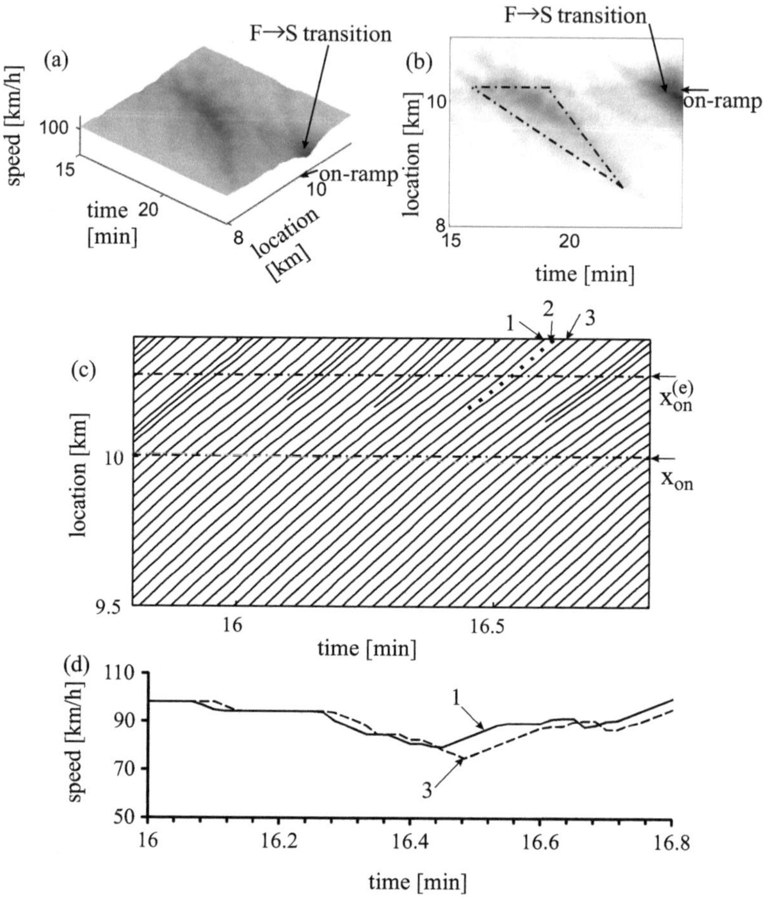

Fig. 6.10 Simulations of dynamics of permanent local speed disturbance at on-ramp bottleneck on single-lane road for traffic flow consisting only of manual driving vehicles: (**a**, **b**) Speed in space and time (**a**) and the same data presented by regions with variable shades of gray (**b**) (shades of gray vary from white to black when the speed decreases from 105 km/h (*white*) to 20 km/h (*black*)); the simulated data is a fragment of the data shown in Fig. 6.3a,b. (**c**) Fragment of vehicle trajectories in space and time related to (**a**, **b**). (**d**) Microscopic vehicle speeds along trajectories as time functions labeled by the same numbers as those in (**c**). In (**b**), *dashed-dotted lines* denote F→S→F transitions. q_{on} = 320 vehicles/h, q_{in} = 2000 vehicles/h. Adapted from [10]

1 and 3 in Fig. 6.8f with speeds of vehicles 1 and 3 in Fig. 6.10d, respectively). Due to a stronger speed reduction at the bottleneck caused by the ACC-vehicles, the probability of traffic breakdown increases at the same flow rates as those in the case of traffic flow consisting only of manual driving vehicles (curve "No ACC" in Fig. 6.7a).

When the percentage $\gamma^{(ACC)}$ of the ACC-vehicles increases, the frequency of large speed disturbances at the bottleneck caused by the ACC-vehicles increases either. This explains why at the given flow rates the probability of traffic breakdown increases up to $\gamma^{(\mathrm{ACC})}_{\mathrm{cr,\ increase}}$ (the value $\gamma^{(\mathrm{ACC})}_{\mathrm{cr,\ increase}} \approx 30\%$ for parameters of ACC-vehicles used in Fig. 6.7b). Only when the percentage of the ACC-vehicles $\gamma^{(\mathrm{ACC})} > \gamma^{(\mathrm{ACC})}_{\mathrm{cr,\ increase}}$, long stable platoons of the ACC-vehicle can occur that lead to a decrease in the breakdown probability. Indeed, free flow consisting of 100% ACC-vehicles is stable.

To understand the physics of the deterioration of the performance of the traffic system through the ACC-vehicles in more details, we should note that in accordance with the hypothesis of the three-phase theory about the existence of 2D-steady states of synchronized flow [6, 7] (see Sect. 5.9.1 for more details), drivers do not control the space gap g to the preceding vehicle when conditions

$$g_{\mathrm{safe}} \leq g \leq G, \tag{6.19}$$

are satisfied, where G and g_{safe} are the synchronization and safe space gaps, respectively. Formula (6.19) is equivalent to formula

$$\tau_{\mathrm{safe}} \leq \tau^{(\mathrm{net})} \leq \tau_{\mathrm{G}}, \tag{6.20}$$

where $\tau^{(\mathrm{net})}$ is a time headway (net time gap) of the vehicle to the preceding vehicle, $\tau_{\mathrm{safe}} = g_{\mathrm{safe}}/v$, $\tau_{\mathrm{G}} = G/v$ as well as it is assumed that $v > 0$.

In contrast to the manual driver behavior (6.19) (or (6.20)), in accordance with the classical ACC-model (6.16), the ACC-vehicle tries to reach an "optimal" space gap given by formula (6.7). This qualitative different dynamic behavior of the ACC-vehicles and manual driving vehicles could explain the occurrence of large disturbances in free flow at the bottleneck. When the space gap between the ACC-vehicle and the merging vehicle is smaller than that given by formula (6.7), the ACC-vehicle decelerates, whereas a manual driving vehicle should not decelerate as long as conditions (6.19) (or (6.20)) are satisfied. The deceleration of the ACC-vehicle is the stronger, the larger the coefficient K_1 in (6.16). This can explain the result of simulations that the larger the coefficient K_1, the more is the shift of the function $P^{(\mathrm{B})}(q_{\mathrm{sum}})$ to the left in the flow rate-axis (curves 1–4 in Fig. 6.7a). We can make the following conclusion.

When dynamics rules of motion of automatic driving vehicles differ considerably from those of human driving vehicles, such automatic driving vehicles can cause the deterioration of the performance of the traffic system. In particular, speed disturbances in mixed traffic flow at road bottlenecks can increase strongly. This can cause the considerable increase in the probability of traffic breakdown.

Naturally, cooperative driving through the use of vehicle-to-vehicle communication (see explanations and references in the review [8, 10]) could reduce this negative effect of automatic driving vehicles on traffic flow.[7] However, it seems better to develop such dynamics rules of motion of automatic driving vehicles that avoid situations at which human drivers can consider automatic driving vehicles as "obstacles".

6.5 Conclusions

Depending on the parameters of automatic driving vehicles and on the percentage of the automatic driving vehicles in mixed traffic flow consisting of a random distribution of automatic driving and manual driving vehicles, the automatic driving vehicles can either decrease or increase the probability of traffic breakdown in mixed traffic flow.

The increase in the probability of traffic breakdown at a bottleneck, i.e., the deterioration of the performance of the traffic system can occur already at a small percentage (about 5%) of automatic driving vehicles. The negative effect of the automatic driving vehicles on traffic flow can be realized, even if any platoon of the automatic driving vehicles satisfies condition for string stability. This effect occurs even at the same flow rates at the bottleneck at which there is no traffic breakdown in free flow consisting of 100% of automatic driving vehicles.

This deterioration of the performance of the traffic system through automatic driving vehicles can also lead to the increase in the frequency of accidents in mixed traffic flow. Indeed, as explained in this chapter, the dynamic behavior of automatic driving vehicles is qualitative different in comparison with that of manual driving vehicles. Therefore, the "unexpected" behavior of automatic driving vehicles in exigent situations can irritate some of the drivers. This can explain results of a recent empirical study made by Schoettle and Sivak [21], in which it has been found that the frequency of accidents caused by manual driving vehicles increases in mixed traffic flow.

References

1. L.C. Davis, Phys. Rev. E. **69**, 066110 (2004)
2. L.C. Davis, Phys. Lett. A **376**, 2658–2662 (2012)
3. L.C. Davis, Physica A **392**, 3755–3764 (2013)
4. L.C. Davis, Physica A **392**, 3798–3805 (2013)

---

[7]In particular, one can expect that cooperative merging could alleviate the problem of large disturbances occurring during vehicle merging at the bottleneck in mixed traffic flow illustrated in Fig. 6.8.

5. L.C. Davis, Physica A **395**, 580 (2014)
6. B.S. Kerner, *The Physics of Traffic* (Springer, Berlin, New York, 2004)
7. B.S. Kerner, *Introduction to Modern Traffic Flow Theory and Control* (Springer, Heidelberg, Dordrecht, London, New York, 2009)
8. B.S. Kerner, Elektrotechnik und Informationstechnik **132**, 417–433 (2015)
9. B.S. Kerner, Phys. Rev. E **92**, 062827 (2015)
10. B.S. Kerner, Physica A **450**, 700–747 (2016)
11. B.S. Kerner, S.L. Klenov, J. Phys. A Math. Gen. **35**, L31–L43 (2002)
12. B.S. Kerner, S.L. Klenov, Phys. Rev. E **68**, 036130 (2003)
13. B.S. Kerner, S.L. Klenov, J. Phys. A Math. Gen. **37**, 8753–8788 (2004)
14. B.S. Kerner, S.L. Klenov, Phys. Rev. E **80**, 056101 (2009)
15. B.S. Kerner, S.L. Klenov, J. Phys. A Math. Theor. **43**, 425101 (2010)
16. C.-Y. Liang, H. Peng, Veh. Syst. Dyn. **32**, 313–330 (1999)
17. C.-Y. Liang, H. Peng, JSME Int. J. Ser. C **43**, 671–677 (2000)
18. W.Levine, M. Athans, IEEE Trans. Automat. Contr. **11**, 355–361 (1966)
19. R. Müller, G. Nöcker, in *Proceedings of the Intelligent Vehicles '92 Symposium*, ed. by I. Masaki (IEEE, Detroit, USA, 1992), pp. 173–178
20. R. Rajamani, *Vehicle Dynamics and Control, Mechanical Engineering Series* (Springer US, Boston, MA, 2012)
21. B. Schoettle, M. Sivak, "A preliminary analysis of real-world crashes involving self-driving vehicles", *Report No. UMTRI-2015-34*,(The University of Michigan, Transp. Res. Institute, Ann Arbor, Michigan, USA, 2015)
22. D. Swaroop, J.K. Hedrick, IEEE Trans. Automat. Contr. **41**, 349–357 (1996)
23. D. Swaroop, J.K. Hedrick, S.B. Choi, IEEE Trans. Veh. Technol. **50**, 150–161 (2001)

Chapter 7
Future Automatic Driving Based on Three-Phase Theory

7.1 Introduction

The deterioration of the performance of the traffic system through the ACC-vehicles discussed in Chap. 6 could be avoided through the use of automatic driving systems in vehicles, which learn from behaviors of drivers in real traffic as incorporated in hypotheses of the three-phase theory.

To understand this statement, we should firstly recall a result of Sect. 6.2 that for each given speed $v > 0$ of a classical ACC-vehicle, there is *only one* operating point of the ACC that is determined by formula $\tau^{(\mathrm{net})} = \tau_{\mathrm{d}}^{(\mathrm{ACC})}$ (6.4): When both the preceding vehicle and the ACC-vehicle move at the same time-independent speed $v = v_{\ell} = const > 0$, then the time headway $\tau^{(\mathrm{net})}$ of the classical ACC vehicle is steered to a given fixed time headway $\tau_{\mathrm{d}}^{(\mathrm{ACC})}$ chosen by a driver that is also called a desired time headway. The operating points of the classical ACC (6.4), (6.5) related to different values of the speed of the classical ACC can be presented by a curve in the gap–speed flow plane (Fig. 6.2).

In contrast with this behavior of the classical ACC-vehicle, from hypothesis 11 of the three-phase theory about 2D-steady states of synchronized flow (Sect. 1.9) [3–8] it follows that drivers do not try to maintain some fixed time headway to the preceding vehicle while moving at a nearly time-independent speed. This behavior of real drivers is confirmed by studies of real field traffic data (Sect. 1.9). Therefore, one of the main features of real drivers that automatic driving vehicles should learn from behaviors of drivers in real traffic is the driver behavior incorporated in hypothesis 11 of the three-phase theory.

The main result of this chapter is a consideration of an ACC based on the three-phase theory. In Sect. 7.2, we show that the ACC based on the three-phase theory is related to hypothesis 11 of the three-phase theory (Sect. 1.9). In contrast with the classical ACC, there is at least one speed range within which for *each given speed* $v > 0$ of the ACC based on the three-phase theory, there are the *infinite number* of the operating points [9–12].

B.S. Kerner, *Breakdown in Traffic Networks*, DOI 10.1007/978-3-662-54473-0_7

Additionally, in Sect. 7.3 we consider some other ideas for the development of future automatic driving vehicles learning from behaviors of drivers in real traffic. Recall that the behaviors of drivers in real traffic that explain the nucleation features of traffic breakdown have been disclosed in the three-phase theory (Sect. 5.15). This explains the title of this chapter.

7.2 Automatic Driving Based on Three-Phase Theory

7.2.1 Infinite Number of Operating Points for Given Speed of Automatic Driving Vehicle

In accordance with hypothesis 11 of the three-phase theory about 2D-steady states of synchronized flow [3–8] (see Sects. 1.9 and 5.9.1 for more details), in 2004 we introduced an ACC based on the three-phase theory [9–12] (see also Fig. 9.18 in Sec. 9.6 of the book [14]). In contrast with the classical model of the ACC-vehicle (6.1), as above-mentioned, for the ACC-vehicle based on the three-phase theory there is *no* fixed desired time headway of the ACC-vehicle to the preceding vehicle. We call automatic driving vehicles in which there is no fixed desired time headway to the preceding vehicle as "automatic driving based on the three-phase theory".

To understand the difference between the classical model of ACC (6.1) and the model of automatic driving based on the three-phase theory [9–12], we assume that the space gap of the ACC-vehicle g to the preceding vehicle satisfies conditions (6.19) that for the simplicity of the discussion we rewrite here:

$$g_{\rm safe} \leq g \leq G, \tag{7.1}$$

where G and $g_{\rm safe}$ are the synchronization and safe space gaps, respectively (Fig. 7.1). We assume that the speed of the ACC-vehicle is higher than zero ($v > 0$). Then, formula (7.1) is equivalent to conditions (6.20) that we also rewrite here:

$$\tau_{\rm safe} \leq \tau^{\rm (net)} \leq \tau_{\rm G}, \tag{7.2}$$

where $\tau_{\rm safe} = g_{\rm safe}/v$, $\tau_{\rm G} = G/v$.

In general, the safe space gap $g_{\rm safe}$ and the synchronization space gap G in (7.1) can be complex functions of the speed of the ACC-vehicle v and the speed of the preceding vehicle v_ℓ; these functions can be different in different driving situations [11–13]. Respectively, the safe time headway $\tau_{\rm safe}$ and the synchronization time headway $\tau_{\rm G}$ of the ACC based on the three-phase theory can be complex speed functions depending on a driving situation.

We consider possible operating points of the ACC based on the three-phase theory. A operating point of the ACC is related to the case, when the speed difference

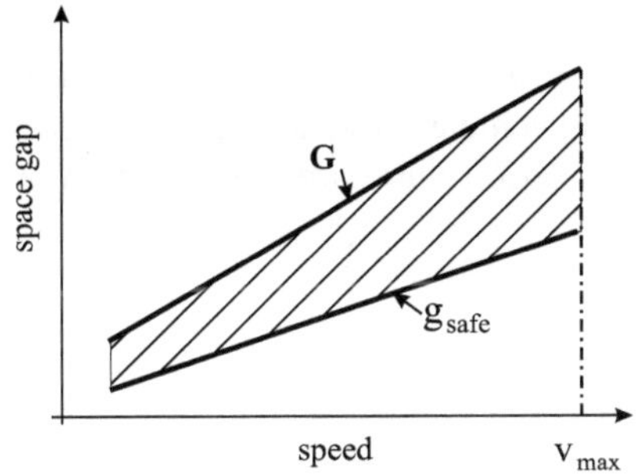

Fig. 7.1 Qualitative explanation of the infinite number of the operating points for the ACC based on the three-phase theory [9–12]: A part of 2D-region for operating points of the ACC in the space-gap–speed plane (*dashed region*) that satisfies conditions (7.1). In contrast, in the classical model of ACC (Sect. 6.2) at the same time-independent speed there is only one operating point of the ACC (see Fig. 6.2)

between the speed of the preceding vehicle and the speed of the ACC-vehicle $\Delta v = v_\ell - v$ is equal to zero (6.5) *and* the preceding vehicle moves at a time-independent speed v_ℓ (where we assume that $v_\ell > 0$). Indeed, in this case in accordance with the definition of an operating point of the ACC-vehicle (Sect. 6.2), the ACC-vehicle moves at the time-independent speed that is equal to the speed of the preceding vehicle: $v = v_\ell$. From conditions (7.2), we get that for the ACC based on the three-phase theory moving at a time-independent speed $v = v_\ell > 0$ an operating point of the ACC should satisfy conditions

$$\tau_{\mathrm{safe}}(v) \le \tau^{(\mathrm{net})} \le \tau_{\mathrm{G}}(v), \tag{7.3}$$

where we have taken into account that in a general case the safe time headway τ_{safe} and/or the synchronization time headway τ_{G} can be speed functions. Additionally, in (7.3) for each given speed v the safe time headway $\tau_{\mathrm{safe}}(v)$ and the synchronization time headway $\tau_{\mathrm{G}}(v)$ are some constant values that satisfy condition

$$\tau_{\mathrm{G}}(v) > \tau_{\mathrm{safe}}(v). \tag{7.4}$$

Conditions (7.3) for the automatic driving vehicles based on the three-phase theory are the same ones as conditions (5.38) for manual driving vehicles following from the hypothesis of the three-phase theory about 2D-steady states of synchronized flow (Sect. 5.9.1).

We can see that for an automatic driving vehicle based on the three-phase theory (Fig. 7.1) [9–12] there should be at least one speed range within which for each given speed that is larger than zero there are the infinite number of different time headway $\tau^{(\mathrm{net})}$ that satisfy conditions (7.3).

We define an automatic driving vehicle based on the three-phase theory [9–12] as the automatic driving vehicle for which there is a speed range within which for each given vehicle speed there are the infinite number of the operating points.

7.2.2 About Dynamic Behavior of Automatic Driving Vehicle Based on Three-Phase Theory

However, the question arises:

- What can be the dynamic behavior of ACC based on the three-phase theory?

We consider an usual case at which the speed of the ACC-vehicle $v(t)$ and the speed of the preceding vehicle $v_\ell(t)$ are time-functions and the speed of the ACC-vehicle is not equal to the speed of the preceding vehicle: $v(t) \neq v_\ell(t)$. When conditions (7.1) (or (7.2)) are satisfied, the space gap of the ACC-vehicle to the preceding vehicle is within a 2D-region in the space-gap–speed plane (dashed region in Fig. 7.2). Then, rather than the classical formula (6.1), the acceleration (deceleration) of automatic driving vehicles based on the three-phase theory is given by formula [9–12]

$$a(t) = K_{\Delta \mathrm{v}} \Delta v(t), \tag{7.5}$$

where $\Delta v(t) = v_\ell(t) - v(t)$, $v(t)$ is the speed of the ACC-vehicle at time instant t, $v_\ell(t)$ is the speed of the preceding vehicle at time instant t, $K_{\Delta \mathrm{v}}$ is a dynamic coefficient of the ACC-vehicle ($K_{\Delta \mathrm{v}} > 0$).

Formula (7.5) means that the ACC-vehicle adapts its speed $v(t)$ to the speed of the preceding vehicle $v_\ell(t)$ without caring, what the precise space gap (time headway) to the preceding vehicle is. Thus, the dynamics of the automatic driving vehicle based on the three-phase theory should not necessarily depend on the time headway to the preceding vehicle $\tau^{(\mathrm{net})}(t)$ as long as conditions (7.2) are satisfied. We can also formulate this conclusion as follows: An automatic driving vehicle based on the three-phase theory [9–12] is the automatic driving vehicle in which there is no fixed desired time headway to the preceding vehicle.

Outside the 2D-region in the space-gap–speed plane (Fig. 7.2) formula (7.5) is not applied [11–13]: At $g > G$, i.e., $\tau^{(\mathrm{net})} > \tau_{\mathrm{G}}$, the ACC-vehicle accelerates (labeled by "acceleration" in Fig. 7.2), whereas at $g < g_{\mathrm{safe}}$, i.e., $\tau^{(\mathrm{net})} < \tau_{\mathrm{safe}}$, the ACC-vehicle decelerates (labeled by "deceleration" in Fig. 7.2).[1]

[1] In real ACC based on the three-phase theory, the synchronization time headway τ_{G}, the safe time headway τ_{safe}, and the coefficient of ACC adaptation $K_{\Delta \mathrm{v}}$ in (7.5) can be complex functions of

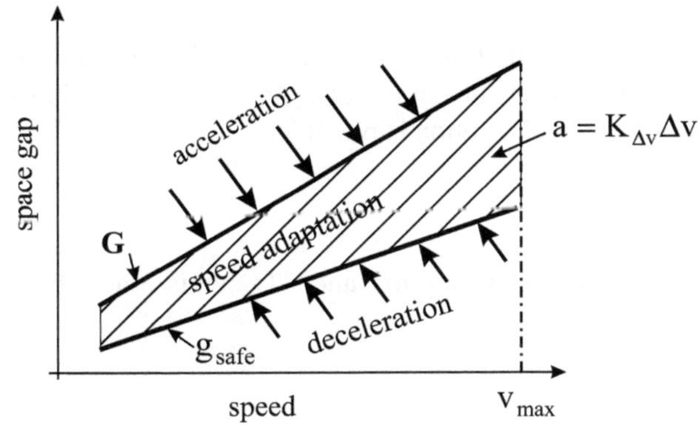

Fig. 7.2 Explanation of ACC based on three-phase theory [9–12]: A part of 2D-region for operating points of the ACC in the space-gap–speed plane (dashed region adapted from Fig. 7.1) within which an ACC-vehicle moves in accordance with Eq. (7.5)

For the automatic driving vehicle based on the three-phase theory there is no fixed time headway to which the automatic driving vehicle should necessarily be steered [9–12].

Thus, in some traffic situations, acceleration (deceleration) of the ACC-vehicle based on the three-phase theory does not depend on the space gap, i.e., on the time headway to the preceding vehicle at all.

In 2007–2008, an ACC based on the three-phase theory, in which formula (7.5) under conditions (7.1), (7.2) was used, was implemented in real ACC-vehicles and proven in test drives by the Daimler Company. It was found that the ACC-vehicle based on the three-phase theory exhibits a more comfortable ACC behavior as well as it leads to the reduction of fuel consumption and CO_2 emissions. To the knowledge of the author, ACC-vehicles based on the three-phase theory [9–13] are currently on the market.

Possible advantages of the ACC-system based on the three-phase theory in comparison with the classical ACC-system (6.1) are as follows.

The prevention of the deterioration of the performance of the traffic system though automatic driving vehicles.

---

the speed of the vehicle and the speed of the preceding vehicle as well as of the current space gap (current time headway). Moreover, these functions can depend on a driving situation. Some examples of these functions can be found in [11–13].

The reducing of a conflict between the dynamic and comfortable ACC behavior. In particular, at the same other ACC dynamic behaviors a much comfortable ACC-vehicle motion is possible.

The reduction of fuel consumption and CO_2 emissions while moving in congested traffic.

Through the use of condition (7.5), automatic driving vehicles can alleviate the problem of large disturbances occurring during vehicle merging at a bottleneck. For this reason, such automatic driving vehicles can prevent traffic breakdown at the bottleneck.

A further development of automatic driving vehicles based on the three-phase theory can be a very interesting task for further investigations.

For this reason, below in Sect. 7.3 we consider some ideas for the further development of automatic driving vehicles based on the three-phase theory that can decrease the probability of traffic breakdown in a mixed traffic flow at network bottlenecks.[2] There are at least two possibilities for this development of automatic driving vehicles:

(i) The dynamic behavior of the automatic driving vehicles should learn from dynamic driver behaviors as found in the three-phase theory.
(ii) Through the use of cooperative driving, automatic driving and manual driving vehicles in mixed traffic flow should prevent the occurrence of dangerous driving situations and decrease the probability of the emergence of nuclei for traffic breakdown at network bottlenecks.

7.3 Driver Behaviors Facilitating Free Flow

As explained in the three-phase theory (Chap. 5), traffic breakdown is associated with a competition between two opposing tendencies occurring within a local disturbance within which the speed is lower and vehicle density is greater than in an

[2]Because a detailed consideration of traffic congestion resulting from traffic breakdown is out of the scope of this book, in the main text of the book we do not consider requirements for the further development of automatic driving vehicles based on the three-phase theory related to dynamic control of congested traffic [14, 15] like the "jam-absorption" effect (see explanations of the "jam-absorption" effect in Sect. A.7.3 of Appendix A).

initial free flow: A tendency towards synchronized flow due to vehicle deceleration associated with the speed adaptation effect and a tendency towards free flow due to vehicle acceleration associated with the over-acceleration effect. Due to a time delay in driver over-acceleration, the probability of over-acceleration is a discontinuous density function (Fig. 5.18). This means that the over-acceleration effect is associated with traffic flow instability in synchronized flow (S→F instability) leading to the local increase in vehicle speed (Sect. 5.13). Thus, if the over-acceleration effect is stronger than the speed adaptation, due to the S→F instability a phase transition from synchronized flow to free flow occurs. Otherwise, if the over-acceleration effect is weaker than the speed adaptation, synchronized flow persists.

In an initial free flow, the competition between the over-acceleration and speed adaptation leads to traffic breakdown in an initially free flow, if a nucleus for traffic breakdown appears in free flow at the bottleneck. Within the nucleus the over-acceleration effect is weaker than the speed adaptation. Therefore, the above-mentioned S→F instability associated with the over-acceleration effect is suppressed. This results in the occurrence of synchronized flow, i.e., in traffic breakdown at the bottleneck.

Thus, we see that changes in driver behaviors for the prevention of traffic breakdown should lead to

- the reduction in the probability of the occurrence of a nucleus for traffic breakdown through the decrease in speed fluctuations in free flow or/and
- the facilitation of the over-acceleration effect that increases the tendency towards free flow.

The decrease in speed fluctuations in free flow in a neighborhood of on- or off-ramp bottleneck can be achieved through the use of cooperative driving in which vehicles moving on the main road in the neighbor lane to on- or off-ramps

- keep larger space gaps for those vehicles that merge from the on-ramp onto the main road or
- change lane to enable vehicles freely merging onto the main road from the on-ramp, or/and
- decrease deceleration (if it is possible).

The facilitation of the over-acceleration effect can be achieved through the use of vehicle systems that help a driver

- to change to a faster lane in a neighborhood of the bottleneck or/and
- to increase space gap (time headway) approaching a local disturbance in free flow at the bottleneck or/and
- to decrease a driver time delay in acceleration at the downstream front of the local disturbance.

The above-mentioned changes in driver behaviors can lead to the interruption of the nucleus growth and, finally, to the dissolution of the nucleus. However, we should mention that the increase in lane changing frequency to a faster lane can lead to the opposite effect—the increase in a local disturbance and, consequently, to the occurrence of a nucleus required for traffic breakdown. Indeed, on the one hand, lane changing to a faster lane can lead to the dissolution of the nucleus in the lane. On the other hand, the lane changing can lead to the occurrence of a nucleus for traffic breakdown, when due to lane changing a vehicle forces the following vehicle in the target lane to decelerate strongly. In more details, such a dual role of lane changing for traffic breakdown has been discussed in Sec. 3.4 of the book [14].

The opposite effect leading to the occurrence of a nucleus for traffic breakdown can also occur through the decrease in a driver time delay in acceleration at the downstream front of a local disturbance in free flow. This is because if a driver accelerates with a too short time delay after the preceding vehicle has just begun to accelerate, then a possible subsequent (and unexpected) deceleration of the preceding vehicle can cause strong deceleration of the following drivers with the subsequent occurrence of the nucleus for traffic breakdown.

We see that very accurate and precise cooperative and well-coordinated changes in individual driver behavior in different road lanes in a neighborhood of a highway bottleneck are required to reach the goal of the prevention of traffic breakdown.

We hope that these dynamic driver behaviors facilitating free flow at network bottlenecks can be effectively used for the future development of automatic driving based on the three-phase theory.

To illustrate dynamic driver behaviors facilitating free flow at network bottlenecks, with the use of simulations based on the Kerner-Klenov microscopic three-phase traffic flow model we show that cooperative driving based on vehicle ad-hoc network can indeed prevent traffic breakdown due to the decrease in speed fluctuations by vehicle merging at an on-ramp bottleneck.

Wireless vehicle-to-vehicle (V2V) communication, which is the basic technology for ad-hoc vehicle networks, is one of the most important scientific fields of future ITS. In general, ad-hoc vehicle networks can be built by a combination of V2V-communication and vehicle-to-infrastructure communication (called V2X-communication). There are many ideas of applications of ad-hoc vehicle networks for cooperative driving in vehicular traffic, including systems for danger warning, adaptive assistance systems, traffic information, improving of traffic flow characteristics, etc. (see, e.g., [1, 2] and references there).

In accordance with driver behaviors facilitating free flow at network bottlenecks discussed above, we can assume that there can be the following two hypothetical possibilities to prevent traffic breakdown at an on-ramp bottleneck through changes in driver behavior of communicating vehicles:

(i) A decrease in the amplitude of disturbances on the main road occurring when vehicles merge from on-ramp onto the right lane of the main road. This decreases the probability of nucleus occurrence for traffic breakdown.
(ii) An increase in the probability of driver over-acceleration.

In simulations (Fig. 7.3), there is an on-ramp bottleneck at location 16 km on two-lane road. If there is no V2V-communication, a general congested pattern (GP) occurs at the bottleneck (Fig. 7.3a). In simulations of "a united network model" that incorporates both a model of ad-hoc network and the Kerner-Klenov three-phase

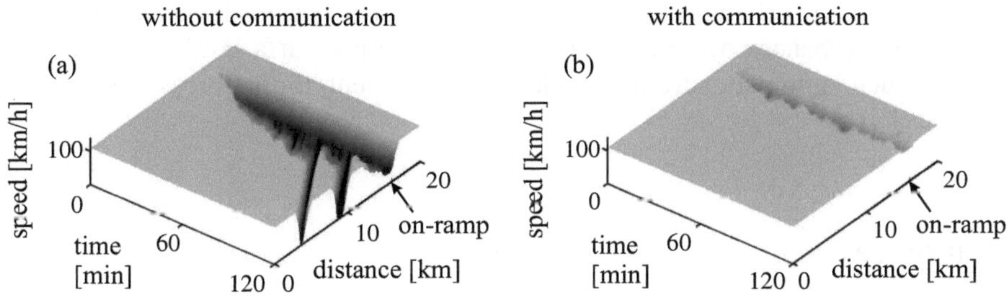

Fig. 7.3 Simulations of prevention of traffic breakdown at on-ramp bottleneck through vehicle communication: Speed in time and space (averaged across the road) without communication (**a**) and with vehicle communication (**b**). The flow rate on the main road upstream of the bottleneck is $q_{\text{in}} = 1827$ vehicles/(h lane); the flow rate to the on-ramp is $q_{\text{on}} = 600$ vehicles/h. In (**a**) and (**b**), we have used the same initial value for random function *rand*() in the model (see Appendix A). Adapted from [16]

microscopic stochastic traffic flow model [16] (for reviews, see [14, 15]), we assume that vehicles moving in the on-ramp lane send a message for neighbor vehicles moving in the right road lane when the vehicle intends to merge from the on-ramp onto the main road. We assume that the following vehicle in the right lane increases a time headway to satisfy a safe gap between the merging vehicle and the following vehicle in the right lane of the main road. Simulations show that in comparison with the case in which no V2V-communication is applied and traffic breakdown occurs (Fig. 7.3a) this change in driver behavior of communicating vehicles decreases disturbances in free flow at the bottleneck. This results in the prevention of traffic breakdown (Fig. 7.3b).

7.4 Conclusions

An automatic driving vehicle based on the three-phase theory [9–12] is the automatic driving vehicle for which there is a speed range within which at each given vehicle speed (that is larger than zero) there are the infinite number of the operating points.

Automatic driving based on the three-phase theory uses automatic driving systems in vehicles, which learn from behaviors of drivers in real traffic as incorporated in hypotheses of the three-phase theory.

The automatic driving based on the three-phase theory can reduce the conflict between the dynamic and comfortable ACC behavior.

The automatic driving based on the three-phase theory can reduce fuel consumption and CO_2 emissions while moving in congested traffic.

The automatic driving based on the three-phase theory can improve the performance of the traffic system considerably.

The automatic driving based on the three-phase theory can alleviate the problem of large disturbances occurring during vehicle merging at a bottleneck. For this reason, such automatic driving vehicles can prevent traffic breakdown at the bottleneck.

References

1. W. Chen (ed.), *Vehicular Communications and Networks* (Woodhead Publishings, Cambridge, 2015)
2. H. Hartenstein, K. Laberteaux (eds.), *VANET—Vehicular Applications and Inter-Networking Technologies* (Wiley, New York, 2010)
3. B.S. Kerner, Phys. Rev. Lett. **81**, 3797–3800 (1998)
4. B.S. Kerner, in *Proceedings of the 3rd Symposium on Highway Capacity and Level of Service*, ed. by R. Rysgaard, Vol. 2 (Road Directorate, Ministry of Transport, Denmark, 1998), pp. 621–642
5. B.S. Kerner, Trans. Res. Rec. **1678**, 160–167 (1999)
6. B.S. Kerner, in *Transportation and Traffic Theory*, ed. by A. Ceder (Elsevier Science, Amsterdam, 1999), pp. 147–171
7. B.S. Kerner. J. Phys. A Math. Gen. **33**, L221–L228 (2000)
8. B.S. Kerner, *The Physics of Traffic* (Springer, Berlin, New York, 2004)
9. B.S. Kerner, Verfahren zur Ansteuerung eines in einem Fahrzeug befindlichen verkehrsadaptiven Assistenzsystems, German patent publication DE 10308256A1 (2004) (https://google.com/patents/DE10308256A1); Patent WO 2004076223A1 (2004) (https://google.com/patents/WO2004076223A1); EU Patent EP 1597106B1 (2006); German patent DE 502004001669D1 (2006)
10. B.S. Kerner, Method for actuating a traffic-adaptive assistance system which is located in a vehicle, USA patent US 20070150167A1 (2007) (https://google.com/patents/US20070150167A1); USA patent US 7451039B2 (2008)
11. B.S. Kerner, Betriebsverfahren für ein fahrzeugseitiges verkehrsadaptives Assistenzsystem, German patent publication DE 102007008253A1 (2007) (https://register.dpma.de/DPMAregister/pat/PatSchrifteneinsicht?docId=DE102007008253A1)
12. B.S. Kerner, Betriebsverfahren für ein fahrzeugseitiges verkehrsadaptives Assistenzsystem, German patent publication DE 102007008257A1 (2007) (https://register.dpma.de/DPMAregister/pat/PatSchrifteneinsicht?docId=DE102007008257A1)
13. B.S. Kerner, Betriebsverfahren für ein fahrzeugseitiges verkehrsadaptives Assistensystem, German patent publication DE 102007008254A1 (2008); H. Eisele, B.S. Kerner, R. Möbus, Verfahren zur Steuerung eines verkehrsadaptiven Assistenzsystems eines Fahrzeugs, DE 102008023704A1 (2009)
14. B.S. Kerner, *Introduction to Modern Traffic Flow Theory and Control* (Springer, Heidelberg, Dordrecht, London, New York, 2009)
15. B.S. Kerner, in [1], pp. 223–254
16. B. S. Kerner, S. L. Klenov, A. Brakemeier, arXiv: 0712.2711 (2007), available at http://arxiv.org/abs/0712.2711; in *Proceedings of IEEE IV 2008*, paper 35; in *Proceedings of the Fourth international Workshop on Vehicle-to-Vehicle Communication V2VCOM 2008*, pp. 57–63 (2008)

Chapter 8
The Reason for Incommensurability of Three-Phase Theory with Classical Traffic Flow Theories

8.1 Introduction

The empirical nucleation nature of traffic breakdown (F→S transition) at a bottleneck is the empirical fundamental of transportation science (Chap. 3). In Chap. 5 we have explained that in the three-phase theory [38–52, 54, 55, 59–63, 65] the nucleation nature of traffic breakdown (F→S transition) at the bottleneck results from the nucleation nature of the S→F instability.

The nucleation nature of the S→F instability revealed in the three-phase theory is the reason for the nucleation nature of traffic breakdown (F→S transition) at the bottleneck.

The three-phase theory explains the empirical fundamental of transportation science through the nucleation nature of the S→F instability.

As shown in Chap. 4, none of the classical traffic flow theories incorporates the metastability of free flow with respect to traffic breakdown (F→S transition) at a road bottleneck. This fact is the main reason for the criticism of the classical traffic and transportation theories made by the author in [38–48, 50–63, 65] and in this book.

However, the author's experience shows that the above critical statement that any classical traffic and transportation theory is inconsistent with the empirical

---

The original version of this chapter was revised. An erratum to the chapter can be found at DOI 10.1007/978-3-662-54473-0_16.

B.S. Kerner, *Breakdown in Traffic Networks*, DOI 10.1007/978-3-662-54473-0_8

fundamental of transportation science could not be accepted by most of the traffic researches. In particular, the following questions arise in the traffic and transportation research community:

- May some of the classical traffic-flow theories be relatively easily adjusted to take into account the empirical evidence of the induced F→S transition at the bottleneck?
- May some of the generally accepted methods for dynamic optimization and traffic control in traffic and transportation networks be consistent with the empirical evidence of the induced F→S transition at the bottleneck?

In Chap. 4, we have already made a negative answer to these questions: None of classical traffic and transportation theories is consistent with the empirical evidence of the induced F→S transition at the bottleneck.

Indeed, this critical conclusion is obvious for the LWR model [91, 106, 114] and its applications [10–13, 91]: Nonc of induced transitions from free flow to congested traffic at the bottleneck is possible in the framework of the LWR model (Sect. 4.3).

However, this criticism is not obvious for two-phase traffic flow models of the GM model class. Indeed, in Sect. 4.4, we have shown that there is the metastability of free flow in the framework of the GM model class. This free flow metastability is the result of the classical traffic flow instability leading to a growing wave of speed reduction in free flow introduced in the GM model. The classical traffic flow instability was discovered in 1958–1961 by Herman, Gazis, Montroll, Potts, Rothery, and Chandler [8, 19, 20, 29] and further incorporated in a huge number of two-phase traffic flow models of the GM model class [1, 2, 4, 9, 17, 18, 22, 23, 28, 35, 52, 88, 90, 93–98, 100–104, 107, 112, 113, 115, 116]. Moreover, as in the three-phase theory (Sect. 5.7), there are the infinity of highway capacities in the concept of stochastic highway capacity in the framework of the classical understanding of highway capacity (Sect. 4.10). Therefore, the following questions arise.

1. Why is this classical traffic flow instability of the GM model class (Sect. 4.4) inconsistent with the empirical evidence of the induced F→S transition at the bottleneck?
2. Why is the classical concept containing the infinity of stochastic highway capacities (Sect. 4.10) inconsistent with the empirical evidence of the induced F→S transition at the bottleneck?

In this chapter, we will try to answer these questions. To reach this goal, we use a concept for the distinguishing of a new paradigm in a scientific field introduced by Kuhn [89]. This concept is mainly based on an analysis whether a new theory is incommensurable with the old one or not. In other words, the term *incommensurability* has been introduced by Kuhn in his classical book to explain a paradigm shift in a scientific field. A critical comparison of three-phase traffic flow models with two-phase traffic flow models and with the classical understanding of highway capacity made in this chapter discloses the *incommensurability* of the three-phase theory with any other classical traffic and transportation theories that is explained as follows.

None of the classical traffic and transportation theories incorporates the S→F instability revealed in the three-phase theory. The nucleation nature of the S→F instability is the basic feature of the three-phase theory. This is because the nucleation nature of the S→F instability explains the empirical fundamental of transportation science.

The three-phase theory is incommensurable with the classical traffic and transportation theories because the nucleation nature of the S→F instability has *no sense* for the classical traffic and transportation theories.

The objective of this chapter is to show that the nucleation nature of the S→F instability has indeed no sense for the classical traffic and transportation theories. As above-mentioned, this is the reason for the incommensurability of the three-phase theory with any other classical traffic and transportation theories.

This chapter can be considered consisting of three parts:

1. A numerical study of the effect of the classical traffic flow instability of Herman, Gazis, Montroll, Potts, Rothery, and Chandler [8, 19, 20, 29] and the S→F instability introduced in the three-phase theory on *spontaneous phase transitions* in traffic flow (Sects. 8.2–8.5).
2. A numerical study of the effect of the classical traffic flow instability and the S→F instability on *critical nuclei* for phase transitions in traffic flow (Sects. 8.6–8.10).
3. A critical consideration of the classical theory of stochastic highway capacity versus the stochastic highway capacity definition made in the three-phase theory (Sect. 8.11).

8.2 Classical Traffic Flow Instability Versus S→F Instability of Three-Phase Theory

The basic difference between the classical traffic flow instability discovered in 1958–1961 by Herman, Gazis, Montroll, Potts, Rothery, and Chandler [8, 19, 20, 29] and further incorporated in a huge number of traffic flow models of the GM model class [1, 2, 9, 17, 18, 22, 23, 28, 35, 52, 88, 90, 93–98, 100–104, 112, 113, 115, 116] (Sect. 4.4.1) and the S→F instability introduced in the three-phase theory [38–40, 52, 62] (Sects. 5.12 and 5.13) is as follows.

The classical traffic flow instability is a growing wave of a local **decrease** in the speed in traffic flow (Fig. 8.1a).

Contrary to classical traffic flow instability, the S→F instability of the three-phase theory is a growing wave of a local **increase** in the speed in synchronized flow (Fig. 8.1b,c).

The classical traffic flow instability [1, 2, 8, 9, 17–20, 22, 23, 28, 29, 35, 88, 90, 93–98, 100–104, 112, 113, 115, 116] should explain traffic breakdown through the driver reaction time (time delay in driver over-deceleration). However, when the classical traffic flow instability occurs in free flow, the instability leads to a phase transition from free flow to a wide moving jam (F→J transition) (Fig. 8.2a) [1, 2, 9, 17, 18, 28, 35, 52, 62, 80, 88, 93–97, 112, 113]. This contradicts real field traffic data in which traffic breakdown is an F→S transition (Chap. 3).

In contrast with the classical traffic flow instability, the S→F instability of the three-phase theory leads to the phase transition from synchronized flow to free flow (S→F transition) (Fig. 8.2b). In its turn, the S→F instability governs the occurrence of an F→S transition at a highway bottleneck (Sects. 5.12 and 5.13). As observed in all known real field traffic data, traffic breakdown is the F→S transition (Chap. 3).

8.3 Moving Jam Emergence in Classical Theories and Three-Phase Theory

In this section, based on results of simulations of two-phase and the three-phase traffic flow models we show that in the relation to the metastability of traffic flow with respect to the moving jam emergence there is no fundamental difference between classical traffic flow theories and the three-phase theory. However, before we consider simulation results, we discuss briefly in Sect. 8.3.1 the metastability of free flow with respect to F→J transition found in real field traffic data.

8.3.1 Empirical Metastability of Free Flow with Respect to F→J Transition

In 1996, Kerner and Rehborn revealed the empirical metastability of free flow with respect to an F→J transition (Fig. 8.3) [83]. In Fig. 8.3, there is a sequence of two moving jams propagating upstream on a road section with a bottleneck.

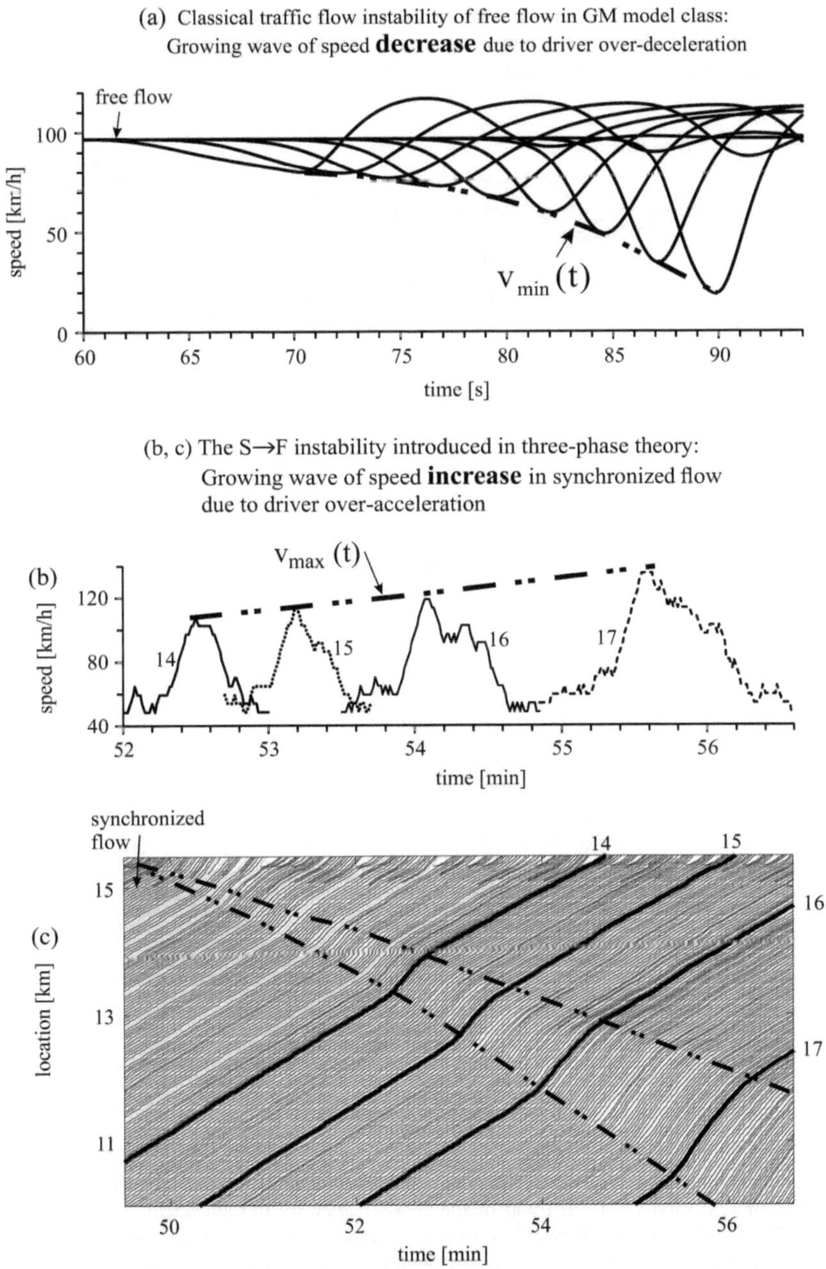

Fig. 8.1 Comparison of classical traffic flow instability of free flow of GM model class (**a**) with the S→F instability introduced in three-phase theory (**b**, **c**): (**a**) Growing wave of speed decrease due to driver over-deceleration (adapted from Fig. 4.5; each second trajectory is shown). (**b**, **c**) Growing wave of speed increase in synchronized flow due to driver over-acceleration as time-function (**b**) and on vehicle trajectories (**c**) (adapted from Figs. 5.24 and 5.28)

(a) Classical traffic flow instability of free flow in GM model class:
Growing wave of speed **decrease** due to driver over-deceleration
leading to the formation of wide moving jam

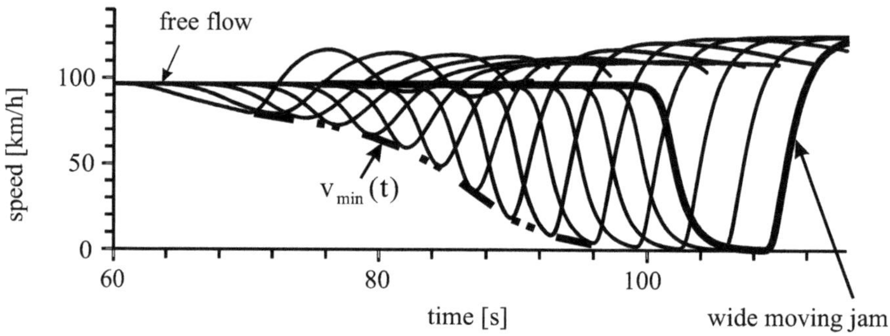

(b) The S→F instability introduced in three-phase theory:
Growing wave of speed **increase** in synchronized flow
due to driver over-acceleration leading to the transition
from synchronized flow to free flow

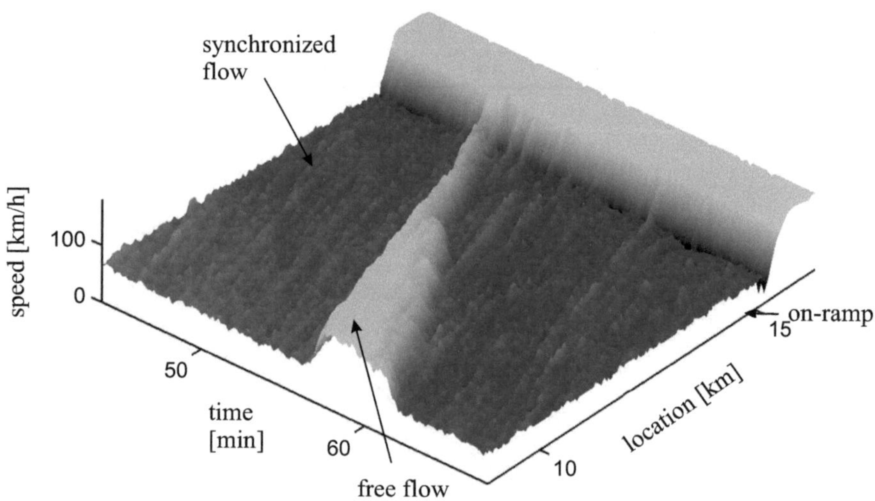

Fig. 8.2 Continuation of Fig. 8.1. Subsequent development of classical traffic flow instability of free flow of GM model class (**a**) and of the S→F instability introduced in three-phase theory (**b**): (**a**) The F→J transition: Emergence of a wide moving jam from the growing wave of speed decrease in free flow due to driver over-deceleration (continuation of Fig. 8.1a). (**b**) The S→F transition: Emergence of free flow from the growing wave of speed increase in synchronized flow due to driver over-acceleration (continuation of Figs. 8.1b,c)

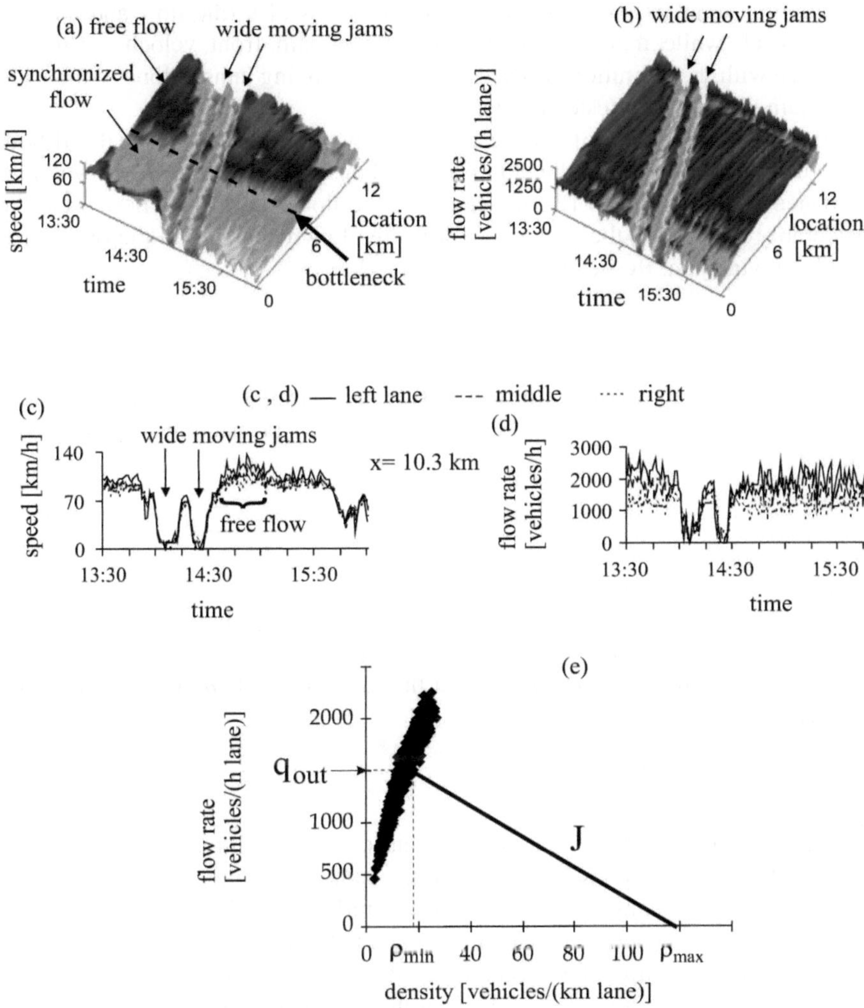

Fig. 8.3 Empirical metastability of free flow with respect to F→J transition: (**a**, **b**) Vehicle speed (**a**) and flow rate averaged across all road lanes (**b**) as functions of time and space. (**c**, **d**) The average vehicle speed (**c**) and flow rate (**d**) in different road lanes measured at location $x = 10.3$ km in (**a**, **b**). (**e**) Free flow (*black diamonds*) and the line J in the flow–density plane; free flow is measured at location $x = 10.3$ km in (**a**, **b**); line J represents the propagation of the downstream front of the downstream jam in the sequence of two wide moving jams in (**a**, **b**). 1-min average data measured by road detectors on the freeway A5-North in Germany on October 09, 1992. Adapted from [83]

These moving jams propagate through different states of traffic flow and through the bottleneck while maintaining the downstream jam front velocity. Thus, in accordance with the definition [J] (Sect. 1.3), these moving jams belong to the wide moving jam phase of congested traffic.

During the propagation of a wide moving jam, free flow can be realized both upstream and downstream of the jam. Empirical observations [83–87] show that in accordance with formula (4.12) of Sect. 4.5.1 the flow rate in free flow q can be considerably larger than the flow rate q_{out} in free flow occurring in the outflow of the wide moving jam. Respectively, condition

$$q \geq q_{\text{out}} \tag{8.1}$$

can be satisfied (Fig. 8.3e). Empirical observations [83–87] confirm that under condition (8.1) the wide moving jam persists while propagating on the road. Contrarily, if

$$q < q_{\text{out}}, \tag{8.2}$$

the wide moving jam dissolves over time.

Due to this empirical evidence, we can state that free flow in which the flow rate q in free flow satisfies condition (8.1) is metastable free flow with respect to an F→J transition. Indeed, under condition (8.1) at the same flow rate q at least two traffic phases can exist:

(i) The free flow phase.
(ii) The wide moving jam phase.

This means that the free flow phase is in a metastable state with respect to the F→J transition: Because two phases can exist at the same flow rate q (8.1) in an initial free flow, there can always be a local speed disturbance in this free flow that is a nucleus for the emergence of a wide moving jam in the free flow.

The important result of the observations of the metastability of free flow with respect to the F→J transition [83–87] is that the threshold flow rate in free flow for induced F→J transition is equal to the jam outflow rate q_{out}.

Indeed, under condition (8.1) a wide moving jam can be induced in free flow. This is because the wide moving jam can exist in free flow. Contrarily, if the opposite condition (8.2) is satisfied, no steadily propagation of a wide moving jam is possible in free flow: Any moving jam dissolves over time. This means that no moving jam that persists over time while propagating in free flow can be induced in free flow. This is because there is no wide moving jam phase in this free flow: Under condition (8.2), free flow is stable with respect to an F→J transition. Thus, under condition

$$q = q_{\text{out}}, \tag{8.3}$$

an F→J transition can still be induced in free flow.

8.3.2 Probability of Spontaneous F→J Transitions at On-Ramp Bottleneck in Two-Phase Model

To show common features of the two-phase and three-phase traffic flow models with respect to the emergence of moving jams, firstly we should discuss features of spontaneous emergence of moving jams in a two-phase model. Indeed, we know that in two-phase traffic flow models of the GM model class, an F→J transition can occur *spontaneously* in the metastable free flow at a highway bottleneck (see Fig. 4.8 of Sect. 4.4) [28, 52, 61–64, 66, 80–82, 96, 97, 113].

It should be noted that formulas (8.1)–(8.3) result from formula (4.12) of Sect. 4.5.1. Formula (4.12) is related to the moving jam emergence in free flow on a homogeneous road without bottlenecks (see Fig. 4.10). Below, we consider simulations of an F→J transition at an on-ramp bottleneck. In this case, the flow rate q in free flow in formula (4.12) should be replaced by the flow rate in free flow at the bottleneck $q_{\rm sum} = q_{\rm in} + q_{\rm on}$. Respectively, the critical flow rate $q_{\rm cr}^{\rm (J)}$ in formula (4.12) should be replaced by another critical flow rate denoted by $q_{\rm cr}^{\rm (B)}$. Consequently, formula (4.12) should be replaced by formula

$$q_{\rm out} \leq q_{\rm sum} < q_{\rm cr}^{\rm (B)}, \tag{8.4}$$

where, as in (4.12), $q_{\rm out}$ is the characteristic flow rate related to free flow in the wide moving jam outflow. Formula (8.4) determines the flow rate range within which free flow is in a metastable state with respect to the moving jam emergence at the bottleneck.

A difference between the critical flow rate $q_{\rm cr}^{\rm (J)}$ in (4.12) and the critical flow rate $q_{\rm cr}^{\rm (B)}$ in (8.4) is as follows. At the critical flow rate $q_{\rm cr}^{\rm (J)}$ in (4.12) free flow on the *homogeneous road without bottlenecks* becomes unstable with respect to the classical traffic flow instability of Herman, Gazis, Montroll, Potts, Rothery, and Chandler [8, 19, 20, 29]. At the critical flow rate $q_{\rm cr}^{\rm (B)}$ in (8.4) free flow at the *bottleneck* becomes unstable with respect to this classical traffic flow instability. In other words, conditions (8.4) are related to the emergence of moving jams due to the spatiotemporal development of local speed disturbances that occur initially at the bottleneck. Simulations show that values $q_{\rm cr}^{\rm (J)}$ and $q_{\rm cr}^{\rm (B)}$ can be different at the same other model parameters. This difference is explained by the effect of the bottleneck on the development of the classical traffic flow instability of Herman, Gazis, Montroll, Potts, Rothery, and Chandler [8, 19, 20, 29].[1]

Simulations of spontaneous F→J transition at the bottleneck with the Nagel-Schreckenberg CA model are shown in Fig. 8.4. As a result of the F→J transition, wide moving jams emerge in the two-phase model (Fig. 8.4). We come also to another well-known result [28, 52, 61–64, 66, 82, 96, 97, 113] found firstly

[1]Because a detailed consideration of the moving jam emergence is out of scope of the book, we do not discuss the effect of the bottleneck on the value of the critical flow rate $q_{\rm cr}^{\rm (B)}$.

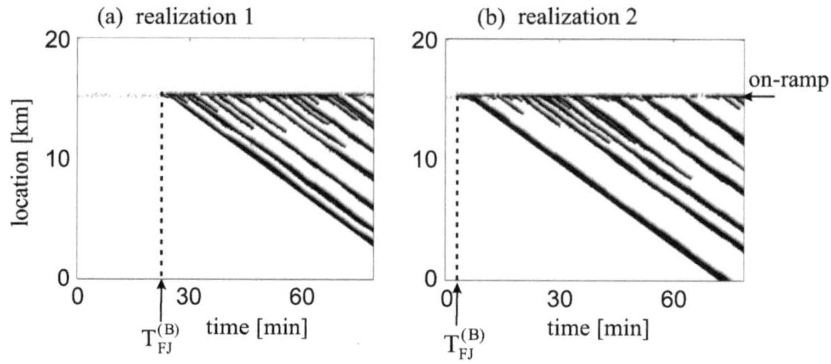

Fig. 8.4 Simulations of spontaneous F→J transition at the on-ramp bottleneck in Nagel-Schreckenberg CA model ("two-phase model"): (**a**, **b**) Speed data for two different simulation realizations presented by regions with variable shades of gray (shades of gray vary from white to black when the speed decreases from 120 km/h (*white*) to zero (*black*)). $q_{in} = 1731$ vehicles/h, $q_{on} = 400$ vehicles/h. Time delays of the F→J transition at the on-ramp bottleneck are: $T_{FJ}^{(B)} = 22$ min in (**a**) and 3 min in (**b**). Adapted from [78]

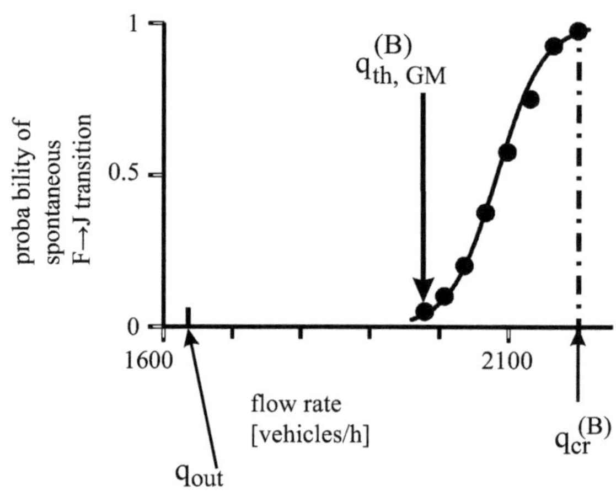

Fig. 8.5 Probabilistic features of spontaneous F→J transition at the on-ramp bottleneck in two-phase model: Dependence of probability of spontaneous F→J transition $P_{GM}^{(B)}(q_{sum})$ on the flow rate $q_{sum} = q_{in} + q_{on}$ at given value of on-ramp inflow rate $q_{on} = 400$ vehicles/h. Adapted from [78]

in [80, 81]: There is a limited flow rate range within which free flow is in a metastable state to respect to the F→J transition. This flow rate rage is given by conditions (8.4).

In [78], the probability of spontaneous F→J transition $P_{GM}^{(B)}(q_{sum})$ as a flow-rate function has been calculated (Fig. 8.5). The probability $P_{GM}^{(B)}(q_{sum})$ is an increasing flow rate function (Fig. 8.5). The probability $P_{GM}^{(B)}(q_{sum})$ is related to the occurrence of the F→J transition at the bottleneck during a given time interval T_{ob} in n_{r} of N_{r} simulation realizations (runs):

$$P_{\mathrm{GM}}^{(\mathrm{B})}(q_{\mathrm{sum}}) = \frac{n_{\mathrm{r}}}{N_{\mathrm{r}}}, \tag{8.5}$$

where N_{r} is number of realizations used for probability calculations.

It has been found [78] that there is a flow rate range

$$q_{\mathrm{th,\,GM}}^{(\mathrm{B})} \leq q_{\mathrm{sum}} < q_{\mathrm{cr}}^{(\mathrm{B})} \tag{8.6}$$

within which the probability $P_{\mathrm{GM}}^{(\mathrm{B})}(q_{\mathrm{sum}})$ of the occurrence of spontaneous F→J transition during a time interval for the observing traffic flow T_{ob} satisfies conditions

$$0 < P_{\mathrm{GM}}^{(\mathrm{B})}(q_{\mathrm{sum}}) < 1. \tag{8.7}$$

Therefore, the critical flow rate $q_{\mathrm{sum}} = q_{\mathrm{cr}}^{(\mathrm{B})}$ in (8.4) and (8.6) has the following meaning: At $q_{\mathrm{sum}} = q_{\mathrm{cr}}^{(\mathrm{B})}$ spontaneous F→J transition at the bottleneck occurs during the time interval T_{ob} with the probability

$$P_{\mathrm{GM}}^{(\mathrm{B})}(q_{\mathrm{cr}}^{(\mathrm{B})}) = 1. \tag{8.8}$$

Two different simulation realizations 1 and 2 shown in Fig. 8.4 are related to $P_{\mathrm{GM}}^{(\mathrm{B})} = 0.775$, i.e., conditions (8.7) are satisfied. It has been found that in different simulation realizations, time delays to the F→J transition denoted by $T_{\mathrm{FJ}}^{(\mathrm{B})}$ are random values. In other words, the F→J transition is a random time-delayed transition from free flow to a wide moving jam at the bottleneck. This conclusion is the result of the metastability of free flow with respect to the F→J transition in traffic flow models of the GM model class.

Because for model parameters used in Fig. 8.4 the breakdown probability $P_{\mathrm{GM}}^{(\mathrm{B})}$ satisfies conditions (8.7), we have indeed found that additionally to simulation realizations in which traffic breakdown has occurred (realizations 1 and 2 in Figs. 8.4a,b), there are simulation realizations in which *no* traffic breakdown occurs at the bottleneck (realization 3 in Fig. 8.6a).

However, as follows from (8.4) under condition condition (8.7) free flow is in a metastable state with respect to the F→J transition. Therefore, if no spontaneous F→J transition occurs during the time interval T_{ob}, as this is realized in simulation realization 3, an F→J transition should be possible to induce in this simulation realization. This conclusion is confirmed in numerical simulations shown in Fig. 8.6b [78].

In (8.6), $q_{\mathrm{th,\,GM}}^{(\mathrm{B})}$ is a threshold flow rate for *spontaneous* moving jam emergence (spontaneous F→J transition) (Fig. 8.5). The threshold flow rate $q_{\mathrm{th,\,GM}}^{(\mathrm{B})}$ has the following physical meaning. Under condition

$$q_{\mathrm{sum}} < q_{\mathrm{th,\,GM}}^{(\mathrm{B})}, \tag{8.9}$$

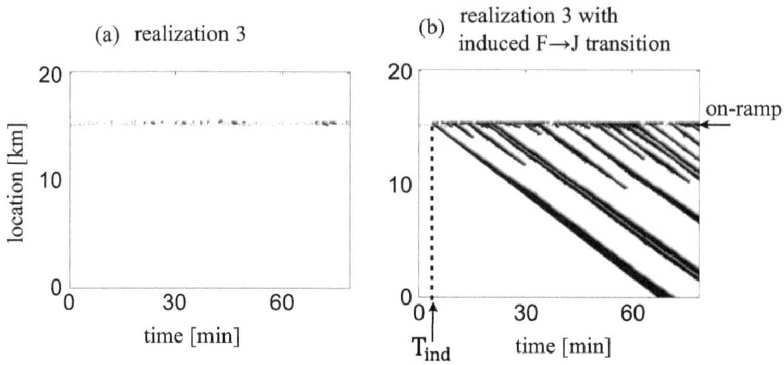

Fig. 8.6 Continuation of Fig. 8.4: (**a**) Simulation realization 3 related to the same set of the flow rates q_{in} and q_{on} as those in Fig. 8.4. (**b**) F→J transition has been induced in realization 3 (**a**) due to the impulse of on-ramp inflow $\Delta q_{on} = 200$ vehicles/h of duration $\Delta t_{ind} = 1$ min that is applied at $T_{ind} = 3$ min (marked by *vertical dashed line*). Speed data presented by regions with variable shades of gray (shades of gray vary from white to black when the speed decreases from 120 km/h (*white*) to zero (*black*))

the probability of spontaneous F→J transition at the bottleneck

$$P^{(B)}_{GM}(q_{sum}) = 0. \tag{8.10}$$

We know that when conditions (8.4) are satisfied, then free flow is in a metastable state with respect to the F→J transition. Under condition (8.9) no spontaneous F→J transition is possible during the time interval T_{ob} and free flow is metastable with respect to the F→J transition. This means that under condition

$$q_{out} \leq q_{sum} < q^{(B)}_{th,\, GM} \tag{8.11}$$

the F→J transition can only be *induced* at the bottleneck (see Sect. 8.7.1 below).

8.3.3 S→J Transition in Two-Phase and Three-Phase Traffic Flow Models

As above-mentioned, simulations show that an F→J transition does not occur spontaneously in a three-phase traffic flow model. In the three-phase model, moving jam emerge spontaneously only in synchronized flow (spontaneous S→J transition). Hypothesis 12 of the three-phase theory (Sect. 1.9) explains the S→J transition as follows (Fig. 8.7).

Synchronized flow states that lie below the line *J* are stable with respect to S→J transitions (labeled by “stable synchronized flow with respect to S→J transition” in Fig. 8.7). Synchronized flow states that lie on or above the line *J* are metastable

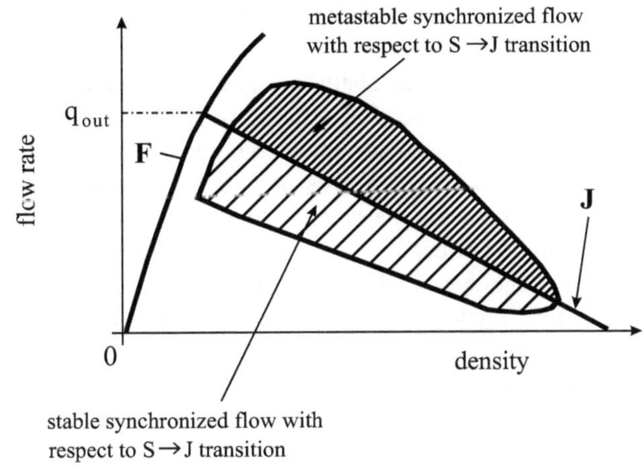

Fig. 8.7 Hypothesis 12 (Sect. 1.9) of the three-phase theory [36–40, 42]: Qualitative presentation of states of free flow (F), steady states of synchronized flow (*2D dashed region*) adapted from Fig. 5.15a, and the line *J* in the flow–density plane

with respect to an S→J transition (labeled by "metastable synchronized flow with respect to S→J transition" in Fig. 8.7).[2]

An S→J transition is explained by a competition between the speed adaptation effect (Sect. 5.9), which describes a tendency towards the metastable synchronized flow with respect to the S→J transition, and an over-deceleration effect (Sect. 4.4.1), which describes a tendency towards a wide moving jam [37, 52, 61–64, 70, 71].

When the tendency towards a wide moving jam overcomes on average the tendency towards synchronized flow, an S→J instability occurs in the metastable synchronized flow. The S→J instability is a growing wave of a local decrease in the speed in synchronized flow. The S→J instability leads to an S→J transition. The over-deceleration effect that leads to the development of the S→J instability is also responsible for the classical traffic flow instability of Herman, Gazis, Montroll, Potts, Rothery, and Chandler [8, 19, 20, 29] (Sect. 4.4.1). For this reason, the metastability of synchronized flow with respect to an S→J transition is associated with the classical traffic flow instability of Herman, Gazis, Montroll, Potts, Rothery, and Chandler.

Simulations show [52, 61–64, 70, 71] that the S→J instability occurs in synchronized flow, when a nucleus for the S→J transition appears in a metastable state of synchronized flow with respect to S→J transition ("metastable synchronized flow with respect to S→J transition" in Fig. 8.7). To understand the nucleus for the S→J transition (wide moving jam emergence in synchronized flow), we consider a local speed disturbance of speed reduction in metastable synchronized flow with respect

[2]A more detailed explanation of the stable and metastable synchronized flow states with respect to the S→J transition can be found in Chap. 6 of the book [52] and Chap. 5 of the book [62].

to the S→J transition: The speed within this local speed disturbance is smaller than outside the disturbance.

The nucleus is a such local speed disturbance of speed reduction in synchronized flow within which the average speed is equal to or smaller than the average speed within a critical nucleus required for the S→J transition. Within a nucleus for the S→J transition the tendency towards a wide moving jam due to the over-deceleration effect overcomes on average the tendency towards synchronized flow due to the speed adaptation effect.

We denote the average speed within the critical nucleus by $v_{\text{cr, SJ}}$. The speed $v_{\text{cr, SJ}}$ within the critical nucleus for the S→J transition is a function of the speed and density in the initial synchronized flow in which the nucleus appears.

Otherwise, when the speed within a local speed disturbance of speed reduction in synchronized flow is larger than the speed $v_{\text{cr, SJ}}$ within the critical nucleus for the S→J transition, the disturbance decays over time. In this case, within the disturbance the tendency towards synchronized flow due to the speed adaptation effect overcomes on average the tendency towards a wide moving jam due to the over-deceleration effect.

The S→J transition and a return J→S transition can be presented by a Z-characteristic for S→J and J→S phase transitions (Fig. 8.8). The Z-characteristic for S→J and J→S phase transitions is a macroscopic characteristic presenting these phase transitions in the speed–flow-rate plane. The Z-characteristic consists of 2D-states of synchronized flow (S), the average (macroscopic) value of the speed $v_{\text{cr, SJ}}$

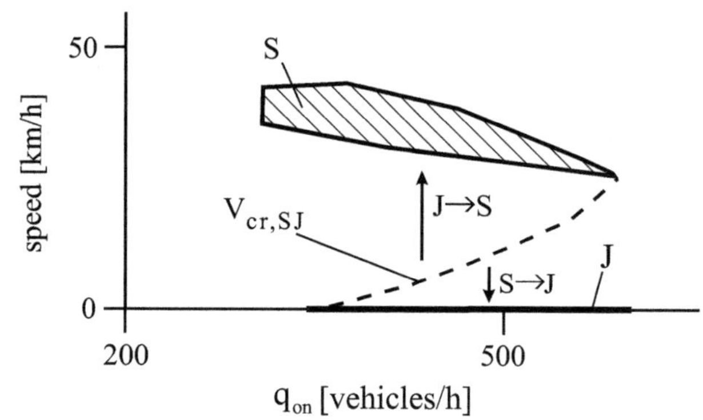

Fig. 8.8 Simulation of Z-characteristic for S→J and J→S phase transitions with the Kerner-Klenov stochastic microscopic three-phase model: Z-characteristic as function of the on-ramp inflow rate q_{on} and given flow rate upstream of the on-ramp bottleneck $q_{\text{in}} = 1850$ vehicles/h. S—synchronized flow phase, J—wide moving jam phase. Arrows S→J and J→S illustrate symbolically the related phase transitions between the phases S and J. The dashed curve shows the flow-rate dependence of the average (macroscopic) speed $v_{\text{cr, SJ}}$ within a critical nucleus required for S→J transition. Adapted from [71]

within the critical nucleus for the S→J transition (dashed curve in Fig. 8.8), and states with the speed $v = 0$ related to a wide moving jam.[3]

The arrow labeled by "S→J" in Fig. 8.8 shows symbolically the S→J transition. As mentioned, the S→J transition occurs only if a nucleus for the S→J transition appears in metastable states of synchronized flow with respect to the S→J transition.

When we have discussed an Z-characteristic for S→J and J→S phase transitions (Fig. 1.9), we have assumed that traffic breakdown has already occurred at an on-ramp bottleneck. As a result of the breakdown, synchronized flow is formed upstream of the bottleneck. In Sect. 4.8, we have explained that there can be "combined" two-phase traffic flow models for which condition (4.25) is satisfied. In this case, as in the LWR-model and the Daganzo's cell transmission model, traffic breakdown at the bottleneck is associated with the formation of synchronized flow at the bottleneck (Figs. 4.12c,d).

It must be emphasized that there is no S→F instability in the synchronized flow of the combined two-phase traffic flow models. Therefore, there is also no metastability of free flow with respect to traffic breakdown in the combined two-phase traffic flow models.

Nevertheless, in the combined two-phase traffic flow models there can be the metastability of synchronized flow with respect to an S→J transition. Indeed, as above-mentioned, the metastability of synchronized flow with respect to the S→J transition is caused by the over-deceleration effect (driver reaction time). The over-deceleration effect that is responsible for the classical instability of Herman, Gazis, Montroll, Potts, Rothery, and Chandler [8, 19, 20, 29] has been incorporated in the combined two-phase traffic flow models (Sect. 4.8). For this reason, as three-phase traffic flow models, combined two-phase traffic flow models can also show the S→J instability occurring in the metastable synchronized flow with respect to the S→J transition (Fig. 4.12e).

Sugiyama et al. [108], Nakayama et al. [99], and Tadaki et al. [109] have made driver experiments in which moving jams emerge spontaneously or have been induced in synchronized flow on a circular road. For example, in [108] $N_{\text{veh}} = 22$ vehicles have moved along a single-lane circle road with $L_{\text{circle}} = 230$ m at the speed of about 30 km/h. There have been no bottlenecks on the single-lane circle road in the driver experiment of [108]. Therefore, in this driver experiment vehicles could not overtake each other and the vehicle speed (30 km/h) has been considerably lower than the speed in free flow (usually higher than 80 km/h for passenger cars). Moreover, the initial density in the initial traffic flow has been $\rho = N_{\text{veh}}/L_{\text{circle}} \approx 95$ vehicles/km. This vehicle density is considerably greater than the maximal density

[3]In reality, a wide moving jam exhibits a complex microscopic spatiotemporal structure that consists of alternations of moving blanks with flow interruption intervals [74–76]. For simplicity, moving blanks within the wide moving jam have been neglected in Fig. 8.8. Through this simplification of the real complex microscopic spatiotemporal structure within wide moving jams, the wide moving jam phase is related to the speed that is equal to zero (labeled by J in Fig. 8.8). A detailed consideration of moving blanks within wide moving jams is out of scope of this book (see [74–76] and Sec. 2.6 of the book [62]).

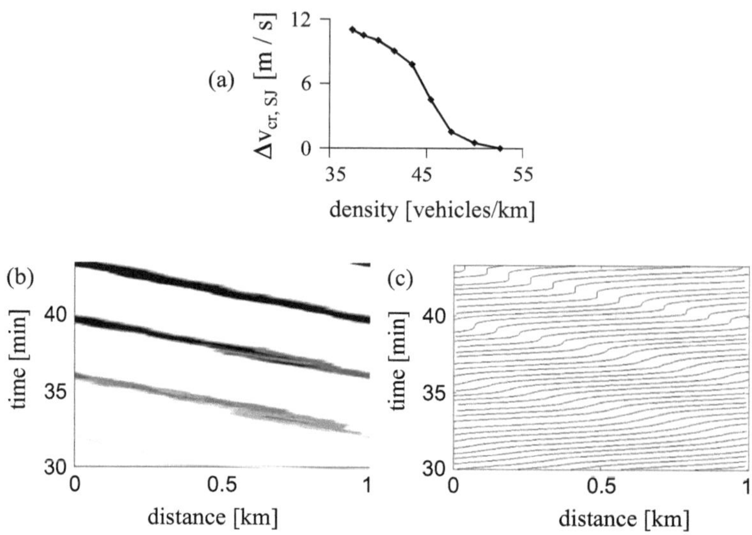

Fig. 8.9 Simulations of the emergence of wide moving jams in synchronized flow on circle road: (**a**) The amplitude of critical nucleus for S→J transition $\Delta v_{cr, SJ}$ as a function of the density at initial synchronized flow speed 40 km/h (for a comparison, the free flow speed $v_{free} = 108$ km/h). (**b, c**) Single-vehicle speed data (**b**) and vehicle trajectories (for each 10th vehicle) (**c**) showing the spontaneous emergence of wide moving jam in synchronized flow in space and time at $N_{veh} = 48$ vehicles, $L_{circle} = 1$ km; vehicles start to move from a vehicle standstill at $t = 0$; resulting flow rate $q = 1210$ vehicles/h. In (**b**), single-vehicle speed data is presented by regions with variable darkness (the lower the speed, the darker the region; in white regions the speed is higher than 20 km/h). Adapted from [60]

observed in free flow, which is usually appreciably smaller than 40 vehicles/(km lane). Therefore, the initial traffic flow in the experiment of [99, 108, 109] is *congested traffic associated with the synchronized flow phase.*

To compare results of the hypothesis of the three-phase theory about the S→J transition discussed above (Fig. 8.7) with the driver experiments of [99, 108, 109], we present simulations of the Kerner-Klenov stochastic microscopic model (Fig. 8.9) [60]. These simulations have been made at approximately the same parameters of synchronized flow as those used in real driver experiments of [99, 108, 109]. We have found that as in the driver experiment [108], in the simulations the spontaneous emergence of wide moving jam(s) in initial synchronized flow (S→J transition) is observed (Fig. 8.9).

To study the metastability of synchronized flow with respect to the emergence of wide moving jams, in simulations (Fig. 8.9) [60] a local speed disturbance of speed reduction has been initiated in an initial homogeneous synchronized flow with the vehicle speed v_{syn}. To initiate the local speed disturbance in this synchronized flow, one of the vehicles has decelerated during a short time interval. We denote the speed within the local disturbance resulting from this vehicle deceleration by $v^{(syn)}_{dis}$. Within the local disturbance, the local decrease in the speed in comparison with the initial speed in synchronized flow is equal to some value $\Delta v_{\rm syn} = v_{\rm syn} - v^{\rm (syn)}_{\rm dis}$, where $\Delta v_{\rm syn} > 0$. It has been found that in accordance with features of the metastability of synchronized flow with respect to an S→J transition discussed above, there is a critical nucleus required for the S→J transition. The critical nucleus is a critical local speed disturbance of speed reduction required for the S→J transition in synchronized flow. The amplitude of the critical nucleus is equal to $\Delta v_{\rm cr,\ SJ} = v_{\rm syn} - v_{\rm cr,\ SJ}$ (Fig. 8.9a).

If $\Delta v_{\rm syn} \geq \Delta v_{\rm cr,\ SJ}$ (Fig. 8.9a), the initial local disturbance in synchronized flow transforms into a growing wave of the local decrease in the speed called a growing narrow moving jam. During the upstream jam propagation, the narrow jam transforms into a wide moving jam, i.e., the S→J transition occurs. If in contrast, $\Delta v_{\rm syn} < \Delta v_{\rm cr,\ SJ}$, then the local speed disturbance decays over time. Simulations show that the nucleus appearance and the associated S→J transition are *only* possible in metastable synchronized flow associated with synchronized flow states that lie on or above the line J in the flow-density plane (Fig. 8.7).

If the density of synchronized flow is great enough, then $\Delta v_{\rm cr,\ SJ} \rightarrow 0$ (Fig. 8.9a). This means that in such an initially homogeneous synchronized flow on a circle road without bottlenecks already small speed (density) disturbances, which occur spontaneously in synchronized flow due to model fluctuations, grow propagating upstream with the subsequent spontaneous emergence of a wide moving jam (Fig. 8.9b,c). Thus, the features of S→J transitions found in the three-phase theory explain empirical results found in the driver experiment of [99, 108, 109].

The consideration of S→J transitions made above allows us to assume that a Z-characteristic for S→J and J→S phase transitions can also be derived with the use of combined two-phase traffic flow models. A possible difference of such a Z-characteristic for S→J and J→S phase transitions derived with combined two-phase traffic flow models in comparison with the Z-characteristic shown in Fig. 8.8 for the three-phase model can be as follows. In the three-phase theory, there are 2D-states of synchronized flow, whereas for known combined two-phase traffic flow models there are 1D-states of synchronized flow. It should be stressed that this difference is not important for the aim of the explanation of the incommensurability of the classical theories with the three-phase theory: There is no fundamental difference between combined two-phase traffic flow models and three-phase traffic flow models in relation to the metastability of traffic flow with respect to the emergence of wide moving jams either in free flow (F→J transition) or in synchronized flow (S→J transition).

It must be noted that the statement that both two-phase and three-phase traffic flow models exhibit common features of the emergence of moving jam in traffic flow does *not* mean that dynamic behavior of congested traffic is qualitatively the same in the two-phase and three-phase traffic flow models. This explains why we use the term "look very similar", when congested traffic patterns simulated by different two-phase traffic flow models are compared with empirical spatiotemporal congested pattern (see Sect. 1.13). A criticism of the description of traffic congestion with two-phase traffic flow models can be found in Sec. 10.3.11 of the book [62].

To study the development of the S→J instability, Jiang et al. [32, 33, 110, 111] have performed a microscopic traffic experiment of the classical traffic flow instability of Herman, Gazis, Montroll, Potts, Rothery, and Chandler [8, 19, 20, 29]. As in the driver experiment of Sugiyama, Nakayama, Tadaki et al. [99, 108, 109], in the driver experiment of Jiang et al. [32, 33, 110, 111] the spontaneous emergence of moving jams has been found when an initial traffic flow is related to the synchronized flow traffic phase (S→J transition). Indeed, the spontaneous emergence of moving jams has been observed only then, when the initial vehicle speed of drivers has been smaller than the speed about 40 km/h [32, 33, 110, 111]. Therefore, as in the driver experiment of [99, 108, 109], the initial vehicle speed of traffic flow in the driver experiment of [32, 33, 110, 111] is related to the synchronized flow traffic phase.

However, in contrast with the driver experiment of Sugiyama, Nakayama, Tadaki et al. [99, 108, 109] that has been done on a circular road, the driver experiment of Jiang et al. [32, 33, 110, 111] has been done under conditions that can simulate traffic flow on an open section of a road as it occurs in real traffic. In the microscopic traffic experiment of [32, 33, 110, 111], car-following behavior in a 25-car-platoon and a 51-car-platoon has been studied. Jiang et al. [32, 33, 110, 111] have discovered that speed disturbances in synchronized flow (nuclei for S→J transitions) grow in a *concave way* along vehicles in the platoon leading the moving jam emergence. In [32, 33, 110, 111] have been shown that two-phase traffic flow models cannot simulate this experimental dynamic behavior of the growing nuclei for S→J transitions. As found in [32–34, 110, 111], to simulate the concave way of the growth of the nuclei for S→J transitions, a traffic flow model should exhibit the speed adaptation effect associated with 2D-states of synchronized flow of the three-phase theory (Sect. 5.9). In other words, both studies of traffic congestion presented in [62] and in [32–34, 110, 111] show that dynamic features of congested traffic can depend qualitatively on the description of synchronized flow states by a traffic flow model. However, as above-mentioned, a detailed consideration of the dynamic behavior of the S→J transition is out of scope of this book.

The reason for this limitation made in the book is that the incommensurability of the classical theories with the three-phase theory has *no* relation to the emergence of moving jams in traffic flow (F→J and S→J transitions) as well as to other features of congested traffic. However, the proof of the incommensurability of the classical theories with the three-phase theory is the objective of this chapter.

Both two-phase and three-phase traffic flow models exhibit common features of the metastability of traffic flow with respect to the emergence of wide moving jams in traffic flow. There is no basic difference between two-phase and three-phase traffic flow models in relation to the metastability of traffic flow with respect to the emergence of wide moving jams either in free flow (F→J transition) or in synchronized flow (S→J transition).

The incommensurability of the classical traffic theories with the three-phase theory is associated with the description of the nucleation nature of traffic breakdown at a highway bottleneck, rather than with the description of congested traffic resulting from traffic breakdown. The proof of the incommensurability of the classical traffic theories with the three-phase theory is the subject of Sects. 8.4 and 8.7–8.10 presented below.

8.4 General Congested Patterns Resulting from Sequence of Two Different Time-Delayed Transitions in Three-Phase Models

8.4.1 F→S→J Transitions

As explained in Chap. 5, traffic breakdown is a random time-delayed F→S transition at a highway bottleneck. An WSP shown in Fig. 8.10a that results from the F→S transition at an on-ramp bottleneck at a single-lane road exhibits qualitatively the same features as already studied in details in Sects. 5.12 and 5.13: In particular, a random time delay $T^{(B)}$ of the F→S transition is governed by the S→F instability introduced in the three-phase theory.

As in Sects. 5.12 and 5.13, when the flow rate q_{in} in free flow upstream of the on-ramp bottleneck is large enough but the on-ramp inflow rate q_{on} is not large, the WSP occurs at the bottleneck (Fig. 8.10a). The flow rate within the WSP is not considerably smaller than the flow rate q_{in} (Fig. 8.11a, right). Respectively, the mean flow rate in congested traffic denoted by q_{cong} that has been measured by a virtual detector in synchronized flow 500 m upstream of the beginning of the on-ramp merging region is only 125 vehicles/h smaller than the flow rate $q_{\text{in}} = 2000$ vehicles/h (Fig. 8.12a). Correspondingly, the flow rate difference (flow rate drop)

$$\Delta q = q_{\text{sum}} - q_{\text{out}}^{(\text{bottle})} \tag{8.12}$$

is about 5.6% of the flow rate q_{sum} (Fig. 8.12b). This corresponds to the empirical findings discussed in Sect. 2.2.2. In (8.12), $q_{\text{sum}} = q_{\text{in}} + q_{\text{on}}$ is the flow rate before traffic breakdown has occurred (pre-discharge flow rate) (measured at time interval $0 < t < T^{(B)}$ in Fig. 8.10) and $q_{\text{out}}^{(\text{bottle})}$ is the mean flow rate after the onset of congestion (discharge flow rate) (measured at $t > T^{(B)}$ in Fig. 8.10); the flow rates q_{sum} and $q_{\text{out}}^{(\text{bottle})}$ have been measured by a virtual detector in free flow downstream of the bottleneck.

When at the same flow rate q_{in} the on-ramp inflow rate q_{on} increases as known (see references in the book [52]) the average speed in synchronized flow decreases. In synchronized flow of a smaller speed, the pinch effect is realized: The vehicle density increases and nuclei for the emergence of wide moving jams (S→J transitions) appear randomly in synchronized flow. The nuclei for S→J transitions

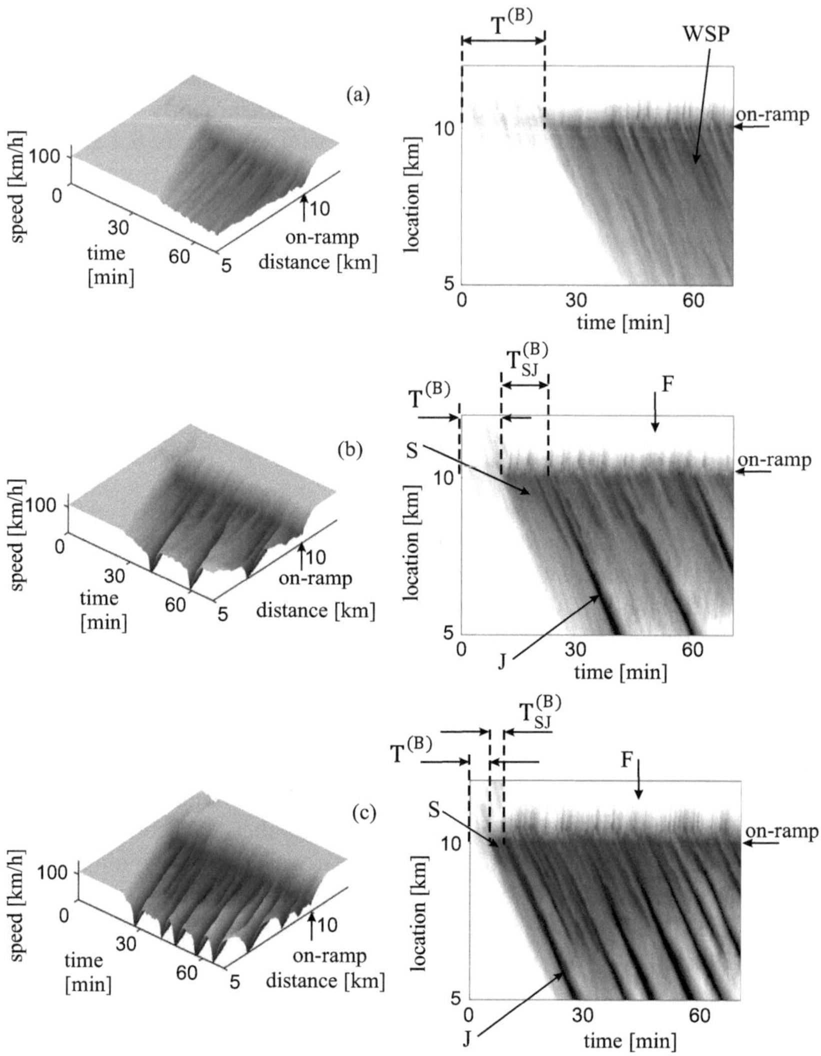

Fig. 8.10 Simulations of the F→S→J transitions at on-ramp bottleneck on single-lane road with the Kerner-Klenov model. Speed in space and time (*left column*) and the same data presented by regions with variable shades of gray (shades of gray vary from white to black when the speed decreases from 105 km/h (*white*) to zero (*black*)) (*right column*). In simulations, the same flow rate in free flow upstream of the on-ramp $q_{\rm in}$ = 2000 vehicles/h and different on-ramp inflow rates have been used: (**a**) $q_{\rm on}$ = 250, (**b**) $q_{\rm on}$ = 350, and (**c**) $q_{\rm on}$ = 550 vehicles/h. Random time delays of traffic breakdown (F→S transition) $T^{\rm (B)}$ and random time delays of S→J transitions $T^{\rm (B)}_{\rm SJ}$ have taken the following values in simulation realizations shown in (**a**–**c**): $T^{\rm (B)}$ = 25 min in (**a**); $T^{\rm (B)}$ = 13 min and $T^{\rm (B)}_{\rm SJ}$ = 15 min in (**b**); $T^{\rm (B)}$ = 7 min and $T^{\rm (B)}_{\rm SJ}$ = 6 min in (**c**). F—free flow, S—synchronized flow, J—wide moving jam, WSP—widening synchronized flow pattern. $x_{\rm on}$ = 10 km and $x^{\rm (e)}_{\rm on}$ = 10.3 km are, respectively, the beginning and the end of the merging region of the on-ramp which which vehicles can merge from the on-ramp onto the main road. Model parameter of vehicle merging from the on-ramp on the main road $\lambda_{\rm b}$ = 0.7; other model parameters are the same as those in Tables A.1 and A.3 of Appendix A

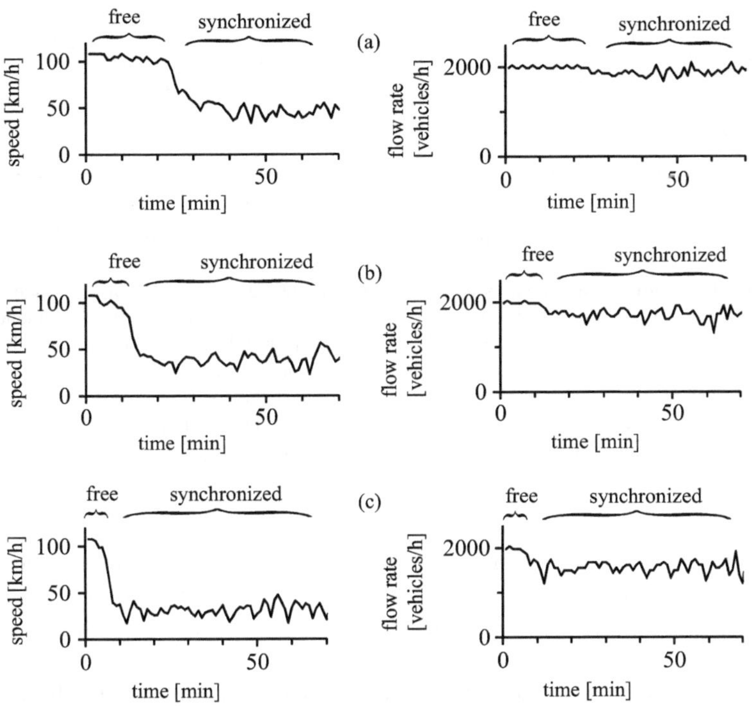

Fig. 8.11 Time-functions of the average speed (*left column*) and the flow rate (*right column*) in free flow and synchronized flow (1-min average data measured by a virtual detector at the location $x = 9.5$ km that is 500 m upstream of the beginning of the merging region of the on-ramp $x_{on} = 10$ km) for traffic patterns shown in Fig. 8.10: (**a**) For WSP in Fig. 8.10a. (**b**) For GP in Fig. 8.10b. (**c**) For GP in Fig. 8.10c

propagate upstream in synchronized flow growing in their amplitude. The growing nuclei have been called growing narrow moving jams [52]. Narrow moving jams transform into wide moving jams resulting in a general congested pattern (GP) that consists of the synchronized flow and wide moving jam phases (Fig. 8.10b,c).[4]

The larger the on-ramp inflow rate q_{on} at a given flow rate q_{in}, the smaller the average speed and the smaller the flow rate within the pinch region of the synchronized flow (Fig. 8.11b,c). For this reason, as well-known [52], the mean flow rate in congested traffic q_{cong} can become considerably smaller than the flow rate q_{in} (points 2 and 3 in Fig. 8.12a). Correspondingly, the flow rate drop $\Delta q = q_{sum} - q_{out}^{(bottle)}$ (8.12) between the pre-discharge flow rate q_{sum} and the discharge flow rate $q_{out}^{(bottle)}$ increases considerably (points 2 and 3 in Fig. 8.12b).

[4]The mean flow rate q_{cong} is measured within the pinch region of synchronized flow of the GP. In a theory of GPs, this flow rate has been denoted by $q^{(pinch)}$ (see Sec. 18.3 of the book [52]). Here we use a general designation q_{cong} used in [60] for the mean flow rate in congested traffic.

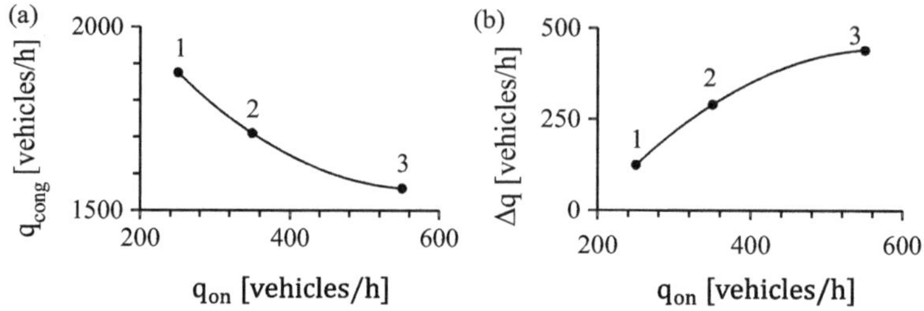

Fig. 8.12 Flow–flow characteristics for traffic patterns shown in Fig. 8.10: (**a**) Dependence of the mean flow rate in congested traffic q_{cong} on the on-ramp inflow rate q_{on}. (**b**) Dependence of the mean flow rate drop due to congested traffic Δq (8.12) in free flow downstream of the bottleneck (at $x = 11$ km) on the on-ramp inflow rate q_{on}. The mean flow rate in congested traffic q_{cong} averaged over 60 min is measured by a virtual detector at the road location $x = 9.5$ km that is 500 m upstream of the beginning of the merging region of the on-ramp $x_{\text{on}} = 10$ km. Point 1 is related to WSP in Fig. 8.10a, point 2 is related to GP in Fig. 8.10b, and point 3 is related to GP in Fig. 8.10c

The F→S transition is a random time-delayed phase transition. Examples of simulations of time-delayed traffic breakdown (F→S transition) are shown in Fig. 8.10. We can see that the F→S transition occurs after a random time delay $T^{(B)}$.

An S→J transition is also a random time-delayed phase transition: There is a time delay denoted by $T^{(B)}_{SJ}$ between the time instant $t = T^{(B)}$ of the F→S transition and a time instance $t = T^{(B)} + T^{(B)}_{SJ}$ of the appearance of a nucleus for the S→J transition in synchronized flow. The development of the nucleus leads to the emergence of the first wide moving jam in synchronized flow (Fig. 8.10b,c). A sequence of two time-delayed F→S and S→J transitions has been called as the F→S→J transitions in the three-phase theory [52] (hypothesis 7 of Sect. 1.9).

In accordance with Sect. 8.3.3, the S→J transition is governed by the classical traffic flow instability of the GM model class. For this reason, the time delay of this phase transition $T^{(B)}_{SJ}$ (Fig. 8.10b,c) has the same physical sense as the time delay of the F→J transition $T^{(B)}_{FJ}$ in traffic flow models of the GM motor class that examples are shown in Fig. 8.4. Thus, the S→J transitions in three-phase traffic flow models and combined two-phase traffic flow models (see Fig. 4.12c,d of Sect. 4.8) as well as the F→J transitions in other two-phase traffic flow models are the same random time-delayed phase transitions governed by the same classical traffic flow instability of Herman, Gazis, Montroll, Potts, Rothery, and Chandler [8, 19, 20, 29]. This explains why the spatiotemporal structure of congested traffic of the GPs in the three-phase model (Fig. 8.10b,c) after traffic breakdown has occurred at the bottleneck (i.e., at $t > T^{(B)}$) looks very similar to congested traffic patterns simulated with different two-phase traffic flow models [28, 112, 113].

As already emphasized in Sect. 1.13, rather than the spatiotemporal structure of congested traffic patterns, the basic difference between the three-phase and two-phase traffic flow models is the existence in the three-phase models of the time-delayed traffic breakdown (F→S transition). The time-delayed traffic breakdown is governed by the S→F instability of the three-phase theory. None of the two-phase traffic flow models incorporates the S→F instability. Therefore, none of the two-phase traffic flow models can be able to show the time-delayed F→S transition associated with the empirical nucleation nature of traffic breakdown at highway bottlenecks—the empirical fundamental of transportation science (Chap. 3).

8.4.2 Complexity of Phase Transitions in Vehicular Traffic

One of the important consequences that follows from the S→F instability introduced in the three-phase theory is illustrated in Fig. 8.13 that can be considered a continuation of simulations in Fig. 8.10 made at the same flow rate $q_{\text{in}} = 2000$ vehicles/h. We have chosen the on-ramp inflow rate $q_{\text{on}} = 300$ vehicles/h at which the probability of traffic breakdown (F→S transition) $P^{(\text{B})}$ during the time interval $T_{\text{ob}} = 30$ min is larger than zero but it is smaller than 1:

$$0 < P^{(\text{B})} < 1. \tag{8.13}$$

For this reason, there are simulation realizations at which no traffic breakdown occurs at the bottleneck at all (realization 1 in Fig. 8.13a). In this case, no moving jams can emerge, because in the three-phase traffic flow model moving jams can emerge spontaneously in synchronized flow only. Due to condition $P^{(\text{B})} > 0$ (8.13), there are also simulation realizations in which traffic breakdown occurs during the time interval T_{ob} (realizations 2–4 in Fig. 8.13b–d).

Due to traffic breakdown (F→S transition), synchronized flow emerges at the bottleneck (realizations 2–4 in Fig. 8.13b–d). Accordingly to results of Sect. 8.3.3, in synchronized flow an S→J transition leading to moving jam emergence can occur spontaneously. Indeed, simulations show that at the chosen flow rates q_{on} and q_{in} the probability of the S→J transition $P_{\text{SJ}}^{(\text{B})}$ during a chosen time interval $T_{\text{ob}}^{(\text{SJ})}$ for the observing of synchronized flow (we have chosen $T_{\text{ob}}^{(\text{SJ})} = 40$ min in Fig. 8.13b–d) is larger than zero but it is smaller than 1:

$$0 < P_{\text{SJ}}^{(\text{B})} < 1. \tag{8.14}$$

For this reason, there are simulation realizations in which after traffic breakdown has occurred at the bottleneck no nucleus for the S→J transition appears within synchronized flow. Therefore, rather than an GP appears, an WSP remains at the bottleneck (realization 2 in Fig. 8.13b).

Due to condition $P_{\text{SJ}}^{(\text{B})} > 0$ (8.14), there are also simulation realizations in which, after synchronized flow has emerged at the bottleneck, nuclei for an S→J transition appear within synchronized flow. This results in the transformation of an WSP into an GP over time (realizations 3 and 4 in Fig. 8.13c,d). Because both the time delay

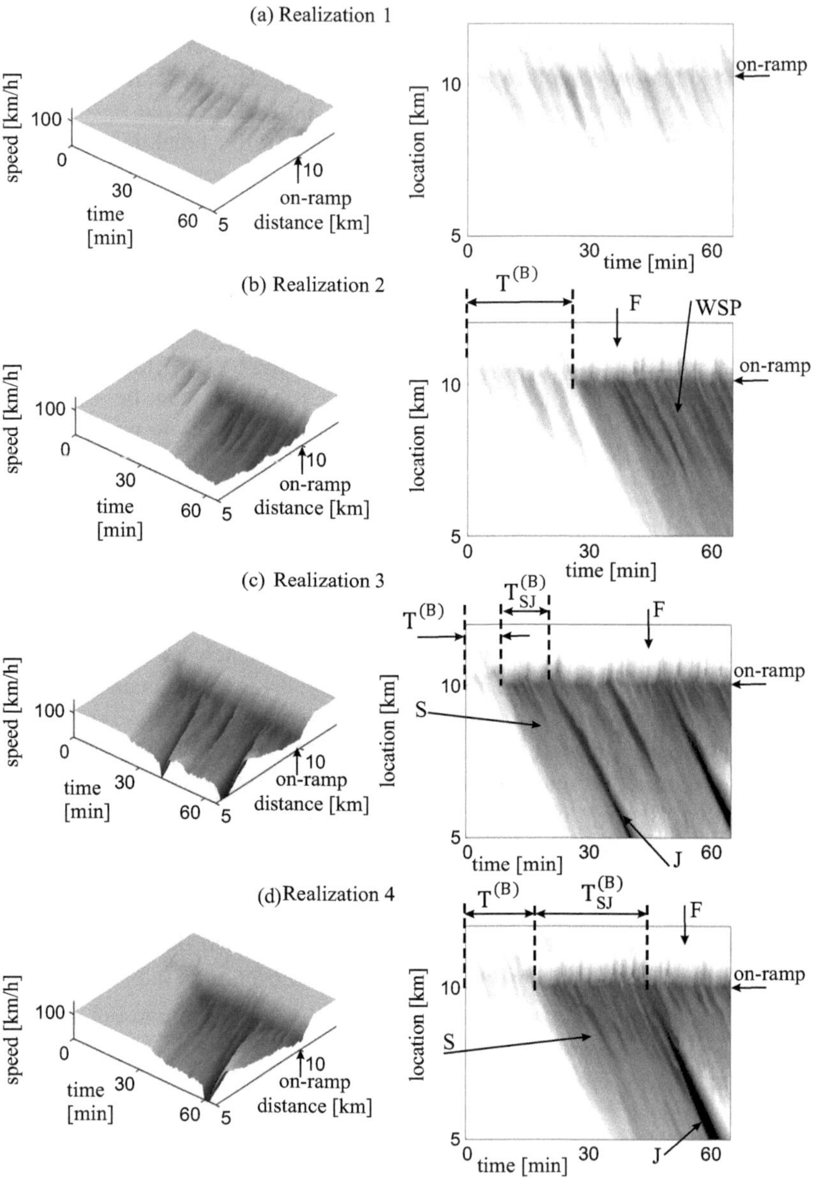

Fig. 8.13 Different simulation realizations 1–4 calculated at a given set of flow rates $q_{\text{in}} = 2000$ vehicles/h and $q_{\text{on}} = 300$ vehicles/h as well as at the same other model parameters (only initial values for random function *rand*() in the Kerner-Klenov model have been different ones for different simulation realizations 1–4): (**a**–**d**) Speed in space and time (*left column*) and the same data presented by regions with variable shades of gray (shades of gray vary from white to black when the speed decreases from 105 km/h (*white*) to zero (*black*)) (*right column*). Random time delays of traffic breakdown $T^{(\text{B})}$ and of the S→J transition $T^{(\text{B})}_{\text{SJ}}$ are as follows: $T^{(\text{B})} = 28$ min in realization 2; $T^{(\text{B})} = 11$ min and $T^{(\text{B})}_{\text{SJ}} = 14$ min in realization 3; $T^{(\text{B})} = 20$ min and $T^{(\text{B})}_{\text{SJ}} = 27$ min in realization 4. Other designations and model parameters are the same as those in Fig. 8.10

$T^{(B)}$ for traffic breakdown and the time delay $T_{SJ}^{(B)}$ for the occurrence of a nucleus for the S→J transition are random values, in different simulation realizations, in which the sequence of F→S→J transitions is observed, we have found very different values of $T^{(B)}$ and $T_{SJ}^{(B)}$. Two of such simulation realizations 3 and 4 are shown in Fig. 8.13c,d.

Simulations of the three-phase model presented in Fig. 8.13 allow us to conclude that there is a variety of diverse traffic phenomena at a highway bottleneck, which are realized at the same set of the flow rates at the bottleneck and at the same other parameters of the three-phase model. In particular, at the same model parameters during the same given time interval T_{ob} the following qualitatively different traffic phenomena can randomly occur at the bottleneck:

1. No traffic breakdown occurs: Free flow persists at the bottleneck (realization 1 in Fig. 8.13a).
2. A random time-delayed traffic breakdown $T^{(B)}$ (F→S transition) occurs. However, no moving jams emerge in synchronized flow resulting from the breakdown (realization 2 in Fig. 8.13b).
3. After a random time delay of traffic breakdown $T^{(B)}$, synchronized flow emerges at the bottlneck. Then, after another random time-delay $T_{SJ}^{(B)}$ a nucleus for an S→J transition appears in the synchronized flow. Finally, an GP emerges at the bottleneck (realizations 3 and 4 in Fig. 8.13c,d).

The F→S instability that governs a random time-delayed F→S transition has no sense for two-phase traffic flow models: The random time-delayed F→S transition is not possible in the two-phase traffic flow models. For this reason, the two-phase traffic flow models cannot show the set of above-mentioned diverse traffic phenomena 1–3 at the bottleneck related to the same set of the flow rates and to the same other parameters of a two-phase traffic flow model. These results confirm critical statements made in Sect. 1.13:

(i) Two-phase traffic flow models *cannot* simulate the nucleation nature of traffic breakdown.
(ii) Two-phase traffic flow models *cannot* simulate the S→F instability that is a fundamental feature of synchronized flow.
(iii) Two-phase traffic flow models are *not* approximations of the three-phase traffic flow models.
(iv) Because the S→F instability has no sense for two-phase traffic flow models, the two-phase models are *incommensurable* with the three-phase traffic flow models.

The incommensurability of the classical theories with the three-phase theory becomes more clear when in Sect. 8.8 we consider the effect of S→F instability on the appearance of nuclei for traffic breakdown (F→S transition) in free flow at a bottleneck. However, before we study this subject, in Sect. 8.5 through simulations of some features of congested traffic we explain the statement of Sect. 1.6 that the understanding of empirical nucleation nature of real traffic breakdown has the fundamental priority for the development of reliable ITS.

8.5 The Fundamental Requirement for Reliability of ITS

In Sect. 8.3, we have shown that there is no basic difference between two-phase and three-phase traffic flow models in relation to the metastability of traffic flow with respect to the emergence of wide moving jams either in free flow or in synchronized flow. Therefore, a question arises:

- Why does the author state that the classical traffic theories and models cannot be applied for reliable traffic control, dynamic traffic assignment, dynamic traffic optimization, and other ITS-applications?

The objective of this section is to show that the fundamental requirement for the reliability of ITS is the consistence of ITS (e.g., ITS for automatic driving, cooperative driving, methods and strategies for traffic control, dynamic traffic assignment as well as dynamic network optimization) with the nucleation nature of traffic breakdown, *not* the consistence of ITS with features of the emergence of moving jams in traffic flow or with other features of congested traffic.

Indeed, after a congested traffic pattern has already developed upstream of a network bottleneck due to traffic breakdown, it is very difficult with the use of ITS to dissolve the congested traffic pattern. This leads to a well-known conclusion that the basic objective of any reliable ITS is to prevent traffic breakdown in a traffic or transportation network. However, to prevent real traffic breakdown, ITS must be consistent with the nucleation nature of traffic breakdown at the bottleneck. However, the classical traffic theories and models are not consistent with the nucleation nature of traffic breakdown at the bottleneck (Chap. 4). This is the reason why the author state that the classical traffic theories and models cannot be applied for reliable traffic control, dynamic traffic assignment, dynamic traffic optimization, and other ITS-applications.

In this section, we prove the above statement that after a congested traffic pattern has already developed upstream of a network bottleneck, it is very difficult with the use of ITS to dissolve the congested traffic pattern. To reach this goal, we consider firstly simulations of the GP presented in Fig. 8.10c. We assume that the on-ramp inflow rate q_{on} cannot be changed. However, through traffic assignment via off-ramp(s) upstream of the on-ramp bottleneck we can reduce the flow rate q_{in} in free flow upstream of the on-ramp bottleneck. Nevertheless, even when we reduce the flow rate q_{in} from $q_{\text{in}} = 2000$ vehicles/h to $q_{\text{in}} = 1700$ vehicles/h, an GP that propagates upstream does occur spontaneously at the bottleneck (Fig. 8.14). Simulations show that to avoid traffic congestion at the bottleneck at the flow rate $q_{\text{on}} = 550$ vehicles/h that has been used in simulation of the GP shown in Fig. 8.10c, we have to reduce the flow rate q_{in} in free flow upstream of the on-ramp bottleneck from $q_{\text{in}} = 2000$ vehicles/h (Fig. 8.10c) to $q_{\text{in}} \approx q_{\text{cong}} = 1560$ vehicles/h.

In general, the larger the on-ramp inflow rate q_{on}, the smaller the mean flow rate in congested traffic q_{cong} (Fig. 8.12a). Therefore, the larger the on-ramp inflow rate q_{on}, the more difficult to dissolve traffic congestion through the reduction of the flow rate q_{in} in free flow upstream of the on-ramp bottleneck. A numerical proof of this statement is presented in Figs. 8.15, 8.16, 8.17.

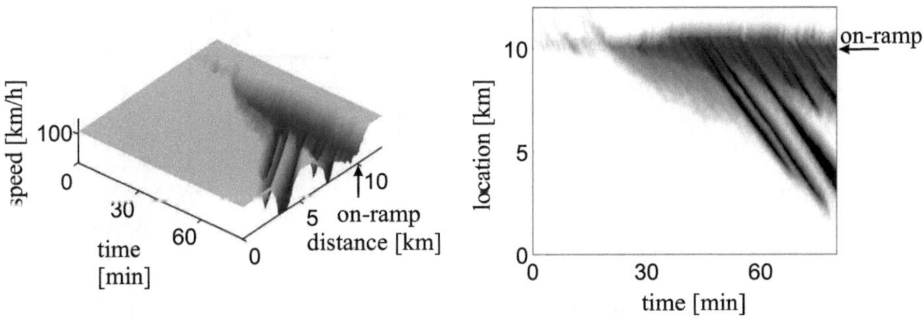

Fig. 8.14 Simulated speed in space and time (*left figure*) and the same data presented by regions with variable shades of gray (shades of gray vary from white to black when the speed decreases from 105 km/h (*white*) to zero (*black*)) (*right figure*). $q_{in} = 1700$ vehicles/h, $q_{on} = 550$ vehicles/h. Other model parameters are the same as those in Fig. 8.10

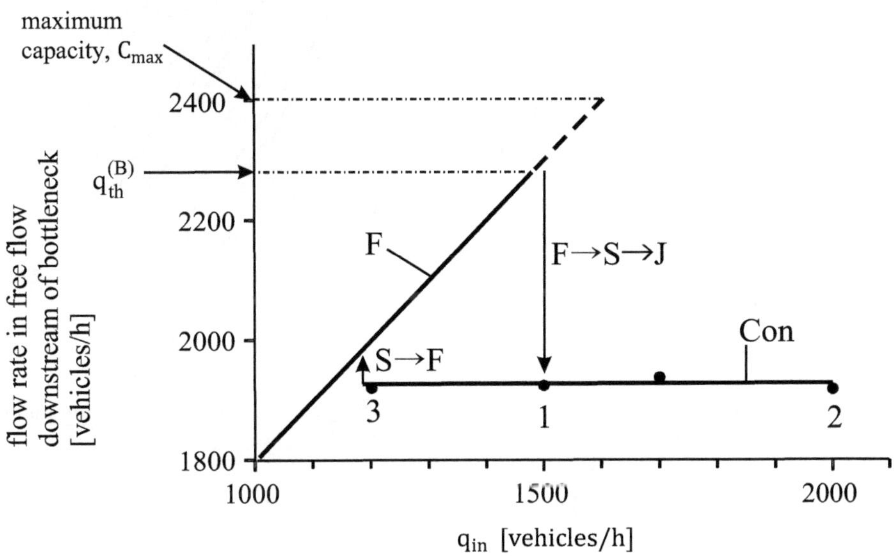

Fig. 8.15 Simulated flow–flow characteristic for the emergence and evolution of congested patterns at on-ramp bottleneck related to a given on-ramp inflow rate q_{on} and a variable value of the flow rate q_{in} in free flow upstream of the congested patterns. $q_{on} = 770$ vehicles/h. Simulations with the Kerner-Klenov model. Curve F is related to free flow. Curve labeled by "Con" is related to the discharge flow rate $q_{out}^{(bottle)}$ from congested patterns measured through the use of a virtual detector at location $x = 11$ km (1 km downstream of the beginning of the on-ramp merging region); averaging time interval of the flow rate $q_{out}^{(bottle)}$ is equal to 60 min. Calculated values $q_{th}^{(B)} = 2280$ vehicles/h, $C_{max} = 2400$ vehicles/h, $q_{out}^{(bottle)} \approx 1930$ vehicles/h. Other model parameters are the same as in Fig. 8.10

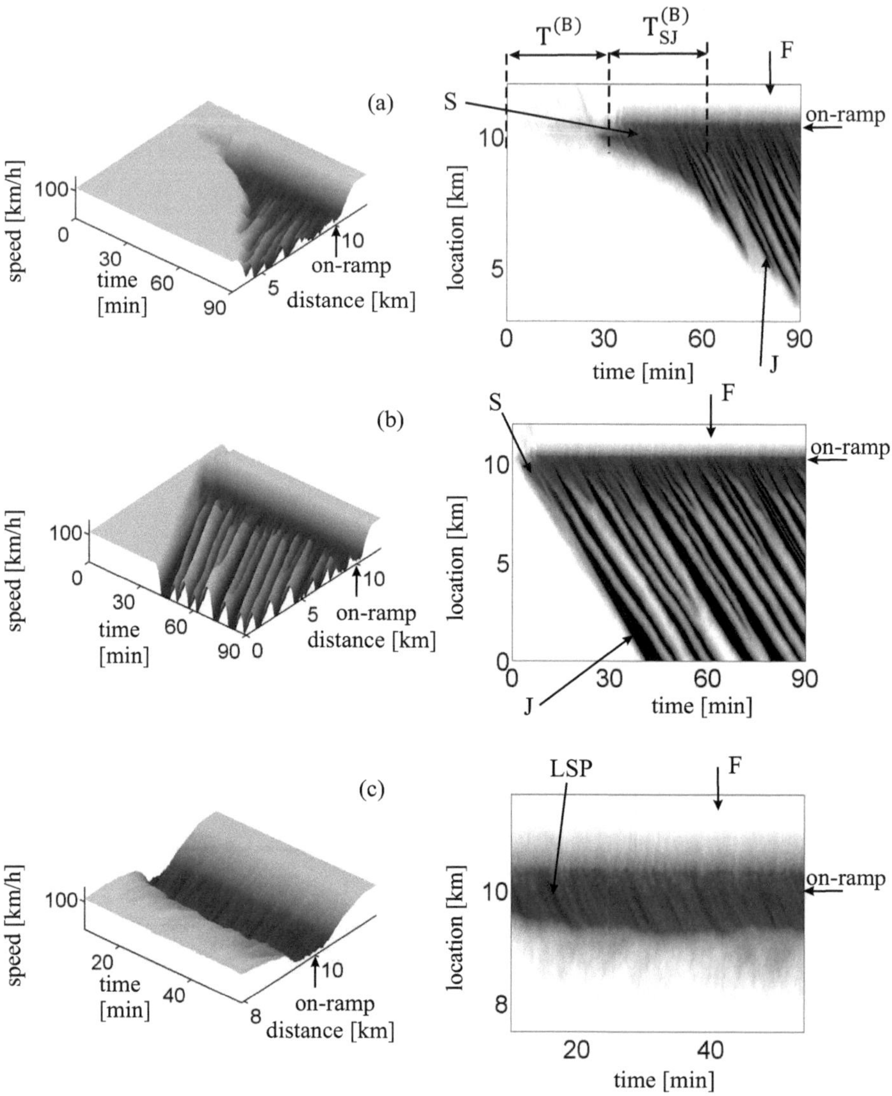

Fig. 8.16 Simulations of congested patterns at on-ramp bottleneck on single-lane road with the Kerner-Klenov model related to points 1,2, and 3 in Fig. 8.15. Speed in space and time (*left column*) and the same data presented by regions with variable shades of gray (shades of gray vary from white to black when the speed decreases from 105 km/h (*white*) to zero (*black*)) (*right column*). Simulations have been made at the same on-ramp inflow rate $q_{\text{on}} = 770$ vehicles/h and different flow rates q_{in} in free flow upstream of congested patterns: (**a**) GP at $q_{\text{in}} = 1500$ vehicles/h (point 1 in Fig. 8.15). (**b**) GP at $q_{\text{in}} = 2000$ vehicles/h (point 2 in Fig. 8.15). (**c**) LSP at $q_{\text{in}} = 1200$ vehicles/h (point 3 in Fig. 8.15). In (**a**), time delay of traffic breakdown (F→S transition) is equal to $T^{(\text{B})} = 29$ min and time delay of S→J transition is equal to $T_{\text{SJ}}^{(\text{B})} = 25$ min. F—free flow, S—synchronized flow, J—wide moving jam, LSP—localized synchronized flow pattern. LSP has been induced due to the application of short-time on-ramp impulse at $t = 0$. Other model parameters are the same as in Fig. 8.10

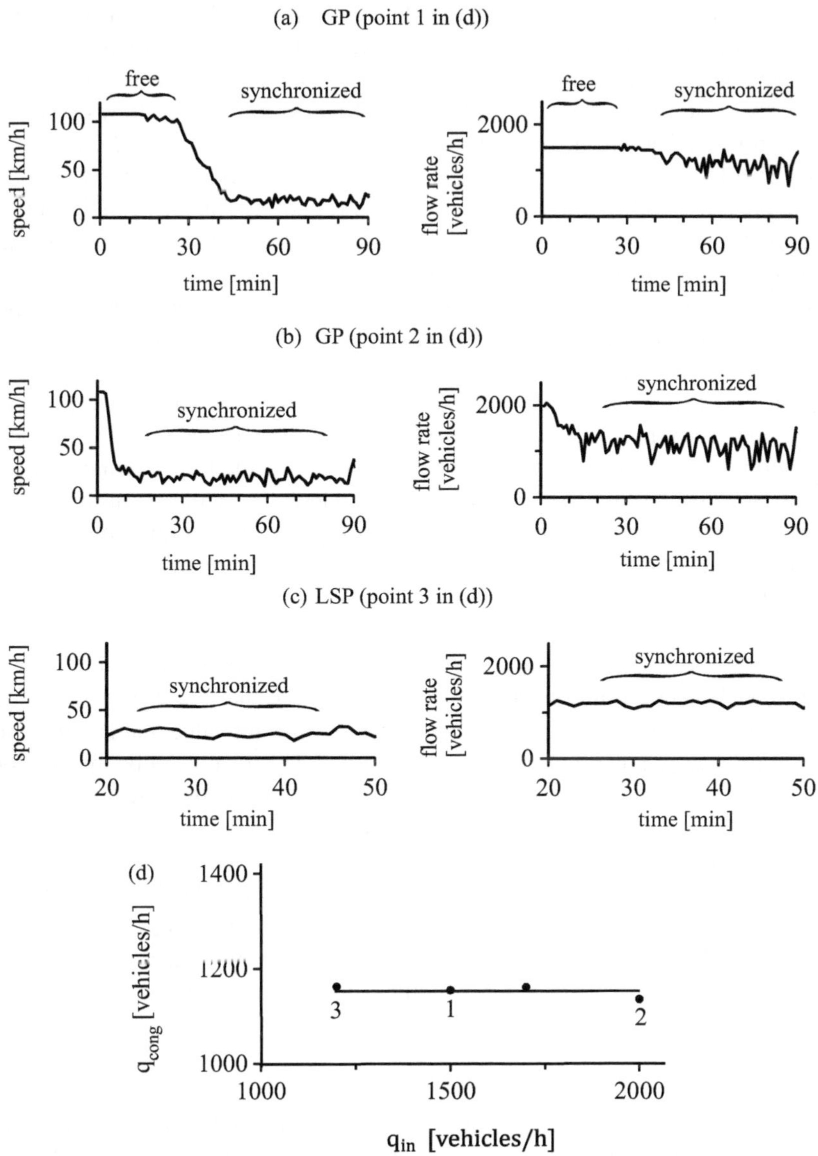

Fig. 8.17 Characteristics of congested patterns shown in Figs. 8.15 and 8.16: (**a–c**) Time-functions of the average speed (*left column*) and the flow rate (*right column*) in free flow and synchronized flow (1-min average data measured by a virtual detector at road location $x = 9.5$ km that is 500 m upstream of the beginning of the merging region of the on-ramp $x_{on} = 10$ km) for traffic patterns shown in Fig. 8.16: (**a**) For GP in Fig. 8.16a. (**b**) For GP in Fig. 8.16b. (**c**) For LSP in Fig. 8.16c. (**d**) Dependence of the mean flow rate q_{cong} within congested traffic patterns shown in Fig. 8.16 on the inflow rate q_{in} in free flow upstream of congested patterns. The mean flow rate q_{cong} that is averaged over 60 min is measured by a virtual detector upstream of the bottleneck at road location $x = 9.5$. In (**d**), calculated value $q_{cong} \approx 1160$ vehicles/h; point 1 is related to GP in Fig. 8.16a, point 2 is related to GP in Fig. 8.16b, point 3 is related to LSP in Fig. 8.16c

In Fig. 8.15, the dependence of the flow rate in free flow downstream of the on-ramp bottleneck on the flow rate q_{in} in free flow upstream of the bottleneck calculated at a given on-ramp inflow rate q_{on} is shown. Curve F is related to free flow at the on-ramp bottleneck. In this case, the flow rate in free flow downstream of the on-ramp bottleneck is equal to $q_{sum} = q_{in} + q_{on}$. A dashed part of curve F in Fig. 8.15 is related to metastable free flow with respect to F→S transition (traffic breakdown) for which the breakdown probability satisfies condition (8.13), where $q_{th}^{(B)}$ is the threshold flow rate for traffic breakdown and C_{max} is the maximum highway capacity of free flow at the bottleneck (see Chap. 5).

In the case under consideration (Figs. 8.15–8.17), under condition (8.13) an GP emerges spontaneously due to a sequence of F→S→J transitions (arrow F→S→J in Fig. 8.15). The vehicle speed and flow rate within the pinch region of the GP (Fig. 8.17a) exhibit qualitatively the same features as those discussed in Sect. 8.4 for the GP shown in Figs. 8.10c and 8.11c. It should be noted that all wide moving jams within the GP shown in Fig. 8.16a dissolve during their upstream propagation. This is because the flow rate $q_{in} = 1500$ vehicles/h is smaller than the flow rate in the jam outflow q_{out} that under chosen model parameter is equal to $q_{out} = 1808$ vehicles/h. Nevertheless, the upstream front of the GP propagates upstream over time (Fig. 8.16a).

A solid curve labeled by "Con" in Fig. 8.15 is related to the dependence of the discharge flow rate $q_{out}^{(bottle)}$ from congested patterns at the bottleneck on the flow rate q_{in} in free flow upstream of the patterns. We can see that the discharge flow rate $q_{out}^{(bottle)}$ (curve "Con" in Fig. 8.15) is nearly some constant value that is independent of the flow rate q_{in} and on pattern characteristics.[5] This result can be understood, if we consider the dependence of the mean flow rate within congested patterns q_{cong} on the flow rate q_{in} (Fig. 8.17d): The mean flow rate within congested patterns q_{cong} is nearly some constant value that is independent on the flow rate q_{in}.

For this reason, if the flow rate q_{in} increases to 2000 vehicles/h, nether the mean flow rate q_{cong} within the pinch region of the GP nor the discharge flow rate $q_{out}^{(bottle)}$ have changed (point 2 on curve "Con" in Fig. 8.15). We observe only that the spatiotemporal structure of the GP becomes more complex (Fig. 8.16b). However,

[5]The chosen on-ramp inflow rate q_{on} in Figs. 8.15–8.17 is equal to a characteristic on-ramp inflow rate $q_{on}^{(strong)}$ at which so-called "weak" congestion transforms into "strong" congestion. A theory of "weak" and "strong" congestion in GPs has been presented in Sec. 18.3 of the book [52]: When the on-ramp inflow rate q_{on} exceeds the characteristic on-ramp inflow rate $q_{on}^{(strong)}$, the mean flow rate within the pinch region of the GP q_{cong} does not depend on the on-ramp inflow rate q_{on} any more reaching a limit value denoted by $q_{cong} = q_{lim}^{(pinch)}$ in Sec. 18.3 of the book [52]. In accordance with results of the theory "weak" and "strong" congestion in GPs, simulations show that if we choose the flow rate q_{on} that is larger than $q_{on}^{(strong)} = 770$ vehicles/h, then neither the discharge flow rate $q_{out}^{(bottle)}$ in Fig. 8.15 nor the flow rate q_{cong} in Fig. 8.17d have on average changed. This is because at $q_{on} > q_{on}^{(strong)}$ congested traffic occurs in the on-ramp lane that limits on average the on-ramp inflow onto the main road to the value $q_{on}^{(strong)} = 770$ vehicles/h.

the average speed and the mean flow rate within the pinch region of synchronized flow of the GPs do not depend on the flow rate q_{in} (Figs. 8.17a,b).

If the flow rate q_{in} decreases to 1200 vehicles/h, no wide moving jams emerge within synchronized flow any more: the GP transforms into an SP. However, also in this case nether the mean flow rate q_{cong} within the pinch region of the SP nor the discharge flow rate $q_{out}^{(bottle)}$ have changed (point 3 on curve "Con" in Fig. 8.15). Because the flow rate q_{in} = 1200 vehicles/h is only about 3% larger than $q_{cong} \approx$ 1160 vehicles/h, the upstream front of the SP propagates upstream extremely slow: The width (in the longitudinal direction) of the SP does not almost change over time. Therefore, we can consider this SP as an LSP (localized SP). However, the speed and the flow rate within the LSP (Fig. 8.17c) remain qualitatively the same as those in the GPs shown in Fig. 8.17a,b.

Thus, in the case under consideration traffic congestion cannot be dissolve even when the flow rate q_{in} in free flow upstream of the bottleneck decreases from 2000 vehicles/h to 1200 vehicles/h. When the flow rate q_{in} becomes smaller than the mean flow rate within the pinch region of synchronized flow $q_{cong} \approx$ 1160 vehicles/h, an S→F transition occurs at the bottleneck (arrow labeled by S→F in Fig. 8.15). At $q_{in} < q_{cong}$, no traffic congestion can occur at the bottleneck.

The above analysis of traffic congestion leads to the conclusion that rather than controlling of traffic congestion in the network, ITS should prevent either traffic breakdown or limit the development of traffic congestion in the network (see Chap. 11 and Sect. 12.6). To prevent traffic breakdown, ITS in the framework of the three-phase theory are needed. This is because none of the classical traffic theories can explain the empirical nucleation nature of real traffic breakdown at highway bottlenecks.

The fundamental requirement for the reliability of ITS is that ITS should be consistent with the empirical nucleation nature of traffic breakdown at a highway bottleneck.

8.6 Methodology of Study of Critical Nuclei Required for Phase Transitions

As emphasized in Sect. 8.3, both the two-phase traffic flow models of the GM model class (see reviews and books [28, 52, 62, 96, 97, 113]) and the three-phase theory [52, 62] show a *metastable free flow* at a highway bottleneck. This free flow metastability leads to a complex dynamics of congested traffic patterns at highway bottlenecks found in the both traffic flow theories. This can explain the highly controversial discussion in the field of traffic science [28, 52, 61–64, 66, 96, 97, 113].

In Sect. 8.4, we have tried to resolve this highly controversial discussion through a consideration of two qualitatively different random time-delayed phase transitions that occur spontaneously in traffic flow. The first of these two random time-delayed transitions is traffic breakdown (F→S transition) at a highway bottleneck. This random time delayed traffic breakdown is governed by the S→F instability introduced in the three-phase theory. The S→F instability has no sense for the classical traffic flow theories. This fact has explained the incommensurability of the three-phase theory and the classical traffic flow theories.

In following sections of this chapter, for a more clear proof of the incommensurability of the three-phase theory and the classical traffic flow theories we use another methodology applied in [78]. Rather than the consideration of two qualitatively different random time-delayed phase transitions that occur spontaneously in traffic flow, we consider simulations of *induced* phase transitions in free flow on a single-lane road with an on-ramp bottleneck. These simulations have been made with stochastic microscopic three-phase and two-phase traffic flow models: The Nagel-Schreckenberg CA model with the slow-to-start-rule [3, 98] that belongs to the GM model class ("two-phase model" for short) and the KKSW CA model [73, 77–79] in the framework of the three-phase theory ("three-phase model" for short). Rules of vehicular motion in the KKSW CA and the Nagel-Schreckenberg CA model can be found in Sect. B.2 and Sect. B.4 of Appendix B, respectively. The both models incorporate the *same mechanism* of the metastability of traffic flow with respect to the moving jam emergence. For this reason, the both models exhibits the same line J in the flow–density plane that represents the downstream front of a wide moving jam in the plane as well as the same value of the flow rate q_{out} in free flow related to the outflow from the wide moving jam (see explanations of the line J in Sect. 4.5.1).

Simulations show that no spontaneous F→J transitions is observed in the three-phase model. Nevertheless, the three-phase model does exhibit the metastability of free flow with respect to an F→J transition. For this reason, there should be a critical nucleus required for the F→J transition in the three-phase model. To study critical nuclei required for phase transitions in free flow, we use the following procedure of numerical simulations (Fig. 8.18).

We consider a possibility of induced phase transitions in metastable free flow at the bottleneck with the use of the two-phase model and the three-phase model. In simulations, to induce a phase transition in metastable free flow at the bottleneck (Fig. 8.18a) we use an impulse of the on-ramp inflow rate Δq_{on} of the duration 1 min. In other words, during time interval $0 \leq t < T_{ind}$ the on-ramp inflow is equal to a given constant value q_{on}. At time instant $T_{ind} = 3$ min the flow rate increases to $q_{on} + \Delta q_{on}$ during the time interval $\Delta t_{ind} = 1$ min. After time instant $t = 4$ min the on-ramp inflow rate is equal to the initial constant value q_{on} (Fig. 8.18b).

For any chosen set of the flow rates q_{in} and q_{on} in free flow at the bottleneck we make simulations with different amplitudes Δq_{on} of the on-ramp inflow impulse (Fig. 8.18b). For any set of the flow rates q_{in} and q_{on}, we try find a *minimum* amplitude of the on-ramp inflow impulse at which induced phase transition in metastable free flow at the bottleneck occurs. The on-ramp inflow impulse with the minimum amplitude can be considered the critical on-ramp inflow impulse or a *critical nucleus* required for a phase transition in metastable free flow at the bottleneck. Therefore, when the on-ramp inflow impulse induces a phase transition in free flow, then the amplitude of the critical nucleus is equal to the minimum

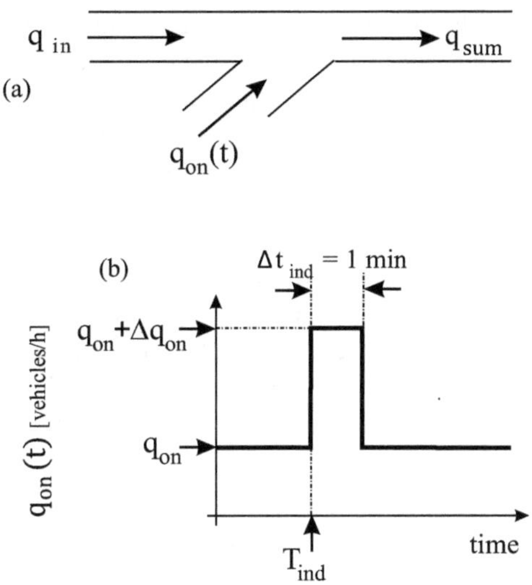

Fig. 8.18 Explanations of methodology of a numerical study of induced phase transitions in two-phase and three-phase models: (**a**) Schema of single-lane road with on-ramp bottleneck. (**b**) Qualitative time-dependence of on-ramp inflow $q_{on}(t)$ showing on-ramp inflow impulse applied to induce a phase transition in free flow at the bottleneck

amplitude of the on-ramp inflow impulse at which the phase transition can still be induced at the bottleneck.

The use of the term "critical nucleus" for the critical on-ramp inflow impulse can be explained as follows. Simulations show that the on-ramp inflow impulse $\Delta q_{on}(t)$ (Fig. 8.18b) is the source for the occurrence of a local disturbance in free flow within which the average speed decreases and the density increases in a neighborhood of the bottleneck.

At the chosen short duration of the on-ramp inflow impulse $\Delta t_{ind} = 1$ min, simulations show the obvious result: The larger the impulse amplitude Δq_{on} (Fig. 8.18b), the smaller the speed within the local speed disturbance. At a given set of the flow rates q_{in} and q_{on}, for each chosen amplitude Δq_{on} of the on-ramp inflow impulse there is a definite value of the minimum average speed within the local speed disturbance.

The critical on-ramp inflow impulse induces a critical local speed disturbance in free flow at the bottleneck required for a phase transition in metastable free flow at the bottleneck. The critical local speed disturbance can be considered a critical nucleus required for the phase transition in free flow at the bottleneck. The difference between the average speed at the bottleneck before the on-ramp inflow impulse is applied and the average speed within the critical local speed disturbance resulting from the on-ramp inflow impulse can be considered the amplitude of the critical nucleus required for the phase transition.[6]

[6]It should be noted that when the local speed disturbance is close to the critical nucleus for a phase transition, the phase transition exhibits a variety of diverse probabilistic features (Chap. 5 and Sect. 8.3.2). However, qualitative conclusions about critical nuclei made in this chapter below with the use of induced phase transitions are independent of probabilistic features of the phase transitions. For this reason, probabilistic features of *induced* phase transitions are not considered in the book.

The peculiarity of the case under consideration is that the critical local speed disturbance is caused by the on-ramp inflow impulse $\Delta q_{\mathrm{on}}(t)$.

In this sense, it is equivalent to consider either the critical on-ramp inflow impulse or the critical local speed disturbance resulting from this impulse as a "critical nucleus" required for a phase transition in metastable free flow at the bottleneck. Clearly, the amplitude of the critical on-ramp inflow impulse is a function of the flow rates q_{in} and q_{on}. Therefore, the amplitude of the critical nucleus required for a phase transition depends also on the flow rates q_{in} and q_{on}. For qualitative explanations of simulation results below, for simplicity, while considering different amplitudes of critical nuclei for different phase transitions in free flow, we denote a critical nucleus of a smaller amplitude as "smaller critical nucleus" and a critical nucleus of a larger amplitude as "larger critical nucleus".

8.7 Induced F→J Transitions in Three-Phase and Two-Phase Traffic Flow Models

We have seen in Sect. 8.3.2 that two-phase traffic flow models of the GM model class show a flow rate range (8.6) in which either spontaneous or induced F→J transition is possible and another flow rate range (8.11) in which only induced F→J transition is possible. In contrast, independent of the flow rate no F→J transitions is observed in the three-phase model. This conclusion becomes clear when one compare Figs. 8.4 and 8.6 for the two-phase model with Figs. 5.11 and 5.12 for the three-phase model [78]: Rather than moving jams (Figs. 8.4 and 8.6), in the three-phase model an F→S transition occurs at the bottleneck (Figs. 5.11 and 5.12). Nevertheless, below we show that an F→J transition can be induced in the three-phase model.

8.7.1 Induced F→J Transition at On-Ramp Bottleneck in Two-Phase Model

When we study the critical on-ramp inflow impulse Δq_{on} at which an induced phase transition in metastable free flow at the bottleneck occurs in the two-phase model (Fig. 8.19a), then we find that this critical impulse induces an F→J transition, i.e., moving jams appears at the bottleneck (Fig. 8.19c). In the two-phase model, *no* other phase transition in free flow is possible to induce with the on-ramp inflow impulse (Fig. 8.19a).

In other words, at a given set of the flow rates q_{on} and q_{in} in the two-phase model there is only one critical on-ramp inflow impulse for a phase transition in free flow at the bottleneck. This critical on-ramp inflow impulse is a critical nucleus required for the F→J transition at the bottleneck.

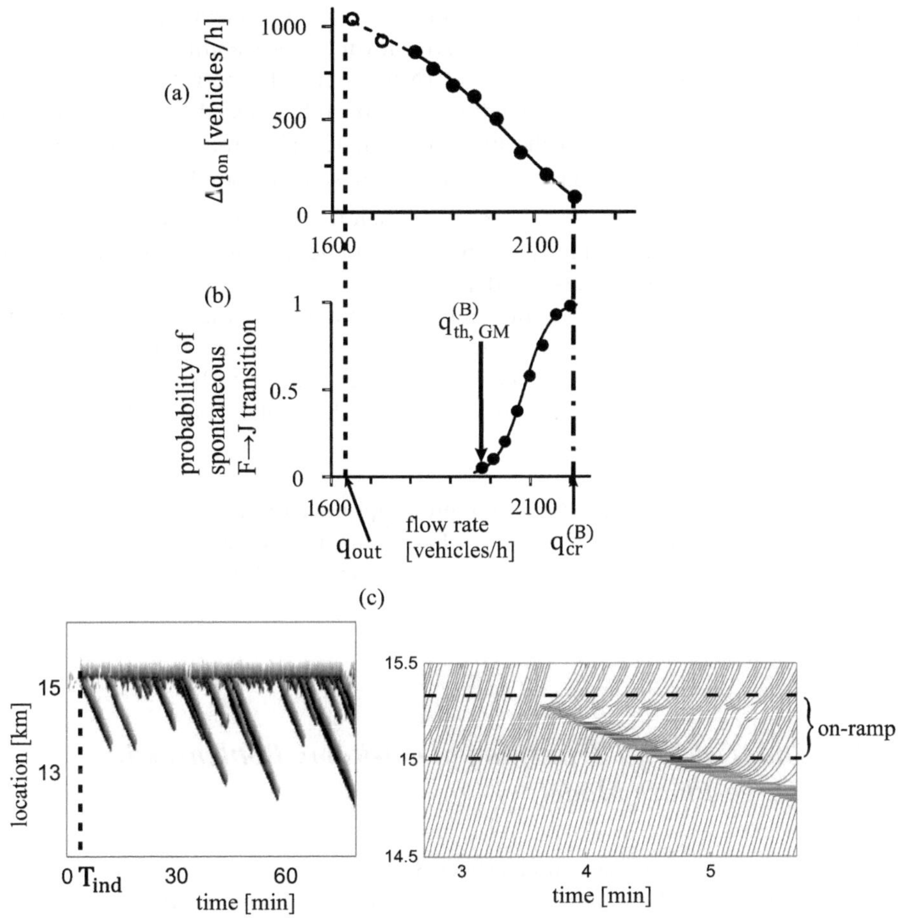

Fig. 8.19 Induced F→J transition in two-phase model (Nagel-Schreckenberg CA model): (**a**) Dependence of the critical on-ramp inflow impulse Δq_{on} on the flow rate downstream of the on-ramp bottleneck $q_{\text{sum}} = q_{\text{in}} + q_{\text{on}}$ at given $q_{\text{on}} = 400$ vehicles/h. (**b**) Probability of spontaneous F→J transition adapted from Fig. 8.5. (**c**) Induced F→J transition under condition (8.11) at $(q_{\text{in}}, \Delta q_{\text{on}}) = (1406, 860)$ vehicles/h. In (**c**), left figure—speed data presented by regions with variable shades of gray (shades of gray vary from white to black when the speed decreases from 120 km/h (*white*) to zero (*black*)); right figure—vehicle trajectories. Dashed part of the curve in (**a**) is related to small values of the flow rate $q_{\text{in}} < q_{\text{out}}$ at which moving jams induced at the bottleneck dissolve during some time interval that is considerably longer than the impulse duration 1 min. $T_{\text{ind}} = 3$ min

The induced F→J transition is possible in the whole flow rate range (8.4) within which free flow is in a metastable state with respect to the F→J transition (Fig. 8.19). The critical amplitude $\Delta q_{\text{on}}(q_{\text{sum}})$ of the on-ramp inflow impulse (critical nucleus) that induces the F→J transition is an decreasing flow rate function (Fig. 8.19a).

To compare the induced F→J transition with the spontaneous F→J transition studied in Sect. 8.3.2, in Fig. 8.19b we have shown the probability of spontaneous F→J transition in the two-phase model (Nagel-Schreckenberg CA model). As mentioned in Sect. 8.3.2, under conditions (8.6) both induced and spontaneous F→J transition are possible in the two-phase model. However, under conditions $q_{\text{out}} \leq q_{\text{sum}} < q^{(\text{B})}_{\text{th, GM}}$ (8.11) the probability of spontaneous F→J transition is equal to zero: $P^{(\text{B})}_{\text{GM}}(q_{\text{sum}}) = 0$ (Fig. 8.19b). Therefore, we have found that under conditions (8.11) in metastable free flow at the bottleneck an F→J transition can only be induced in the two-phase model.

We have also found that at the flow rate at the bottleneck, which is only slightly larger than the flow rate q_{out}, to induce the wide moving jam at the bottleneck, we should apply a large time-limited speed disturbance: The disturbance should force all drivers to stop during some seconds in a neighborhood of the bottleneck location.

In the two-phase model, the minimum amplitude of the on-ramp inflow impulse (critical nucleus) at which a phase transition in metastable free flow occurs at the bottleneck is related to an F→J transition (Fig. 8.19).

8.7.2 *Induced F→J Transition at On-Ramp Bottleneck in Three-Phase Model*

In contrast with the two-phase model (Sect. 8.7.1), as we will show in Sect. 8.8.2, in the three-phase model the *minimum amplitude* of the on-ramp inflow rate impulse at which a phase transition can occur at the bottleneck is related to an F→S transition at the bottleneck, *not* to the F→J transition. Nevertheless, it is possible to induce an F→J transition in the three-phase model. To reach this goal, we have to choose the on-ramp inflow rate impulse of a large enough amplitude (Fig. 8.20a) at which moving jams are induced at the bottleneck (Fig. 8.20b). As we will find in Sect. 8.8.2, the amplitude of this on-ramp inflow rate impulse (Fig. 8.20a) is larger than the minimum amplitude of the on-ramp inflow rate impulse at which the F→S transition occurs at the same set of the flow rates q_{on} and q_{in} at the bottleneck.

It should also be noted that in the two-phase model, the F→J transition can be induced in the whole range of the flow rate (8.4) in which free flow is in a metastable state with respect to the F→J transition. For the three-phase model, there is the same range of the flow rate (8.4) in which free flow is in a metastable state with respect to the F→J transition. However, in contrast with the two-phase model, the F→J transition can be induced only in a relatively small part of the flow rate range (8.4) (curve in Fig. 8.20a). The reason for this behavior of the three-phase model will also be explained in Sect. 8.8.2.

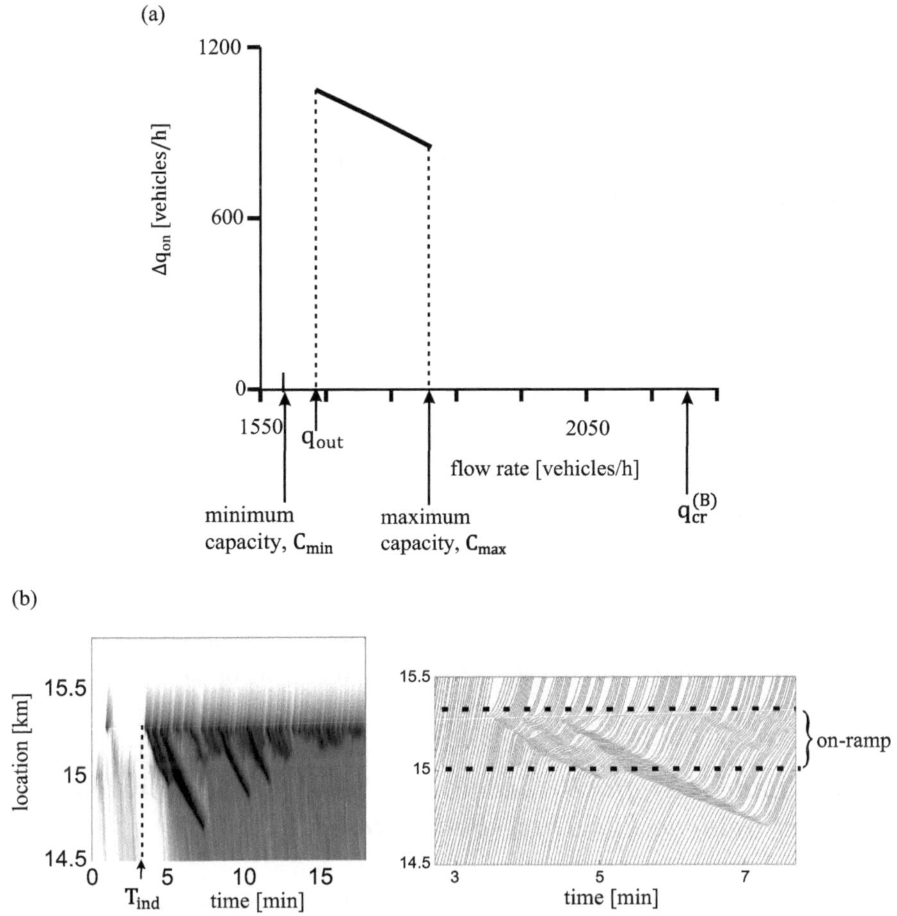

Fig. 8.20 Induced F→J transition in three-phase model (KKSW CA model) (Appendix B): (**a**) Flow-rate–dependence of critical on-ramp inflow impulse $\Delta q_{on}(q_{sum})$ required for induced F→J transition. (**b**) Congested pattern occurring through the F→J transition induced in the three-phase model at the same values of on-ramp inflow rate q_{on} = 400 vehicles/h and the same critical amplitude of impulse of on-ramp inflow Δq_{on} as those values, respectively, used in Fig. 8.19c for the two-phase model: Left figure—speed data presented by regions with variable shades of gray (shades of gray vary from white to black when the speed decreases from 120 km/h (*white*) to zero (*black*)), right figure—vehicle trajectories. As in Fig. 8.19c, $(q_{in}, \Delta q_{on}) = (1406, 860)$ vehicles/h. T_{ind} = 3 min. Calculated minimum capacity C_{min} = 1585 vehicles/h. Calculated maximum capacity C_{max} = 1810 vehicles/h. Adapted from [78]

We see that for the three-phase model there are very special simulation conditions at which moving jams can be induced in free flow at the bottleneck (Fig. 8.20). As we will show in Sect. 8.8.2, these conditions cannot be realized in real traffic flow. For this reason, the emergence of moving jams does not observe in real free flow at the bottleneck (Chap. 3). The objective of the hypothetical numerical study presented in Fig. 8.20 is to prove that the F→J transition can indeed be induced in the three-phase model.

8.8 Effect of S→F Instability on Nuclei for Traffic Breakdown at Bottleneck

In Sect. 8.3.3, we have mentioned that the metastability of free flow states (states F at the flow rate $q \geq q_{\text{out}}$ in Fig. 8.7) with respect to an F→J transition (formation of wide moving jams in free flow) [81] is a common result for the two-phase and three-phase traffic flow models. The empirical evidence that in real field traffic data the F→J transition is not observed in the metastable states of free flow is explained by the three-phase theory [52, 61–64, 70, 71] through hypothesis 10 of this theory (Sect. 1.9) as follows:

- At any flow rate within a range of the flow rate in free flow at which either an F→J transition or an F→S transition is possible, a critical nucleus required for the F→S transition is considerably smaller than a critical nucleus required for the F→J transition.
- For this reason, the probability of the spontaneous occurrence of the F→J transition is *negligible* in comparison with the probability of the F→S transition.

8.8.1 Induced Traffic Breakdown (Induced F→S Transition) at Bottleneck in Three-Phase Model

In contrast with the two-phase model (Fig. 8.19), when we study the *minimum amplitude* of the on-ramp inflow impulse Δq_{on} at which induced phase transition in metastable free flow at the bottleneck occurs in the three-phase model (Fig. 8.21a), we find that this critical on-ramp inflow impulse (critical nucleus) results in an F→S transition (traffic breakdown) (Fig. 8.21c), *not* in the F→J transition shown in Fig. 8.20.

This result remains in the whole range of the flow rate $C_{\min} \leq q_{\text{sum}} < C_{\max}$ (5.1), in which free flow is in a metastable state with respect to traffic breakdown (F→S transition) at the bottleneck. As explained in Sect. 5.4, in the range of the flow rate q_{sum} (5.1), either spontaneous or induced traffic breakdown is possible at the bottleneck.

> Contrarily to the two-phase model (Sect. 8.7.1), for the three-phase model the minimum amplitude of the on-ramp inflow impulse (critical nucleus) at which a phase transition in metastable free flow occurs at the bottleneck is related to an F→S transition (Fig. 8.21).

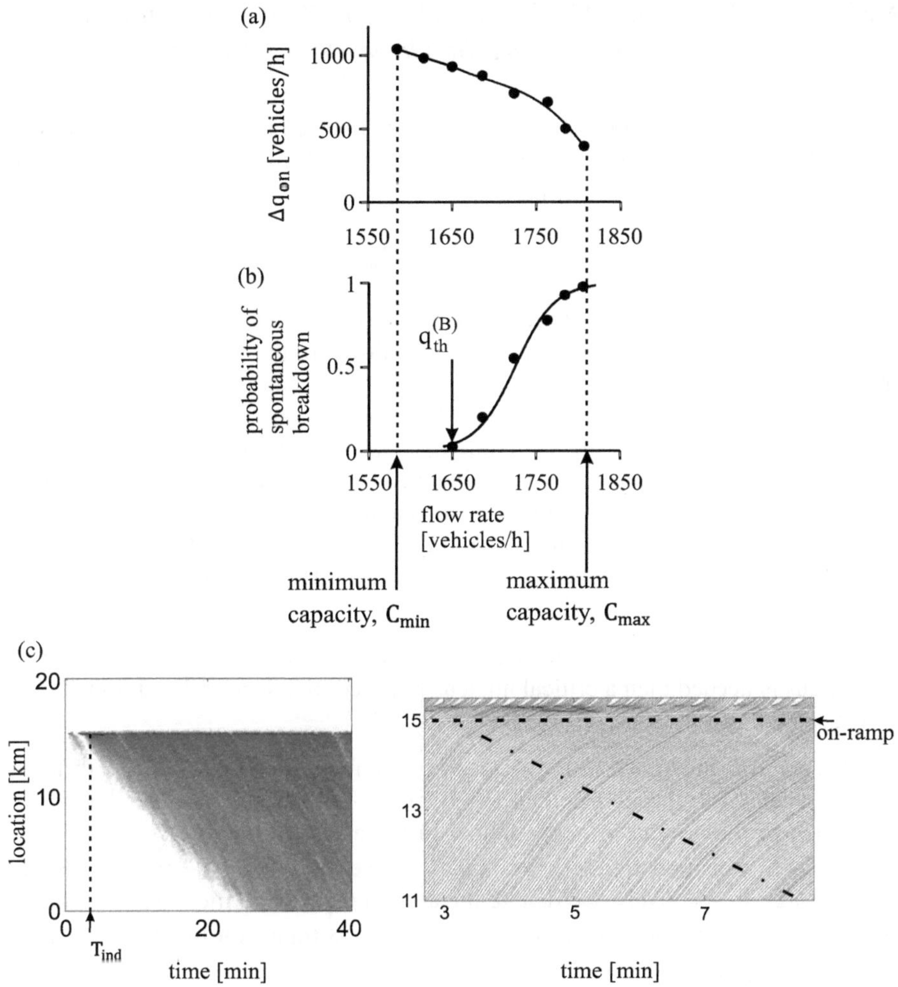

Fig. 8.21 Induced traffic breakdown (induced F→S transition) at on-ramp bottleneck on single lane road in three-phase model (KKSW CA model) (Appendix B): (**a**) Dependence of critical amplitude of impulse $\Delta q_{on}(q_{sum})$ required for induced traffic breakdown on the flow rate $q_{sum} = q_{on} + q_{in}$ downstream of bottleneck at a variable value q_{in} and a given value q_{on}. (**b**) Flow rate dependence of the probability of spontaneous traffic breakdown (spontaneous F→S transition) $P^{(B)}(q_{sum})$ adapted from Fig. 5.9. (**c**) WSP resulting from induced traffic breakdown at the bottleneck for $(q_{in}, \Delta q_{on}) = (1406, 380)$ vehicles/h. In (**c**), left figure—speed data presented by regions with variable shades of gray (shades of gray vary from white to black when the speed decreases from 120 km/h (*white*) to zero (*black*)), right figure—vehicle trajectories; dashed-dotted line (*right figure*) marks the propagation of the upstream fronts of the WSP (*left figure*). $q_{on} = 400$ vehicles/h. Calculated minimum capacity $C_{min} = 1585$ vehicles/h. Calculated maximum capacity $C_{max} = 1810$ vehicles/h. The beginning and the end of the merging region of the on-ramp are $x_{on} = 15$ km and $x_{on}^{(e)} = 15.3$, respectively. Adapted from [78]

8.8.2 *Two Different "Critical Nuclei" for Phase Transitions in Free Flow at Bottleneck in Three-Phase Theory*

As shown in Fig. 8.20, an F→J transition can be induced in metastable free flow at the bottleneck in the three-phase model. This is possible in the three-phase model within the range of the flow rate q_{sum}

$$q_{\text{out}} \leq q_{\text{sum}} < C_{\max}. \tag{8.15}$$

This is because within the flow rate range (8.15) free flow at the bottleneck is in a metastable state with respect to an F→J transition. In the three-phase model, the induced F→J transition can occur at the same flow rates $(q_{\text{in}},\ q_{\text{on}})$ at which the induced F→S transition is possible at the bottleneck (in both Figs. 8.20b and 8.21c the set of the flow rates is the same one $(q_{\text{in}},\ q_{\text{on}}) = (1406,\ 400)$ vehicles/h). A limitation for the flow rate q_{sum} for the induced F→J transition associated with the right-hand condition (8.15) is caused by a spontaneous F→S transition that occurs in the three-phase model with the probability $P^{(\text{B})} = 1$ at $q_{\text{sum}} \geq C_{\max}$ (Sect. 5.4).

The fundamental difference between the induced F→J and F→S transitions in the three-phase model is as follows.

- To induce an F→S transition at the bottleneck, a considerably smaller critical nucleus is needed then a critical nucleus required to induce an F→J transition at the same flow rates $(q_{\text{in}},\ q_{\text{on}})$ at the bottleneck.

This result is illustrated in Fig. 8.22. We have found that there are two different critical nuclei for these different F→J and F→S transitions. We denote the amplitude of the critical on-ramp inflow impulse that causes the critical nucleus for an F→S transition by $\Delta q_{\text{on}} = \Delta q_{\text{on}}^{(\text{FS})}$ (curve $\Delta q_{\text{on}}(q_{\text{sum}})$ labeled by "F→S" in Fig. 8.22). Respectively, we denote the amplitude of the critical on-ramp inflow impulse that causes the critical nucleus for an F→J transition by $\Delta q_{\text{on}} = \Delta q_{\text{on}}^{(\text{FJ})}$ (curve $\Delta q_{\text{on}}(q_{\text{sum}})$ labeled by "F→J" in Fig. 8.22).

For example, to induce the F→S transition leading to the WSP shown in Fig. 8.21c, the critical on-ramp inflow impulse $\Delta q_{\text{on}}^{(\text{FS})} = 380$ vehicles/h. However, to induce the F→J transition leading to moving jams shown in Fig. 8.20b, the critical on-ramp inflow impulse $\Delta q_{\text{on}}^{(\text{FJ})} = 860$ vehicles/h. Therefore, at the same flow rate q_{sum} the value $\Delta q_{\text{on}}^{(\text{FJ})}$ is considerably larger than the value $\Delta q_{\text{on}}^{(\text{FS})}$.

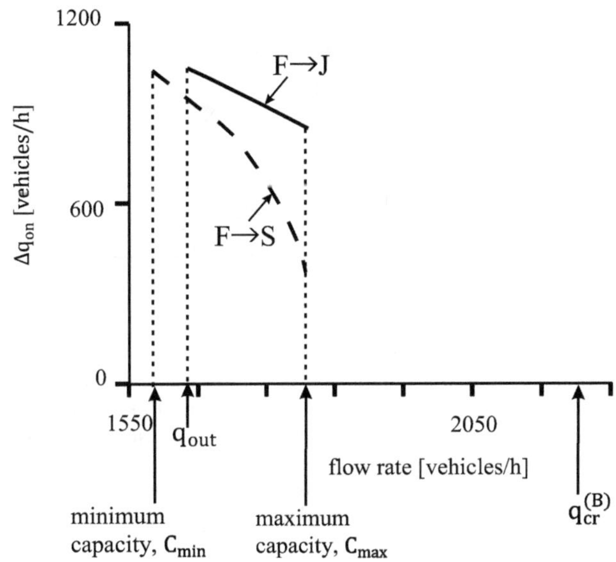

Fig. 8.22 Comparison of induced F→J and F→S transitions in three-phase model: Flow-rate–dependencies of critical on-ramp inflow impulse $\Delta q_{\mathrm{on}}(q_{\mathrm{sum}})$ that are required for induced F→J transition (solid curve labeled by "F→J") and that are required for induced F→S transition (dashed curve labeled by "F→S"). *Solid curve* labeled by "F→J" has been taken from Fig. 8.19a. *Dashed curve* labeled by "F→S" has been taken from Fig. 8.21a

This result remains for any flow rate q_{sum} (8.15) at which induced F→J transition in the three-phase model is possible (Fig. 8.22):

$$\Delta q_{\mathrm{on}}^{(\mathrm{FJ})}(q_{\mathrm{sum}}) > \Delta q_{\mathrm{on}}^{(\mathrm{FS})}(q_{\mathrm{sum}}). \tag{8.16}$$

It turns out that for the range of the flow rate q_{sum} in which moving jams can be induced in free flow at the bottleneck, the flow-rate dependence of the amplitude of the critical on-ramp inflow impulse $\Delta q_{\mathrm{on}}^{(\mathrm{FJ})}(q_{\mathrm{sum}})$ for the F→J transition in the two-phase model (Fig. 8.19a) is *identical one* with the flow-rate dependence of the amplitude of the critical on-ramp inflow impulse $\Delta q_{\mathrm{on}}^{(\mathrm{FJ})}(q_{\mathrm{sum}})$ for the F→J transition in the three-phase model (curve labeled by "F→J" in Fig. 8.22). We have also found that at the same flow rate at the bottleneck q_{sum}, which is only slightly larger than the flow rate in free flow in the jam outflow q_{out}, the amplitude of the critical on-ramp inflow impulse $\Delta q_{\mathrm{on}}^{(\mathrm{FJ})}$ required to induce moving jams in simulations either with the two-phase model or with the three-phase model should be very large. This means that to induce a moving jam in free flow at the bottleneck, the local speed disturbance at the bottleneck caused by the on-ramp inflow impulse should force some drivers to stop during some seconds in a neighborhood of the bottleneck location.

Thus, simulations presented in Fig. 8.22 show that at any chosen set of the flow rates $(q_{\mathrm{on}},\ q_{\mathrm{in}})$ at which either an F→S transition or an F→J transition could

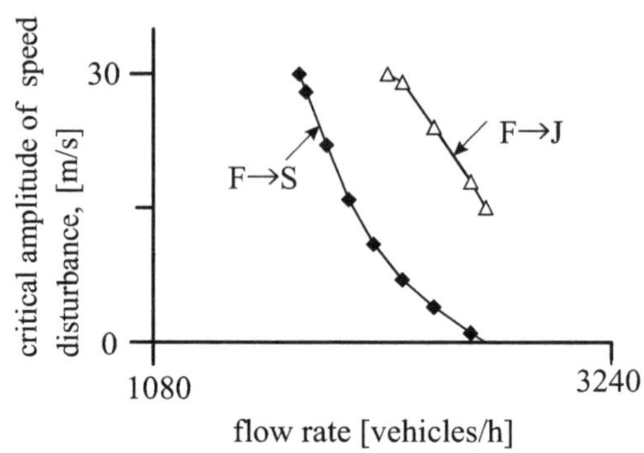

Fig. 8.23 Simulations of the critical amplitude of a local speed disturbance in free flow (the amplitude of the critical nucleus for F→S transition) required for F→S transition (curve labeled by "F→S") and of the critical amplitude of a local speed disturbance in this free flow (the amplitude of the critical nucleus for F→J transition) required for F→J transition (curve labeled by "F→J") as a function of the flow rate. Simulations of the Kerner-Klenov microscopic stochastic three-phase traffic flow model. Curve labeled by "F→S" is taken from Fig. 5.6. Adapted from [70]

be possible at the bottleneck, to induce the F→S transition at the bottleneck, a considerably smaller critical nucleus is needed in comparison with a critical nucleus required to induce the F→J transition at the bottleneck. This conclusion has already been derived (with the use of a different simulation approach) in first simulations of the Kerner-Klenov stochastic microscopic three-phase model (Fig. 8.23) [70]. From Fig. 8.23 we can see that the amplitude of the critical nucleus required for the F→S transition (curve labeled by "F→S") is considerably smaller than the amplitude of the critical nucleus required for an F→J transition (curve labeled by "F→J"). In other words, at any flow rate in free flow, at which either an F→J transition or an F→S transition is possible, the critical nucleus for the F→S transition is considerably smaller than the critical nucleus required for the F→J transition. For this reason, during the same time interval of traffic observations T_{ob} the probability of the spontaneous occurrence of the F→J transition is *negligible* in comparison with the probability of the F→S transition.

The microscopic theory of the S→F instability (Sects. 5.12 and 5.13) explains this result as follows. The S→F instability tries to prevent the decrease in the speed within a local speed disturbance in free flow at the bottleneck. In other words, the S→F instability tries to destroy any local speed decrease in free flow at the bottleneck. Because the S→F instability exhibits the nucleation nature (Sect. 5.13), the S→F instability destroys an initial local speed decrease in free flow at the bottleneck only if the speed within this local speed disturbance is not very small. Otherwise, when the speed within the local speed disturbance is small enough, the S→F instability does not occur. In this case, the local speed disturbance is a nucleus for the F→S transition (traffic breakdown). This explains the mechanism of the occurrence of the nucleus for the F→S transition.

In the three-phase theory, the S→F instability is associated with a discontinuity of the density-dependence of the probability of driver over-acceleration caused by a time-delay in driver over-acceleration (Sect. 5.10). This discontinuity of the density-dependence of the probability of driver over-acceleration leads to the metastability of synchronized flow with respect to an S→F transition. This metastability of synchronized flow results from the nucleation nature of the S→F instability. In its turn, the nucleation nature of the S→F instability is the origin the nucleation nature of traffic breakdown (F→S transition) at a highway bottleneck.

None of the classical traffic flow theories incorporates an S→F instability: The S→F instability has no sense for two-phase traffic flow models. For this reason, the classical traffic flow theories cannot explain the empirical fundamental of transportation science. This is the basic qualitative difference between the classical theories and the three-phase theory. This explains also that the three-phase theory is incommensurable with the classical traffic flow theories.

In the three-phase theory, at any flow rate at which either induced F→S transition or induced F→J transition is possible, there are two qualitatively different critical nuclei for two qualitatively different phase transitions in free flow, respectively. These two qualitatively different critical nuclei for two qualitatively different phase transitions in free flow at a highway bottleneck in the three-phase theory are as follows:

1. The first critical nucleus is the critical nucleus required for an F→J transition. The occurrence of the critical nucleus is governed by the classical traffic flow instability of Herman, Gazis, Montroll, Potts, Rothery, and Chandler [8, 19, 20, 29].
2. The second critical nucleus is the critical nucleus required for an F→S transition. The occurrence of the critical nucleus required for an F→S transition is governed by the S→F instability introduced in the three-phase theory.

In the three-phase theory, at any flow rate at which either an F→S transition or an F→J transition is possible, the critical nucleus required for the F→J transition is considerably larger than the critical nucleus required for the F→S transition (traffic breakdown) at the bottleneck.

In the same free flow, the probability of the occurrence of a local speed disturbance with a small critical amplitude (small critical nucleus) is considerably larger than the probability of the occurrence of a local speed disturbance with a large critical amplitude (large critical nucleus). The results of numerical simulations of the three-phase model presented above (Figs. 8.22 and 8.23) confirm the hypothesis of the three-phase theory formulated in 1998–1999 [36–39] that is as follows:

- At any flow rate at which either an F→S transition or an F→J transition is possible in the three-phase model, the probability of the F→S transition is considerably larger than the probability of the F→J transition at the bottleneck.

This explains why rather than an F→J transition, only the F→S transition is observed at highway bottlenecks in real field traffic data (Chap. 3).

The S→F instability has no sense for the classical traffic flow models like the LWR model and Daganzo's cell-transmission model, traffic flow models of the GM model class, combined two-phase models that incorporate features of the LWR model and models of the GM model (Sect. 4.8) as well as for any other classical traffic flow models and theories.

In the three-phase theory, the nucleation nature of the S→F instability governs the nucleation nature of the F→S transition (traffic breakdown) at the bottleneck. For this reason, the nucleation nature of the F→S transition at the bottleneck has no sense for the classical traffic flow models and theories.

This explains that and why the three-phase theory is incommensurable with all earlier traffic flow theories and models.

In its turn, the empirical nucleation nature of the F→S transition (traffic breakdown) at the bottleneck is the empirical fundamental of transportation science. The three-phase theory explains the empirical fundamental of transportation science.

For this reason, the three-phase theory can be considered the theoretical fundamental of transportation science.

8.9 Basic Requirement for Three-Phase Traffic Flow Models

The empirical nucleation nature of the F→S transition (traffic breakdown) at the bottleneck is the empirical fundamental of transportation science. The nucleation nature of the F→S transition (traffic breakdown) results from the nucleation nature of the S→F instability.

For this reason, a three-phase traffic flow model should show the nucleation nature of the S→F instability.

The theory of the S→F instability presented in Sects. 5.12 and 5.13 is a *microscopic* theory. In general, there can be a variety of microscopic, macroscopic, and other mathematical approaches to mathematical traffic flow models in the framework of the three-phase theory. Therefore, a question arises:

- Is there a *macroscopic* characteristic that shows the effect of the nucleation nature of the S→F instability on traffic breakdown at a bottleneck?

In Sect. 5.4.2, we have shown that such a macroscopic characteristic is the Z-characteristic for traffic breakdown (Fig. 1.8): Both the existence of the critical nucleus required for traffic breakdown and the flow-rate dependence of the averaged speed within the critical nucleus $v^{(B)}_{\mathrm{cr,\ FS}}(q_{\mathrm{sum}})$ result from the nucleation nature of the S→F instability. For this reason, a three-phase traffic flow model should either show or incorporate the Z-characteristic for traffic breakdown (F→S transition).

Indeed, the Z-characteristic for traffic breakdown is realized due to the hypothesis of the three-phase theory about the discontinuous character of over-acceleration probability (Fig. 8.24b). As explained in Chap. 5, the discontinuous character of over-acceleration probability leads to the nucleation nature of the S→F instability. The nucleation nature of the S→F instability explains the nucleation nature of traffic breakdown (F→S transition) at a bottleneck. In its turn, the Z-characteristic for traffic breakdown is a macroscopic characteristic that presents the nucleation nature of traffic breakdown (F→S transition) in the speed–flow-rate plane (Fig. 8.24a).

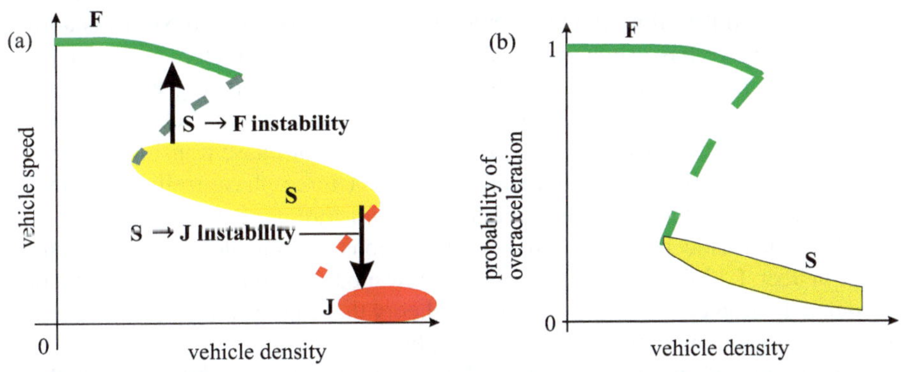

Fig. 8.24 Qualitative illustrations of 2Z-characteristic for phase transitions (**a**) and discontinuous character of over-acceleration probability made in three-phase theory (**b**) [38–48, 50–52, 54, 55, 59–63, 65]. 2Z-characteristic for phase transitions consists of two different Z-characteristics. The first Z-characteristic is a Z-characteristic for traffic breakdown that results from the nucleation nature of the S→F instability introduced in the three-phase theory (labeled by "S→F instability" in (**a**)) [38–40, 52, 62]. The second Z-characteristic for the S→J and J→S transitions results from the nucleation nature of the S→J instability (labeled by "S→J instability" in (**a**)) associated with the classical traffic flow instability of Herman, Gazis, Montroll, Potts, Rothery, and Chandler [8, 19, 20, 29]

A three-phase traffic flow model should either show or incorporate a Z-characteristic for traffic breakdown at the bottleneck.

In accordance with Hypothesis 9 of Sect. 1.9, there are two different traffic flow instabilities in the three-phase theory:

1. The S→F instability (labeled by "S→F instability" in Fig. 8.24a). The S→F instability occurs due to a time delay in driver over-acceleration. The S→F instability leads to a growing wave of the local *increase* in the speed in synchronized flow. The S→F instability has been introduced in the three-phase theory [38–40, 52, 62]. The S→F instability governs the nucleation nature of traffic breakdown (F→S transition). The S→F instability has been considered in Sects. 5.12 and 5.13.
2. The S→J instability (labeled by "S→J instability" in Fig. 8.24a). The S→J instability occurs due to the driver over-deceleration effect (driver reaction time). The S→J instability leads to a growing wave of the local *decrease* in the speed in synchronized flow resulting in an S→J transition. The S→J instability is associated with the classical traffic flow instability of Herman, Gazis, Montroll, Potts, Rothery, and Chandler [8, 19, 20, 29]. The S→J instability has been considered in Sect. 8.3.3.

Thus, additionally to the Z-characteristic for traffic breakdown (F→S transition), a three-phase traffic flow model should show or incorporate the S→J instability. In Sect. 8.3.3, we have shown that such a macroscopic characteristic of the S→J instability is the Z-characteristic for S→J and J→S transitions (Fig. 8.8).

The Z-characteristic for traffic breakdown together with the Z-characteristic for S→J and J→S transitions build an 2Z-characteristic for phase transitions in traffic flow (Fig. 8.24a). Therefore, the statement that a three-phase traffic flow model should show or incorporate the Z-characteristic for traffic breakdown together with the Z-characteristic for S→J and J→S transitions is equivalent to the statement made in Sect. 1.12 that a three-phase traffic flow model is a mathematical traffic flow model in the framework of the three-phase theory that can show an 2Z-characteristic for phase transitions in traffic flow (Fig. 8.24a).

Simulations of an 2Z-characteristic for phase transitions in traffic flow with the Kerner-Klenov model (Fig. 8.25)[7] confirm general features of the 2Z-characteristic

[7]It should be noted that when qualitative features of an Z-characteristic for traffic breakdown as a function of the flow rate q_{sum} (Fig. 5.5) has been discussed (Sect. 5.3), we have considered the metastability of the free flow phase (F) with respect to an F→S transition only, rather than the metastability of the synchronized flow phase (S) with respect to an S→F transition. The reason for this limitation of the consideration of the Z-characteristic for traffic breakdown made in Sect. 5.3 is as follows. In the phase F, the sum of the flow rates $q_{\text{in}} + q_{\text{on}}$ is equal to the flow rate $q_{\text{sum}} = q_{\text{in}} + q_{\text{on}}$ in free flow downstream of the on-ramp bottleneck. For this reason, we can consider the Z-characteristic for traffic breakdown as a function of the flow rate q_{sum} (Fig. 5.5). One of the reasons

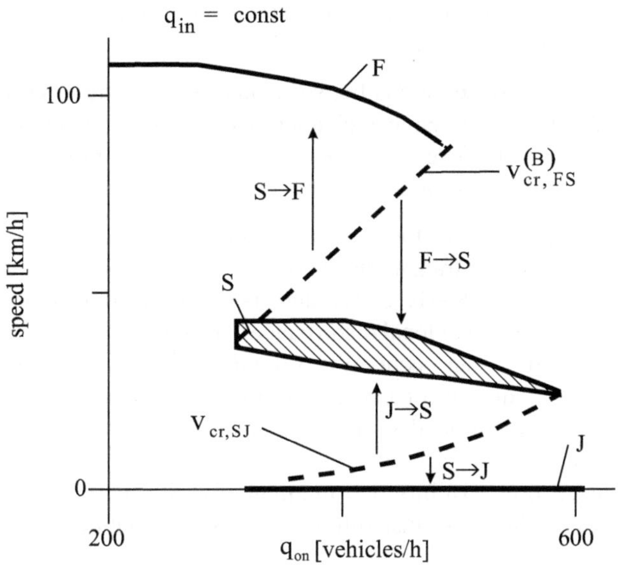

Fig. 8.25 Simulation of 2Z-characteristic for phase transitions with the Kerner-Klenov stochastic microscopic three-phase model: 2Z-characteristic for phase transitions as function of the on-ramp inflow rate q_{on} and given flow rate upstream of the on-ramp bottleneck $q_{in} = 1850$ vehicles/h. The first Z-characteristic between the phases F and S is the same one as that shown in Fig. 5.8. The second Z-characteristic between the phases S and J is the same one as that shown in Fig. 8.8. F—free flow phase, S—synchronized flow phase, J—wide moving jam phase. Arrows F→S, S→F, S→J, and J→S illustrate qualitatively the related phase transitions between the three traffic phase. Adapted from [71]

for phase transitions found in the three-phase theory (Fig. 8.24a). This explains the statement that the basic requirement for a three-phase traffic flow model is that the model should either show or incorporate an 2Z-characteristic for phase transitions in traffic flow.[8] Thus, we can conclude that the 2Z-characteristic for phase transitions in traffic flow should exhibit the following features:

---

for such a Z-characteristic is that it shows the minimum highway capacity C_{min} and maximum highway capacity C_{max} of free flow at the bottleneck. However, it must be stressed that strictly speaking the presentation of the Z-characteristic for phase transitions in Fig. 5.5 is valid for the F→S transition *only*. Contrarily, when the synchronized flow phase (S) is at a highway bottleneck, then in a general case the flow rate within the synchronized flow is not equal to the flow rate q_{in} in free flow upstream of the bottleneck. In this case, the flow rate in free flow downstream of the bottleneck is not equal to the sum $q_{in} + q_{on}$ (see Sects. 8.4.1 and 8.5). For this reason, we present the Z-characteristic for S→J and J→S phase transitions (Fig. 8.8) as well as the 2Z-characteristic (Fig. 8.25) as a function of the on-ramp inflow rate q_{on} at a given flow rate q_{in}.

[8]An 2Z-characteristic can also be shown by three-phase traffic flow models that do not satisfy hypothesis of the three-phase theory about 2D-steady states of synchronized flow (Sect. 5.9.1). In other words, 2D-steady states of synchronized flow is not a necessarily condition for a three-phase traffic flow model [72].

(i) The 2Z-characteristic for phase transitions in traffic flow consists of two different Z-characteristics:

- The first Z-characteristic is between states of the traffic phases F and S in Fig. 8.25. This is the same Z-characteristic as that shown in Fig. 5.8.
- The second Z-characteristic is between states of the traffic phases S and J in Fig. 8.25. This is the same Z-characteristic as that shown in Fig. 8.8.

(ii) There is a finite range of the flow rate within which each of the F→S, S→F, S→J, and J→S phase transitions is possible.

(iii) Each of the F→S, S→F, S→J, and J→S phase transitions exhibits the nucleation nature: Small local disturbances in an initial traffic phase decay, i.e., no phase transition can occur. When the disturbance amplitude in the initial traffic phase is equal to or exceeds a critical value, a phase transition does occur; as above-mentioned, such a local disturbance is called a nucleus for the phase transition.

(iv) Synchronized flow exhibits two qualitatively different kinds of traffic flow instabilities: 1. The S→F instability. 2. The S→J instability.

The basic requirement for a three-phase traffic flow model is that the model should either show or incorporate a 2Z-characteristic for phase transitions in traffic flow.

8.10 Basic Difference Between Three-Phase and Two-Phase Traffic Flow Models

The basic difference between three-phase and two-phase traffic flow models becomes more obvious when we compare the physical sense of the minimum highway capacity C_{min} with the physical sense of threshold flow rate q_{out} (Fig. 8.26).[9]

---

[9]At model parameters chosen in the three-phase model that simulations presented in Fig. 8.26, condition

$$C_{min} < q_{out} \tag{8.17}$$

is satisfied. However, under conditions

$$C_{min} \leq q_{sum} < q_{out} \tag{8.18}$$

no F→J transition is possible. Therefore, under condition (8.18) only the F→S transition can occur at the bottleneck in the three-phase model. In contrast, in the two-phase model at any flow rate $q_{sum} < q_{out}$ no phase transition can be induced in free flow at the bottleneck. We have found

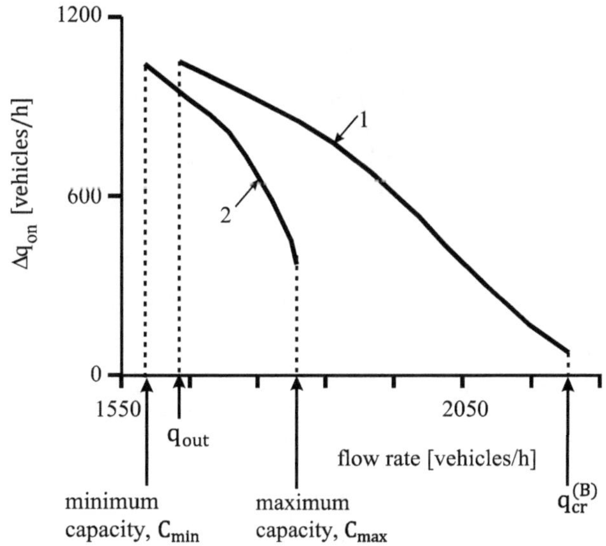

Fig. 8.26 Comparison of flow-rate–dependence of the critical on-ramp inflow impulse $\Delta q_{on}(q_{sum})$ that is required for induced F→J transition in the two-phase model (Nagel-Schreckenberg CA model) (curve 1) with flow-rate–dependence of the critical on-ramp inflow impulse $\Delta q_{on}(q_{sum})$ that is required for induced F→S transition in the three-phase model (KKSW CA model) (curve 2). Curve 1 is adapted from Fig. 8.19a. Curve 2 is adapted from Fig. 8.21a. It should be noted that the part of curve 1 within the flow rate range (8.15) coincides with curve F→J for induced F→J transition in the three-phase model (Fig. 8.22). This is because the Nagel-Schreckenberg CA model (two-phase model) and the KKSW CA model (three-phase model) (Appendix B) used in simulations exhibit the same parameters of free flow metastability with respect to the F→J transition; in particular, the flow rate q_{out} takes the same value in the both models: $q_{out} = 1636$ vehicles/h. Adapted from [78]

In both the three-phase and two-phase models, the characteristic flow rate q_{out} called also as the jam outflow rate is the minimum flow rate q_{sum} downstream of the on-ramp bottleneck at which an F→J transition can still be induced (Figs. 8.19, 8.26, and 8.27a). For this reason, the jam outflow rate q_{out} is the threshold flow rate for the existing of a moving jam in free flow (Sect. 4.5.1). This physical meaning of the jam outflow rate q_{out} is the same in the two-phase and three-phase traffic flow models.

In the three-phase theory, the minimum highway capacity C_{min} is the minimum flow rate q_{sum} at the bottleneck (Fig. 8.27b) at which an F→S transition (traffic breakdown) can still be induced (Sect. 5.4.3.2). For this reason, in the three-phase theory the threshold flow rate for the existing of synchronized flow at the bottleneck is equal to the minimum highway capacity C_{min}.

that the minimum highway capacity C_{min} can depend considerably on the value q_{on}. In particular, at other model parameters than those used in Fig. 8.26 we can also find that $C_{min} > q_{out}$.

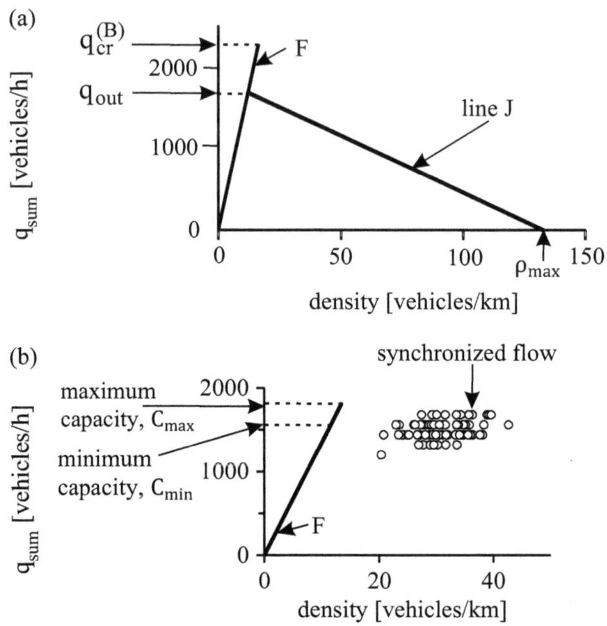

Fig. 8.27 Threshold and critical flow rates at the bottleneck in three-phase and two-phase traffic flow theories. Metastable free flow in the flow–density plane for the two-phase model (**a**) and for the three-phase model (**b**). Synchronized flow (1-min averaged data measured by a virtual detector at location 15.32 km) is related to WSP shown in Fig. 8.21c. F—free flow. Line J is the line *J* (Sect. 4.5.1). Traffic flow characteristics q_{out}, $q_{cr}^{(B)}$, C_{min}, and C_{max} have the same meaning as those in Fig. 8.26, respectively. In (**a**), $\rho_{max} = 1000/d$ is the vehicle density within a wide moving jam. Adapted from [78]

In contrast with the jam outflow rate q_{out} that has the same physical sense in the two-phase and three-phase models, the minimum highway capacity C_{min} has a physical sense *only* for the three-phase model. Indeed, in two-phase traffic flow models no F→S transition that exhibits the nucleation nature is possible. Therefore, the minimum highway capacity C_{min} of the three-phase theory has a sense for *none* of the two-phase traffic flow models. This emphasizes the fundamental physical difference between threshold values C_{min} and q_{out}.

We have found that the maximum highway capacity C_{max} of the three-phase model used for simulations (KKSW CA model) is always smaller than the critical flow rate $q_{cr}^{(B)}$ for spontaneous F→J transition of the two-phase model (Nagel-Schreckenberg CA model) (Fig. 8.26):

$$C_{max} < q_{cr}^{(B)}. \tag{8.19}$$

Thus, the critical flow rate $q_{cr}^{(B)}$ for spontaneous F→J transition cannot be reached in simulations with the three-phase model. This is because, as shown above, in the three-phase theory at any flow rate q_{sum} at which either an F→J transition or an F→S transition is possible, the probability of the F→S transition at the bottleneck is considerably larger than that for the F→J transition.

The metastability of free flow with respect to the F→J transition resulting from the classical traffic flow instability of Herman, Gazis, Montroll, Potts, Rothery, and Chandler [8, 19, 20, 29] can be represented in the flow–density plane by the line J that intersects the curve for free flow in the threshold point q_{out} (Fig. 8.27a). However, this theoretical result of the Kerner-Konhäuser theory of wide moving jams [81], which now is a theoretical basis of a huge number of further theoretical studies of traffic flow (see references in [28, 52, 61–64, 66, 80–82, 96, 97, 113]), has *no relation* to the metastability of free flow observed in real measured traffic data.

Indeed, as explained in Chaps. 3 and 5, the metastability of free flow at a highway bottleneck observed in real measured traffic data results from the metastability of free flow with respect to F→S transition at the bottleneck. Characteristics of the metastability of free flow with respect to F→S transition at the bottleneck are the minimum highway capacity C_{min} and the maximum highway capacity C_{max} of the three-phase theory. In Fig. 8.27b, we have illustrated these characteristics of the metastability of free flow with respect to F→S transition at the bottleneck in the flow–density plane together with synchronized flow points related to the flow rate and density within synchronized flow at the bottleneck.

The existence in the three-phase theory of the minimum highway capacity C_{min} at which traffic breakdown (F→S transition) can still be induced at a highway bottleneck has *no sense* for the classical traffic and transportation theories.

We can conclude the main results of this analysis as follows.

1. There are some common features of two-phase and three-phase traffic flow models:
   (i) In both two-phase and three-phase traffic flow models, an induced F→J transition is possible.
   (ii) In accordance with item (i), in both the two-phase and three-phase traffic flow models there is a threshold flow rate in free flow at which a moving jam can still be induced in free flow at the bottleneck (induced F→J transition).
   (iii) This threshold flow rate related to the induced F→J transition at the bottleneck is the same in both the two-phase and three-phase traffic flow models: This threshold flow rate is equal to the flow rate in free flow formed in the outflow from a wide moving jam q_{out} as discussed in Sect. 4.5.1 (see Fig. 4.9).
   (iv) As explained in Sect. 4.5.1, in two-phase traffic flow models of the GM model class the existence of the threshold flow rate q_{out} is associated

with the classical traffic flow instability of Herman, Gazis, Montroll, Potts, Rothery, and Chandler [8, 19, 20, 29].

2. There are some qualitatively different features of the two-phase and three-phase traffic flow models that lead to the conclusion about the incommensurability of the three-phase theory with all earlier traffic flow theories:

    (i) In contrast with the two-phase traffic flow models, in the three-phase theory, additionally to the classical traffic flow instability, there is also another traffic flow instability, an S→F instability. The S→F instability governs an F→S transition at the bottleneck (Sect. 5.12). In other words, there are two qualitatively different instabilities in the three-phase theory.

    (ii) The S→F instability determines the existence a minimum highway capacity C_{min}: In the three-phase theory, the threshold flow rate in free flow at which traffic breakdown (F→S transition) can still be induced at a highway bottleneck is equal to the minimum highway capacity C_{min}.

    (iii) In the three-phase theory, the threshold flow rate q_{out} and the minimum highway capacity C_{min} are totally different physical threshold characteristics of free flow at the bottleneck:

        - The threshold flow rate q_{out} determines the threshold for the F→J transition.
        - The minimum highway capacity C_{min} determines the threshold for the F→S transition at the bottleneck.

    The minimum highway capacity C_{min} has no sense for the classical traffic flow theories.

The S→F instability of the three-phase theory has no sense for the classical traffic flow theories. The minimum highway capacity C_{min} that determines the threshold for the F→S transition at the bottleneck results from features of the S→F instability. For this reason, the minimum highway capacity C_{min} has also no sense for the classical traffic flow theories. This shows *the incommensurability* of the three-phase theory with all earlier traffic flow theories.

8.11 Stochastic Highway Capacity: Classical Theory Versus Three-Phase Theory

As explained in Sect. 4.10, in the classical understanding of stochastic highway capacity $C(t)$, free flow is stable under condition $q_{sum}(t) < C(t)$ (4.30). This means that no traffic breakdown can occur or be induced at the bottleneck at long as the flow rate in free flow at the bottleneck is smaller than the stochastic highway capacity. This contradicts to the empirical fact that traffic breakdown can be induced at the bottleneck due to the upstream propagation of a localized congested pattern (Sect. 3.6).

We have also already mentioned in Sect. 4.10 that in contrast with the classical understanding of stochastic highway capacity, the evidence of the empirical induced traffic breakdown (induced F→S transition) means that free flow is in a metastable state with respect to the breakdown (F→S transition). The metastability of free flow at the bottleneck with respect to the F→S transition (traffic breakdown) should exist for all flow rates at which traffic breakdown can be induced at the bottleneck as observed in real traffic (Fig. 3.16a). The evidence of the empirical metastability of free flow at the bottleneck with respect to traffic breakdown contradicts fundamentally the concept of the Brilon's stochastic highway capacity (Sect. 4.10), in which free flow is stable under condition (4.30). This explains why the generally accepted classical understanding of stochastic highway capacity [5–7, 14–16, 21, 24–27, 30, 31, 92, 93, 105] is invalid for real traffic.

The classical understanding of stochastic highway capacity is based on the assumption that the empirical probability of traffic breakdown is determined by the capacity distribution function, i.e., condition (4.29) is valid. In contrast, the assumption of the three-phase theory about the metastability of traffic breakdown with respect to traffic breakdown (5.7) is based on the evidence that empirical traffic breakdown can be induced at a bottleneck. In both the classical theory and three-phase traffic theory there are the infinite number of stochastic highway capacities. However, in the classical understanding of stochastic highway capacity at a *given time instant* there is *only one* value of capacity (Fig. 4.13) that we do not know because capacity is a stochastic value.

Contrarily, in the three-phase theory the existence of an infinite number of highway capacities *at any time instant* means that highway capacity is stochastic. Indeed, in the three-phase theory at any given time instant there are the infinite number of stochastic capacities within some capacity range between minimum $C_{\rm min}^{\rm (stoch)}$ and maximum highway capacities $C_{\rm max}^{\rm (stoch)}$ (Sects. 5.1 and 5.7). We do not know values of $C_{\rm min}^{\rm (stoch)}$ and $C_{\rm max}^{\rm (stoch)}$ because $C_{\rm max}^{\rm (stoch)}(t)$ and $C_{\rm min}^{\rm (stoch)}(t)$ are stochastic values. Due to the evidence of the possibility of empirical induced traffic breakdown, we know only that $C_{\rm max}^{\rm (stoch)}(t) > C_{\rm min}^{\rm (stoch)}(t)$ (Fig. 5.14).

Thus, the observation of empirical induced breakdowns proves that condition (4.29) of Brilon's stochastic highway capacity [6, 7, 14, 15, 21] cannot be valid for real traffic. However, the following question arises:

- What are the consequences of this controversial understanding of the nature of traffic breakdown?

With the use of Fig. 8.28, we can qualitatively illustrate the basic difference between the classical understanding of stochastic highway capacity and the understanding of the infinite number of stochastic highway capacities made in the three-phase theory. In the classical understanding of stochastic capacity (4.29), for

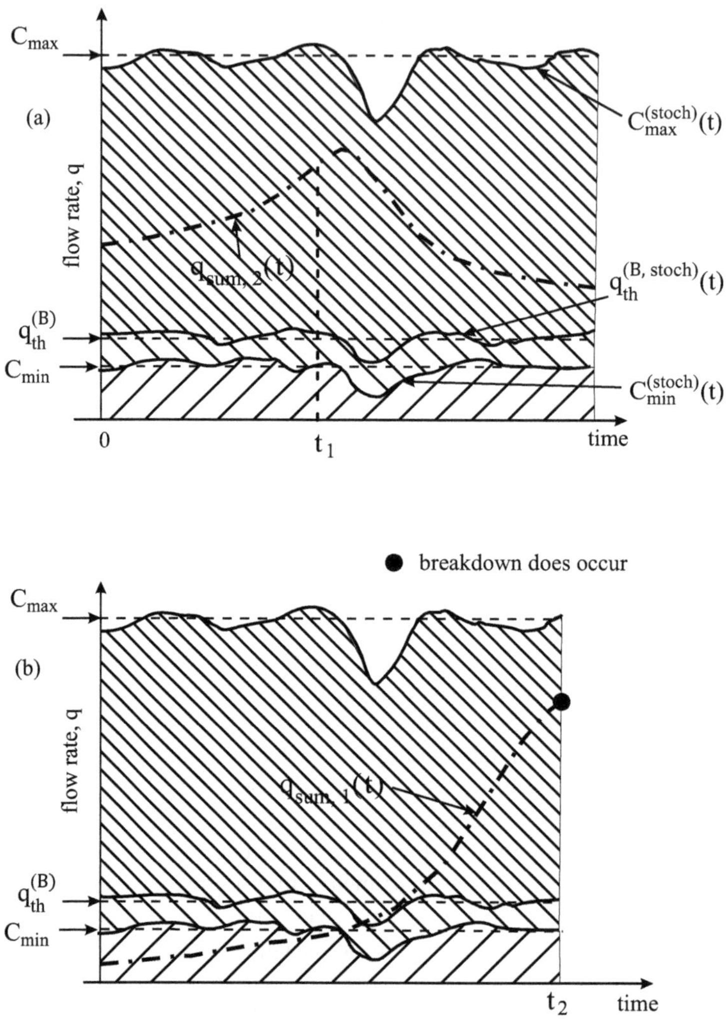

Fig. 8.28 Qualitative explanation of traffic breakdown with the use of the infinite number of capacities of free flow at a highway bottleneck of the three-phase theory. Hypothetical time-functions $C^{(stoch)}_{max}(t)$, $C^{(stoch)}_{min}(t)$, and $q^{(B,\ stoch)}_{th}(t)$ are adapted from Fig. 5.14. Hypothetical time functions of the flow rates $q_{sum,\ 2}(t)$ in (**a**) and $q_{sum,\ 1}(t)$ in (**b**) as well as time instant t_1 in (**a**) are, respectively, the same as those in Fig. 4.14. Adapted from [67–69]

the hypothetical time dependence of the flow rate $q_{sum,\ 2}(t)$ shown in Fig. 4.14b, traffic breakdown has occurred at time instant t_1 at which condition $q_{sum,\ 2}(t) = C(t)$ is satisfied, i.e., when the flow rate is equal to the capacity value. In contrast, in accordance with the three-phase theory for the same time dependence of the flow rate $q_{sum,\ 2}(t)$, for which conditions (5.16) are satisfied, no breakdown should be necessarily occur both at time instant t_1 and for a later time interval (Fig. 8.28a).

In the classical understanding of stochastic capacity (4.29), for the hypothetical time dependence of the flow rate $q_{\text{sum},\,1}(t)$ shown in Fig. 4.14a, traffic breakdown could not occur because for all time instants condition $q_{\text{sum}}(t) < C(t)$ (4.30) is satisfied. In contrast, in accordance with the three-phase theory for the same time dependence of the flow rate $q_{\text{sum},\,1}(t)$ traffic breakdown can occur spontaneously as this is shown for time instant t_2 in Fig. 8.28b.

Because the classical understanding of stochastic highway capacity (4.28), (4.29) contradicts the empirical nucleation nature of real traffic breakdown, the understanding of stochastic highway capacity made in [5–7, 14, 15, 21] cannot be used for reliable highway design and highway operations.

This critical conclusion is also related to a diverse variety of applications of the classical understanding of highway capacity. An example is the term *capacity drop* that criticism has been given in Sect. 4.11.2. Other invalid terms of the classical traffic flow theory associated with the classical understanding of highway capacity are *critical density*, *critical flow rate*, and *critical occupancy*. We have already briefly critically discuss these terms in Sect. 4.9.2. These terms follow either from the LWR theory or from a study of traffic flow models of the GM model class. However, as we have explained in Sect. 4.7 as well as in this chapter above, this understanding of critical density, critical occupancy, and critical flow rate contradicts fundamentally the empirical nucleation nature of traffic breakdown at the bottleneck (Chap. 3). Indeed, there are neither some critical density, nor critical occupancy, nor else critical flow rate within the flow rate range between the minimum and maximum highway capacities of the three-phase theory: At any density, any occupancy, or any flow rate related to the flow rate range in free flow (5.3) of Sect. 5.1, traffic breakdown can occur with some finite probability at the bottleneck. In particular, this explain why well-known methods for dynamic traffic control (like on-ramp metering and speed limit control) based on these assumptions of the classical traffic flow theories are invalid for real applications. This critical conclusion is also related to many other free flow control methods developed by traffic researchers in accordance with conclusions of the classical traffic flow theories (see references in Sect. 4.9).

8.12 Conclusions

(i) The incommensurability of the classical traffic theories (two-phase traffic flow models) with the three-phase theory is associated with the description of the nucleation nature of traffic breakdown at a highway bottleneck, rather than with the description of congested traffic resulting from traffic breakdown.

(ii) There is *no* basic difference between the two-phase and three-phase traffic flow models in relation to the metastability of traffic flow with respect to the emergence of moving jams either in free flow (F→J transition) or in synchronized flow (S→J transition). This conclusion is explained as follows.

- The metastability of free flow and metastability of synchronized flow with respect to moving jam emergence is associated with the classical traffic flow instability of Herman, Gazis, Montroll, Potts, Rothery, and Chandler [8, 19, 20, 29] that has been incorporated in the two-phase and three-phase models.

(iii) Contrarily to the two-phase traffic flow models, additionally to the metastability of free flow with respect to the F→J transition, three-phase traffic flow models exhibit the metastability of free flow with respect to a transition from free flow to synchronized flow (F→S transition).

(iv) This metastability of free flow with respect to the F→S transition of the three-phase theory explains the empirical nucleation nature of traffic breakdown at a highway bottleneck—the empirical fundamental of transportation science.

(v) The metastability of free flow with respect to the F→S transition introduced in the three-phase theory has a sense for *none* of the classical traffic flow models and theories.

(vi) The metastability of free flow with respect to the F→S transition is governed by the S→F instability introduced in the three-phase theory.

(vii) The nucleation nature of the S→F instability introduced in the three-phase theory explains the empirical nucleation nature of traffic breakdown at the bottleneck that is the empirical fundamental of transportation science. The S→F instability has no sense for the classical traffic flow models and theories. This shows the incommensurability of the classical traffic flow models and theories with the three-phase theory.

(viii) The S→F instability introduced in the three-phase theory is also responsible for the following features of traffic flow:

1. At any flow rate at which either an F→S transition or an F→J transition is possible in a three-phase traffic flow model, the critical nucleus required for the F→J transition is considerably larger than the critical nucleus required for the F→S transition (traffic breakdown) at the bottleneck.
2. A consequence of the latter conclusion is as follows: At any flow rate at which either an F→S transition or an F→J transition is possible in the three-phase model, the probability of the F→S transition is considerably larger than the probability of the F→J transition at the bottleneck. This explains why rather than the F→J transition, only the F→S transition is observed in real field traffic data (Chap. 3).
3. In the three-phase theory, the S→F instability determines the existence of the minimum highway capacity $C_{\min}$. The flow rate at which an F→S transition can still occur at the bottleneck is equal to the minimum highway capacity $C_{\min}$.
4. In the three-phase theory, the threshold flow rate q_{out} and the minimum highway capacity $C_{\min}$ are totally different physical threshold characteristics of free flow at the bottleneck:

- The threshold flow rate q_{out} determines the threshold for the F→J transition at the bottleneck. The threshold flow rate q_{out} has the same meaning in the two-phase and three-phase traffic flow models.
- The minimum highway capacity C_{min} determines the threshold for the F→S transition at the bottleneck. In contrast with the three-phase theory, the minimum highway capacity C_{min} has no sense for the classical traffic flow theories and models.

(ix) The classical understanding of stochastic highway capacity contradicts the empirical nucleation nature of real traffic breakdown. For this reason, the classical understanding of stochastic highway capacity cannot be used for reliable highway design and highway operations. This critical conclusion is also related to a diverse variety of applications of the classical understanding of stochastic highway capacity for dynamic traffic control methods.

The incommensurability of the classical traffic theories (two-phase traffic flow models) with the three-phase theory is associated with the description of the nucleation nature of traffic breakdown at a highway bottleneck, rather than with the description of congested traffic resulting from traffic breakdown.

The empirical nucleation nature of traffic breakdown at a highway bottleneck is the empirical fundamental of transportation science. The S→F instability governs the nucleation nature of traffic breakdown at the bottleneck. The basic feature of the three-phase theory is the nucleation nature of the S→F instability. The S→F instability has no sense for the classical traffic flow models and theories. Therefore, the classical traffic models and theories are *not* consistent with the empirical fundamental of transportation science. This explains why ITS-applications of the classical traffic models and theories have failed in the real world.

The S→F instability introduced in the three-phase theory is needed to explain the empirical nucleation nature of traffic breakdown at a highway bottleneck that is the empirical fundamental of transportation science. Because the S→F instability has *no* sense for the classical traffic flow models and theories, the three-phase theory is incommensurable with the classical models and theories. The term *incommensurability* has been introduced by Kuhn [89] to explain a paradigm shift in a scientific field.

References

1. A. Aw, M. Rascle, SIAM J. Appl. Math. **60**, 916–938 (2000)
2. M. Bando, K. Hasebe, A. Nakayama, A. Shibata, Y. Sugiyama, Phys. Rev. E **51**, 1035–1042 (1995)
3. R.Barlović, L. Santen, A. Schadschneider, M. Schreckenberg, Eur. Phys. J. B **5**, 793–800 (1998)
4. R. Borsche, M. Kimathi, A. Klar, Comp. Math. Appl. **64**, 2939–2953 (2012)
5. W. Brilon, J. Geistefeldt, M. Regler, in *Traffic and Transportation Theory* ed. by H.S. Mahmassani (Elsevier Science, Amsterdam, 2005), pp. 125–144
6. W. Brilon, M. Regler, J. Geistefeld Straßenverkehrstechnik, Heft **3**, 136 (2005)
7. W. Brilon, H. Zurlinden Straßenverkehrstechnik, Heft **4**, 164 (2004)
8. R.E. Chandler, R. Herman, E.W. Montroll, Oper. Res. **6**, 165–184 (1958)
9. D. Chowdhury, L. Santen, A. Schadschneider, Phys. Rep. **329**, 199 (2000)
10. C.F. Daganzo, Trans. Res. B **28**, 269–287 (1994)
11. C.F. Daganzo, Transp. Res. B **29**, 79–93 (1995)
12. C.F. Daganzo, *Fundamentals of Transportation and Traffic Operations* (Elsevier Science, New York, 1997)
13. C.F. Daganzo, M.J. Cassidy, R.L. Bertini, Transp. Res. A **33**, 365–379 (1999)
14. L. Elefteriadou, *An Introduction to Traffic Flow Theory, Springer Optimization and Its Applications*, vol. 84 (Springer, Berlin 2014)
15. L. Elefteriadou, A. Kondyli, W. Brilon, F.L. Hall, B. Persaud, S. Washburn, J. Transp. Eng. **140**, 04014003 (2014)
16. L. Elefteriadou, R.P. Roess, W.R. McShane Transp. Res. Rec. **1484**, 80–89 (1995)
17. N.H. Gartner, C.J. Messer, A. Rathi (eds.), *Traffic Flow Theory* (Transportation Research Board, Washington, D.C., 2001)
18. D.C. Gazis, *Traffic Theory* (Springer, Berlin, 2002)
19. D.C. Gazis, R. Herman, R.B. Potts, Oper. Res. **7**, 499–505 (1959)
20. D.C. Gazis, R. Herman, R.W. Rothery, Oper. Res. **9**, 545–567 (1961)
21. J. Geistefeldt, W. Brilon, in *Transportation and Traffic Theory 2009*, ed. by W.H.K. Lam, S.C. Wong, H.K. Lo (Springer, Dordrecht, Heidelberg, London, New York, 2009), pp. 583–602
22. P.G. Gipps, Transp. Res. B **15**, 105–111 (1981)
23. P.G. Gipps, Transp. Res. B. **20**, 403–414 (1986)
24. F.L. Hall, Transp. Res. A **21**, 191–201 (1987)
25. F.L. Hall, K. Agyemang-Duah Transp. Res. Rec. **1320**, 91–98 (1991)
26. F.L. Hall, V.F. Hurdle, J.H. Banks Transp. Res. Rec. **1365**, 12–18 (1992)
27. F.L. Hall, M.A. Gunter, Transp. Res. Rec. **1091**, 1–9 (1986)
28. D. Helbing, Rev. Mod. Phys. **73**, 1067–1141 (2001)
29. R. Herman, E.W. Montroll, R.B. Potts, R.W. Rothery, Oper. Res. **7**, 86–106 (1959)
30. *Highway Capacity Manual 2000* (National Research Council, Transportation Research Board, Washington, D.C., 2000)
31. *Highway Capacity Manual 2010* (National Research Council, Transportation Research Board, Washington, D.C., 2010)
32. R. Jiang, M.-B. Hu, H.M. Zhang, Z.-Y. Gao, B. Jia, Q.-S. Wu, Transp. Res. B **80**, 338–354 (2015)
33. R. Jiang, M.-B. Hu, H.M. Zhang, Z.-Y. Gao, B. Jia, Q.-S. Wu, B. Wang, M. Yang, PLOS One **9**, e94351 (2014)
34. R. Jiang, C.-J. Jin, H.M. Zhang, Y.-X. Huang, J.-F. Tian, W. Wang, M.-B. Hu, H. Wang, B. Jia, in *Proc. Int. Symposium on Traffic and Transportation Theory ISTTT22* (2017) (accepted)
35. R. Jiang, Q.S. Wu, Z.J. Zhu, Phys. Rev. E **64**, 017101 (2001)
36. B.S. Kerner, in *Proceedings of the 3rd Symposium on Highway Capacity and Level of Service*, ed. by R. Rysgaard, vol. 2. (Road Directorate, Ministry of Transport, Denmark, 1998), pp. 621–642

37. B.S. Kerner, Phys. Rev. Lett. **81**, 3797–3800 (1998)
38. B.S. Kerner, Trans. Res. Rec. **1678**, 160–167 (1999)
39. B.S. Kerner, in *Transportation and Traffic Theory*, ed. by A. Ceder (Elsevier Science, Amsterdam, 1999), pp. 147–171
40. B.S. Kerner, Physics World **12**, 25–30 (1999)
41. B.S. Kerner, J. Phys. A Math. Gen. **33**, L221–L228 (2000)
42. B.S. Kerner, in *Traffic and Granular Flow '99: Social, Traffic and Granular Dynamics*, ed. by D. Helbing, H.J. Herrmann, M. Schreckenberg, D.E. Wolf (Springer, Heidelberg, Berlin, 2000), pp. 253–284
43. B.S. Kerner, Transp. Res. Rec. **1710**, 136–144 (2000)
44. B.S. Kerner, in *2001 IEEE Intelligent Transp. Systems Conference Proc.* (Oakland, USA, 2001), pp. 88–93
45. B.S. Kerner, Netw. Spat. Econ. **1**, 35–76 (2001)
46. B. S. Kerner, Transp. Res. Rec. **1802**, 145–154 (2002)
47. B. S. Kerner, in *Traffic and Transportation Theory in the 21st Century*, ed. by M.A.P. Taylor (Elsevier Science, Amsterdam, 2002), pp. 417–439
48. B.S. Kerner, Phys. Rev. E. **65**, 046138 (2002)
49. B. S. Kerner, Math. Comput. Modell. **35**, 481–508 (2002)
50. B.S. Kerner, in *Traffic and Granular Flow' 01*, ed. by M. Schreckenberg, Y. Sugiyama, D. Wolf (Springer, Berlin, 2003), pp. 13–50
51. B.S. Kerner, Physica A **333**, 379–440 (2004)
52. B.S. Kerner, *The Physics of Traffic* (Springer, Berlin, New York, 2004)
53. B.S. Kerner, Physica A **355**, 565–601 (2005)
54. B.S. Kerner, IEEE Trans. ITS **8**, 308–320 (2007)
55. B.S. Kerner, Transp. Res. Rec. **1999**, 30–39 (2007)
56. B.S. Kerner, Traffic Eng. Contr. **48**, 28–35 (2007)
57. B.S. Kerner, Traffic Eng. Contr. **48**, 68–75 (2007)
58. B.S. Kerner, Traffic Eng. Contr. **48**, 114–120 (2007)
59. B.S. Kerner, Transp. Res. Rec. **2088**, 80–89 (2008)
60. B.S. Kerner, J. Phys. A Math. Theor. **41**, 215101 (2008)
61. B.S. Kerner, in *Transportation Research Trends*, ed. by P.O. Inweldi (Nova Science Publishers, New York, 2008), pp. 1–92
62. B.S. Kerner, *Introduction to Modern Traffic Flow Theory and Control* (Springer, Heidelberg, Dordrecht, London, New York, 2009)
63. B.S. Kerner, in *Encyclopedia of Complexity and System Science*, ed. by R.A. Meyers (Springer, Berlin, 2009), pp. 9302–9355
64. B.S. Kerner, in *Encyclopedia of Complexity and System Science*, ed. by R.A. Meyers (Springer, Berlin, 2009), pp. 9355–9411
65. B.S. Kerner, Phys. Rev. E. **85**, 036110 (2012)
66. B.S. Kerner, Physica A **392**, 5261–5282 (2013)
67. B.S. Kerner, in *Vehicular Communications and Networks*, ed. by W. Chen (Woodhead Publishings, Cambridge, 2015), pp. 223–254
68. B.S. Kerner, Elektrotechnik Informationstechnik **132**, 417–433 (2015)
69. B.S. Kerner, Physica A **450**, 700–747 (2016)
70. B.S. Kerner, S.L. Klenov, J. Phys. A Math. Gen. **35**, L31–L43 (2002)
71. B.S. Kerner, S.L. Klenov, Phys. Rev. E **68**, 036130 (2003)
72. B.S. Kerner, S.L. Klenov, J. Phys. A Math. Gen. **39**, 1775–1809 (2006)
73. B.S. Kerner, S.L. Klenov, G. Hermanns, M. Schreckenberg, Physica A **392**, 4083–4105 (2013)
74. B.S. Kerner, S.L. Klenov, A. Hiller, J. Phys. A Math. Gen. **39**, 2001–2020 (2006)
75. B.S. Kerner, S.L. Klenov, A. Hiller, Non. Dynam. **49**, 525–553 (2007)
76. B.S. Kerner, S.L. Klenov, A. Hiller, H. Rehborn, Phys. Rev. E **73**, 046107 (2006)
77. B.S. Kerner, S.L. Klenov, M. Schreckenberg, Phys. Rev. E **84**, 046110 (2011)
78. B.S. Kerner, S.L. Klenov, M. Schreckenberg, Phys. Rev. E **89**, 052807 (2014)
79. B.S. Kerner, S.L. Klenov, D.E. Wolf, J. Phys. A Math. Gen. **35**, 9971–10013 (2002)

80. B.S. Kerner, P. Konhäuser, Phys. Rev. E **48**, 2335–2338 (1993)
81. B.S. Kerner, P. Konhäuser, Phys. Rev. E **50**, 54–83 (1994)
82. B.S. Kerner, P. Konhäuser, M. Schilke, Phys. Rev. E **51**, 6243–6246 (1995)
83. B.S. Kerner, H. Rehborn, Phys. Rev. E **53**, R4275–R4278 (1996)
84. B.S. Kerner, H. Rehborn, Phys. Rev. E **53**, R4275–R4278 (1996)
85. B.S. Kerner, H. Rehborn, Phys. Rev. Lett. **79**, 4030–4033 (1997)
86. B.S. Kerner, H. Rehborn, Internat. Verkehrswesen **50**, 196–203 (1998)
87. B.S. Kerner, H. Rehborn, Internat. Verkehrswesen **50**, 347–348 (1998)
88. S. Krauß, P. Wagner, C. Gawron, Phys. Rev. E **55**, 5597–5602 (1997)
89. T.S. Kuhn, *The Structure of Scientific Revolutions*, 4th edn. (The University of Chicago Press, Chicago, London, 2012)
90. J.A. Laval, C.S. Toth, Y. Zhou, Transp. Res. B **70**, 228–238 (2014)
91. M.J. Lighthill, G.B. Whitham, Proc. Roy. Soc. A **229**, 281–345 (1955)
92. M. Lorenz, L. Elefteriadou, Transp. Res. Circ. **E-C018**, 84–95 (2000)
93. A.D. May, *Traffic Flow Fundamentals* (Prentice-Hall, New Jersey, 1990)
94. T. Nagatani, Physica A **261**, 599–607 (1998)
95. T. Nagatani, Phys. Rev. E **59**, 4857–4864 (1999)
96. T. Nagatani, Rep. Prog. Phys. **65**, 1331–1386 (2002)
97. K. Nagel, P. Wagner, R. Woesler, Oper. Res. **51**, 681–716 (2003)
98. K. Nagel, M. Schreckenberg, J. Phys. (France) I **2**, 2221–2229 (1992)
99. A. Nakayama, M. Fukui, M. Kikuchi, K. Hasebe, K. Nishinari, Y. Sugiyama, S-i. Tadaki, S. Yukawa, New J. Phys. **11**, 083025 (2009)
100. G.F. Newell, Oper. Res. **9**, 209–229 (1961)
101. G.F. Newell, in *Proc. Second Internat. Sympos. on Traffic Road Traffic Flow* (OECD, London 1963), pp. 73–83
102. G.F. Newell, Transp. Res. B **36**, 195–205 (2002)
103. H.J. Payne, in *Mathematical Models of Public Systems*, ed. by G.A. Bekey, vol. 1 (Simulation Council, La Jolla, 1971), pp. 51–60
104. H.J. Payne, Transp. Res. Rec. **772**, 68 (1979)
105. B.N. Persaud, S. Yagar, R. Brownlee, Transp. Res. Rec. **1634**, 64–69 (1998)
106. P.I. Richards, Oper. Res. **4**, 42–51 (1956)
107. A. Schadschneider, D. Chowdhury, K. Nishinari, *Stochastic Transport in Complex Systems* (Elsevier Science, New York, 2011)
108. Y. Sugiyama, M. Fukui, M. Kikuchi, K. Hasebe, A. Nakayama, K. Nishinari, S.-I. Tadaki, S. Yukawa, New J. Phys. **10**, 033001 (2008)
109. S. Tadaki, M. Kikuchi, M. Fukui, A. Nakayama, K. Nishinari, A. Shibata, Y. Sugiyama, T. Yoshida, S. Yukawa, New J. Phys. **15**, 103034 (2013)
110. J.-F. Tian, R. Jiang, B. Jia, Z.-Y. Gao, S.F. Ma, Transp. Res. B **93**, 338–354 (2016)
111. J.-F. Tian, R. Jiang, G. Li, M. Treiber, B. Jia, C.Q. Zhu, Transp. Rec. F **41**, 55–65 (2016)
112. M. Treiber, A. Hennecke, D. Helbing, Phys. Rev. E **62**, 1805–1824 (2000)
113. M. Treiber, A. Kesting, *Traffic Flow Dynamics* (Springer, Berlin, 2013)
114. G.B. Whitham, *Linear and Nonlinear Waves* (Wiley, New York, 1974)
115. G.B. Whitham, Proc. R. Soc. Lond. A **428**, 49 (1990)
116. R. Wiedemann, *Simulation des Verkehrsflusses* (University of Karlsruhe, Karlsruhe, 1974)

Chapter 9
Time-Delayed Breakdown at Traffic Signal in City Traffic

9.1 Introduction—When Can Classical Traffic Flow Theories Be Considered Special Cases of Three-Phase Theory?

As explained in Chap. 8, the basic feature of the three-phase theory is the nucleation nature of the S→F instability. This is because the S→F instability governs the nucleation nature of traffic breakdown (F→S transition) at a highway bottleneck. The nucleation nature of traffic breakdown at the bottleneck explains the empirical fundamental of transportation science. However, the S→F instability has no sense for the classical traffic flow theories. Therefore, the classical theories are *not* consistent with the empirical fundamental of transportation science.

In particular, under "classical traffic flow theories and models" we understand traffic flow models of the GM model class and the LWR model as well as all their applications. For simplicity, to distinguish the classical traffic flow theories from the three-phase theory, we have also used a generic term "two-phase traffic flow models" (Sect. 1.7) for the LWR model and traffic flow models of the GM model class as well as for any combinations of these traffic flow models.

As shown in Sect. 8.10, the nucleation nature of the S→F instability results in the existence of the minimum highway capacity $C_{\min}$. In the three-phase theory, the minimum highway capacity $C_{\min}$ is the minimum flow rate $q_{\rm sum}$ at a highway bottleneck at which an F→S transition (traffic breakdown) can still be induced (Fig. 8.27b). For this reason, in the three-phase theory the threshold flow rate for the existing of synchronized flow at the bottleneck is equal to the minimum highway capacity $C_{\min}$.

We have mentioned above that the S→F instability has no sense for the classical traffic flow theories. For this reason, a two-phase traffic flow model cannot show the metastability of free flow with respect to an F→S transition at the bottleneck. On the one hand, this is the reason for the incommensurability of two-phase and three-phase theories: None of two-phase traffic flow models can explain the empirical nucleation features of traffic breakdown at a highway bottleneck (F→S transition) that is the empirical fundamental of transportation science. On the other hand, this

B.S. Kerner, *Breakdown in Traffic Networks*, DOI 10.1007/978-3-662-54473-0_9

fact explains why the three-phase theory, which is devoted to the explanation of the empirical nucleation features of traffic breakdown at a highway bottleneck (F→S transition), can be considered a new paradigm in traffic and transportation science.

The empirical nucleation nature of traffic breakdown (F→S transition) at a highway bottleneck is the empirical fundamental of transportation science. The empirical fundamental of transportation science can be explained by *none* of classical traffic flow theories. To these classical theories belong two-phase traffic flow models, in particular, traffic flow models of the GM model class and the LWR model.

Contrarily to the classical theories, the basic feature of the three-phase theory is the metastability of free flow with respect to an F→S transition at a highway bottleneck. The three-phase theory explains the empirical nucleation nature of traffic breakdown. This is the reason for the paradigm change in traffic and transportation science.

As known from the history of science (see the classical book by Kuhn [33]), for some applications almost any old theory (old paradigm in a scientific field) can be considered a special case of a new theory (new paradigm): The old theory explains some of the phenomena exactly as the new one. A prominent example is the Newtonian mechanics that in a number of applications can be considered either a special case of relativistic mechanics or a special case of quantum mechanics [33].

Thus, we can expect that there should be a least one kind of traffic breakdown at a road bottleneck whose qualitative explanation is the same with the classical traffic flow theories and with the three-phase theory. When such a kind of traffic breakdown exists, then for the analysis of this phenomenon a classical traffic flow theory can be considered a special case of the three-phase theory. In this chapter, we show that under some conditions discussed below this kind of traffic breakdown can occur at traffic signal in city traffic. As well-known, due to traffic breakdown at traffic signal a transition from under–saturated to over-saturated traffic is realized [2–4, 8, 12, 14, 17, 18, 22, 32, 35–42, 44, 45, 47, 49]. Therefore, the term *traffic breakdown at traffic signal* used below is synonym of the term *transition from under–saturated to over-saturated traffic*.

To study conditions at which classical traffic flow theories can be considered a special case of the three-phase theory, we present a theory of time-delayed traffic breakdown at traffic signal in city traffic developed by the author recently [26, 27]. The term *time-delayed traffic breakdown at the signal* used below is synonym of the term *time-delayed transition from under-saturated to over-saturated traffic at the signal*.

In accordance with the classical theory of traffic at the signal [2–4, 8, 12, 14, 17, 18, 22, 32, 35–42, 44, 45, 47, 49], traffic breakdown (transition from under-traffic to over-saturated traffic) at the signal should occur without any delay when the average arrival flow rate at the signal exceeds a classical signal capacity. In contrast with the classical model of traffic breakdown at the signal, in a theory of traffic at the signal [26, 27], it has been found that at the same value of the average arrival flow rate at the signal that exceeds the classical signal capacity, there can be two states of traffic at the signal: (i) A metastable state of under-saturated traffic. (ii) A state of over-saturated traffic. Due to the existence of the metastable under-saturated traffic, the transition from under- to over-saturated traffic at the signal is a probabilistic effect that exhibits the nucleation nature. For this reason, there can a random time delay before the transition from under-saturated traffic to over-saturated traffic occurs at the signal [26, 27]. The time delay of the transition from under- to over-saturated traffic at the signal can be equal to several signal cycles.

We have found that both traffic flow models of the GM model class and three-phase traffic flow models can describe the time-delayed traffic breakdown at the signal [26, 27]. This is because the signal introduces an extremely large disturbance in traffic flow while forcing all vehicles come to a stop at the signal during the red signal phase. We will also find that when the three-phase model is used for the description of city traffic, then moving synchronized flow patterns (MSP) occurring in under-saturated city traffic effect on *quantitative characteristics* of time-delayed traffic breakdown at the signal only. Therefore, for the description of time-delayed traffic breakdown at the signal, the two-phase models of the GM model class can be considered some approximations of the three-phase models.

In city traffic vehicles have to come to a stop at the signal during the red signal phase. We will show that for this reason both traffic flow models of the GM model class and three-phase traffic flow models can describe the time-delayed traffic breakdown at the signal. For the description of time-delayed traffic breakdown at the signal, the models of the GM model class can be considered some approximations of the three-phase models.

However, we have also found that the classical model of the transition from under- to over-saturated traffic (traffic breakdown) at the signal [2–4, 8, 12, 14, 17, 18, 22, 32, 35–42, 44, 45, 47, 49] can be considered a special case of the theory of time-delayed traffic breakdown at the signal [26, 27] only in a non-realistic case, when all vehicles arrive the signal during the red signal phase (called a "red wave" (RW)) [28].

We will find that the classical model of traffic breakdown at the signal [2–4, 8, 12, 14, 17, 18, 22, 32, 35–42, 44, 45, 47, 49] can be considered a special case of the theory of time-delayed traffic breakdown at the signal [26, 27] only in a non-realistic case, when all vehicles arrive the signal during the red signal phase [28].

The chapter is organized as follows. In Sect. 9.2, we review briefly the classical theory of traffic breakdown at the signal [2–4, 8, 12, 14, 17, 18, 22, 32, 35–42, 44, 45, 47, 49]. An overview of features of the time-delayed traffic breakdown at the signal is given in Sect. 9.3.

An application of the phenomenon of the time-delayed traffic breakdown at the signal to a theory of the breakdown of a green wave (GW) in city traffic is the subject of Sect. 9.4. In Sect. 9.5, we study the effect of time-dependence of the arrival flow rate on probabilistic characteristics of traffic breakdown at the signal.

Based on the theory of the breakdown of the GW in city traffic, in Sect. 9.6 we compare time-delayed traffic breakdown in a three-phase traffic flow model versus a two-phase traffic flow model of the GM model class. We will explain why the time-delayed traffic breakdown at the signal can be shown by both two-phase models of the GM model class and the three-phase model. However, we will explain that due to the existence of the synchronized flow phase in the three-phase model, there is also a difference in results of simulations of the time-delayed traffic breakdown at the signal with two-phase models of the GM model class and the three-phase model [26, 27]. The importance of a consideration of the synchronized flow phase in city traffic follows also from theoretical and empirical studies of over-saturated city traffic made recently in [20, 21, 29, 31].

An analysis of possible reasons for the metastability of under-saturated traffic and the signal as well as a consideration of the three-phase theory versus classical continuum theory of city traffic will be made in Sect. 9.7. In Sect. 9.8, we show that the classical theory of traffic signal shows exactly the same features of traffic breakdown at the signal as that resulting from the three-phase theory *only* for the non-realistic case of "red wave" (RW) in city traffic.

9.2 Traffic Breakdown at Signal in Classical Theory of City Traffic

Results of this section are based on a brief consideration of the classical Webster's model of traffic at the signal [49] as well as the further development of the Webster's model made by several generations of traffic researchers (see, e.g., [2–18, 22, 32, 35–42, 44–47] and references three).

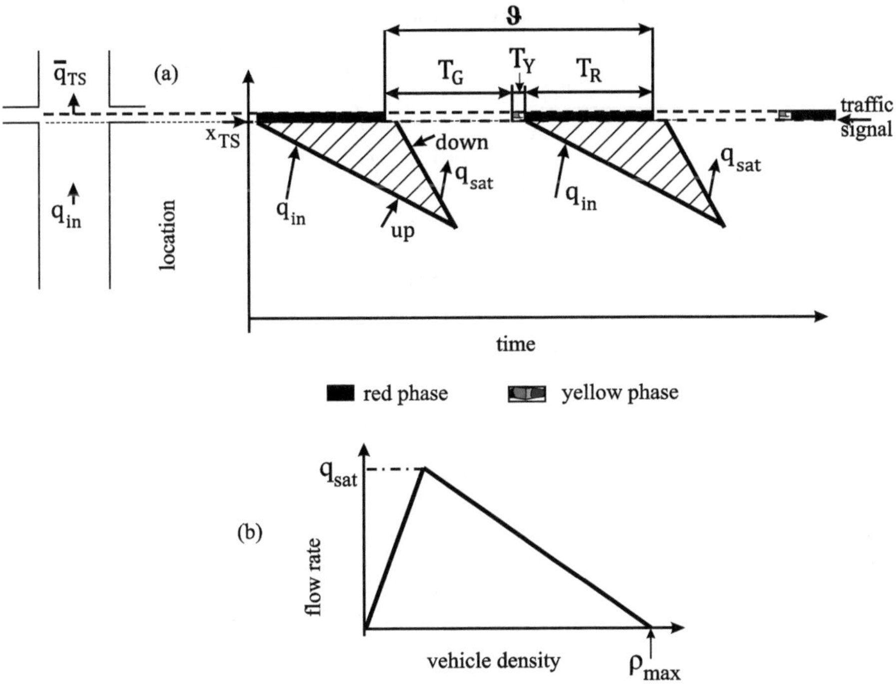

Fig. 9.1 Qualitative explanation of traffic at signal in classical theory of city traffic: (**a**) Queue patterns (*dashed regions*) in under-saturated traffic. (**b**) Fundamental diagram of traffic flow whose maximum point is given by the saturation flow rate q_{sat}. The arrival flow rate q_{in} is assumed to be time-independent. $\bar{q}_{TS}$ is the average flow rate downstream of traffic signal. x_{TS} is the signal location. T_G, T_R, and T_Y are, respectively, the durations of the green, red, and yellow signal phases. $\vartheta = T_G + T_Y + T_R$ is the duration of the signal cycle. Qualitative results that are well-known in the classical theory of traffic at signal (e.g., [2, 4, 14, 18, 22, 32, 35, 37–42, 44, 45, 47, 49])

9.2.1 Vehicle Queue at Signal Versus Wide Moving Jam in Highway Traffic

An important characteristic of the classical theory of traffic at traffic signal is the saturation flow rate q_{sat} (Fig. 9.1). All vehicles that arrive at the signal during the red signal phase must stop at signal location (location x_{TS} in Fig. 9.1a). This results in a vehicular queue within which vehicles are motionless (dashed region in Fig. 9.1a). For simplicity, we assume in this section that the arrival flow rate at the signal q_{in} is time-independent. In this case, the velocity of the upstream front of the queue, at which vehicles decelerate to a stop within the queue, is also time-independent. The related increase in the queue length at the signal is presented by a line labeled by "up" in Fig. 9.1a.

During the green signal phase T_G, vehicles accelerate at the downstream front of the queue at the signal (queue front labeled by "down" in Fig. 9.1a). As a result, the

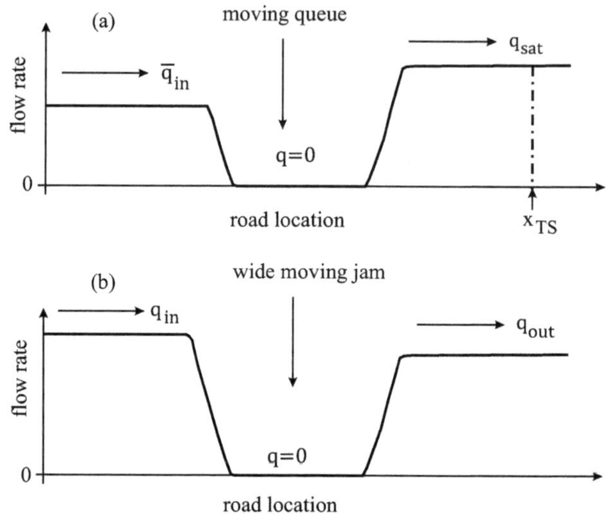

Fig. 9.2 Qualitative explanation of the characteristic flow rates q_{sat} and q_{out}: Qualitative space-dependencies of the flow rates at a fixed time instant for a moving queue during the green signal phase at traffic signal (signal location is labeled by x_{TS}) in city traffic (**a**) and for a wide moving jam in highway traffic (**b**)

downstream front of the queue moves upstream from the signal location over time (Fig. 9.1a). Thus, during the green phase both upstream and downstream fronts of the queue of motionless vehicles move upstream of the signal location. We call the queue of motionless vehicles whose both upstream and downstream fronts moving in the upstream direction as "moving queue". We assume that at the downstream front of the queue all vehicles can freely accelerate to a free flow speed downstream of the signal. In this case, empirical observations show that when the downstream front of the moving queue is far enough from the signal location and almost identical vehicles are within the queue, then during the green signal the average flow rate out of the queue is on average a time-independent value. The average flow rate downstream of the moving queue is called the saturation flow rate q_{sat} (Fig. 9.1) [2, 4, 14, 18, 22, 32, 35, 37–42, 44, 45, 47, 49].

It must be noted that the saturation flow rate q_{sat} of classical theory of city traffic at traffic signal is equal to the flow rate out of a wide moving jam in highway traffic q_{out} (Fig. 9.2):

$$q_{sat} = q_{out}. \tag{9.1}$$

Formula (9.1) is only valid when the following conditions are satisfied:

(i) all vehicles within the wide moving jam are motionless,
(ii) the vehicles accelerate at the downstream front of the wide moving jam to the same free flow speed as the free flow speed downstream of the signal, and

(iii) both the queue length and the jam width in the longitudinal direction (the jam width has the same sense as the length of the moving queue) are long enough.[1]

Thus, in (9.1) we assume that at the downstream front of the moving queue and at the downstream front of the wide moving jam vehicles accelerate from a standstill to the same free flow speed (Fig. 9.2). This explains formula (9.1): Under the above-mentioned conditions (i)–(iii), the effect of vehicle acceleration at the downstream fronts of the moving queue and the wide moving jam is absolutely the same. This means that there is no qualitative difference in the propagation of the downstream front of the moving queue upstream of the signal location during the green signal phase and the propagation of the downstream front of the wide moving jam in highway traffic.

However, there are also basic differences between moving queues at the signal and a wide moving jam in highway traffic. These differences are associated with different reasons of the emergence of these congested patterns. The moving queue occurs usually due to the stop of vehicles occurring during the red signal phase (Fig. 9.1a). The vehicles that have moved initially in free flow must stop at the signal. In other words, the signal "induces" the queue at the signal. Therefore, the queue emergence can be considered an induced F→J transition. In contrast, in highway traffic a wide moving jam occurs usually due to a spontaneous phase transition from the synchronized flow phase to the wide moving jam phase (spontaneous S→J transition; see Sect. 8.3.3).

9.2.2 "Lost Time" and Effective Green Phase Duration at Signal

Another important characteristic of traffic at the signal is so-called "lost time" at the signal δt (see references in the review [14]) (Fig. 9.3). To explain this characteristic, firstly we discuss another traffic flow characteristic, the saturated gross time headway between vehicles $\tau_{\rm sat}^{\rm (gross)}$. Note that a gross time headway $\tau^{\rm (gross)}$ is a gross time gap between two vehicles: At a free flow speed $v_{\rm free}$, the difference between gross time headway and time headway (net time gap) $\tau^{\rm (net)}$ is equal to $d/v_{\rm free}$, where d is the vehicle length:

$$\tau^{\rm (gross)} = \tau^{\rm (net)} + d/v_{\rm free}. \qquad (9.2)$$

The saturated gross time headway $\tau_{\rm sat}^{\rm (gross)}$ is the mean gross time headway in free flow that is formed by the discharge from a moving queue during the green phase of the signal. The flow rate in this free flow is equal to the saturated flow rate $q_{\rm sat}$.

[1] The queue length and the jam width should be considerably larger than the lengths of the upstream and downstream fronts of the moving queue and the wide moving jam, respectively.

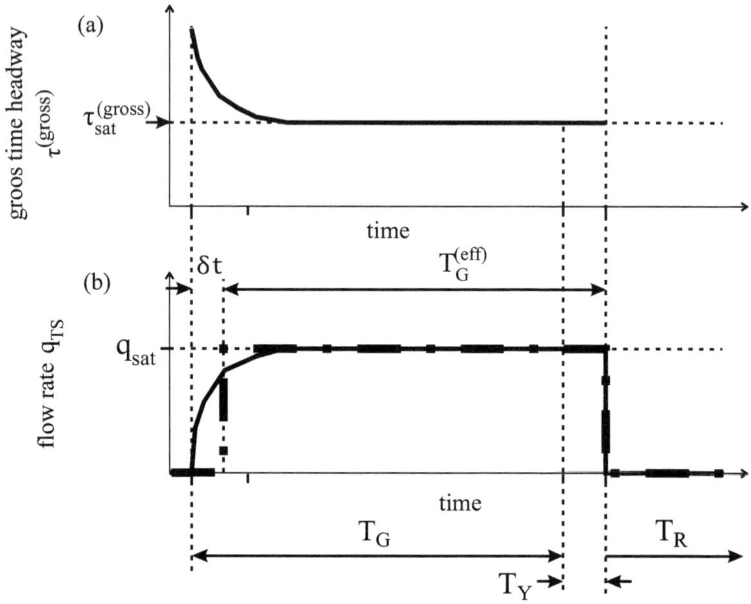

Fig. 9.3 Qualitative explanation of "lost time" δt at the signal: (**a**, **b**) Qualitative dependence of the gross time headway $\tau^{(\text{gross})}$ as time-function at the signal location (**a**) and related time-dependence of the flow rate downstream of traffic signal $q_{\text{TS}}(t)$ (**b**). *Bold dashed-dotted curve* shows the saturated flow rate q_{sat} within the effective green phase time $T_{\text{G}}^{(\text{eff})}$ that corresponds to condition (9.6)

Therefore, the saturated gross time headway $\tau_{\text{sat}}^{(\text{gross})}$ determines the saturation flow rate in accordance with the obvious formula

$$q_{\text{sat}} = 3600/\tau_{\text{sat}}^{(\text{gross})} \quad \text{vehicles/h}. \tag{9.3}$$

Simulations presented in Fig. 9.4 explain formula (9.3) as follows. We consider over-saturated traffic with long enough queues at the signal at a given signal cycle time ϑ for two different values of the green phase duration T_{G} (left and right panels in Fig. 9.4).

To find q_{sat}, we calculate a time-dependence of a gross time headway $\tau^{(\text{gross})}(t)$ for traffic passing the signal during the green phase. As in empirical observations [14, 19], in simulations the gross time headway $\tau^{(\text{gross})}$ between two vehicles following each other is measured at the signal location (stop line of the signal). For each of the signal cycles, we set vehicle running numbers $i = 1, 2 \ldots$, while beginning from vehicle number $i = 1$ for the first vehicle passing the signal, number $i = 2$ for the second vehicle, and so on. Because of model fluctuations in the queue discharge, we average gross time headway $\bar{\tau}^{(\text{gross})}$ separately for each of the vehicles with the same number i in the vehicle sequence passing the signal in different signal cycles (Fig. 9.4b).

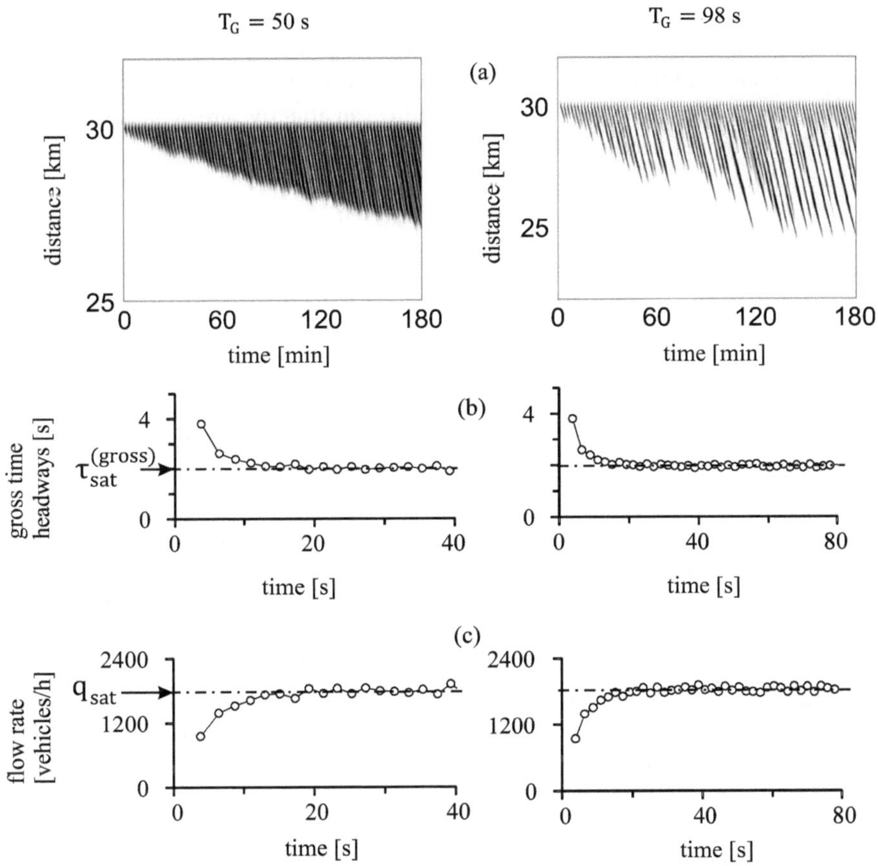

Fig. 9.4 Simulations of saturation flow rate and lost time in over-saturated traffic for different durations of green signal phase T_G at the same signal cycle $\vartheta = 120$ s and the yellow phase $T_Y = 2$ s: Figures in left panel are related to $T_R = 68$ s, i.e., $T_G = 50$ s. Figures in right panel are related to $T_R = 20$ s, i.e., $T_G = 98$ s. (**a**) Speed data presented by regions with variable shades of gray (shades of gray vary from white to black when the speed decreases from 50 km/h (*white*) to zero (*black*)). (**b**) Time-dependence of average gross time headway $\bar{\tau}^{(gross)}$ for vehicles passing the signal; gross time headway is averaged separately for each of the vehicles with the same number in different vehicle sequences found in 900 subsequent signal cycles. (**c**) Time-dependence of the flow rate $q_{TS}(t)$ of traffic passing the signal during the green phase associated with (**b**). Time headway and the flow rate $q_{TS}(t)$ are measured at the signal location $x_{TS} = 30$ km. Simulations of the Kerner-Klenov model for city traffic (Sect. A.12 of Appendix A). Adapted from [30]

From these simulations, we see that a time-function of the gross time headway $\tau^{(gross)}(t)$ decreases firstly from a long gross time headway $\tau^{(gross)}$ related to the first pair of vehicles that have passed the signal after the beginning of each green signal phase (Fig. 9.4b). After a few seconds, there is a saturation of the time-function $\tau^{(gross)}(t)$ to a saturated gross time headway $\tau_{sat}^{(gross)}$. This saturated gross time headway $\tau_{sat}^{(gross)}$ determines the saturation flow rate q_{sat} downstream of a moving queue accordingly to formula (9.3) (Fig. 9.4c).

In the classical theory of traffic at the signal, to take into account this initial decrease in values $\tau^{(\mathrm{gross})}$ at the beginning of each green signal phase (Fig. 9.4b), a lost time δt has been introduced (see references in review [14]). Figures 9.4b and 9.3 illustrate the lost time δt as follows. We introduce an effective green phase time $T_{\mathrm{G}}^{(\mathrm{eff})}$ (Fig. 9.3b) through formula [14]

$$T_{\mathrm{G}}^{(\mathrm{eff})} = \vartheta - T_{\mathrm{R}} - \delta t, \tag{9.4}$$

where

$$\vartheta = T_{\mathrm{G}} + T_{\mathrm{Y}} + T_{\mathrm{R}} \tag{9.5}$$

is the duration of the cycle time of traffic signal that is assumed to be time-independent; T_{G}, T_{Y}, and T_{R} are durations of the green, yellow, and red phases of traffic signal, respectively.[2] In accordance with formula (9.4), the effective green phase time $T_{\mathrm{G}}^{(\mathrm{eff})}$ is the portion of the cycle time during which vehicles are assumed to pass traffic signal at constant rate q_{sat} [14]. This is illustrated in Fig. 9.3b by a bold dashed-dotted curve. In empirical observations, $\delta t \approx 3$–4 s [14, 19].

The physical sense of the lost time can be explained as follows [14]. Vehicles accelerate from a standstill within the queue. After the beginning of the green phase, the flow rate of vehicles passing the signal $q_{\mathrm{TS}}(t)$, which is measured at the signal location x_{TS}, increases during some time interval from zero to q_{sat} (solid dependence of the flow rate $q_{\mathrm{TS}}(t)$ in Fig. 9.3b). Therefore, we can effective consider real complex time-dependence of the flow rate $q_{\mathrm{TS}}(t)$ (solid dependence of the flow rate $q_{\mathrm{TS}}(t)$ in Fig. 9.3b) as a constant value of the saturated flow rate q_{sat} during some effective green phase $T_{\mathrm{G}}^{(\mathrm{eff})}$ and zero outside the time interval of the effective green phase (bold dashed-dotted curve in Fig. 9.3b). This consideration is related to formula

$$q_{\mathrm{sat}} T_{\mathrm{G}}^{(\mathrm{eff})} = \vartheta \bar{q}_{\mathrm{TS}}, \tag{9.6}$$

where for the average flow rate $\bar{q}_{\mathrm{TS}}$ we get

$$\bar{q}_{\mathrm{TS}} = (Z\vartheta)^{-1} \int_0^{Z\vartheta} q_{\mathrm{TS}}(t)dt, \tag{9.7}$$

Z is the number of signal cycles within which the flow rate q_{TS} has been averaged ($Z \gg 1$).

To calculate the lost time δt (Fig. 9.4), we use simulations with the Kerner-Klenov stochastic microscopic three-phase traffic flow model for city traffic (Sect. A.12 of Appendix A). The Kerner-Klenov model for city traffic is also used

[2]The cycle time of traffic signal and the durations of the green, yellow, and red phases of traffic signal are often called signal parameters or signal control settings.

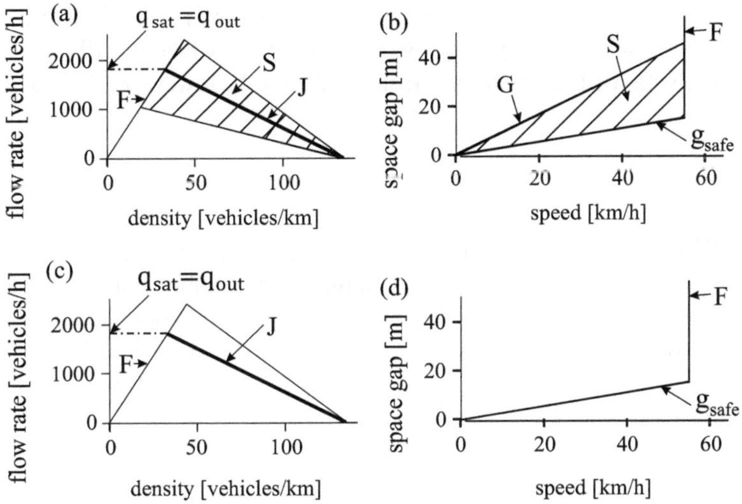

Fig. 9.5 Steady speed states of the Kerner-Klenov microscopic stochastic three-phase traffic flow model (**a**, **b**) and a two-phase traffic flow model of the GM model class (**c**, **d**) in the flow–density (**a**, **c**) and space-gap–speed planes (**b**, **d**). In (**a**, **c**), line J is the line *J* associated with the wide moving jam traffic phase, which corresponds to a moving queue in city traffic. F—free flow, S—synchronized flow

for most of other simulation results presented in this chapter. We find a time interval Δt between the beginning of the green phase and a time instant at which the average gross time headway $\bar{\tau}^{(\mathrm{gross})}$ can be considered as time-independent. This time-independent value $\bar{\tau}^{(\mathrm{gross})}$ is equal to the saturated gross time headway $\tau_{\mathrm{sat}}^{(\mathrm{gross})}$ (Fig. 9.4b). The lost time is found from condition [14, 19]

$$\delta t = \Delta t - n_{\Delta t}\tau_{\mathrm{sat}}^{(\mathrm{gross})}, \tag{9.8}$$

where $n_{\Delta t}$ is the number of vehicles passing the stop line of the signal during the time interval Δt. The calculations have been repeated for several different simulation realizations (runs).[3] We find that with the accuracy about 5% both q_{sat} and δt remain to be approximately equal to (Fig. 9.5)

$$q_{\mathrm{sat}} \approx 1808 \text{ vehicles/h}, \quad \delta t \approx 3.2 \text{ s}. \tag{9.9}$$

[3]Different simulation realizations (runs) are made at the same chosen q_{in} and signal parameters (signal control settings). The only difference between these runs is different *initial conditions* for model fluctuations in the different runs. To reach this goal, initial values for random function *rand*() in the Kerner-Klenov model (Appendix A) have been different ones for different simulation realizations.

As above-mentioned, the saturated gross time headway $\tau_{\mathrm{sat}}^{(\mathrm{gross})}$ determines the saturation flow rate q_{sat} (Fig. 9.4c) in accordance with formula (9.3). The mean value of the saturated net time gap (saturated time headway) $\tau_{\mathrm{sat}}^{(\mathrm{net})}$ is determined by formula

$$\tau_{\mathrm{sat}}^{(\mathrm{net})} = \tau_{\mathrm{del}}^{(\mathrm{acc})}(0), \tag{9.10}$$

where $\tau_{\mathrm{del}}^{(\mathrm{acc})}(0)$ is the mean time delay in vehicle acceleration from a standstill within a moving queue to free flow downstream of the moving queue. For parameters of the Kerner-Klenov stochastic microscopic model used in simulations of city traffic, the value $\tau_{\mathrm{del}}^{(\mathrm{acc})}(0) = 1.5\,\mathrm{s}$; therefore, accordingly to (9.10), we get $\tau_{\mathrm{sat}}^{(\mathrm{net})} = 1.5\,\mathrm{s}$. Thus, as follows from formula (9.2), in which we use $d = 7.5\,\mathrm{m}$ and $v_{\mathrm{free}} = 55\,\mathrm{km/h}$, in formula (9.3) the value $\tau_{\mathrm{sat}}^{(\mathrm{gross})}$ is equal to

$$\tau_{\mathrm{sat}}^{(\mathrm{gross})} \approx 1.991\ \mathrm{s}. \tag{9.11}$$

It should be stressed that as long as $T_{\mathrm{G}} \gg \delta t$, neither q_{sat} nor δt depends on the green phase duration T_{G} (Fig. 9.4b,c). This is exactly the result of the classical theory [2, 4, 14, 18, 22, 32, 35, 37–42, 44, 45, 47, 49].

9.2.3 Classical Signal Capacity

Traffic breakdown at the signal is the transition from under- to over-saturated traffic. In under-saturated traffic at the signal, all vehicles that have come to a stop at the signal during the red signal phase can pass the signal location during the green signal phase. In other words, the queue of vehicles dissolves fully during the green signal phase (Figs. 9.1a and 9.6a).

In contrast, in over-saturated traffic at the signal, the queue of vehicles built during the red signal phase does not dissolve fully during the green signal phase (Fig. 9.6c). As a result, at least one of the vehicles must wait at the signal longer than one signal cycle to pass the signal. In other words, some of the vehicles that have come to a stop at the signal during the red signal phase cannot pass the signal location during the green signal phase.

In the classical theories of city traffic it is assumed [2, 4, 14, 18, 22, 32, 35, 37–42, 44, 45, 47, 49] that there is a particular value of the signal capacity given by formula

$$C_{\mathrm{cl}} = q_{\mathrm{sat}} T_{\mathrm{G}}^{(\mathrm{eff})} / \vartheta. \tag{9.12}$$

We call C_{cl} (9.12) as classical signal capacity.

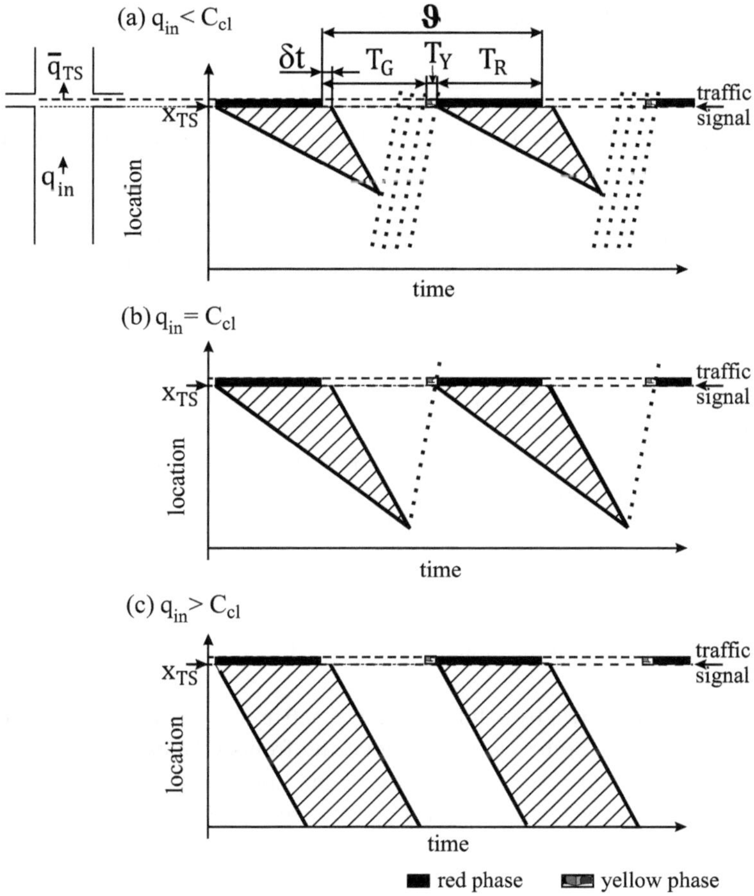

Fig. 9.6 Qualitative explanation of traffic breakdown at signal in classical theory of city traffic at a time independent arrival flow rate q_{in}: (**a–c**) Queue patterns (*solid lines*) in under-saturated traffic at $q_{\mathrm{in}} < C_{\mathrm{cl}}$ (**a**), $q_{\mathrm{in}} = C_{\mathrm{cl}}$ (**b**) and in well-developed over-saturated traffic at $q_{\mathrm{in}} > C_{\mathrm{cl}}$ (**c**); dotted lines—vehicle trajectories. In (**a**), δt is the lost time, x_{TS} is the signal location, $\bar{q}_{\mathrm{TS}}$ is the average flow rate downstream of traffic signal (9.7). Results are adapted from [2, 4, 14, 18, 22, 32, 35, 37–42, 44, 45, 47, 49] (figures have been modified from the originals presented in [2, 4, 14, 18, 22, 32, 35, 37–42, 44, 45, 47, 49])

In the classical theory of traffic at the signal it is assumed that traffic breakdown occurs at the signal when the average arrival flow rate at the signal $\bar{q}_{\mathrm{in}}$ (called as the average arrival traffic rate on the approach [14])

$$\bar{q}_{\mathrm{in}} = \vartheta^{-1} \int_0^{\vartheta} q_{\mathrm{in}}(t)dt \tag{9.13}$$

is larger than the classical signal capacity:

$$\bar{q}_{\rm in} > C_{\rm cl}. \tag{9.14}$$

In (9.13), $q_{\rm in}(t)$ is the arrival flow rate at the signal that can be a complex function of time.

In the classical theory (see, e.g. [2, 4, 14, 18, 22, 32, 35, 37–42, 44, 45, 47, 49]), under-saturated traffic exists at the signal only then, when the average arrival flow rate at the signal $\bar{q}_{\rm in}$ is smaller than the classical capacity, i.e., when

$$\bar{q}_{\rm in} < C_{\rm cl}. \tag{9.15}$$

In this case, on average the queue has fully dissolved and the last vehicle of this queue passes the signal location before the next red phase begins (Fig. 9.6a).

When fluctuations in the arrival flow rate and in the saturated flow rate are neglected (the effect of flow fluctuations will be discussed in Sect. 9.8), in the classical theory it is assumed that in under-saturated traffic under condition

$$\bar{q}_{\rm in} = C_{\rm cl} \tag{9.16}$$

the rate of traffic passing the signal $q_{\rm TS}$ is fully determined by the queue discharge (Fig. 9.6b).

Queue patterns shown by solid lines in Fig. 9.6 are related to a time-independent arrival flow rate $q_{\rm in}$. In the classical theory, the fronts of these queue patterns are calculated with the use of the velocities of the upstream and downstream queue fronts given, respectively, by the well-known formulas [2, 4, 14, 18, 22, 32, 35, 37–42, 44, 45, 47, 49]

$$v_{\rm up} = \frac{q_{\rm in}}{(q_{\rm in}/v_{\rm free}) - \rho_{\rm max}} \tag{9.17}$$

and

$$v_{\rm down} = \frac{q_{\rm sat}}{(q_{\rm sat}/v_{\rm free}) - \rho_{\rm max}}, \tag{9.18}$$

where $\rho_{\rm max} = 1/d$ is the density within the vehicle queue.

Under condition (9.14) there are on average vehicles at the queue end that must stop at the signal during the next red phase. This means that under condition (9.14) the queue grows, i.e., over-saturated traffic must occur at the signal [2, 4, 14, 18, 22, 32, 35, 37–42, 44, 45, 47, 49]. Associated queue patterns at the signal location for well-developed over-saturated traffic are shown by solid lines in Fig. 9.6c calculated with formula $v_{\rm up} = v_{\rm down}$ (9.18).

These results of the classical theory are associated with the assumption that the saturation flow rate $q_{\rm sat}$ is the maximum rate of flow on a street [14]. In turn, this classical assumption is agreed with a fundamental diagram of the classical theories of traffic flow, in particular, with the LWR theory (Fig. 9.1b) [34, 43].

In the classical theory, it is assumed that the presented theory of traffic breakdown at the signal is valid independent of the time-dependence of the arrival flow rate $q_{\rm in}(t)$: When the average arrival flow rate $\bar{q}_{\rm in}$ exceeds the classical signal capacity $C_{\rm cl}$ (9.14), then the transition from under-saturated traffic to over saturated traffic, i.e., the breakdown at the signal does occur.

9.3 Time-Delayed Breakdown at Signal in Two-Phase and Three-Phase Traffic Flow Models: An Overview

Contrarily to the classical theory of traffic breakdown presented in Sect. 9.2, in the theory of traffic breakdown developed by the author recently [26, 27], traffic breakdown at the signal is a time-delayed transition from under-saturated traffic to over-saturated traffic at the signal (time-delayed traffic breakdown at the signal). It has also be found in this theory of traffic breakdown at the signal [26, 27] that features of the transition from under-saturated traffic to over-saturated traffic at the signal depend qualitatively on the time-dependence of the arrival flow rate $q_{\rm in}(t)$.

9.3.1 Metastability of Under-Saturated Traffic at Signal

It turns out that there is a finite range of the average arrival flow rate that exceeds the classical signal capacity (9.14) within which there can be two states of traffic at the signal:

- A metastable state of under-saturated traffic.
- A state of over-saturated traffic.

Due to the existence of the metastable under-saturated traffic, the transition from under- to over-saturated traffic at the signal exhibits the nucleation nature. Due to the nucleation nature of traffic breakdown at the signal, under condition (9.14) within a finite range of the average arrival flow rate discussed below there is a time delay to the transition from under- to over-saturated traffic at the signal. In this section, we make an overview of general features of the time-delayed traffic breakdown at the signal [26, 27].

The general features of the time-delayed traffic breakdown at the signal are as follows [26, 27].

(i) There should not be necessarily traffic breakdown at the signal in metastable under-saturated traffic at the signal: Under condition (9.14), a metastable state of under-saturated traffic can persist at the signal over a long time interval that can be considerably longer than the signal cycle ϑ.

(ii) In the metastable state of the under-saturated traffic, traffic breakdown (transition from under-saturated traffic to over-saturated traffic) at the signal occurs after a time delay denoted by $T^{(B)}$.
(iii) This time delay of traffic breakdown at the signal $T^{(B)}$ is a random value.
(iv) The time delay to the breakdown $T^{(B)}$ can take many signal cycles.
(v) The above-listed features of the time-delayed traffic breakdown at the signal mean that at the same given average arrival flow rate as well as given signal parameters under condition (9.14) metastable under-saturated traffic can exist during several signal cycles.

These features of traffic breakdown at the signal explains why we call the model of traffic breakdown at the signal as the *model of random time-delayed traffic breakdown at the signal*. The model of random time-delayed traffic breakdown at the signal introduced firstly in [26] has been further developed for different parameters of the arrival flow rate and traffic signal in [27, 28, 30].

The model of random time-delayed traffic breakdown at the signal [26, 27] predicts that under-saturated traffic at the signal can exist during a long time interval even under condition (9.14), i.e., when the average arrival flow rate in under-saturated traffic is larger than the classical signal capacity.

9.3.2 General Characteristics of Time-Delayed Traffic Breakdown at Signal

In [26, 27], it has been shown that there are some features of random time-delayed traffic breakdown at the signal, which are general ones in the two-phase models of the GM model class and the three-phase theory. These general features of random time-delayed transition from under-saturated to over-saturated traffic at the signal are as follows:

1. *At any time instant*, there are the infinite number of signal capacities C that are between a minimum signal capacity $C_{\rm min}$ and a maximum signal capacity $C_{\rm max}$:

$$C_{\rm min} \leq C \leq C_{\rm max}, \tag{9.19}$$

   where $C_{\rm min} < C_{\rm max}$. The existence of an infinite number of signal capacities at any time instant means that signal capacity is stochastic.
2. The minimum signal capacity $C_{\rm min}$ is equal to the classical signal capacity $C_{\rm cl}$ (9.12):

$$C_{\rm min} = C_{\rm cl}. \tag{9.20}$$

3. Within a range of the average arrival flow rate $\bar{q}_{in}$

$$C_{min} \le \bar{q}_{in} < C_{max}, \tag{9.21}$$

under-saturated traffic is in a metastable state with respect to the transition from metastable under-saturated traffic to over-saturated traffic (traffic breakdown) at the signal: Any average arrival flow rate $\bar{q}_{in}$ that satisfies (9.21) is a signal capacity.
4. In metastable under-saturated traffic, the probability $P^{(B)}(\bar{q}_{in})$ of spontaneous traffic breakdown at the signal during the observation time T_{ob} is an increasing function of the average arrival flow rate $\bar{q}_{in}$ (Fig. 9.7).
5. There can be two different regions of the metastable under-saturated traffic within which traffic breakdown exhibits different probabilistic features:

   - Under condition

$$q_{th}^{(B)} \le \bar{q}_{in} < C_{max}, \tag{9.22}$$

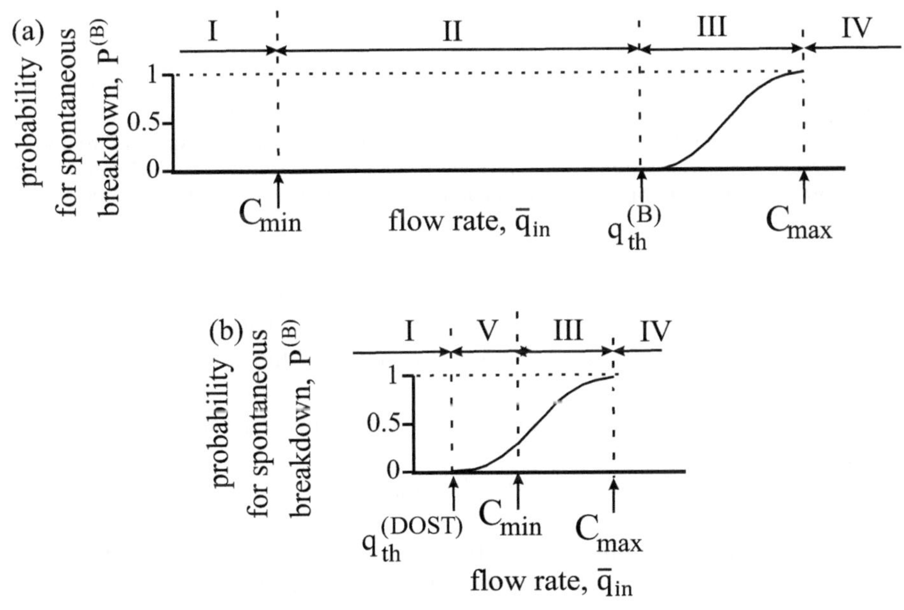

Fig. 9.7 Probabilistic characteristics of random time-delayed traffic breakdown at the signal: Two types (**a**) and (**b**) (see explanations in text) of qualitative dependencies of probability $P^{(B)}(\bar{q}_{in})$ of spontaneous traffic breakdown at the signal during the observation time T_{ob} as function of the average arrival flow rate $\bar{q}_{in}$. The flow rate range I is related to stable under-saturated traffic, II—metastable under-saturated traffic, III—metastable under-saturated traffic in which spontaneous breakdown can occur, IV—unstable under-saturated traffic, and region V (in (**b**))—dissolving over-saturated traffic (DOST). Adapted from [28]

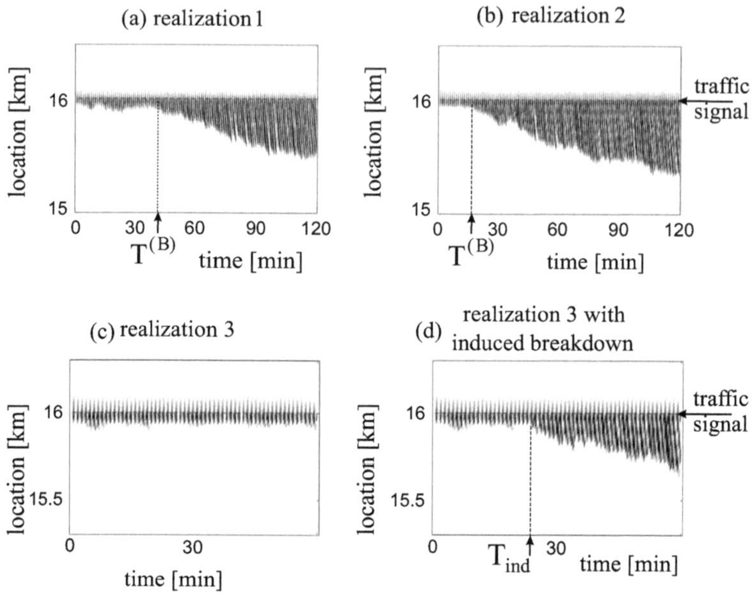

Fig. 9.8 Traffic breakdown at signal in three-phase theory of city traffic: Simulations of different realizations of metastable under-saturated traffic under condition (9.14). Speed data in space-time plane are shown presented by regions with variable shades of gray (shades of gray vary from white to black when the speed decreases from 50 km/h (*white*) to zero (*black*)). $\vartheta = 1$ min, $T^{(B)} =$ 40 min (**a**) and 20 min (**b**). Simulations of the Kerner-Klenov model for city traffic (Sect. A.12 of Appendix A). Adapted from [26]

during a chosen observation time T_{ob} a time-delayed spontaneous traffic breakdown occurs with some probability (region III in Fig. 9.7a). In (9.22), $q_{th}^{(B)}$ is a threshold average arrival flow rate. Within this region III of the average arrival flow rate (Fig. 9.7a) the probability of spontaneous traffic breakdown at the signal during the observation time T_{ob} satisfies condition

$$0 < P^{(B)}(\bar{q}_{in}) < 1. \tag{9.23}$$

This means that in region III of metastable under-saturated traffic (Figs. 9.7a,b) either spontaneous traffic breakdown or induced traffic breakdown can occur (Fig. 9.8).

- Under condition

$$C_{min} \le \bar{q}_{in} < q_{th}^{(B)} \tag{9.24}$$

(region II in Fig. 9.7a) probability for spontaneous traffic breakdown at the signal during the observation time T_{ob} satisfies condition

$$P^{(B)} = 0. \tag{9.25}$$

This means that in region II of metastable under-saturated traffic (Fig. 9.7a) only induced traffic breakdown is possible.

6. At $\bar{q}_{in} = C_{max}$, traffic breakdown occurs spontaneously with probability $P^{(B)} = 1$ during the observation time T_{ob}. This means that under condition

$$\bar{q}_{in} \geq C_{max} \tag{9.26}$$

under-saturated traffic can be considered as unstable under-saturated traffic (region IV in Fig. 9.7).

7. Under condition

$$\bar{q}_{in} < C_{min}, \tag{9.27}$$

features of a random time-delayed transition from under-saturated traffic to over-saturated traffic depend qualitatively on the time-dependence of the arrival flow rate $q_{in}(t)$. The latter result explains two different dependencies of the breakdown probability $P^{(B)}$ on the average arrival flow rate $\bar{q}_{in}$ presented in Figs. 9.7a and 9.7b.

8. When the arrival flow rate $q_{in}(t)$ during the green signal phase is considerably larger than it is during the red signal phase, as this occurs often in a green wave (GW) in a city (see next Sect. 9.4), then the function $P^{(B)}(\bar{q}_{in})$ is given by that shown in Fig. 9.7a. In this case, under-saturated traffic is stable under condition (9.27) (region I in Fig. 9.7a).
9. Contrarily, if the arrival flow rate $q_{in}(t)$ during the green and red signal phases are almost the same, then the function $P^{(B)}(\bar{q}_{in})$ is given by that shown in Fig. 9.7b (Sect. 9.5).

   - In the case shown in Fig. 9.7b, under condition (9.27) dissolving over-saturated traffic (DOST) can occur (physical features of the DOST will be considered in Sect. 9.5.3). In dissolving over-saturated traffic, random emergence and subsequent random dissolution of a growing queue at traffic signal follows each other randomly (Fig. 9.9).
   - The DOST occurs under conditions

$$q_{th}^{(DOST)} \leq \bar{q}_{in} < C_{min} \tag{9.28}$$

   (region V in Fig. 9.7b). In (9.28), the flow rate $q_{th}^{(DOST)}$ is a threshold flow rate for the occurrence of the DOST.
   - When condition (9.28) for the occurrence of the DOST is satisfied, then condition for stable under-saturated traffic, in which no DOST appears, is given by

$$\bar{q}_{in} < q_{th}^{(DOST)} \tag{9.29}$$

   (region I in Fig. 9.7b).

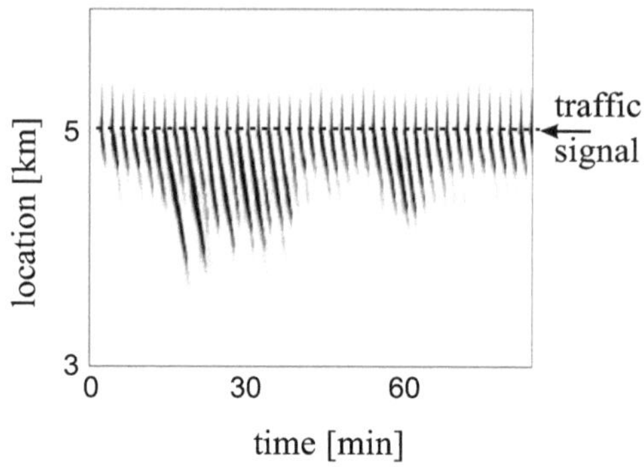

Fig. 9.9 Traffic breakdown at signal in three-phase theory of city traffic: Simulations of dissolving over-saturated traffic (DOST). Speed data in space-time plane are shown presented by regions with variable shades of gray (shades of gray vary from white to black when the speed decreases from 50 km/h (*white*) to zero (*black*)). Simulations of the Kerner-Klenov model for city traffic (Sect. A.12 of Appendix A). Adapted from [28]

It should be noted that there is a characteristic feature of spontaneous traffic breakdown (region III in Figs. 9.7a) that is as follows. As above-mentioned, the spontaneous breakdown occurs after a random time interval $T^{(B)}$ that can considerably exceed the signal cycle ϑ (Figs. 9.8a,b). Because in region III (Fig. 9.7) breakdown probability $P^{(B)} < 1$, in some of the simulation realizations no traffic breakdown occurs during the time interval T_{ob} (Fig. 9.8c). However, the breakdown can be induced by the application of a large enough local speed disturbance in under-saturated traffic (Fig. 9.8d).

9.3.3 *Effect of Large Fluctuations in Under-Saturated Traffic on Time-Delayed Traffic Breakdown at Signal*

For the same reasons as explained in Sect. 5.6, the larger the fluctuations in under-saturated city traffic, the larger the probability that a large nucleus for traffic breakdown occurs spontaneously at the signal during the observation time T_{ob}. The larger nuclei can occur spontaneously in traffic flow, the smaller the difference between the threshold flow rate $q^{(B)}_{th}$ and the minimum signal capacity C_{min}. In a particular limit case of very large traffic flow fluctuations, we should assume that

$$C_{min} = q^{(B)}_{th}. \tag{9.30}$$

Therefore, under conditions (9.21) and (9.30), either spontaneous or induced traffic breakdown can occur in the metastable under-saturated traffic at the signal. In other words, within the flow rate range (9.21) there is a finite probability that traffic breakdown occurs spontaneously during the time interval T_{ob} at the signal. Thus, in this particular under-saturated traffic with large fluctuations (9.30), the flow-rate range II (9.24) disappears and region III (9.22) is determined by conditions (9.21).

9.3.4 Stochastic Minimum and Maximum Signal Capacities

As for highway traffic (Sect. 5.7), the maximum signal capacity $C_{\min}$, the minimum signal capacity $C_{\max}$ as well as the values $q_{\rm th}^{(\rm B)}$ and $q_{\rm th}^{(\rm DOST)}$ depend on traffic parameters. These traffic parameters are weather, mean driver's characteristics (e.g., mean driver reaction time), share of long vehicles, etc. In real city traffic, these traffic parameters change over time. For this reason, the values of the minimum signal capacity $C_{\min}$, the maximum signal capacity $C_{\max}$ as well as the values $q_{\rm th}^{(\rm B)}$ and $q_{\rm th}^{(\rm DOST)}$ change also over time. Moreover, in real city traffic, the traffic parameters are stochastic time functions. Therefore, in real city traffic we should consider some stochastic minimum signal capacity $C_{\min}^{(\rm stoch)}(t)$, stochastic maximum signal capacity $C_{\max}^{(\rm stoch)}(t)$ as well as stochastic threshold flow rates $q_{\rm th}^{(\rm B,\ stoch)}(t)$ and $q_{\rm th}^{(\rm DOST,\ stoch)}(t)$ whose time dependencies are determined by stochastic characteristics of traffic parameters. A hypothetical fragment of time-functions $C_{\min}^{(\rm stoch)}(t)$, $C_{\max}^{(\rm stoch)}(t)$, and $q_{\rm th}^{(\rm B,\ stoch)}(t)$ related to the case shown in Fig. 9.7a is qualitatively the same as that presented in Fig. 5.14 for highway traffic.

We can also generalize the definition of stochastic highway capacity (Sect. 5.7) for stochastic signal capacity at the signal as follows:

At any time instant t, there are the infinite number of stochastic signal capacities C of under-saturated traffic at the signal. The range of these capacities is limited by a minimum stochastic signal capacity $C_{\min}^{(\rm stoch)}(t)$ and a maximum stochastic signal capacity $C_{\max}^{(\rm stoch)}(t)$:

$$C_{\min}^{(\rm stoch)}(t) \leq C \leq C_{\max}^{(\rm stoch)}(t), \tag{9.31}$$

where $C_{\min}^{(\rm stoch)}(t) < C_{\max}^{(\rm stoch)}(t)$.

It should be noted that stochastic functions $C_{\max}^{(\rm stoch)}(t)$, $C_{\min}^{(\rm stoch)}(t)$, $q_{\rm th}^{(\rm B,\ stoch)}(t)$, and $q_{\rm th}^{(\rm DOST,\ stoch)}(t)$ cannot be measured in empirical observations. Only their mean values, respectively, $C_{\max}$, $C_{\min}$, $q_{\rm th}^{(\rm B)}$, and $q_{\rm th}^{(\rm DOST)}$ can be found in empirical studies of measured traffic data. In particular, the mean values $C_{\max}$ and $q_{\rm th}^{(\rm B)}$ can be found from an empirical study of the breakdown probability $P^{(\rm B)}(\bar{q}_{\rm in})$ as a flow rate function (see empirical examples of the dependence of the breakdown probability $P^{(\rm B)}(\bar{q}_{\rm in})$ at the signal in Sect. 9.5.2).

Qualitative features of ranges I–V of the average arrival flow rate $\bar{q}_{\rm in}$ in under-saturated traffic at the signal considered in Sect. 9.3.2 can be generalized for stochastic functions $C_{\max}^{(\rm stoch)}(t)$, $C_{\min}^{(\rm stoch)}(t)$, $q_{\rm th}^{(\rm B,\ stoch)}(t)$, and $q_{\rm th}^{(\rm DOST,\ stoch)}(t)$ as it has been made for highway traffic. For example, formulas (5.31), (5.32), and (5.33) of Sect. 5.7 are also valid for stochastic functions $C_{\max}^{(\rm stoch)}(t)$, $C_{\min}^{(\rm stoch)}(t)$, and $q_{\rm th}^{(\rm B,\ stoch)}(t)$

related to the signal, if we replace in these formulas $q_{\rm sum}$ by $\bar{q}_{\rm in}$. For the occurrence of the DOST, conditions (9.28) should be replaced by conditions $q_{\rm th}^{(\rm DOST,\ stoch)}(t) \leq \bar{q}_{\rm in} < C_{\rm min}^{(\rm stoch)}(t)$.

9.4 Breakdown of Green Wave (GW) in City Traffic in Framework of Three-Phase Theory

In Sect. 9.3 we have mentioned that when the arrival flow rate $q_{\rm in}(t)$ during the green signal phase is considerably larger than it is during the red signal phase, then the flow-rate function of the breakdown probability at the signal $P^{(\rm B)}(\bar{q}_{\rm in})$ is related to that shown in Fig. 9.7a. This case is related to a GW in a city. Results of this section are based on a theory of the breakdown of the GW developed in [27, 28].

9.4.1 Model of GW

In a GW model, a GW propagates on a single-lane road through a sequence of traffic signals with the same signal parameters ϑ, $T_{\rm G}$, $T_{\rm Y}$, and $T_{\rm R}$ (Fig. 9.10). For simplicity, we assume that all signals are at equidistant locations $x = x_{\rm TS}^{(u)}$ with a distance between them $\Delta x_{\rm TS}$, where $u = 1, 2, 3, \ldots, U$; U is the number of signals. For the GW existence, there is should be a time shift $\Delta T_{\rm G}$ between the beginning of the green phases for different signals. For the choice of $\Delta T_{\rm G}$, we assume that all

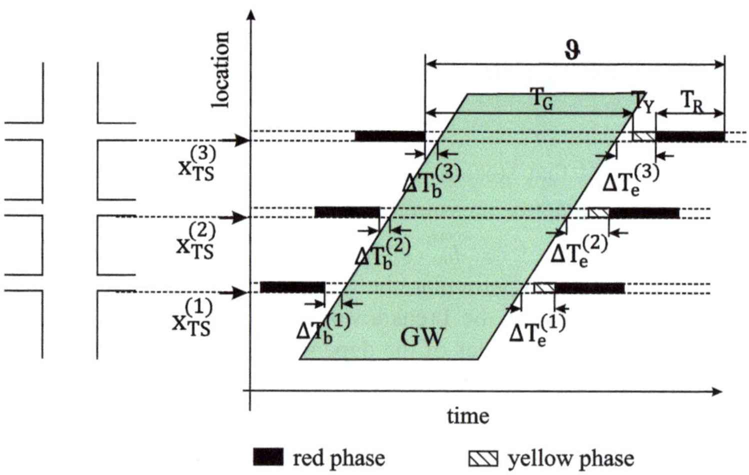

Fig. 9.10 GW model: Model of GW propagating through several city intersections

vehicles within the GW move at a maximum free flow speed v_{free}. Therefore,

$$\Delta T_G = \Delta x_{TS}/v_{free}. \tag{9.32}$$

However, in reality there are speed disturbances within a GW occurring during GW propagation. For this reason, to organize GWs in a city, the GW duration (GW band width) is chosen shorter than the green phase T_G [14]. There are two time intervals [14]:

(i) A time interval ΔT_b between the end of the red phase and the beginning of the GW.
(ii) A time interval ΔT_e between the end of the GW and the beginning of the next red phase.

Values $\Delta T_b^{(u)}$ and $\Delta T_e^{(u)}$ can be different for different signals $u = 1, 2, 3, \ldots, U$ (Fig. 9.10).

When a GW propagates through several traffic signals (Fig. 9.10), the probability that a spontaneous GW breakdown occurs in *at least* at one of the signals is larger

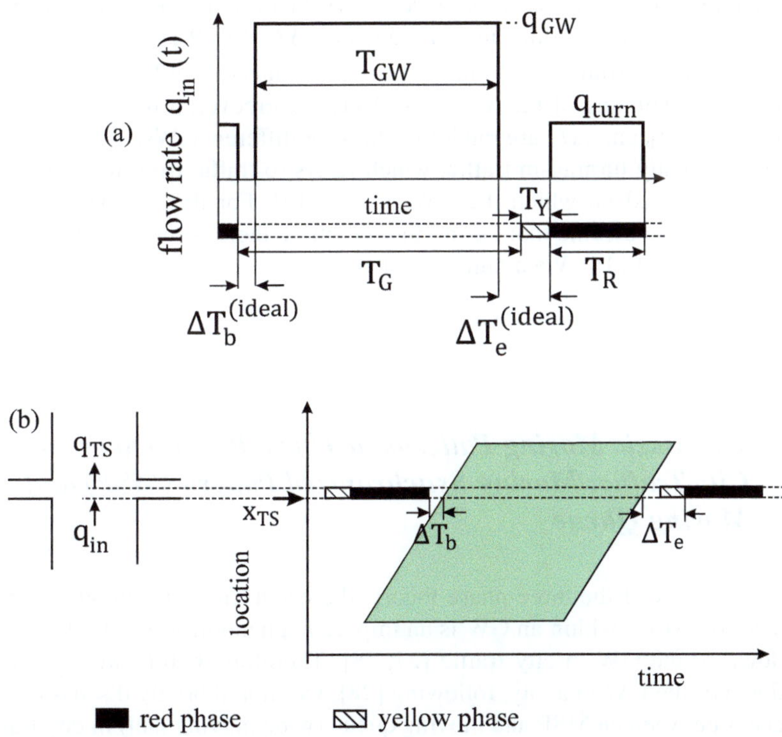

Fig. 9.11 GW model: (**a**) Time-dependence of the arrival flow rate $q_{in}(t)$ during one of the signal cycles $\vartheta = T_G + T_Y + T_R$ related to the beginning of the road $x = x_b$; q_{GW} is the flow rate within GW and q_{turn} is the flow rate of turning-in traffic. (**b**) Model of GW propagation through an isolated signal at single city intersection at $q_{turn} = 0$

than probability that the breakdown occurs *only* at a chosen signal. Therefore, to reveal the physics of under-saturated city traffic, firstly we limit a study by a consideration of traffic an *isolated* signal at location $x = x_{\text{TS}}$ (Fig. 9.11). As shown in Sect. 9.4.6, general physical features of under-saturated city traffic found in this idealized model of the GW propagation through the signal (Fig. 9.11) govern complex spatiotemporal traffic phenomena of self-organization occurring when GWs travel through a sequence of the signals (Fig. 9.10).

Model features of GW propagation through the isolated signal (Fig. 9.11) are as follows. In Fig. 9.11a, $\Delta T_{\text{b}}^{\text{(ideal)}}$ is a given parameter of the GW that determines an "ideal" time interval between the end of the red phase and the beginning of an GW at $x = x_{\text{TS}}$ related to an undisturbed vehicle motion at the free flow speed v_{free}. Respectively,

$$\Delta T_{\text{e}}^{\text{(ideal)}} = T_{\text{G}} + T_{\text{Y}} - T_{\text{GW}} - \Delta T_{\text{b}}^{\text{(ideal)}} \quad (9.33)$$

is an "ideal" time interval between the end of the GW and the beginning of the red phase related to vehicle motion at the free flow speed v_{free}; T_{GW} is the duration of the undisturbed GW.

Simulations of GW propagation have shown that due to speed disturbances in under-saturated traffic real time intervals ΔT_{b} and ΔT_{e} (Fig. 9.11b) can considerably differ, respectively, from ideal values of the time intervals $\Delta T_{\text{b}}^{\text{(ideal)}}$ and $\Delta T_{\text{e}}^{\text{(ideal)}}$ chosen at the beginning of the road (Fig. 9.11a). Moreover, it turns out that these time intervals ΔT_{b} and ΔT_{e} are random values for different GWs.

There is usually turning-in traffic, which refers to traffic from the cross streets that enters the road on which the GW travels [14]. For this reason, in the GW model [27, 28] we assume that during the red phase there is turning-in traffic with the rate q_{turn} (Fig. 9.11a). We assume also that

$$q_{\text{turn}} < q_{\text{GW}}. \quad (9.34)$$

9.4.2 *Two Basic Moving Patterns in Three-Phase Theory of City Traffic: Moving Synchronized Pattern (MSP) and Moving Queue*

In the framework of the three-phase theory, the occurrence of a moving synchronized pattern (MSP) within an GW is an important phenomenon that effects on the breakdown of the GW in city traffic [27, 28]. Therefore, before we consider the breakdown of the GW in a city, following [28], we should briefly discuss the basic differences between an MSP and moving queue (wide moving jam) in city traffic.

In Sect. 9.2.1, we have shown that the main characteristic parameters of a wide moving jam propagating in free flow on a highway are the same as those for the propagation of a moving queue in over-saturated traffic far enough upstream of the signal location. For this reason, to understand the basic difference between the

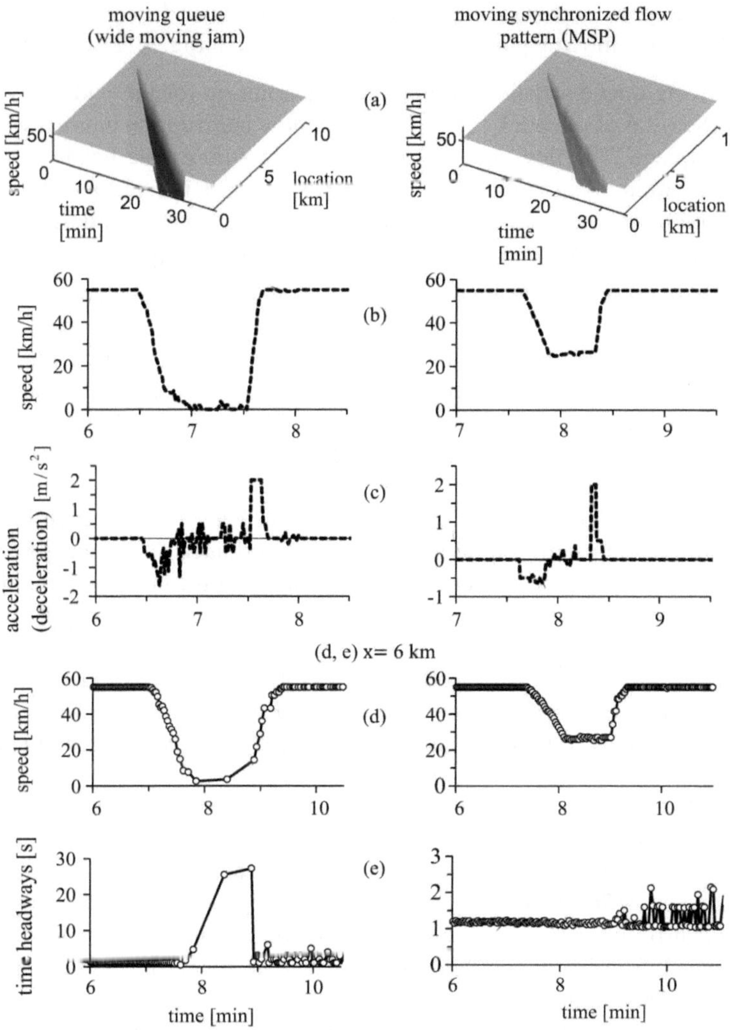

Fig. 9.12 Simulations of moving queue (wide moving jam) (*left panel*) and moving synchronized flow pattern (MSP) (*right panel*) on homogeneous single-lane road without bottlenecks: (**a**) Speed in time and space. (**b**, **c**) Single vehicle speed (**b**) and acceleration (deceleration) (**c**) for a vehicle moving through the moving queue (*left*) or MSP (*right*). (**d**, **e**) Single vehicle speed (**d**) and time headway (**e**) of vehicles measured at a virtual detector at location $x = 6$ km. $q_{\text{in}} = 2156$ vehicles/h. Simulations of the Kerner-Klenov model for city traffic (Sect. A.12 of Appendix A). Adapted from [27, 28]

MSP and a moving queue in city traffic as well as to disclose the physics of MSP propagation within a GW discussed in Sect. 9.4.3 below, it is sufficient to consider a hypothetical case of the emergence and subsequent propagation of a wide moving jam (Fig. 9.12, left panel) and an MSP (Fig. 9.12, right panel) within the same free flow on a homogeneous road without bottlenecks. The basic differences between the MSP and the wide moving jam (moving queue) shown in Fig. 9.12 are as follows.

1. The moving queue satisfies the microscopic criterion for the wide moving jam phase (Sec. 2.6 of the book [25]): There is a flow interruption interval within the moving queue. The flow interruption interval is a long time headway between vehicles within the moving queue (about 27 s in Fig. 9.12e, left panel). The flow interruption interval is considerably longer than the mean time delay of vehicle acceleration $\tau_{\rm del}^{\rm (acc)}(0)$ from a standstill within the moving queue to free flow that is downstream of the queue ($\tau_{\rm del}^{\rm (acc)}(0) = 1.5$ s in the Kerner-Klenov model used for simulations).
2. In contrast with the moving queue, there is no flow interruption within the MSP (Fig. 9.12e, right panel)—the microscopic criterion for the moving queue is not satisfied. Accordingly to the microscopic criterion for traffic phases in congested traffic (Sec. 2.6 of the book [25]), the MSP belongs to the synchronized flow phase (S).
3. The wide moving jam (Fig. 9.12a, left panel) has been induced in an initial free flow through the use of a time-limited local disturbance of the speed reduction: One of the vehicles has decelerated during a limited time interval in free flow. To induce the wide moving jam in free flow (F→J transition) the amplitude of this local disturbance should be very large: In simulations (Fig. 9.12a, left panel), to induce the local disturbance leading to the formation of the wide moving jam, the decelerating vehicle should reduce its speed as low as zero. Only in this case, the F→J transition with the resulting emergence of the wide moving jam could be possible.
4. An MSP (Fig. 9.12a, right panel) has also been induced in an initial free flow (F→S transition) through the use of a time-limited local disturbance of the speed reduction. However, in contrast with the emergence of the wide moving jam, to induce an MSP in free flow (Fig. 9.12a, right panel), a considerably smaller local disturbance in free flow is required: The decelerating vehicle leading to the local speed disturbance should decelerate to some synchronized flow speed that is considerably higher than zero.
5. Thus, simulations show that at each given flow rate in free flow at which either an F→S transition (the emergence of the MSP) or an F→J transition (the emergence of the wide moving jam) is possible, the speed within the critical local disturbance required for the F→J transition (critical nucleus for F→J transition) is considerably lower than the speed within the critical local disturbance required for the F→S transition (critical nucleus for F→S transition). This result of the three-phase theory has already been discussed in Sect. 8.8.2.

The above results of simulations of the emergence and propagation of the MSP and the wide moving jam (Fig. 9.12a) mean that there is a wide range of the amplitudes of local speed disturbances in free flow, which do not lead to the emergence of moving queues in under-saturated city traffic. However, the same local speed disturbances do lead to the emergence of MSPs in under-saturated city traffic.

This result is confirmed by simulations of under-saturated city traffic made in [28]. Some of the simulations of the emergence of MSPs in under-saturated city traffic will be considered below.

Under conditions mentioned in Sect. 9.2.1, the saturation flow rate $q_{\rm sat}$ related to a moving queue is equal to the flow rate in the jam outflow $q_{\rm out}$, i.e., condition $q_{\rm out} = q_{\rm sat}$ (9.1) is valid. As for a steadily propagation of a wide moving jam in free flow (Sect. 4.5.1.2), the flow rate $q_{\rm in}$ in free flow is a threshold flow rate for a steadily propagation of a moving queue, when the flow rate $q_{\rm in}$ is equal to the flow rate in free flow formed in the queue outflow – the saturation flow rate $q_{\rm sat}$:

$$q_{\rm in} = q_{\rm sat}. \tag{9.35}$$

Under condition (9.35), the flow rates upstream and downstream of the moving queue are equal to each other. Respectively, the mean downstream velocity of the moving queue (wide moving jam) $v_{\rm g}$ is equal to the mean upstream velocity of the moving queue $v_{\rm up}$:

$$v_{\rm g} = v_{\rm up}. \tag{9.36}$$

In the threshold case (9.35), (9.36) the initial mean length of the queue (the mean width of a wide moving jam) does not depend on time, when the moving queue (wide moving jam) propagates upstream on the road.

Simulations show that as for the moving queue, there is a threshold flow rate in free flow for the existence of an MSP (MSP propagation in homogeneous free flow). The threshold flow rate for the existence of an MSP is equal to the flow rate in free flow in the MSP outflow denoted by $q_{\rm out}^{\rm (MSP)}$ (called also as the outflow rate of the MSP or the discharge flow rate from the MSP). At the threshold flow rate for the existence of an MSP, the flow rate in free flow upstream of the MSP $q_{\rm in}$ is equal to the outflow rate of the MSP[4]:

$$q_{\rm in} = q_{\rm out}^{\rm (MSP)}. \tag{9.37}$$

Under condition (9.37), the mean velocity of the downstream front of the MSP denoted by $v_{\rm down}^{\rm (MSP)}$ is equal to the mean velocity of the upstream front of the MSP denoted by $v_{\rm up}^{\rm (MSP)}$:

$$v_{\rm down}^{\rm (MSP)} = v_{\rm up}^{\rm (MSP)}. \tag{9.38}$$

In the threshold case for the MSP existence (9.37), (9.38), the initial mean width (in the longitudinal direction) of the MSP does not depend on time, when the MSP propagates upstream on the road.

[4]Note that in Sec. 5.2.7 of the book [24] we have explained that the flow rate $q_{\rm out}^{\rm (MSP)}$ determines the threshold flow rate $q_{\rm in} = q_{\rm th}$ for the F→S transition (traffic breakdown) on a hypothetical *homogeneous* road without bottlenecks: $q_{\rm th} = q_{\rm out}^{\rm (MSP)}$.

When the flow rate $q_{\rm in}$ in free flow upstream of the MSP satisfies condition

$$q_{\rm in} < q_{\rm min}^{\rm (MSP)}, \tag{9.39}$$

the absolute value of the mean velocity of the downstream front of the MSP $| v_{\rm down}^{\rm (MSP)} |$ is larger than the absolute value of the mean velocity of the upstream front of the MSP $| v_{\rm up}^{\rm (MSP)} |$:

$$| v_{\rm down}^{\rm (MSP)} | > | v_{\rm up}^{\rm (MSP)} | . \tag{9.40}$$

Respectively, the initial MSP dissolves over time: No steadily propagation of an MSP is possible. We will call such an MSP as *a dissolving MSP*.

Contrarily, when the flow rate $q_{\rm in}$ in free flow upstream of the MSP satisfies condition

$$q_{\rm in} > q_{\rm min}^{\rm (MSP)}, \tag{9.41}$$

the absolute value of the mean velocity of the downstream front of the MSP $| v_{\rm down}^{\rm (MSP)} |$ is smaller than the absolute value of the mean velocity of the upstream front of the MSP $| v_{\rm up}^{\rm (MSP)} |$:

$$| v_{\rm down}^{\rm (MSP)} | < | v_{\rm up}^{\rm (MSP)} | . \tag{9.42}$$

Respectively, the mean width of the MSP increases over time. The case (9.41), (9.42) is shown in Fig. 9.12a, right panel.

We denote the mean time headway (mean net time gap) between vehicles that discharge from the MSP by $\tau_{\rm out}^{\rm (MSP)}$. For parameters of the Kerner-Klenov model used for simulations, we have found that

$$\tau_{\rm out}^{\rm (MSP)} \approx 1.3 \ {\rm s}. \tag{9.43}$$

This value is considerably shorter than the average time headway (net time gap) between vehicles that discharge from the moving queue $\tau_{\rm sat}^{\rm (net)} = 1.5$ s. Taken into account formula (9.2), in which we use $d = 7.5$ m and $v_{\rm free} = 55$ km/h, for the average gross time gap between vehicles that discharge from the MSP during the green signal phase we get

$$\tau_{\rm MSP}^{\rm (gross)} \approx 1.791 \ {\rm s}. \tag{9.44}$$

Using a well-known formula for the flow rate

$$q = 3600/\tau^{\rm (gross)} \ {\rm vehicles/h}, \tag{9.45}$$

we get

$$q_{\rm out}^{\rm (MSP)} \approx 2010 \quad {\rm vehicles/h}. \tag{9.46}$$

Accordingly to formulas (9.3) and (9.11) for a moving queue, the saturation flow rate $q_{\rm sat}$ is equal to $q_{\rm sat} = 1808$ vehicles/h. Therefore, we find that

$$q_{\rm out}^{\rm (MSP)} > q_{\rm sat}. \quad (9.47)$$

Thus, the threshold flow rate for the steadily MSP propagation in free flow $q_{\rm in} = q_{\rm out}^{\rm (MSP)}$ (9.37) is appreciably larger than the threshold flow rate $q_{\rm in} = q_{\rm sat}$ (9.35) of the steadily propagation of a moving queue in free flow. It should be emphasized that under-saturated city traffic far enough upstream of the signal exhibits the same features of the propagation of the MSP as those of MSP propagation in free flow on a road without traffic signals. For this reason, the result (9.47) can be used for a further analysis of MSP emergence and propagation in under-saturated city traffic.

The threshold flow rate for the steadily MSP propagation in under-saturated traffic is appreciably larger than the threshold flow rate of the steadily propagation of a moving queue in under-saturated traffic far enough upstream of the signal.

9.4.3 Physics of GW Breakdown at Signal

A propagation of GWs through the signal is associated with under-saturated traffic at the signal. However, in Fig. 9.13a we can observe that this GM propagation, i.e., under-saturated traffic is realized during some time interval $T^{\rm (B)}$ only. This time interval $T^{\rm (B)} = 30$ min is considerably longer than the signal cycle that in the example shown in Fig. 9.13 is equal to $\vartheta = 2$ min. At $t > T^{\rm (B)}$, the breakdown of the GW occurs resulting in over saturated traffic at the signal. This time interval $T^{\rm (B)}$ is a time delay to the GW breakdown. We have found that the time delay to the GW breakdown is a random value: In different simulation realizations (runs) made at the same model parameters we have found very different values of $T^{\rm (B)}$ (see Fig. 9.8a,b). Thus, the breakdown of the GW is a time-delayed traffic breakdown.

The physics of the random time-delayed GW breakdown shown in Fig. 9.13a is as follows. Turning-in traffic leads to a queue formation during the red phase. During the green signal phase, the discharge of this queue forces vehicles moving within a GW to decrease their speeds in a neighborhood of the signal location in comparison with the free flow speed $v_{\rm free}$ within the GW far away upstream of the signal. This effect of turning-in traffic on the GW results in a speed disturbance within a GW that appears at the beginning of each of the GWs when the GW approaches the signal. The speed disturbance causes the emergence of a moving synchronized flow pattern (MSP) within each of the GWs. The MSP propagates upstream through the GW ($\rm MSP_1$-$\rm MSP_5$ in Figs. 9.13b, 9.14, and 9.15).

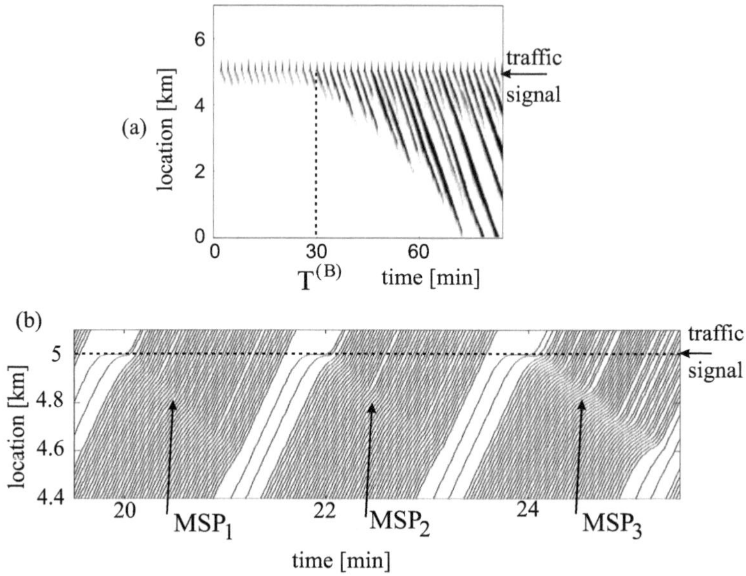

Fig. 9.13 Simulations of MSPs within GW at single intersection: (**a**) Speed data presented by regions with variable shades of gray (shades of gray vary from white to black when the speed decreases from 50 km/h (*white*) to zero (*black*)). (**b**) Vehicle trajectories. $\vartheta = 120$ s, $T_R = 20$ s, $T_Y = 2$ s, $T_{GW} = 90$ s, and $\Delta T_b^{(ideal)} = 3$ s. $q_{turn} = 400$ vehicles/h, $q_{GW} = 2000$ vehicles/h. Simulations of the Kerner-Klenov model for city traffic (Sect. A.12 of Appendix A). Adapted from [27, 28]

The transition from a speed disturbance to the MSP that occurs within each of the GWs can be considered an F→S transition. Due to upstream MSP propagation through the GW as well as due to turning-in-traffic, time intervals ΔT_b and ΔT_e (Fig. 9.11b) are, respectively, longer and shorter than $\Delta T_b^{(ideal)}$ and $\Delta T_e^{(ideal)}$ for an undisturbed GW. Parameters of MSPs that occur in different cycles are random values. Consequently, within time interval $0 < t < 24$ min the time intervals ΔT_b and ΔT_e change randomly for different GWs (Fig. 9.13).

The vehicle speed within an MSP is lower than in free flow (Fig. 9.14c). For this reason, when vehicles move through the MSP, vehicle delays occur. The vehicle delays can cause the formation of a longer vehicle queue in comparison with the initial queue caused by the flow rate q_{turn}. Indeed, due to vehicle delays occurring due to the vehicle motion through the MSP, one or several vehicles at the end of a GW have to stop at the signal (see Fig. 9.14b, where the stopped vehicles are related to bold trajectories 4 and 5, i.e., the queue at the signal increases from two vehicles associated with turning-in traffic to four vehicles).

In its turn, the growth of the queue leads to the decrease in the speed disturbance occurring at next signal cycle, when a new GW approaches the signal location. Because a random number of vehicles have to come to stop at the end of the subsequent green signal phases, the increase in the resulting queue length is also a random

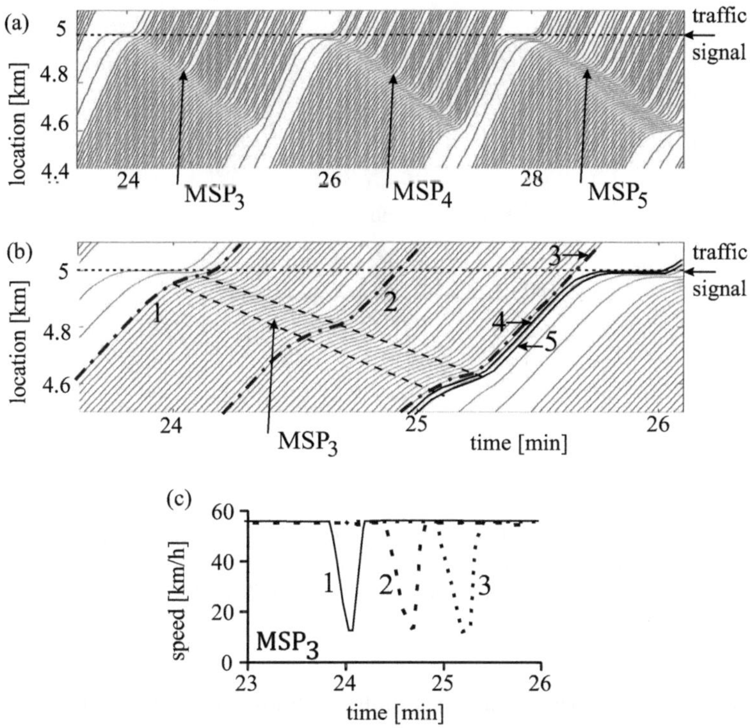

Fig. 9.14 Simulations of the development of MSPs within GW at single intersection (continuation of Fig. 9.13): (**a, b**) Vehicle trajectories. (**c**) Microscopic vehicle speed along trajectories marked by the same numbers in (**b**). Simulations of the Kerner-Klenov model for city traffic (Sect. A.12 of Appendix A). Adapted from [27, 28]

value. Therefore, random values of the speed decrease within speed disturbances at the beginning of the GWs following each other at the subsequent signal cycles occur. The random decreases in the speed within the speed disturbances at the beginning of the GWs cause the random decreases in the speed within MSPs propagating through these GWs at next signal cycles (MSP_5 in Fig. 9.15). For this reason, after a random time interval $T^{(B)}$ (Fig. 9.13a) that is considerably longer than the signal cycle the queue growth at the signal results in the GW breakdown at $t = T^{(B)} = 30$ min. The GW breakdown can be considered an S→J transition: A moving queue is formed (moving queue in Fig. 9.15a) resulting in over-saturated traffic at the signal.

Thus, the GW breakdown can be considered resulting from a sequence of the F→S→J transitions at the signal: (i) Firstly, due to an F→S transition, an MSP emerges within the GW (MSP_5 in Fig. 9.15a). (ii) Due to MSP propagation through the GW, a moving queue (J) (moving queue in Fig. 9.15a) occurs at the signal.

However, it should be noted that there is a qualitative difference of the F→S→J transitions at the signal and the F→S→J transitions in highway traffic discussed in Sect. 8.4.1. This difference is as follows. In highway traffic, firstly an F→S tran-

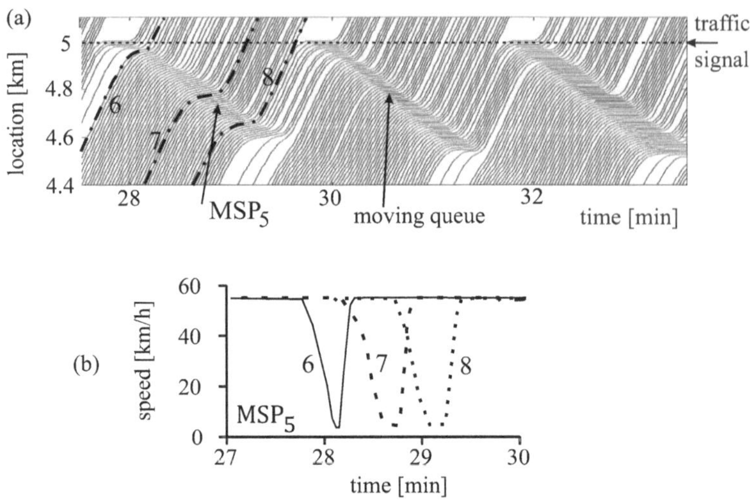

Fig. 9.15 Simulations of the GW breakdown at single intersection (continuation of Fig. 9.14): (**a**) Vehicle trajectories. (**b**) Microscopic vehicle speed along trajectories marked by the same numbers in (**a**). Simulations of the Kerner-Klenov model for city traffic (Sect. A.12 of Appendix A). Adapted from [27, 28]

sition (traffic breakdown) occurs at a highway bottleneck. This traffic breakdown results in synchronized flow at the bottleneck. Later and usually at another road location, an S→J transition(s) occurs within the synchronized flow leading to the emergence of wide moving jam(s).

Contrarily to highway traffic, at the signal in city traffic an F→S transition does *not* lead to the breakdown at the signal directly. Rather than the breakdown, an MSP occurs in under-saturated traffic only. No transition from under- to over-saturated traffic is usually realized during the MSP formation. The MSP formation can however lead to a random increase in the vehicle queue at the end of the green signal phase. When this increase in the vehicle queue is large enough, then at *next signal cycle(s)* the transition from under- to over-saturated traffic (traffic breakdown) can randomly occur. This transition results in moving queue formation upstream of the signal. Because the main characteristics of moving queues propagating upstream of the signal are the same as those of wide moving jams in highway traffic (Sect. 9.2.1), we have denoted a moving queue as the wide moving jam phase by the letter "J". The emergence of MSP(s) in under-saturated traffic increases the probability of the transition from under- to over-saturated traffic (traffic breakdown) at the signal. Therefore, we can call the whole sequence of the events leading to the breakdown at the signal as the F→S→J transitions at the signal.

The emergence of MSPs in under-saturated traffic increases the probability of the transition from under- to over-saturated traffic (traffic breakdown) at the signal. We can call the whole sequence of the events leading to traffic breakdown at the signal as the F→S→J transitions at the signal.

9.4.4 Probability of GW Breakdown at Signal

Within the flow rate range (9.22), GW breakdown occurs at the signal during the time interval $T_{ob} = 60$ min with some probability $P^{(B)} < 1$ *only*.[5] This means that in some of the different simulation runs the GW breakdown does occur, however, in other runs the GW breakdown does not occur (see Fig. 9.8a–c). The probability of the GW breakdown $P^{(B)}(\bar{q}_{in})$ is an increasing flow rate function (Fig. 9.16), which can be fitted with function [27, 28]

$$P^{(B)}(\bar{q}_{in}) = [1 + \exp[\beta(q_P - \bar{q}_{in})]]^{-1}, \tag{9.48}$$

where β and q_P depend on q_{turn} and signal parameters.

The GW breakdown is a time-delayed traffic breakdown in under-saturated traffic at the signal. When the averaged arrival flow rate is larger than the minimum signal capacity but it is smaller than the maximum signal capacity, the probability of the GW breakdown during a given observation time interval (that is considerably longer than the signal cycle) is larger than zero but it is smaller than 1: In some simulation runs, the GW breakdown occurs, however, in some other simulation runs no GW breakdown is realized.

Simulations confirm general features of traffic breakdown at the signal reviewed in Sect. 9.3 (compare Fig. 9.7a with Fig. 9.16). In particular, for the GW breakdown we have found that the smaller the flow rate q_{turn} (Fig. 9.11a), consequently, the smaller the initial speed disturbance at the beginning of each of the GWs, the larger the maximum signal capacity C_{max}. The largest maximum signal capacity C_{max} has been found for the limit case of $q_{turn} = 0$.

[5]For the calculation of the probability of GW breakdown we have used $N_r = 40$ different simulation realizations (runs) at the same chosen $q_{in}(t)$ and the same signal parameters. The only difference between these runs is different initial conditions for random function *rand*() used in the Kerner-Klenov stochastic microscopic model at time $t = 0$ (see Appendix A).

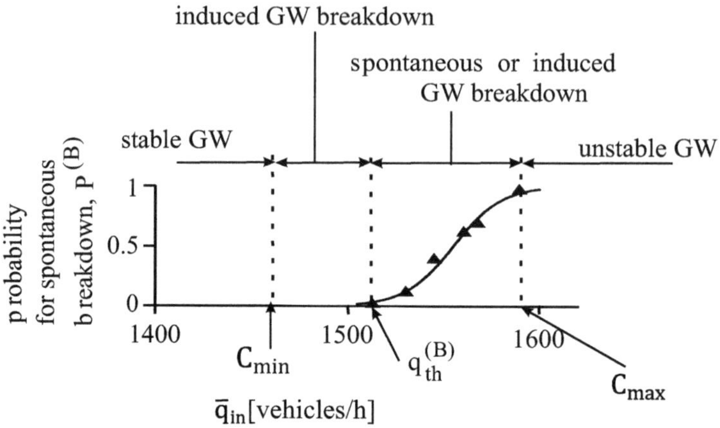

Fig. 9.16 Flow-rate function for probability of GW breakdown $P^{(B)}(\bar{q}_{in})$ for turning-in traffic used in Fig. 9.13. Simulations of the Kerner-Klenov model for city traffic (Sect. A.12 of Appendix A). Adapted from [27, 28]

9.4.5 Flow–Flow Characteristic of GW Breakdown at Signal

The metastability of GWs with respect to the GW breakdown can also be understood with the use of a flow–flow characteristic $\bar{q}_{TS}(\bar{q}_{in})$ (Fig. 9.17a), where $\bar{q}_{TS} = (Z\vartheta)^{-1}\int_0^{Z\vartheta} q_{TS}(t)dt$ (9.7) is the average flow rate downstream of the signal, i.e., in the signal outflow; Z is the number of signal cycles within which the flow rate q_{TS} has been averaged ($Z \gg 1$).

If the average arrival flow rate $\bar{q}_{in}$ increases beginning from small values, we get $\bar{q}_{TS} = \bar{q}_{in}$ (branch U for under-saturated traffic in Fig. 9.17a). Under-saturated traffic associated with GWs can exist even when $\bar{q}_{in} > C_{min}$. However, at $\bar{q}_{in} = C_{max}$ during the time interval T_{ob} with probability $P^{(B)} = 1$ the GW breakdown does occur: The GW is destroyed resulting in the decrease in the average outflow rate from $\bar{q}_{TS} = C_{max}$ to $\bar{q}_{TS} = C_{min}$ (arrow from branch U to branch O for over-saturated traffic) caused by the GW breakdown.

Before GW breakdown occurs, the number of vehicles passing the signal during the signal cycle is almost time-independent ($t < T^{(B)}$ in Fig. 9.17b). After the GW breakdown, it exhibits a very complex time behavior ($t > T^{(B)}$).

After over-saturated traffic has occurred, the presented theory shows the well-known result of the classical theory [2, 14]: When over-saturated traffic exists at the signal (branch O in Fig. 9.17b), a large decrease in $\bar{q}_{in}$ to $\bar{q}_{in} < C_{min}$ is needed for the return transition to under-saturated traffic.

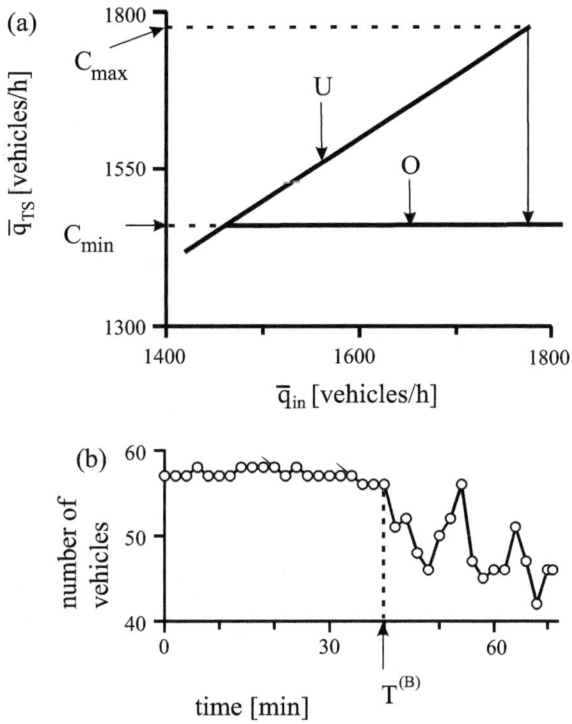

Fig. 9.17 Characteristics of GW breakdown: (**a**) Flow–flow characteristic $\bar{q}_{TS}(\bar{q}_{in})$. (**b**) Dependence of number of vehicles passing traffic signal during the signal cycle for GWs in one of the simulation realizations at $q_{GW} = 2316$ vehicles/h, $q_{turn} = 0$. Other model parameters are the same as those in Fig. 9.13. Simulations of the Kerner-Klenov model for city traffic (Sect. A.12 of Appendix A). Adapted from [28]

9.4.6 Spatiotemporal Interaction of MSPs Induced by GW Propagation Though Sequence of City Intersections

In this Sect. 9.4.6, based on the model for GW propagation of Sect. 9.4.1, we study GW propagation through several city intersections (Fig. 9.10). To understand the physics of this GW propagation, we consider a limit case $q_{turn} = 0$, i.e., when there is no turning-in traffic (Figs. 9.18 and 9.19). We have revealed the following traffic phenomena [28]:

(i) During GW propagation through several city intersections (Fig. 9.10), in a neighborhood of *each of* the signals an MSP can occur. Because vehicles move through MSPs occurring at different signals, the mean values of random time intervals ΔT_b and ΔT_e depend on signal location.
(ii) Features of GW breakdown are the same as those found for the isolated signal (Sects. 9.4.3–9.4.5). In particular, there is a random time delay to the GW breakdown (Fig. 9.18a).
(iii) However, there is a stochastic *dynamic competition* in moving queue formation at different signals during GW breakdown. In particular, it often turns out that although a moving queue appears firstly at one of the signals, this queue dissolves later and GW breakdown is finally realized at another signal.

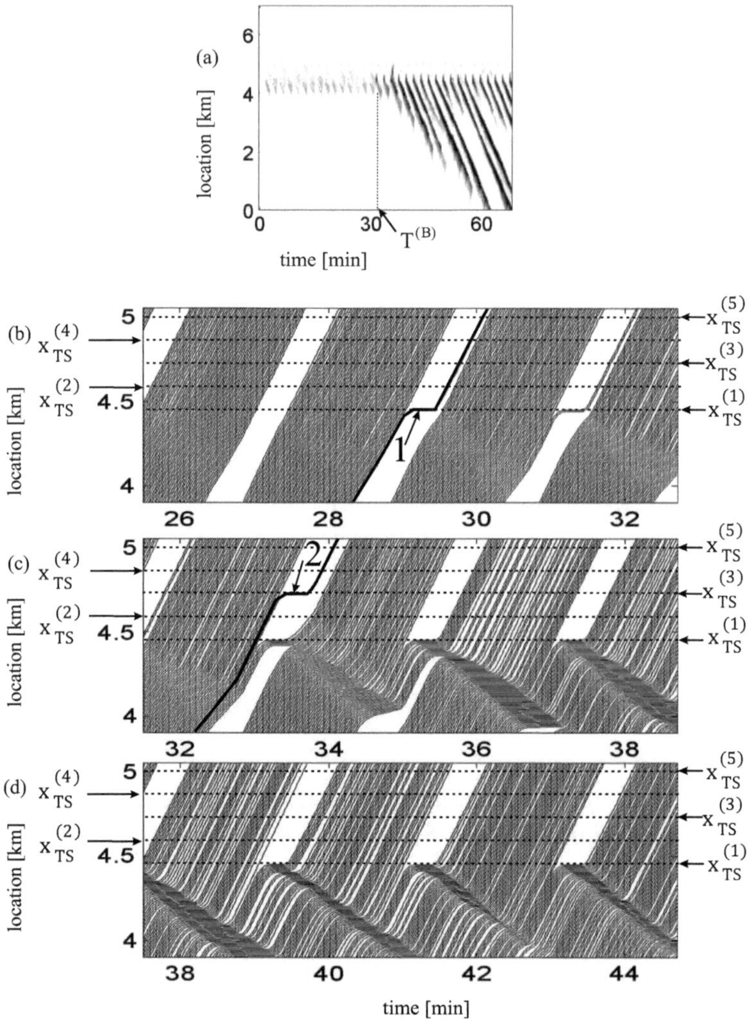

Fig. 9.18 GW breakdown at five signals at $\vartheta = 120$ s, $T_R = 20$ s, $T_Y = 2$ s, $T_{GW} = 90$ s, and $\Delta T_b^{(\text{ideal})} = 3$ s: (**a**) Speed data in space-time plane presented by regions with variable shades of gray (shades of gray vary from white to black when the speed decreases from 50 km/h (*white*) to zero (*black*)). (**b–d**) Vehicle trajectories. $\Delta x_{TS} = 137.5$ m. $q_{GW} = 2252$ vehicles/h, $q_{turn} = 0$. Simulations of the Kerner-Klenov model for city traffic (Sect. A.12 of Appendix A). Adapted from [28]

Characteristic features of spatiotemporal competition in queue formation at different signals depend on values $\Delta T_b^{(\text{ideal})}$ and Δx_{TS} (Sect. 9.4.1) as follows.

If Δx_{TS} is considerably shorter than the distance that an MSP propagates to the GW end (about 400 m in simulations), then with the largest probability GW breakdown occurs at one of the upstream signals (Fig. 9.18).

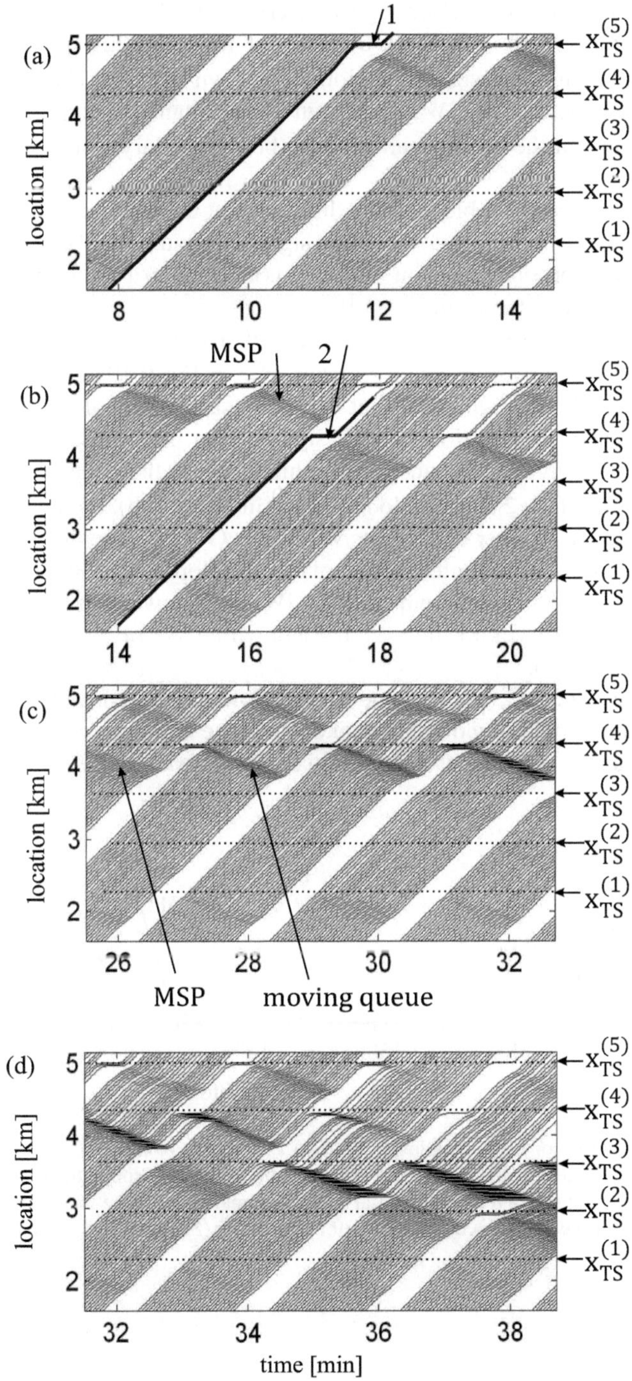

Fig. 9.19 GW breakdown at five signals: Vehicle trajectories. $\Delta x_{TS} = 687.5$ m. Other model parameters are the same as those in Fig. 9.18. Adapted from [28]

Even when Δx_{TS} is short enough and GW breakdown occurs at the farthest upstream signal (see bold vehicle trajectory 1 and resulting queue development at the upstream signal at signal location $x_{TS}^{(1)}$ shown in Fig. 9.18b–d), nevertheless during the development of GW breakdown a queue can also emerge randomly and dissolve at some of the downstream signals (bold trajectory 2 at signal location $x_{TS}^{(3)}$ in Fig. 9.18c).

When Δx_{TS} is comparable with the distance that the MSP propagates to the GW end, a MSP that occurs at the downstream signal can cause MSP emergence within a GW at the neighborhood upstream signal. This can lead to a complex spatiotemporal traffic phenomena as shown in Fig. 9.19.

In Fig. 9.19, the stop of the last vehicle of a GW occurs at the downstream signal (bold trajectory 1 at signal location $x_{TS}^{(5)}$ in Fig. 9.19a). This results in the emergence of MSPs within GWs at the downstream signal. However, no GW breakdown occurs at this downstream signal. This is because one of the MSPs (MSP marked in Fig. 9.19b) causes the vehicle stop at the neighborhood upstream signal (bold trajectory 2 at signal location $x_{TS}^{(4)}$ in Fig. 9.19b). In its turn, this vehicle stop decreases the number of vehicles within the GW approaching the downstream signal. This results in the interruption of the development of the GW breakdown at this signal: The development of the GW breakdown that has started at the downstream signal leads to the GW breakdown at one or a few of the upstream signals (Fig. 9.19c,d).

As can be expected, at the same arrival flow rate $q_{in}(t)$ and the same signal parameters, the probability of GW breakdown at the sequence of signals (curve 1 in Fig. 9.20) is larger than that at the isolated signal (curve 2 in Fig. 9.20).

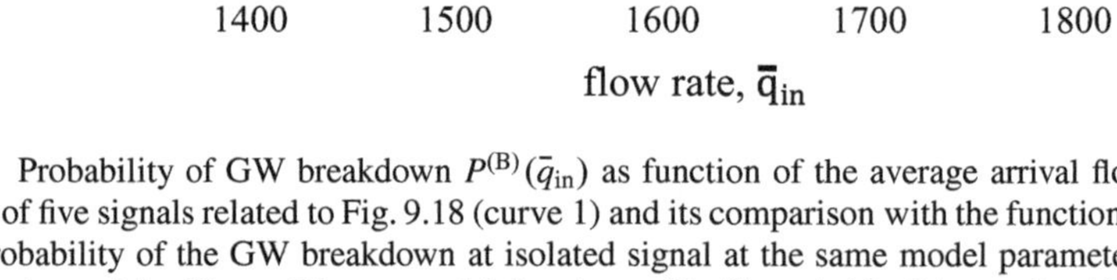

Fig. 9.20 Probability of GW breakdown $P^{(B)}(\bar{q}_{in})$ as function of the average arrival flow rate at sequence of five signals related to Fig. 9.18 (curve 1) and its comparison with the function $P^{(B)}(\bar{q}_{in})$ for the probability of the GW breakdown at isolated signal at the same model parameters (curve 2). Simulations of the Kerner-Klenov model for city traffic (Sect. A.12 of Appendix A). Adapted from [28]

There can be a complex spatiotemporal dynamics of MSPs occurring within an GW at different signals in city intersections. Due to this spatiotemporal dynamics of the MSPs, the development of the GW breakdown that has started at one of the downstream signals can lead to the GW breakdown at one or a few of the upstream signals.

The probability of GW breakdown at the sequence of signals is larger than that at the isolated signal.

9.5 Effect of Time-Dependence of Arrival Flow Rate on Traffic Breakdown at Signal

It should be noted that in metastable under-saturated traffic (9.21), two different cases should be distinguished:

(i) There is at least one time interval during the green signal phase, when the arrival flow rate is larger than the saturation flow rate:

$$q_{\text{in}}(t) > q_{\text{sat}}. \tag{9.49}$$

(ii) At any time instant, the arrival flow rate is smaller than the saturation flow rate

$$q_{\text{in}}(t) < q_{\text{sat}}. \tag{9.50}$$

Condition (9.49) contradicts assumptions of the classical theory (Fig. 9.1b). Condition (9.49) can be satisfied, for example, when an GW is organized in city traffic (Sect. 9.4). For the GW model of Sect. 9.4.1 the flow rate q_{in} during the green phase is equal to q_{GW} (Fig. 9.11b); therefore, in this case condition (9.49) is equivalent to condition

$$q_{\text{GW}} > q_{\text{sat}}. \tag{9.51}$$

In contrast, condition (9.50) does not contradict assumptions of the classical theory. Physical reasons for the existence of a metastable under-saturated traffic at the signal under conditions (9.21) can be different when either condition (9.49) or condition (9.50) is satisfied. These different reasons for metastable under-saturated traffic at the signal will be considered in Sect. 9.7.

In this Sect. 9.5, we show that the probabilistic features of a time-delayed traffic breakdown (time-delayed transition from under- to over-saturated traffic) at the signal remain qualitatively the same when either condition (9.49) or condition (9.50) is satisfied. Moreover, we have found that the probabilistic features of traffic breakdown do not also qualitatively change when the arrival flow rate $q_{\rm in}(t)$ is a complex time-function. However, we will find that the minimum signal capacity and the maximum signal capacity as well as congested patterns occurring due to traffic breakdown at the signal can depend considerably on the time-function of the arrival flow rate $q_{\rm in}(t)$ and on either condition (9.49) or condition (9.50) is satisfied.

9.5.1 Characteristics of Probability of Traffic Breakdown at Signal

Formulas (9.21)–(9.25) as well as qualitative features of the metastability of under-saturated traffic at the signal discussed above for GWs remain also when there are *no* GWs in city traffic. To show this result, we consider simulations of traffic breakdown at the isolated signal on a single-lane road for different periodic functions

$$q_{\rm in}(t) = q_{\rm in}(t + \vartheta) \tag{9.52}$$

of the arrival flow rate (Fig. 9.21a). The simulations show [28] that traffic breakdown exhibits qualitatively the same features and characteristics (Fig. 9.21b,c) as those for the GW breakdown at the isolated signal.

In particular, a flow-rate dependence of the breakdown probability $P^{(\rm B)}(\bar{q}_{\rm in})$ (Fig. 9.21b) exhibits qualitatively the same features as that for GW breakdown. The dependence of the probability $P^{(\rm B)}(\bar{q}_{\rm in})$ on the average arrival flow rate $\bar{q}_{\rm in}$ is determined by the same formula (9.48) as that for the GW.

In a diagram of traffic breakdown (Fig. 9.21c) found for periodic functions (9.52) (Fig. 9.21a), diagram regions I–V have the same meaning as those discussed in Sect. 9.3. In the diagram shown in Fig. 9.21c,

$$R = q_{\rm in}^{(\rm green)}/q_{\rm in}^{(\rm red)} \tag{9.53}$$

is the ratio between average rates of traffic arriving the signal during different signal phases:

$$q_{\rm in}^{(\rm red)} = T_{\rm R}^{-1}\int_{\vartheta - T_{\rm R}}^{\vartheta} q_{\rm in}(t)dt, \tag{9.54}$$

$$q_{\rm in}^{(\rm green)} = (\vartheta - T_{\rm R})^{-1}\int_{0}^{\vartheta - T_{\rm R}} q_{\rm in}(t)dt. \tag{9.55}$$

In city traffic without GWs, there can be many traffic signals at which the arrival flow rates during the green and the red signal phases do not differ considerably, i.e.,

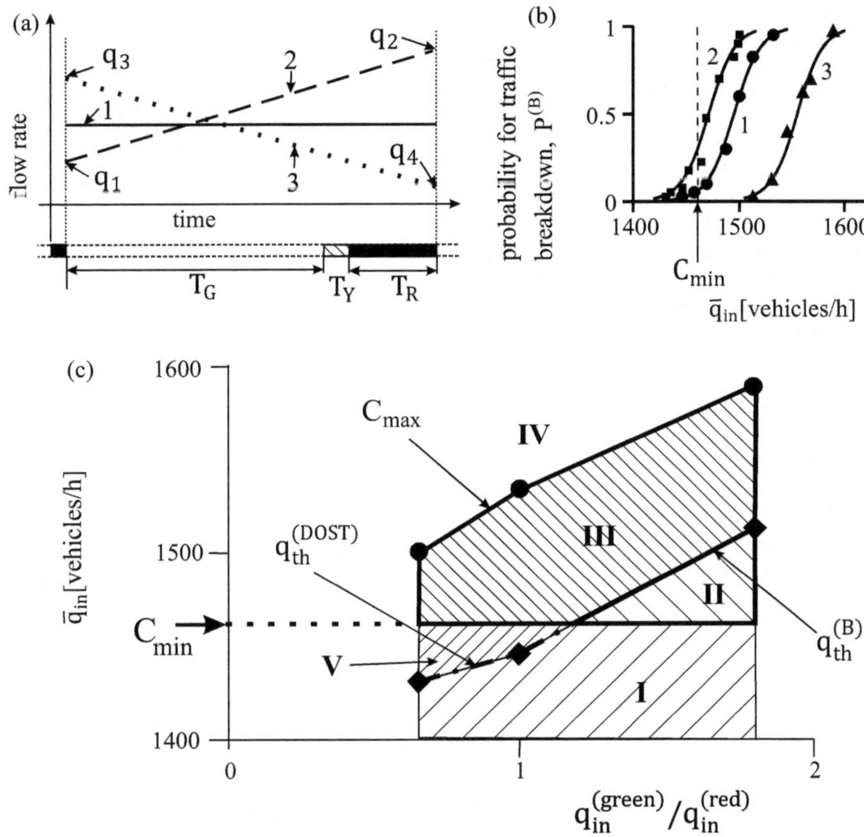

Fig. 9.21 Characteristics of traffic breakdown at the isolated signal on a single-lane road: (**a**) Different periodic functions $q_{\text{in}}(t)$ associated with a time-independent flow rate (curve 1), an increase in $q_{\text{in}}(t)$ during the signal cycle (curve 2), and a decrease in $q_{\text{in}}(t)$ (curve 3); $q_1 = 0.35q_2$, $q_4 = 0.35q_3$. (**b**) Probabilities of traffic breakdown for the periodic functions $q_{\text{in}}(t)$ in (**a**) labeled by the same numbers in (**a**) and (**b**). (**c**) Diagram of traffic breakdown at signal associated with functions $P^{(\text{B})}(\bar{q}_{\text{in}})$ shown by curves 1–3 in (**b**). In (**c**), diagram region I is related to stable under-saturated traffic, II—metastable under-saturated traffic, III—metastable under-saturated traffic in which spontaneous breakdown can occur, IV—unstable under-saturated traffic, and region V—dissolving over-saturated traffic (DOST). $C_{\min} = 1461$ vehicles/h. $T_{\text{ob}} = 60$ min. Simulations of the Kerner-Klenov model for city traffic (Sect. A.12 of Appendix A). Adapted from [28]

in (9.53) value $R \approx 1$. Nevertheless, as above-mentioned, under-saturated traffic can also be in a metastable state with respect to traffic breakdown.

An example is shown in Fig. 9.22a, which has been simulated under conditions (9.14), (9.50): Metastable under-saturated traffic persists during a time interval $T^{(\text{B})} \gg \vartheta$. After this random time-interval, a time-delayed traffic breakdown from under- to over-saturated traffic occurs.

We find that the evolution of metastable under-saturated traffic does not also qualitatively depend on whether a GW is organized in city traffic with the flow rate

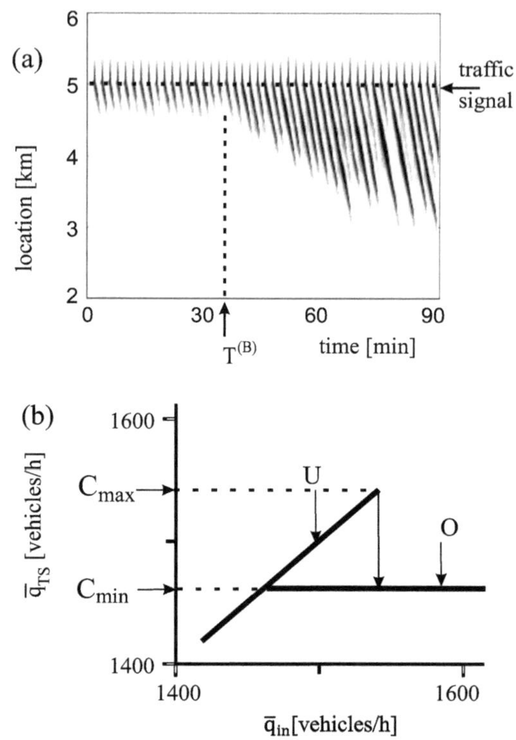

Fig. 9.22 Random time-delayed traffic breakdowns under conditions (9.14), (9.50) for durations of green phase $T_G = 98$ s and time independent arrival flow rate $q_{in} = 1500$ vehicles/h: (**a**) Speed data presented by regions with variable shades of gray (shades of gray vary from white to black when the speed decreases from 50 km/h (*white*) to zero (*black*)). $\vartheta = 120$ s, $T_Y = 2$ s. $T_R = 20$ s. (**b**) Flow–flow characteristic of traffic breakdown at the signal shown in (**a**). Minimum signal capacity $C_{min} = 1458$ vehicles/h. Simulations of the Kerner-Klenov model for city traffic (Sect. A.12 of Appendix A)

$q_{GW} > q_{sat}$ (9.51), i.e., condition (9.49) is satisfied, or the arrival flow rate is almost the same during in the green phase and the red signal phase, when the opposite condition (9.50) is satisfied. In particular, we find qualitatively the same hysteresis effect in the flow–flow plane when under condition (9.49) the GW breakdown occurs (Fig. 9.17a) and when under the opposite condition (9.50) a random time-delayed breakdown occurs at time-independent arrival flow rate (Fig. 9.22b).

Probabilistic characteristics of a time-delayed traffic breakdown at the signal remain qualitatively the same for a diverse variety of time-functions of the arrival flow rate.

9.5.2 Empirical Probability of Traffic Breakdown at Signal

Based on a study of different days, at which traffic breakdown at traffic signal in city Düsseldorf (Germany) has been observed, we have found [29] that the empirical probability of traffic breakdown measured at the detector as function of the flow

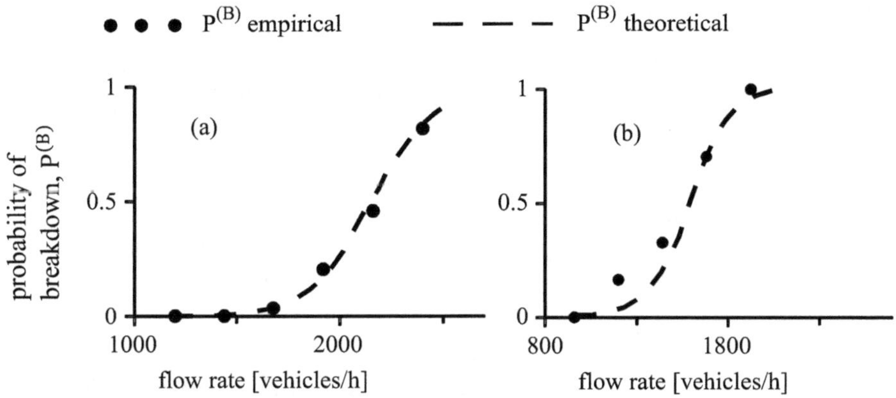

Fig. 9.23 Comparison of empirical (*black points*) and theoretical (*dashed curves*) probabilities of traffic breakdown measured at detector on 219 days for 06:30–11:00 (mornings) (**a**) and on 60 days for 18:00–20:30 (evenings) (**b**). Adapted from [29]

rate (black points in Fig. 9.23) is well fitted with a theoretical one given by formula (9.48) with fitting parameters $(q_{\rm P},\ \beta^{-1})\ =\ (2160,\ 148.8)$ vehicles/h for Fig. 9.23a and (1590, 111.1) vehicles/h for Fig. 9.23b.

As follows from simulations made in [28], $q_{\rm P}$ and β in formula (9.48) should depend on the red phase duration $T_{\rm R}$. In empirical data, the road computer increases $T_{\rm R}$ from 34 s (mornings) to 39 s (evenings) at constant signal cycle $\vartheta\ =\ 70$ s and yellow phase $T_{\rm Y}\ =\ 4$ s. In accordance with the physics the probabilistic transition from metastable under-saturated traffic to oversaturated traffic at the signal (Sect. 9.3), the increase in $T_{\rm R}$ results in a decrease in the minimum signal capacity. This can explain the shift of function $P^{(\rm B)}(\bar{q}_{\rm in})$ in Fig. 9.23a to smaller flow rates $\bar{q}_{\rm in}$ in Fig. 9.23b. This is related to the decrease in $q_{\rm P}$ from 2160 vehicles/h for Fig. 9.23a to 1590 vehicles/h for Fig. 9.23b found in theoretical formula (9.48).

9.5.3 *Physical Reason for Dissolving Over-Saturated Traffic at Signal*

In Sect. 9.3, we have already mentioned that under condition (9.27) dissolving over-saturated traffic (DOST) can occur. In DOST, random emergence and subsequent random dissolution of a growing queue at traffic signal follows each other randomly (Fig. 9.9). To understand the physics and features of the DOST, we consider simulations of the number of vehicles passing the signal and the flow rate passing the signal $q_{\rm TS}(t)$ associated with DOST (Fig. 9.24).

In the case shown in Fig. 9.24a, conditions $q_{\rm th}^{(\rm DOST)} \leq \bar{q}_{\rm in} < C_{\rm min}$ (9.28) of DOST occurrence are satisfied (region V in Figs. 9.7b and 9.21c). Under condition $\bar{q}_{\rm in} < C_{\rm min}$ that is satisfied in Fig. 9.24a, no over-saturated traffic can occur *on average* at the signal. However, for a *finite time interval* that can however be considerably

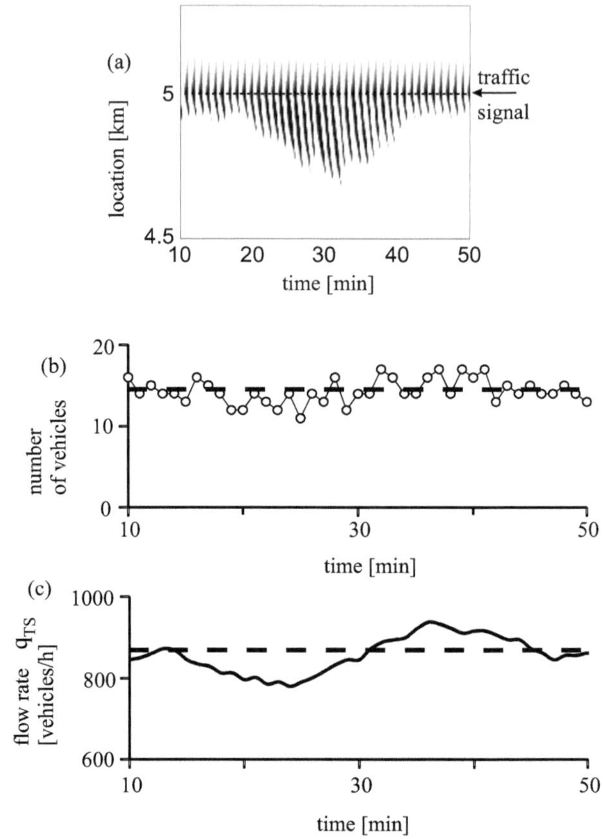

Fig. 9.24 Simulations of a fragment of dissolving over-saturated traffic (DOST) (**a**) with the related values of the number of vehicles $N_{\text{signal}}(t)$ passing the signal during the signal cycle (**b**) and the flow rate passing the signal $q_{\text{TS}}(t)$ (**c**). $\vartheta = 60$ s, $T_{\text{R}} = 28$ s, $T_{\text{Y}} = 2$ s, time-independent arrival flow rate $q_{\text{in}} = 857$ vehicles/h. Calculated value $C_{\min} = 868$ vehicles/h. *Dashed lines* in (**b**, **c**) show the mean values of N_{signal} (**b**) and q_{TS} (**c**), respectively. Simulations of the Kerner-Klenov model for city traffic (Sect. A.12 of Appendix A)

longer than the signal cycle over-saturated traffic does occur, when the arrival flow rate is very close to the minimum signal capacity (Fig. 9.24a), i.e., when

$$\bar{q}_{\text{in}} < C_{\min}, \qquad \frac{C_{\min} - \bar{q}_{\text{in}}}{C_{\min}} \ll 1. \tag{9.56}$$

The physics of this DOST is explained by large fluctuations in the number of vehicles $N_{\text{signal}}(t)$ passing the signal during different signal cycles (Fig. 9.24b). In its turn, the large fluctuations in $N_{\text{signal}}(t)$ result from fluctuations in the discharge rate from the queue built during the red phase at the signal. The large fluctuations in $N_{\text{signal}}(t)$ can lead to a random decrease in the flow rate $q_{\text{TS}}(t)$ with the subsequent random increase in the flow rate $q_{\text{TS}}(t)$. When this sequence of the decrease with the subsequent increase in the flow rate $q_{\text{TS}}(t)$ takes many signal cycles as it occurs in Fig. 9.24c, DOST appears at the signal.

9.6 Two-Phase Models of GM Model Class Versus Three-Phase Theory

In Sects. 9.1 and 9.3, we have already mentioned that a time-delayed traffic breakdown at the signal can be found in simulations of both two-phase traffic flow models of the GM model class and the three-phase traffic flow models. In this Sect. 9.6 we explain this fact and consider some differences in the description of the time-delayed traffic breakdown at the signal with the two-phase and the three-phase models.

To reach this goal, we derive a two-phase traffic flow model of the GM model class from the Kerner-Klenov stochastic three-phase traffic flow model used above for the simulations of the time-delayed traffic breakdown at the signal (see Sect. A.12.3 of Appendix A). Such a two phase traffic flow model results from the Kerner-Klenov stochastic microscopic three-phase model, if we remove the description of driver behaviors associated with the three-phase theory [24]. In particular, we should remove the 2D-region of synchronized flow states (dashed region S in Fig. 9.5a,b), the speed adaptation effect within these states as well as the mathematical description of the over-acceleration effect. Steady states of the resulting two-phase traffic flow model are related to a fundamental diagram (Fig. 9.5c,d). The features of the two-phase model are qualitatively the same as well-known for many other two-phase models of the GM model class: In these two-phase models, traffic breakdown at a highway bottleneck and at the signal is governed by an F→J transition. Characteristics of a moving queue in the three-phase and two-phase models are *identical* (line J in Fig. 9.5a,c).

Nevertheless, we have found that at the same flow rate $\bar{q}_{\text{in}} = q_{\text{GW}}(T_{\text{GW}}/\vartheta)$ for GWs without turning-in traffic $q_{\text{turn}} = 0$ and other model parameters as those used in above simulations of GWs with the three-phase model in *none* of simulation realizations made with the two-phase model the GW breakdown can occur (Fig. 9.25a). To understand this, note that at any given flow rate $\bar{q}_{\text{in}}$ the probability of a sequence of F→S→J transitions occurring at the signal in the three-phase model (curve labeled by F→S→J in Fig. 9.26) is considerably larger than probability of an F→ J transition that occurs in the two-phase model (curve labeled by F→J in Fig. 9.26). Indeed, although in simulations with the two-phase model there are also initial speed disturbances at the beginning of the GWs (Fig. 9.25b), however, *no* MSPs can emerge in the two-phase model.

This is because in contrast with the three-phase model, there is no synchronized flow phase in the two-phase model. As a result, in the two-phase model the speed within the initial speed disturbances increases during disturbance propagation through the GWs (Fig. 9.25c).

This explains why the GW breakdown in the two-phase model occurs at considerably larger flow rates $\bar{q}_{\text{in}}$ than those in the three-phase model. At these large flow rates, the initial disturbance with a considerably lower speed occurs causing long vehicle delays. The subsequent development of the transition from under- to over-saturated traffic (traffic breakdown) at the signal in the two-phase

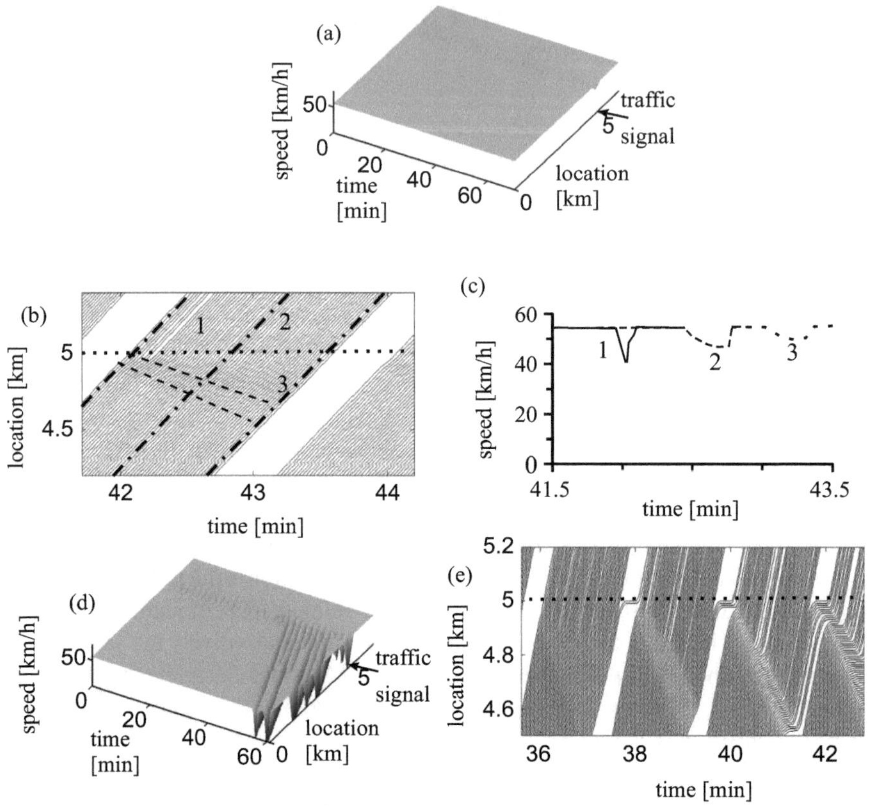

Fig. 9.25 Simulations of GW breakdown at signal with a two-phase traffic flow model of the GM model class: (**a**) Speed in space and time simulated with two-phase model of the GM model class. (**b, c**) Vehicle trajectories (**b**) and associated microscopic speeds along vehicle trajectories 1–3 (**c**) related to (**b**). (**d, e**) Speed in space and time and trajectories for GW breakdown simulated with two-phase model at greater flow rate $\bar{q}_{\text{in}}$ than that in (**a–c**). $q_{\text{GW}} = 2316$ (**a–c**) and 2446 vehicles/h (**d, e**). $\vartheta = 120$ s, $T_{\text{R}} = 20$ s, $T_{\text{Y}} = 2$ s, $T_{\text{GW}} = 90$ s, and $\Delta T_{\text{b}}^{(\text{ideal})} = 3$ s, $q_{\text{turn}} = 0$. Simulations of the two-phase traffic flow model of Sect. A.12.3 of Appendix A

model (Figs. 9.25d,e) is qualitatively similar to that explained above with the use of the three-phase model. Because no MSPs appear in under-saturated traffic in simulations with the two-phase model, we denote the transition from under- to over-saturated traffic at the signal as an F→J transition ("F" denotes under-saturated traffic, "J" denotes moving queue that propagates upstream of the signal in over-saturated traffic).

The result that at a given flow rate $\bar{q}_{\text{in}}$ probability of a sequence of F→S→J transitions (three-phase model) is considerably larger than probability of an F→J transition (two-phase model) remains also when $\bar{q}_{\text{in}}$ is smaller than the threshold flow rate for the MSP existence in the three-phase model $q_{\text{out}}^{(\text{MSP})}$ (9.37) (Sect. 9.4.2).

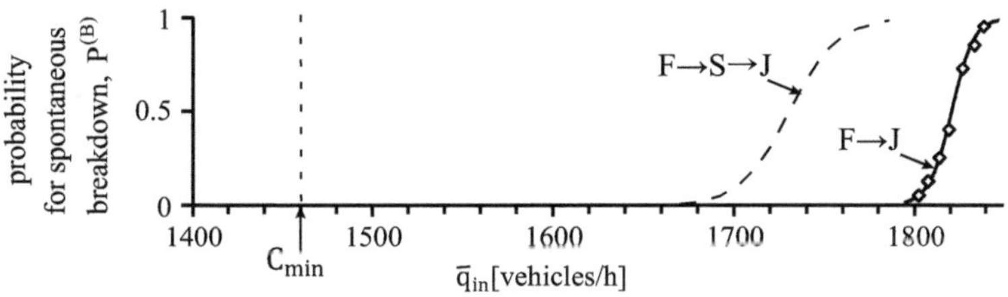

Fig. 9.26 Comparison of flow-rate functions for the probability of GW breakdown $P^{(B)}(\bar{q}_{in})$ at $q_{turn} = 0$ for two-phase model (*solid curve* labeled by F→J) and for three-phase model (*dashed curve* labeled by F→S→J) simulated at the same signal parameters. The three-phase model is the Kerner-Klenov model for city traffic (Sect. A.12 of Appendix A). The two-phase model is considered in Sect. A.12.3 of Appendix A. Adapted from [27]

This is explained in Fig. 9.27 through a consideration of the dissolution of a wide moving jam on a homogeneous road without traffic signals and other bottlenecks in the three-phase model (left panel in Fig. 9.27) and the two-phase model (right panel in Fig. 9.27). In the three-phase model, after the jam has dissolved, a dissolving MSP occurs (see explanations of dissolving MSP in Sect. 9.4.2). The MSP dissolves slowly (figures in left panel in Fig. 9.27c,e). This slow dissolution of the MSP causes the associated slow dissolution of a local region of a lower speed within the dissolving MSP. In contrast, in the two-phase model no synchronized flow can appear. Therefore, local disturbances in free flow dissolve much quicker (right panel in Fig. 9.27c,e) than those in the three-phase model (left panel in Fig. 9.27c,e).

It should be noted that there can be many two-phase models of the GM model class that are able to simulate the time-delayed traffic breakdown (time-delayed transition from under- to over-saturated traffic) at the signal revealed firstly in [26, 27]. In particular, some two-phase models of the GM model class that show the time-delayed traffic breakdown at the signal have been developed by Jiang et al. [23] and Wang and Chen [48].

In both two-phase models of the GM model class and the three-phase theory a metastable state of under-saturated traffic can exist when the average arrival flow rate at the signal exceeds the classical signal capacity. The existence of the metastable under-saturated traffic results in time-delayed traffic breakdown at the signal revealed in [26, 27] based on simulations with the two-phase and three-phase traffic flow models.

The two-phase traffic flow models that can show a time-delayed transition from under- to over-saturated traffic (traffic breakdown) at the signal can be considered a special case of the three-phase theory of city traffic at the signal.

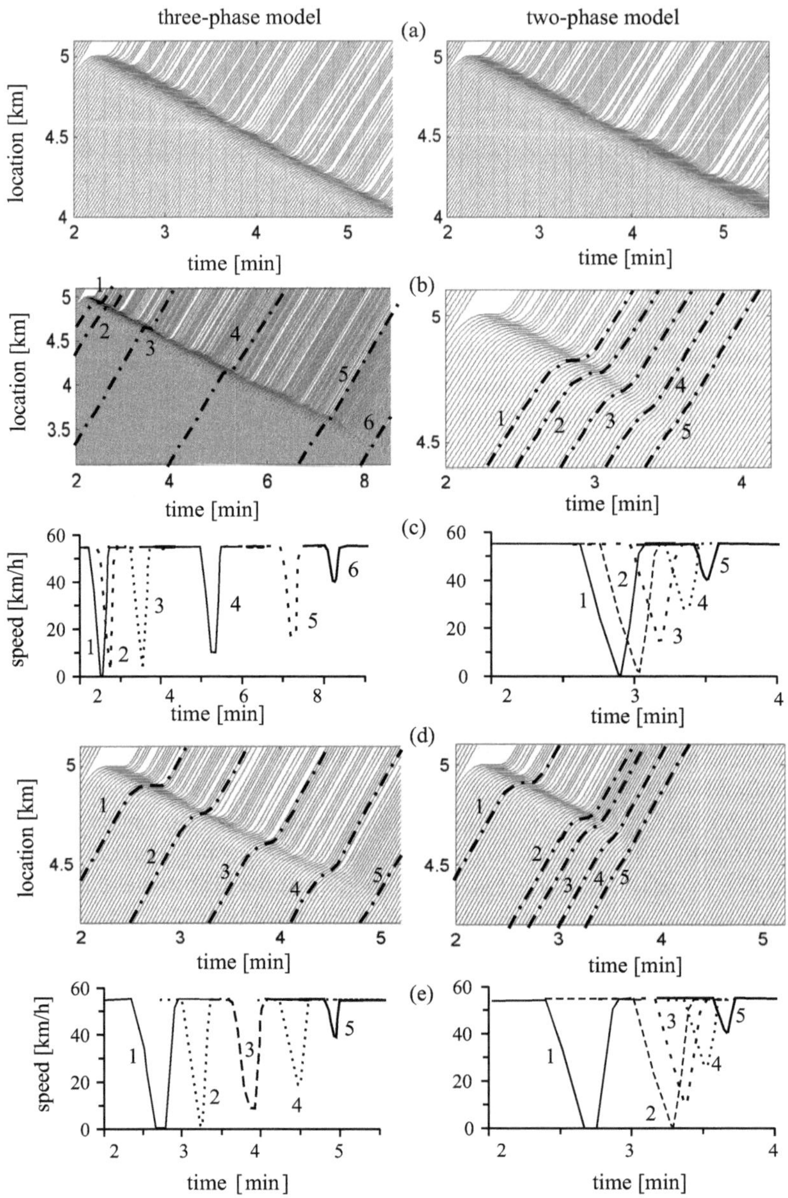

Fig. 9.27 Features of dissolution of wide moving jam (queue dissolution) in three-phase model (*left panel*) and two-phase model (*right panel*): (**a**, **b**, **d**) Vehicle trajectories for wide moving jam propagation without jam dissolution at $q_{in} = q_{out}$ (**a**) and under jam dissolution occurring at $q_{in} < q_{out}$ (**b**, **d**). (**c**, **e**) Vehicle speed along trajectories related to (**b**, **d**), respectively. Simulations of traffic flow on homogeneous road without traffic signal and other bottlenecks. To induced initial wide moving jam, one of the vehicles comes to a stop for 10 s; after this time interval, the vehicle accelerates in accordance with model rules of vehicle motion. q_{in} = 1808 (**a**), 1800 (**b**, **c**) 1565 vehicles/h (**d**, **e**). The three-phase model is the Kerner-Klenov model for city traffic (Sect. A.12 of Appendix A). The two-phase model is considered in Sect. A.12.3 of Appendix A

9.7 Reasons for Metastable Under-Saturated Traffic at Signal

As mentioned in Sect. 9.3, a time-delayed traffic breakdown at the signal occurs because under under conditions (9.21) under-saturated traffic is in a metastable state with respect to traffic breakdown (transition from under- to over-saturated traffic) at the signal. In this section, we discuss some of the physical reasons for this metastability of under-saturated traffic at the signal. The importance of this study is as follows.

Contrarily to the model of a time-delayed traffic breakdown in metastable under-saturated city traffic [27, 28], in the classical theory of traffic at the signal under conditions (9.21) no metastable under-saturated traffic can exist: When the average arrival flow rate exceeds the classical signal capacity $\bar{q}_{\rm in} > C_{\rm cl}$ (9.14), the transition from under- to over-saturated traffic (traffic breakdown) must occur at the signal (see, e.g., the review [14]). Recall that in the model of the time-delayed traffic breakdown at the signal the classical signal capacity is equal to the minimum signal capacity $C_{\rm cl} = C_{\rm min}$ (9.20).

Before we discuss the physical reasons for metastability of under-saturated traffic at the signal, it must be emphasized that in the Kerner-Klenov stochastic microscopic model for identical vehicles and identical drivers moving on a single-lane road used for simulations in this chapter, during the green signal phase the mean time-delay in vehicle acceleration from a standstill within the queue at the signal to free flow downstream of the signal $\tau^{\rm (acc)}_{\rm del}(0)$ does not depend on the length of the queue at the signal. In accordance with formulas (9.3), (9.10), and (9.11) this means that the saturation flow rate $q_{\rm sat}$ does not also depend on the length of the queue at the signal: The value of $q_{\rm sat}$ remains constant during the queue discharge regardless of how long the queue is. This result, which is the same as in the classical theory of traffic breakdown at the signal, has been shown in simulations presented in Fig. 9.4. Therefore, a question arises:

- While the saturation flow rate $q_{\rm sat}$ is assumed to be constant during the whole time interval of the queue discharge during the green signal phase, what is the reason for the occurrence of under-saturated traffic under conditions (9.21)?

Below we show that there can be different reasons for the existence of a metastable under-saturated traffic at the signal. The reasons for the metastable under-saturated traffic at the signal under conditions (9.21) depend on either condition (9.49) or condition (9.50) is satisfied. Therefore, we analysis these possible reasons for these two cases separately each other. The analysis can explain in more details a qualitative difference between the model of the time-delayed traffic breakdown at the signal [26, 27] and the classical model of traffic breakdown at the signal [2–4, 8, 12, 14, 17, 18, 22, 32, 35–42, 44, 45, 47, 49].

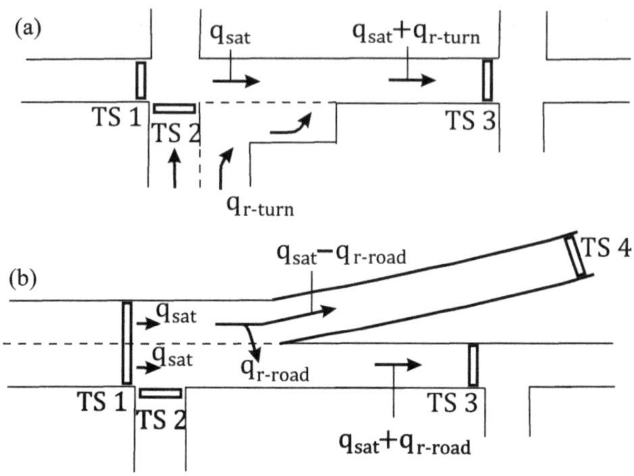

Fig. 9.28 Qualitative explanations of condition (9.49)

9.7.1 Arrival Flow Rate Exceeds Saturation Flow Rate During Green Signal Phase

In this Sect. 9.7.1, we consider traffic phenomena at the signal for the case when condition (9.49) is satisfied.

9.7.1.1 Explanation of Condition (9.49) for Real City Traffic

Under condition (9.49), there is at least one time interval during the green signal phase, when the arrival flow rate $q_{in}(t)$ is larger than the saturation flow rate q_{sat}. This condition contradicts the assumption of the classical theory that the saturation flow rate is the maximum flow rate on a road [14]. However, there can be different reasons for condition (9.49) in real city traffic. Two examples are explained in Fig. 9.28.

In Fig. 9.28a, the flow rate downstream of the signal TS 1 is assumed to be equal to q_{sat} during a time interval within the green phase. Due to right turn traffic with the rate q_{r-turn} from a crossing street that is allowed during the red phase of the signal TS 2 (only left turn and straight traffic must stop at this signal), the arrival flow rate at the signal TS 3 increases to $q_{sat} + q_{r-turn} > q_{sat}$, i.e., condition (9.49) can be satisfied.

In another example (Fig. 9.28b), a two-lane road diverges into two single-lane roads downstream. In this case, we can also assume that the flow rate on each of the lanes downstream of the signal TS 1 is equal to q_{sat} during a time interval within the green phase. We assume also that some traffic with the rate q_{r-road}, which is initially in the left lane, should follow the right road. Therefore, the associated vehicles have to change from the left road to the right road. For this reason, the arrival flow rate at the signal TS 3 increases to $q_{\mathrm{sat}} + q_{\mathrm{r-road}} > q_{\mathrm{sat}}$, i.e., condition (9.49) can be satisfied.

9.7.1.2 Duration of Green Phase and Metastability of Under-Saturated Traffic

Figures 9.13–9.15 in which the GW breakdown has been presented above are related to condition (9.49). However, in these figures the duration of the green signal phase has been chosen much longer than the duration of the red signal phase. Therefore, the question arises:

- Do results discussed in Sect. 9.4 remain qualitatively, when an opposite case of a short green phase duration in comparison with the duration of the red phase is considered?

We have found that when the green phase T_{G} is relatively short in comparison with the cycle ϑ, a time-delayed traffic breakdown occurs (Fig. 9.29a). Moreover, the GW breakdown exhibits qualitatively the same stages as found for a considerably longer green phase T_{G} (Figs. 9.13–9.15): An MSP appears during each of the GWs. Randomly at some of the cycles, one (or several) vehicle(s) at the GW end stops (bold vehicle trajectory 1 in Fig. 9.29b). The synchronized flow speed within the MSP occurring within a GW decreases at the next cycles. This leads to stop of more vehicles during the red phase, and so on. Finally, a growing moving queue occurs, i.e., traffic breakdown is realized (growing moving queue in Fig. 9.29c).

Thus, we see that a random time-delayed breakdown is realized, even if the green phase is considerably shorter than the red phase (Figs. 9.29 and 9.30). Indeed, as shown in Fig. 9.30, there are two different solutions for traffic at the signal at chosen model parameters: One of the solutions is under-saturated traffic (Fig. 9.30a). The second solution is related to over-saturated traffic (Fig. 9.30b). This is in contrast with the classical theory in which under condition (9.14) only over-saturated traffic is possible. Only the solution for over-saturated traffic remains the same in both theories (compare Fig. 9.6c with Fig. 9.30b).

9.7.1.3 Qualitative Explanation of Metastability of Under-Saturated City Traffic

Here, we give a qualitative explanation of the metastability of under-saturated traffic under condition (9.49). In all realistic cases, there is at least one time interval during the green and yellow signal phases, when the arrival flow rate $q_{\mathrm{in}}(t) \neq 0$. In Sect. 9.3, we have shown that for this usual case in the theory of a time-delayed traffic breakdown (time-delayed transition from under- to over-saturated traffic) at the signal [26, 27] there are the infinite number of signal capacities between the minimum capacity C_{min} and maximum capacity C_{max}. When the average arrival flow rate $\bar{q}_{\mathrm{in}}$ exceeds the minimum signal capacity but it is smaller than the maximum

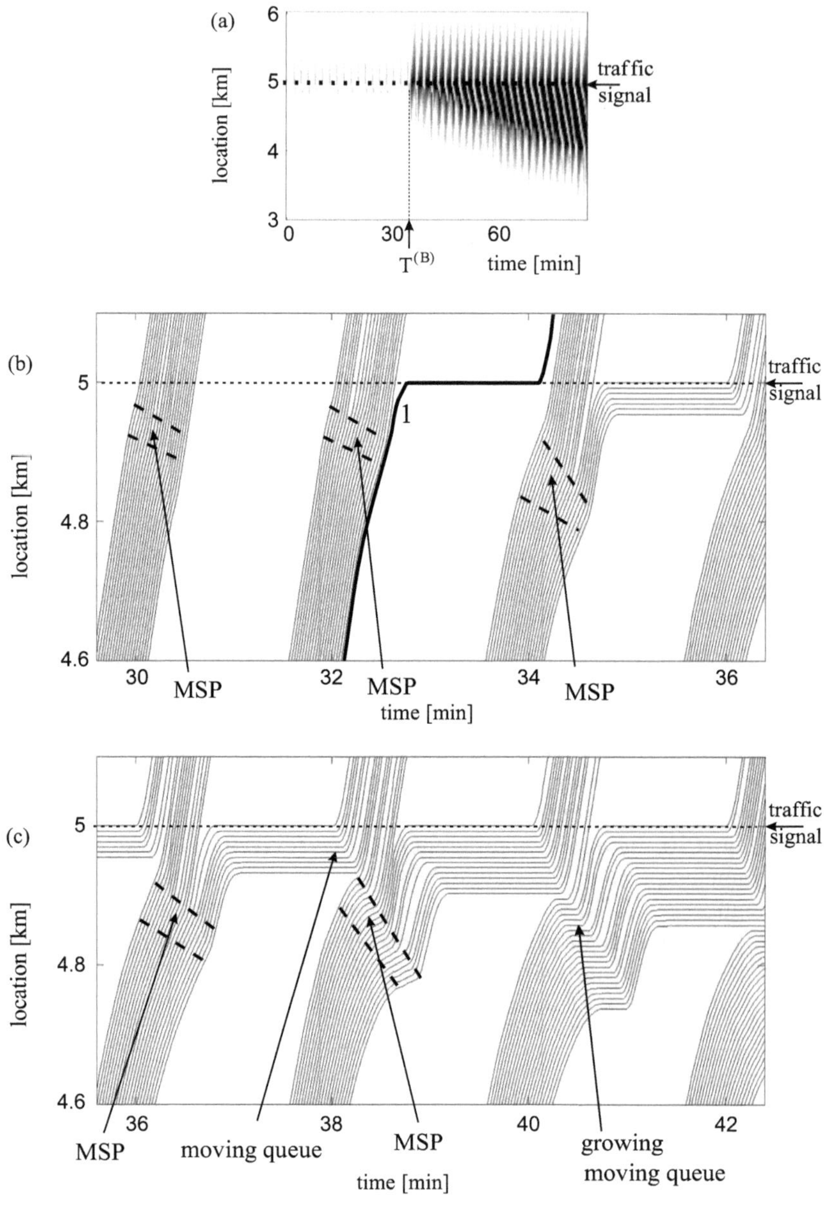

Fig. 9.29 Random time-delayed GW breakdown under condition (9.49) for a short duration of green signal phase $T_G = 38$ s in comparison with a long green signal phase $T_G = 98$ s used in Figs. 9.13–9.15 that have been simulated at the same value $\vartheta = 120$ s: (**a**) Speed data presented by regions with variable shades of gray (shades of gray vary from white to black when the speed decreases from 50 km/h (*white*) to zero (*black*)). (**b**, **c**) Vehicle trajectories related to (**a**); MSPs are marked by dashed curves. $q_{GW} = 2250$ vehicles/h. Minimum signal capacity C_{min} (classical signal capacity) is equal to $C_{min} = 554$ vehicles/h (in Figs. 9.13–9.15, $C_{min} = 1458$ vehicles/h). The average arrival flow rate is equal to $\bar{q}_{in} = 638$ vehicles/h. Simulations of the Kerner-Klenov model for city traffic (Sect. A.12 of Appendix A). Adapted from [30]

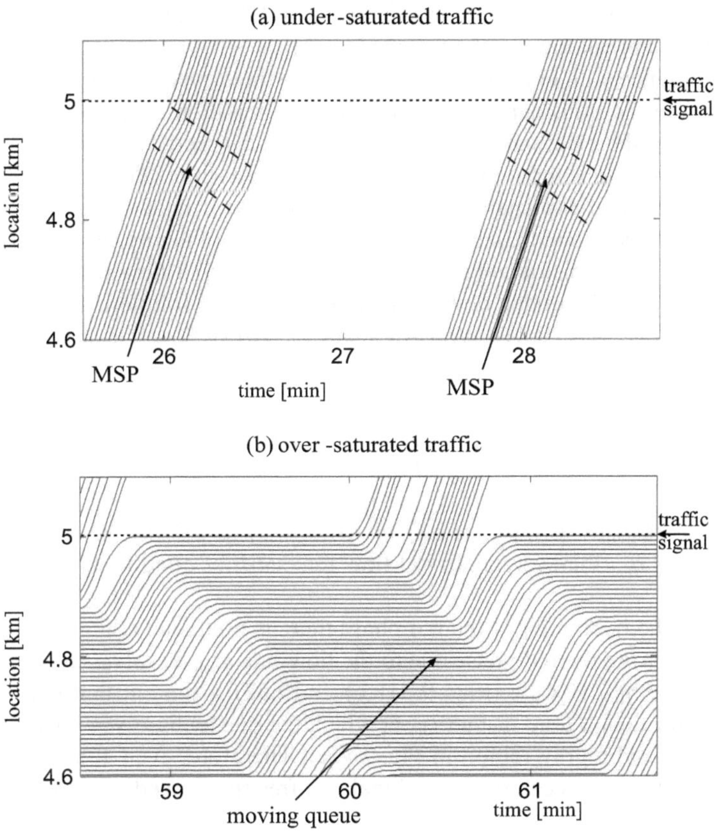

Fig. 9.30 Continuation of Fig. 9.29: Existence of either under-saturated traffic (**a**) or over-saturated traffic (**b**) at the same traffic flow and signal parameters under condition (9.14). Adapted from [30]

signal capacity, under-saturated traffic is in a metastable sate with the respect to traffic breakdown at the signal. The metastable under-saturated traffic can persist many signal cycles without the transition from this under-saturated to over-saturated traffic at the signal. When fluctuations in under-saturated traffic and in the discharge flow rate from the queue at the signal are small enough, then under conditions $C_{\rm min} < \bar{q}_{\rm in} < C_{\rm max}$ *no transition* from under-saturated to over-saturated traffic at the signal should be necessarily occur.

In contrast, in the classical theory of city traffic there cannot be a time-delayed traffic breakdown at the signal [2, 14]: When the average arrival flow rate $\bar{q}_{\rm in}$ exceeds the classical signal capacity $C_{\rm cl}$, traffic breakdown (transition from under- to over-saturated traffic) must occur at the signal without any delay (Sect. 9.2). Recall that the classical signal capacity $C_{\rm cl}$ is equal to the minimum signal capacity $C_{\rm min}$ (9.20) of the theory of a time-delayed traffic breakdown at the signal [27] (Sect. 9.3). Here, we give qualitative explanations of this crucial difference between results of the classical theory of traffic breakdown at the signal [2, 14] and the theory of the time-delayed traffic breakdown at the signal [26, 27].

In accordance with empirical data [24], in the three-phase theory the arrival flow rate $q_{\rm in}(t)$ can be considerably larger than the saturation flow rate $q_{\rm sat}$ (Fig. 9.5). In contrast, in the classical continuum theory of city traffic it is assumed that the saturation flow rate is the maximum rate of flow on a street (Fig. 9.1b) [14].

For a qualitative explanation of the metastability of under-saturated traffic, we use the above-mentioned assumption of the three-phase theory that the arrival flow rate $q_{\rm in}(t)$ can be considerably larger than the saturation flow rate $q_{\rm sat}$. However, we describe both the queue development and queue dissolution with the classical continuum theory (Fig. 9.31). We assume that the arrival flow rate $q_{\rm in}(t) = q_{\rm in}(t+\vartheta)$ is given by formula

$$q_{\rm in}(t) = \begin{cases} q_{\rm in1} \ \text{at} \ t_1 \leq t < t_2 \\ q_{\rm in2} \ \text{at} \ t_2 \leq t < t_3, \end{cases} \tag{9.57}$$

where $q_{\rm in1} < C_{\min} = C_{\rm cl}$, $q_{\rm in2} > q_{\rm sat}$, $t_3 - t_1 = \vartheta$ (Fig. 9.31a). We assume also that due to $q_{\rm in2} > q_{\rm sat}$ the average arrival flow rate satisfies conditions $\bar{q}_{\rm in} > C_{\min}$ and $\bar{q}_{\rm in} < C_{\max}$.

When at the beginning of the red phase there is no queue at the signal, then the queue built during the red phase due to the flow rate $q_{\rm in1}$ dissolves during the green phase because $q_{\rm in1} < C_{\min}$ (Figs. 9.31b,c). In (9.57), a time instant t_2 is chosen to be a few later than a time instant $t_{\rm dis}$ at which the queue dissolves fully (solid lines in Fig. 9.31b and a gray region in Fig. 9.31c). Therefore, even at a large flow rate $q_{\rm in2}$ (9.57) all vehicles can pass the signal at a free flow speed, i.e., under-saturated traffic is realized.

However, let us now assume that at the time instant $t = t_1$ there is already a long enough queue at the signal (time instant $t = t_1$ on dashed lines in Fig. 9.31b and on a hatched region in Fig. 9.31d). In this case, at the time instant t_2 the queue cannot fully dissolve. Therefore, vehicles that arrive at the signal with the rate $q_{\rm in2}$ (9.57) must stop at the end of this queue (dashed lines in Fig. 9.31b and a hatched region in Fig. 9.31d). As a result, over-saturated traffic occurs. Thus depending on an initial vehicle queue at the signal, either under-saturated traffic or over-saturated traffic occurs at the same arrival flow rate and signal parameters. This explains the metastability of city traffic under conditions (9.21), if we take into account the empirical fact that the arrival flow rate can be considerably larger than the saturation flow rate.

9.7.2 Arrival Flow Rate Is Smaller Than Saturation Flow Rate

In this Sect. 9.7.2, we consider traffic phenomena at the signal for the case when condition (9.50) is satisfied.

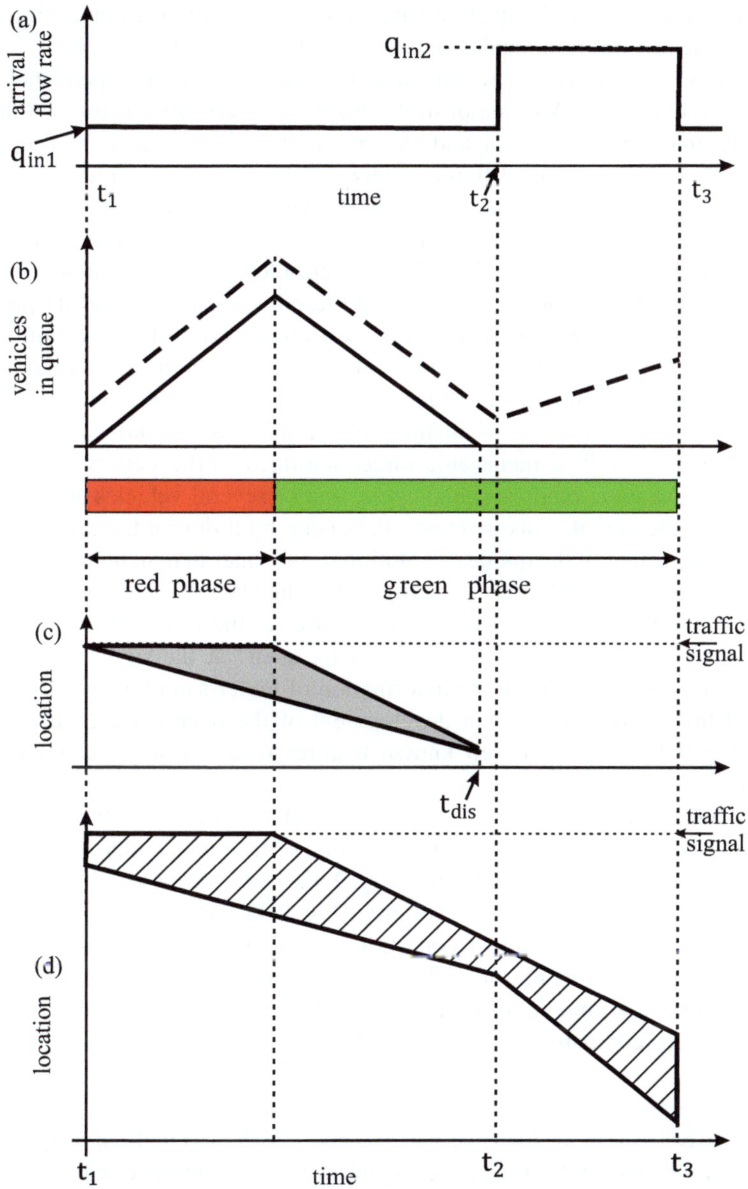

Fig. 9.31 Qualitative explanation of metastability of city traffic: (**a**) Time dependence of the arrival flow rate (9.57). (**b**) Number of vehicles in queue in under-saturated traffic (*solid curve*) and in over-saturated traffic (*dashed curve*). (**c, d**) Queue patterns with motionless vehicles in under-saturated traffic (**c**) and over-saturated traffic (**d**). Yellow signal phase is neglected

We have found that although there are quantitative differences in spatiotemporal characteristics of traffic breakdown, under condition (9.50) qualitative features of the time-delayed breakdown remain independent of the time-dependence of the arrival flow rate $q_{in}(t)$. We demonstrate this general conclusion for two different cases presented in Figs. 9.22a and 9.32a. In the both cases, conditions (9.14) and (9.50) are satisfied. The difference between these two examples is the duration of the green signal phase: In the example shown in Fig. 9.32a the duration of the green phase is considerably shorter than a considerably longer duration of the green phase in the example shown in Fig. 9.22a. Nevertheless, we find qualitatively similar results of traffic breakdown at the signal in these two different cases. In particular, in the both cases metastable under-saturated traffic persists during a time interval $T^{(B)} \gg \vartheta$. After this random time-interval, a time-delayed traffic breakdown, i.e., the time-delayed transition from under- to over-saturated traffic occurs.

We have also found that as in the case of the GW breakdown discussed in Sect. 9.4, at $t < T^{(B)}$ a metastable under-saturated traffic persists at the signal (Fig. 9.32b). At some random signal cycle, one or several vehicles at the GW end must stop at the signal. This increases the queue built during the red phase. As a result, a time instant of the queue dissolution occurs later than in the previous signal cycles. Due to longer delays of vehicles within the GW, the number of vehicles at the end of the GW that come to a stop increases at the next signal cycle(s). This results in the formation of a growing moving queue at the signal, i.e., in traffic breakdown associated with the transformation of under-saturated traffic into over-saturated traffic. The subsequent development of the over-saturated traffic at the signal (Fig. 9.32c) shows a well known features revealed in the classical theory (Fig. 9.6c).

Thus, when condition (9.14) is satisfied, in the theory of the time-delayed breakdown at the signal even under condition (9.50) either under-saturated traffic (Fig. 9.32b) or over-saturated traffic (Fig. 9.32c) can occur *at the same signal parameters and the same arrival flow rate* (Fig. 9.32). This is in contrast with the classical theory, in which under condition (9.14) only over-saturated traffic can exist (Fig. 9.6c). For this reason, only for over-saturated traffic the classical theory and the theory of a time-delayed traffic breakdown at the signal show qualitatively the same result (compare Figs. 9.6c and 9.32c).

If the average arrival flow rate exceeds the classical signal capacity, in the classical theory of traffic at the signal *only* over-saturated traffic can be realized at the signal. Contrarily to the classical theory of traffic breakdown at the signal, if the average arrival flow rate exceeds the classical signal capacity, in the theory of a time-delayed traffic breakdown at the signal at the same signal parameters and the same average arrival flow rate there can be two different states of city traffic at the signal: (i) metastable under-saturated traffic and (ii) over-saturated traffic.

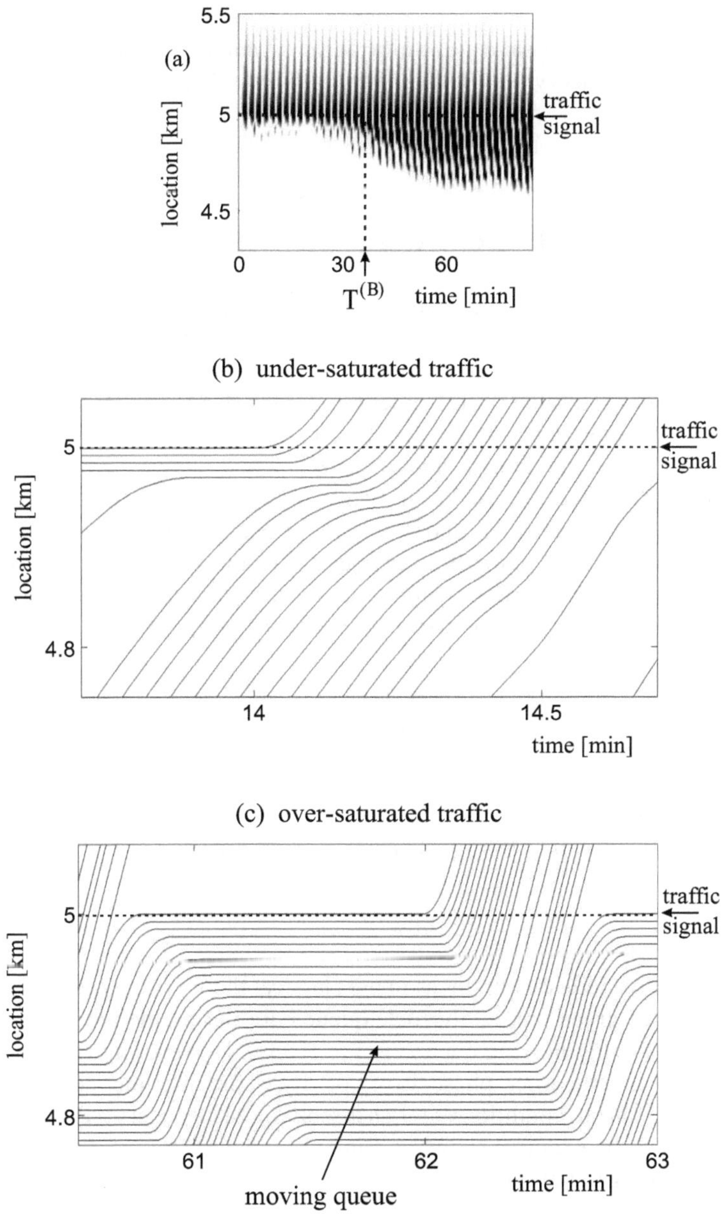

Fig. 9.32 Existence of either under-saturated traffic or over-saturated traffic at the same traffic flow and signal parameters under conditions (9.14) and (9.50): (**a**) Speed data presented by regions with variable shades of gray (shades of gray vary from white to black when the speed decreases from 50 km/h (*white*) to zero (*black*)). (**b, c**) vehicle trajectories for under-saturated traffic ($t < T^{(B)}$) (**b**) and over-saturated traffic $t > T^{(B)}$ (**c**) related to (**a**). $\vartheta = 120$ s, $T_Y = 2$ s. $T_R = 80$ s. $q_{GW} = 1440$ vehicles/h, $q_{turn} = 225$ vehicles/h. Minimum capacity C_{min} is equal to $C_{min} = 554$ vehicles/h, the average arrival flow rate is equal to $\bar{q}_{in} = 584$ vehicles/h. Simulations of the Kerner-Klenov model for city traffic (Sect. A.12 of Appendix A). Adapted from [30]

Only features of over-saturated traffic at the signal are qualitatively the same in the classical theory and in the theory of a time-delayed traffic breakdown at the signal.

9.7.2.1 Dissolving MSP in Under-Saturated Traffic

The existence of metastable under-saturated traffic under conditions (9.14) and (9.50) presented in Fig. 9.32b is associated with MSP occurrence in under-saturated traffic at the signal. To prove this, we make here a microscopic analysis of metastable under-saturated traffic under conditions (9.14) and (9.50). We will see that due to MSP emergence at the signal the rate of traffic passing the signal $q_{\rm TS}(t)$ becomes larger than the saturation flow rate $q_{\rm sat}$.

In Fig. 9.32 related to condition (9.50), we observe qualitatively the same stages of GW breakdown as those shown in Figs. 9.13–9.15 for the opposite condition (9.49): In each of the signal cycles during the green phase, due to turning-in traffic, firstly, a dissolving moving queue occurs ("dissolving moving queue" in Fig. 9.33a). The moving queue dissolves because the inflow rate $q_{\rm GW}$ within the GW satisfies condition

$$q_{\rm GW} < q_{\rm sat}. \tag{9.58}$$

After the moving queue has dissolved, a dissolving MSP occurs ("dissolving MSP" in Fig. 9.33a). This is because a threshold flow rate required for stationary propagation of an MSP $q_{\rm out}^{\rm (MSP)}$ (Sect. 9.4.2) is larger than $q_{\rm GW}$:

$$q_{\rm GW} < q_{\rm out}^{\rm (MSP)}. \tag{9.59}$$

Thus, we can conclude that firstly as in the classical theory (Fig. 9.6a) [14], after the beginning of the green phase, a vehicle queue built during the red phase begins to dissolve ("dissolving moving queue" in Fig. 9.33a). However, in contrast with the classical theory, in the three-phase theory of city traffic after the moving queue has fully dissolved, due to condition (9.59), a dissolving MSP occurs ("dissolving MSP" in Fig. 9.33). In the example shown in Fig. 9.33a, the MSP does not fully dissolve while propagating to the GW end.

9.7.2.2 Compression of Under-Saturated Traffic at Signal Due to Formation of Dissolving MSP

The effect of the dissolving MSP on features of the metastable under-saturated traffic is demonstrated in Fig. 9.34: As follows from a consideration of the MSP

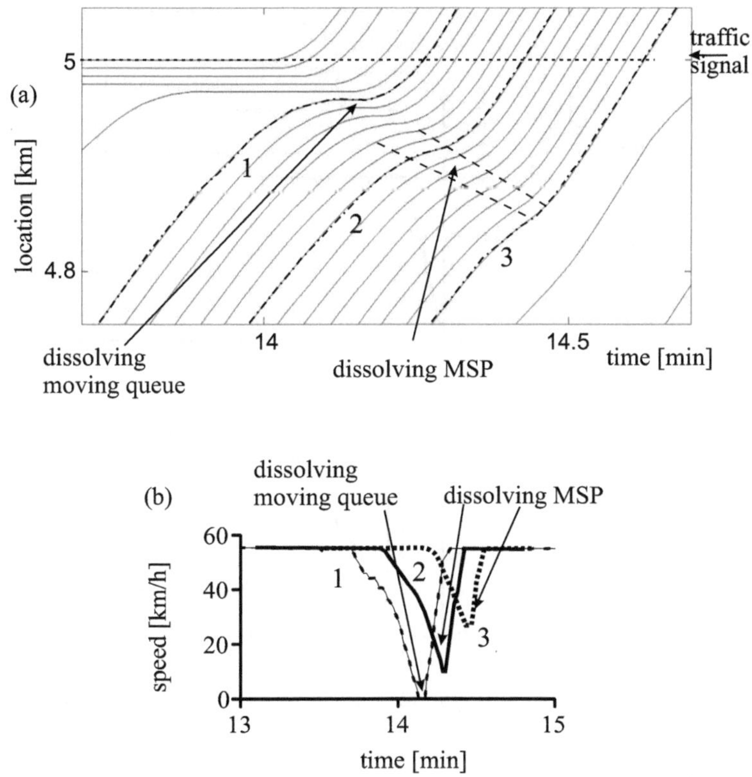

Fig. 9.33 Formation of dissolving moving queues and dissolving MSPs during green signal phase in example of time-delayed traffic breakdown shown in Fig. 9.32a: (**a**) Vehicle trajectories during the signal cycle adapted from Fig. 9.32b. (**b**) Single-vehicle speeds along vehicle trajectories labeled by the same numbers in (**a**)

propagation discussed in Sect. 9.4.2, the average time headway (average net time gap) $\tau_{\rm out}^{\rm (MSP)} \approx 1.3$ s between vehicles that discharge from the MSP during the green signal phase is considerably shorter than the average time headway between vehicles that discharge from the moving queue $\tau_{\rm sat}^{\rm (net)} = 1.5$ s (Fig. 9.34b). Using formulas (9.2), (9.45), and (9.47), we can find a qualitative time-dependence of the rate of traffic passing the signal $q_{\rm TS}(t)$ (Fig. 9.34a).

After some time interval from the beginning of the green signal phase (Sect. 9.2.2), as in the classical theory, the value $q_{\rm TS}(t)$ during a time interval related to the queue discharge is equal to

$$q_{\rm TS} = q_{\rm sat}, \tag{9.60}$$

where for model parameters used in simulations accordingly to formula (9.9) $q_{\rm TS} \approx$ 1808 vehicles/h. Later, $q_{\rm TS}(t)$ is determined by the MSP discharge (Sect. 9.4.2). This

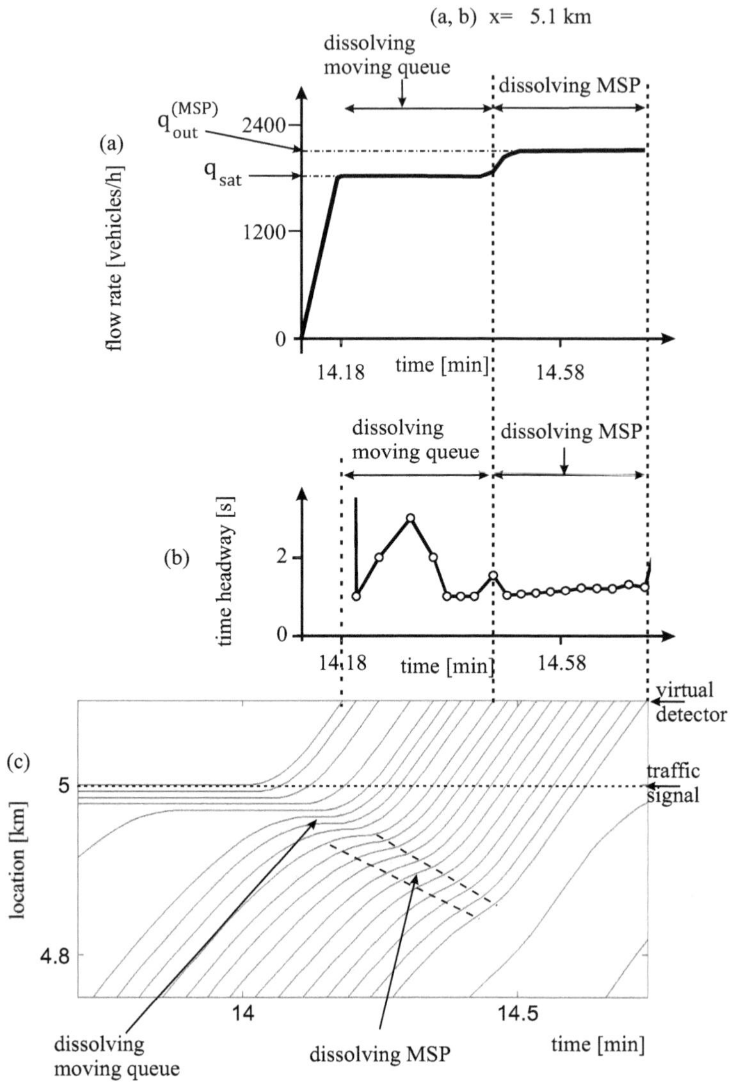

Fig. 9.34 Compression of under-saturated traffic at signal due to the formation of dissolving MSP shown in Fig. 9.33a: (**a**) Qualitative time-dependence of the rate of traffic passing the signal $q_{TS}(t)$ calculated with time-function of time headway in (**b**). (**b**) Time-function of time headway between vehicles. (**c**) Vehicle trajectories adapted from Fig. 9.33a. Time headway between the first vehicle and the beginning the green phase has not been shown. We measure time headway and the related flow rate $q_{TS}(t)$ at a virtual detector that is at location 100 m downstream of the signal location at which vehicles passing the signal have already accelerated to the speed in free flow v_{free}. Adapted from [30]

flow rate is equal to

$$q_{\mathrm{TS}} = q_{\mathrm{out}}^{(\mathrm{MSP})}, \tag{9.61}$$

where for model parameters used in simulations $q_{\mathrm{TS}} = q_{\mathrm{out}}^{(\mathrm{MSP})} \approx 2010$ vehicles/h. Thus, during the green signal phase, the flow rate $q_{\mathrm{TS}}(t)$ increases over time from the classical saturation flow rate $q_{\mathrm{sat}} \approx 1808$ vehicles/h to a larger flow rate $q_{\mathrm{out}}^{(\mathrm{MSP})} \approx$ 2010 vehicles/h that is determined by the MSP discharge.

Therefore, during the green signal phase, the flow rate q_{TS} increases over time from the saturation flow rate to a larger flow rate associated with the MSP discharge. This is because the MSP causes a "compression" of under-saturated traffic at signal:

(i) The vehicle density can increase within the MSP considerably in comparison with the initial density in the arrival flow.
(ii) In accordance with formula (9.47), the rate of the MSP discharge can be much larger than the arrival flow rate.

Thus, we can conclude that the importance of a consideration of the dissolving MSP is as follows: Due to MSP emergence in under-saturated traffic at the signal, there is a compression of under-saturated traffic. For this reason, the rate of traffic passing the signal $q_{\mathrm{TS}}(t)$ becomes larger than the saturation flow rate q_{sat} during the existence of the dissolving MSP in under-saturated traffic (Figs. 9.34a and 9.35a). This result contradicts the classical theory (Fig. 9.1a) [14], in which the rate of traffic passing the signal $q_{\mathrm{TS}}(t)$ cannot exceed the saturation flow rate q_{sat}.

In the classical theory of traffic at the signal, the rate of traffic passing the signal $q_{\mathrm{TS}}(t)$ cannot exceed the saturation flow rate q_{sat}.

In contrast with the classical theory, in the theory of a time-delayed traffic breakdown at the signal, due to MSP emergence in under-saturated traffic at the signal the rate of traffic passing the signal $q_{\mathrm{TS}}(t)$ can become larger than the saturation flow rate q_{sat}.

9.7.2.3 Reason for Metastable Under-Saturated Traffic Under Condition (9.50)

We have shown that under condition (9.50) under-saturated traffic can exist even if condition (9.14) is satisfied. The cause of the existence of this metastable under-saturated traffic is as follows (Fig. 9.36).

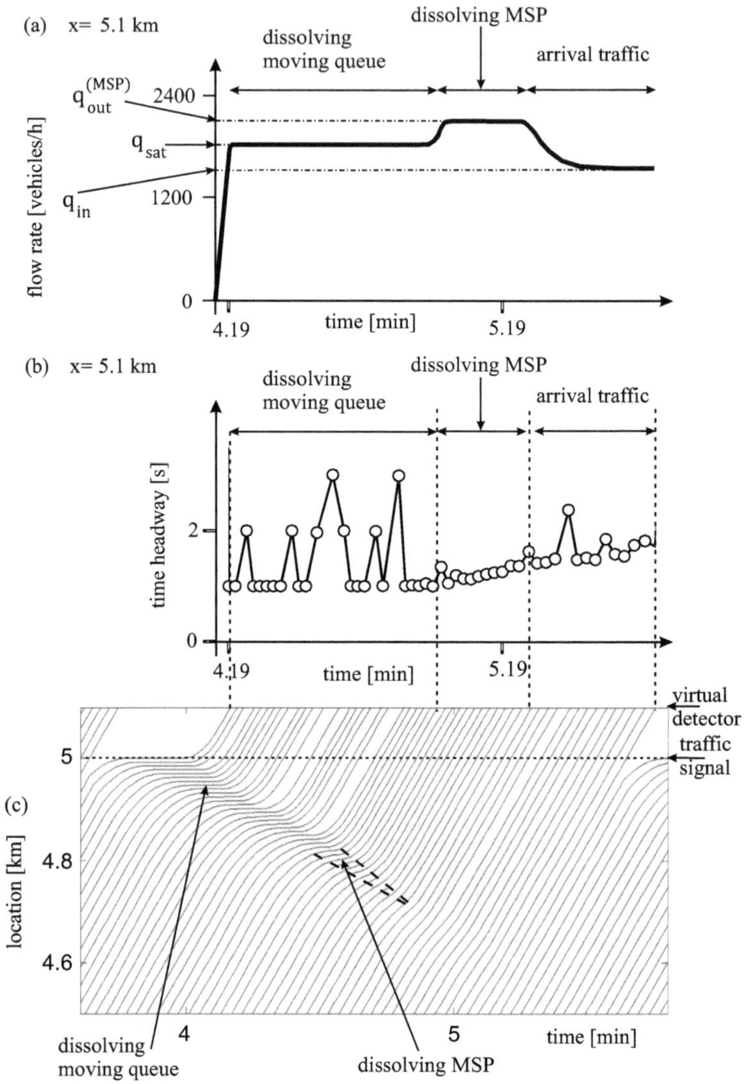

Fig. 9.35 Compression of under-saturated traffic at signal due to the formation of dissolving MSP occurring in example of time-delayed traffic breakdown at the signal shown in Fig. 9.22a: (**a**) Qualitative time-dependence of the rate of traffic passing the signal $q_{TS}(t)$ calculated with time-function of time headway in (**b**). (**b**) Time-function of time headway between vehicles. (**c**) Vehicle trajectories in under-saturated traffic in a time interval related to $t < T^{(B)}$ in Fig. 9.22a. Time headway between the first vehicle and the beginning the green phase has not been shown. We measure time headway and the related flow rate $q_{TS}(t)$ at a virtual detector that is at location 100 m downstream of the signal location at which vehicles passing the signal have already accelerated to the speed in free flow v_{free}. Adapted from [28, 30]

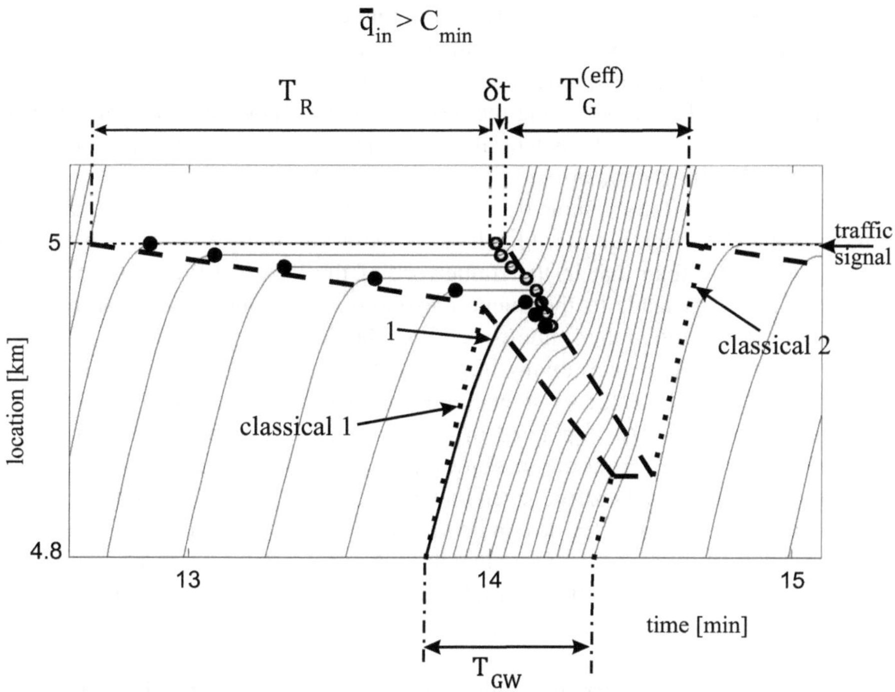

Fig. 9.36 Explanation of cause for metastable under-saturated traffic under condition (9.50). Vehicle trajectories adapted from Fig. 9.33a. *Black points* on vehicle trajectories mark co-ordinates at which vehicles come to a stop at the upstream queue front. *Circle points* on vehicle trajectories mark co-ordinates at which vehicles begin to accelerate at the downstream queue front. *Dashed lines* show queue patterns calculated with formulas (9.17) and (9.18) of the classical continuum theory of city traffic; *dotted lines* show vehicle trajectories associated with the queue patterns of the classical theory. T_{GW} is GW duration (see Fig. 9.11b)

While approaching a queue of motionless vehicles at the signal, a vehicle decelerates within a finite road length. This road length determines the length of the upstream front of the queue. As long as the queue length is considerably longer than the length of the upstream queue front, the length of this front can be neglected. However, this is not the case in under-saturated traffic at the end of the queue dissolution when the queue length becomes short enough.

This is because it takes some time for vehicle deceleration from a free flow speed to a standstill within the queue. When the length of a moving queue of motionless vehicles is short enough, then during this finite time interval of vehicle deceleration the queue can dissolve fully before the decelerating vehicle should come to a stop. This effect leads to queue dissolution at an earlier time instant in comparison with a time instant of the queue dissolution that is found in the classical theory.

Indeed, in the framework of the classical theory, the first vehicle at the GW beginning (dotted line labeled by “classical 1” at the beginning of time interval T_{GW} in Fig. 9.36) stops at the end of the moving queue shown by dashed lines in

Fig. 9.36. Because at the chosen arrival flow rate condition (9.14) is satisfied, the queue dissolves fully only after the end of the effective green time. This means that the last vehicle of this queue (dotted line labeled by "classical 2" in Fig. 9.36) comes to the signal location during the next red phase. Therefore, the vehicle must stop at the signal. This explains why in the framework of the classical theory *only* over-saturated traffic can occur under condition (9.14).

In reality, due to finite vehicle deceleration at the upstream queue front, the first vehicle at the GW beginning (bold vehicle trajectory 1 in Fig. 9.36) comes to stop considerably *later* than that in the framework of the classical theory (dotted line "classical 1"). This leads to the queue dissolution at a considerably earlier time instant than that found in the framework of the classical theory.

For this reason, after the moving queue has fully dissolved, there is still a long enough time interval for the existence and propagation of the dissolving MSP in under-saturated traffic. As explained in Sect. 9.7.2.2, during this time interval the flow rate q_{TS} increases from the saturation flow rate to the larger rate of the MSP discharge (Fig. 9.34a). Therefore, even under condition (9.50), condition (9.14) for metastable under-saturated traffic can be satisfied. The finite vehicle deceleration at upstream queue front leads to a finite length of the upstream queue front. Therefore, it is equivalent to consider "finite vehicle deceleration at the upstream queue front" or "finite length of the upstream queue front" as a cause for the metastability of under-saturated traffic at the signal.

In contrast with the theory of a time-delayed traffic breakdown at the signal [26–28] presented in this chapter, in the classical continuum theory the length of the upstream queue front is neglected (dotted and dashed lines in Fig. 9.36) [14]. This explains why in the classical theory there is no metastability of under-saturated traffic at the signal.

Therefore, we can conclude that the finite value of vehicle deceleration at the upstream front of a moving queue leads to queue dissolution during the green phase at an earlier time instant in comparison with a time instant of the queue dissolution that is found in the framework of the classical theory (Fig. 9.36). As a result, after the moving queue has fully dissolved, there is still a long enough time interval for the existence and propagation of a dissolving MSP in under-saturated traffic (Figs. 9.34c, and 9.35c). During this time interval the rate of traffic passing the signal $q_{TS}(t)$ increases from the saturation flow rate to the larger rate of MSP discharge (Figs. 9.34a, and 9.35a).

This result can explain results of empirical studies in which an increasing time-function $q_{TS}(t)$ has been found (see Fig. 4 of [1]). To understand this conclusion, we should note that in [1] (as well as in all known for us other empirical studies of the flow rate downstream of the signal $q_{TS}(t)$) vehicle trajectories upstream of the signal have not been measured. Additionally, there are no traffic measurement in [1], which can conclude whether over-saturated traffic or metastable under-saturated traffic is realized in this empirical study made at peak time period of the data collection. This means that it is *not* possible from known empirical data to conclude whether the increasing function $q_{TS}(t)$ is determined by the saturation flow rate as assumed by the authors of [1] or by the rate of MSP discharge as found in our theoretical study.

Thus, for the answer on the question about the reason of the increasing function $q_{TS}(t)$ found in empirical data [1], *both* the flow rate $q_{TS}(t)$ *and* single vehicle data upstream of the signal should be measured.

9.8 "Red Wave" in City Traffic: Classical Theory of Traffic at Signal as Special Case of Three-Phase Theory

We have shown above that the theory of a time-delayed traffic breakdown at the signal [26–28] leads to crucially different features of city traffic in comparison with the classical theory of traffic at the signal [2, 4, 14, 18, 22, 32, 35, 37–42, 44, 45, 47, 49].

However, there is one non-realistic traffic situation in which results for the transition from under- to over-saturated traffic (traffic breakdown) at the signal coincide exactly each other for the classical theory and the theory of a time-delayed traffic breakdown at the signal. This case is a "red wave" (RW) in a city. The RW means that all vehicles arrive the signal only during the red signal phase: During the green and yellow phases the arrival flow rate is equal to zero. In accordance with formula (9.53), for a RW (Fig. 9.37a) we get

$$R = q_{\rm in}^{\rm (green)}/q_{\rm in}^{\rm (red)} = 0. \tag{9.62}$$

Simulations of traffic breakdown at the signal in the case of the RW show that the difference $C_{\max} - C_{\min}$ becomes very small (Fig. 9.37b). Therefore, the transition from under- to over-saturated traffic occurs on average at

$$\bar{q}_{\rm in} = C_{\min} \tag{9.63}$$

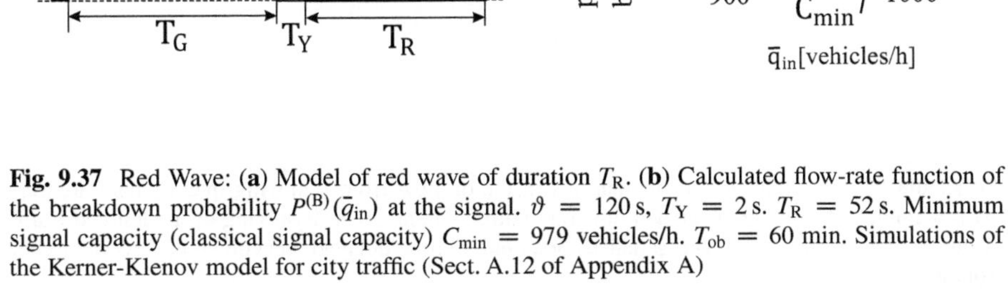

Fig. 9.37 Red Wave: (**a**) Model of red wave of duration T_R. (**b**) Calculated flow-rate function of the breakdown probability $P^{(B)}(\bar{q}_{\rm in})$ at the signal. $\vartheta = 120$ s, $T_Y = 2$ s. $T_R = 52$ s. Minimum signal capacity (classical signal capacity) $C_{\min} = 979$ vehicles/h. $T_{\rm ob} = 60$ min. Simulations of the Kerner-Klenov model for city traffic (Sect. A.12 of Appendix A)

as stated in the classical traffic flow theories [2, 4, 14, 18, 22, 32, 35, 37–42, 44, 45, 47, 49]. A smoothing character of function $P^{(B)}(\bar{q}_{in})$ around $\bar{q}_{in} = C_{min}$ for the RW found in simulations (Fig. 9.37b) is caused by the effect of model fluctuations on the queue discharge outflow.

Thus, only for the RW one can determine signal capacity based on the classical capacity definition (9.12). This emphasizes that and why in all realistic cases, in which $q_{in}(t) \neq 0$ during at least one time interval within the green and yellow phases, there are the infinite number of signal capacities between the minimum signal capacity C_{min} and maximum signal capacity C_{max} (Sect. 9.3).

We have found the following results for the case of a RW (Figs. 9.38 and 9.39):

(i) When $\bar{q}_{in} = C_{min}$, i.e., condition (9.16) is satisfied, then as in the classical theory, the last vehicle of the queue of motionless vehicles at the signal can just pass on average the signal location before the next red signal phase begins (Fig. 9.6b). Due to model fluctuations, some of the vehicles in the queue outflow must stop at the signal in some of the cycles (vehicle trajectory 1 in Fig. 9.38c). This increase in the vehicle queue at the signal can lead to over-saturated traffic but it should not necessarily lead to over-saturated traffic. This is because during the next signal cycle the vehicle queue can fully dissolve during the green phase. Therefore, *on average* no growth of the vehicle queue should necessarily occur at the signal, when a lot of signal cycles are taken into account.[6]

(ii) Even when $\bar{q}_{in}$ exceeds C_{min} only slightly, i.e., condition (9.14) is just satisfied, then as in the classical theory, over-saturated traffic is developing over time (Fig. 9.39).

An explanation of these results is as follows. In the case of the RW, the rate of traffic passing the signal is determined *only* by the queue discharge, i.e., by the saturation flow rate (Figs. 9.38 and 9.39). As in the classical theory, in the Kerner-Klenov stochastic microscopic three-phase model used for simulations the saturation flow rate does not depend on the green phase duration. This explains why in the case of the RW the classical capacity separates on average under- and over-saturated traffic in the both theories of city traffic.

As mentioned, the RW is a non-realistic case of city traffic. In all realistic cases of city traffic, there is at least one time interval during the green and yellow phases, when $q_{in}(t) \neq 0$. Therefore, for physical reasons explained above, the theory of a time-delayed traffic breakdown at the signal and the classical theory of city traffic predict qualitatively different features of traffic breakdown at the signal.

---

[6] It should be noted that due to fluctuations in the outflow from a moving queue, there are very narrow ranges of the average arrival flow rate within which either dissolving over-saturated traffic occurs (when condition (9.28) is satisfied) or a random time-delayed breakdown is observed (when condition (9.21) is satisfied).

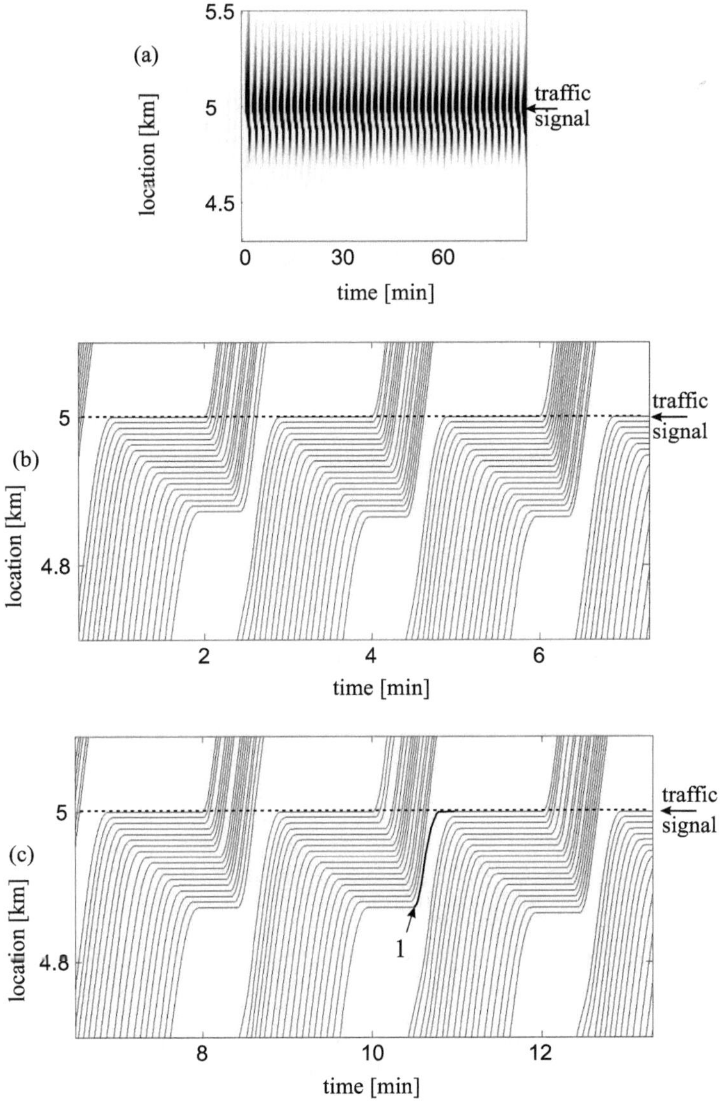

Fig. 9.38 Simulations of traffic breakdown in the case of "red wave" (RW) at $\bar{q}_{\text{in}} = C_{\min}$: (**a**) Speed data presented by regions with variable shades of gray (shades of gray vary from white to black when the speed decreases from 50 km/h (*white*) to zero (*black*)). (**b, c**) Vehicle trajectories. $\vartheta = 120$ s, $T_{\text{Y}} = 2$ s. $T_{\text{R}} = 80$ s. $\bar{q}_{\text{in}} = C_{\min} = 554$ vehicles/h. Simulations of the Kerner-Klenov model for city traffic (Sect. A.12 of Appendix A)

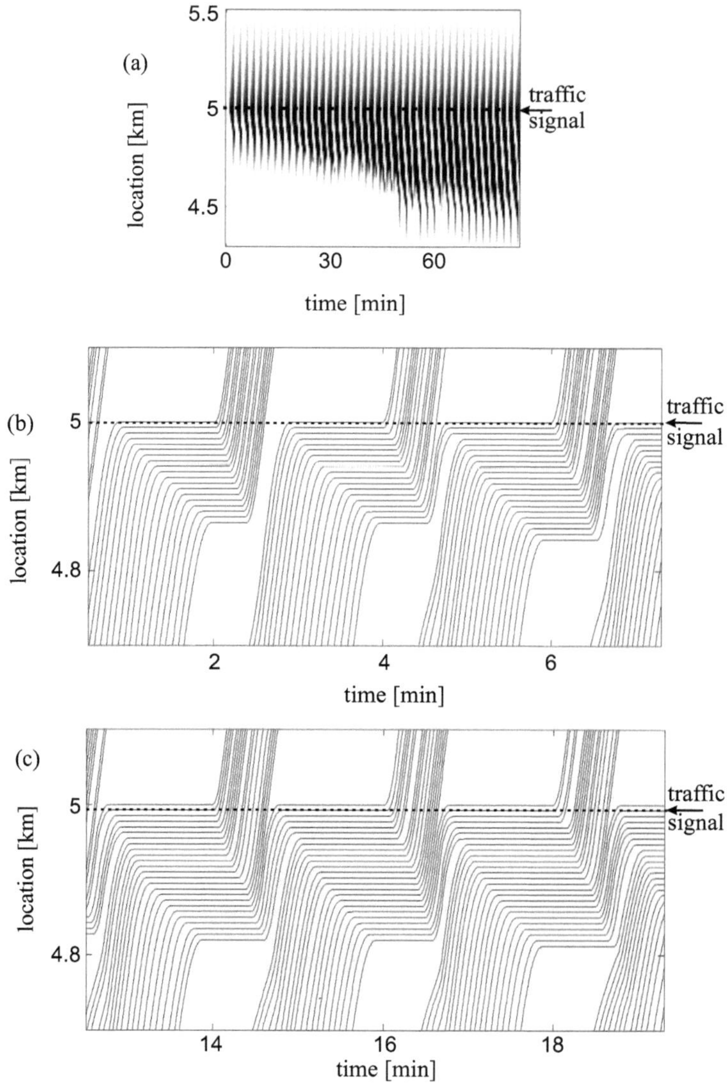

Fig. 9.39 Simulations of traffic breakdown in the case of RW at $\bar{q}_{\text{in}} > C_{\text{min}}$: (**a**) Speed data presented by regions with variable shades of gray (shades of gray vary from white to black when the speed decreases from 50 km/h (*white*) to zero (*black*)). (**b**, **c**) Vehicle trajectories. $\vartheta = 120$ s, $T_{\text{Y}} = 2$ s. $T_{\text{R}} = 80$ s. $\bar{q}_{\text{in}} = 572$ vehicles/h. $C_{\text{min}} = 554$ vehicles/h. Simulations of the Kerner-Klenov model for city traffic (Sect. A.12 of Appendix A)

The theory of a time-delayed traffic breakdown at the signal [26–28] leads to crucially different features of city traffic in comparison with the classical theory [2, 4, 14, 18, 22, 32, 35, 37–42, 44, 45, 47, 49].

For a "red wave" (RW) in city traffic, results of the classical theory and the theory of a time-delayed traffic breakdown at the signal coincide each other. The RW means that during the green and yellow signal phases the arrival flow rate is equal exactly to zero. However, in all realistic cases of city traffic, there is at least one time interval during the green and yellow phases, when the arrival flow rate $q_{\text{in}}(t) \neq 0$. Therefore, the RW is a non-realistic case of city traffic.

In all realistic situations in city traffic, the theory of a time-delayed traffic breakdown at the signal predicts qualitatively different features of traffic breakdown at the signal in comparison with the classical theory of city traffic.

9.9 Conclusions

(i) Under-saturated traffic at the signal in a city traffic can persist during many signal cycles even when the average arrival flow rate in this under-saturated traffic exceeds the classical signal capacity (9.12).

(ii) When the average arrival flow rate in under-saturated traffic exceeds the classical signal capacity (9.12) (item (i)), the under-saturated traffic is in a metastable state with respect to a transition from under- to over-saturated traffic at the signal. The transition from under- to over-saturated traffic is called as traffic breakdown at the signal.

(iii) After a time delay that can be considerably longer than the signal cycle, traffic breakdown occurs in the metastable under-saturated traffic at the signal. The time delay of the breakdown at the signal is a random value. Thus, in a general case traffic breakdown at the signal in a city traffic is a time delayed transition (a time delayed transition from under- to over-saturated traffic at the signal) [26, 27].

(iv) In under-saturated traffic, moving synchronized flow patterns (MSP) can emerge. The MSPs can effect considerably on probabilistic characteristics of a time delayed transition from under- to over-saturated traffic at the signal.

(v) In under-saturated traffic at the signal, a dissolving MSP propagating upstream in the under-saturated traffic can emerge. Within the dissolving MSP, there is a compression of traffic flow. Due to this flow compression, the flow rate in the MSP discharge can be considerably larger than the saturation flow rate. For this reason, the flow rate passing the signal during the green signal phase can be larger than the saturation flow rate determined by the queue discharge.

(vi) A finite vehicle deceleration at the upstream queue front leads to a finite length of the upstream queue front. This is one of the reasons for the metastability of under-saturated traffic at the signal.

(vii) The signal forces all vehicles to stop during the red signal phase. Undisturbed under-saturated traffic during the green signal phase can be considered free flow (F). Therefore, the effect of the red signal phase on the under-saturated traffic can be considered an induced F→J transition at the signal (J—a vehicle queue built at the signal during the red signal phase). Two-phase traffic flow models of the GM model class can explain and simulate the F→J transition. For this reason, at least some of the two-phase traffic flow models of the GM model class can explain and simulate a time-delayed transition from under- to over-saturated traffic (time-delayed traffic breakdown) at the signal [26, 27]. Nevertheless, we have found that the three-phase theory leads to sometimes qualitatively different results for a time-delayed traffic breakdown at the signal in comparison with two-phase models of the GM model class. This difference is associated with the occurrence of MSPs in under-saturated traffic in a neighborhood of the signal.

(viii) Only in a case of the "red wave" (RW) in city traffic, when all vehicles arrive the signal location only during the red signal phase, the classical theory of city traffic [2, 4, 14, 18, 22, 32, 35, 37–42, 44, 45, 47, 49] is a special case of the three-phase theory. However, the RW is a non-realistic case of city traffic. In all realistic cases of city traffic, there is at least one time interval during the green and yellow phases, when the arrival flow rate $q_{\text{in}}(t) \neq 0$. Thus, the classical theory of city traffic [2, 4, 14, 18, 22, 32, 35, 37–42, 44, 45, 47, 49] and the theory of a time-delayed transition from under- to over-saturated traffic (time-delayed traffic breakdown) at the signal [26–28] predict qualitatively different features of traffic breakdown at the signal.

References

1. M.S. Chaudhry, P. Ranjitkar, in *TRB 92nd Annual Meeting Compendium of Papers*, TRB Paper 13-3396 (Transportation Research Board, Washington D.C., 2013)
2. F. Dion, H. Rakha, Y.S. Kang, Transp. Rec. B **38**, 99–122 (2004)
3. N.H. Gartner, Transp. Sci. **6**, 88–93 (1972)
4. N.H. Gartner, Transp. Res. Rec. **906**, 75–81 (1983)
5. N.H. Gartner, Transp. Res. A **19**, 369–373 (1985)
6. N.H. Gartner, M. Al-Malik, Transp. Res. Rec. **1554**, 27–35 (1996)

7. N.H. Gartner, S.F. Assmann, F. Lasaga, D.L. Hou, Transp. Res. Rec. **1287**, 212–222 (1990)
8. N.H. Gartner, S.F. Assmann, F. Lasaga, D.L. Hou, Transp. Res. B **25**, 55–74 (1991)
9. N.H. Gartner, J.D.C. Little, Transp. Res. Rec. **531**, 58–69 (1975)
10. N.H. Gartner, J.D.C. Little, H. Gabbay, Transp. Sci. **9**, 321–343 (1975)
11. N.H. Gartner, J.D.C. Little, H. Gabbay, Transp. Res. Rec. **596**, 6–15 (1976)
12. N.H. Gartner, Ch. Stamatiadis, Math. Comp. Model. **35**, 657–671 (2002)
13. N.H. Gartner, Ch. Stamatiadis, J. Intel. Transp. Sys. **8**, 77–86 (2004)
14. N.H. Gartner, Ch. Stamatiadis, in *Encyclopedia of Complexity and System Science*, ed. by R.A. Meyers (Springer, Berlin, 2009), pp. 9470–9500
15. N.H. Gartner, Ch. Stamatiadis, P.J. Tarnoff, Transp. Res. Rec. **1494**, 98–105 (1995)
16. N.H. Gartner, P.J. Tarnoff, C.M. Andrews, Transp. Res. Rec. **1324**, 105–114 (1991)
17. D.L. Gerlough, M.J. Huber, *Traffic Flow Theory Special Report 165* (Transp. Res. Board, Washington D.C., 1975)
18. R.B. Grafton, G.F. Newell, in *Vehicular Traffic Science*, ed. by L.C. Edie, R. Herman, R. Rothery (Elsevier, New York, 1967), pp. 239–257
19. B.D. Greenshields, D. Schapiro, E.L. Ericksen, *Technical Report No.1* (Yale Bureau of Highway Traffic, Yale University, 1947)
20. P. Hemmerle, M. Koller, H. Rehborn, B.S. Kerner, M. Schreckenberg, IET Intell. Transp. Syst. **10**, 122–129 (2016)
21. G. Hermanns, P. Hemmerle, H. Rehborn, M. Koller, B.S. Kerner, M. Schreckenberg, Transp. Res. Rec. **2490**, 47–55 (2015)
22. P.B. Hunt, D.I. Robertson, R.D. Bretherton, R.I. Winton, *TRRL Report No. LR 1014* (Transp. and Road Res. Lab., Crowthorne, UK, 1981)
23. R. Jiang, M.-B. Hu, B. Jia, Z.-Y. Gao, Comp. Phys. Comm. **185**, 1439 (2014)
24. B.S. Kerner, *The Physics of Traffic* (Springer, Berlin, New York, 2004)
25. B.S. Kerner, *Introduction to Modern Traffic Flow Theory and Control* (Springer, Berlin, New York, 2009)
26. B.S. Kerner, Phys. Rev. E **84**, 045102(R) (2011)
27. B.S. Kerner, Europhys. Lett. **102**, 28010 (2013)
28. B.S. Kerner, Physica A **397**, 76–110 (2014)
29. B.S. Kerner, P. Hemmerle, M. Koller, G. Hermanns, S.L. Klenov, H. Rehborn, M. Schreckenberg, Phys. Rev. E **90**, 032810 (2014)
30. B.S. Kerner, S.L. Klenov, M. Schreckenberg, J. Stat. Mech. P03001 (2014)
31. B.S. Kerner, S.L. Klenov, G. Hermanns, P. Hemmerle, H. Rehborn, M. Schreckenberg, Phys. Rev. E **88**, 054801 (2013)
32. W.R. McShane, R.P. Roess, *Traffic Engineering* (Prentice Hall, Englewood Cliffs, NJ, 1990)
33. T.S. Kuhn, in *The Structure of Scientific Revolutions*, 4th edn. (The University of Chicago Press, Chicago, London, 2012)
34. M.J. Lighthill, G.B. Whitham, Proc. Roy. Soc. A **229**, 281–345 (1955)
35. J.D.C. Little, Oper. Res. **14**, 568–594 (1966)
36. J.D.C. Little, M.D. Kelson, N.H. Gartner, Transp. Res. Rec. **795**, 40–46 (1981)
37. P.G. Michalopoulos, G. Stephanopoulos, V.B. Pisharody, Transp. Sci. **14**, 9–41 (1980)
38. P.G. Michalopoulos, G. Stephanopoulos, G., Stephanopoulos, Transp. Res. B **15**, 35–51 (1981)
39. J.T. Morgan, J.D.C. Little, Oper. Res. **12**, 896–912 (1964)
40. G.F. Newell, Ann. Math. Stat. **31**, 589–597 (1960)
41. G.F. Newell, SIAM Rev. **575**, 223–240 (1965)
42. G.F. Newell, *Applications of Queuing Theory* (Chapman Hall, London, 1982)
43. P.I. Richards, Oper. Res. **4**, 42–51 (1956)
44. D.I. Robertson, *TRRL Report No. LR 253* (Transportation and Road Research Laboratory, Crowthorne, UK, 1969)
45. D.I. Robertson, in *Proceedings of the International Symposium on Traffic Control Systems*, pp. 262–288 (1979)
46. Ch. Stamatiadis, N.H. Gartner, Transp. Res. Rec. **1554**, 9–17 (1996)

47. G. Stephanopoulos, P.G. Michalopoulos, Transp. Res. A **13**, 295–307 (1979)
48. Y. Wang, Y.-Y. Chen, Physica A **463**, 12–24 (2016)
49. F.V. Webster, *Road Research Technical Paper No. 39* (Road Research Laboratory, London, UK, 1958)

Chapter 10
Theoretical Fundamental of Transportation Science—Breakdown Minimization (BM) Principle

10.1 Introduction—Motivation for BM Principle

Real traffic and transportation networks consist of many road links. Each of the network link can have a bottleneck. Therefore, there can be a lot of network bottlenecks in a traffic network. At each of the network bottlenecks traffic breakdown can occur, if the flow rate at the bottleneck is large enough. Obviously, each of the traffic control or optimization methods should try to prevent traffic breakdown in a traffic network.

Indeed, users of traffic and transportation networks expect that through the use of dynamic traffic assignment and control in the networks traffic breakdown can be prevented in a network. This is because travel costs like travel time and fuel consumption increase considerably in congested traffic in comparison with travel costs in free flow. Therefore, any approach for dynamic traffic assignment and control in the traffic and transportation networks should be consistent with the empirical fundamental of transportation science—the empirical nucleation nature of traffic breakdown at a network bottleneck (Chap. 3).

As shown in Chap. 4 as well as we will study in Chap. 13, generally accepted methods for traffic control, dynamic traffic assignment, or other types of network optimization do not take into account the empirical nucleation nature of traffic breakdown at network bottlenecks. For example, in the standard dynamic traffic assignment and network optimization methodology travel times or/and other travel costs on network links are considered to be self-evident traffic characteristics for objective functions used for optimization transportation problems. The main aim of associated generally accepted classical approaches is to minimize travel times or/and other travel costs in a traffic or transportation network. This standard methodology is related to the state-of-the-art in traffic and transportation research.

However, when the total network inflow rate increases, the minimization of travel times in the network based on the standard dynamic traffic assignment methodology leads to considerably larger increases in the flow rates on short network routes (paths) in comparison with increases in the flow rates on long network routes.

B.S. Kerner, *Breakdown in Traffic Networks*, DOI 10.1007/978-3-662-54473-0_10

Due to the increase in the flow rates on short network routes, the probability of traffic breakdown can increase considerably on these short network routes. Therefore, traffic breakdown can randomly occur in the network. In other words, the minimization of travel times and/or other travel costs in a traffic network based on the state-of-the-art in traffic and transportation research can deteriorate the traffic system considerably, while provoking heavy traffic congestion in urban networks (see a detailed discussion of this critical statement in Chap. 13).

To prevent traffic breakdown in a traffic or transportation network and to satisfy the empirical nucleation nature of traffic breakdown, in 2011 the author introduced a breakdown minimization principle (BM principle) [1]. The application of the BM principle should lead to a minimization of the probability of traffic breakdown in a traffic or transportation network.

The motivation for the BM principle is as follows. Methods for traffic control, dynamic traffic assignment, or other types of network optimization based on the BM principle should ensure a *minimum probability of traffic congestion* in the network.

The necessity of the application of the BM principle for future dynamic traffic assignment and control in traffic and transportation networks will be explained in Sec. 14.2.

10.2 Definition of BM Principle

The breakdown minimization (BM) principle for dynamic traffic assignment, the optimization and control in traffic and transportation networks is defined as follows [1]:

- The BM principle states that the optimum of a traffic or transportation network with N bottlenecks (where $N > 1$) is reached, when dynamic traffic assignment, optimization and/or control are performed in the network in such a way that the probability $P_{\rm net}$ for the occurrence of traffic breakdown in at least one of the network bottlenecks during a given time interval $T_{\rm ob}$ for observing traffic flow reaches the minimum possible value.
- The BM principle is equivalent to the maximization of the probability that during the time interval $T_{\rm ob}$ traffic breakdown occurs at *none* of the network bottlenecks.

The objective of any application of the BM principle is to find the distinct assignment of network link inflow rates in a network and the set of control parameters of network bottlenecks at which the probability $P_{\rm net}$ that traffic breakdown occurs during the time interval $T_{\rm ob}$ in at least at one of the network bottleneck exhibits a minimum. For short, we call the probability $P_{\rm net}$ as the "probability of traffic breakdown in the network".

10.3 Model of Traffic and Transportation Networks

As explained above in this book (see Sec. 1.2 and Chap. 3), in known real field (empirical) traffic data, traffic breakdowns occur usually at some road locations of a traffic network called network bottlenecks. Network bottlenecks are caused, for example, by on- and off-ramps, road gradients, road-works, a decrease in the number of road lines (in the flow direction), traffic signals in city traffic, etc. A bottleneck introduces a speed disturbance localized in a neighborhood of the bottleneck. As a result, in the empirical data at the same flow rate the probability of traffic breakdown at a bottleneck on a network link is considerably larger than on the link outside the bottleneck.

We consider a traffic or transportation network with N network bottlenecks, where $N > 1$. We assume that there are M network links (where $M > 1$) for which the inflow rates q_m, $m = 1, 2, \ldots, M$, can be adjusted; q_m is the link inflow rate for a link with index m.

At the network boundaries, there are I links for the network inflow rates $q_i^{(\mathrm{o})}(t)$, $i = 1, 2, \ldots, I$ (called "origins", for short $\mathrm{O_i}$, $i = 1, 2, \ldots, I$), where $I \geq 1$. At the network boundaries, there are also J links for the network outflow rates $q_j^{(\mathrm{d})}(t)$, $j = 1, 2, \ldots, J$ (called "destinations", for short $\mathrm{D_j}$, $j = 1, 2, \ldots, J$), where $J \geq 1$. The network inflow rates $q_i^{(\mathrm{o})}(t)$ and the total network inflow rate $Q(t)$ are determined by the network inflow rates $q_{ij}^{(\mathrm{o})}(t)$ of vehicles moving from origin $\mathrm{O_i}$ to destination $\mathrm{D_j}$ (called as origin-destination pair $\mathrm{O_i}$–$\mathrm{D_j}$ of the network) through the well-known formulas

$$q_i^{(\mathrm{o})}(t) = \sum_{j=1}^{J} q_{ij}^{(\mathrm{o})}(t), \tag{10.1}$$

$$Q(t) = \sum_{i=1}^{I} q_i^{(\mathrm{o})}(t). \tag{10.2}$$

In this book, the network inflow rates $q_{ij}^{(\mathrm{o})}(t)$ (origin-destination matrix) are assumed to be known time-functions.

For each of the network bottlenecks $k = 1, 2, \ldots, N$, the following probabilistic characteristics of traffic breakdown can be distinguished (Sects. 5.4.9 and 9.3):

- A minimum capacity $C_{\mathrm{min}}^{(k)} = C_{\mathrm{min}}^{(k)}(\alpha_k, \mathbf{R}_k)$.
- A threshold flow rate for a spontaneous traffic breakdown $q_{\mathrm{th}}^{(\mathrm{B},k)} = q_{\mathrm{th}}^{(\mathrm{B},k)}(\alpha_k, \mathbf{R}_k)$.
- A maximum capacity $C_{\mathrm{max}}^{(k)} = C_{\mathrm{max}}^{(k)}(\alpha_k, \mathbf{R}_k)$.

- The probability $P^{(\mathrm{B},k)} = P^{(\mathrm{B},k)}(q^{(k)}_{\mathrm{sum}}, \ \alpha_k, \mathbf{R}_k)$ that during the time interval T_{ob} spontaneous traffic breakdown[1] occurs at network bottleneck k, where $q^{(k)}_{\mathrm{sum}}$ is the flow rate in free flow at bottleneck k.
- When a network bottleneck is due to traffic signal (Sec. 9.3), the flow rate $q^{(k)}_{\mathrm{sum}}$ in free flow at bottleneck k is related to the average arrival flow rate in under-saturated traffic $\bar{q}^{(k)}_{\mathrm{in}}$ and the probability of traffic breakdown at signal is $P^{(\mathrm{B},k)} = P^{(\mathrm{B},k)}(\bar{q}^{(k)}_{\mathrm{in}}, \ \alpha_k, \mathbf{R}_k)$.

Here, α_k is the set of control parameters of bottleneck k, $\mathbf{R}_k$ is a matrix of percentages of vehicles with different vehicle (and/or driver) characteristics that takes into account that dynamic traffic assignment at bottleneck k is possible to perform individually for each of the vehicles.

The matrix $\mathbf{R}_k$ can exhibit two attributes. Each element of this matrix is a percentage of vehicles in traffic flow with two attributes: (i) a type of vehicles and (ii) a set of driver characteristics. As mentioned, the matrix $\mathbf{R}_k$ takes into account that dynamic assignment is possible to perform individually for each of the vehicles. For example, different assignment can be done for long vehicles (trucks) and passenger vehicles as well as for usual vehicles and electric vehicles. Different vehicle types can be driven by drivers with different characteristics, like different driver reaction times. Therefore, percentages of vehicles with different vehicle (and/or driver) characteristics that influence on the breakdown probability at a bottleneck can be important control parameters by the application of the BM principle for the optimization of real networks [2–6].

As mentioned, α_k is the set of control parameters of bottleneck k. Examples of possible control parameters of network bottlenecks are the on-ramp inflow $q_{\mathrm{on},\,k}$ for the case when bottleneck k is an on-ramp bottleneck as well as the flow rate of vehicles $q_{\mathrm{off},\,k}$ leaving the main road to an off-ramp for the case when bottleneck k is an off-ramp bottleneck. Control parameters of a bottleneck caused by traffic signal in a city network are signal parameters (Sec. 9.5). It can turn out that signal parameters should be changed to satisfy the minimum probability of traffic breakdown in the network found with the BM principle.

10.4 A Mathematical Formulation of BM Principle

The definition of the BM principle (Sec. 10.2) can formally be written as follows [1]:

$$\min_{q_1,q_2,\ldots,q_{\mathrm{M}},\alpha_1,\alpha_2,\ldots,\alpha_N,\mathbf{R}_1,\mathbf{R}_2,\ldots,\mathbf{R}_{\mathrm{N}}} \{P_{\mathrm{net}}(q_1, q_2, \ldots, q_{\mathrm{M}}, \alpha_1, \alpha_2, \ldots, \alpha_N, \mathbf{R}_1, \mathbf{R}_2, \ldots, \mathbf{R}_{\mathrm{N}})\}. \tag{10.3}$$

[1] In future applications of the BM principle for real traffic and transportation networks, the flow-rate dependence of the probability $P^{(\mathrm{B},k)}$ of spontaneous traffic breakdown at network bottleneck k should be found from empirical observations. See explanations in footnote 1 of Sec. 5.2.

where it is assumed that the probability P_{net} depends on variables q_m, $m = 1, 2, \ldots, M$ as well as α_k and $\mathbf{R}_k$, $k = 1, 2, \ldots, N$.

The BM principle (10.3) is equivalent to the maximization of the probability that traffic breakdown occurs at *none* of the network bottlenecks [1]:

$$\max_{q_1, q_2, \ldots, q_M, \alpha_1, \alpha_2, \ldots, \alpha_N, \mathbf{R}_1, \mathbf{R}_2, \ldots, \mathbf{R}_N} \{P_{\mathrm{C,\ net}}(q_1, q_2, \ldots, q_M, \alpha_1, \alpha_2, \ldots, \alpha_N, \mathbf{R}_1, \mathbf{R}_2, \ldots, \mathbf{R}_N)\}, \quad (10.4)$$

where

$$P_{\mathrm{C,\ net}} = 1 - P_{\mathrm{net}}. \quad (10.5)$$

In (10.3) and (10.4), it is assumed that traffic dynamic assignment is performed between some alternative network routes only. The definition of the alternative network routes applied for the BM principle will be made in Sec. 10.5.

For a complete formulation of the optimization principle (10.3) (or (10.4)), link inflow rates q_{m}, $m = 1, 2, \ldots, M$ should be connected with network inflow rates. To reach this goal in a general case of dynamic traffic assignment in a traffic network, one should use a dynamic traffic flow model. This dynamic model should calculate spatiotemporal dynamics of vehicular traffic variables within the network under given network inflow rates that can be time-functions. However, network inflow rates can also be considered inflow rates for some network links at the network boundaries. Usually, network inflow rates are given, i.e., they cannot be adjusted based on some traffic optimization. This explains why network links $m = 1, 2, \ldots, M$ in (10.3) (or (10.4)) include only those network links for which link inflow rates can be adjusted with the use of dynamic traffic assignment in the network.

In general, the BM principle (10.3) (or (10.4)) is devoted to the optimization of large scale, complex vehicular traffic networks consisting of a great number of links ($M \gg 1$). The BM principle (10.3) (or (10.4)) can be applied as long as free flow conditions remain at the network bottlenecks $k = 1, 2, \ldots, N$. When traffic breakdown has already occurred at some of the network bottlenecks, a combination of the use of the BM principle for the network parts that are not effected by traffic congestion together with applications of dynamic control for congestion dissolution at these network bottlenecks should be applied (see Sec. 12.6).

10.5 Constrain "Alternative Network Routes"

In contrast with the standard dynamic traffic assignment and network optimization methodology that is the state-of-the-art in traffic and transportation research (see Chap. 13), the BM principle for dynamic traffic assignment, control, and optimization of traffic and transportation networks does *not* include travel times (or other "travel costs") on different network routes (paths).

To avoid the use of network routes with considerably longer travel times in comparison with the shortest routes, the BM principle (10.3) or (10.4) is applied for some "alternative network routes (paths)" (for short, alternative routes) *only*. This means that there is a constrain for alternative network routes (paths) by any application of the BM principle. The constrain for the alternative routes prevents the use of too long routes in dynamic traffic assignment with the use of the BM principle. To determine the constrain for the alternative routes, we consider free flow conditions in the network related to a very small network inflow flow rate ($Q \to 0$) at which travel time on each of the network routes can be considered to be equal to its minimum possible value.

Each of origin-destination pair $\mathrm{O}_i - \mathrm{D}_j$ of the network is connected by a set of paths (routes) through the network. This set is denoted by $\mathscr{S}_{ij}$. Path (route) s connects origin O_i and destination D_j ($s \in \mathscr{S}_{ij}$). $T_{ij}^{(s)}$ represents the travel time on path s under free flow conditions at the small enough total network inflow rate ($Q \to 0$). The travel time on path $s = s_{\text{short}}$ ($s_{\text{short}} \in \mathscr{S}_{ij}$) with the shortest travel time from origin O_i to destination D_j is denoted by $T_{ij}^{(s)} = T_{ij}^{(s_{\text{short}})}$. A difference between the travel time $T_{ij}^{(s)}$ on path s and the shortest travel time $T_{ij}^{(s_{\text{short}})}$ on path $s = s_{\text{short}}$ is denoted by $\Delta T_{ij}^{(s)} = T_{ij}^{(s)} - T_{ij}^{(s_{\text{short}})}$.

We define the constrain "alternative network routes (paths)" as follows. The set of alternative network routes denoted by $\mathscr{A}_{ij}$ (where $\mathscr{A}_{ij} \subset \mathscr{S}_{ij}$) is related to possible different network routes (paths) from origin O_i to destination D_j for which the difference between route travel times $\Delta T_{ij}^{(s)}$ does not exceed a chosen threshold value denoted by Λ_{ij}:

$$\Delta T_{ij}^{(s)} \leq \Lambda_{ij}. \tag{10.6}$$

Condition (10.6) should be proven for all routes (paths) $s \in \mathscr{S}_{ij}$. A set of routes (paths) that satisfies (10.6) is the set of alternative network routes $\mathscr{A}_{ij}$. A set of all other network routes (paths) from origin O_i to destination D_j that do not satisfy condition (10.6) will be called as the set of "non-applicable network routes (paths)" denoted by $\mathscr{N}_{ij}$ (where $\mathscr{N}_{ij} \subset \mathscr{S}_{ij}$). Clearly, that $\mathscr{A}_{ij} \cup \mathscr{N}_{ij} = \mathscr{S}_{ij}$.

For each of the origin-destination pairs $\mathrm{O}_i - \mathrm{D}_j$ of the network, the set of alternative network routes $\mathscr{A}_{ij}$ (where $i = 1, 2, \ldots, I$, $j = 1, 2, \ldots, J$) should be found. Therefore, condition (10.6) should be applied for each of the origin-destination pairs $\mathrm{O}_i - \mathrm{D}_j$ of the network separately each other. Threshold values Λ_{ij} in formula (10.6) can be different for different origin-destination pairs $\mathrm{O}_i - \mathrm{D}_j$, $i = 1, 2, \ldots, I, j = 1, 2, \ldots, J$ of the network.

Therefore, in the methodology of dynamic traffic assignment in a network with the BM principle there are two subsequent stages:

(i) In accordance with condition (10.6), the set of the alternative routes $\mathscr{A}_{ij}$ should be found. The set of the non-applicable network routes $\mathscr{N}_{ij}$ is excluded from dynamic traffic assignment in the network. This procedure should be performed for each of the origin-destination pairs $O_i - D_j$ of the network. The BM principle (10.3) (or (10.4)) is *not* used at this stage.

(ii) After the sets of alternative routes $\mathscr{A}_{ij}$, $i = 1, 2, \ldots, I$, $j = 1, 2, \ldots, J$ have been determined, dynamic traffic assignment is performed through an application of the BM principle (10.3) (or (10.4)) with the use of these sets of alternative routes $\mathscr{A}_{ij}$ only. This means that dynamic traffic assignment in the network made through the use of the BM principle (10.3) (or (10.4)) does *not* depend on current values of the route travel times in the network.

10.6 Basic Applications of BM Principle

The objective of any application of the BM principle is to find the distinct assignment of network link inflow rates q_m in a network and the set of control parameters of network bottlenecks α_k, $\mathbf{R}_k$ at which the probability of traffic breakdown in the network P_{net} exhibits the minimum possible value denoted by

$$P_{\text{net}} = P_{\text{net}}^{(\text{min})}. \tag{10.7}$$

At a small enough total network inflow rate Q (10.2) we can expect that no traffic breakdown can occur at the network bottlenecks. In this case the minimum probability of traffic breakdown in the network $P_{\text{net}} = P_{\text{net}}^{(\text{min})} = 0$, i.e.,

$$P_{\text{net}} = 0. \tag{10.8}$$

A different situation can be expected, when the total network inflow rate Q (10.2) is large enough. Then although the application of the BM principle reduces the probability of traffic breakdown in the network to the minimum possible value $P_{\text{net}} = P_{\text{net}}^{(\text{min})}$ (10.7), this minimum probability of traffic breakdown in the network $P_{\text{net}} = P_{\text{net}}^{(\text{min})} > 0$. This means that the probability of traffic breakdown in the network resulting from the application of the BM principle remains larger than zero:

$$P_{\text{net}} > 0. \tag{10.9}$$

In other words, there can be two different applications of the BM principle:

1. **The network throughput maximization approach related to the case (10.8).** When the total network inflow rate Q (10.2) is small enough, no application of the BM principle is required. However, when the total network inflow rate Q increases, there can be an application of the BM principle for dynamic traffic assignment in the network that keeps condition (10.8) while maximizing the network throughput. Such an application of the BM principle for "zero breakdown probability" revealed in [7] permits to find the distinct assignment of network

link inflow rates q_m for the case $P_{\text{net}} = 0$ (10.8). The application of the BM principle for "zero breakdown probability" (10.8) that maximizes the network throughput has been called *network throughput maximization approach* [7].

With the use of the network throughput maximization approach, general physical conditions for the maximization of the network throughput at which free flow conditions are ensured in the whole traffic or transportation network can be found. It turns out that there is a physical measure of the network called a *network capacity* that characterizes general features of the network with respect to the maximization of the network throughput [7, 8]. The network throughput maximization approach and the network capacity will be considered in Chap. 11.

2. **The application of the BM principle related to the case (10.9)**. This application of the BM principle is related to a large enough value of the total network inflow rate Q (10.2), when the probability of traffic breakdown in the network cannot be decreased to zero. The application of the BM principle permits the distinct assignment of network link inflow rates q_m in the network while minimizing the probability of traffic breakdown in the network [1–6]. This application of the BM principle will be considered in Chap. 12.

10.7 Conclusions

(i) Methods for traffic control, dynamic traffic assignment, or other types of network optimization that perform in accordance with the BM principle should ensure a *minimum probability of traffic congestion* in the network.

(ii) The BM principle states that the optimum of a traffic or transportation network with N bottlenecks is reached, when dynamic traffic assignment, optimization and/or control are performed in the network in such a way that the probability for occurrence of traffic breakdown in at least one of the network bottlenecks during a given observation time interval reaches the minimum possible value.

(iii) The BM principle is equivalent to the maximization of the probability that traffic breakdown occurs at *none* of the network bottlenecks.

(iv) There can be two applications of the BM principle:

- At a smaller total network inflow rate, the application of the BM principle should ensure that the probability of traffic breakdown in the network remains zero, while maximizing the network throughput. This application of the BM principle for "zero breakdown probability" is called the network throughput maximization approach.
- At a larger total network inflow rate, when the probability of traffic breakdown in the network cannot be decreased to zero, the application of the BM principle should ensure the minimum probability of traffic breakdown in the network.

References

1. B.S. Kerner, J. Phys. A Math. Theor. **44**, 092001 (2011)
2. B.S. Kerner, Phys. Rev. E **84**, 045102(R) (2011)
3. B.S. Kerner, Traffic Eng. Contr. **52**, 379–386 (2011)
4. B.S. Kerner, in *Proc. of 2011 IEEE Forum on Integrated and Sustainable Transp. Systems* (Austria, Vienna, 2011), pp. 196–201
5. B.S. Kerner, in *Proc. of the 19th ITS World Congress* (Austria, Vienna, 2012), Paper No. EU-00190
6. B.S. Kerner, Physica A **397**, 76–110 (2014)
7. B.S. Kerner, Eur. Phys. J. B **89**, 199 (2016)
8. B.S. Kerner, Physica A **466**, 626–662 (2017)

Chapter 11
Maximization of Network Throughput Ensuring Free Flow Conditions in Network

11.1 Introduction

As mentioned in Chap. 10, the BM principle [3–7] permits to find the distinct assignment of network link inflow rates q_m while minimizing the probability P_{net} of traffic breakdown in a traffic or transportation network. In real traffic and transportation networks the total network inflow rate $Q(t)$ (10.2) increases from very small values (at night) to larger values during rash hour. Obviously that at small enough network inflow rates, traffic breakdown cannot occur in the network, i.e., the probability of traffic breakdown in the network $P_{\text{net}} = 0$ (10.8).

When the total network inflow rate $Q(t)$ increases subsequently, as mentioned in Sect. 10.6, there is an application of the BM principle that maintains condition $P_{\text{net}} = 0$ (10.8) (called the BM principle for "zero breakdown probability") while maximizing the network throughput. Through the maximization of the network throughput no traffic breakdown can occur in the traffic network: This network throughput maximization allows us to ensure free flow conditions in the whole network. Such an application of the BM principle for "zero breakdown probability" (10.8) introduced in [10] has been called *the network throughput maximization approach* [10, 11].

In this chapter, we consider the network throughput maximization approach. As mentioned, this approach to dynamic traffic assignment and control in traffic and transportation networks permits the maximization of the network throughput at which free flow conditions are ensured in the whole network. We introduce a physical measure of a traffic or transportation network called *network capacity*. The network capacity characterizes general features of the network with respect to the maximization of the network throughput at which free flow conditions are ensured in the network. Results of this chapter are based on [10, 11].

B.S. Kerner, *Breakdown in Traffic Networks*, DOI 10.1007/978-3-662-54473-0_11

Before we start with the consideration of the network throughput maximization approach and the network capacity, we would like to recall that traffic breakdown exhibits the nucleation nature. As explained in Sects. 5.4.9 and 9.3, each of network bottlenecks is characterized by a minimum capacity $C_{\rm min}$ that separates stable and metastable states of free flow at the bottleneck: At

$$q_{\rm sum} < C_{\rm min}, \tag{11.1}$$

free flow is stable at the bottleneck, where $q_{\rm sum}$ is the flow rate in free flow at the bottleneck. Therefore, under condition (11.1) no traffic breakdown can occur at the bottleneck. Contrarily, at

$$q_{\rm sum} \geq C_{\rm min} \tag{11.2}$$

free flow is metastable with respect to traffic breakdown at the bottleneck. Therefore, under condition (11.2) traffic breakdown can occur at the bottleneck.

We consider the network model of Sect. 10.3. In real traffic and transportation networks, the total network inflow rate Q (10.2) and, therefore, the flow rates $q^{(k)}_{\rm sum}$, $k = 1, 2, \ldots, N$ in the network increase from very small values (at night) to large values during daytime related to rush hours in urban networks. At small initial values Q and $q^{(k)}_{\rm sum}$ condition (11.1) is valid for each of the network bottlenecks:

$$q^{(k)}_{\rm sum} < C^{(k)}_{\rm min}, \quad k = 1, 2, \ldots, N. \tag{11.3}$$

In accordance with condition (11.1), conditions (11.3) mean that free flow is stable with respect to traffic breakdown at each of the network bottlenecks. Therefore, no traffic breakdown can occur in the network.

Results of the applications of the BM principle presented in Chaps. 11–13 do not depend on road bottleneck types in a network. In particular, when a network bottleneck is due to traffic signal, then the physical sense of formulas (11.1) and (11.2) remains also for the signal. However, in accordance with Sects. 9.3 and 10.3, the flow rate $q_{\rm sum}$ in formulas (11.1)–(11.3) as well as in all other formulas presented below in Chaps. 11–13 should be replaced by the average arrival flow rate in under-saturated traffic $\bar{q}^{(k)}_{\rm in}$. Respectively, the terms "stable free flow" and "metastable free flow" at a network bottleneck should be replaced for network bottleneck due to traffic signal, respectively, by the terms "stable under-saturated traffic" and "metastable under-saturated traffic".

11.2 Network Throughput Maximization Approach: The Maximization of Network Throughput by Prevention of Breakdown in Network

As mentioned, conditions (11.3) mean that free flow is stable with respect to traffic breakdown at each of the network bottlenecks. Therefore, no traffic breakdown can occur in the network. This means that in the BM principle (10.3) the probability of traffic breakdown in the network $P_{\rm net} = 0$ (10.8). Following [7], in this section we consider the network throughput maximization approach. This approach for dynamic traffic assignment allows us to find the distinct dynamic traffic assignment of the link inflow rates q_m that maximizes the network throughput at which the probability of traffic breakdown in the network $P_{\rm net}$ remains zero (10.8).

When the total network inflow rate Q (10.2) is small enough, no dynamic traffic assignment is needed. Due to the increase in Q over time, the flow rate $q^{(k)}_{\rm sum}$ at least for one of the network bottlenecks $k = k^{(1)}_1$ becomes very close to $C^{(k)}_{\rm min}$. Therefore, in formula

$$q^{(k)}_{\rm sum} + \epsilon^{(k)} = C^{(k)}_{\rm min} \text{ for } k = k^{(1)}_1 \tag{11.4}$$

a positive value $\epsilon^{(k)} = C^{(k)}_{\rm min} - q^{(k)}_{\rm sum}$ becomes very small:

$$\epsilon^{(k)}/C^{(k)}_{\rm min} \ll 1 \text{ for } k = k^{(1)}_1. \tag{11.5}$$

When the total network inflow rate Q (10.2) increases subsequently, we begin to use the network throughput maximization approach for dynamic traffic assignment in the network that is as follows [7]. We maintain condition (11.4) at the expense of the increase in q_m on other alternative routes. As a result, condition $q^{(k)}_{\rm sum} + \epsilon^{(k)} = C^{(k)}_{\rm min}$ at $\epsilon^{(k)}/C^{(k)}_{\rm min} \ll 1$ is satisfied for another bottleneck $k = k^{(1)}_2$. When Q increases further, we repeat the above procedure for other network bottlenecks. Consequently, we get

$$\begin{aligned} q^{(k)}_{\rm sum} + \epsilon^{(k)} &= C^{(k)}_{\rm min} \text{ for } k = k^{(1)}_z \\ q^{(k)}_{\rm sum} &< C^{(k)}_{\rm min} \text{ for } k \neq k^{(1)}_z, \\ z = 1, 2, \ldots, Z_1, \ \text{where } Z_1 \geq 1,\ Z_1 &\leq N;\ k = 1, 2, \ldots, N; \end{aligned} \tag{11.6}$$

values $\epsilon^{(k)}$ satisfy conditions

$$0 < \epsilon^{(k)}/C^{(k)}_{\rm min} \ll 1 \tag{11.7}$$

for bottlenecks $k = k_z^{(1)}$, $z = 1, 2, \ldots, Z_1$. Thus, when the network inflow rate Q increases, we maintain conditions (11.6) at the expense of the increase in q_m on other alternative routes.[1]

However, due to the constrain "alternative routes", the number Z_1 of bottlenecks satisfying (11.6) is limited by some value Z (where $Z \leq N$). All values $\epsilon^{(k)}$ in (11.6) accordingly to (11.7) are positive ones. Therefore, under conditions (11.6), conditions (11.3) are satisfied. Thus, due to the application of the network throughput maximization approach free flow remains to be stable with respect to traffic breakdown at each of the network bottlenecks. For this reason, no traffic breakdown can occur in the network.

This means that conditions (11.6) in the limit case $Z_1 = Z$ and $\epsilon^{(k)} \to 0$ are related to *the maximization of network throughput* at which free flow conditions are still ensured in the whole network. This is because at the subsequent increase in the total network inflow rate Q the constrain "alternative routes" (Sect. 10.5) does not allow us to maintain conditions (11.6) at the expense of the increase in q_m on other alternative routes: At least for one of the network bottlenecks condition (11.2) is satisfied. Therefore, traffic breakdown can occur in the network.

The network throughput maximization approach is an approach to dynamic traffic assignment and control in a traffic or transportation network. The network throughput maximization approach maximizes the network throughput while keeping free flow conditions in the whole network. The network throughput maximization approach maintains condition $P_{\text{net}} = 0$, where P_{net} is the probability of traffic breakdown in the network. Therefore, the network throughput maximization approach is also called the BM principle for "zero breakdown probability". Through the maximization of the network throughput with the network throughput maximization approach no traffic breakdown can occur in the network.

11.3 A Physical Measure of Traffic and Transportation Networks—Network Capacity

Within a steady-state analysis of traffic and transportation networks (see footnote 9 of Sect. 1.7.3), the limit case $Z_1 = Z$ in conditions (11.6) allows us to define a network measure (or "metric")—*network capacity* denoted by C_{net} as follows [7]. The network capacity C_{net} is the maximum total network inflow rate Q (10.2) at

[1]This increase in q_m on other alternative routes maintaining conditions (11.6) can be obtained with the use of different routing methodologies including those that are based on the Wardrop's UE or SO.

which conditions

$$q_{\rm sum}^{(k)} = C_{\rm min}^{(k)} \text{ for } k = k_z^{(1)}, \qquad (11.8)$$
$$q_{\rm sum}^{(k)} < C_{\rm min}^{(k)} \text{ for } k \neq k_z^{(1)},$$
$$z = 1, 2, \ldots, Z;\ Z \geq 1,\ Z \leq N;\ k = 1, 2, \ldots, N$$

are satisfied.[2] From a comparison of conditions (11.6) and (11.8) we can see that the total network inflow rate reaches the network capacity[3]

$$Q = C_{\rm net} \qquad (11.10)$$

---

[2]In accordance with results of Sects. 5.7 and 9.3.4, the minimum capacity $C_{\rm min}^{(k)}$ of a network bottleneck is a stochastic value. Therefore, the network capacity $C_{\rm net}$ (11.8) that depends on the values of the minimum capacities of the network bottlenecks is also a stochastic value. A consideration of the effect of stochastic features of minimum capacities $C_{\rm min}^{(k)}$, $k = 1, 2, \ldots, N$ of the network bottlenecks on the application of the network throughput maximization approach and on the network capacity $C_{\rm net}$ is out of scope of this book. This can be an interesting task for future investigations.

[3]In city traffic, there is a hypothetical case of "red wave", when all vehicles approach a traffic signal during the red signal phase *only* (Sect. 9.8). In this case, the definition of the network capacity $C_{\rm net}$ through conditions (11.8) remains. However, when at one of the network bottlenecks due to the signal "red wave" is realized, then the minimum capacity of the signal $C_{\rm min}$ is equal to the classical signal capacity $C_{\rm cl}$ (Sect. 9.8). Therefore, in (11.8) the value of $C_{\rm min}$ for this signal should be replaced by $C_{\rm cl}$. To explain this, we note that in the case of "red wave" the classical theory of traffic at the signal is a special case of the three-phase theory (Sect. 9.8): When the average arrival flow rate (flow rate at a bottleneck due to the signal) exceeds $C_{\rm min} = C_{\rm cl}$, then traffic breakdown, i.e., the transition from under-saturated traffic to over-saturated (congested) traffic occurs at the signal without time delay. In a non-realistic case, when all network bottlenecks are due to the signals at which red wave conditions are realized, then the network capacity $C_{\rm net}$ is found from conditions (11.8) as follows:

$$q_{\rm sum}^{(k)} = C_{\rm cl}^{(k)} \text{ for } k = k_z^{(1)}, \qquad (11.9)$$
$$q_{\rm sum}^{(k)} < C_{\rm cl}^{(k)} \text{ for } k \neq k_z^{(1)},$$
$$z = 1, 2, \ldots, Z;\ Z \geq 1,\ Z \leq N;\ k = 1, 2, \ldots, N,$$

where bottlenecks $k = k_z^{(1)}$ and value Z are found in accordance with the constrain "alternative routes" as described in Sect. 10.5.

As explained in Sec. 12.4 of review [11], formula (11.9) is also applicable in the framework of the classical traffic flow theories in which it is assumed that there is a particular value of capacity for any network bottleneck. In particular, the measure "network capacity" introduced in [10], which is discussed in this section, has the same physical sense in the classical theory as that in the three-phase traffic theory. However, as emphasized in Sect. 4.10, the classical understanding of stochastic highway capacity at a bottleneck is inconsistent with the empirical nucleation nature of traffic breakdown at the bottleneck. For this reason, in the book we will not consider the network capacity under classical understanding of stochastic highway capacity at a network bottleneck (see such a consideration in Sec. 12.4 of [11]).

when the limit case $Z_1 = Z$ in (11.6) is realized *and* all values $\epsilon^{(k)}$ in (11.6) are set to zero: $\epsilon^{(k)} = 0$. In accordance with the definition of the network capacity (11.8) and condition (11.2), at $Q = C_{\rm net}$ (11.10) free flow becomes the metastable one with respect to traffic breakdown at network bottleneck(s) $k = k_z^{(1)}$, $z = 1, 2, \ldots, Z$ (11.8). This means that under conditions (11.8) traffic breakdown can occur in the network.

When

$$Q < C_{\rm net}, \qquad (11.11)$$

as explained above, conditions (11.6) are satisfied. Under conditions (11.6), free flow is stable with respect to traffic breakdown in the whole network. Therefore, no traffic breakdown can occur in the network. This explains the physical sense of the network capacity $C_{\rm net}$: Dynamic traffic assignment in the network in accordance with the network throughput maximization approach maximizes the network throughput at which traffic breakdown cannot occur in the whole network.

Respectively, when

$$Q > C_{\rm net}, \qquad (11.12)$$

due to the constrain "alternative routes" of the network throughput maximization approach at least for one of the network bottlenecks the flow rate in free flow at the bottleneck $q^{(k)}_{\rm sum}$ becomes larger than the minimum bottleneck capacity $C^{(k)}_{\rm min}$. We denote network bottlenecks k for which condition $q^{(k)}_{\rm sum} > C^{(k)}_{\rm min}$ is satisfied by $k = k_w^{(2)}$, where $w = 1, 2, \ldots, W$; $W \geq 1$, $W \leq N$. Thus, rather than conditions (11.8) under condition (11.12) we get[4]

$$\begin{aligned} q^{(k)}_{\rm sum} &> C^{(k)}_{\rm min} \ \text{at} \ k = k_w^{(2)}, \qquad (11.13) \\ q^{(k)}_{\rm sum} &\leq C^{(k)}_{\rm min} \ \text{at} \ k \neq k_w^{(2)}, \\ w = 1, 2, \ldots, W;\ W \geq 1,\ &W \leq N;\ k = 1, 2, \ldots, N. \end{aligned}$$

Therefore, in accordance with (11.2), free flow is the metastable one with respect to traffic breakdown at the related network bottleneck(s). This means that under conditions (11.13) traffic breakdown can occur in the network.

To explain the term *maximum total network inflow rate* in the definition of the network capacity $C_{\rm net}$ in more details, we note that in some cases the minimum capacity $C_{\rm min}$ of a network bottleneck can be a function of link inflow rates q_m. For example, for an on-ramp bottleneck $C_{\rm min}$ can be a function of the on-ramp inflow rate $q_{\rm on}$, for a bottleneck due to traffic signal $C_{\rm min}$ is a function of signal parameters

[4] As the total network inflow rate Q in (11.12) is subsequently increased, the number W of bottlenecks for which conditions (11.13) are satisfied, can change.

(signal control settings). For this reason, to find the network capacity $C_{\rm net}$, the assignment of the flow rates q_m as well as control of bottleneck parameters should ensure the maximum total network inflow rate Q at which conditions (11.8) are satisfied. A development of a general procedure for the calculation of the network capacity for traffic networks has not been made up to now; this will be an interesting task for further investigations. Examples of calculations $C_{\rm net}$ for simple network models are presented in Sect. 12.4.

Note that the network capacity $C_{\rm net}$ depends on the network inflow rates $q_{ij}^{(\rm o)}(t)$ (Sect. 10.3) and the set of the alternative routes $\mathscr{A}_{ij}$ (Sect. 10.3), which are assumed to be given. For the same day, there can be also different values $C_{\rm net}$ at different time instants.

The network capacity $C_{\rm net}$ is a measure (or "metric") of a traffic or transportation network. The network capacity allows us to formulate a general physical condition for the maximization of the network throughput at which free flow does persist in the whole network: Under application of the network throughput maximization approach, as long as the total network inflow rate is smaller than the network capacity traffic breakdown cannot occur in the network. The network capacity is a stochastic value.

11.4 The Maximization of Network Throughput in Non-Steady State of Network

As above-mentioned, the definition of the network capacity (11.8), (11.10) is valid only within a steady-state analysis of the networks. Contrarily, conditions (11.6) at $Z_1 = Z$ and $\epsilon^{(k)} \to 0$, for which free flow is stable with respect to traffic breakdown at each of the network bottlenecks, are applicable even when free flow distribution in the network cannot be considered a steady one. Indeed, in conditions (11.6) only local flow rates $q_{\rm sum}^{(k)}$ in free flow at network bottlenecks are used.

Therefore, conditions (11.6) at which free flow conditions persist in the whole network can be used for dynamic traffic assignment in real traffic and transportation networks under real non-steady state conditions,[5] i.e., without involving an analysis of the network capacity [10, 11]. We can see that the basic objective of the network

[5]Under non-steady state conditions in the network, even when free flow is realized in the whole network, nevertheless, at a given time instant the total network inflow rate should not be necessarily equal to the the total network outflow rate: The total network inflow rate is equal to the total network outflow rate *on average* only.

throughput maximization approach is to guarantee that condition (11.2) is satisfied at *none* of the network bottlenecks.[6]

11.5 Behavior of Probability of Traffic Breakdown in Traffic and Transportation Networks

As discussed above, when the total network inflow rate Q (10.2) increases subsequently, condition $Q > C_{\rm net}$ (11.12) can be satisfied. Under this condition, free flow at some of the network bottlenecks is in a metastable state with respect to traffic breakdown. For this reason, the network throughput maximization approach to dynamic traffic assignment in the network cannot be applied any more. Therefore, a question arises: What is the procedure of the application of the BM principle in this case? To answer this question, firstly we should consider features of possible critical traffic phenomena that can occur in metastable free flow at network bottlenecks. Important features of the critical phenomena that determine the application of the BM principle are conditions for either induced or spontaneous traffic breakdown at network bottlenecks. In its turn, in accordance with empirical results (Chap. 3) these conditions depend on the value of fluctuations of traffic variables (the flow rate, the speed, and the density) in free flow at network bottlenecks.

11.5.1 Fluctuations in Metastable Free Flow and Spontaneous Traffic Breakdown at Network Bottlenecks

Accordingly Sects. 5.1 and 9.3.2, within the flow rate range

$$C_{\rm min} \leq q_{\rm sum} < C_{\rm max} \tag{11.14}$$

free flow is in a metastable state with respect to traffic breakdown at a bottleneck (Fig. 11.1).

As shown in Chap. 5, when fluctuations in free flow are large enough, then either induced or spontaneous traffic breakdown can occur in metastable free flow (11.14) (Fig. 11.1a). Contrarily, when fluctuations in free flow are small enough, then in accordance with results of Sects. 5.4 and 9.3.2 two different flow rate ranges in

[6] It should be noted that when the network throughput maximization approach is applied under non-steady state conditions in the network, then the set of the network bottlenecks $k = k_z^{(1)}$ (where $z = 1, 2, \ldots, Z_1$) in conditions (11.6) can be different from the set of the network bottlenecks $k = k_z^{(1)}$ (where $z = 1, 2, \ldots, Z$) in conditions (11.8). Respectively, the limit number of network bottlenecks $Z_1 = Z$ in (11.6) should not necessarily coincide with the number of network bottlenecks Z in (11.8).

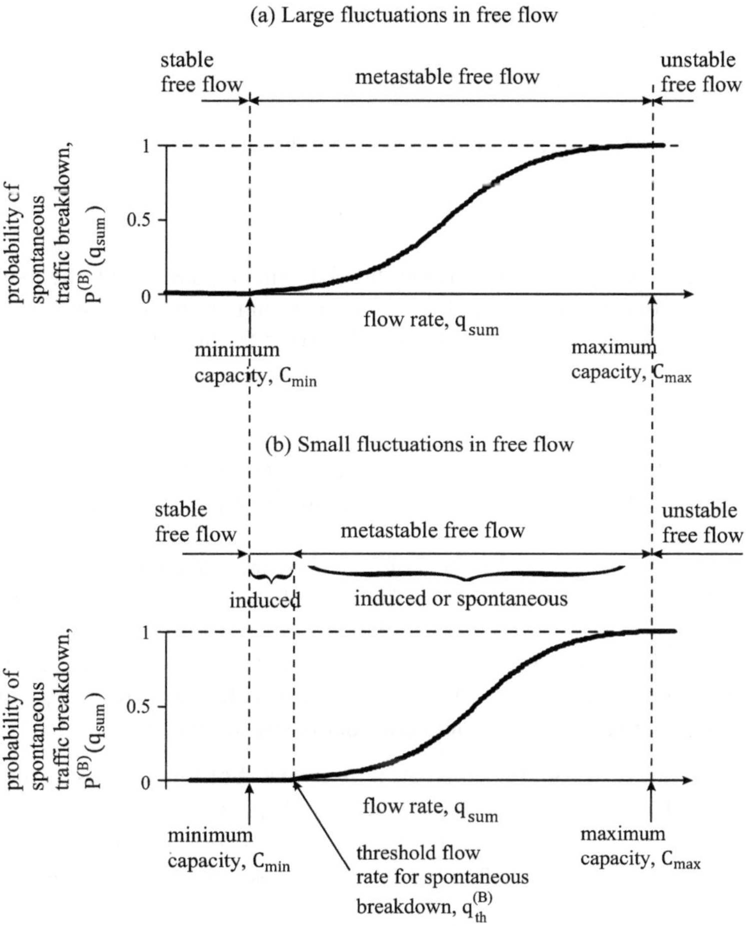

Fig. 11.1 Qualitative flow-rate dependencies of probability $P^{(B)}(q_{sum})$ of traffic breakdown at bottleneck [1, 2, 13]: (**a**) Large fluctuations in free flow: Either induced or spontaneous traffic breakdown can occur in metastable free flow (11.14) (adapted from Fig. 5.5c). (**b**) Small fluctuations in free flow at which there are two different ranges of the flow rate in metastable free flow (11.14) (adapted from Fig. 5.10c): (i) under conditions (11.15) induced traffic breakdown is only possible at bottleneck; (ii) under conditions (11.17) either induced or spontaneous traffic breakdown can occur at bottleneck

metastable free flow (11.14) should be distinguished: In the flow rate range defined by conditions

$$C_{min} \leq q_{sum} < q_{th}^{(B)} \tag{11.15}$$

only induced traffic breakdown is possible at a network bottleneck. This is because under conditions (11.15) the probability of spontaneous traffic breakdown at the bottleneck during the time interval T_{ob} is zero[7]:

$$P^{(B)}(q_{sum}) = 0. \tag{11.16}$$

Contrarily, in the flow rate range defined by conditions

$$q_{th}^{(B)} \leq q_{sum} < C_{max} \tag{11.17}$$

either induced or spontaneous traffic breakdown is possible at the bottleneck (Fig. 11.1b). This means that traffic breakdown can occur spontaneously at the bottleneck during the time interval T_{ob} for observing traffic flow with the probability

$$0 < P^{(B)}(q_{sum}) < 1. \tag{11.18}$$

Formulas (11.15)–(11.18) explain why the characteristic flow rate $q_{sum} = q_{th}^{(B)}$ is called a threshold value of the flow rate q_{sum} for spontaneous traffic breakdown at the bottleneck.

Therefore, when at least at one of the network bottlenecks free flow is in a metastable state with respect to traffic breakdown, i.e., conditions (11.14) are satisfied, the probability P_{net} of traffic breakdown in the network should depend on the value of fluctuations of traffic variables (the flow rate, the speed, the density) in free flow at network bottlenecks. Thus, to discuss the probability P_{net} of traffic breakdown in the network we should consider two different cases:

(i) Large free flow fluctuations at a network bottleneck. As explained in Sects. 5.6 and 9.3.3, in this case the probability of spontaneous traffic breakdown at the network bottleneck during the time interval T_{ob} becomes larger than zero, when the flow rate q_{sum} reaches the minimum capacity C_{min} of the bottleneck:

$$\begin{aligned} P^{(B)}(q_{sum}) &= 0 \quad \text{at} \quad q_{sum} < C_{min}, \\ P^{(B)}(q_{sum}) &> 0 \quad \text{at} \quad q_{sum} \geq C_{min}. \end{aligned} \tag{11.19}$$

(ii) Small free flow fluctuations at a network bottleneck. As explained in Sects. 5.4.7 and 9.3.2, in this case the probability of spontaneous traffic breakdown at the network bottleneck during the time interval T_{ob} becomes larger than zero,

[7]In numerical calculations of the breakdown probability $P^{(B)}(q_{sum})$ [12] only a finite number N_r of simulation realizations (runs) can be made for each given flow rate q_{sum}. As shown in Sect. 5.4.5, for the smallest value of the breakdown probability at the flow rate $q_{sum} = q_{th}^{(B)}$ we get $P^{(B)}\,|_{q_{sum}=q_{th}^{(B)}} = \frac{1}{N_r}$. This formula determines the accuracy of the calculation of the threshold flow rate $q_{th}^{(B)}$ related to the flow rate ranges (11.15) and (11.17). In other words, the larger the number N_r, the more exactly the threshold flow rate $q_{sum} = q_{th}^{(B)}$ can be calculated.

when the flow rate $q_{\rm sum}$ reaches the threshold flow rate $q_{\rm th}^{\rm (B)}$ of the bottleneck:

$$P^{\rm (B)}(q_{\rm sum}) = 0 \quad \text{at} \quad q_{\rm sum} < q_{\rm th}^{\rm (B)}, \qquad (11.20)$$
$$P^{\rm (B)}(q_{\rm sum}) > 0 \quad \text{at} \quad q_{\rm sum} \geq q_{\rm th}^{\rm (B)}.$$

11.5.2 Probability of Traffic Breakdown in Network Under Large Free Flow Fluctuations

It is clear that the probability of traffic breakdown in the network $P_{\rm net}$ depends on the probabilities for spontaneous traffic breakdown at the network bottlenecks $P^{({\rm B},k)}(q_{\rm sum}^{(k)})$, $k = 1, 2, \ldots, N$ during the time interval $T_{\rm ob}$. In particular, when free flow is stable with respect to traffic breakdown for each of the network bottlenecks, we get

$$P^{({\rm B},k)}(q_{\rm sum}^{(k)}) = 0, \ k = 1, 2, \ldots, N. \qquad (11.21)$$

In Sect. 11.2 we have shown that under increase in the total network inflow rate Q (10.2) as long as conditions (11.6) are satisfied, free flow is stable with respect to traffic breakdown at each of the network bottlenecks. This means that Eqs. (11.21) are valid. Respectively, the probability of traffic breakdown in the network $P_{\rm net} = 0$ (10.8).

In this Sect. 11.5.2, we discuss a case when fluctuations of free flow at all network bottleneck can be considered large enough (Fig. 11.1a). In this case, accordingly formula (11.19), the probability of spontaneous traffic breakdown at a network bottleneck becomes larger than zero, when the flow rate $q_{\rm sum}$ reaches the minimum capacity of the bottleneck $C_{\rm min}$.

For this reason, the probability $P_{\rm net}$ of traffic breakdown in the network should also become larger than zero, when conditions (11.6) cannot be satisfied any more. In this case, free flow becomes metastable with respect to traffic breakdown at least at one of the network bottlenecks. This conclusion results in the following conditions for the probability $P_{\rm net}$ of traffic breakdown in the network:

$$P_{\rm net} = 0 \quad \text{at} \quad q_{\rm sum}^{(k)} < C_{\rm min}^{(k)}, \ k = 1, 2, \ldots, N, \qquad (11.22)$$

$$P_{\rm net} > 0 \quad \text{at} \quad \begin{cases} q_{\rm sum}^{(k)} \geq C_{\rm min}^{(k)} \ \text{at} \ k = k_g^{(3)}, \\ q_{\rm sum}^{(k)} < C_{\rm min}^{(k)} \ \text{at} \ k \neq k_g^{(3)}, \end{cases} \qquad (11.23)$$
$$g = 1, 2, \ldots, G; \ G \geq 1, \ G \leq N; \ k = 1, 2, \ldots, N.$$

Conditions (11.22) are related to the case, when conditions (11.6) can be satisfied. This is because free flow is stable with respect to traffic breakdown at each of the network bottlenecks. Contrarily, conditions (11.23) are related to the case, when conditions (11.6) cannot be satisfied any more. In this case, free flow becomes metastable with respect to traffic breakdown at least at some of the network bottlenecks denoted by $k = k_g^{(3)}$, $g = 1, 2, \ldots, G$ in (11.23). As explained in Sect. 11.4, formulas (11.22) and (11.23) are also valid under real non-steady state conditions in a traffic or transportation network.

Under a steady state analysis of a traffic network, as shown in Sects. 11.2 and 11.3, the network measure called the network capacity $C_{\rm net}$ can be used. In this case, we can find the probability $P_{\rm net}(Q)$ of traffic breakdown in the network as a function of the total network inflow rate Q (10.2). In accordance with conditions (11.22) and (11.23), the probability $P_{\rm net}(Q)$ becomes larger than zero, when the network inflow rate reaches the network capacity:

$$P_{\rm net}(Q) = 0 \quad \text{at} \quad Q < C_{\rm net}, \tag{11.24}$$

$$P_{\rm net}(Q) > 0 \quad \text{at} \quad Q \geq C_{\rm net}. \tag{11.25}$$

The physical sense of this flow-rate behavior of the probability $P_{\rm net}(Q)$ of traffic breakdown in the network is the same as that explained for formulas (11.22) and (11.23). When $Q < C_{\rm net}$, free flow is stable with respect to traffic breakdown at each of the network bottlenecks. Contrarily, at $Q \geq C_{\rm net}$ free flow becomes metastable with respect to traffic breakdown at least at some of the network bottlenecks.

11.6 Effect of Fluctuations on Prevention of Spontaneous Traffic Breakdown in Networks

Contrarily to the case of large enough fluctuations in free flow discussed in Sect. 11.5.2, in this Sect. 11.6 we consider cases, when fluctuations in free flow are small enough. Small free flow fluctuations can be realized either in the whole network or at least at some of network bottlenecks. We have found some important features of the application of the BM principle depending on the value of fluctuations in free flow at the network bottlenecks.

In a hypothetical case, in which free flow fluctuations are small enough in the whole network, formulas (11.15)–(11.18) are valid for each of the network bottlenecks $k = 1, 2, \ldots, N$. Accordingly formulas (11.15) and (11.16), under conditions

$$C_{\rm min}^{(k)} \leq q_{\rm sum}^{(k)} < q_{\rm th}^{({\rm B},k)}, \ k = 1, 2, \ldots, N \tag{11.26}$$

the probability of spontaneous traffic breakdown at each of the network bottlenecks during the time interval $T_{\rm ob}$ is zero: $P^{({\rm B},k)}(q_{\rm sum}^{(k)}) = 0, \ k = 1, 2, \ldots, N$.

Nevertheless, in a general case we cannot state that the probability of traffic breakdown in the network $P_{\text{net}} = 0$ (10.8). This is because under conditions (11.26) free flow is in a metastable state with respect to traffic breakdown for each of the network bottlenecks. Therefore, as explained in Sect. 11.5.1, traffic breakdown can be induced at a network bottleneck (Fig. 11.1b). For this reason, firstly we should consider a possible effect of induced traffic breakdown on the probability of traffic breakdown in the network P_{net}.

11.6.1 Empirical Induced and Spontaneous Traffic Breakdowns in Networks

To understand the effect of possible induced traffic breakdown on the probability P_{net} of traffic breakdown in the network during the time interval T_{ob}, we consider empirical results presented in Fig. 11.2. In empirical data measured up to now, real traffic scenarios of empirical induced traffic breakdown (labeled "induced breakdown" in Fig. 11.2) have usually been as follows. Firstly, spontaneous traffic breakdown has occurred at a downstream bottleneck (labeled "spontaneous breakdown" in Fig. 11.2). Then due to the subsequent upstream propagation of a congested traffic pattern to an upstream bottleneck, the pattern reaches the bottleneck while inducing traffic breakdown at the bottleneck.

When a congested pattern that induces traffic breakdown at a network bottleneck has initially occurred at one of the downstream bottlenecks of the network, then there has earlier been spontaneous traffic breakdown at the downstream bottleneck of the network leading to the formation of this congested pattern. This is *only* possible if the probability of traffic breakdown in the network $P_{\text{net}} > 0$ (10.9). This means that when the probability P_{net} of traffic breakdown in the network satisfies condition $P_{\text{net}} = 0$ (10.8), *no* induced traffic breakdown can occur at the network bottlenecks if the flow rate $q^{(k)}_{\text{sum}}$ satisfies conditions (11.26). The physical sense of this result is as follows. As long as no spontaneous traffic breakdown has already occurred at one of the network bottlenecks, no induced traffic breakdown can occur at other network bottlenecks.

Thus, we can make the following conclusion. In the case of small enough fluctuations in free flow in the whole network, the metastability of free flow at network bottlenecks for which conditions (11.15) are satisfied does not affect on the probability P_{net} of traffic breakdown in the network: The probability of traffic breakdown in the network $P_{\text{net}} = 0$ (10.8).

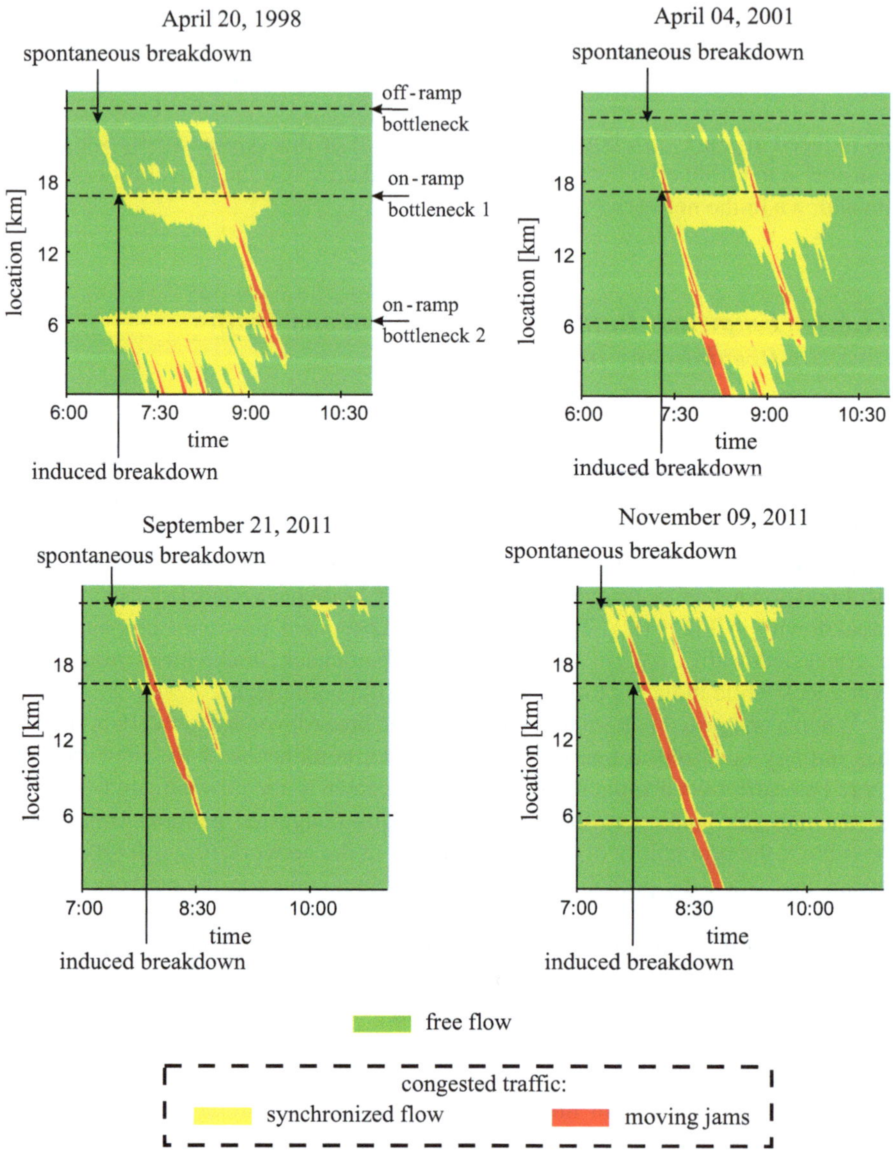

Fig. 11.2 Examples of empirical induced traffic breakdown at highway bottlenecks measured on four different days on freeway A5-South in Germany (1 min averaged field data): Speed data measured with road detectors installed along road section; data is presented in space and time with averaging method described in Sec. C.2 of [14]. Freeway section schema is shown in Fig. 2.1 of the book [1]

11.6.2 Network Throughput Maximization Preventing Spontaneous Breakdown Under Small Free Flow Fluctuations in Networks

The consideration made in Sect. 11.6.1 shows that when fluctuations in free flow are small enough at each of the network bottlenecks, the probability of traffic breakdown in the network $P_{\rm net} = 0$ (10.8) as long as conditions (11.26) are satisfied. Therefore, for the case of small enough fluctuations in free flow in a traffic network we can repeat the same analysis of the maximization of the network throughput as made in Sect. 11.2. In this case, the network throughput maximization approach is devoted to the prevention of *spontaneous* traffic breakdown at one of the network bottlenecks maximizing the network throughput.

In the case of small enough fluctuations in free flow at the network bottlenecks, the network throughput maximization approach is as follows. When the initial value Q (10.2) and, therefore, values of the flow rates $q^{(k)}_{\rm sum}$, $k = 1, 2, \ldots, N$ are not large enough, we can assume that the following conditions are valid for each of the network bottlenecks:

$$q^{(k)}_{\rm sum} < q^{({\rm B},k)}_{\rm th}, \ k = 1, 2, \ldots, N. \tag{11.27}$$

In accordance with the discussion of Sects. 5.5 and 11.5.1, conditions (11.27) mean that although free flow can be metastable at least at some of the network bottlenecks, nevertheless, due to small fluctuations no spontaneous traffic breakdown can occur at the network bottlenecks during the time interval $T_{\rm ob}$. Therefore, no spontaneous traffic breakdown can occur in the network. Due to the increase in the total network inflow rate Q (10.2), the flow rate $q^{(k)}_{\rm sum}$ at least at one of the network bottlenecks $k = k^{(4)}_1$ becomes very close to $q^{({\rm B},k)}_{\rm th}$. Therefore, in formula

$$q^{(k)}_{\rm sum} + \epsilon^{(k)}_{\rm th} = q^{({\rm B},k)}_{\rm th} \ \text{for} \ k = k^{(4)}_1 \tag{11.28}$$

a positive value $\epsilon^{(k)}_{\rm th} = q^{({\rm B},k)}_{\rm th} - q^{(k)}_{\rm sum}$ becomes very small:

$$\epsilon^{(k)}_{\rm th}/q^{({\rm B},k)}_{\rm th} \ll 1 \ \text{for} \ k = k^{(4)}_1. \tag{11.29}$$

When the total network inflow rate Q (10.2) increases subsequently, we maintain condition (11.28) at the expense of the increase in q_m on other alternative routes. As a result, condition $q^{(k)}_{\rm sum} + \epsilon^{(k)}_{\rm th} = q^{({\rm B},k)}_{\rm th}$ at $\epsilon^{(k)}_{\rm th}/q^{({\rm B},k)}_{\rm th} \ll 1$ can be satisfied for another bottleneck $k = k^{(4)}_2$. When Q increases further, we repeat the above procedure for other network bottlenecks. Consequently, we get

$$\begin{gathered} q^{(k)}_{\rm sum} + \epsilon^{(k)}_{\rm th} = q^{({\rm B},k)}_{\rm th} \ \text{for} \ k = k^{(4)}_f \\ q^{(k)}_{\rm sum} < q^{({\rm B},k)}_{\rm th} \ \text{for} \ k \neq k^{(4)}_f, \\ f = 1, 2, \ldots, F_1, \ \text{where} \ F_1 \geq 1, \ F_1 \leq N; \ k = 1, 2, \ldots, N; \end{gathered} \tag{11.30}$$

values $\epsilon_{\rm th}^{(k)}$ satisfy conditions

$$0 < \epsilon_{\rm th}^{(k)} / q_{\rm th}^{({\rm B},k)} \ll 1 \tag{11.31}$$

for bottlenecks $k = k_f^{(4)}, f = 1, 2, \ldots, F_1$ in (11.30). Thus, when the total network inflow rate Q (10.2) increases, we maintain conditions (11.30) at the expense of the increase in q_m on other alternative routes.

However, due to the constrain "alternative routes" (Sect. 10.5), the number F_1 of bottlenecks satisfying (11.30) is limited by some value F (where $F \leq N$). All values $\epsilon_{\rm th}^{(k)}$ in (11.30) accordingly to (11.31) are positive ones. Therefore, under conditions (11.30), conditions (11.27) are satisfied. This means that the probability of spontaneous traffic breakdown at each of the network bottlenecks during the time interval $T_{\rm ob}$ is equal to zero: $P^{({\rm B},k)}(q_{\rm sum}^{(k)}) = 0, \ k = 1, 2, \ldots, N$. Thus, under conditions (11.30) related to small enough fluctuations in free flow, the probability of traffic breakdown in the network $P_{\rm net} = 0$ (10.8). In other words, no spontaneous traffic breakdown can occur in the network.

This means that conditions (11.30) in the limit case $F_1 = F$ and $\epsilon_{\rm th}^{(k)} \to 0$ are related to *the maximization of the network throughput* at which spontaneous traffic breakdown is still prevented in the whole network. This is because at the subsequent increase in the total network inflow rate Q (10.2) the constrain "alternative routes" does not allow us to maintain conditions (11.30) at the expense of the increase in q_m on other alternative routes: At least at one of the network bottlenecks the flow rate at the bottleneck becomes equal to or larger than the threshold flow rate:

$$q_{\rm sum} \geq q_{\rm th}^{(\rm B)}. \tag{11.32}$$

In this case, at the network bottlenecks for which condition (11.32) is satisfied the probability of spontaneous traffic breakdown during the time interval $T_{\rm ob}$ becomes larger than zero.

11.6.3 Probability of Traffic Breakdown in Network Under Small Free Flow Fluctuations

As explained in Sect. 11.5.1, when the flow rate $q_{\rm sum}$ at a network bottleneck becomes equal to or larger than the threshold flow rate $q_{\rm th}^{(\rm B)}$, the probability of spontaneous traffic breakdown at the network bottleneck becomes larger than zero (11.20). Consequently, in this case the probability of traffic breakdown in the network $P_{\rm net}$ should also become larger than zero. Indeed, the probability $P_{\rm net}$ of traffic breakdown in the network becomes larger than zero, when conditions (11.30) cannot be satisfied any more. We denote network bottlenecks k for which condition $q_{\rm sum}^{(k)} \geq q_{\rm th}^{({\rm B},k)}$ is satisfied by $k = k_d^{(5)}$, where $d = 1, 2, \ldots, D$; $D \geq 1$, $D \leq N$.

Consequently, we get:

$$P_{\text{net}} = 0 \quad \text{at} \quad q_{\text{sum}}^{(k)} < q_{\text{th}}^{(\text{B},k)}, \ k = 1, 2, \ldots, N, \tag{11.33}$$

$$P_{\text{net}} > 0 \quad \text{at} \quad \begin{cases} q_{\text{sum}}^{(k)} \geq q_{\text{th}}^{(\text{B},k)} \ \text{at} \ k = k_d^{(5)}, \\ q_{\text{sum}}^{(k)} < q_{\text{th}}^{(\text{B},k)} \ \text{at} \ k \neq k_d^{(5)}, \end{cases} \tag{11.34}$$

$$d = 1, 2, \ldots, D; \ D \geq 1, \ D \leq N; \ k = 1, 2, \ldots, N.$$

It should be noted that conditions (11.30) at $F_1 = F$ and $\epsilon_{\text{th}}^{(k)} \to 0$, at which the probability of traffic breakdown in the network is still equal to zero (i.e., $P_{\text{net}} = 0$ (10.8)), are applicable even under real non-steady state conditions in the network. Indeed, in conditions (11.30) only local flow rates $q_{\text{sum}}^{(k)}$ in free flow at network bottlenecks are used. For this reasons, formulas (11.33) and (11.34) are also valid under non-steady state conditions in the network.

11.6.4 Network Capacity Under Small Free Flow Fluctuations

As in Sect. 11.3, within a steady-state analysis of traffic and transportation networks, the limit case $F_1 = F$ and $\epsilon_{\text{th}}^{(k)} \to 0$ in conditions (11.30) allows us to define network capacity C_{net} for the case of small enough fluctuations in free flow at the network bottlenecks. The network capacity C_{net} is the maximum total network inflow rate Q at which conditions[8]

$$q_{\text{sum}}^{(k)} = q_{\text{th}}^{(\text{B},k)} \ \text{for} \ k = k_f^{(4)}, \tag{11.35}$$

$$q_{\text{sum}}^{(k)} < q_{\text{th}}^{(\text{B},k)} \ \text{for} \ k \neq k_f^{(4)},$$

$$f = 1, 2, \ldots, F; \ F \geq 1, \ F \leq N; \ k = 1, 2, \ldots, N$$

are satisfied. From a comparison of conditions (11.30) and (11.35) we can see that the total network inflow rate reaches the network capacity $Q = C_{\text{net}}$, when the limit case $F_1 = F$ in (11.30) is realized *and* all values $\epsilon_{\text{th}}^{(k)}$ in (11.30) are set to zero: $\epsilon_{\text{th}}^{(k)} = 0$.

In accordance with the definition of the network capacity (11.35) and condition (11.20), at $Q = C_{\text{net}}$ the probability of spontaneous traffic breakdown at the bottlenecks $k = k_f^{(4)}$, $f = 1, 2, \ldots, F$ during the time interval T_{ob} is larger than zero. Consequently, the probability of traffic breakdown in the network P_{net} becomes also larger than zero: $P_{\text{net}} > 0$ (10.9).

[8] As in Sect. 11.3, the network capacity C_{net} defined through conditions (11.35) depends on the network inflow rates $q_{ij}^{(\text{o})}(t)$ and the set of the alternative routes, which are assumed to be given. For the same day, there can be also different values C_{net}.

When $Q < C_{\text{net}}$, conditions (11.30) are satisfied. Therefore, the probability of traffic breakdown in the network $P_{\text{net}} = 0$. This explains the physical sense of the network capacity C_{net} for the case of small fluctuations in free flow at the network bottlenecks: Dynamic traffic assignment in the network in accordance with the network throughput maximization approach maximizes the network throughput at which spontaneous traffic breakdown is prevented in the whole network.

Respectively, when $Q > C_{\text{net}}$, at least at one of the network bottlenecks the flow rate in free flow at the bottleneck $q_{\text{sum}}^{(k)}$ becomes larger than the threshold flow rate $q_{\text{th}}^{(\text{B},k)}$. We denote network bottlenecks k for which condition $q_{\text{sum}}^{(k)} > q_{\text{th}}^{(\text{B},k)}$ is satisfied by $k = k_y^{(6)}$, where $y = 1, 2, \ldots, Y$; $Y \geq 1$, $Y \leq N$. Thus, rather than conditions (11.35), under condition $Q > C_{\text{net}}$ we get[9]

$$\begin{aligned} q_{\text{sum}}^{(k)} &> q_{\text{th}}^{(\text{B},k)} \text{ at } k = k_y^{(6)}, \\ q_{\text{sum}}^{(k)} &\leq q_{\text{th}}^{(\text{B},k)} \text{ at } k \neq k_y^{(6)}, \\ y = 1, 2, \ldots, Y;\ Y \geq 1,\ Y &\leq N;\ k = 1, 2, \ldots, N. \end{aligned} \tag{11.36}$$

Therefore, the probability of spontaneous traffic breakdown at some of the network bottlenecks during the time interval T_{ob} is larger than zero. For this reason, the probability of traffic breakdown in the network P_{net} becomes also larger than zero: $P_{\text{net}} > 0$ (10.9).

These results show that formulas (11.24), (11.25), which describe the probability $P_{\text{net}}(Q)$ of traffic breakdown in the network as a function of the total network inflow rate Q (10.2), are also valid for the case of small fluctuations in free flow of the network. However, in the case of small fluctuations, the network capacity C_{net} in formulas (11.24), (11.25) is defined by conditions (11.35).

11.6.5 Heterogeneous Free Flow Fluctuations in Networks

In a real traffic or transportation network fluctuations of the flow rate, the vehicle speed, and the vehicle density can be very different in different parts of the network with N bottlenecks. Therefore, there can be S network bottlenecks $k = 1, 2, \ldots, S$ (where $S \leq N$) for which fluctuations in free flow at these bottlenecks should be considered as larger one. For these S network bottlenecks, the characteristic parameter of the bottleneck in the network throughput maximization approach (Sect. 11.2) is the minimum capacity $C_{\min}^{(k)}$. For the remaining $N - S$ network bottlenecks $k = S{+}1, S{+}2, \ldots, N$, fluctuations in free flow at these bottlenecks should be considered as small one. For these $N - S$ network bottlenecks, the characteristic parameter of the bottleneck in the network throughput maximization approach (Sect. 11.6.2) is

[9] As the total network inflow rate Q is subsequently increased, the number Y of bottlenecks for which conditions (11.36) are satisfied can change.

the threshold flow rate $q_{\mathrm{th}}^{(\mathrm{B},k)}$. Thus, the network throughput maximization approach in this case of non-homogeneous free flow fluctuations in network is a combination of the approaches of Sect. 11.2 and Sect. 11.6.2 that is as follows.

When the total network inflow rate Q (10.2) increases beginning from small initial values, we maintain conditions (11.6) at the expense of the increase in q_m on other alternative routes for network bottlenecks for which fluctuations in free flow at these bottlenecks should be considered as larger one; in conditions (11.3) and (11.6) we should consider only bottlenecks $k = 1, 2, \ldots, S$, i.e., number N of bottlenecks should be replaced by number S. Respectively, we maintain conditions (11.30) at the expense of the increase in q_m on other alternative routes for network bottlenecks for which fluctuations in free flow at these bottlenecks should be considered as small one; in conditions (11.27) and (11.30) we should consider only bottlenecks $k = S + 1, S + 2, \ldots, N$, i.e., in conditions (11.27) and (11.30) number N of bottlenecks should be replaced by number $N - S$.

As explained in Sect. 11.2, due to the constrain "alternative routes" the number Z_1 of bottlenecks satisfying conditions (11.6) is limited by some value Z, where $Z \leq S$; correspondingly Sect. 11.6.2, the number F_1 of bottlenecks satisfying conditions (11.30) is limited by some value F, where $F \leq N - S$. These conditions in the limit case $Z_1 = Z$, $F_1 = F$ and $\epsilon^{(k)} \to 0$, $\epsilon_{\mathrm{th}}^{(k)} \to 0$ are related to the maximization of the network throughput at which during the time interval T_{ob} spontaneous traffic breakdown is still prevented in the whole network. In this case, the probability of traffic breakdown in the network is still equal to zero: $P_{\mathrm{net}} = 0$ (10.8).

At the subsequent increase in the total network inflow rate Q (10.2) the constrain "alternative routes" does not allow us to maintain conditions (11.6) and/or (11.30) at the expense of the increase in q_m on other alternative routes. For this reason, conditions (11.6) and/or (11.30) cannot be satisfied any more: At least at one of the network bottlenecks the probability of spontaneous traffic breakdown during the time interval T_{ob} becomes larger than zero. This leads to $P_{\mathrm{net}} > 0$ (10.9).

We see that within a steady state analysis of a network we can define the network capacity C_{net} as the maximum possible total network inflow rate Q (10.2) at which the following conditions are satisfied:

$$\begin{aligned} q_{\mathrm{sum}}^{(k)} &= C_{\mathrm{min}}^{(k)} \text{ for } k = k_z^{(1)}, \\ q_{\mathrm{sum}}^{(k)} &< C_{\mathrm{min}}^{(k)} \text{ for } k \neq k_z^{(1)}, \\ &\text{where } z = 1, 2, \ldots, Z, \\ Z \geq 1,\ Z \leq S,\ k &= 1, 2, \ldots, S; \end{aligned} \tag{11.37}$$

$$\begin{aligned} q_{\mathrm{sum}}^{(k)} &= q_{\mathrm{th}}^{(\mathrm{B},k)} \text{ for } k = k_f^{(4)}, \\ q_{\mathrm{sum}}^{(k)} &< q_{\mathrm{th}}^{(\mathrm{B},k)} \text{ for } k \neq k_f^{(4)}, \\ &\text{where } f = 1, 2, \ldots, F, \\ F \geq 1,\ F \leq N - S,\ k &= S + 1, S + 2, \ldots, N. \end{aligned} \tag{11.38}$$

11.6.6 "Non-Isolated" Traffic Networks

Applications of the BM principle discussed above are related to the case of an isolated traffic network. Only in this case, we can assume that network bottlenecks for which conditions (11.15) are satisfied do not affect on the probability P_{net} of traffic breakdown in the network.

There can be the following exclusion from this conclusion. We can assume that a traffic network under consideration is a part of a larger traffic network. In other words, the network should be considered as a non-isolated one from other traffic networks connected with it. In this case, a congested pattern can occur outside the network. If the congested pattern propagates through the network boundaries, the pattern can reach one of the network bottlenecks for which conditions (11.15) are satisfied. In this case, the pattern induces traffic breakdown at this bottleneck.

Therefore, in this particular case of a non-isolated traffic network, even when fluctuations in free flow in the network are small enough, the maximization of the network throughput at which no traffic breakdown occurs in the network is determined by conditions (11.6) in the limit case $Z_1 = Z$ and $\epsilon^{(k)} \to 0$ (Sect. 11.2). Respectively, the definition of the network capacity C_{net} follows from formula (11.8).

11.6.7 Prevention of Dissolving Over-Saturated Traffic at Traffic Signals in City Networks

In a theory of time-delayed traffic breakdown at the signal in city traffic [4, 8], it has been found that in free flow at traffic signals short-living over-saturated traffic that dissolves always during several cycles of the signal can occur spontaneously at the signal. The related pattern has been called "dissolving over-saturated traffic" (DOST) [9] (Sect. 9.3). The dissolving over-saturated traffic can occur with some probability during the time interval T_{ob} within the flow rate range $q_{\text{th}}^{(\text{DOST})} \le \bar{q}_{\text{in}} < C_{\text{min}}$ (9.28), where $q_{\text{th}}^{(\text{DOST})}$ is a threshold flow rate for the occurrence of dissolving over-saturated traffic at the signal, $\bar{q}_{\text{in}}$ is the average arrival flow rate at the signal.

At $\bar{q}_{\text{in}} < q_{\text{th}}^{(\text{DOST})}$, the probability for the occurrence of dissolving over-saturated traffic is equal to zero. To prevent the occurrence of dissolving over-saturated traffic at traffic signals of a network, we can use the application of the network throughput maximization approach of Sect. 11.6.5 as follows. We limit a consideration by the case of large enough fluctuations in free flow in the whole network.

In city network with N bottlenecks, there can be H network bottlenecks $k = 1, 2, \ldots, H$ (where $H \le N$) at which no dissolving over-saturated traffic can occur; this case is realized, for example, for highway bottlenecks. For these H network bottlenecks, the characteristic parameter of the bottleneck in the network throughput maximization approach is the minimum capacity $C_{\text{min}}^{(k)}$. For the remaining $N - H$

network bottlenecks $k = H + 1, H + 2, \ldots, N$ due to traffic signals we assume that condition (9.28) is valid. For these $N - H$ network bottlenecks, to prevent the occurrence of dissolving over-saturated traffic, the characteristic parameter of the bottleneck in the network throughput maximization approach is the threshold flow rate $q_{\text{th}}^{(\text{DOST},k)}$ for the occurrence of dissolving over-saturated traffic at the signal.

We can see that formally the application of the network throughput maximization approach for this network heterogeneity is qualitatively the same as explained in Sect. 11.6.5 for the network heterogeneity in free flow fluctuations: All explanations and formulas remain the same as those derived in Sect. 11.6.5, if we replace the number of the bottlenecks S by H as well as the threshold flow rate $q_{\text{th}}^{(\text{B},k)}$ by the threshold flow rate $q_{\text{th}}^{(\text{DOST},k)}$.

11.7 Conclusions

1. A general approach for the maximization of the network throughput introduced in [10] is considered. This network maximization throughput approach allows us to find general conditions for the *maximization* of the network throughput at which traffic breakdown cannot occur in the whole network. Under these conditions, free flow is ensured in the network. It is shown that the network maximization throughput approach is the application of the breakdown minimization (BM) principle for the case "zero breakdown probability" (10.8) in the network.
2. The network throughput maximization approach allows us to find a distinct dynamic assignment of link inflow rates in a traffic or transportation network maximizing the network throughput at which no traffic breakdown can still occur in the network.
3. The network throughput maximization approach can be applied for dynamic traffic assignment even when non-steady state conditions are realized in the network.
4. Under steady state conditions in the network, based on the network throughput maximization approach, a physical measure of traffic and transportation networks called *network capacity* introduced in [10] has been considered.
5. The network capacity allows us to formulate a general physical condition for the maximization of the network throughput at which free flow does persist in the whole network: As long as the total network inflow rate is smaller than the network capacity, under application of the network throughput maximization approach traffic breakdown cannot occur in the network.
6. The network capacity can considerably depend on the amplitude of fluctuations (the flow rate, the vehicle speed, and the vehicle density) as well as on the degree of the heterogeneity of free flow at network bottlenecks [11].
7. When the total network inflow rate increases continuously, the probability of traffic breakdown in the network becomes larger than zero, when the total network inflow rate reaches the network capacity.

The network throughput maximization approach to dynamic traffic assignment and control in traffic and transportation networks introduced in [10] permits the maximization of the network throughput at which free flow conditions are ensured in the whole network: No traffic breakdown occurs in the network.

The network throughput maximization approach is the application of the BM principle for "zero breakdown probability": The network throughput maximization approach guarantees that under increasing in the total network inflow rate the probability of traffic breakdown in the whole network remains zero [10].

The network throughput maximization approach is applicable even in the realistic case in a traffic or transportation network when non-steady state conditions are realized in the network [10].

An application of the network throughput maximization approach in a steady state theoretical analysis of a traffic or transportation network allows us to introduce a physical measure of the network called *network capacity*. The network capacity characterizes general features of the network with respect to the maximization of the network throughput: Under application of the network throughput maximization approach, as long as the total network inflow rate is smaller than the network capacity traffic breakdown cannot occur in the network [10].

References

1. B.S. Kerner, *The Physics of Traffic* (Springer, Berlin, New York, 2004)
2. B.S. Kerner, *Introduction to Modern Traffic Flow Theory and Control* (Springer, Heidelberg, Dordrecht, London, New York, 2009)
3. B.S. Kerner, J. Phys. A Math. Theor. **44**, 092001 (2011)
4. B.S. Kerner, Phys. Rev. E **84**, 045102(R) (2011)
5. B.S. Kerner, Traffic Eng. Contr. **52**, 379–386 (2011)
6. B.S. Kerner, in *Proc. of 2011 IEEE Forum on Integrated and Sustainable Transp. Systems* (Austria, Vienna, 2011), pp. 196–201

7. B.S. Kerner, in *Proc. of the 19th ITS World Congress* (Austria, Vienna, 2012), Paper No. EU-00190
8. B.S. Kerner, Europhys. Lett. **102**, 28010 (2013)
9. B.S. Kerner, Physica A **397**, 76–110 (2014)
10. B.S. Kerner, Eur. Phys. J. B **89**, 199 (2016)
11. B.S. Kerner, Physica A **466**, 626–662 (2017)
12. B.S. Kerner, S.L. Klenov, D.E. Wolf, J. Phys. A Math. Gen. **35**, 9971–10013 (2002)
13. B.S. Kerner, M. Koller, S.L. Klenov, H. Rehborn, M. Leibel, Physica A **438**, 365–397 (2015)
14. B.S. Kerner, H. Rehborn, R.-P. Schäfer, S.L. Klenov, J. Palmer, S. Lorkowski, N. Witte, Physica A **392**, 221–251 (2013)

Chapter 12
Minimization of Traffic Congestion in Networks

12.1 Introduction

In Chap. 11, we have considered the application of the BM principle for "zero breakdown probability" $P_{\text{net}} = 0$ (10.8), where P_{net} is the probability of traffic breakdown in a traffic or transportation network during a given time interval for observing of traffic flow T_{ob}. This application of the BM principle called the network throughput maximization approach allows us to prevent traffic breakdown in the whole network while keeping condition (10.8).

The objective of this chapter is to study the case of larger values of the total network inflow rate Q (10.2) related to rush hours in urban networks at which condition $P_{\text{net}} = 0$ (10.8) cannot be satisfied through the application of the BM principle. This means that under application of the BM principle the probability P_{net} that traffic breakdown occurs in the network becomes larger than zero: $P_{\text{net}} > 0$ (10.9). In other words, at such large values of the total network inflow rate Q related to rush hours in urban networks, the application of the BM principle can only reduce the probability of the occurrence of traffic breakdown in the network during the time interval T_{ob} to a minimum value of the probability:

$$P_{\text{net}} = P_{\text{net}}^{(\text{min})} > 0. \tag{12.1}$$

Based on results of the analysis of the BM principle made in [21] as well as applications of "congested pattern control approach" [14–18] to the dissolution of traffic congestion at network bottlenecks, in this chapter we analysis traffic assignment and control in a traffic or transportation network under condition (12.1) [23–25, 27, 31, 32]. We will also show that even if under application of the BM principle

B.S. Kerner, *Breakdown in Traffic Networks*, DOI 10.1007/978-3-662-54473-0_12

traffic breakdown occurs at some of the network bottlenecks, through control of traffic breakdown the development of traffic congestion in the network can be prevented.

12.2 An Explicit Formulation for BM Principle

To find the minimum probability $P_{\rm net} = P_{\rm net}^{(\rm min)}$ of the occurrence of traffic breakdown in the network during the time interval $T_{\rm ob}$, we should use an explicit formulation for the probability $P_{\rm net}$ of traffic breakdown in the network as a function of characteristics of traffic breakdown at network bottlenecks. Characteristics of network bottlenecks have already been formulated in Sect. 10.3. A theory of traffic breakdown at network bottlenecks has been presented in Chaps. 5 and 9. In this chapter, we will use results of this theory.

By the explicit formulation for the probability $P_{\rm net}$ of traffic breakdown in the network, we assume that the probability $P_{\rm net}$ is determined only[1] by the probabilities of spontaneous traffic breakdown at the network bottlenecks during a given time interval $T_{\rm ob}$ for observing traffic flow. We assume also that at different bottlenecks spontaneous traffic breakdown occurs independently each other. Under these assumptions, we can use a formulation for the probability $P_{\rm net}$ of traffic breakdown in the network during the time interval $T_{\rm ob}$ following from classical probability theory [9]:

$$P_{\rm net} = 1 - \prod_{k=1}^{N}(1 - P^{({\rm B},k)}). \tag{12.2}$$

To explain formula (12.2), note that in accordance with the classical probability theory [9], the probability $P_{\rm C,\,net} = 1 - P_{\rm net}$ (10.5) that during the time interval $T_{\rm ob}$ traffic breakdown occurs at none of the network bottlenecks is equal to

$$P_{\rm C,\,net} = \prod_{k=1}^{N} P_{\rm C}^{({\rm B},k)}, \tag{12.3}$$

where $P_{\rm C}^{({\rm B},k)}$ is the probability that traffic breakdown does not occur at bottleneck k during the time interval $T_{\rm ob}$, i.e., the probability that during this time interval free

[1]This means that applications of the BM principle under condition (12.1) for the case of non-isolated traffic networks (Sect. 11.6.6) will not be studied in this book.

flow persists at bottleneck k. Using formula

$$P_{\mathrm{C}}^{(\mathrm{B},k)} = 1 - P^{(\mathrm{B},k)}, \tag{12.4}$$

we can rewrite formula (12.3) as follows:

$$P_{\mathrm{C,\ net}} = \prod_{k=1}^{N} (1 - P^{(\mathrm{B},k)}). \tag{12.5}$$

In accordance with (10.5) and (12.5), we get formula (12.2) for the probability for spontaneous occurrence of traffic breakdown in at least one of the network bottlenecks P_{net} during the time interval T_{ob}.

Using formula (12.2), an explicit formulation of the BM principle (10.3) can be written as [21, 22]:

$$\min_{q_1, q_2, \ldots, q_{\mathrm{M}}, \alpha_1, \alpha_2, \ldots, \alpha_N, \mathbf{R}_1, \mathbf{R}_2, \ldots, \mathbf{R}_{\mathrm{N}}} \{1 - \prod_{k=1}^{N} (1 - P^{(\mathrm{B},k)}(q_{\mathrm{sum}}^{(k)}, \ \alpha_k, \mathbf{R}_k))\}. \tag{12.6}$$

Respectively, using (12.5), for an equivalent explicit formulation of the BM principle (10.4) we get

$$\max_{q_1, q_2, \ldots, q_{\mathrm{M}}, \alpha_1, \alpha_2, \ldots, \alpha_N, \mathbf{R}_1, \mathbf{R}_2, \ldots, \mathbf{R}_{\mathrm{N}}} \prod_{k=1}^{N} (1 - P^{(\mathrm{B},k)}(q_{\mathrm{sum}}^{(k)}, \ \alpha_k, \mathbf{R}_k)). \tag{12.7}$$

Either the explicit formulation of the BM principle (12.6) or (12.7) can be used for traffic assignment and control resulting in the minimization of the probability of traffic breakdown in traffic and transportation networks. The connection between the flow rate $q_{\mathrm{sum}}^{(k)}$ in free flow at bottleneck k that is located on network link m and the link inflow rate q_m depends on the bottleneck type. For example, for an on-ramp bottleneck we get $q_{\mathrm{sum}}^{(k)} = q_m + q_{\mathrm{on}}^{(k)}$, where $q_{\mathrm{on}}^{(k)}$ is the on-ramp inflow rate of the on-ramp bottleneck.[2]

As mentioned in Sect. 10.3, in general, the probability for traffic breakdown at a network bottleneck depends on the flow rate $q_{\mathrm{sum}}^{(k)}$, a set of bottleneck parameters α_k, and a matrix of percentages of vehicles with different vehicle (and/or driver) characteristics $\mathbf{R}_k$: $P^{(\mathrm{B},k)} = P^{(\mathrm{B},k)}(q_{\mathrm{sum}}^{(k)}, \ \alpha_k, \mathbf{R}_k)$. However, following [21] to disclose the physics of the BM principle, we limit the further analysis of the BM principle by a consideration of the dependence of the breakdown probability

[2]It should be noted that in some networks the on-ramp inflow rate $q_{\mathrm{on}}^{(k)}$ can be equal to the inflow rate for another network link; in this case, $q_{\mathrm{sum}}^{(k)}$ depends on the inflow rates for two different network links.

on the flow rate $q_{\rm sum}^{(k)}$ *only*. In this case, for the flow-rate dependence of the probability for traffic breakdown $P^{({\rm B},k)}(q_{\rm sum}^{(k)})$ during the time interval $T_{\rm ob}$ we can use formulas (5.12) and (9.48) [35]:

$$P^{({\rm B},k)}(q_{\rm sum}^{(k)}) = \frac{1}{1 + \exp[\beta^{(k)}(q_{\rm P}^{(k)} - q_{\rm sum}^{(k)})]}, \qquad (12.8)$$

where $\beta^{(k)}$ and $q_{\rm P}^{(k)}$ are parameters[3] of bottleneck k ($k = 1, 2, \ldots, N$). Respectively, from (12.6) and (12.8) we can write a simplified explicit formulation of the BM principle as follows [21]:

$$\min_{q_1, q_2, \ldots, q_{\rm M}} \left\{ 1 - \prod_{k=1}^{N} \left(1 - \frac{1}{1 + \exp[\beta^{(k)}(q_{\rm P}^{(k)} - q_{\rm sum}^{(k)})]} \right) \right\}. \qquad (12.9)$$

It should be noted that recently Guo, Cao, Zhang, Niyato, Seshadri, and Fastenrath [10, 11] have applied the explicit formulation of the BM principle (12.9) [21] for the development of a new distributed multi-vehicle routing algorithm. The works [10, 11] are the first ones in which the BM principle (12.9) [21] has been applied for large scale road networks.

12.3 Empirical Spontaneous Traffic Breakdowns as Independent Events in Network

The explicit formulation of the BM principle (12.6) (or (12.7)) is only valid, when traffic breakdowns at neighborhood bottlenecks can be considered independent events. The latter is usually the case in real traffic networks as explained below. In Fig. 12.1, we show three different empirical examples in which spontaneous traffic breakdowns have occurred at off- and on-ramp bottlenecks. We see that in each of these empirical examples, there are two spontaneous traffic breakdowns 1 and 2 (labeled, respectively, by "spontaneous breakdown 1" and "spontaneous breakdown 2" in Fig. 12.1). In each of the three days at which real field traffic data was measured, these two spontaneous traffic breakdowns 1 and 2 occur at different bottlenecks at different time instants.

In other words, in each of the empirical examples, after spontaneous traffic breakdown 1 has already occurred at one of the bottlenecks, there has been

[3]Recall that as mentioned in Sect. 11.1, when network bottleneck k is due to traffic signal, for this bottleneck k the flow rate $q_{\rm sum}^{(k)}$ in all formulas should be replaced by the average arrival flow rate at the approach $\bar{q}_{\rm in}^{(k)}$ (see Sect. 9.3).

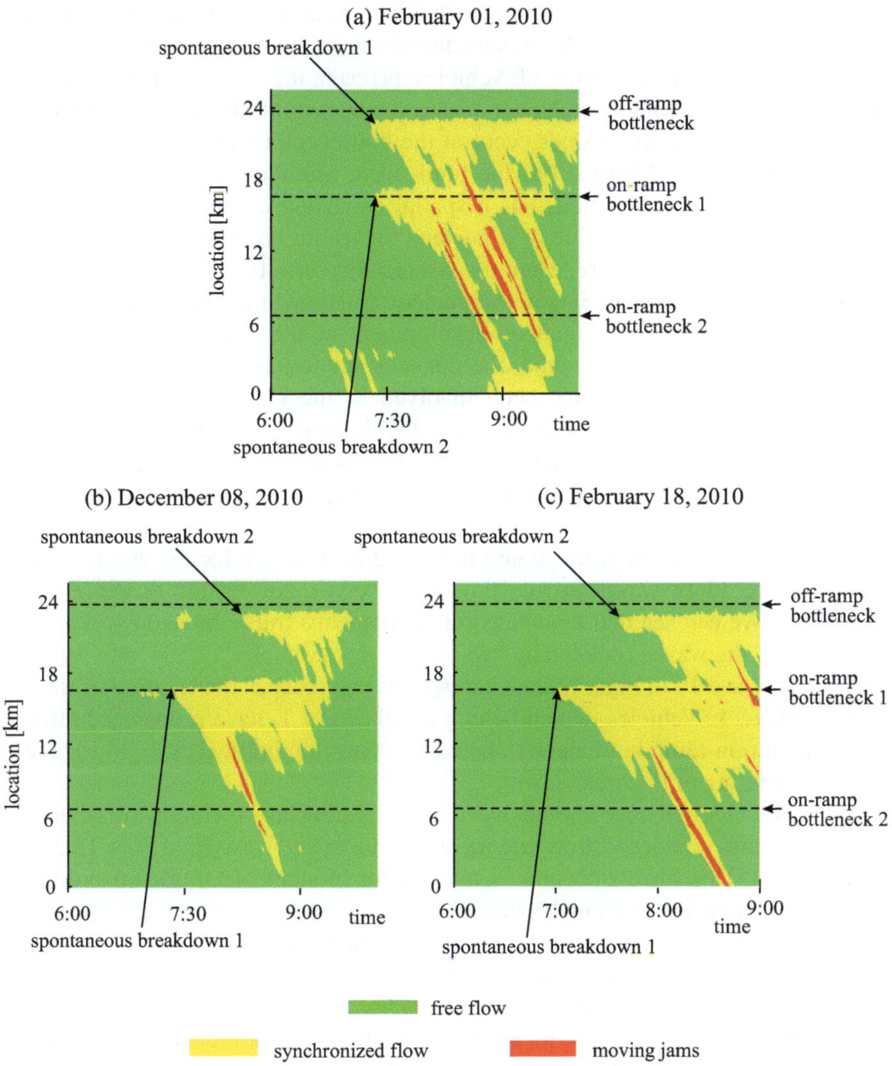

Fig. 12.1 Empirical examples showing independence of traffic breakdowns (spontaneous F→S transitions) at different highway bottlenecks. Real field traffic data measured by road detectors on freeway A5-South in Germany on February 01, 2010 (**a**), December 08, 2010 (**b**), and February 18, 2010 (**c**). Speed data is presented in space and time with averaging method described in Sec. C.2 of [37]. Off-ramp bottleneck is related to road intersection "Nordwestkreuz Frankfurt"; on-ramp bottleneck labeled by "on-ramp bottleneck 1" is related to intersection "Bad Homburger Kreuz"; on-ramp bottleneck labeled by "on-ramp bottleneck 2" is related to intersection "Friedberg" (see explanations for these road intersections in Fig. 2.1 of the book [14])

free flow at another bottleneck at which later spontaneous traffic breakdown 2 occurs. Moreover, it should be noted that spontaneous traffic breakdown results from some dynamic behavior of vehicles propagating through the bottleneck. When this vehicle dynamic behavior causes the occurrence of a nucleus for traffic breakdown, the breakdown does occur at the bottleneck (Sect. 3.4.4). The nucleus is localized in a neighborhood of the bottleneck. Simulations and empirical data show [14, 36] that the nucleus has a spatial extension about or less 500 m (see explanations of the occurrence of a nucleus for traffic breakdown at a bottleneck in Chaps. 3 and 5). Therefore, when the distance between two neighbor bottlenecks is considerably larger than 500 m, the nucleus emergence at one of the bottleneck does not influence on the nucleus development at the neighbor bottleneck. In other words, when two neighbor bottlenecks are on a large enough distance each other, vehicles that have been involved in the emergence of the nucleus that is responsible for spontaneous traffic breakdown 1 at a bottleneck, have no effect on the behavior of vehicles that have been involved in the emergence of the nucleus that is responsible for spontaneous traffic breakdown 2 at another bottleneck.

The latter conclusion is illustrated in Fig. 12.2, in which for the empirical data shown in Fig. 12.1a waves in free flow have been made visible as explained in Sect. 3.4.1. We can see that a nucleus for spontaneous traffic breakdown 1 occurs, when one of the waves (labeled by "wave 1 of slow vehicles", "flow rate wave 1", and "speed wave 1" in Figs. 12.2a,b,c, respectively) reaches the location of the off-ramp bottleneck. Vehicles, which belong to this wave 1, have propagated through the upstream on-ramp bottleneck (labeled by "on-ramp bottleneck 1" in Fig. 12.2) without initiating of the breakdown.

Other vehicles belong to another wave 2 propagating later on highway (labeled by "wave 2 of slow vehicles", "flow rate wave 2", and "speed wave 2" in Figs. 12.2a,b,c, respectively). This wave 2 becomes a nucleus for spontaneous traffic breakdown 2, when wave 2 reaches on-ramp bottleneck 1.

Thus, we see that spontaneous traffic breakdowns occurring in free flow at different highway bottlenecks in real free flow can indeed be considered independent events. However, this general conclusion is only valid as long as the distance between the effective locations of two neighbor bottlenecks is larger than some critical distance (about 500 m) at which the formation of a nucleus at one of the bottlenecks can effect on the formation of another nucleus at the neighbor bottleneck.

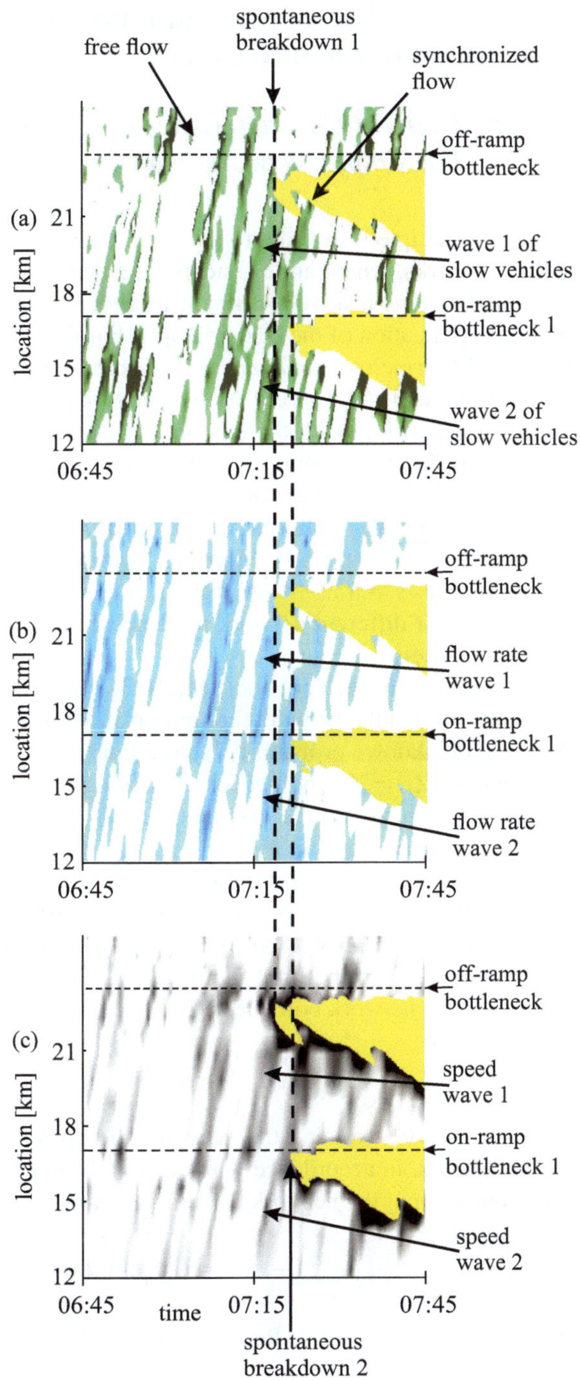

Fig. 12.2 Empirical nucleation of independent spontaneous traffic breakdowns due to waves propagation through off- and on-ramp bottlenecks associated with empirical data in Fig. 12.1a: (**a**) Waves of $\Delta\psi_{\rm wave}$ are presented by regions with variable shades of gray (green in the on-line version). (**b**) Waves of $\Delta q_{\rm wave}$ are presented by regions with variable shades of gray (blue in the on-line version). (**c**) Waves of $\Delta v_{\rm wave}$ are presented by regions with variable shades of gray. Real field traffic data measured on freeway A5-South in Germany on February 01, 2010. Wave presentation has been explained in Sect. 3.4.1

12.4 Simulations of Minimum Probability of Traffic Breakdown in Networks

12.4.1 General Characteristics of Applications of BM Principle for Simple Network Model

The BM principle is devoted to the optimization of large scale, complex vehicular traffic networks, which are of interest in engineering applications. However, under such applications it is very difficult to understand the physical effects associated with the application of the BM principle. For this reason, to understand the physics of the BM principle, in *numerical simulations* of the applications of the BM principle we will study a very simple network model.

In the network model, there is only one origin and one destination. In general, there are N alternative routes from the origin to the destination in the model, where $N > 1$. We assume that for all alternative routes condition (10.6) is satisfied. This means that all N alternative routes satisfy the constrain "alternative network routes (paths)" (Sect. 10.5). Moreover, we assume that at each of the alternative network routes $s = 1, 2, \ldots, N$ there is only one bottleneck. In this case, index $k = 1, 2, \ldots, N$ for different network bottlenecks is the same as index $s = 1, 2, \ldots, N$ for different network routes: $s = k$.

For a general traffic network (Sect. 10.3), due to the constrain "alternative routes" the application of the network throughput maximization approach (the BM principle for "zero breakdown probability") has a limit. This limit case is associated with the limit number $Z_1 = Z$ of bottlenecks satisfying conditions (11.6) (Sect. 11.2): For the general traffic network, the limit number $Z_1 = Z$ can be less than the total number of the network bottlenecks N.

Contrarily to the general network model, for the simple network model under consideration, all alternative network routes satisfy the constrain "alternative routes" for the application of the BM principle. Therefore, for the simple network model, in conditions (11.6) the limit number $Z_1 = Z$ of bottlenecks is equal to the total number of the network bottlenecks:

$$Z = N. \tag{12.10}$$

Formula (12.10) explains the sense of the consideration of the simple network model. Indeed, in accordance with (12.10), from conditions (11.6) following from the application of the network throughput maximization approach, we get

$$q_{\text{sum}}^{(k)} + \epsilon^{(k)} = C_{\text{min}}^{(k)} \text{ for } k = 1, 2, \ldots, N, \tag{12.11}$$

where $0 < \epsilon^{(k)}/C_{\text{min}}^{(k)} \ll 1$. Respectively, conditions (11.8) for the network capacity reduce to

$$q_{\text{sum}}^{(k)} = C_{\text{min}}^{(k)} \text{ for } k = 1, 2, \ldots, N. \tag{12.12}$$

Under a steady-state analysis of the simple network model, the total network inflow rate Q (10.2) is equal to

$$Q = \sum_{k=1}^{N} q_{\rm sum}^{(k)}. \tag{12.13}$$

For the simple network model, from (12.12) and (12.13) we get that the network capacity is found from the following formula:

$$C_{\rm net} = \sum_{k=1}^{N} C_{\rm min}^{(k)}. \tag{12.14}$$

In accordance with results of Sect. 11.5.1, formulas (12.11), (12.12), and (12.14) are related to large free flow fluctuations in the whole network.

Contrarily, when free flow fluctuations in the whole network are small enough, accordingly to results of Sect. 11.6, for the simple network model conditions (11.30) following from the application of the network throughput maximization approach reduce to

$$q_{\rm sum}^{(k)} + \epsilon_{\rm th}^{(k)} = q_{\rm th}^{({\rm B},k)} \ \text{for}\ k = 1, 2, \ldots, N, \tag{12.15}$$

where $0 < \epsilon_{\rm th}^{(k)}/q_{\rm th}^{({\rm B},k)} \ll 1$. For the case of small free flow fluctuations, conditions (11.35) for the network capacity reduce to

$$q_{\rm sum}^{(k)} = q_{\rm th}^{({\rm B},k)} \ \text{for}\ k = 1, 2, \ldots, N. \tag{12.16}$$

Respectively, instead of formula (12.14), in the case of small free flow fluctuations in the whole simple network under consideration for the network capacity we get the following formula:

$$C_{\rm net} = \sum_{k=1}^{N} q_{\rm th}^{({\rm B},k)}. \tag{12.17}$$

12.4.2 Two-Route and Three-Route Simple Network Models

At a large number N of alternative routes in the simple model of traffic network discussed in Sect. 12.4.1, simulations of the applications of the BM principle can nevertheless become too complex for the understanding of traffic phenomena. Therefore, following [21], in the book we limit simulations by the analysis of the applications of the BM principle for traffic assignment in two simple network models related to the network model discussed in Sect. 12.4.1: (i) a network model with $N = 2$, i.e., two-route network[4] with two on-ramp bottlenecks (Fig. 12.3a,b) and (ii) a network model with $N = 3$, i.e., three-route network with with three on-ramp bottlenecks (Fig. 12.4a).

One of the reasons of the consideration of two networks (Figs. 12.3a and 12.4a) is to show that the increase in the number of routes at which traffic breakdown can occur in a network does not change the main feature of the BM principle: The BM principle allows us to find the assignment of the link inflow rates q_m at which breakdown probability $P_{\rm net}$ (12.2) for the whole network exhibits a minimum. The consideration of two different examples of simple networks is also useful for a study of probabilistic features of traffic breakdown at network bottlenecks under application of the BM principle.

For simulations of the applications of the BM principle, we use the Kerner-Klenov stochastic three-phase traffic flow model [33, 34] (see Appendix A), in which the maximum speed of vehicles in free flow is a function of the space gap g between vehicles (Fig. 12.5) given by formula:

$$v_{\rm free}(g) = \max(v_{\rm free}^{(\rm min)},\ v_{\rm free}^{(\rm max)}(1 - \kappa d/(d + g))), \tag{12.18}$$

where in simulations presented below we have used the following model parameters: $v_{\rm free}^{(\rm min)} = 90$ km/h, $v_{\rm free}^{(\rm max)} = 150$ km/h, $d = 7.5$ m (vehicle length), and $\kappa = 1.73$.

In both networks, vehicles move from origin O to destination D (Figs. 12.3a and 12.4a). Each of the network routes is a two-lane road with an on-ramp bottleneck (Fig. 12.3b). In two-route network (Fig. 12.3a), there are two alternative routes 1 and 2 with lengths L_1 and L_2 (with $L_2 > L_1$). In three-route network (Fig. 12.4a), there are three alternative routes 1, 2, and 3 with lengths L_1, L_2, and L_3 (with $L_2 > L_3 > L_1$).

Due to a complex dynamics of local permanent speed disturbances existing in free flow at the on-ramp bottlenecks (Sects. 3.4 and 5.12) as well as a decreasing-dependence of the vehicle speed in free flow on the vehicle density (12.18), route travel times denoted by T_s (where $s = 1, 2$ for two-route network and $s = 1, 2, 3$ for three-route network) depend considerably on the link inflow rates q_m (Figs. 12.3a and 12.4a), even when free flow persists at the bottlenecks. Moreover, due to possible traffic breakdowns at the bottlenecks, route travel times T_s can become complex time-functions, even when the on-ramp inflow rates $q_{\rm on}^{(k)}$ are given constants ($k = 1, 2$ in Fig. 12.3a and $k = 1, 2, 3$ in Fig. 12.4a).

[4]Two-route network models are often used for studies of other traffic phenomena (e.g., [1–8, 38–43]).

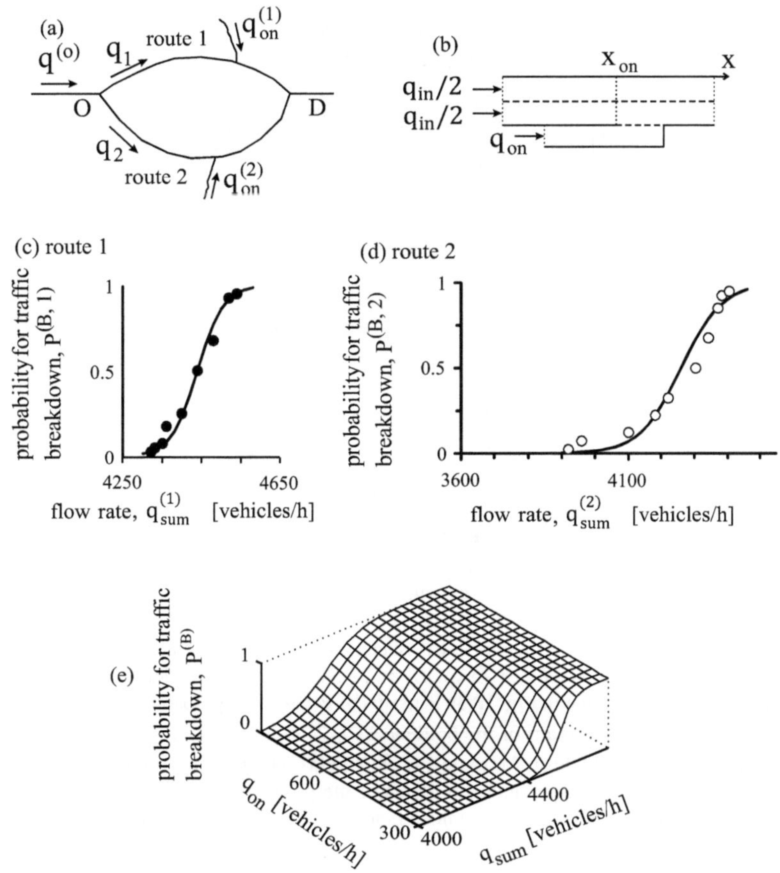

Fig. 12.3 Model of two-route network: (**a**) Sketch of network with two routes and two on-ramp bottlenecks; $L_1 = 20$ km, $L_2 = 25$ km. (**b**) Model of on-ramp bottleneck on a two-lane road link, the beginning of the on-ramp merging region is at $x_{on} = 15$ km, q_{in} is the flow rate in free flow upstream of the bottleneck (in all simulations presented below, the vehicle speed averaged across two-lanes of a road link is shown). (**c**, **d**) Probabilities $P^{(B,k)}(q^{(k)}_{sum})$ of spontaneous traffic breakdown at on-ramp bottlenecks $k = 1$ (**c**) and $k = 2$ (**d**) as functions of the total flow rate downstream of the bottlenecks $q^{(k)}_{sum}$ for given on-ramp inflow rates $q^{(k)}_{on} = 400$ vehicles/h for route 1 ($k = 1$) and 700 vehicles/h for route 2 ($k = 2$); calculated points of the probabilities $P^{(B,k)}(q^{(k)}_{sum})$ are fitted with curves given by formula (12.8). $T_{ob} = 30$ min, number of different simulation realizations $N_r = 40$. (**e**) Probability of traffic breakdown at on-ramp bottleneck as a function of the on-ramp inflow rate q_{on} and the flow rate downstream of the bottleneck $q_{sum} = q_{in} + q_{on}$

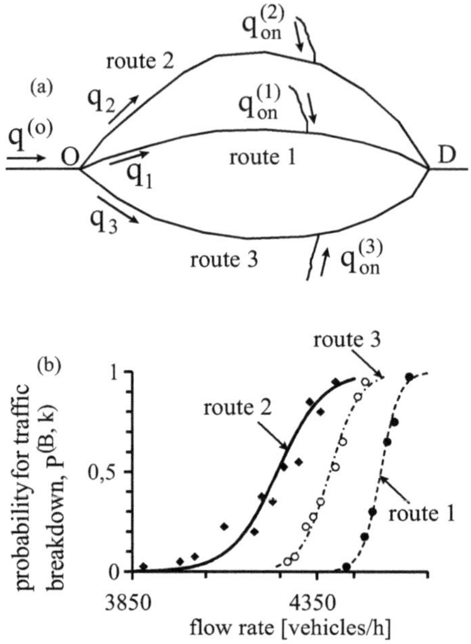

Fig. 12.4 Model of three-route network: (**a**) Sketch of network with three routes and three on-ramp bottlenecks; $L_1 = 20$ km, $L_2 = 25$ km, $L_3 = 22.5$ km; the same model of on-ramp bottleneck on a two-lane road link as that in Fig. 12.3b is used. (**b**) Breakdown probabilities $P^{(\mathrm{B},k)}$ as functions of the total flow rate downstream of the bottlenecks $q_{\mathrm{sum}}^{(k)}$ (where $k = 1, 2, 3$) for given on-ramp inflow rates $q_{\mathrm{on}}^{(k)} = 300$ vehicles/h for route 1 ($k = 1$), 800 vehicles/h for route 2 ($k = 2$), and 500 vehicles/h for route 3 ($k = 3$); calculated points of the probabilities $P^{(\mathrm{B},k)}$ are fitted with curves given by formula (12.8). Other model parameters are the same as those in Fig. 12.3

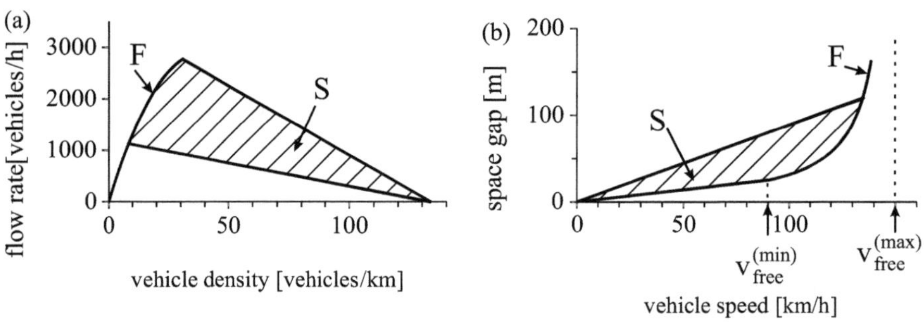

Fig. 12.5 Model steady states in Kerner-Klenov model: (**a**) Steady states in the flow-density plane. (**b**) Steady states in the space-gap-speed plane. F—free flow. S—synchronized flow

12.4.3 Probabilistic Features of Traffic Breakdown in Networks

To disclose the physics of traffic assignment in traffic and transportation networks with the BM principle, the assignment of the link inflow rates q_m (Fig. 12.3a and 12.4a), which can depend on time, is performed at time-independent network inflow rates at the network boundaries (origin of the network "O" and on-ramps in Figs. 12.3a and 12.4a, respectively). Therefore, the total network inflow rate for two-route network $Q = Q_2$, where

$$Q_2 = \sum_{m=1}^{2} q_m + \sum_{k=1}^{2} q_{\rm on}^{(k)}, \tag{12.19}$$

and the total network inflow rate for three-route network $Q = Q_3$, where

$$Q_3 = \sum_{m=1}^{3} q_m + \sum_{k=1}^{3} q_{\rm on}^{(k)}, \tag{12.20}$$

do not depend on time.

Under chosen parameters of the Kerner-Klenov model used for simulations presented in this chapter as well as in the next Chap. 13, there is a flow rate range (11.15) in metastable free flow at the network bottlenecks within which the probability of spontaneous traffic breakdown at the bottlenecks during the time interval $T_{\rm ob}$ is equal to zero (11.16). In other words, the parameters of the Kerner-Klenov model are related to *small fluctuations* in free flow at the network bottlenecks (Sect. 11.5.1). For this case, the network capacity is given by formula (12.17) in which $N = 2$ for two-route network and $N = 3$ for three-route network.

Simulations show that at chosen network parameters for two-route network we have found values $q_{\rm th}^{({\rm B},k)} = 4320$ for $k = 1$, 3920 vehicles/h for $k = 2$. For three-route network, we have found values $q_{\rm th}^{({\rm B},k)} = 4430$ for $k = 1$, 3880 for $k = 2$, and 4270 vehicles/h for $k = 3$. Accordingly to (12.17) we get $C_{\rm net} = 8240$ vehicles/h for two-route and $C_{\rm net} = 12580$ vehicle/h for three-route networks, respectively. As long as $Q < C_{\rm net}$, where the network capacity $C_{\rm net}$ is given by (12.17), simulations show that under application of the network throughput maximization approach of Sect. 11.6.2, no spontaneous traffic breakdown (F→S transition) occurs at the network bottlenecks during the time interval $T_{\rm ob}$.

When the total network inflow rate Q exceeds the network capacity $C_{\rm net}$ given by Eq. (12.17), the network throughput maximization approach cannot be applied any more. This is because simulations show that the probability of traffic breakdown in the network $P_{\rm net} > 0$: Spontaneous traffic breakdown does occur during the time interval $T_{\rm ob}$ at least at one of the network bottlenecks. To calculate the minimum value of this probability $P_{\rm net} = P_{\rm net}^{(\rm min)}$ (10.9), we use the BM principle (12.6). For two-route network (Fig. 12.3a), from (12.6) we get

$$\min_{q_1,q_2}\{1-\prod_{k=1}^{2}(1-P^{(\mathrm{B},k)}(q_{\mathrm{sum}}^{(k)}))\}, \tag{12.21}$$

where q_1 and q_2 correspond to (12.19) at a given value $Q = Q_2$. For three-route network (Fig. 12.4a), from (12.6) we get

$$\min_{q_1,q_2,q_3}\{1-\prod_{k=1}^{3}(1-P^{(\mathrm{B},k)}(q_{\mathrm{sum}}^{(k)}))\}, \tag{12.22}$$

where q_1, q_2, and q_3 correspond to (12.20) at a given value $Q = Q_3$.

The flow rate-dependence of the breakdown probability $P^{(\mathrm{B},k)}(q_{\mathrm{sum}}^{(k)})$ in (12.21) or (12.22) is given by formula (12.8) (curves for different routes in Fig. 12.3c,d and Fig. 12.4b).

For each given total network inflow rate Q, the flow rates q_m (q_1, q_2 in Fig. 12.3a for two-route network and q_1, q_2, q_3 in Figs. 12.4c,d for three-route network) that satisfy the BM principle (12.21) (for two-route network) or (12.22) (for three-route network) are initially found in numerical simulations with the use of the flow-rate dependence of the breakdown probability $P^{(\mathrm{B},k)}(q_{\mathrm{sum}}^{(k)})$ (12.8). The connection between the flow rates $q_{\mathrm{sum}}^{(k)}$ in free flow at the network bottlenecks in (12.8) and the link inflow rates q_m is as follows. The link inflow rates q_m determines the flow rates in free flow upstream of the bottlenecks; therefore, the flow rates in free flow at the bottlenecks

$$q_{\mathrm{sum}}^{(k)} = q_m + q_{\mathrm{on}}^{(k)}, \tag{12.23}$$

where $m = k$, $k = 1, 2, \ldots, N$; $N = 2$ for two-route network (Fig. 12.3a) and $N = 3$ for three-route network (Fig. 12.4a). Because the breakdown probabilities $P^{(\mathrm{B},k)}(q_{\mathrm{sum}}^{(k)})$ of spontaneous traffic breakdown at the network bottlenecks during the time interval T_{ob} depend on the on-ramp inflow rate (12.23), the flow-rate dependencies of the breakdown probabilities related to different $q_{\mathrm{on}}^{(k)}$ are shifted each other in the flow rate axis (Figs. 12.3c,d and 12.4b).

It should be emphasized that the breakdown probability at any single bottleneck exhibits no minimum: The breakdown probability is always a monotonously increasing flow rate function (Figs. 12.3c,d and 12.4b). This result is independent of the on-ramp inflow rate as shown in Fig. 12.3e. For this reason, the minimization of breakdown probability $P^{(\mathrm{B})}$ for a single bottleneck is not possible. However, the minimization of the probability P_{net} of traffic breakdown in a *traffic network* is possible, as formulated in the BM principle (10.3).

To understand this conclusion, in Fig. 12.6 we present calculations of flow rate dependencies of breakdown probability P_{net} (12.2) for two-route network (Fig. 12.3a). At a given value $Q = Q_2$ (12.19), through the use of traffic assignment

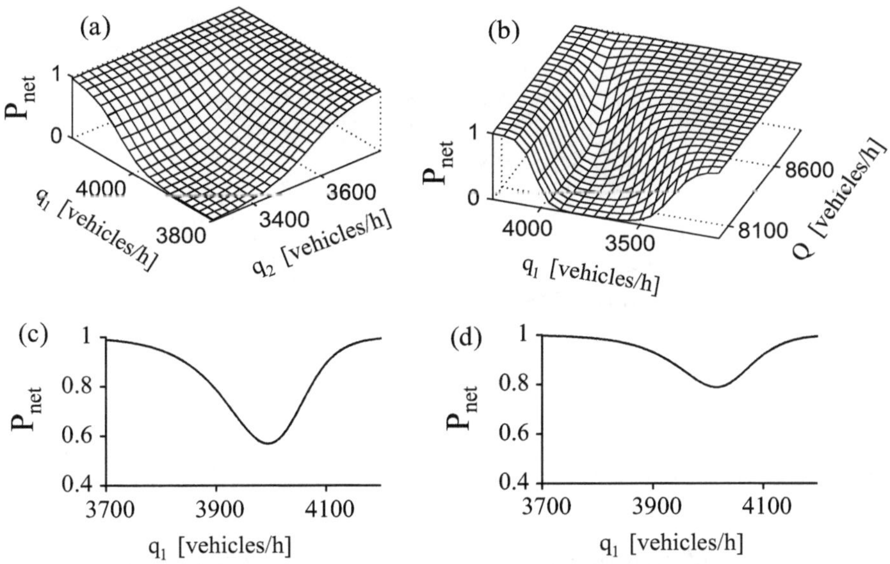

Fig. 12.6 Breakdown probability $P_{\rm net}$ (12.2) for two-route network shown in Fig. 12.3a as a function of the flow rates q_1 and q_2 (**a**) and a function of the flow rate q_1 and the total network inflow rate Q (12.19) (**b**). (**c, d**) $P_{\rm net}$ (12.2) as a function of q_1 for different given values $Q = 8640$ (**c**) and 8720 vehicles/h (**d**)

at the origin of the network (Fig. 12.3a) we can increase, for example, the flow rate q_1 upstream of bottleneck 1 and decrease the flow rate q_2 upstream of another bottleneck 2. This traffic assignment in the network changes breakdown probabilities for both bottlenecks: The breakdown probability at bottleneck 1 increases, whereas the breakdown probability at bottleneck 2 decreases.

There should be an optimum in the assignment of the link flow rates q_1 and q_2 (Fig. 12.3a) at which probability for spontaneous occurrence of traffic breakdown in at least one of these two network bottlenecks $P_{\rm net}$ (12.2) during the time interval $T_{\rm ob}$ reaches a minimum (Fig. 12.6b–d) at a given value $Q = Q_2$. In other words, probability that during the time interval $T_{\rm ob}$ traffic breakdown occurs at none of these two network bottlenecks reaches a maximum. Thus, although breakdown probability for each of the bottlenecks separately has no minimum (Fig. 12.3c,d), there is a minimum in breakdown probability $P_{\rm net}$ (12.2) for the network at $Q = Q_2 = {\rm const}$ (Fig. 12.6c,d).

The same conclusion is illustrated in Fig. 12.7 in which the probability $P_{\rm net}$ (12.2) of traffic breakdown in three-route network (Fig. 12.4a) has been calculated as a function of link inflow rates q_1 and q_2 at a given value of the total network inflow rate $Q = Q_3$ (12.20).

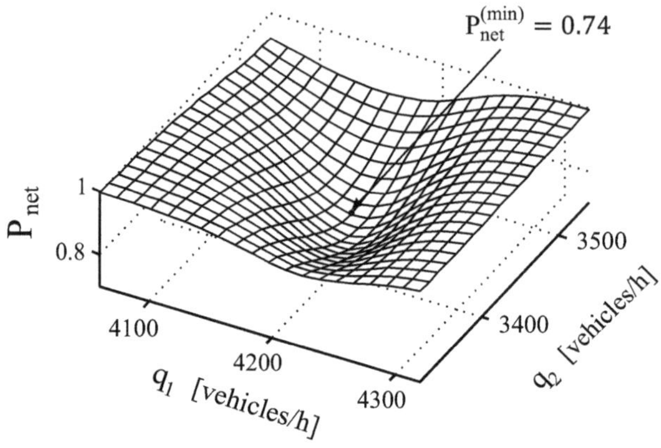

Fig. 12.7 Breakdown probability P_{net} (12.2) for three-route network shown in Fig. 12.4a as a function of link inflow rates q_1 and q_2 at constant value $Q = 13090$ vehicles/h

12.5 Effect of Application of BM Principle on Random Traffic Breakdown at Network Bottlenecks

The application of the BM principle for traffic assignment ensures the minimum of the probability $P_{net} = P_{net}^{(min)}$ that during a given observation time T_{ob} traffic breakdown occurs in at least one of the bottlenecks of a network. We consider the total network inflow rates Q (10.2) at which

$$0 < P_{net}^{(min)} < 1. \tag{12.24}$$

Under condition (12.24), there is at least one bottleneck k in the network at which during the time interval T_{ob} traffic breakdown can occur.

For example, in the network consisting of two routes (Fig. 12.3a) at the total network inflow rate $Q = Q_2 = 8640$ vehicles/h as follows from the analysis of the BM principle (12.21) together with (12.19), the breakdown probability P_{net} exhibits a minimum value that is equal to $P_{net} = P_{net}^{(min)} = 0.57$ (Fig. 12.6c). This minimum of the breakdown probability in the network is related to the optimum assignment between of the inflow rates $q_1 = 3994$ vehicles/h and $q_2 = 3546$ vehicles/h.

For these flow rates, the probabilities of traffic breakdown during the same observation time $T_{ob} = 30$ min are equal to $P^{(B,\,1)} = 0.23$, $P^{(B,\,2)} = 0.44$. We have found that under application of the BM principle there are simulation realizations in which traffic breakdown (F→S transition) occurs at none of the bottlenecks during the observation time $T_{ob} = 30$ min (Fig. 12.8a), simulation realizations in which traffic breakdown occurs at bottleneck 1 only (Fig. 12.8b), simulation realizations in which traffic breakdown occurs at bottleneck 2 only (Fig. 12.8c), simulation realizations in which traffic breakdown occurs at both bottlenecks 1 and 2 (Fig. 12.8d). Because a time delay for traffic breakdown at a bottleneck is a random

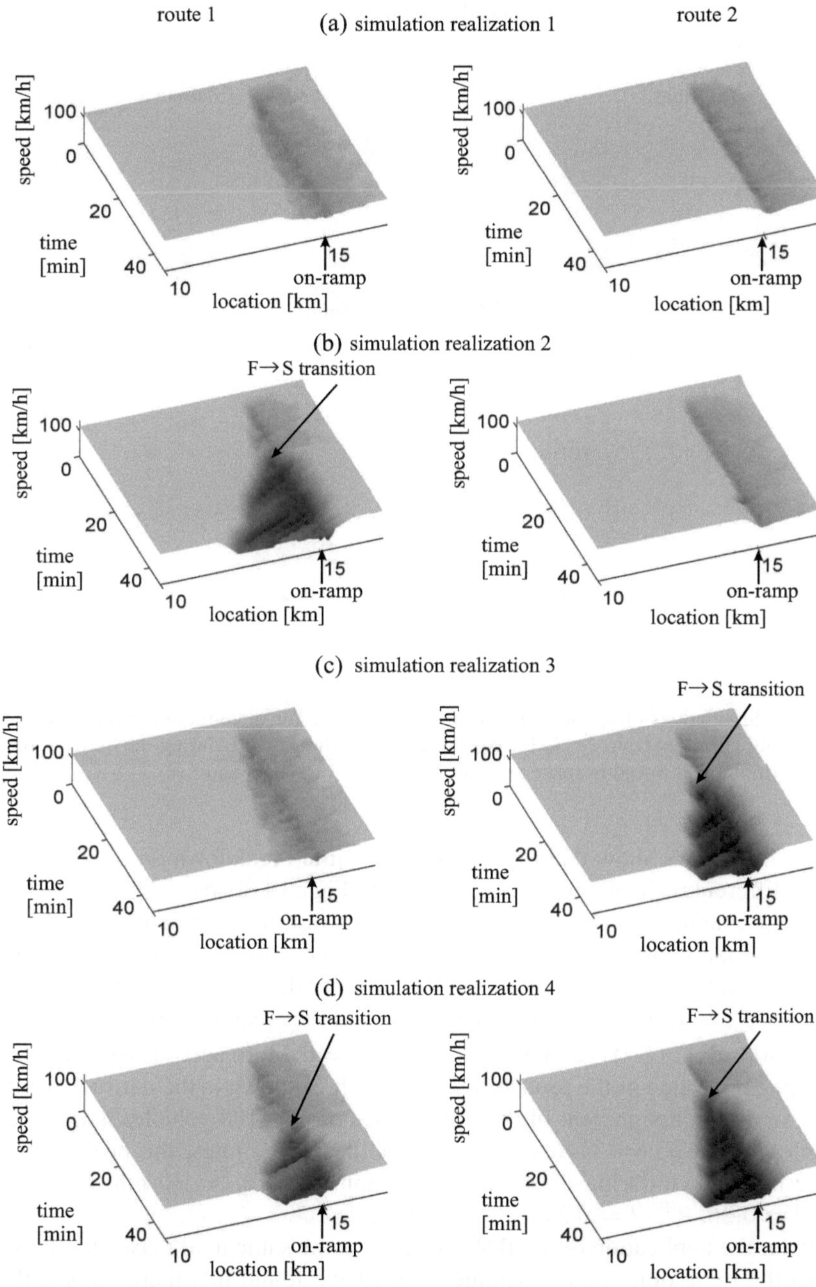

Fig. 12.8 Simulations of applications of the BM principle for two-route network (Fig. 12.3a): Speed in space and time on route 1 (*left panel*) and route 2 (*right panel*) for four different realizations (runs) simulated at the same model parameters. $Q = 8640$ vehicles/h. $P^{(\mathrm{B},k)} = 0.23$ ($k = 1$) and 0.44 ($k = 2$). $P_{\mathrm{net}}^{(\mathrm{min})} = 0.57$. Traffic breakdown at network bottleneck is labeled by "F→S transition"

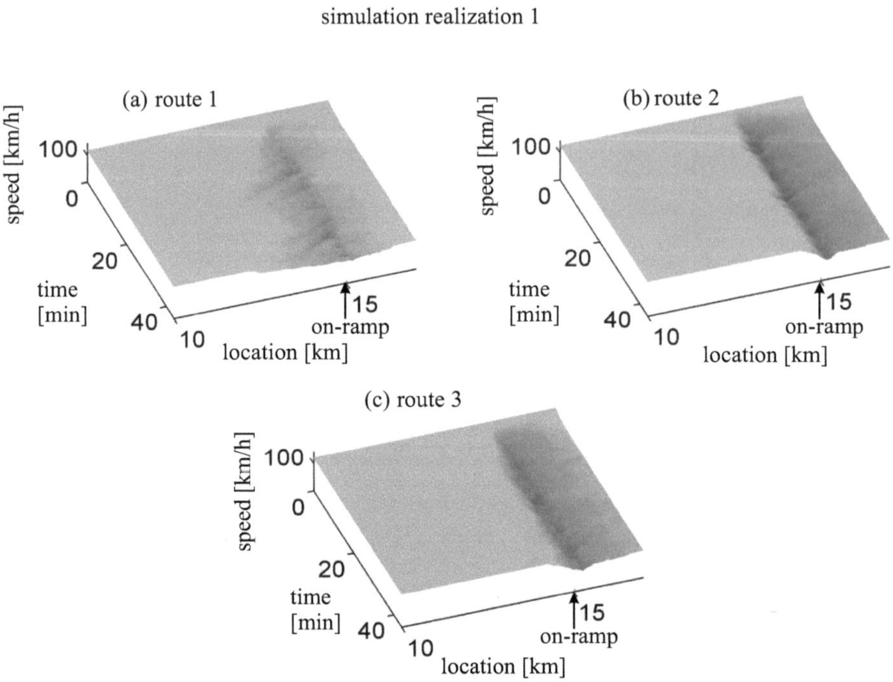

Fig. 12.9 Simulations of applications of the BM principle for three-route network (Fig. 12.4a). Simulation realization (run) 1: No breakdown occurs in the network during the observation time $T_{\rm ob} = 30$ min. (**a–c**) Speed in space and time on routes 1 (**a**), 2 (**b**), and 3 (**c**)

value, simulations show that time delays for traffic breakdown on different routes and in different realizations vary considerably (Figs. 12.8b–d).

The probabilistic behavior of a traffic network becomes considerably more complex if the number of network routes, for which condition (12.24) is satisfied, increases. Indeed, for three-route network (Fig. 12.4a) at the total network inflow rate $Q = Q_3 = 13090$ vehicles/h the minimum of the breakdown probability $P_{\rm net}$ for the network (Fig. 12.7) is equal to $P_{\rm net} = P^{\rm (min)}_{\rm net} = 0.74$. In accordance with the BM principle, this value of the probability of traffic breakdown in the network is related to the optimum assignment of the inflow rates $q_1 = 4188$ vehicles/h, $q_2 = 3856$ vehicles/h, and $q_3 = 3446$ vehicles/h. For these flow rates, the probabilities of traffic breakdown during the same observation time $T_{\rm ob} = 30$ min are equal to $P^{\rm (B,\,1)} = 0.26$, $P^{\rm (B,\,2)} = 0.513$, $P^{\rm (B,\,3)} = 0.328$.

Under the application of the BM principle, at the same total network inflow rate Q and the same other model parameters we have found that there are simulation realizations in which traffic breakdown (F→S transition) occurs at none of the bottlenecks during the observation time $T_{\rm ob} = 30$ min (Fig. 12.9), simulation realizations in which traffic breakdown occurs randomly at one of the bottlenecks only (Figs. 12.10, 12.11, and 12.12), simulation realizations in which traffic breakdown

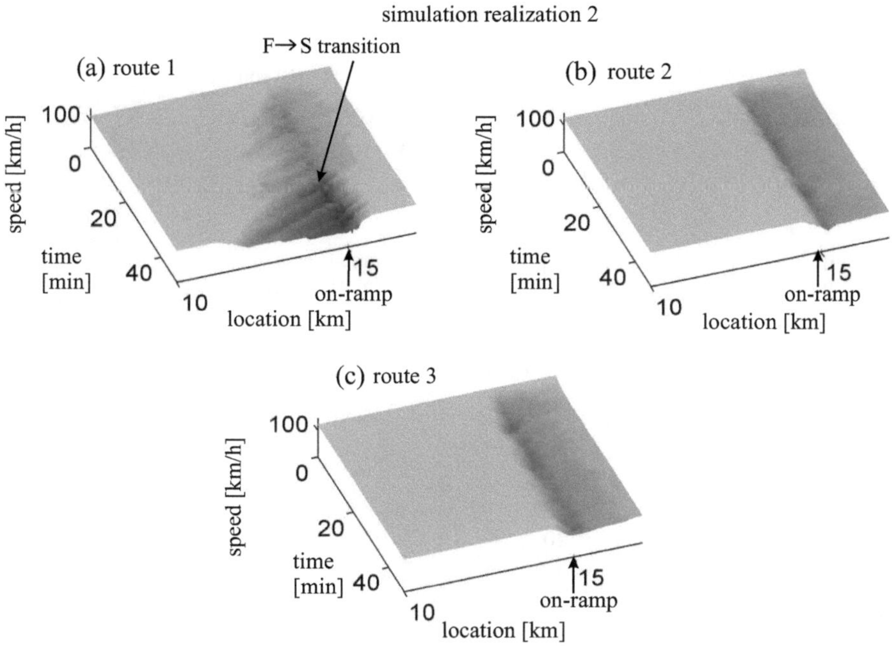

Fig. 12.10 Continuation of Fig. 12.9. Simulation realization (run) 2: Traffic breakdown occurs on route 1 only. Traffic breakdown at network bottleneck is labeled by "F→S transition"

occurs randomly at two of the bottlenecks only (Figs. 12.13, 12.14, and 12.15) as well as simulation realizations in which traffic breakdown occurs at all three bottlenecks 1, 2, and 3 (Fig. 12.16). In accordance with the three-phase theory (Sect. 5.4.6) [29, 30], simulations show that time delays for traffic breakdowns on different routes and in different realizations are random values that can differ each other considerably (Figs. 12.10 12.16).

12.6 Traffic Control in Framework of Three-Phase Theory

12.6.1 Congested Pattern Control Approach

The BM principle can be applied only when there is free flow at bottlenecks of a network. Therefore, a question arises.

- Can the BM principle be applied when due to traffic breakdown at some of the bottlenecks in a network, there is traffic congestion in a part(s) of the network?

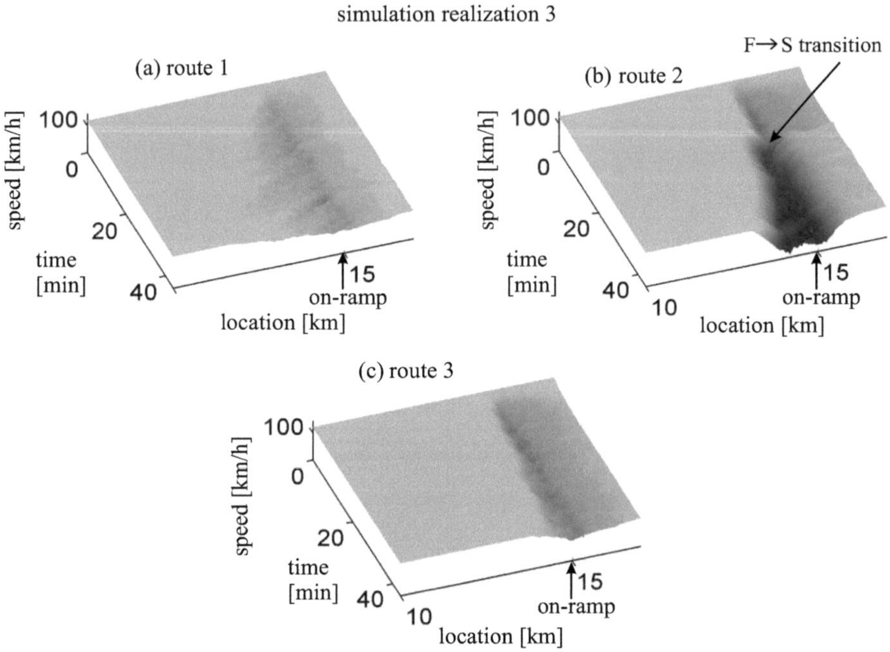

Fig. 12.11 Continuation of Fig. 12.9. Simulation realization (run) 3: Traffic breakdown occurs on route 2 only

As introduced in [21], if traffic breakdown has already occurred at a network bottleneck, then network optimization with the BM principle can consist of the stages:

(i) The dissolution of congestion due to control of traffic breakdown at congested bottlenecks. A consideration of congestion dissolution at network bottlenecks is the objective of this Sect. 12.6.
(ii) When values of the total network inflow rate Q are large enough, this congestion dissolution is not always possible to ensure. In this case, spatial limitation of congestion growth at network bottlenecks should be a target of traffic control. To reach this goal, firstly a detailed analysis of nonlinear features of the synchronized flow phase and wide moving jam phase in congested traffic is required. However, such an analysis is out of the scope of this book (see the book [14] and the three-phase theory for traffic congestion at heavy bottlenecks [19]).
(iii) The minimization of traffic breakdown probability with the BM principle in the remaining network, i.e., the network part that is not influenced by congestion.

Here, we show that the application of both the BM principle *and* control of traffic breakdown allows us to maintain free flow in the network during almost all time, even when traffic breakdowns occur randomly at network bottlenecks.

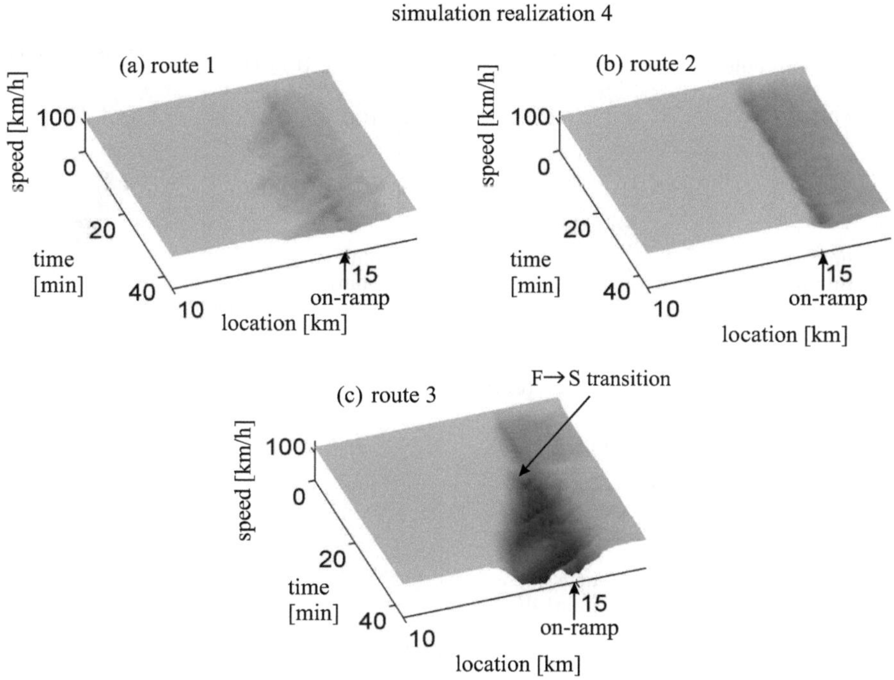

Fig. 12.12 Continuation of Fig. 12.9. Simulation realization (run) 4: Traffic breakdown occurs on route 3 only

To reach this goal, we will apply a "congested pattern control approach" [14–18]. In this approach, *no control* of traffic flow at a network bottleneck is realized as long as free flow is realized at the bottleneck. This means that the occurrence of random traffic breakdown is permitted to occur at the bottleneck: Congestion at the bottleneck is allowed to set in. The main idea of this approach is as follows. Only after traffic breakdown has occurred, traffic control starts. In this approach, two control scenarios are possible:

1. Due to the congested pattern control approach, congestion dissolves and free flow recovers at the bottleneck. In this case, the congested pattern control approach can also be called "control of traffic breakdown at bottleneck".
2. Due to the congested pattern control approach, congestion at the bottleneck is maintained at a minimum possible level.

Thus, in the congested pattern control approach only after traffic breakdown has occurred at a network bottleneck, traffic control is automatically switched on. The main aim of the congested pattern control approach is congestion dissolution. However, if congestion dissolution is not possible to achieve, then through the use of the congested pattern control approach the maintenance of congestion at an on-ramp bottleneck should be realized at a minimum possible level. The "minimum

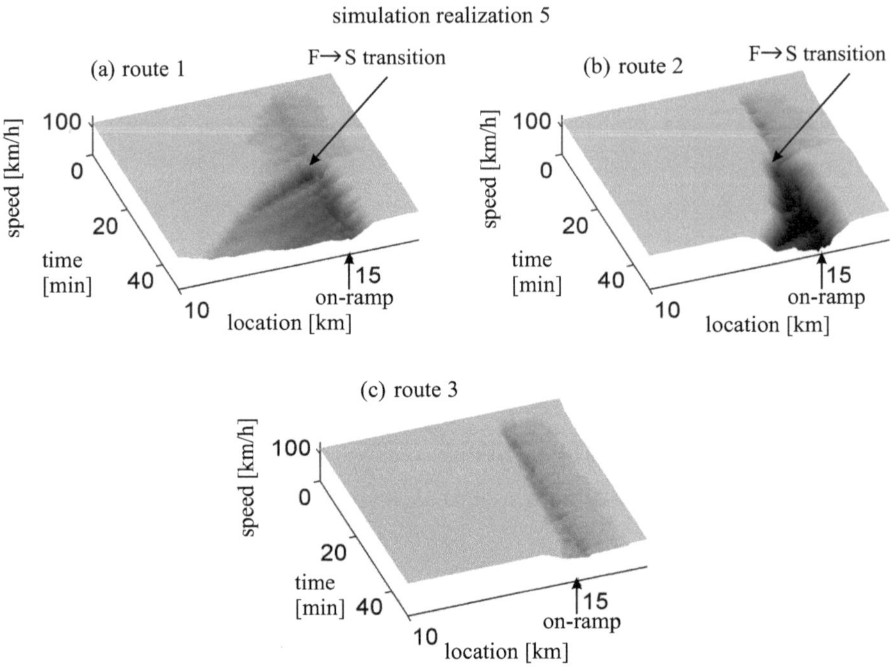

Fig. 12.13 Continuation of simulations shown in Fig. 12.9. Simulation realization 5: Traffic breakdown occurs on routes 1 and 2 only

possible level" of congestion means that traffic congestion should be localized at the bottleneck: Continuous congestion propagation upstream of the bottleneck should be prevented in the congested pattern control approach [14–18]. In other words, rather than propagating upstream, the congested pattern should be localized at the bottleneck within a relatively small stretch of the main road.

12.6.2 ANCONA On-Ramp Metering

When the congested pattern control approach is applied for on-ramp metering, it has been called "**a**utomatic o**n**-ramp **c**ontrol of c**on**gested p**a**tterns" (ANCONA) [14–18].

To register a random traffic breakdown with the subsequent formation of synchronized flow pattern (SP) at the bottleneck, in ANCONA there is feedback detector that location is upstream of the effective location of the bottleneck

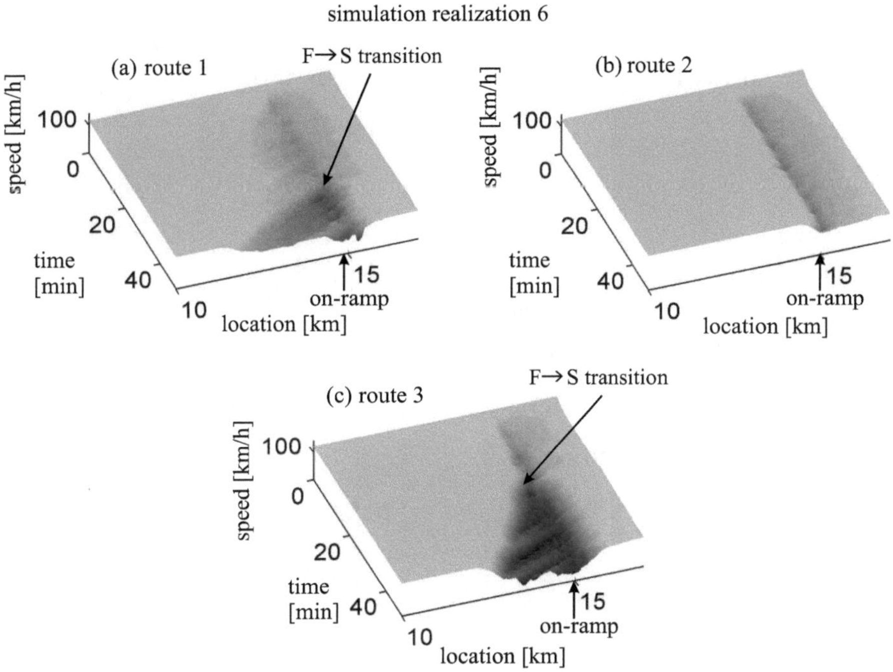

Fig. 12.14 Continuation of simulations shown in Fig. 12.9. Simulation realization 6: Traffic breakdown occurs on routes 1 and 3 only

(Fig. 12.17).[5] In accordance with the idea of the congested pattern approach, ANCONA does not restrict on-ramp inflow as long as free flow is measured at the bottleneck: There is no on-ramp control as long as free flow is measured at the bottleneck. This means that the flow rate of vehicles denoted by $q_{\text{on}}^{(\text{cont})}$, which can merge from the on-ramp onto the main road, is equal to

$$q_{\text{on}}^{(\text{cont})} = q_{\text{on}}, \tag{12.25}$$

where q_{on} is the flow rate to the on-ramp (Fig. 12.17).

Only after an SP due to random traffic breakdown is set in at the bottleneck, ANCONA begins to control this SP through the feedback decrease in on-ramp inflow. Due to traffic breakdown, the average speed denoted by $v^{(\text{det})}$ measured at the feedback control detectors upstream of the bottleneck (Fig. 12.17) decreases. If the

[5]In [17] it has been found that the closer the feedback detector location to the effective location of the bottleneck, the more efficient is ANCONA application. However, in any case the feedback detector location in ANCONA must be *upstream* of the effective bottleneck location. For simplicity, we study here only a case [14, 15] in which a feedback detector location in ANCONA is upstream of the bottleneck as shown in Fig. 12.17.

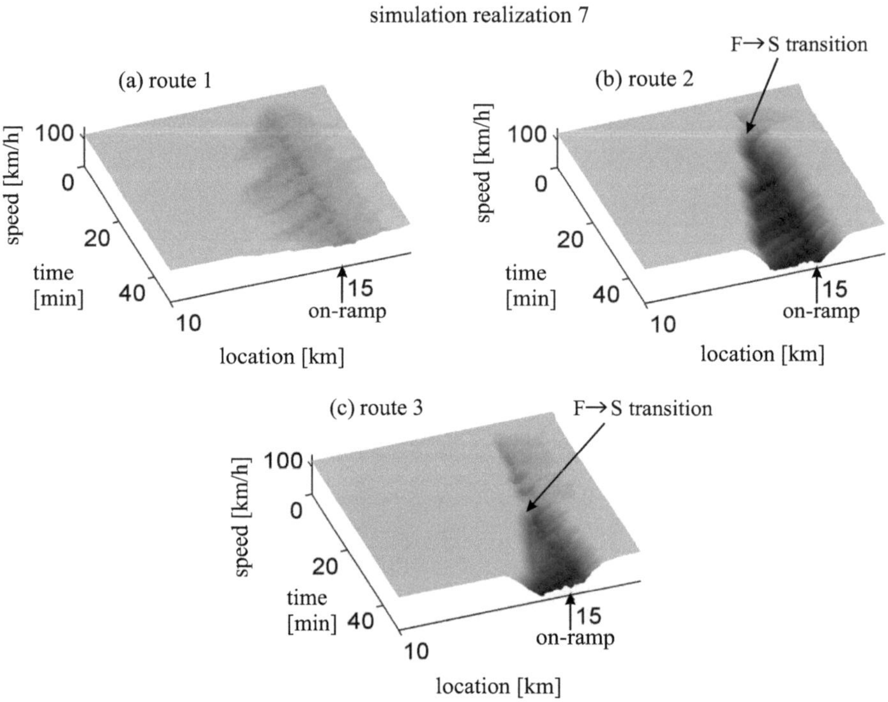

Fig. 12.15 Continuation of simulations shown in Fig. 12.9. Simulation realization 7: Traffic breakdown occurs on routes 2 and 3 only

average speed $v^{(\text{det})}$ is equal to or drops below a chosen "congestion speed" denoted by $v_{\text{cong, 1}}$, i.e.,

$$v^{(\text{det})} \leq v_{\text{cong, 1}}, \tag{12.26}$$

feedback on-ramp metering via ANCONA begins to perform. We have applied the following feedback control procedure: Through the use of traffic signal in the on-ramp line, ANCONA reduces the flow rate $q_{\text{on}}^{(\text{cont})}$ to a smaller flow rate denoted by $q_{\text{on 1}}$:

$$q_{\text{on}}^{(\text{cont})} = q_{\text{on 1}} < q_{\text{on}}. \tag{12.27}$$

Due to this decrease in the flow rate (12.27), ANCONA tries to achieve a return phase transition from synchronized flow to free flow (S→F transition) at the bottleneck. As a result of this S→F transition, the initial SP dissolves and the average speed at the detector $v^{(\text{det})}$ increases above some chosen value of the speed $v_{\text{cong, 2}}$:

$$v^{(\text{det})} > v_{\text{cong, 2}}. \tag{12.28}$$

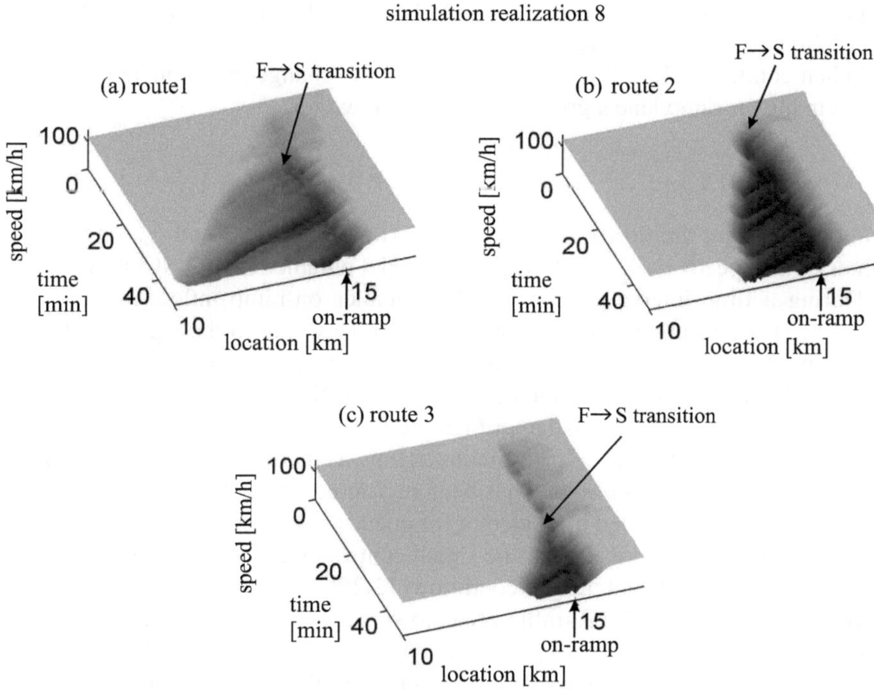

Fig. 12.16 Continuation of simulations shown in Fig. 12.9. Simulation realization 8: Traffic breakdown occurs on all routes 1, 2, and 3

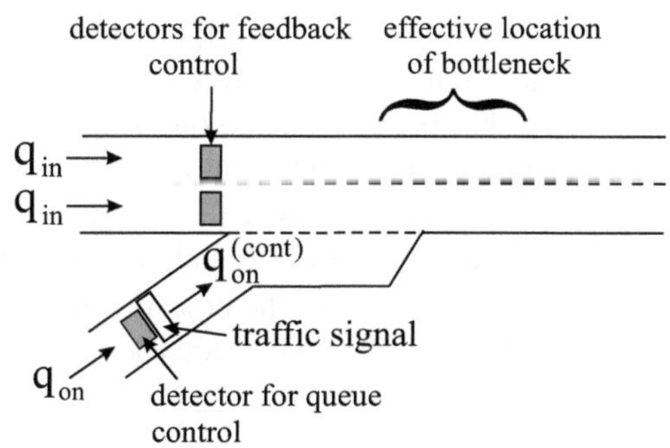

Fig. 12.17 Qualitative scheme of on-ramp bottleneck with upstream location of feedback control detectors for feedback on-ramp metering as well as with a detector in the on-ramp lane for queue control

In general, a value of the speed $v_{\text{cong},\,1}$ in (12.26) and a value of the speed $v_{\text{cong},\,2}$ in (12.28) can be chosen different ones.

When condition (12.28) is satisfied, through the change in parameters of traffic signal in the on-ramp lane a greater on-ramp inflow rate

$$q_{\text{on}}^{(\text{cont})} = q_{\text{on}\,2}, \tag{12.29}$$

is allowed, where the flow rate $q_{\text{on}\,2} > q_{\text{on}\,1}$ (flow rates $q_{\text{on}\,1}$ and $q_{\text{on}\,2}$ as well as values of the speeds $v_{\text{cong},\,1}$ and $v_{\text{cong},\,2}$ are model parameters of ANCONA).

During a time interval within which a smaller on-ramp inflow rate $q_{\text{on}\,1} < q_{\text{on}}$ (12.27) is permitted to merge onto the main road due to the signal in the on-ramp lane (Fig. 12.17), a vehicle queue is built at the signal in the on-ramp lane. To dissolve the queue under condition (12.28), ANCONA tries to apply the on-ramp inflow rate $q_{\text{on}\,2}$ (12.29) that is larger than the flow rate q_{on} upstream of the queue. The on-ramp flow rate $q_{\text{on}\,2} > q_{\text{on}}$ remains as long as there is still a vehicle queue at the signal. The queue at the signal has been registered through the use of a detector in the on-ramp lane (labeled by "detector for queue control" in Fig. 12.17). After the queue has dissolved, there is no traffic signal control in the on-ramp lane any more, i.e., ANCONA is switched off and condition (12.25) is satisfied.

If under condition (12.25) traffic breakdown occurs at the bottleneck later once more, an incipient SP begins to propagate upstream of the bottleneck. As a result, the speed at the feedback control detector decreases. Therefore, condition (12.26) is satisfied once more. This leads to a new decrease in $q_{\text{on}}^{(\text{cont})}$, and so on.

Results of ANCONA application are presented in Fig. 12.18.[6] We see that traffic congestion shown in Fig. 12.8 dissolves through the use of ANCONA. In Fig. 12.18, accordingly to formula (12.27) after the F→S transition has been registered on the main road at one of the bottlenecks k ($k = 1, 2$), due to traffic signal in the on-ramp lane, the on-ramp inflow rate decreases to a value $q_{\text{on}\,1}^{(k)} = q_{\text{on}}^{(k)} - \Delta q_{\text{on}}^{(\text{FS},k)}$. After a return S→F transition has been registered on the main road, accordingly to formula (12.29) due to a change in traffic signal parameters, the on-ramp inflow rate increases to a value $q_{\text{on}\,2}^{(k)} = q_{\text{on}}^{(k)} + \Delta q_{\text{on}}^{(\text{SF},k)}$. When in the queue at traffic signal only one vehicle remains, traffic signal is switched off and, therefore, the on-ramp inflow rate returns to the initial value $q_{\text{on}}^{(k)}$. When traffic breakdown occurs later once more at a network bottleneck, ANCONA dynamic feedback control of the breakdown starts automatically again as explained above.

[6]In simulations presented in Fig. 12.18, the location of detectors for feedback control is upstream of the effective bottleneck location (Fig. 12.17). On-ramp inflow bottleneck control starts *only* after traffic breakdown (F→S transition) is registered by the detectors for feedback control installed on the main road 100 m upstream of bottleneck location $x_{\text{on}} = 15$ km. Additionally to the detectors for feedback control installed on the main road, there is a detector in the on-ramp lane. Due to the detector in the on-ramp lane, the queue length of vehicles at traffic signals in the on-ramp lane is measured.

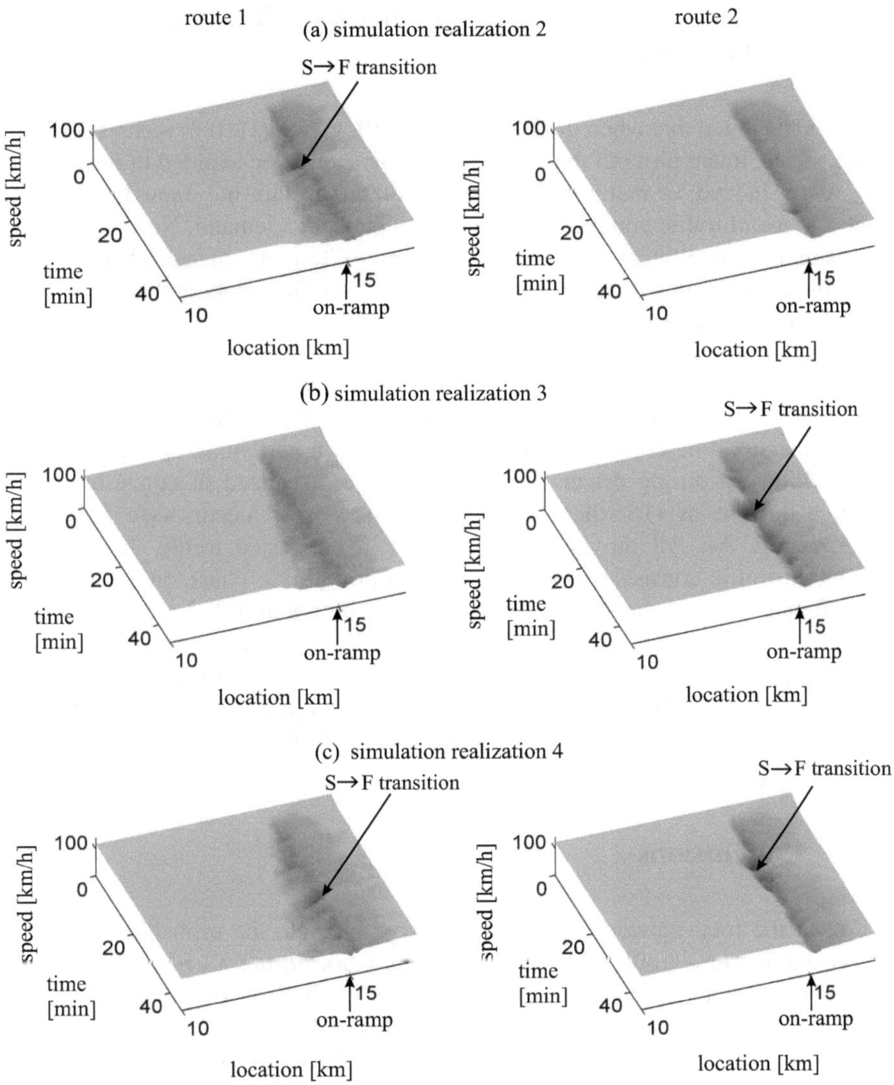

Fig. 12.18 Control of traffic breakdown in two-route network with the use of a combination of the BM principle and ANCONA. Simulations of the dissolution of traffic congestion shown in Fig. 12.8b–d through the use of on-ramp control ANCONA: Speed in space and time on route 1 (*left panel*) and route 2 (*right panel*) for different realizations 2–4 in Fig. 12.8 for which parameters of on-ramp control ANCONA are as follows: $q^{(k)}_{\rm on\ 1} = q^{(k)}_{\rm on} - \Delta q^{({\rm FS},\ k)}_{\rm on}$, $q^{(k)}_{\rm on\ 2} = q^{(k)}_{\rm on} + \Delta q^{({\rm SF},\ k)}_{\rm on}$, where $k = 1, 2$, $(\Delta q^{({\rm FS},\ 1)}_{\rm on}, \Delta q^{({\rm SF},\ 1)}_{\rm on}) = (73, 114)$ vehicles/h; $(\Delta q^{({\rm FS},\ 2)}_{\rm on}, \Delta q^{({\rm SF},\ 2)}_{\rm on}) = (250, 200)$ vehicles/h. $v_{\rm cong,\ 1} = v_{\rm cong,\ 2} = 72$ km/h. $Q = 8640$ vehicles/h. $P^{(\rm min)}_{\rm net} = 0.57$. Adapted from [32]

12.6.3 Enforcing Synchronized Flow Under Heavy Traffic Congestion

It should be noted that when the total network inflow rate Q (10.2) increases to very large values, it can turn out that congested traffic cannot be avoided in the network any more. Indeed, in many real traffic networks there are not enough alternative routes to avoid traffic congestion at large enough traffic demand. Nevertheless, the application of intelligent transportation systems (ITS) can change characteristics of traffic congestion with the objective to increase traffic safety and comfort while moving in heavy congested traffic.

To reach this goal, features of real congested traffic patterns that can be influenced by ITS should be understood. As follows from the analysis of spatiotemporal features of traffic congested patterns made in [14, 19, 20, 26, 28], these features can been understood based on the three-phase theory. In particular, as found in [26], if conditions of strong driver speed adaptation are realized in congested traffic through the use of ITS, then the jam-absorption effect occurs (see Sect. A.7.3 of Appendix A): All moving jams dissolve in congested traffic. In this case, congested traffic consists of the synchronized flow traffic phase only [26]. It has been found [12, 13] that fuel consumption can be about 1.2–2 times smaller in synchronized flow than that in congested traffic in which moving jams propagate. A development of effective methods of traffic control in networks in the case of extremely large total network inflow rates at which congestion propagates through the whole network will be a very interesting task for further investigations.

12.7 Conclusions

1. When the total network inflow rate is large enough, the probability of traffic breakdown in a traffic or transportation network during a given time interval becomes larger than zero. In this case, the application of the BM principle minimizes the probability of traffic breakdown in the network.
2. Complex probabilistic effects of random traffic breakdown at different network bottlenecks can occur, when the probability of traffic breakdown in the network is larger than zero.
3. If under application of the BM principle, nevertheless, traffic breakdown occurs at some of the network bottlenecks, through the application of congested pattern control approach the development of traffic congestion in the network can be prevented. In particular, we have shown that through the use of control of traffic breakdown (ANCONA) in a neighborhood of the bottleneck the dissolution of traffic congestion can be achieved.

References

1. B. Chen, W. Tong, W. Zhang, X. Sun, B. Wang, EPL **97**, 14001 (2012)
2. B. Chen, W. Tong, W. Zhang, X. Sun, B. Wang, Comput. Phys. Comm. **183**, 2081–2088 (2012)
3. R. Claes, T. Holvoet, and D. Weyns, IEEE Trans. ITS **12**, 364–373 (2011)
4. L.C. Davis, Physica A **388**, 4459 (2009)
5. L.C. Davis, Physica A **389**, 3588 (2010)
6. C. Dong, X. Ma, B. Wang, Phys. Lett. A **374**, 1326–1331 (2010)
7. C. Dong, X. Ma, B. Wang, Sci. China-Infor. Sci. **53**, 2265–2271 (2010)
8. C. Dong, X. Ma, G. Wang, B. Wang, X. Sun, Physica A **389**, 3274–3281 (2010)
9. Ch.M. Grinstead, J.L. Snell *Introduction to Probability*, 2nd edn. (American Mathematical Society, Rhode Island, 1997)
10. H. Guo, Z. Cao, J. Zhang, D. Niyato, U. Fastenrath, in *17th Inter. IEEE Conference on ITS (ITSC)*, (China, Qingdao, 2014), pp. 2180–2187
11. H. Guo, Z. Cao, J. Zhang, D. Niyato, M. Seshadri, U. Fastenrath, IEEE Trans. Emerg. Top. Comp. Intell. (2017). doi:10.1109/TETCI.2017.2665592
12. P. Hemmerle, M. Koller, H. Rehborn, B.S. Kerner, M. Schreckenberg, IET Intell. Transp. Syst. **10**, 122–129 (2016)
13. G. Hermanns, P. Hemmerle, H. Rehborn, M. Koller, B.S. Kerner, M. Schreckenberg, Transp. Res. Rec. **2490**, 47–55 (2015)
14. B.S. Kerner, *The Physics of Traffic* (Springer, Berlin, New York, 2004)
15. B.S. Kerner, Physica A **355**, 565–601 (2005)
16. B.S. Kerner, in *Traffic and Transportation Theory*, ed. by H.S. Mahmassani (Elsevier Science, Amsterdam, 2005), pp. 181–203
17. B.S. Kerner, IEEE Trans. ITS **8**, 308–320 (2007)
18. B.S. Kerner, Transp. Res. Rec. **1999**, 30–39 (2007)
19. B.S. Kerner, J. Phys. A Math. Theor. **41**, 215101 (2008)
20. B.S. Kerner, *Introduction to Modern Traffic Flow Theory and Control* (Springer, Heidelberg, Dordrecht, London, New York, 2009)
21. B.S. Kerner, J. Phys. A Math. Theor. **44**, 092001 (2011)
22. B.S. Kerner, Phys. Rev. E **84**, 045102(R) (2011)
23. B.S. Kerner, Traffic Eng. Contr. **52**, 379–386 (2011)
24. B.S. Kerner, in *Proc. of 2011 IEEE Forum on Integrated and Sustainable Transp. Systems* (Austria, Vienna, 2011), pp. 196–201
25. B.S. Kerner, in *Proc. of the 19th ITS World Congress* (Austria, Vienna, 2012), Paper No. EU-00190
26. B.S. Kerner, Phys. Rev. E **85**, 036110 (2012)
27. B.S. Kerner, Physica A **397**, 76–110 (2014)
28. B.S. Kerner, Phys. Rev. E **92**, 062827 (2015)
29. B.S. Kerner, Elektrotechnik Informationstechnik **132**, 417–433 (2015)
30. B.S. Kerner, Physica A **450**, 700–747 (2016)
31. B.S. Kerner, Eur. Phys. J. B **89**, 199 (2016)
32. B.S. Kerner, Physica A **466**, 626–662 (2017)
33. B.S. Kerner, S.L. Klenov, J. Phys. A Math. Gen. **35**, L31–L43 (2002)
34. B.S. Kerner, S.L. Klenov, Phys. Rev. E **80**, 056101 (2009)
35. B.S. Kerner, S.L. Klenov, D.E. Wolf, J. Phys. A Math. Gen. **35**, 9971–10013 (2002)
36. B.S. Kerner, M. Koller, S.L. Klenov, H. Rehborn, M. Leibel, Physica A **438**, 365–397 (2015)
37. B.S. Kerner, H. Rehborn, R.-P. Schäfer, S.L. Klenov, J. Palmer, S. Lorkowski, N. Witte, Physica A **392**, 221–251 (2013)
38. K. Lee, P. Hui, B. Wang, N. Johnson, J. Phys. Soc. Jpn. **70**, 3507–3510 (2001)
39. K. Tobita, T. Nagatani, Physica A **391**, 6137–6145 (2012)

40. J. Wahle, A.L.C. Bazzan, F. Klugl, M. Schreckenberg, Physica A **287**, 669 (2000)
41. J. Wahle, A. Bazzan, F. Klugl, M. Schreckenberg, Transp. Res. C **10**, 399–417 (2002)
42. W. Wang, B. Wang, W. Zheng, C. Yin, T. Zhou, Phys. Rev. E **72**, 066702 (2005)
43. X.-m. Zhao, D.-f. Xie, Q. Li, Comput. Phys. Comm. **187**, 106–114 (2015)

Chapter 13
Deterioration of Traffic System Through Standard Dynamic Traffic Assignment in Networks

13.1 Introduction—Wardrop's User Equilibrium (UE) and System Optimum (SO)

To find traffic optima and make an effective traffic control in traffic and transportation networks, a huge number of models for dynamic traffic assignment as well as other advanced traffic models have been introduced (see, e.g., [1, 2, 4–18, 21–33, 36–56, 59, 60, 62–74, 83, 85, 86, 88, 90–98, 100–125, 127–131, 133–152] and conference proceedings [3, 19, 20, 61, 84, 87, 89, 99, 126]).

Travel times and/or other travel costs in a network seems to be the most important parameters of traffic and transportation networks. Therefore, most of the traffic researchers and practitioners are strongly believe that the minimization of travel times and/or other travel costs in the network should be the explicit objective of any reasonable principle for dynamic control, optimization, and assignment in traffic and transportation networks. In other words, travel times or/and other travel costs a traffic or transportation network have been generally accepted in classical traffic and transportation theories to be self-evident traffic characteristics for objective functions used for the optimization of transportation networks, like dynamic traffic assignment. The main aim of the classical approaches is to minimize travel times or/and other travel costs in the network. The minimization of network travel times (or/and other travel costs) is the state-of-the-art in dynamic traffic assignment and control in traffic and transportation networks (see, e.g., [1, 2, 4–16, 24–27, 29–33, 36, 40–53, 55, 59, 60, 62, 63, 65–67, 69, 70, 72–74, 83, 85, 86, 91, 93, 95, 96, 98, 100–105, 107, 109–112, 114–121, 124, 127, 133–139, 142, 143, 148, 150–152] and conference proceedings [3, 19, 20, 61, 84, 87, 89, 99, 126]).

B.S. Kerner, *Breakdown in Traffic Networks*, DOI 10.1007/978-3-662-54473-0_13

The main objective of this chapter is to show that applications of this standard methodology of the minimization of network travel times deteriorates traffic system basically while provoking heavy traffic congestion in the network.

To show that applications of the standard methodology of the minimization of network travel times provokes heavy traffic congestion in a traffic or transportation network, we consider below the most famous approach to an analysis of traffic and transportation networks that is based on the Wardrop's user equilibrium (UE) and system optimum (SO) equilibrium introduced by Wardrop in 1952 [132]. The Wardrop's UE and SO are also called the Wardrop's principles or the Wardrop's equilibria.

Wardrop's UE Traffic on a network distributes itself in such a way that the travel times on all routes used from any origin to any destination are equal, while all unused routes have equal or greater travel times.

Wardrop's SO The network-wide travel time should be a minimum.

The Wardrop's principles reflect either the wish of drivers to reach their destinations as soon as possible (UE) or the wish of network operators to reach the minimum network-wide travel time (SO).

During last 50 years a huge number of theoretical works to dynamic traffic assignment and control in a traffic or transportation network based on the Wardrop's UE and SO have been made by several generations of researchers that are devoted to the minimization of route travel times (or/and other travel costs) in the network, in particular, in works by Merchant and Nemhauser [103, 104], Daganzo and Sheffi [29, 119], Daganzo et al. [30–33, 53], Bernstein and Friesz [41], Bell et al. [7–12, 14, 24, 25, 136–139], Mahmassani et al. [1, 26, 65, 98, 100–102, 105, 110, 111, 114, 143, 150, 151], Friesz et al. [40, 42–48, 55, 124, 127, 133] and by many other scientific groups (see, e.g., [2, 4–6, 13, 15, 16, 27, 36, 49–52, 59, 60, 62, 63, 66, 67, 69, 70, 72–74, 83, 85, 86, 91, 93, 95, 96, 107, 109, 112, 115–118, 120, 121, 134, 135, 142, 148, 152] and conference proceedings [3, 19, 20, 61, 84, 87, 89, 99, 126]).

However, as already explained in Sect. 1.7.4, network optimization approaches based on these classical traffic and transportation theories have failed by their applications in the real world. Even several decades of a very intensive effort to improve and validate network optimization models have no success. Indeed, there can be found no examples where on-line implementations of the network optimization models based on these fundamentals and methodologies could reduce congestion in real traffic and transportation networks.

In this chapter, we do not review other research that challenges the standard dynamic traffic assignment and optimization methodology. Such reviews can be found, for example, in [13, 41, 98, 112, 114, 115, 119]. Instead, we make a critical analysis of basic assumptions of this standard methodology that is the state-of-the-art in traffic and transportation research (see, e.g., [1, 2, 4–16, 24–

27, 29–33, 36, 40–53, 55, 59, 60, 62, 63, 65–67, 69, 70, 72–74, 83, 85, 86, 91, 93, 95, 96, 98, 100–105, 107, 109–112, 114–121, 124, 127, 133–139, 142, 143, 148, 150–152] and conference proceedings [3, 19, 20, 61, 84, 87, 89, 99, 126]). We show that applications of this standard dynamic traffic assignment for the minimization of network travel times (or/and other travel costs) provokes heavy traffic congestion in the network.

The chapter is organized as follows. Firstly, in Sect. 13.2 we make a general proof of the deterioration of the traffic system in an arbitrary traffic network occurring when dynamic traffic assignment is performed in the network based on the classical Wardrop's UE and SO. This general analysis is independent of a mathematical approach to the dynamic traffic assignment in which either the Wardrop's UE or SO is used. Then, in Sects. 13.3 and 13.4 we support results of this general analysis with numerical simulations of two simple two-route and three-rote networks that have already been used for a numerical study of the applications of the BM principle in Sect. 12.4. In Sect. 13.5 we consider the effect of traffic control on traffic breakdown at network bottlenecks under applications of the Wardrop's UE and ANCONA method of Sect. 12.6 at the same total network inflow rate Q (10.2) as that used in Fig. 12.18 associated with the application of the BM principle.

13.2 BM Principle Versus Wardrop's Equilibria: General Results

We consider the network model of Sect. 10.3. We assume that a steady state analysis of a traffic or transportation network can be applied (see footnote 9 of Sect. 1.7.3). The steady state analysis of the networks is a standard analysis of most application of the Wardrop's equilibria (e.g., reviews and books [13, 41, 98, 112, 114, 115, 119]). Under a steady state of a network, we can apply the measure of the network called the network capacity $C_{\rm net}$ (Sect. 11.3). In this Sect. 13.2, we consider only the case when the total network inflow rate Q (10.2) is smaller than the network capacity $C_{\rm net}$, i.e., condition (11.11) is satisfied.

In other words, a general analysis of the BM principle versus the Wardrop's equilibria is performed below *only* for the case $Q < C_{\rm net}$ (11.11). As we have shown in Sect. 11.3, in this case the application of the BM principle for "zero breakdown probability, i.e., the network throughput maximization approach (11.6) ensures that traffic breakdown cannot occur in the network. One of the benefits of this analysis is that we can find a general explanation of the deterioration of traffic system through application of the Wardrop's equilibria that is independent of network characteristics.

In Sect. 11.2 it has been shown that conditions (11.6) in the limit case $Z_1 = Z$ and $\epsilon^{(k)} \to 0$ are related to the maximization of the network throughput at which free flow conditions are still ensured in the whole network: Free flow is stable with respect to traffic breakdown at each of the network bottlenecks, i.e., no traffic breakdown can occur in the network. The physical sense of the limit case (in particular, the meaning of number Z of bottlenecks in a network related to the limit case of the application of the network throughput maximization approach) has been explained in Sect. 11.2. In its turn, in the limit case $Z_1 = Z$ and $\epsilon^{(k)} \to 0$ in a steady state analysis of the network, condition (11.11) is still valid that as above-mentioned is used below in this Sect. 13.2.

A real traffic or transportation network consists of alternative routes with very different lengths. Therefore, at small enough link inflow rates q_m (Sect. 10.3), there are routes with short travel times ("short routes") and routes with longer travel times ("long routes"). When the total network inflow rate Q and, consequently, values q_m increase, the minimization of travel times in the network with the use of dynamic traffic assignment based on the Wardrop's UE or SO leads to considerably larger increases in the flow rates on short routes in comparison with increases in the flow rates on long routes (see, e.g., [1, 2, 4–16, 24–27, 29–33, 36, 40–53, 55, 59, 60, 62, 63, 65–67, 69, 70, 72–74, 83, 85, 86, 91, 93, 96, 98, 100–105, 107, 109–112, 114–121, 124, 127, 133–139, 142, 143, 148, 150–152]).

Thus, through the application of the Wardrop's UE or SO, the flow rate on a short route should be larger than the flow rate on a long route. This is independent of values $C^{(k)}_{\rm min}$ for bottlenecks on the route. Contrarily, in accordance with conditions (11.6) resulting from the network throughput maximization approach, the flow rate at any network bottleneck $q^{(k)}_{\rm sum}$ should be smaller than the minimum capacity $C^{(k)}_{\rm min}$ of the bottleneck. This does not depend on whether a bottleneck is on a short route or the bottleneck is on a long route.

Therefore, at $Q \to C_{\rm net}$ (however, the total network inflow rate is still smaller than the network capacity: $Q < C_{\rm net}$ (11.11)), rather than conditions (11.6), any application of the Wardrop's UE or SO leads to conditions

$$q^{(k)}_{\rm sum} < C^{(k)}_{\rm min} \tag{13.1}$$

for some of the bottlenecks on long routes, whereas for some of the bottlenecks on short routes, we get

$$q^{(k)}_{\rm sum} > C^{(k)}_{\rm min}. \tag{13.2}$$

Condition (13.2) means that free flow becomes metastable one with respect to the breakdown at these bottlenecks. Therefore, in accordance with condition (13.2), traffic breakdown can occur at network bottlenecks on the short routes.

One of the consequences of this general conclusion is that already at relatively small total network inflow rates $Q < C_{\rm net}$ the application of the Wardrop's equilibria leads to the occurrence of congestion in urban networks. Even if the metastability of free flow with respect to traffic breakdown at network bottlenecks has been taken into account in travel time costs, the performance of applications of the Wardrop's equilibria with respect to the prevention of traffic breakdown in a network could *not* be achieved. We illustrate this conclusion in Sects. 13.3 and 13.4, in which simulations of the application of the Wardrop's equilibria to dynamic traffic assignment in two-route and three-route network models are shown: These simulations have been made with the Kerner-Klenov microscopic stochastic three-phase traffic flow model that shows the metastability of free flow with respect to an F→S transition (traffic breakdown) at network bottlenecks.

The above conclusion that at $Q \to C_{\rm net}$ any application of the Wardrop's equilibria results in the occurrence of the metastable free flow at some of the network bottlenecks can additionally be explained through a consideration of the following hypothetical network: In the network with different lengths of alternative routes, values of the minimum capacity $C_{\rm min}$ are assumed the same for any network bottleneck. We assume that our statement about the metastability of free flow at some of the network bottlenecks might be incorrect: An application of the Wardrop's UE or SO might result in conditions (11.6) at which free flow is stable with respect to traffic breakdown at each of the network bottlenecks. However, at $C^{(k)}_{\rm min} = C_{\rm min}$ and $\epsilon^{(k)} \to 0$ in (11.6), on network routes with different lengths the flow rates $q^{(k)}_{\rm sum}$ must be the same for all bottlenecks for which conditions (11.6) are satisfied. This contradicts the sense of any application of the Wardrop's UE or SO: On average, the flow rates should be larger on short routes then those on long routes.

Thus, due to the application of the Wardrop's UE or SO, already at $Q < C_{\rm net}$ (11.11) it turns out that for some of the network bottlenecks $q^{(k)}_{\rm sum} > C^{(k)}_{\rm min}$ (13.2). Therefore, traffic breakdown can occur at these bottlenecks. In this case, it is not possible to predict the time instant at which traffic breakdown occurs at these bottlenecks. This is because the time delay to the breakdown $T^{\rm (B)}$ is a random value (Sect. 5.4.4).

After traffic breakdown has occurred at a network bottleneck, we can apply the Wardrop's UE or SO for dynamic traffic assignment without any delay. This is possible because the speed decrease due to the breakdown at the network bottleneck can be measured. However, after the assignment has been made, there is always a time delay in the change of traffic variables at the bottleneck location. This time delay is caused by travel time from the beginning of a link to the bottleneck location on the link. Therefore, it is not possible to avoid congested traffic occurring due to traffic breakdown at the network bottleneck. A control method can only effect on a spatiotemporal distribution of congestion in the network. From this general analysis, we can see that the main benefit of the network throughput maximization approach in comparison with the Wardrop's equilibria is as follows.

The physics of traffic breakdown in traffic and transportation networks revealed in Sects. 11.2 and 11.3 shows that the wish of humans to use shortest routes of a network contradicts fundamentally another wish of humans to drive under free flow conditions in the network. Therefore, the use of the classical Wardrop's equilibria results basically in the occurrence of congestion

(continued)

in urban networks. This can explain why approaches to dynamic traffic assignment based on the classical Wardrop's equilibria (see, e.g., [1, 2, 4–16, 24–27, 29–33, 36, 40–53, 55, 59, 60, 62, 63, 65–67, 69, 70, 72–74, 83, 85, 86, 91, 93, 96, 98, 100–105, 107, 109–112, 114–121, 124, 127, 133–139, 142, 143, 148, 150–152] and conference proceedings [3, 19, 20, 61, 84, 87, 89, 99, 126]) have failed by their applications in the real world.

In Sects. 13.3 and 13.4, for the case $Q < C_{\rm net}$ (11.11) we confirm general results of Sect. 13.2 with numerical simulations of the models of two-route and three-route networks discussed in Sect. 12.4 (Figs. 12.3a,b and 12.4a). A comparison of the performance of the Wardrop's UE versus the BM principle for the case $Q > C_{\rm net}$ will be considered in Sect. 13.5.

13.3 Facilitation of Traffic Breakdown in Networks Through the Use of Wardrop's UE

13.3.1 Wardrop's UE in Simple Network Models

In a general consideration made in Sect. 13.2, we have shown that the Wardrop's equilibria should deteriorate the traffic system while leading to traffic congestion in the network already at the minimum possible theoretical value of the network capacity $C_{\rm net}$ (11.8). This minimum possible theoretical value of the network capacity $C_{\rm net}$ (11.8) is reached when free flow fluctuations in the whole network are large enough (Sect. 11.5.1). The term "minimum possible theoretical value of the network capacity" means that at $Q < C_{\rm net}$ free flow in the whole network is stable; therefore, no traffic breakdown can occur, *even when free flow fluctuations are very large*. For networks in Fig. 12.3a,b and Fig. 12.4a, the minimum possible theoretical value of the network capacity $C_{\rm net}$ is given by formula (12.14) in which $N = 2$ for two-route network and $N = 3$ for three-route network.

For two-route network (Fig. 12.3a) we get $C^{(k)}_{\rm min} = 3980$ for $k = 1$, 3760 vehicles/h for $k = 2$; for three-route network (Fig. 12.4a) we get $C^{(k)}_{\rm min} = 4000$ for $k = 1, 3740$ for $k = 2$, and 3860 vehicles/h for $k = 3$. Therefore, in accordance with Eq. (12.14) we get $C_{\rm net} = 7740$ vehicles/h for two-route and $C_{\rm net} = 11600$ vehicle/h for three-route networks, respectively.[1]

[1]In Sect. 12.4 we have mentioned that in the Kerner-Klenov model used for all simulations in this chapter conditions (5.13) are satisfied and, therefore, the threshold for spontaneous traffic breakdown at a bottleneck is determined by the threshold flow rate $q^{({\rm B},k)}_{\rm th}$ of bottleneck k. For this reason, in simulations with the Kerner-Klenov model the network capacity $C_{\rm net}$ is given by formula (12.17). The threshold flow rate $q^{({\rm B},k)}_{\rm th}$ of on-ramp bottleneck k is larger than the minimum

We consider the same parameters of the models of two-route and three-route networks as those mentioned in captions to Figs. 12.3 and 12.4. In particular, the density dependence of the free flow speed in the Kerner-Klenov stochastic microscopic three-phase traffic flow model (Appendix A) used for all simulations presented in this chapter is determined by formula (12.18); the same values of the on-ramp inflow rates at network bottlenecks are also used.

In many typical calculations of dynamic traffic assignment through the use of Wardrop's UE or SO traffic flow is optimized based on an assumption about a *monotonic* dependence of the vehicle speed on the flow rate on each link of the network (see, e.g., [13, 98, 112, 114, 115, 119]). However, as we have emphasized in Sects. 1.12 and 8.9, rather than a monotonic one, a dependence of the vehicle speed on the flow rate exhibits a *discontinuity*: The discontinuity is caused by the metastability of free flow at a bottleneck on a network link with respect to an F→S transition (traffic breakdown) at the bottleneck (Chap. 5). The Kerner-Klenov model used in all simulations below shows this discontinuity.

The Wardrop's UE for two-route network shown in Fig. 12.3a results in

$$T_1(q_1, q_{\rm on}^{(1)}) = T_2(q_2, q_{\rm on}^{(2)}), \quad q^{(o)} = \sum_{m=1}^{2} q_m, \tag{13.3}$$

where T_s, $s = 1, 2$ are travel times on routes $s = 1$ and $s = 2$, respectively. For three-route network (Fig. 12.4a) under conditions $q_m > 0$, $m = 1, 2, 3$ that are realized in all simulations presented below, we get

$$T_1(q_1, q_{\rm on}^{(1)}) = T_2(q_2, q_{\rm on}^{(2)}) = T_3(q_3, q_{\rm on}^{(3)}), \quad q^{(o)} = \sum_{m=1}^{3} q_m, \tag{13.4}$$

where T_s, $s = 1, 2, 3$ are travel times on routes 1,2, and 3, respectively.

To disclose the physics of the dynamic traffic assignment with the Wardrop's UE under condition $Q < C_{\rm net}$ (11.11), we consider and compare both a hypothetical case of a time-independent inflow rate $q^{(o)}$ at the origin of the network (Figs. 13.1, 13.2, and 13.3) with a more realistic case of a time-dependent inflow rate $q^{(o)}(t)$ (Figs. 13.4, 13.5, 13.6). In Figs. 13.4 and 13.5, in accordance with almost all empirical observations we simulate a morning (or evening) rush hour in which the inflow rate $q^{(o)}(t)$ firstly increases and then decreases over daytime (Figs. 13.4a and 13.5a).

capacity $C_{\rm min}^{(k)}$ of the bottleneck. This explains why the minimum possible theoretical value of the network capacity $C_{\rm net}$ (12.14) is considerably smaller than the network capacity $C_{\rm net}$ (12.17) following from simulations with the Kerner-Klenov model.

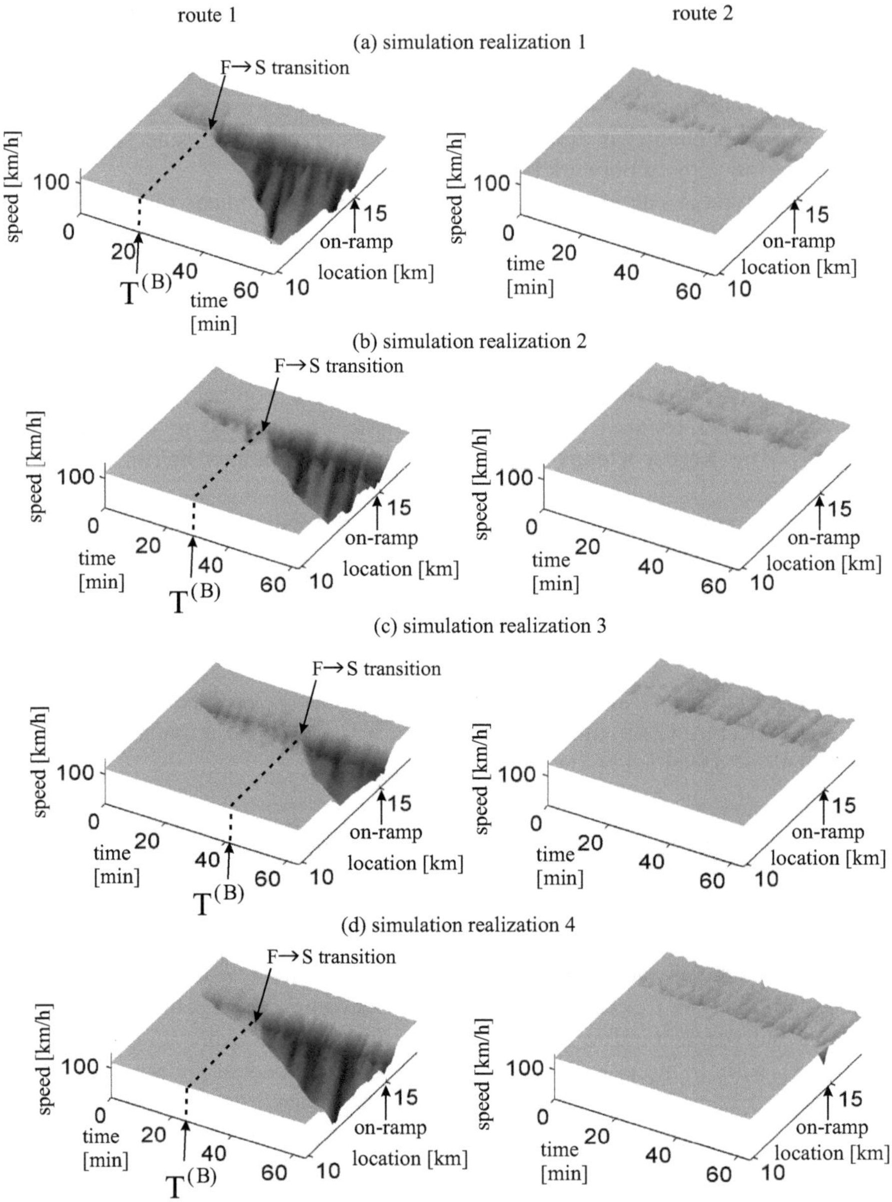

Fig. 13.1 Simulations of static traffic assignment with the use of the Wardrop's UE for network in Fig. 12.3a. (**a–d**) Speed in space and time on routes 1 (*left panel*) and 2 (*right panel*) for four different simulation realizations (runs) with time delays of the breakdown $T^{(B)} = 20$ min (realization 1), 29 min (realization 2), 42 min (realization 3), 24 min (realization 4). $Q = 7000$ vehicles/h. In accordance with (13.3), $q_1 = 4060$ vehicles/h, $q_2 = 1840$ vehicles/h. Arrows F→S show time instants of F→S transitions

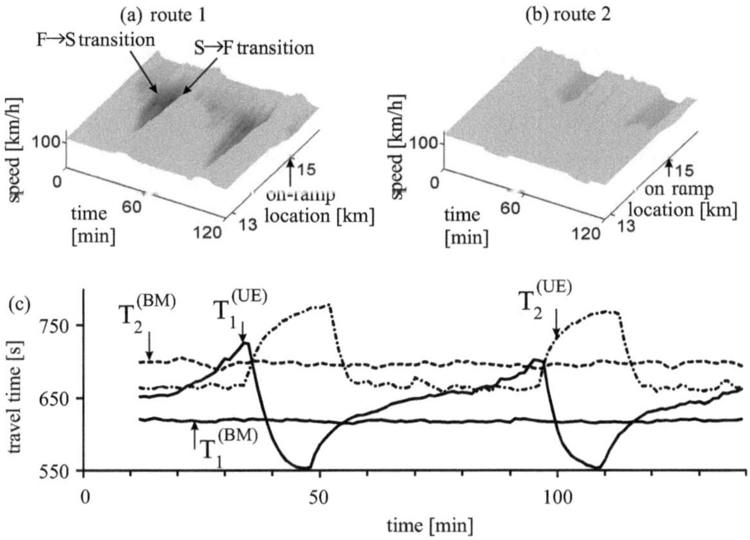

Fig. 13.2 Simulations of dynamic traffic assignment for two-route network (Fig. 12.3a) at time-independent inflow rate $q^{(o)}$. (**a, b**) Speed in space and time on routes 1 (**a**) and 2 (**b**) under application of the Wardrop's UE (13.3) and congested pattern control approach (Sect. 13.3.2) at $\Delta q = 1600$ vehicles/h. (**c**) Travel times $T_1 = T_1^{(UE)}(t)$, $T_2 = T_2^{(UE)}(t)$ for the Wardrop's UE. Travel times $T_1 = T_1^{(BM)}$, $T_2 = T_2^{(BM)}$ are related to application of the network throughput maximization approach (BM principle for "zero breakdown probability" $P_{net} = 0$ (10.8)) for which $q_1 = 3560$ vehicles/h, $q_2 = 2340$ vehicles/h. $Q = 7000$ vehicles/h. Arrows F→S and S→F show time instants of F→S and S→F transitions, respectively

However, before we consider *dynamic* traffic assignment in Sects. 13.3.2–13.3.4 below, here we discuss results of simulations of *static* traffic assignment in which the flow rate $q^{(o)}$ is time-independent and link inflow rates q_m that satisfy the Wardrop's UE at time instant $t = 0$ do not change over time (Fig. 13.1). For two route network (Fig. 12.3a), route 1 is shorter than route 2. To satisfy (13.3), the flow rate q_1 should be larger than the flow rate q_2. We have chosen the total network inflow rate $Q = 7000$ vehicles/h that is considerably smaller than the network capacity $C_{net} = 7740$ vehicles/h. In other words, condition $Q < C_{net}$ (11.11) is satisfied. Under this condition, from the Wardrop's UE (13.3), we get

$$q_1 = 4060 \text{ vehicles/h}, \; q_2 = 1840\text{vehicles/h} \tag{13.5}$$

at $q^{(o)}$ =5900 vehicles/h. Respectively, from (13.5) for the flow rates at network bottlenecks we get $q_{sum}^{(1)} = q_1 + q_{on}^{(1)}$ =4460 vehicles/h and $q_{sum}^{(2)} = q_2 + q_{on}^{(2)}$ =2540 vehicles/h, where we have taken into account that $q_{on}^{(k)} = 400$ (k =1) and 700 (k =2) vehicles/h (see caption to Fig. 12.3a).

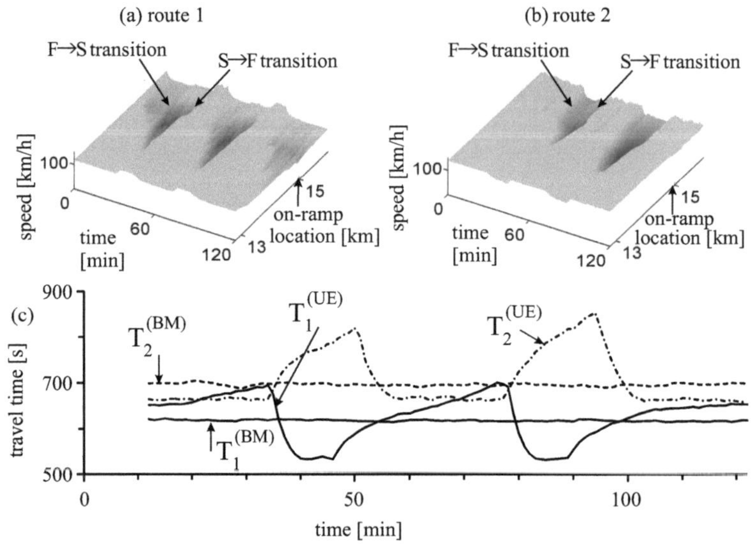

Fig. 13.3 Simulations of dynamic traffic assignment for two-route network (Fig. 12.3a) at time-independent inflow rate $q^{(o)}$. (**c**) Travel times $T_1 = T_1^{(UE)}(t)$, $T_2 = T_2^{(UE)}(t)$ under application of the Wardrop's UE (13.3) and congested pattern control approach (Sect. 13.3.2) at $\Delta q = 2000$ vehicles/h. As in Fig. 13.2, travel times $T_1 = T_1^{(BM)}$, $T_2 = T_2^{(BM)}$ are related to application of the network throughput maximization approach (BM principle for "zero breakdown probability" (10.8)) for which $q_1 = 3560$ vehicles/h, $q_2 = 2340$ vehicles/h. $Q = 7000$ vehicles/h. Arrows F→S and S→F show time instants of F→S and S→F transitions, respectively

Therefore, for the chosen total network inflow rate $Q = 7000$ vehicles/h and the network capacity $C_{net} = 7740$ vehicles/h we get

$$q_{sum}^{(1)} > C_{min}^{(1)}, \tag{13.6}$$

$$q_{sum}^{(2)} < C_{min}^{(2)}, \tag{13.7}$$

where we have taken into account that $q_{sum}^{(1)} = 4460$ vehicles/h, $C_{min}^{(1)} = 3980$ vehicles/h and $q_{sum}^{(2)} = 2540$ vehicles/h, $C_{min}^{(2)} = 3760$ vehicles/h.

Condition (13.6) is related to a general conclusion (13.2) of any application of the Wardrop's equilibria: At the total network inflow rate Q that is considerably smaller than the network capacity C_{net} the application of the Wardrop's UE can provoke traffic congestion the network. Indeed, we have found that the probability of traffic breakdown at bottleneck 1 is equal to 0.59. Therefore, under application of static traffic assignment with the Wardrop's UE we have found a random time-delayed traffic breakdown (F→S transition) at bottleneck 1 leading to traffic congestion (Fig. 13.1).

It should be noted that results of simulations presented in Fig. 13.1 satisfy the Wardrop's UE (13.3) *only* as long as no change in route travel times occurs over

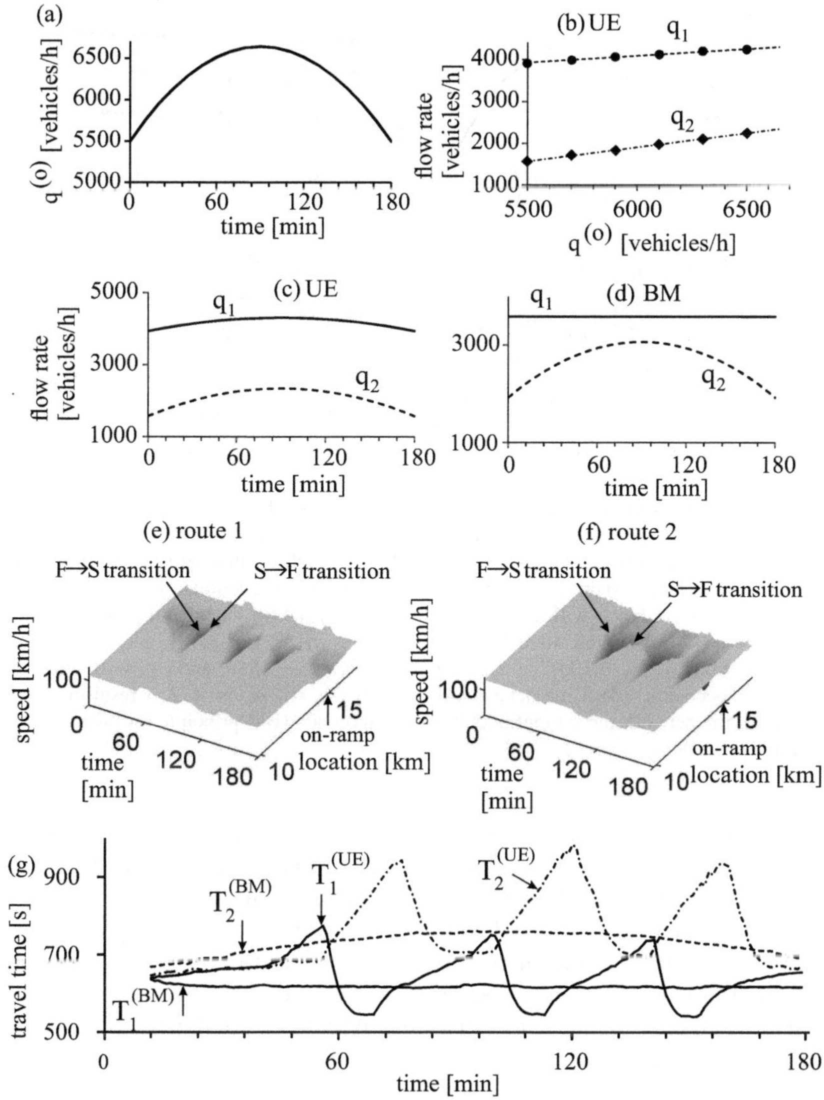

Fig. 13.4 Simulations of dynamic traffic assignment for two-route network (Fig. 12.3a) at time-dependent inflow rate $q^{(o)}(t)$ resulting in the maximum total network inflow rate $Q = 7736$ vehicles/h that satisfies condition (11.11) (**a**). (**b, c**) Flow rates q_m, $m = 1, 2$ for the Wardrop's UE as functions of $q^{(o)}$ (**b**) and on time (**c**) calculated in accordance with (13.3) (see footnote 4); in (**b**), calculated points for the flow rates q_m, $m = 1, 2$ are well fitted by associated lines $q_1(q^{(o)})$ and $q_2(q^{(o)})$ that are given by equations $q_1 = 0.33q^{(o)} + 2170$ and $q_2 = 0.67q^{(o)} - 2170$ vehicles/h, respectively. (**d**) Time-dependent flow rates $q_m(t)$, $m = 1, 2$ for the network throughput maximization approach. (**e, f**) Speed in space and time on routes 1 (**e**) and 2 (**f**) under application of the Wardrop's UE (13.3) and congested pattern control approach (Sect. 13.3.2) at $\Delta q = 2000$ vehicles/h. (**g**) $T_s = T_s^{(UE)}(t)$ on different routes $s = 1, 2$ for the Wardrop's UE. In (**g**), travel times $T_s = T_s^{(BM)}(t)$ on different routes $s = 1, 2$ are related to application of the network throughput maximization approach (BM principle for "zero breakdown probability" (10.8)) at which no traffic breakdown occurs at the bottlenecks

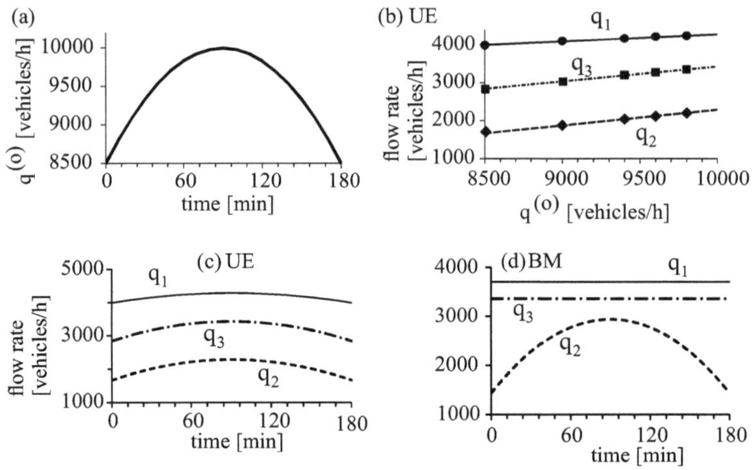

Fig. 13.5 Simulations of distribution of flow rates between different routes through dynamic traffic assignment in three-route network (Fig. 12.4a) at time-dependent inflow rate $q^{(o)}(t)$ resulting in the maximum total network inflow rate $Q = 11594$ vehicles/h that satisfies condition (11.11) (**a**). (**b, c**) Flow rates q_m, $m = 1, 2, 3$ for the Wardrop's UE as functions of $q^{(o)}$ (**b**) and on time (**c**) calculated in accordance with (13.4) (see footnote 4); in (**b**), calculated points for the flow rates q_m, $m = 1, 2, 3$ are well fitted by associated lines $q_1(q^{(o)})$, $q_2(q^{(o)})$, and $q_3(q^{(o)})$ that are given by equations $q_1 = 0.19q^{(o)} + 2345$, $q_2 = 0.41q^{(o)} - 1850$, and $q_3 = 0.39q^{(o)} - 495$ vehicles/h, respectively. (**d**) Time-dependent flow rates $q_m(t)$, $m = 1, 2, 3$ resulting from the application of the network throughput maximization approach (BM principle for "zero breakdown probability" (10.8))

time. This is because in the case of *static* traffic assignment the flow rates q_1 and q_2 that satisfy the Wardrop's UE (13.3) do not change over time. In Fig. 13.1, this is approximately valid as long as no traffic breakdown occurs at network bottlenecks: When traffic breakdown has occurred at a network bottleneck, the associated route travel time increases over time due to congestion propagation. Thus, after traffic breakdown has occurred at one of the network bottlenecks, rather than static traffic assignment, to satisfy the Wardrop's equilibria, dynamic traffic assignment is needed.

13.3.2 Dynamic Traffic Assignment with Congested Pattern Control Approach

We have proven different control methods to decrease congestion in the network occurring due to dynamic traffic assignment with the Wardrop's UE or SO: As explained above, we have found that regardless of control method it is not possible to avoid congestion fully.

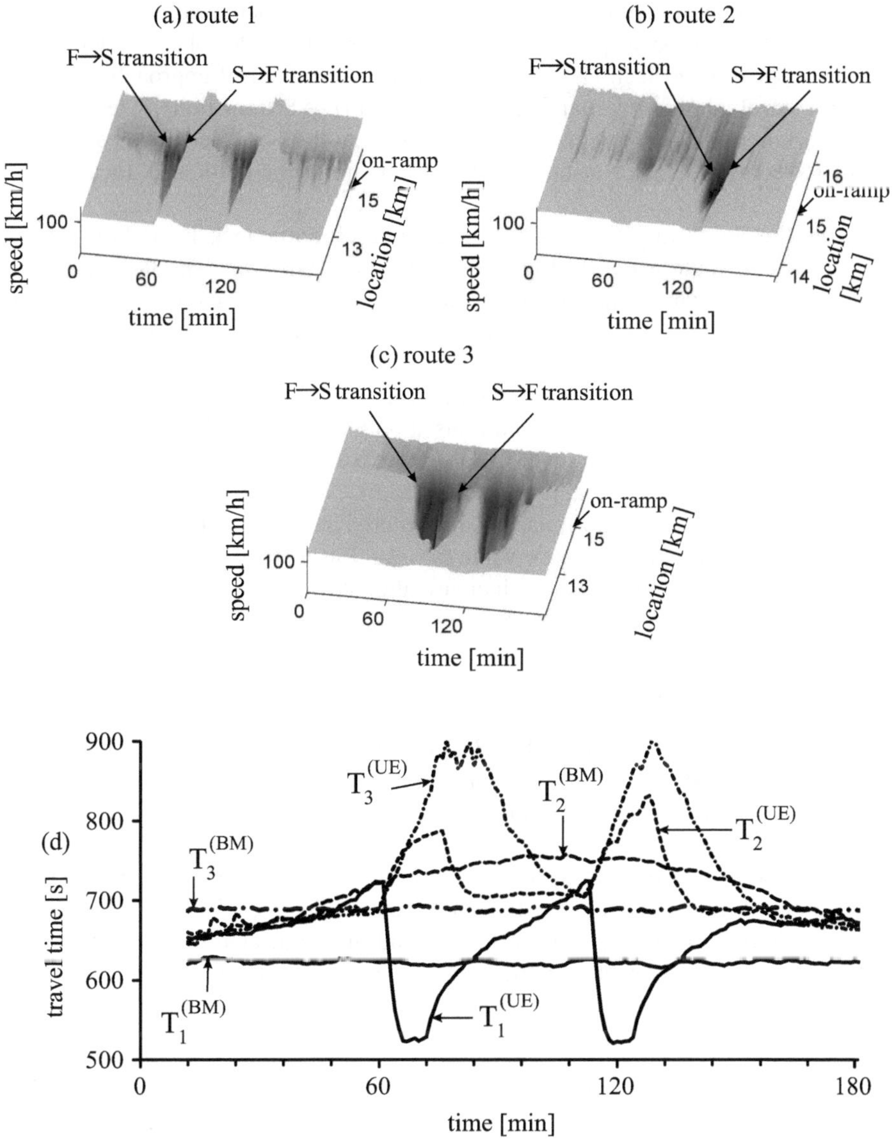

Fig. 13.6 Simulations of dynamic traffic assignment in three-route network accordingly to the flow rates on different routes that calculations are shown in Fig. 13.5: (**a**–**c**) Speed in space and time on routes 1 (**a**), 2 (**b**), and 3 (**c**) under application of the Wardrop's UE (13.4) and congested pattern control approach (Sect. 13.3.2) at $\Delta q = 2400$ vehicles/h. (**d**) $T_s = T_s^{(\mathrm{UE})}(t)$ on different routes ($s = 1, 2, 3$) for the Wardrop's UE. In (**d**), travel times $T_s = T_s^{(\mathrm{BM})}(t)$ ($s = 1, 2, 3$) are related to application of the network throughput maximization approach (BM principle for "zero breakdown probability" (10.8)) at which no traffic breakdown occurs at the bottlenecks

Some of the best results shown in Figs. 13.2–13.6 are achieved with "congested pattern control approach" [75–79] (see explanations of the idea of congested pattern control approach in Sect. 12.6). In congested pattern control approach[2] feedback traffic control starts *only* after traffic breakdown (F→S transition) is registered at a network bottleneck. To register the breakdown at the bottleneck, we have used a feedback control detector installed 500 m upstream of the bottleneck location $x = x_{\mathrm{on}}$.

After the F→S transition has been registered on bottleneck 1, the flow rates q_1 and q_2 of two-route network (Fig. 12.3a) have been changed: The flow rate q_1 on route 1 decreases on a value Δq and the flow rate q_2 on route 2 increases on the same value Δq (Figs. 13.2, 13.3, and 13.4). In other words, the flow rate q_1 decreases to $q_1 - \Delta q$ and the flow rate q_2 increases to $q_2 + \Delta q$. After a return S→F transition has been registered at bottleneck 1, the flow rates q_m, $m = 1, 2$ return to their values found from (13.3) for free flow conditions at both network bottlenecks.

For three-route network (Fig. 12.4a), we have used a similar application of congested pattern control approach (Fig. 13.6). When an F→S transition has occurred at bottleneck 1 and it has been registered at the feedback control detector, the flow rate q_1 decreases to a new value $q_1 - \Delta q$. Respectively, the flow rates q_2 and q_3 on other routes (routes 2 and 3) increase altogether on the value Δq; new values of q_2 and q_3 are found from equations

$$T_2(q_2, q_{\mathrm{on}}^{(2)}) = T_3(q_3, q_{\mathrm{on}}^{(3)}), \; q_2 + q_3 = q^{(\mathrm{o})} - (q_1 - \Delta q)$$

following from the Wardrop's UE (13.4) for these routes for free flow conditions. After a return S→F transition has been registered at bottleneck 1, the flow rates q_m, $m = 1, 2, 3$ return to their values found from (13.4) for free flow conditions at all network bottlenecks.

When an F→S transition occurs on bottleneck 1 later once more, traffic breakdown is registered by the feedback control detector. This leads to a new decrease in the flow rate q_1 on a value Δq and the associated increase in the flow rates on other network routes on the same value Δq, and so on.[3]

---

[2]A comparison of the application of congested pattern control approach used in the main text of the book with a step-by-step method for the searching of the Wardrop's UE can be found in Appendix C.

[3]In the case under consideration, the location of bottleneck 1 is $x = x_{\mathrm{on}} = 15$ km. An F→S transition is registered by the feedback control detector when the speed at the detector $v \leq 80$ km/h (for two-route network) and $v \leq 70$ km/h (for three route network) during 3 min. A return S→F transition is registered by the feedback control detector when the speed at the detector $v \geq 92$ km/h during 2 min.

13.3.3 *Dynamic Traffic Assignment Under Time-Independent Total Network Inflow Rate*

In Fig. 13.2, we have applied *dynamic* traffic assignment of the flow rates q_1 and q_2 in (13.3) based on congested pattern control approach as explained in Sect. 13.3.2. In this dynamic feedback control method, bottleneck control starts *only* after a random traffic breakdown is registered by the road detector. In our case, after delay time $T_1^{(B)} = 20$ min traffic breakdown has occurred at bottleneck 1 (labeled by "F→S transition" in Fig. 13.2a). A congested pattern resulting from the breakdown has been registered by the detector. Without any delay, the flow rate q_1 decreases to a value $q_1 - \Delta q$ and the flow rate q_2 increases to the value $q_2 + \Delta q$.

In Fig. 13.2, we have used a value Δq at which congested patterns have been dissolved at bottleneck 1 and no traffic breakdown is realized at bottleneck 2. Thus, after traffic breakdown has occurred at bottleneck 1, due to the decrease in the flow rate on route 1 to $q_1 - \Delta q$, the congested pattern at bottleneck 1 has dissolved over time (labeled by "S→F transition" in Fig. 13.2a). As a result, free flow has returned at bottleneck 1. Therefore, in accordance with the control method the flow rates q_1 and q_2 return to their values found from Eq. (13.3). Then, after another random time delay traffic breakdown has occurred at bottleneck 1 once more (Fig. 13.2a). A congested pattern resulting from this second breakdown has been registered by the detector. Without any delay, the the flow rate q_1 decreases to a value $q_1 - \Delta q$ and the flow rate q_2 increases to the value $q_2 + \Delta q$. Over time the second congested pattern has dissolved (Fig. 13.2a) and free flow has returned at bottleneck 1, and so on.

Because traffic breakdown occurs at the bottleneck after a random value of delay time $T^{(B)}$, the common result of the application of the Wardrop's UE principle for dynamic traffic assignment is a *random process* of the congested pattern emergence with the subsequent dissolution of the congested pattern due to the application of congested pattern control approach. For this reason, in different simulation realizations we have found different sequences of the congested pattern emergence and dissolution.

In Fig. 13.3, in comparison with the value Δq used in simulations shown in Fig. 13.2, we have increased Δq to provide a faster congestion dissolution. Then, the mean duration of congested traffic at bottleneck 1 becomes shorter. However, traffic breakdown with resulting congested pattern formation is realized at bottleneck 2. In accordance with congested pattern control approach, the flow rates q_1 and q_2 return to their values found from Eq. (13.3) when free flow is measured at bottleneck 1, and so on. In both cases (Figs. 13.2 and 13.3), the random process of a random traffic breakdown with the subsequent dissolution of traffic congestion due to the application of dynamic traffic assignment leads to large oscillations of travel times $T_1^{(UE)}$ and $T_2^{(UE)}$ (Figs. 13.2c and 13.3c).

Thus, we can conclude that the application of the Wardrop's UE for dynamic traffic assignment results in a *random process* of the congested pattern emergence due to an F→S transition with the subsequent dissolution of the pattern due to a return S→F transition, and so on (Figs. 13.2 and 13.3): In different simulation realizations, we have found different sequences of the congested pattern emergence and dissolution. As in many other applications of the Wardrop's UE and other approaches to the minimization of route travel times [34, 35, 88, 129, 130], this random process leads to large oscillations of travel times $T_1^{(UE)}$ and $T_2^{(UE)}$ (Figs. 13.2c and 13.3c).

In contrast with the application of the Wardrop's UE for dynamic traffic assignment (Figs. 13.2 and 13.3), no traffic breakdown and, therefore, no traffic congestion occurs under application of the network throughput maximization approach (BM principle for "zero breakdown probability" (10.8)). For this reason, travel times $T_1^{(BM)}$ and $T_2^{(BM)}$ for dynamic traffic assignment with BM principle are almost time-independent (Figs. 13.2c and 13.3c).

13.3.4 Dynamic Traffic Assignment Under Time-Dependent Total Network Inflow Rate

13.3.4.1 Application of Wardrop's UE

In Fig. 13.4, the flow rate at the origin of the network models (Figs. 12.3a and 12.4a) depends on time: $q^{(o)} = q^{(o)}(t)$. Therefore, the total network inflow rate $Q(t)$ (10.2) depends also on time.[4] We consider only such time-dependencies of the total network inflow rate $Q(t)$ for which the flow rate $Q(t)$ does not exceed the minimum possible theoretical value of the network capacity C_{net} (12.14) that for two-route network is equal to $C_{net} = 7740$ vehicles/h.

We can see (Fig. 13.4e,f) that even in this case dynamic traffic assignment under application of the Wardrop's UE together with the congested pattern control approach leads to qualitatively the same random process of sequences of the congested pattern emergence and dissolution with large oscillations of route travel times.

In Fig. 13.5a, simulations of dynamic traffic assignment under application of the Wardrop's UE together with the congested pattern control approach for three-route network (Fig. 12.4a) are presented. In this case, the total network inflow

---

[4]The dynamic traffic assignment of the link inflow rates q_m ($m = 1, 2$ in Fig. 12.3a and $m = 1, 2, 3$ in Fig. 12.4a) under time-dependent flow rate $q^{(o)}(t)$ (Figs. 13.4–13.6) is performed with the use of standard methods [13, 98, 112, 114, 115, 119]: For some flow rates $q^{(o)}$, values q_m have been found in accordance with the Wardrop's UE (points in Figs. 13.4b and 13.5b). Simulations show that these points are well fitted with lines $q_m(q^{(o)})$ given in captions to Figs. 13.4 and 13.5. These lines have further been used for calculations of q_m under time-dependent flow rate $q^{(o)}(t)$.

rate $Q(t)$ depends also on time. The total network inflow rate $Q(t)$ does not also exceed the minimum possible theoretical value of the network capacity $C_{\rm net}$ (12.14) that for three-route network (Fig. 12.4a) is equal to $C_{\rm net} = 11600$ vehicles/h.

A more complex network structure (Fig. 12.4a) results in a more sophisticated spatiotemporal distribution of congestion between different routes (Fig. 13.6a–c). Nevertheless, we find qualitatively the same random process of sequences of the congested pattern emergence and dissolution on different routes with oscillations of route travel times (Fig. 13.6d) as that in a simpler two-route network (Fig. 13.4).

13.3.4.2 Application of Network Throughput Maximization Approach

Due to the application of the network throughput maximization approach (Sect. 11.2), condition $q^{(1)}_{\rm sum} + \epsilon^{(1)} = C^{(1)}_{\rm min}$ (where $0 < \epsilon^{(1)}/C^{(1)}_{\rm min} \ll 1$, see formulas (11.6) and (11.7)) is satisfied for bottleneck 1 only. Contrarily, for bottleneck 2 condition $q^{(2)}_{\rm sum} < C^{(2)}_{\rm min}$ is valid. Therefore, the increase in $q^{(\rm o)}(t)$ leads to the increase in q_2 on the alternative route 2 only. For this reason, from Fig. 13.4d,g we can see that as long as the inflow rate $q^{(\rm o)}$ increases over time the flow rate q_1 and travel time $T^{(\rm BM)}_1$ does not depend on time, whereas q_2 and $T^{(\rm BM)}_2$ increase over time.

When the network throughput maximization approach is applied for three-route network, then for the flow rate $q^{(\rm o)}(t)$ used in Fig. 13.5a conditions $q^{(k)}_{\rm sum} + \epsilon^{(k)} = C^{(k)}_{\rm min}$ are satisfied for bottlenecks k=1 and k=3 only. Contrarily, for bottleneck 2 condition $q^{(2)}_{\rm sum} < C^{(2)}_{\rm min}$ is satisfied. Therefore, the whole increase in $q^{(\rm o)}(t)$ over time (Fig. 13.5a) leads to the increase in q_2 on the alternative route 2 only (Fig. 13.5d). Respectively, from Figs. 13.5d and 13.6d we can see that as long as the inflow rate $q^{(\rm o)}(t)$ increases over time, the flow rates q_1, q_3 and travel times $T^{(\rm BM)}_1$, $T^{(\rm BM)}_3$ do not depend on time, whereas q_2 and $T^{(\rm BM)}_2$ increase over time.

13.4 Facilitation of Traffic Breakdown in Networks Through the Use of Wardrop's SO

To study dynamic traffic assignment with the use of Wardrop's SO principle, we denote link travel times, respectively, as follows (Fig. 13.7): For links $m = 1, 3$ on route 1 by $T_{1,1}$, $T_{3,1}$; for links $m = 2, 4$ on route 2 by $T_{2,2}$, $T_{4,2}$ (Fig. 13.7). Travel times on routes 1 and 2 are $T_1 = T_{1,1} + T_{3,1}$ and $T_2 = T_{2,2} + T_{4,2}$, respectively.

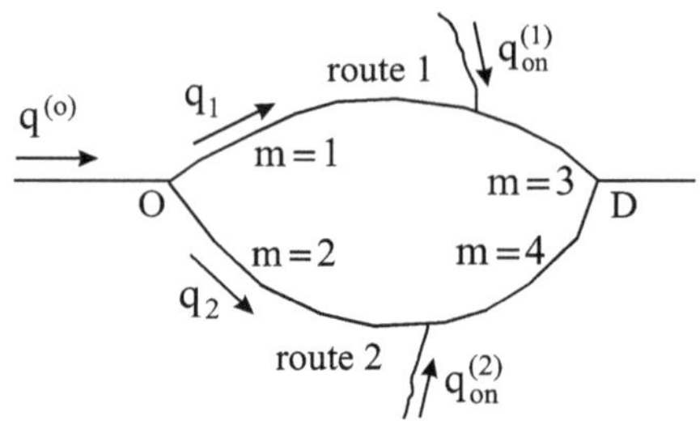

Fig. 13.7 Model of two-route network adapted from Fig. 12.3a,b in which different network links $m = 1, 2, 3, 4$ are denoted as required for the application of the Wardrop SO

Wardrop's SO for two-route network (Fig. 13.7) can be written as follows:

$$\min_{q_1, q_2}\{q_1 T_{1,1} + (q_1 + q_{\text{on}}^{(1)})T_{3,1} + q_2 T_{2,2} + \\ +(q_2 + q_{\text{on}}^{(2)})T_{4,2}\}, \; q^{(o)} = \sum_{m=1}^{2} q_m. \quad (13.8)$$

We have found qualitatively the same results of the application of Wardrop's SO as those found in Sect. 13.3 under the application of Wardrop's UE. In particular, we have chosen the total network inflow rate $Q = 7300$ vehicles/h that is considerably smaller than the network capacity $C_{\text{net}} = 7740$ vehicles/h. In other words, condition $Q < C_{\text{net}}$ (11.11) is satisfied. Under this condition, from the Wardrop's SO (13.8), we get

$$q_1 = 3980 \text{ vehicles/h}, \; q_2 = 2220\text{vehicles/h}. \quad (13.9)$$

Respectively, from (13.9) for the flow rates at network bottlenecks we get $q_{\text{sum}}^{(1)} = q_1 + q_{\text{on}}^{(1)}$ =4380 vehicles/h and $q_{\text{sum}}^{(2)} = q_2 + q_{\text{on}}^{(2)}$ =2920 vehicles/h, where we have taken into account that $q_{\text{on}}^{(k)} = 400$ (k =1) and 700 (k =2) vehicles/h (see caption to Fig. 12.3a).

Therefore, for the chosen total network inflow rate $Q = 7300$ vehicles/h that satisfies condition $Q < C_{\text{net}}$ (11.11) we get

$$q_{\text{sum}}^{(1)} > C_{\text{min}}^{(1)}, \quad (13.10)$$

$$q_{\text{sum}}^{(2)} < C_{\text{min}}^{(2)}, \quad (13.11)$$

where we have taken into account that $C_{\text{min}}^{(1)}$ =3980 vehicles/h and $C_{\text{min}}^{(2)} = 3760$ vehicles/h. Condition (13.10) is related to a general conclusion (13.2) of any application of the Wardrop's equilibria: At the total network inflow rate Q that is smaller than the network capacity C_{net} the application of the Wardrop's SO can provoke traffic congestion the network. Indeed, we have found that the application of the Wardrop's SO causes the increase in the probability of traffic breakdown at bottleneck 1. Simulations show that the probabilities of traffic breakdown at network bottlenecks under application of the Wardrop's SO principle (13.8) are

$$P^{(B,1)} = 0.21, \; P^{(B,2)} = 0 \text{ at } Q = 7300 \text{ vehicles/h}. \tag{13.12}$$

Therefore, under static dynamic assignment with the use of the Wardrop's SO (13.8), we have found a random time-delayed traffic breakdown (F→S transition) at bottleneck 1 (Fig. 13.8). Because the time delay to the breakdown is a random value, it is not possible to predict the time instant at which traffic breakdown occurs. As explained above for the case of the Wardrop's UE, for static traffic assignment results of simulations presented in Fig. 13.8 satisfy the Wardrop's SO (13.8) *only* as long as no change in route travel times occurs over time. Thus, after traffic breakdown has occurred at one of the network bottlenecks, rather than static traffic assignment, to satisfy the Wardrop's SO, dynamic traffic assignment is needed.

If we apply a control method for dynamic traffic assignment with the use of the Wardrop's SO (13.8), then regardless of the control method we have applied, it is not possible to fully avoid congested traffic occurring due to the breakdown: After traffic breakdown has been registered and even when the initial values of the flow rates q_1 and q_2 (13.9) has been changed (the flow rate q_1 decreases and the flow rate q_2 increases in accordance with a control method, for example, as explained in Sect. 13.3.2) without any time delay, there is always a time delay caused by travel time from the origin to the bottleneck location. Therefore, the decrease in the flow rate q_1 can begin to reduce congestion at bottleneck 1 with the time delay only. The control method can only effect on the distribution of congestion between two routes in the network. This conclusion is illustrated in Figs. 13.9 and 13.10. Traffic control methods (Sect. 13.3.2) used by dynamic traffic assignment in Figs. 13.9 and 13.10 are the same as those used in Figs. 13.2 and 13.3, respectively.

When the total network inflow rate is smaller than the network capacity, under application of the network throughput maximization approach no traffic breakdown occurs in a network: Free flow persists in the whole network.

In contrast with the application of the network throughput maximization approach, even when the total network inflow rate is still smaller than the network capacity, the common result of the application of the Wardrop's UE or SO for dynamic traffic assignment is a random sequence of the congested pattern emergence with the subsequent dissolution of the congested pattern.

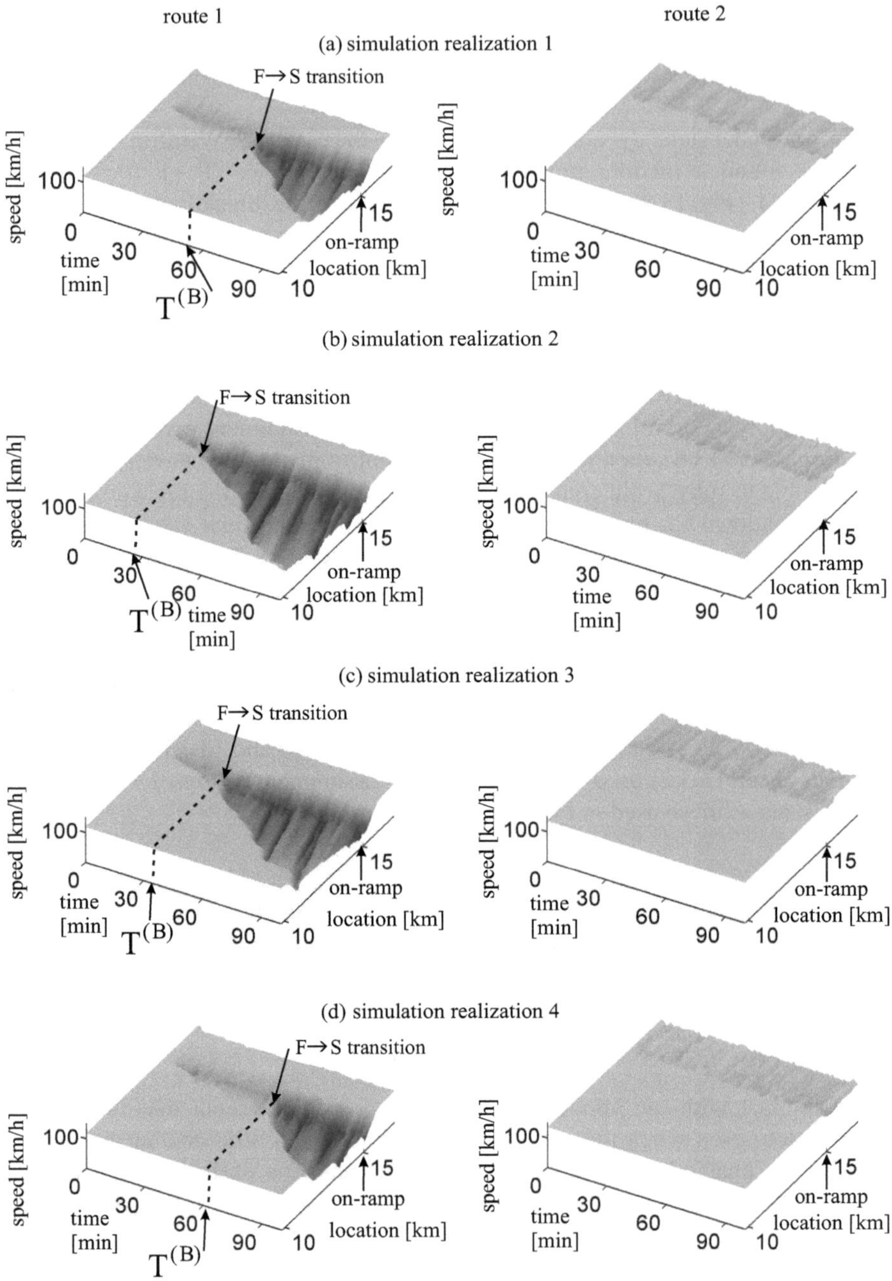

Fig. 13.8 Simulations of applications of Wardrop's SO (13.8) for static traffic assignment for two-route network (Fig. 13.7). (**a–d**) Speed in space and time on routes 1 (*left panel*) and 2 (*right panel*) for four different simulation realizations (runs) with time delays of traffic breakdown $T^{(B)}$ = 54 min (realization 1), 28 min (realization 2), 34 min (realization 3), 61 min (realization 4). q_1 = 3980 vehicles/h and q_2 = 2220 vehicles/h. Q = 7300 vehicles/h

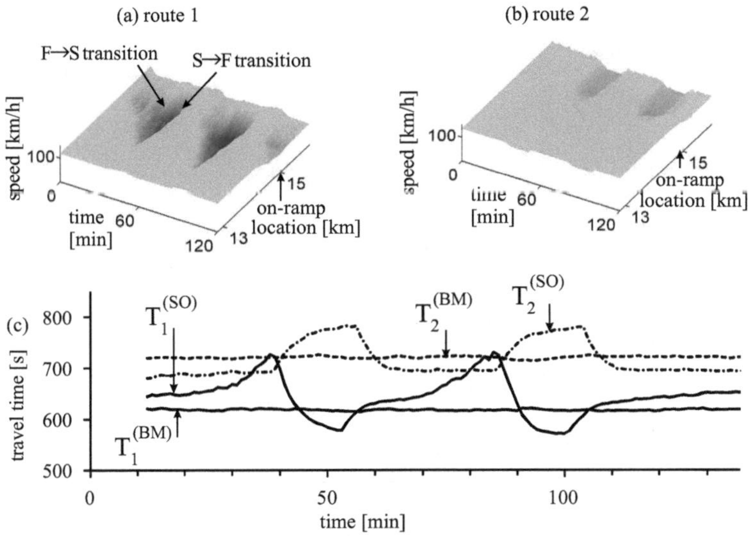

Fig. 13.9 Simulations of dynamic traffic assignment of the flow rates q_1 and q_2 in (13.8) for two-route network (Fig. 13.7) with the use of the Wardrop's SO and congested pattern control approach (Sect. 13.3.2) at $\Delta q = 1600$ vehicles/h. (**a, b**) Speed in space and time on routes 1 (**a**) and 2 (**b**). (**c**) Travel times T_1 and T_2 respectively for route 1 and 2 as time-functions $T_1 = T_1^{(UE)}(t)$, $T_2 = T_2^{(UE)}(t)$ for the Wardrop's SO. In (**c**), $T_1 = T_1^{(BM)}$, $T_2 = T_2^{(BM)}$ are related to application of the network throughput maximization approach (BM principle for "zero breakdown probability" (10.8)) for which $q_1 = 3560$ vehicles/h, $q_2 = 2640$ vehicles/h. Arrows F→S and S→F show time instants of F→S and S→F transitions, respectively. $Q = 7300$ vehicles/h

13.5 Control of Traffic Breakdown in Networks: Wardrop's UE Versus BM Principle

Up to now in this chapter we have discussed the case when the total network inflow rate Q is smaller than the network capacity C_{net}: $Q < C_{net}$ (11.11) . We have found that the use of the BM principle leads to the maintenance of free flow conditions in the whole network. In contract, the use of the Wardrop's UE deteriorates of the performance of traffic system: Rather than free flow conditions, complex congested patterns occur in the network.

As we have found in Sect. 12.5, when the total network inflow rate Q exceeds considerably the network capacity C_{net}, traffic breakdown (F→S transition) can occur at network bottlenecks, even if the BM principle is applied (Fig. 12.8). In the latter case, we has also found that through the application of dynamic control of this traffic breakdown with ANCONA on-ramp metering, a return S→F transition can be forced at the bottleneck (see Fig. 12.18 in Sect. 12.6). It is important to emphasize that due to this on-ramp metering control applied in the combination with the use of the BM principle, no vehicle queue remains in the on-ramp line of the on-ramp bottleneck (Fig. 12.18). The total network inflow rate $Q = 8640$ vehicles/h used

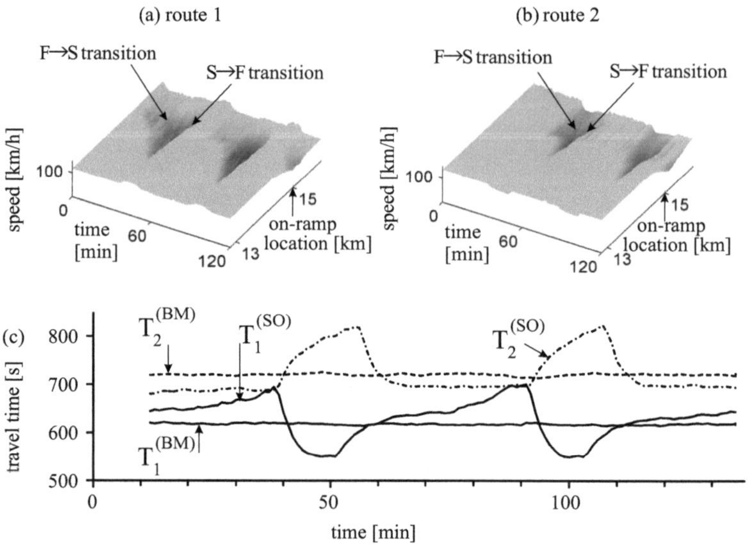

Fig. 13.10 Simulations of dynamic traffic assignment of the flow rates q_1 and q_2 in (13.8) for two-route network (Fig. 13.7) with the use of the Wardrop's SO and congested pattern control approach (Sect. 13.3.2) at $\Delta q = 2000$ vehicles/h. (**a, b**) Speed in space and time on routes 1 (**a**) and 2 (**b**). (**c**) Travel times T_1 and T_2 respectively for route 1 and 2 as time-functions $T_1 = T_1^{(UE)}(t)$, $T_2 = T_2^{(UE)}(t)$ for the Wardrop's SO. In (**c**), $T_1 = T_1^{(BM)}$, $T_2 = T_2^{(BM)}$ are related to application of the network throughput maximization approach (BM principle for "zero breakdown probability" (10.8)) adapted from Fig. 13.9. Arrows F→S and S→F show time instants of F→S and S→F transitions, respectively. $Q = 7300$ vehicles/h

in simulations of two-route network (Figs. 12.8 and 12.18) has been considerably larger than the network capacity for two-route network $C_{net} = 7740$ vehicles/h. Therefore, a question arises:

- Is it possible to force a return S→F transition at the bottleneck under application of the Wardrop's UE principle at the same value of the network inflow rate Q as that studied in Sect. 12.6 (Fig. 12.18)?

Under application of the Wardrop's UE (13.3), already at considerably smaller on-ramp inflow rate Q than the value $Q = 8640$ vehicles/h used in Figs. 12.8 and 12.18, the probability of traffic breakdown at bottleneck 1 is equal to 1. Respectively, when we use the Wardrop's UE (13.3) at the total network inflow rate $Q = 8640$ vehicles/h, we have found that traffic breakdown at bottleneck 1 occurs without any delay (Fig. 13.11a).[5]

[5]It should be noted that a free assignment of the flow rates q_1 and q_2 is only possible as long as free flow is at the origin of the network. Therefore, simulation results presented in Fig. 13.11 has only the sense as long as a contested pattern occurring in the network has not reached the origin ($x = 0$ in Figs. 13.11b, right and c, left).

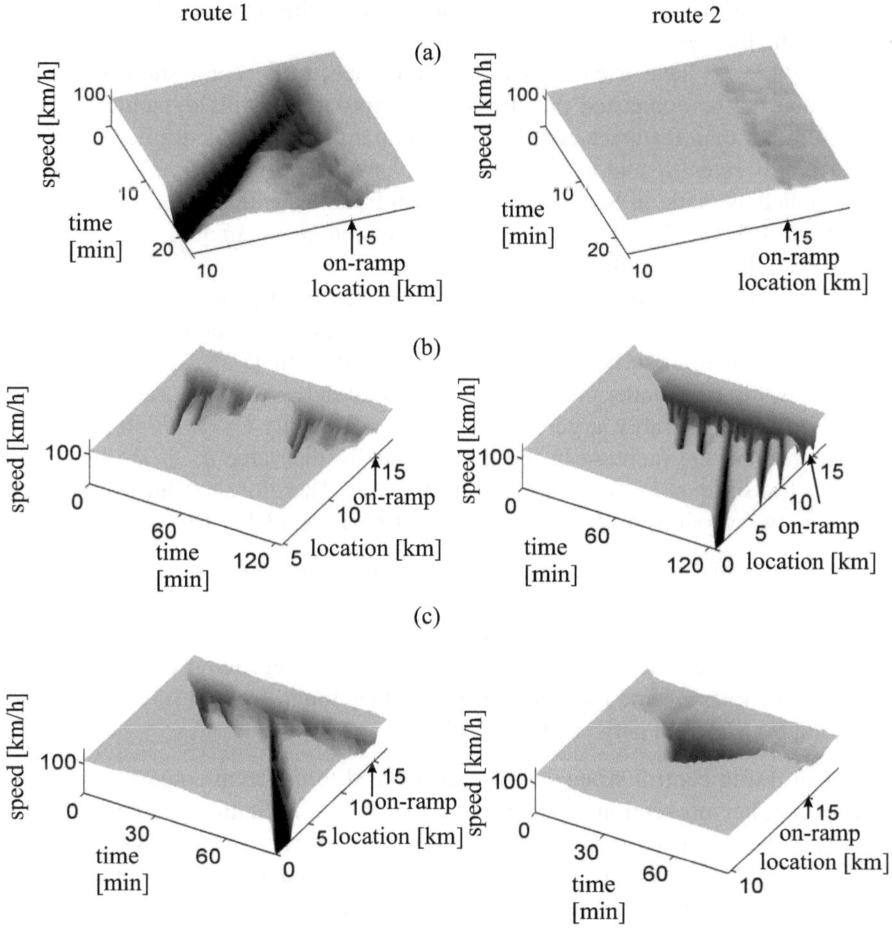

Fig. 13.11 Simulations of the Wardrop's UE (13.3) for two-route network (Fig. 12.3a) at the same network model parameters and the same total network inflow rate $Q = 8640$ vehicles/h as those in Figs. 12.8 and 12.18: Speed in space and time on route 1 (*left panel*) and route 2 (*right panel*). (**a**) Effect of the use of the Wardrop's UE without any control. (**b**, **c**) Effect of dynamic traffic assignment with the Wardrop's UE principle with congested pattern control described in the text, $\Delta q = 1000$ vehicles/h

We have tried to dissolve or reduce traffic congestion occurring due to this traffic breakdown with the use of ANCONA approach, as it has be done in Fig. 12.18 under applications of the BM principle. We have no success: To force a return S$\rightarrow$F transition at bottleneck 1, we have had to reduce the on-ramp inflow through the use of ANCONA to very small value $q_{\text{on}}^{(\text{cont})}$ (see Fig. 12.17) at which a growing queue has formed in the on-ramp lane of bottleneck 1. There can be found no values $q_{\text{on}}^{(\text{cont})}$ in ANCONA on-ramp metering (Fig. 12.17) at which both the growth of traffic congestion on the main road and the growth of the queue in the on-ramp lane of bottleneck 1 can be limited: We have found only an unlimited growth of congestion either on the main road or in the on-ramp lane.

Therefore, we have tried to dissolve or reduce traffic congestion due to the application of the congested pattern control approach. In this dynamic feedback control, rather than feedback control of on-ramp inflow, dynamic traffic assignment of the inflow rates q_1 and q_2 at the origin of two-route network (Fig. 12.3a) has been applied as made in Sect. 13.3.2: After an F→S transition has been registered at bottleneck 1, the flow rate q_1 decreases to a value $q_1 - \Delta q$ and the flow rate q_2 increases to the value $q_2 + \Delta q$.

As in Sects. 13.3.3 and 13.3.4, the value Δq has been chosen to force a return S→F transition at bottleneck 1. However, in Sects. 13.3.3 and 13.3.4 we have considered dynamic traffic assignment under condition $Q < C_{\rm net}$ (11.11). In contrast, in the case under consideration, the total network inflow rate $Q = 8640$ vehicles/h is considerably larger than the network capacity $C_{\rm net} = 7740$ vehicles/h. For this reason, the increase in the flow rate q_2 to the value $q_2 + \Delta q$ has led to an F→S transition at bottleneck 2 (Figs. 13.11b,c). Therefore, we have studied the following two versions of dynamic traffic control (Figs. 13.11b,c):

(i) In Fig. 13.11b, after a return S→F transition has been registered on bottleneck 1, the flow rates q_1 and q_2 return to their values found from (13.3) for free flow conditions at both network bottlenecks.
(ii) In Fig. 13.11c, after an F→S transition has been registered on bottleneck 2, the flow rates q_1 and q_2 return to their values found from (13.3) for free flow conditions at both network bottlenecks.

Although traffic control strategies (i) and (ii) lead to different congested patterns, there is neither prevention of traffic congestion nor limitation of the congested growth under application of dynamic traffic assignment with the Wardrop's UE (Fig. 13.11b,c). In contrast with the Wardrop's UE, under the same total network inflow rate a combination of applications of the BM principle together with congested pattern control approach can lead to the dissolution of traffic congestion in the network (Fig. 12.18).

13.6 Conclusions

Based on the physics of the network capacity, we have shown that any application of the classical Wardrop's equilibria for dynamic traffic assignment in traffic and transportation networks deteriorates basically the traffic system: Even when the total network inflow rate is smaller than the network capacity, the dynamic traffic assignment with the Wardrop's equilibria leads to the occurrence of traffic congestion in networks.

In other words, applications of the classical Wardrop's equilibria and other approaches for dynamic traffic assignment related to the state-of-the-art in traffic and transportation research (see, e.g., [1, 2, 4–16, 24–27, 29–33, 36, 40–53, 55, 59, 60, 62, 63, 65–67, 69, 70, 72–74, 83, 85, 86, 91, 93, 96, 98, 100–105, 107,

109–112, 114–121, 124, 127, 133–139, 142, 143, 148, 150–152] and conference proceedings [3, 19, 20, 61, 84, 87, 89, 99, 126]) deteriorate basically the traffic system while provoking heavy traffic congestion in urban networks.

In contrast with applications of the classical Wardrop's equilibria and other approaches for dynamic traffic assignment related to the state-of-the-art in traffic and transportation research, the application of the network throughput maximization approach (BM principle for "zero breakdown probability" (10.8)) during free flow conditions can increase considerably the time interval of congestion-less traffic in the network.

Empirical analysis of real field microscopic traffic data shows that fuel consumption and, therefore, CO_2 emission can be about 3–3.5 times larger in congested traffic than those in free flow (e.g., [57, 58] and references there). Thus, the maintenance of free flow conditions in urban areas of industrial countries through the application of the BM principle can contribute to the environment protection against the increase in CO_2 emission in the world.

Even when the total network inflow rate Q approaches the network capacity C_{net}, the network throughput maximization approach (BM principle for "zero breakdown probability" (10.8)) does ensure free flow conditions at which traffic breakdown cannot occur in the whole network. Contrarily, at $Q \to C_{net}$ the application of the Wardrop's equilibria does lead to congested traffic in the network.

Standard dynamic traffic assignment in a traffic or transportation network related to the state-of-the-art in traffic and transportation research deteriorates the traffic system provoking heavy traffic congestion in the network.

In contrast with standard dynamic traffic assignment, the application of the network throughput maximization approach for dynamic traffic assignment during free flow conditions can lead to the maintenance of free flow in the network during almost the whole day time, including a large interval of rush hour.

The BM principle [80–82] is the fundamental of transportation science that permits to maximize the network throughput preventing traffic breakdown in the network.

References

1. K. Abdelghany, D. Valdes, A. Abdelfatah, H.S. Mahmassani, Transp. Res. Rec. **1667**, 67–76 (1999)
2. S.D. Ahipas, R. Meskarian, T.L. Magnanti, K. Natarajan, Transp. Res. B **81**, 333–354 (2015)
3. R.E. Allsop, M.G.H. Bell, B.G. Heydecker (eds.), *Transportation and Traffic Theory 2007* (Elsevier Science, Amsterdam, 2007)
4. M. Amirgholy, E.J. Gonzales, Transp. Res. B **92**, 234–252 (2016)
5. D. Atmani, J.-P. Lebacque, N. Bhouri, H. Haj-Salem, Transp. Res. Procedia **3**, 895–904 (2014)
6. H. Bar-Gera, Transp. Sci. **36**, 398–417 (2002)
7. M.G.H. Bell, Transp. Res. B **29**, 125–137 (1995)
8. M.G.H. Bell, Transp. Res. B **29**, 287–295 (1995)
9. M.G.H. Bell, Transp. Res. B **34**, 533–545 (2000)
10. M.G.H. Bell, Ch. Cassir, in *2nd Int. Conf. on Traffic and Transportation Studies (ICTTS), Beijing, China*, ed. by K.C. Wang, G. Xiao, J. Ji (American Society of Civil Engineers, USA, 2000), pp. 9–16
11. M.G.H. Bell, Ch. Cassir, *Reliability of Transport Networks* (Research Studies Press, Baldock, Hertfordshire, 2000)
12. M.G.H. Bell, Ch. Cassir, Transp. Res. B **36**, 671–681 (2002)
13. M.G.H. Bell, Y. Iida, *Transportation Network Analysis* (Wiley, Hoboken, 1997)
14. M.G.H. Bell, C.M. Shielda, F. Busch, G. Kruse, Transp. Res. C **5**, 197–210 (1997)
15. D. Boyce, B. Ralevic-Dekic, H. Bar-Gera, J. Transp. Eng. **130**, 49–55 (2004)
16. M. Carey, P. Humphreys, M. McHugh, R. McIvor, Transp. Res. B **65**, 90–104 (2014)
17. M. Carey, D. Watling, Transp. Res. B. **46**, 634–648 (2012)
18. E. Cascetta, M. Gallo, B. Montella, Ann. Oper. Res. **144**, 301–328 (2006)
19. M.J. Cassidy, A. Skabardonis (eds.), in *Papers selected for the 19th Int. Sym. on Transp. and Traffic Theory*. Procedia Soc. Behav. Sci. **17**, 1–716 (2011)
20. A. Ceder (ed.), *Transportation and Traffic Theory* (Elsevier Science, Oxford, 1999)
21. B. Chen, W. Tong, W. Zhang, X. Sun, B. Wang, EPL **97**, 14001 (2012)
22. B. Chen, W. Tong, W. Zhang, X. Sun, B. Wang, Comput. Phys. Comm. **183**, 2081–2088 (2012)
23. X. Chen, C. Xiong, X. He, Z. Zhu, L. Zhang, Transp. Res. C **63**, 71–95 (2016)
24. H. Ceylan, M.G.H. Bell, J. Adv. Transp. **38**, 291–321 (2004)
25. H. Ceylan, M.G.H. Bell, Transp. Res. B **39**, 169–185 (2005)
26. Y.-C. Chiu, H.S. Mahmassani, Transp. Res. Rec. **1783**, 89–97 (2002)
27. B.D. Chung, H.-J. Cho, T.L. Friesz, H. Huang, T. Yao, Netw. Spat. Econ. **14**, 183–207 (2014)
28. R. Claes, T. Holvoet, D. Weyns, IEEE Trans. ITS **12**, 364–373 (2011)
29. C.F. Daganzo, Y. Sheffi, Transp. Sci. **11**, 253–274 (1977)
30. C.F. Daganzo, Transp. Res. **11**, 385–389 (1977)
31. C.F. Daganzo, Transp. Res. B **14**, 229–239 (1980)
32. C.F. Daganzo, Transp. Sci. **32**, 3–11 (1998)
33. C.F. Daganzo, Transp. Res. B **55**, 98–117 (2013)
34. L.C. Davis, Physica A **388**, 4459–4474 (2009)
35. L.C. Davis, Physica A **389**, 3588–3599 (2010)
36. K. Doan, S.V. Ukkusuri, Transp. Res. C **51**, 41–65 (2015)
37. C. Dong, X. Ma, B. Wang, Phys. Lett. A **374**, 1326–1331 (2010)
38. C. Dong, X. Ma, B. Wang, Sci. China-Infor. Sci. **53**, 2265–2271 (2010)
39. C. Dong, X. Ma, G. Wang, B. Wang, X. Sun, Physica A **389**, 3274–3281 (2010)
40. T.L. Friesz, Transp. Res. A **19**, 413–427 (1985)
41. T.L. Friesz, D. Bernstein, in *Foundations of Network Optimization and Games*, Complex Networks and Dynamic Systems, vol. 3 (Springer, New York, Berlin, 2016)
42. T. L. Friesz, D. Bernstein, T.E. Smith, R.L. Tobin, B.W. Wie, Oper. Res. **41**, 179–191 (1993)
43. T.L. Friesz, D. Bernstein, T.E. Smith R.L. Tobin, B.W. Wie, Oper. Res. **41**, 179–191 (1993)

44. T.L. Friesz, D. Bernstein, N.J. Mehta, R.L. Tobin, S. Ganjalizadeh, Oper. Res. **42**, 1120–1136 (1994)
45. T.L. Friesz, H.-J. Cho, N.J. Mehta, R.L. Tobin, G. Anandalingam, Transp. Sci. **26**, 18–26 (1992)
46. T.L. Friesz, K. Han, P.A. Neto, A. Meimand, T. Yao, Transp. Res. B **47**, 102–126 (2013)
47. T.L. Friesz, J. Luque, R.L. Tobin, B.-W. Wie, Oper. Res. **37**, 893–901 (1989)
48. T.L. Friesz, R.L. Tobin, T.E. Smith, P.T. Harker, J. Reg. Sci. **23**, 337–359 (1983)
49. N.H. Gartner, Transp. Sci. **14**, 174–191 (1980)
50. N.H. Gartner, M. Al-Malik, Transp. Res. Rec. **1554**, 27–35 (1996)
51. N.H. Gartner, Ch. Stamatiadis, Transp. Res. Rec. **1644**, 150–156 (1998)
52. Y.E. Ge, B.R. Sun, H.M. Zhang, W.Y. Szeto, X. Zhou, Netw. Spat. Econ. **15**, 583–598 (2015)
53. E.J. Gonzales, C.F. Daganzo, Transp. Res. B. **46**, 1519–1534 (2012)
54. M. Haghani, Z. Shahhoseini, M. Sarvi, J. Traffic Transp. Eng. (English Edition) **3**, 181–191 (2016)
55. K. Han, T.L. Friesz, T. Yao, Transp. Res. B **53**, 17–30 (2013)
56. C. Hauck, Y. Sun, I. Timofeyev, Stoch. Dyn. **14**, 1350022 (2014)
57. P. Hemmerle, M. Koller, H. Rehborn, B.S. Kerner, M. Schreckenberg, IET Intell. Transp. Syst. **10**, 122–129 (2016)
58. P. Hemmerle, M. Koller, G. Hermanns, H. Rehborn, B.S. Kerner, M. Schreckenberg, Collective Dynamics **1**, A7:1–27 (2016)
59. B.G. Heydecker, J.D. Addison, Transp. Sci. **39**, 39–57 (2005)
60. B.G. Heydecker, J.D. Addison, in *Transportation Networks: Recent Methodological Advances. Selected Proceedings of the 4th EURO Transportation Meeting*, ed. by M.G.H. Bell (Elsevier, Oxford, 1999), pp. 35–49
61. S.P. Hoogendoorn, V.L. Knoop, H. van Lint (eds.), in *20th International Symposium on Transportation and Traffic Theory (ISTTT 2013)*. Procedia Soc. Behav. Sci. **80**, 1–996 (2013)
62. S.P. Hoogendoorn, V.L. Knoop, H.J. Van Zuylen, J. Adv. Transp. **42**, 357–377 (2008)
63. H.-J. Huang, W.H.K. Lam, Transp. Res. B **36**, 253–273 (2002)
64. H.-J. Huang, Z.-C. Li, Eur. J. Oper. Res. **176**, 1464–1477 (2007)
65. N. Huynh, H.S. Mahmassani, H. Tavana, Transp. Res. Rec. **1783**, 55–65 (2002)
66. T. Iryo, Netw. Spat. Econ. **15**, 599–616 (2015)
67. B. Janson, C. Zozaya-Gorostiza, Transp. Res. B **21**, 299–310 (1987)
68. X. Ji, C. Shao, B. Wang, Procedia Engineering **137**, 620–629 (2016)
69. Y. Jiang, S.C. Wong, H.W. Ho, P. Zhang, R. Liu, A. Sumalee, Transp. Res. B **4**, 343–363 (2011)
70. W.-L. Jin, Netw. Spat. Econ. **15**, 617–634 (2015)
71. I. Juran, J.N. Prashker, S. Bekhor, I. Ishai, Transp. Res. C. **17**, 240–258 (2009)
72. P. Kachroo, S. Sastry, IEEE Trans. ITS **17**, 1438–1447 (2016)
73. G. Kalafatos, S. Peeta, in *Transportation and Traffic Theory 2009*, ed. by W.H.K. Lam, S.C. Wong, H.K. Lo. (Springer, Dordrecht, Heidelberg, London, New York, 2009), pp. 541–558
74. A. Karoonsoontawong, D.-Y. Lin, Netw. Spat. Econ. **15**, 1011–1048 (2015)
75. B.S. Kerner, *The Physics of Traffic* (Springer, Berlin, Heidelberg, New York, 2004)
76. B.S. Kerner, Physica A **355**, 565–601 (2005)
77. B.S. Kerner, in *Traffic and Transportation Theory*, ed. by H.S. Mahmassani (Elsevier Science, Amsterdam, 2005), pp. 181–203
78. B.S. Kerner, IEEE Trans. ITS **8**, 308–320 (2007)
79. B.S. Kerner, Transp. Res. Rec. **1999**, 30–39 (2007)
80. B.S. Kerner, J. Phys. A Math. Theor. **44**, 092001 (2011)
81. B.S. Kerner, Eur. Phys. J. B **89**, 199 (2016)
82. B.S. Kerner, Physica A **466**, 626–662 (2017)
83. D. Kupsizewska, D. Van Vliet, in *Transport Planning Methods* (Proc. PTRC Planning and Transport Summer Ann. Meeting, Univ. Sussex, 1999)
84. M. Kuwahara, H. Kita, Y. Asakura (eds.), in *21st International Symposium on Transportation and Traffic Theory* Transp. Res. Procedia **7**, 1–704 (2015)
85. W.H.K. Lam, K.S. Chan, Z.-C. Li, M.G.H. Bell, J. Adv. Transp. **44**, 219–230 (2010)

86. W.H.K. Lam, Hai-Jun Huang, Transp. Res. B **26**, 275–287 (1992)
87. W.H.K. Lam, S.C. Wong, H.K. Lo (eds.), *Transportation and Traffic Theory 2009* (Springer, Dordrecht, Heidelberg, London, New York, 2009)
88. K. Lee, P. Hui, B. Wang, N. Johnson, J. Phys. Soc. Jpn. **70**, 3507–3510 (2001)
89. J.-B. Lesort (ed.), *Transportation and Traffic Theory* (Elsevier Science, Oxford, 1996)
90. F. Leurent, H. Boujnah, Transp. Res. C. **47**, 28–46 (2014)
91. H. Liu, D.Z.W. Wang Transp. Res. B **72**, 20–39 (2015)
92. J. Long, W.Y. Szeto, Z. Gao, H.-J. Huang, Q. Shi, Transp. Res. B **83**, 179–206 (2016)
93. J. Long, W.Y. Szeto, Q. Shi, Z. Gao, H.-J. Huang, Transportmetrica A **11**, 388–419 (2015)
94. C.-C. Lu, H.S. Mahmassani, X. Zhou, Transp. Res. C. **16**, 371–389 (2008)
95. L. Lu, Y. Xu, C. Antoniou, M. Ben-Akiva, Transp. Res. C **51**, 149–166 (2015)
96. P. Luathep, A. Sumalee, W.H.K. Lam, Z-Ch. Li, H.K. Lo, Transp. Res. B **45**, 808–827 (2011)
97. M. Lujak, S. Giordani, S. Ossowski, Neurocomputing **151**, 449–460 (2015)
98. H.S. Mahmassani, Netw. Spat. Econ. **1**, 267–292 (2001)
99. H.S. Mahmassani (ed.), *Traffic and Transportation Theory* (Elsevier Science, Amsterdam, 2005)
100. H.S. Mahmassani, G.L. Chang. Transp. Sci. **21**, 89–99 (1987)
101. H.S. Mahmassani, R. Herman, Transp. Sci. **18**, 362–384 (1984)
102. H.S. Mahmassani, S. Peeta, Transp. Res. Rec. **1408**, 83–93 (1993)
103. D.K. Merchant, G.L. Nemhauser, Transp. Sci. **12**, 187–199 (1978)
104. D.K. Merchant, G.L. Nemhauser, Transp. Sci. **12**, 200–207 (1978)
105. E.D. Miller-Hooks, H.S. Mahmassani, Transp. Sci. **34**, 198–215 (2000)
106. A. Nagurney, J. Dong, Transp. Res. B **36**, 445–469 (2002)
107. Y. Nesterov, A. de Palma, Netw. Spat. Econ. **3**, 371–395 (2003)
108. Y. Nie, H.M. Zhang, Netw. Spat. Econ. **10**, 49–71 (2010)
109. M. Patriksson, *The Traffic Assignment Problem–Models and Methods* (VSP, Utrecht, 1994)
110. S. Peeta, H.S. Mahmassani, Transp. Res. C **3**, 83–98 (1995)
111. S. Peeta, H.S. Mahmassani, Ann. Oper. Res. **60**, 81–113 (1995)
112. S. Peeta, A.K. Ziliaskopoulos, Netw. Spat. Econ. **1**, 233–265 (2001)
113. Z. Qian, R. Rajagopal, Transp. Res. C **48**, 468–490 (2014)
114. H. Rakha, A. Tawfik, in *Encyclopedia of Complexity and System Science*, ed. by R.A. Meyers (Springer, Berlin, 2009), pp. 9429–9470
115. B. Ran, D. Boyce, *Modeling Dynamic Transportation Networks* (Springer, Berlin, 1996)
116. B. Ran, D.E. Boyce, L.J. LeBlanc, Oper. Res. **41**, 192–202 (1993)
117. R.A. Reiss, N.H. Gartner, S.L. Cohen, Transp. Res. A **25**, 267–276 (1991)
118. H. Shao, W.H.K. Lam, M.L. Tam, Netw. Spat. Econ. **6**, 173–204 (2006)
119. Y. Sheffi, *Urban Transportation Networks: Equilibrium Analysis with Mathematical Programming Methods* (Prentice-Hall, New Jersey, 1984)
120. W. Shen, H.M. Zhang, Transp. Res. B **65**, 1–17 (2014)
121. V.I. Shvetsov, Autom. Remote Control **70**, 1728–1736 (2009)
122. A. Sopasakis, M.A. Katsoulakis, SIAM J. Appl. Math. **66**, 921–944 (2006)
123. Y. Sun, I. Timofeyev Phys. Rev. E **89**, 052810 (2014)
124. Ch. Suwansirikul, T.L. Friesz, R.L. Tobin, Transp. Sci. **21**, 254–263 (1987)
125. Z. Tan, H. Yang, R.-Y. Guo, Transp. Res. C **61**, 87–105 (2015)
126. M.A.P. Taylor (ed.), *Transportation and Traffic Theory in the 21st Century* (Elsevier Science, Amsterdam, 2002)
127. R.L. Tobin, T.L. Friesz, Transp. Sci. **22**, 242–250 (1988)
128. K. Tobita, T. Nagatani, Physica A **391**, 6137–6145 (2012)
129. J. Wahle, A.L.C. Bazzan, F. Klugl, M. Schreckenberg, Physica A **287**, 669 (2000)
130. J. Wahle, A. Bazzan, F. Klugl, M. Schreckenberg, Transp. Res. C **10**, 399–417 (2002)
131. W. Wang, B. Wang, W. Zheng, C. Yin, T. Zhou, Phys. Rev. E **72**, 066702 (2005)
132. J.G. Wardrop, in *Proc. of Inst. of Civil Eng. II.* **1**, 325–378 (1952)
133. B.-W. Wie, T.L. Friesz, R.L. Tobin, Transp. Res. B **24**, 431–442 (1990)
134. X. Xu, A. Chen, S. Kitthamkesorn, H. Yang, H.K. Lo, Transp. Res. B **81**, 686–703 (2015)

135. J. Yang, G. Jiang, Transp. Res. C **47**, 168–178 (2014)
136. H. Yang, M.G.H. Bell, Transp. Res. B **31**, 303–314 (1997)
137. H. Yang, M.G.H. Bell, Transp. Res. B **32**, 539–545 (1998)
138. H. Yang, M.G.H. Bell, Transp. Rev. **18**, 257–278 (1998)
139. H. Yang, M.G.H. Bell, Q. Menga, Transp. Res. B **34**, 255–275 (2000)
140. H. Yang, X. Zhang, Q. Meng, Transp. Res. B **38**, 191–213 (2004)
141. Q. Yu, D. Fang, W. Du, Eur. J. Oper. Res. **239**, 112–118 (2014)
142. C. Zhang, X. Chen, A. Sumalee, Transp. Res. B **45**, 534–552 (2011)
143. K. Zhang, H.S. Mahmassani, C.-C. Lu, Transp. Res. Rec. **2085**, 86–94 (2008)
144. K. Zhang, H.S. Mahmassani, C.-C. Lu, Transp. Res. C **27**, 189–204 (2013)
145. X. Zhang, H. Yang, Transp. Res. B **38**, 517–537 (2004)
146. X. Zhang, H. Yang, H.-J. Huang, Eur. J. Oper. Res. **189**, 146–158 (2008)
147. C. Zhao, B. Fu, T. Wang, Transp. Res. E **61**, 135–141 (2014)
148. C.-L. Zhao, H.-J. Huang, Transp. Res. C **68**, 22–37 (2016)
149. X.-M. Zhao, D.-f. Xie, Q. Li, Comput. Phys. Comm. **187**, 106–114 (2015)
150. X. Zhou, H.S. Mahmassani, K. Zhang, Transp. Res. C **16**, 167–186 (2008)
151. A.K. Ziliaskopoulos, D. Kotzinos, H.S. Mahmassani, Transp. Res. C **5**, 95–107 (1997)
152. A.K. Ziliaskopoulos, Transp. Sci. **34**, 37–49 (2000)

Chapter 14
Discussion of Future Dynamic Traffic Assignment and Control in Networks

14.1 Introduction

A discussion of a future dynamic traffic assignment and control in traffic and transportation networks that we make in this chapter is based on results presented in Chaps. 10–13.

In particular, in Chap. 11 we have presented the application of the BM principle for "zero breakdown probability" $P_{\rm net} = 0$ (10.8), where $P_{\rm net}$ is the probability of traffic breakdown in a traffic or transportation network. This application of the BM principle called the network throughput maximization approach allows us to prevent traffic breakdown in the whole network while keeping condition (10.8).

We have also shown in Chap. 11 that there is a network measure called the network capacity $C_{\rm net}$. The sense of the network capacity is as follows: As long as the total network inflow rate Q (10.2) is smaller than the network capacity $C_{\rm net}$, under application of the network throughput maximization approach traffic breakdown cannot occur in the network.

In the case of larger values of the total network inflow rate Q related to rush hours in urban networks condition $P_{\rm net} = 0$ (10.8) cannot be satisfied through the application of the BM principle. As we have shown in Chap. 12, at such large values of the total network inflow rate, the application of the BM principle can only reduce the probability of the occurrence of traffic breakdown in the network during a given time interval $T_{\rm ob}$ for observing traffic flow to a minimum value of this probability $P_{\rm net} = P^{(\rm min)}_{\rm net}$ (12.1) that is larger than zero. In other words, at such large values of the total network inflow rate even if the BM principle is applied, traffic breakdown can occur at some of the network bottlenecks during the time interval $T_{\rm ob}$. We have shown in Sect. 12.6 that dynamic feedback control of traffic breakdown at network bottlenecks based on the application of congested pattern control approach can prevent the development of traffic congestion in the network.

B.S. Kerner, *Breakdown in Traffic Networks*, DOI 10.1007/978-3-662-54473-0_14

Contrarily to the applications of the BM principle (Chaps. 11 and 12), in Chap. 13 we have shown that any application of the classical Wardrop's equilibria for dynamic traffic assignment in traffic and transportation networks deteriorates basically the traffic system: Even when the total network inflow rate is smaller than the network capacity, the dynamic traffic assignment with the Wardrop's equilibria leads to the occurrence of traffic congestion in networks. In other words, applications of standard dynamic traffic assignment related to the state-of-the-art in traffic and transportation research deteriorate basically the traffic system while provoking heavy traffic congestion in urban networks.

Thus, a consideration of different applications the BM principle (Chaps. 10–12) and a critical analysis of standard dynamic traffic assignment based on the classical Wardrops' equilibria (Chap. 13) allow us to make the conclusion that the assignment procedure with the BM principle should give a better performance than assignment procedures designed through the use of the Wardrop's UE or SO. However, the following questions arise:

- Even when the BM principle exhibits the better performance in comparison with applications of the Wardrop's equilibria, whether this better performance of the BM principle does lead to *large enough benefits* that justify to use this approach instead of applications of the Wardrop's equilibria?

Indeed, it seems that the use of the BM principle exhibits considerable disadvantages: Some of the drivers should use longer routes.

The objective of this chapter is a discussion of benefits of the application of the BM principle for future traffic organization in traffic and transportation networks in comparison with standard dynamic traffic assignment and control. The chapter is organized as follows. An explanation of the necessity of applications of BM principle for future traffic organization in networks will be discussed in Sect. 14.2. Benefits of the application of BM principle will be additionally considered in Sect. 14.3. In Sect. 14.4 we make a discussion of a reasonable choice of the threshold difference between route travel times in a network for the constrain "alternative network routes (paths)" (Sect. 10.5) used in applications of the BM principle. Finally, in Sect. 14.5 we make a qualitative discussion of possible applications of the BM principle for real traffic and transportation networks.

14.2 The Necessity of Applications of BM Principle

Thus, we have found that the network throughput maximization approach (BM principle for "zero breakdown probability" (10.8)) maintains free flow in the network up to the network inflow rate $Q \rightarrow C_{\mathrm{net}}$. In contrast, the use of the Wardrop's UE or SO for dynamic traffic assignment in traffic and transportation networks leads to traffic congestion in the network already at the total network inflow rate Q that is smaller than the network capacity C_{net}. In other words, the applications of the Wardrop's UE or SO causes the deterioration of the performance of traffic system while provoking heavy traffic congestion in urban networks.

Nevertheless, the author's experience shows that many of theoreticians working in traffic and transportation research would give a statement:

- "Clearly it is not interesting to apply the network throughput maximization approach during free flow conditions: It is not necessary during free flow and not possible when the probability of traffic breakdown is positive".

An answer to this expected statement of the theoreticians is the following:

- The alternative to the application of the network throughput maximization approach is heavy traffic congestion observed almost each day during rush hour on all important roads in industrial countries.

Indeed, we can expect that if the network throughput maximization approach is applied, then free flow conditions can be realized in the whole traffic network during almost the whole day time, including a large interval of rush hour. This is because under conditions (11.6) following from the network throughput maximization approach, no traffic breakdown can occur in the network. Only for a short time interval during rush hour, when the total network inflow rate Q increases to very large values (for a steady state in the network, when Q exceeds the network capacity $C_{\rm net}$), free flow becomes metastable with respect to traffic breakdown at some of the network bottlenecks.

Thus, the application of the network throughput maximization approach during free flow conditions can increase considerably the time interval of congestion-less traffic in the network. This can result in the reduction of traffic congestion in the network.

The wish of humans to use shortest routes of a network contradicts fundamentally another wish to drive under free flow conditions in the network (see Sect. 13.2). The classical Wardrop's equilibria are based on the wish of humans to use shortest routes in the network. As we have explained in Chap. 13, the use of the classical Wardrop's equilibria for standard dynamic traffic assignment in traffic and transportation networks, which is the state-of-the-art in traffic and transportation research, results *basically* in the occurrence of heavy traffic congestion.

The application of the network throughput maximization approach can reduce or even avoid heavy traffic congestion in urban networks. For this reason, the network throughput maximization approach is needed to apply during free flow conditions for future dynamic traffic optimization in urban networks. The question how drivers could be encouraged to follow the rules of dynamic traffic assignment based on the network throughput maximization approach will be discussed in next Sects. 14.3 and 14.4.

14.3 Benefits of Applications of BM Principle

In the methodology to dynamic traffic assignment and control in a traffic network based on the application of the BM principle, route travel times determine only the constrain "alternative routes" that prevents the use of too long routes. After all sets of alternative routes have been determined with the use of this constrain (Sect. 10.5), the BM principle is applied for these sets of alternative routes only. For this reason, conditions (11.6) (or/and (11.30)) for dynamic traffic assignment resulting from the network throughput maximization approach (Sects. 11.2 and 11.6) do not depend on route travel times.

Thus, it seems that the use of the BM principle in real traffic and transportation networks has considerable disadvantages in comparison with applications of the Wardrop's equilibria: Some of the drives should use longer routes to avoid congestion in the network.

However, possible benefits of the application of the BM principlc for thc future organization of traffic and transportation networks can overcome these disadvantages. Congested traffic in a network occurs through the use of short routes in a traffic network. On the one hand, as mentioned in Sect. 13.5, empirical analysis of real field microscopic traffic data shows that fuel consumption and, therefore, CO_2 emission can be about 3–3.5 times larger in congested traffic than those in free flow. On the other hand, travel times can considerably increase due to congestion on short routes (Figs. 13.2, 13.3, 13.4, and 13.6). Therefore, due to traffic congestion the use of short routes can lead to a considerably increase in travel costs in real networks.

To maintain free flow conditions in urban networks for such an environment protection, there are technical possibilities already now.[1] Through communication of GPS vehicle data to a traffic control center, the center can provide appropriate information to the drivers as well as organize an efficient network organization with the network throughput maximization approach:

(i) Traffic center can store characteristics of traffic breakdown at network bottlenecks found from measurements of traffic variables with road detectors, video cameras and/or GPS probe vehicles (FCD—floating car data).
(ii) Traffic center can inform drivers individually about an eco-route.

The application of the BM principle results in a *dynamic system optimum* in a traffic network. In contrast with the Wardrop's SO or UE that demands some minimization of travel times (travel costs) in the network, in the dynamic system optimum resulting from applications of the BM principle *no* travel times (or other travel costs) should be minimized. The dynamic system optimum resulting from the network throughput maximization approach allows us the maximization of the total

[1] While working at the Daimler Company, the author was lucky to take part in the development of systems for traffic information used in vehicle routing systems, which are on the market.

network inflow rate at which free flow conditions can be maintained in the whole network.

When drivers move in free flow, there is a considerable dynamic user advantage in travel times. Therefore, when the BM principle is applied, in some sense *a dynamic user optimum* is achieved in the network, at which drivers can move in free flow even at a relatively large total network inflow rate, which in some cases can be related to rush hours in urban networks.

14.4 Choice of Threshold for Constrain "Alternative Network Routes (Paths)" in Applications of BM Principle

We have already mentioned that in the methodology of dynamic traffic assignment and control in a traffic network based on the application of the BM principle, route travel times in the network determine only the constrain "alternative routes" that prevents the use of too long network routes (Sect. 10.5). In the constrain "alternative routes", for each origin-destination pair $O_i - D_j$ we should determine the set of the alternative routes through the choice of a threshold value Λ_{ij} for the difference between travel times for different routes under free flow conditions related to a small enough total network inflow rate ($Q \to 0$) (Sect. 10.5): After all sets of the alternative routes have been found, dynamic traffic assignment is performed with the use of the BM principle for these alternative routes only. For this reason, dynamic traffic assignment based on the BM principle does not depend on network route travel times.

For example, in a simple two-route network (Fig. 12.3a) we have used the following length of routes 1 and 2: $L_1 = 20$ km and $L_2 = 25$ km. At small enough link inflow rates, this corresponds to travel times $T_1 = 8$ min and $T_2 = 10$ min. Respectively, the difference between route travel times is equal to $\Delta T = T_2 - T_1 = 2$ min. This means that in dynamic traffic assignment with the use of the BM principle we have assumed that this difference between travel times does not exceed a threshold value Λ.

Now we assume that the length of route 2 is $L_2 = 250$ km. Then $T_2 = 100$ min and the difference between travel times is equal to $\Delta T = 92$ min. Clearly, such a large difference between travel times should be considerably larger than a chosen threshold value Λ that can still be acceptable for drivers. In this case, through the application of the constrain "alternative routes", route 2 is excluded from dynamic traffic assignment. This means that rather than a two-route network, there is only one route from origin to destination: The BM principle cannot be applied in this case.

However, the author has often dealt with the following questions:

- How to apply a dynamic system optimum in real traffic while users are selfish? No one would like to spend a longer time to let the system performance better.

- What is a condition for the choice of the threshold value Λ_{ij} in the constrain "alternative routes"?

One of the possibilities to facilitate the use of alternative routes associated with the network throughput maximization approach of Sects. 11.2, 11.4, and 11.6 is the use of electronic road charge. Through the use of electronic road charge, drivers that would like to use shorter routes in a traffic network should pay a larger road charge than the drivers that agree to use longer routes in the network. Indeed, electronic road charge systems based on GPS vehicle data can facilitate the use of alternative routes associated with the network throughput maximization approach for the maintenance of conditions (11.6) (or/and (11.30)).

We can assume that these is a maximum road charge that is still acceptable for a large enough amount of the drivers at which conditions (11.6) can be satisfied. This maximum road charge can be used for the choice of the threshold value Λ_{ij} in the constrain "alternative routes". We can assume that under free flow conditions the larger the difference between travel times on the shortest alternative route in comparison with another alternative route from origin O_i to destination D_j, the larger should be the road charge for the use of the shortest route. However, when the road charge is chosen too large one, conditions (11.6) cannot be satisfied for the shortest route: There are not enough drivers that are willing to pay this large road charge. Therefore, the threshold value Λ_{ij} can be determined by the maximum road charge that is still acceptable for a large enough amount of the drivers at which conditions (11.6) can be satisfied for the shortest route. A calculation of the threshold value Λ_{ij} for large scale traffic networks can be an interesting task for future investigations.

14.5 Possible Applications of BM Principle for Real Traffic and Transportation Networks

In Chaps. 10–12, we have found that the optimum of a traffic or transportation network is reached, when dynamic traffic assignment, traffic optimization and/or control are performed in the network in such a way that the *probability* for the occurrence of traffic breakdown in at least one of the network bottlenecks during a given time interval for observing traffic flow reaches the *minimum possible value*. Rather than travel time (or some other travel costs), in the BM principle the objective function that should be minimized is the probability of traffic breakdown in the network.

The BM principle demands the minimization of the probability of traffic congestion in the network.

Thus, the objective function in the BM principle that should be minimized depends neither on travel time nor other travel costs. The BM principle demands the minimization of the probability of the occurrence of congestion in the network. Under great traffic demand, the application of the BM principle should result in relatively small travel time (small travel costs) associated with free flow in a network. Here a question arises:

- What is a procedure of dynamic traffic assignment and control in a traffic or transportation network, when the network inflow rate increases continuously during daytime?

In this section, we discuss qualitatively possible applications of the BM principle under subsequent increase in the total network inflow rate Q (10.2) from initially small values to larger ones related to rush hours in urban networks.

14.5.1 Applications of Network Throughput Maximization Approach

As mentioned in Sect. 10.6, when the total network inflow rate Q (10.2) in a traffic or transportation network is very small, no dynamic traffic assignment with the use of the BM principle is required. This changes when under increase in the total network inflow rate Q condition (11.6) (at large enough fluctuations in free flow) or condition (11.30) (at small fluctuations in free flow) is satisfied. Then, the network throughput maximization approach should be applied for dynamic traffic assignment in the network as described in Chap. 11.

In the network throughput maximization approach, route travel times determine only the constrain "alternative routes" that prevents the use of too long routes in the network. After all sets of alternative routes have been determined (Sect. 10.5), dynamic traffic assignment is performed in the network in accordance with conditions (11.6) (at large enough fluctuations in free flow) or conditions (11.30) (at small fluctuations in free flow) for the alternative routes only.

At relatively small values of the network inflow rate, dynamic traffic assignment and control in a traffic or transportation network should be performed through the application of the BM principle called the network throughput maximization approach. The network throughput maximization approach maximizes the network throughput ensuring free flow condition in the whole network: The probability of traffic breakdown in the network remains zero.

The BM principle is the fundamental of transportation science that permits to maximize the network throughput preventing traffic breakdown in the network.

14.5.2 Possible Applications of BM Principle Under Subsequent Increase in Total Network Inflow Rate

Under a large enough total network inflow rate $Q(t)$ (10.2), formula (11.6) or formula (11.30) cannot be satisfied: The probability of the breakdown in the network $P_{\rm net}$ during the time interval $T_{\rm ob}$ becomes larger than zero. Thus, the network throughput maximization approach cannot be used any more.

Through the application of the BM principle (12.6), the probability of the occurrence of traffic breakdown $P_{\rm net}$ in the network during the time interval $T_{\rm ob}$ can be reduced to the minimum possible value that remains larger than zero: $P_{\rm net} = P_{\rm net}^{(\rm min)} > 0$ (12.1). As described in Sect. 12.6, when traffic breakdown has occurred at one of the network bottlenecks, control of traffic breakdown at the bottleneck can recover free flow at this bottleneck.

At larger values of the total network inflow rate related to rush hours, it is often the case that the probability of traffic breakdown in the network cannot remain zero. Then, dynamic traffic assignment and control in a traffic or transportation network should be performed through another application of the BM principle that minimizes the probability of traffic breakdown in the network.

If traffic breakdown has already occurred at a network bottleneck, then dynamic traffic assignment, network optimization and control can consist of the stages:

(i) The dissolution of traffic congestion due to dynamic control of traffic breakdown in a neighborhood of the bottleneck. Through the use of control of traffic breakdown in a neighborhood of the bottleneck the dissolution of traffic congestion can be achieved.
(ii) The minimization of traffic breakdown probability with the BM principle in the remaining network, i.e., the network part that is not influenced by congestion.

14.5.3 About Future Control of Heavy Traffic Congestion in Networks

As follows from the BM principle (12.6), the minimum possible value of the probability of traffic breakdown the network $P_{\rm net} = P_{\rm net}^{(\rm min)}$ satisfies conditions

$$P_{\rm net}^{(\rm min)} \geq P^{(\rm B,\mathit{k})}, \ k = 1, 2, \ldots, N. \tag{14.1}$$

Therefore, if there are two or more network bottlenecks for which $P^{(\rm B,\, \mathit{k})} > 0$, then the value $P_{\rm net}^{(\rm min)}$ is larger than each of the values $P^{(\rm B,\mathit{k})}$. For this reason, when the total network inflow rate $Q(t)$ increases and the number of the network bottlenecks for which $P^{(\rm B,\, \mathit{k})} > 0$ becomes large enough, the value $P_{\rm net}^{(\rm min)}$ can quickly reach 1. This means that during the time interval $T_{\rm ob}$ traffic breakdown does occur in at least one of the network bottlenecks.

When the total network inflow rate Q (10.2) increases further, it can turn out that congested traffic cannot be avoided in the network any more. Nevertheless, due to the application of the congested pattern control approach (Sect. 12.6) we can try to maintain congestion conditions at the minimum possible level at which congested traffic is localized in some parts of the network only. However, in many real traffic networks there are not enough alternative routes to avoid traffic congestion at large enough traffic demand. Thus, at large enough traffic demand even "perfect" methods of traffic control and dynamic network optimization could not prevent traffic breakdown resulting in heavy traffic congestion in a traffic network.

Due to heavy traffic congestion, travel time and fuel consumption as well as other travel costs can increase considerably in the network. For this reason, users of traffic and transportation networks would expect that even if traffic congestion cannot be avoided in traffic networks, nevertheless, the application of intelligent transportation systems (ITS) can change characteristics of traffic congestion with the objective to increase traffic safety and comfort while moving in congested traffic.

Therefore the following questions arise:

1. What are empirical features of congested patterns that can be influenced for reliable spatial limitation of congestion growth and/or for congestion dissolution?
2. How can driver behavior change features of congested patterns with the aim to increase safety and comfortable driving in congested traffic?
3. What are vehicle systems for automatic driving and other ITS-applications that can help to increase safety and comfort while driving in congested traffic?

Studies of congested traffic made in the framework of the three-phase theory have answered these questions. For example, as mentioned in Sect. 12.6.3, if through the use of ITS conditions of strong driver speed adaptation are realized in congested traffic, all moving jams dissolve in congested traffic: Congested traffic consists only of the synchronized flow traffic phase in which fuel consumption can be considerably smaller than that in congested traffic consisting of synchronized flow and wide moving jams. However, a detailed consideration of traffic phenomena occurring under heavy traffic congestion is out of scope of this book.

14.6 Conclusions

1. In contrast with standard dynamic traffic assignment and control in traffic and transportation networks related to the state-of-the-art in traffic and transportation research (Chap. 13), the application of the network throughput maximization approach for dynamic traffic assignment (Chap. 11) can lead to the maintenance of free flow conditions in the network during almost the whole day time, including a large interval of rush hour.
2. A procedure of dynamic traffic assignment and control in a traffic or transportation network with the use of the BM principle under subsequent increase in the total network inflow rate can be as follows.

   (i) When the total network inflow rate is not large, the network throughput maximization approach (Chap. 11) shown be applied.
   (ii) When the total network inflow rate is large enough and the probability of traffic breakdown in the network during a given time interval becomes larger than zero, the application of the BM principle that minimizes the probability of the occurrence of traffic breakdown in the network should be applied (Chap. 12).
   (iii) If traffic breakdown has already occurred at a network bottleneck, then network optimization and control can consist of the stages:

   - The dissolution of traffic congestion due to dynamic feedback control of traffic breakdown in a neighborhood of the bottleneck. We have shown that through the use of dynamic feedback control of traffic breakdown at the bottleneck (ANCONA) the dissolution of traffic congestion can indeed be achieved (Sect. 12.6).
   - The minimization of traffic breakdown probability with the BM principle in the remaining network, i.e., the network part that is not influenced by congestion.

3. The BM principle can be applied, even when non-steady state conditions are realized in a traffic or transportation network.

Chapter 15
Conclusions and Outlook

Main conclusions of this book are as follows:

1. Empirical features of nuclei for traffic breakdown at highway bottlenecks confirm the basic assumption of the three-phase theory about the nucleation nature of traffic breakdown (F→S transition) at highway.
2. The empirical nucleation nature of traffic breakdown (F→S transition) at a highway bottleneck can be considered an empirical fundamental of transportation science.
3. Driver behavioral characteristics assumed in the three-phase theory explain the nucleation nature of traffic breakdown at highway bottlenecks.
4. The three-phase theory is the theoretical fundamental of transportation science that explain the empirical fundamental of transportation science.
5. The BM principle is the theoretical fundamental of transportation science that minimizes traffic congestion in a traffic or transportation network.
6. There is a general approach for the maximization of the network throughput. This network maximization throughput approach allows us to find general conditions for the *maximization* of the network throughput at which traffic breakdown cannot occur in the whole network.
7. Although there are great achievements of classical traffic flow theories and models in the understanding of many important empirical traffic phenomena, the classical theories and traffic flow models cannot explain and show the empirical fundamental of transportation science—the nucleation nature of traffic breakdown at a highway bottleneck. The metastability of free flow with respect to the F→S transition at the bottleneck has no sense for the classical traffic theories. This explains the failure of applications of the classical traffic and transportation theories and models in the real world.

B.S. Kerner, *Breakdown in Traffic Networks*, DOI 10.1007/978-3-662-54473-0_15

8. Applications of the classical traffic flow models for an analysis of the effect of ITS (intelligent transportation systems) on traffic flow lead to invalid conclusions about the ITS performance in real traffic.
9. The three-phase theory is incommensurable with the classical traffic-flow theories.

Below we consider these book's conclusions in more details.

15.1 Empirical Fundamental of Transportation Science

In this book, we have shown that there is an empirical traffic phenomenon that is an fundamental empirical basis of transportation science. This empirical phenomenon is the metastability of free flow with respect to a phase transition from free flow to synchronized flow (F→S transition) at a highway bottleneck. This empirical metastability leads to the nucleation character of the F→S transition (traffic breakdown) at the bottleneck observed in real field traffic data.

The empirical nucleation nature of traffic breakdown (F→S transition) at a highway bottleneck can be considered an empirical fundamental of transportation science.

Some of the main empirical features of the phenomenon "nucleation nature of traffic breakdown at the bottleneck" studied in the book are as follows:

(i) In real field traffic data, a nucleus for empirical spontaneous traffic breakdown at a highway bottleneck occurs through an interaction of one of the waves in free flow with a permanent speed disturbance localized at the bottleneck. When the wave reaches the location of the disturbance at the bottleneck, spontaneous traffic breakdown, i.e., a phase transition from free flow to synchronized flow (F→S transition) occurs.
(ii) To initiate empirical spontaneous traffic breakdown at the bottleneck, i.e., to be a nucleus for the breakdown, a wave in free flow should be equal to or a larger one than a critical wave. Small enough waves in free flow cannot be nuclei for traffic breakdown.
(iii) In contrast with empirical *spontaneous* traffic breakdown, any localized congested pattern becomes usually a nucleus for empirical *induced* traffic breakdown: When the congested pattern reaches the location of the bottleneck, the pattern induces the breakdown at the bottleneck.
(iv) Rather than the nature of traffic breakdown, the terms *empirical spontaneous* and *empirical induced* traffic breakdowns at a bottleneck distinguish different *sources* of a nucleus that occurrence leads to traffic breakdown at the bot-

tleneck. In particular, a nucleus for empirical spontaneous traffic breakdown appears usually due to the interaction of a wave in free flow with the permanent local speed disturbance at the bottleneck (items (i) and (ii)), whereas a nucleus for empirical induced traffic breakdown occurs due to the propagation of a congested traffic pattern to the bottleneck location. The source of a nucleus for traffic breakdown is the solely difference between empirical spontaneous and induced traffic breakdowns at the bottleneck. Empirical spontaneous and induced traffic breakdowns (F→S transitions) at a highway bottleneck exhibit the same nucleation nature.

(v) Qualitative features of empirical synchronized flow at a highway bottleneck resulting from either empirical induced traffic breakdown or empirical spontaneous traffic breakdown are identical ones.

(vi) There is a limited range of the flow rate in free flow at a highway bottleneck at which free flow is in metastable state with respect to traffic breakdown (F→S transition) at the bottleneck. At small enough values of the flow rate in free flow no traffic breakdown is observed at the bottleneck; this free flow is stable with respect to traffic breakdown (F→S transition) at the bottleneck.

(vii) At any time instant, there are the infinite number of highway capacities of free flow at a highway bottleneck. The capacities are related to different flow rates at which free flow is in a metastable state with respect to traffic breakdown (F→S transition) at the bottleneck. The existence of an infinite number of highway capacities *at any time instant* means that highway capacity is stochastic.

> Empirical features of nuclei for traffic breakdown at highway bottlenecks confirm the basic assumption of the three-phase theory about the nucleation nature of traffic breakdown (F→S transition) at highway bottlenecks.

15.2 Theoretical Fundamentals of Transportation Science

15.2.1 The Three-Phase Traffic Theory

Driver behavioral characteristics assumed in the three-phase theory explain the empirical fundamental of transportation science—the nucleation nature of traffic breakdown at highway bottlenecks as follows.

(i) The nucleation nature of traffic breakdown is explained by a competition between two opposing tendencies occurring within a random local disturbance at a bottleneck in which the speed is lower and vehicle density is larger than in an initial free flow:

- A tendency towards synchronized flow due to vehicle deceleration associated with the speed adaptation effect.
- A tendency towards the initial free flow due to vehicle acceleration associated with the over-acceleration effect.

(ii) The speed adaptation effect is driver adaptation of the vehicle speed to the speed of the preceding vehicle at any space gap between the vehicles that is within a space gap range between a synchronization space gap and a safe space gap.

(iii) The over-acceleration effect is driver maneuver leading to a higher speed from initial car-following at a lower speed.

(iv) There is a time delay in driver over-acceleration.

(v) At the same density, driver over-acceleration is more probable to occur in free flow than in synchronized flow: At a given density, there is a drop in probability of over-acceleration, when free flow transforms into synchronized flow: A density function of the probability of over-acceleration exhibits a discontinuity.

(vi) In synchronized flow (S), a traffic flow instability can occur resulting in a growing speed wave of the local increase in the speed that propagates upstream within synchronized flow and leads to free flow (F) at the bottleneck. This S→F instability is caused by a finite time delay in driver over-acceleration. The S→F instability exhibits the nucleation nature: Only large enough local disturbances of speed increase within synchronized flow can be nuclei for the S→F instability; contrarily, small disturbances in synchronized flow decay.

(vii) The S→F instability governs traffic breakdown (F→S transition) at a highway bottleneck as follows. When an F→S transition begins to develop, the S→F instability can randomly occur within the emergent synchronized flow region. Due to the S→F instability, free flow returns at the bottleneck. We call this effect as the sequence of F→S→F transitions. A sequence of F→S→F transitions interrupts the formation of a congested pattern at the bottleneck.

(viii) As long as the S→F instability interrupts a former F→S transition, no congested pattern persists at the bottleneck. At some random time instant, the S→F instability is not able to interrupt a former F→S transition. This results in the formation of a congested pattern that persists at the bottleneck. In its turn, the random time instant at which the S→F instability is not able to interrupt the former F→S transition determines a random time delay to traffic breakdown. For this reason, traffic breakdown is called a time-delayed traffic breakdown (time-delayed F→S transition) at the bottleneck.

The three-phase theory is the theoretical fundamental of transportation science that explains the empirical nucleation nature of traffic breakdown.

Reasons for this conclusion are as follows.

- The three-phase theory explains the empirical fundamental of transportation science —the nucleation nature of traffic breakdown at the bottleneck.
- Therefore, the three-phase theory can be considered the theoretical basis for the development of reliable dynamic management, control, assignment, and organization of traffic and transportation networks.

15.2.2 The Breakdown Minimization (BM) Principle

The BM principle states that the optimum of a traffic network with N bottlenecks is reached, when dynamic traffic assignment, optimization and/or control are performed in the network in such a way that the probability for the occurrence of traffic breakdown in at least one of the network bottlenecks during a given time interval for observing traffic flow reaches the minimum possible value. The BM principle is equivalent to the maximization of the probability that during the given time interval traffic breakdown occurs at none of the network bottlenecks.

With the use of the BM principle, the following results have been obtained in the book:

1. There is a general approach for the maximization of the network throughput. This network maximization throughput approach allows us to find general conditions for the *maximization* of the network throughput at which traffic breakdown cannot occur in the whole network. Under these conditions, free flow is ensured in the whole network. It is shown that the network maximization throughput approach is an application of the BM principle for the case "zero breakdown probability" in the network.
2. Under steady state conditions in the network, based on the network throughput maximization approach, a physical measure of traffic and transportation networks called *network capacity* can be introduced. The network capacity allows us to formulate a general physical condition for the maximization of the network throughput at which free flow does persist in the whole network: As long as the total network inflow rate is smaller than the network capacity, under application of the network throughput maximization approach traffic breakdown cannot occur in the network.
3. At larger values of the total network inflow rate related to rush hours, it is often the case that the probability of traffic breakdown in the network cannot remain zero. Then, dynamic traffic assignment and control in a traffic or transportation network performed through the application of the BM principle minimize the probability of the occurrence of traffic breakdown in the network during a given time interval for observing traffic flow.
4. If traffic breakdown has already occurred at a network bottleneck, then dynamic traffic assignment, network optimization and control can consist of the stages:

(i) The dissolution of traffic congestion due to dynamic feedback control of traffic breakdown in a neighborhood of the bottleneck. Through the use of dynamic feedback control of traffic breakdown (ANCONA) in a neighborhood of the bottleneck the dissolution of traffic congestion can be achieved.
(ii) The minimization of traffic breakdown probability with the BM principle in the remaining network, i.e., the network part that is not influenced by congestion.

5. In contrast with the network throughput maximization approach that guarantees free flow conditions in the whole network when the total network inflow rate is smaller than the network capacity, standard approaches to dynamic traffic assignment in the network related to the state-of-the-art in traffic and transportation research deteriorate the traffic system provoking heavy traffic congestion in the network.

For this reason, the BM principle is the theoretical fundamental of transportation science that permits to maximize the network throughput preventing traffic breakdown in the whole network.

The BM principle can be applied, even when non-steady state conditions are realized in a traffic or transportation network.

The BM principle is the theoretical fundamental of transportation science that minimizes traffic congestion in a traffic or transportation network.

15.3 Failure of Classical Traffic and Transportation Theories

Although there are great achievements of classical traffic flow theories and models in the understanding of many important empirical traffic phenomena, the classical theories and traffic flow models cannot explain and show the empirical fundamental of transportation science—the nucleation nature of traffic breakdown at a highway bottleneck. This explains the failure of applications of the classical traffic and transportation theories and models in the real world.

In particular, the failure of the LWR theory and its applications as well as traffic flow models of the GM model class is explained as follows:

(i) The LWR model and its applications failed because the LWR model cannot show induced traffic breakdown at a highway bottleneck observed in real field traffic data.
(ii) Traffic flow models of the GM model class failed because the models are not able to show the metastability of free flow with respect to an F→S transition at a highway bottleneck. In other words, traffic flow models of the GM class are not able to show the empirical nucleation nature of traffic breakdown (F→S transition) at the bottleneck observed in real field traffic data.

Classical traffic flow models are not consistent with empirical nucleation nature of traffic breakdown (F→S transition) at the bottleneck. This explains the failure of the classical traffic flow models for the description of real traffic breakdown at highway bottlenecks.

Applications of the classical traffic flow models for an analysis of the effect of ITS (intelligent transportation systems) on traffic flow lead to invalid conclusions about the ITS performance in real traffic.

This criticism is related to all ITS that affect traffic flow, for example, on-ramp metering and many other ITS-applications, like the effect of cooperative driving with the use of vehicle-two-vehicle (V2V) communication on traffic flow characteristics, simulations of the performance of dynamic traffic assignment and optimization in traffic and transportation networks, the analysis of the effect of adaptive cruise control (ACC) and automatic driving vehicles (self-driving vehicles) on traffic flow.

Based on the physics of the network capacity, we have shown that any application of the classical Wardrop's equilibria for dynamic traffic assignment in traffic and transportation networks deteriorates basically the traffic system: Even when the total network inflow rate is smaller than the network capacity, the dynamic traffic assignment with the Wardrop's equilibria leads to the occurrence of traffic congestion in networks.

Standard approaches for dynamic traffic assignment related to the state-of-the-art in traffic and transportation research deteriorate the traffic system while provoking heavy traffic congestion in urban networks.

15.4 Paradigm Shift in Transportation Science

None of the classical traffic flow theories incorporates the metastability of free flow with respect to an F→S transition (traffic breakdown) at a highway bottleneck.

The metastability of free flow with respect to the F→S transition (traffic breakdown) at the bottleneck has no sense for the classical traffic theories.

The metastability of free flow with respect to an F→S transition at a highway bottleneck is the main feature of the three-phase theory.

This shows the incommensurability of the three-phase theory and the classical traffic-flow theories. The term *incommensurability* has been introduced by Kuhn in his book "The structure of scientific revolutions" to explain a paradigm shift in a scientific field.

15.5 Challenges for Transportation Science

As explained in this book, classical traffic and transportation theories that are the state-of-the-art in traffic and transportation research failed by their applications in the real world. For this reason, we can make the following conclusion.

The three-phase theory and the BM principle can be considered the theoretical fundamentals of transportation science.

Therefore, there are at least the following challenging areas for transportation science in which the three-phase theory and the BM principle should be used in the future:

1. The development of new traffic flow models in the framework of the three-phase theory should be intensified. This is because only there-phase traffic flow models that incorporates the metastability of free flow with respect to an F→S transition at a highway bottleneck can be used for reliable analysis of the performance of ITS in traffic flow.

2. Congested pattern control approach based on the three-phase theory should be further developed and implemented to the market.
3. The three-phase theory should be also a theoretical basis for the future development of effective methods of traffic control in networks in the case of an extremely large total network inflow rate at which congestion propagates through the whole network. Indeed, in many real traffic networks there are not enough alternative routes to avoid traffic congestion at large enough traffic demand. Thus, at large enough traffic demand even "perfect" methods of traffic control and dynamic network optimization could not prevent traffic breakdown. Nevertheless, the application of ITS can change characteristics of traffic congestion with the objective to increase traffic safety and comfort while moving in congested traffic.
4. A further theoretical development of the network throughput maximization approach (the BM principle for the case "zero breakdown probability") as well as other applications of the BM principle together with congested pattern control approach should be made.
5. Traffic control centers as well as traffic guidance systems in vehicles should be developed that permit dynamic traffic assignment in traffic networks based on the applications of the BM principle.
6. Studies of microscopic (single vehicle) empirical features of traffic breakdown at network bottlenecks and resulting congested patterns should be intensified. In particular, the empirical studies should give the answer to questions about features of congested traffic that can effectively be influenced through the use of ITS-applications for the spatial limitation of congestion growth and/or for the dissolution of traffic congestion.
7. The development of assistant systems in vehicles as well as automatic driving (self-driven) vehicles that can influence on spatial limitation of congestion growth and/or on congestion dissolution in traffic and transportation networks should be made.
8. The development of automatic driving vehicles whose dynamic behavior is automatically adaptable to dynamic behavior of manual drivers should be performed.
9. The network throughput maximization approach should be applied for the design of future urban areas.
10. Features of the nucleation nature of traffic breakdown and applications of the BM principle for dynamic traffic assignment and control in traffic and transportation networks discussed in this book together with spatiotemporal features of the propagation of traffic congestion should be used for development of realistic models of travel decision behavior and transportation demand.

Erratum to: The Reason for Incommensurability of Three-Phase Theory with Classical Traffic Flow Theories

Boris S. Kerner

Erratum to:
Chapter 8 in: B.S. Kerner, *Breakdown in Traffic Networks*, DOI 10.1007/978-3-662-54473-0_8

Unfortunately, there was an error in Fig. 8.28(b) in the original publication of the chapter. The original chapter has been corrected. The correct version of the Figure is shown below.

The updated online version of this chapter can be found at
http://dx.doi.org/10.1007/978-3-662-54473-0_8

B.S. Kerner, *Breakdown in Traffic Networks*, DOI 10.1007/978-3-662-54473-0_16

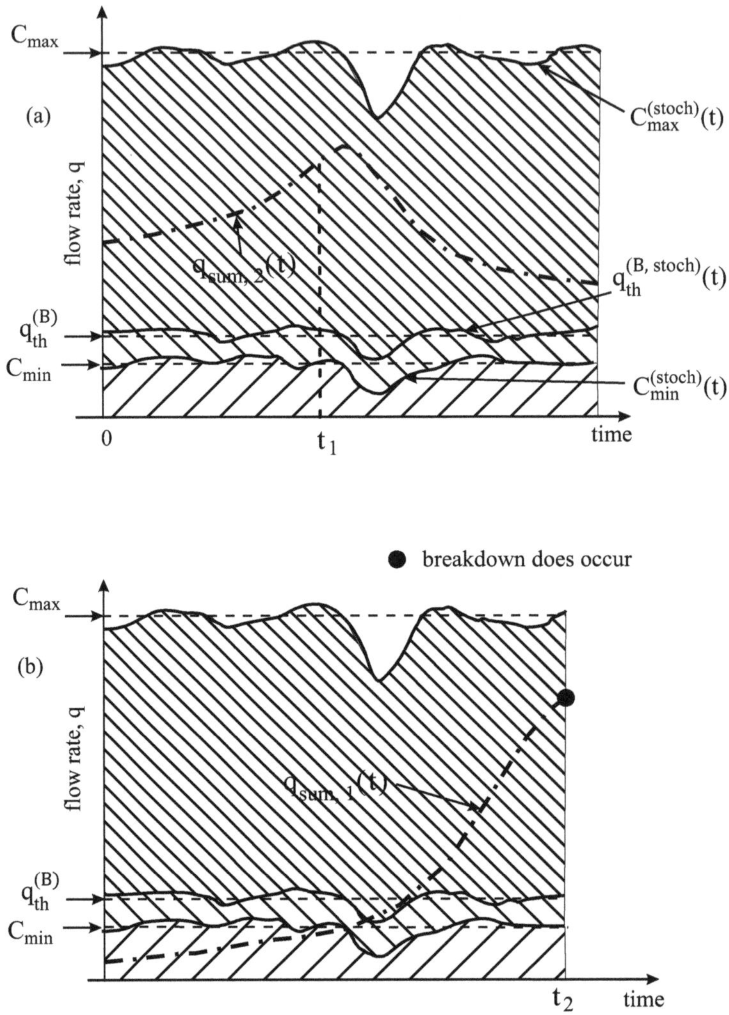

Fig. 8.28 Qualitative explanation of traffic breakdown with the use of the infinite number of capacities of free flow at a highway bottleneck of the three-phase theory. Hypothetical time-functions $C_{\mathrm{max}}^{(\mathrm{stoch})}(t)$, $C_{\mathrm{min}}^{(\mathrm{stoch})}(t)$, and $q_{\mathrm{th}}^{(\mathrm{B,\ stoch})}(t)$ are adapted from Fig. 5.14. Hypothetical time functions of the flow rates $q_{\mathrm{sum},\ 2}(t)$ in (**a**) and $q_{\mathrm{sum},\ 1}(t)$ in (**b**) as well as time instant t_1 in (**a**) are, respectively, the same as those in Fig. 4.14. Adapted from [67–69]

Appendix A
Kerner-Klenov Stochastic Microscopic Model in Framework of Three-Phase Theory

Additional List of Symbols Used in Appendices A and B

δx	A discretization space interval
τ	A discretization time interval
δv	A discretization vehicle speed interval, $\delta v = \delta x/\tau$
δa	A discretization vehicle acceleration (deceleration) interval, $\delta a = \delta v/\tau$
t_{n}	The discrete time, $t_{\mathrm{n}} = \tau n,\ n = 0, 1, \ldots$. Time is measured in values of τ, respectively, value τ in all formulas is assumed below to be the dimensionless value $\tau = 1$
n	The subscript n marks in variables and functions time step corresponding to the discrete time t_{n}
v_n	The vehicle speed at time step n
x_n	The vehicle coordinate at time step n
ξ_n	Model speed fluctuations at time step n
$\tilde{v}_n$	A vehicle speed without speed fluctuations at time step n
$v_{\mathrm{s},n}$	A safe vehicle speed at time step n
g_n	A space gap between two vehicles following each other at time step n, $g_n = x_{\ell,n} - x_n - d$
ℓ	The subscript ℓ marks variables related to the preceding vehicle; for example, $v_{\ell,n}$ is the speed of the preceding vehicle at time step n
d	Vehicle length. The vehicle length includes the mean space gap between vehicles that are in a standstill within a wide moving jam or within a vehicle queue at traffic signal
G_n	The synchronization space gap between two vehicles following each other at time step n
$g_{\mathrm{safe},n}$	The safe space gap between two vehicles following each other at time step n
S_n	The state of vehicle motion at time step n

B.S. Kerner, *Breakdown in Traffic Networks*, DOI 10.1007/978-3-662-54473-0

Δv_n	The difference between the speed of the preceding vehicle and the vehicle speed at time step n: $\Delta v_n = v_{\ell,n} - v_n$
A_{n+1}	The vehicle acceleration (deceleration): $A_{n+1} = (v_{n+1} - v_n)/\tau$
a	The maximum vehicle acceleration
$a_{\max}$	The maximum vehicle acceleration in model of city traffic
$v_{\text{free}}(g)$	The space-gap dependence of free flow speed
v_{free}	Vehicle speed in free flow under assumption that the free flow speed does not depend on the space gap
$v_{\text{free}}^{(\min)}$	A parameter of the space-gap dependence of free flow speed
$v_{\text{free}}^{(\max)}$	A parameter of the space-gap dependence of free flow speed
$v_{\text{free, long}}^{(\max)}$	The maximum speed of long vehicles (trucks)
$v_{\ell}^{(a)}$	An anticipation (predicted) speed of the preceding vehicle
a_n	A random vehicle acceleration at time step n
b_n	A random vehicle deceleration at time step n
p_0	A probability of random vehicle acceleration related to a_n
p_1	A probability of random vehicle deceleration related to b_n
p_2	A probability of random vehicle deceleration related to b_n
ξ_{b}	A random source for vehicle over-deceleration when the vehicle decelerates
ξ_{a}	A random source for vehicle over-acceleration when the vehicle accelerates
$\xi^{(0)}$	A random source for vehicle deceleration and acceleration when the vehicle maintains its speed
p_{b}	A probability of random vehicle deceleration
p_{a}	A probability of random vehicle acceleration
$\tau_{\text{del}}^{(\text{acc})}$	The mean time delay in vehicle acceleration
$\tau_{\text{del}}^{(\text{acc})}(0)$	The mean time delay in vehicle acceleration at the vehicle speed that is equal to zero (at the downstream front of a wide moving jam or a moving queue at the signal)
$\tau_{\text{del, syn}}^{(\text{acc})}$	The mean time delay in acceleration at downstream front of synchronized flow
$R \to L$	Lane changing from the right lane to the left lane
$L \to R$	Lane changing from the left lane to the right lane
$+$	A superscript in variables, parameters, and functions that denotes the preceding vehicle in the "target" (neighboring) lane
$-$	A superscript in variables, parameters, and functions that denotes the trailing vehicle in the "target" (neighboring) lane
$\hat{v}_n$	A vehicle speed at time step n used in lane changing rules and rules of vehicle merging at bottlenecks
$\hat{v}_n^{+}$	A speed of the preceding vehicle in the "target" (neighboring) lane at time step n used in lane changing rules and rules of vehicle merging at bottlenecks
$x_n^{(\text{m})}$	A coordinate at time step n used in lane changing rules and rules of vehicle merging at bottlenecks

$g_{\text{target}}^{(\min)}$	A space gap used in lane changing rules and rules of vehicle merging at bottlenecks
L_{m}	The length of the merging region of on- and off-ramp bottlenecks
L_{r}	The length of a part of the on- and off-ramp lanes, respectively, upstream and downstream of the merging region
x_{on}	The beginning of the merging region of the on-ramp
$v_{\text{free on}}$	The maximum speed in the on-ramp lane
x_{off}	The beginning of the merging region of the off-ramp
$v_{\text{free off}}$	The maximum speed in the off-ramp lane
L_{c}	The length of the main road upstream of the off-ramp within which vehicles going to the off-ramp have to change from the left lane to the right lane of the main road
η_{off}	A percentage of the flow rate of vehicles that go to the off-ramp
v_{M}	The maximum speed of a moving bottleneck
L_{M}	The length of the merging region of the moving bottleneck
p_{c}	A probability of lane changing
v_{pinch}	A parameter of the KKSW CA model (Appendix B) that defines a range of speeds in synchronized flow within which wide moving jams occur spontaneously with a larger probability
$j = 1, 2, \ldots, H$	A superscript in variables, parameters, and functions that identifies vehicles and drivers with different parameters in heterogeneous traffic flow
d_{ℓ}	The length of the preceding vehicle (in heterogeneous traffic flow)
$r = \text{rand}(0, 1)$	A random number uniformly distributed between 0 and 1
$\lfloor z \rfloor$	The largest integer less than or equal to z
$\lceil z \rceil$	The smallest integer greater than or equal to z
$\Theta(z)$	$\Theta(z) = 0$ at $z < 0$ and $\Theta(z) = 1$ at $z \geq 0$

A.1 Motivation

Almost each traffic flow model can describe some real traffic features. Each traffic flow model has limitations for the description of some real traffic features.

After the author has formulated the three-phase theory [10–18], we wanted to develop a *mathematical* microscopic traffic flow model in the framework of the three-phase theory. The model should satisfy the following requirements:

(i) The model should show a 2Z-characteristic for phase transitions between the three traffic phases F, S, and J (see Figs. 1.9 and 8.25). For this reason, we have developed a *three-phase* traffic flow model (see explanations in Sects. 1.12 and 8.9).

(ii) The model should reproduce known empirical spatiotemporal features of traffic flow patterns as close as possible to all known empirical results obtained form studies of real field traffic data measured in different countries (see Chap. 3 as well as results of empirical studies of real spatiotemporal features of traffic patterns presented in Chap. 2 and Part II of the book [19]).

(iii) Each of the driver characteristics should be simulated independent of each other. For this reason, we have developed a *microscopic* traffic flow model.

(iv) The model should simulate the effect of fluctuations on traffic flow phenomena. For this reason, we have developed a *stochastic* microscopic traffic flow model.

(v) The model should be able to simulate all known and new different ITS-applications as close as possible to the reality.

To explain item (v), we should note that during a detailed study of different traffic simulation tools that are on the market, we have understood that there are no traffic simulation tools developed in the engineering traffic community that can be used for a reliable analysis of a variety of ITS-applications (see explanations of this criticism in Chap. 4). This explains why we could use none of the existing traffic simulation tools for reliable simulations of the ITS-applications.

One of the most complex requirements is that the model should simulate each of the driver characteristics independent of each other (item (iii)). To explain the problem, we should note that in different driving situations drivers exhibit usually quantitative different driver time delays in acceleration and deceleration. Therefore, we have decided to use as many model parameters as necessarily to have a possibility for the choice of *different values of driver time delays in acceleration and deceleration* that can be chosen independent of each other in a diverse variety of driving situations.

The choice of model parameters and robustness of simulations against changes of these parameters are as follows. Values of model parameters used in the simulations have been obtained and optimized through a comparison of simulated phase transitions and resulting spatiotemporal features of congested patterns with empirical data [19]. It has turned out that one of the most important empirical features of highway traffic, which have had to be carefully incorporated in the model, is the empirical fundamental of transportation science—the nucleation nature of traffic breakdown (F→S transition) at highway bottlenecks. Time durations of traffic breakdown and other phase transitions, velocity of fronts separating different traffic phases in space and time as well as spatiotemporal distributions of the flow rates, speeds, and densities within different congested patterns, which have been observed in real traffic data measured over years on different highways (between 1995 and 2016) have been used for the validation of model variables and parameters. We have found that results of simulations are robust against changes of the model parameters: For each of the model parameters no fine-tuning within some limited (but not small) parameter range is required.

In this chapter, we consider a *stochastic* microscopic three-phase traffic flow model, in which a variety of driver time delays is simulated through the use of model fluctuations [28, 29]. This is because simulations of the *deterministic* microscopic three-phase traffic flow model derived by Kerner and Klenov [31] have not been considered in this book.

It must be noted that in addition to the incorporation of the hypotheses of the three-phase theory [10–18], in the Kerner-Klenov stochastic microscopic three-phase traffic flow model many ideas about simulations of the driver behavior introduced in earlier two-phase traffic flow models have also been used. In particular, simulations of the over-deceleration effect due to a driver time delay firstly introduced in 1958–1961 by Herman, Gazis, Montroll, Potts, Rothery, and Chandler [2, 6, 7, 9], simulations of driver time delays through the use of models fluctuations introduced by Nagel and Schreckenberg [43], simulations of slow-to-start-rules introduced by Takayasu and Takayasu [45] as well as Barlović et al. [1, 3], simulations of lane changing by Nagel et al [42] as well as simulations of a safe speed in traffic flow introduced by Gipps [8] and Krauß et al. [41] are also very important elements of the Kerner-Klenov stochastic microscopic three-phase traffic flow model (the Kerner-Klenov model for short).

A.2 Discrete Model Version

The Kerner-Klenov model developed in 2002 was a continuum in space model [28]. This model has been further developed and applied for simulations of traffic flow dynamics up to 2008 [20, 29, 30, 35–37]. The continuum in space Kerner-Klenov stochastic microscopic three-phase model has been considered in details in Sects. 16.3 and 20.2 of the book [19].

In 2009, we developed a model version that is discrete in space and time [32]. Although both continuum and discrete model versions exhibit almost the same traffic features at highway bottlenecks, we use the discrete model version for the following reasons:

1. In contrast with the continuum model version [28, 29], in which there are no model fluctuations when the speed difference between two following each other vehicles is zero, in the discrete model version (as in real traffic) there are always speed fluctuations. As shown in [32], this model feature leads to a more accurate simulation of traffic breakdown in some traffic scenarios.
2. Simulations with the discrete model version are quicker than that with the continuous-in space model version.

For these reasons, after 2009 for simulations with the Kerner-Klenov stochastic microscopic three-phase model we use only model versions that are discrete in space and time [22, 24–26, 33, 34]. The objective of this chapter is to consider the Kerner-Klenov stochastic microscopic three-phase model that is discrete in space and time.

In a discrete model version of the Kerner-Klenov stochastic microscopic three-phase model, rather than the continuum space co-ordinate [28, 29], a discretized space co-ordinate with a small enough value of the discretization space interval δx is used. Consequently, the vehicle speed and acceleration (deceleration) discretization intervals are $\delta v = \delta x/\tau$ and $\delta a = \delta v/\tau$, respectively, where τ is time step.

Because in the discrete model version discrete (and dimensionless) values of space coordinate, speed and acceleration are used, which are measured respectively in values δx, δv and δa, and time is measured in values of τ, value τ in all formulas is assumed below to be the dimensionless value $\tau = 1$.

A choice of $\delta x = 0.01$ m made in the model determines the accuracy of vehicle speed calculations *in comparison* with the initial continuum in space stochastic model of [28]. We have found that the discrete model exhibits similar characteristics of phase transitions and resulting congested patterns at highway bottlenecks as those in the continuum model at δx that satisfies the conditions

$$\delta x/\tau^2 \ll b,\ a,\ a^{(\mathrm{a})},\ a^{(\mathrm{b})},\ a^{(0)}, \tag{A.1}$$

where model parameters for driver deceleration and acceleration b, a, $a^{(\mathrm{a})}$, $a^{(\mathrm{b})}$, $a^{(0)}$ will be explained below.

A.3 Update Rules of Vehicle Motion in Road Lane in Model of Identical Drivers and Vehicles

Update rules of vehicle motion in the discrete model for identical drivers and identical vehicles moving in a road lane are as follows:

$$v_{n+1} = \max(0, \min(v_{\mathrm{free}}, \tilde{v}_{n+1} + \xi_n, v_n + a\tau, v_{\mathrm{s},n})), \tag{A.2}$$

$$x_{n+1} = x_n + v_{n+1}\tau, \tag{A.3}$$

where the index n corresponds to the discrete time $t_{\mathrm{n}} = \tau n$, $n = 0, 1, \ldots$; v_n is the vehicle speed at time step n, a is the maximum acceleration, $\tilde{v}_n$ is the vehicle speed without speed fluctuations ξ_n:

$$\tilde{v}_{n+1} = \min(v_{\mathrm{free}}, v_{\mathrm{s},n}, v_{\mathrm{c},n}), \tag{A.4}$$

$$v_{\mathrm{c},n} = \begin{cases} v_n + \Delta_n & \text{at } g_n \le G_n \\ v_n + a_n\tau & \text{at } g_n > G_n, \end{cases} \tag{A.5}$$

$$\Delta_n = \max(-b_n\tau, \min(a_n\tau,\ v_{\ell,n} - v_n)), \tag{A.6}$$

$$g_n = x_{\ell,n} - x_n - d, \tag{A.7}$$

the subscript ℓ marks variables related to the preceding vehicle, $v_{\mathrm{s},n}$ is a safe speed at time step n, v_{free} is the free flow speed in free flow, ξ_n describes speed fluctuations; g_n is a space gap between two vehicles following each other; G_n is the synchronization space gap; all vehicles have the same length d. The vehicle length d includes the mean space gap between vehicles that are in a standstill within a wide moving jam.

Values $a_n \geq 0$ and $b_n \geq 0$ in (A.5), (A.6) restrict changes in speed per time step when the vehicle accelerates or adjusts the speed to that of the preceding vehicle.

A.3.1 Synchronization Space Gap and Hypothetical Steady States of Synchronized Flow

Equations (A.5), (A.6) describe the adaptation of the vehicle speed to the speed of the preceding vehicle, i.e., the speed adaptation effect in synchronized flow (Sect. 5.9). This vehicle speed adaptation takes place within the synchronization gap G_n: At

$$g_n \leq G_n \tag{A.8}$$

the vehicle tends to adjust its speed to the speed of the preceding vehicle. This means that the vehicle decelerates if $v_n > v_{\ell,n}$, and accelerates if $v_n < v_{\ell,n}$ [28].

In (A.5), the synchronization gap G_n depends on the vehicle speed v_n and on the speed of the preceding vehicle $v_{\ell,n}$:

$$G_n = G(v_n, v_{\ell,n}), \tag{A.9}$$

$$G(u, w) = \max(0, \lfloor k\tau u + a^{-1}u(u - w)\rfloor), \tag{A.10}$$

where $k > 1$ is constant; $\lfloor z \rfloor$ denotes the integer part of z.

The speed adaptation effect within the synchronization distance is related to the hypothesis of the three-phase theory: Hypothetical steady states of synchronized flow cover a 2D region in the flow–density (see Fig. A.1a). Boundaries F, L, and U of this 2D-region shown in Fig. A.1a are, respectively, associated with the free flow speed in free flow, a synchronization space gap G, and a safe space gap g_{safe}. A speed-function of the safe space gap $g_{\text{safe}}(v)$ is found from the equation

$$v = v_{\text{s}}(g_{\text{safe}},\ v). \tag{A.11}$$

Respectively, as for the continuum model (see Sec. 16.3 of the book [19]), for the discrete model hypothetical steady states of synchronized flow cover a 2D-region in the flow–density plane (Fig. A.1a,b). However, because the speed v and space gap g are integer in the discrete model, the steady states do not form a continuum in the flow–density plane as they do in the continuum model. The inequalities

$$v \leq v_{\text{free}}, \quad g \leq G(v,\ v), \quad g \geq g_{\text{safe}}(v), \tag{A.12}$$

define a 2D-region in the space-gap–speed plane (Fig. A.1c) in which the hypothetical steady states exist for the discrete model, when all model fluctuations are neglected.

In (A.12), we have taken into account that in the hypothetical steady states of synchronized flow vehicle speeds and space gaps are assumed to be time-independent and the speed of each of the vehicles is equal to the speed of the

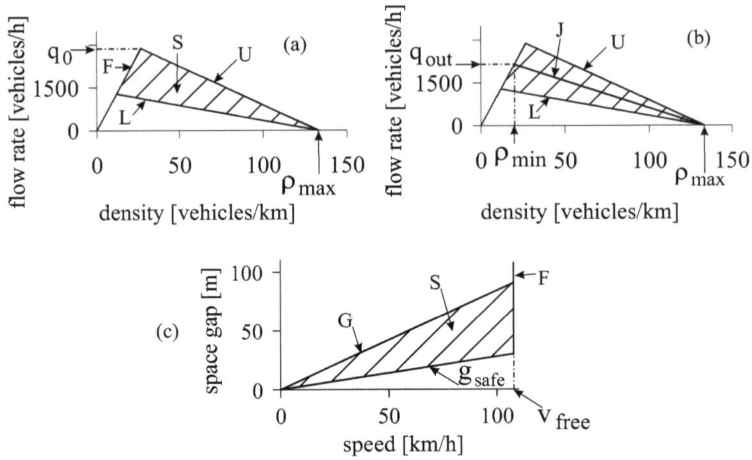

Fig. A.1 Steady speed states for the Kerner-Klenov traffic flow model in the flow–density (**a**, **b**) and in the space-gap–speed planes (**c**). In (**a, b**), L and U are, respectively, lower and upper boundaries of 2D-regions of steady states of synchronized flow. In (**b**), J is the line J whose slope is equal to the characteristic mean velocity v_g of a wide moving jam; in the flow–density plane, the line J represents the propagation of the downstream front of the wide moving jam with time-independent velocity v_g. F—free flow, S—synchronized flow

associated preceding vehicle: $v = v_\ell$. However, due to model fluctuations, steady states of synchronized flow are destroyed, i.e., they do not exist in simulations; this explains the term "hypothetical" steady states of synchronized flow. Therefore, rather than steady states some non-homogeneous in space and time traffic states occur. In other words, steady states are related to a hypothetical model fluctuation-less limit of homogeneous in space and time vehicle motion that does not realized in real simulations. Driver time delays are described through model fluctuations. Therefore, any application of the Kerner-Klenov stochastic microscopic three-phase traffic flow model without model fluctuations has *no sense*. In other words, for the description of real spatiotemporal traffic flow phenomena, model speed fluctuations incorporated in this model are needed.

A.3.2 Model Speed Fluctuations

In the model, random vehicle deceleration and acceleration are applied depending on whether the vehicle decelerates or accelerates, or else maintains its speed:

$$\xi_n = \begin{cases} \xi_{\text{a}} & \text{if } S_{n+1} = 1 \\ -\xi_{\text{b}} & \text{if } S_{n+1} = -1 \\ \xi^{(0)} & \text{if } S_{n+1} = 0. \end{cases} \tag{A.13}$$

Table A.1 Model parameters of vehicle motion in road lane often used in simulations

$\tau_{\text{safe}} = \tau = 1$ s, $d = 7.5$ m$/\delta x$,
$\delta x = 0.01$ m, $\delta v = 0.01$ ms^{-1}, $\delta a = 0.01$ ms^{-2},
$v_{\text{free}} = 30$ ms$^{-1}/\delta v$, $b = 1$ ms$^{-2}/\delta a$, $a = 0.5$ ms$^{-2}/\delta a$,
$k = 3, p_1 = 0.3, p_b = 0.1, p_a = 0.17, p^{(0)} = 0.005$,
$p_0(v_n) = 0.575 + 0.125 \min(1, v_n/v_{01})$,
$p_2(v_n) = 0.48 + 0.32\Theta(v_n - v_{21})$,
$v_{01} = 10$ ms$^{-1}/\delta v$, $v_{21} = 15$ ms$^{-1}/\delta v$,
$a^{(0)} = 0.2a, a^{(a)} = a^{(b)} = a$

State of vehicle motion S_{n+1} in (A.13) is determined by formula

$$S_{n+1} = \begin{cases} -1 & \text{if } \tilde{v}_{n+1} < v_n \\ 1 & \text{if } \tilde{v}_{n+1} > v_n \\ 0 & \text{if } \tilde{v}_{n+1} = v_n. \end{cases} \tag{A.14}$$

In (A.13), ξ_{b}, $\xi^{(0)}$, and ξ_{a} are random sources for deceleration and acceleration that are as follows:

$$\xi_{\text{b}} = a^{(\text{b})}\tau\Theta(p_{\text{b}} - r), \tag{A.15}$$

$$\xi^{(0)} = a^{(0)}\tau \begin{cases} -1 & \text{if } r < p^{(0)} \\ 1 & \text{if } p^{(0)} \le r < 2p^{(0)} \quad \text{and } v_n > 0 \\ 0 & \text{otherwise,} \end{cases} \tag{A.16}$$

$$\xi_{\text{a}} = a^{(\text{a})}\tau\Theta(p_{\text{a}} - r), \tag{A.17}$$

p_{b} is probability of random vehicle deceleration, p_{a} is probability of random vehicle acceleration, $p^{(0)}$ and $a^{(0)} \le a$ are constants, $r = \text{rand}(0, 1)$, $\Theta(z) = 0$ at $z < 0$ and $\Theta(z) = 1$ at $z \ge 0$, $a^{(\text{a})}$ and $a^{(\text{b})}$ are model parameters (see Table A.1), which in some applications can be chosen as speed functions $a^{(\text{a})} = a^{(\text{a})}(v_n)$ and $a^{(\text{b})} = a^{(\text{b})}(v_n)$ (see Sect. A.8 and Table A.4).

A.3.3 Stochastic Time Delays of Acceleration and Deceleration

To simulate time delays either in vehicle acceleration or in vehicle deceleration, a_n and b_n in (A.6) are taken as the following stochastic functions

$$a_n = a\Theta(P_0 - r_1), \tag{A.18}$$

$$b_n = a\Theta(P_1 - r_1), \tag{A.19}$$

$$P_0 = \begin{cases} p_0 & \text{if } S_n \neq 1 \\ 1 & \text{if } S_n = 1, \end{cases} \tag{A.20}$$

$$P_1 = \begin{cases} p_1 & \text{if } S_n \neq -1 \\ p_2 & \text{if } S_n = -1, \end{cases} \tag{A.21}$$

$r_1 = \text{rand}(0, 1)$, p_1 is constant, $p_0 = p_0(v_n)$ and $p_2 = p_2(v_n)$ are speed functions (see Table A.1).

A.3.4 Simulations of Slow-to-Start Rule

In the model, simulations of the well-known effect of the driver time delay in acceleration at the downstream front of synchronized flow or a wide moving jam known as a slow-to-start rule [1] are made as a collective effect through the use of Eqs. (A.5), (A.6), and a random value of vehicle acceleration (A.18). Eq. (A.18) with $P_0 = p_0 < 1$ is applied only if the vehicle did not accelerate at the former time step ($S_n \neq 1$); in the latter case, a vehicle accelerates with some probability p_0 that depends on the speed v_n; otherwise $P_0 = 1$ (see formula (A.20)).

The mean time delay in vehicle acceleration is equal to

$$\tau_{\text{del}}^{(\text{acc})}(v_n) = \frac{\tau}{p_0(v_n)}. \tag{A.22}$$

From formula (A.22), it follows that the mean time delay in vehicle acceleration from a standstill within a wide moving jam (i.e., when in formula (A.22) the speed $v_n = 0$) is equal to

$$\tau_{\text{del}}^{(\text{acc})}(0) = \frac{\tau}{p_0(0)}. \tag{A.23}$$

The mean time delay in vehicle acceleration from a standstill within a wide moving jam determines the parameters of the line J in the flow–density plane (Fig. A.1b).

Probability $p_0(v_n)$ in (A.20) is chosen to be an increasing speed function (see Sect. A.8.3 below). Because the speed within synchronized flow is larger than zero, the mean time delay in vehicle acceleration at the downstream front of synchronized flow that we denote by

$$\tau_{\text{del, syn}}^{(\text{acc})} = \tau_{\text{del}}^{(\text{acc})}(v_n), \quad v_n > 0 \tag{A.24}$$

is shorter than the mean time delay in vehicle acceleration at the downstream front of the wide moving jam $\tau_{\text{del}}^{(\text{acc})}(0)$:

$$\tau_{\text{del, syn}}^{(\text{acc})} < \tau_{\text{del}}^{(\text{acc})}(0). \tag{A.25}$$

A.3.5 Safe Speed

In the model, the safe speed $v_{s,n}$ in (A.2) is chosen in the form

$$v_{s,n} = \min\,(v_n^{(\mathrm{safe})}, g_n/\tau + v_\ell^{(a)}), \tag{A.26}$$

$v_\ell^{(a)}$ is an "anticipation" speed of the preceding vehicle that will be considered below, the function

$$v_n^{(\mathrm{safe})} = \lfloor v^{(\mathrm{safe})}(g_n,\ v_{\ell,n}) \rfloor \tag{A.27}$$

in (A.26) is related to the safe speed $v^{(\mathrm{safe})}(g_n,\ v_{\ell,n})$ in the model by Krauß et al. [41], which is a solution of the Gipps's equation [8]

$$v^{(\mathrm{safe})}\tau_{\mathrm{safe}} + X_{\mathrm{d}}(v^{(\mathrm{safe})}) = g_n + X_{\mathrm{d}}(v_{\ell,n}), \tag{A.28}$$

where τ_{safe} is a safe time gap, $X_{\mathrm{d}}(u)$ is the braking distance that should be passed by the vehicle moving first with the speed u before the vehicle can come to a stop.

The condition (A.28) enables us to find the safe speed $v^{(\mathrm{safe})}$ as a function of the space gap g_n and speed $v_{\ell,n}$ provided $X_{\mathrm{d}}(u)$ is a known function. In the case when the vehicle brakes with a constant deceleration b, the change in the vehicle speed for each time step is $-b\tau$ except the last time step before the vehicle comes to a stop. At the last time step, the vehicle decreases its speed at the value $b\tau\beta$, where β is a fractional part of $u/b\tau$. According to formula (A.3) for the displacement of the vehicle for one time step, the braking distance $X_{\mathrm{d}}(u)$ is [41]

$$X_{\mathrm{d}}(u) = \tau\big(u - b\tau + u - 2b\tau + \ldots + \beta b\tau\big). \tag{A.29}$$

From (A.29), it follows [41]

$$X_{\mathrm{d}}(u) = b\tau^2\left(\alpha\beta + \frac{\alpha(\alpha-1)}{2}\right), \tag{A.30}$$

$\alpha = \lfloor u/b\tau \rfloor$ is an integer part of $u/b\tau$.

The safe speed $v^{(\mathrm{safe})}$ as a solution of equation (A.28) at the distance $X_{\mathrm{d}}(u)$ given by (A.30) and at $\tau_{\mathrm{safe}} = \tau$ has been found by Krauß et al. [41]

$$v^{(\mathrm{safe})}(g_n, v_{\ell,n}) = b\tau(\alpha_{\mathrm{safe}} + \beta_{\mathrm{safe}}), \tag{A.31}$$

where

$$\alpha_{\mathrm{safe}} = \left\lfloor \sqrt{2\frac{X_{\mathrm{d}}(v_{\ell,n}) + g_n}{b\tau^2} + \frac{1}{4}} - \frac{1}{2} \right\rfloor, \tag{A.32}$$

$$\beta_{\text{safe}} = \frac{X_{\text{d}}(v_{\ell,n}) + g_n}{(\alpha_{\text{safe}} + 1)b\tau^2} - \frac{\alpha_{\text{safe}}}{2}. \tag{A.33}$$

The safe speed in the model by Krauß et al. [41] provides collision-less motion of vehicles if the time gap g_n/v_n between two vehicles is greater than or equal to the time step τ, i.e., if $g_n \geq v_n\tau$ [40]. In the model, it is assumed that in some cases, mainly due to lane changing or merging of vehicles onto the main road within the merging region of bottlenecks, the space gap g_n can become less than $v_n\tau$. In these critical situations, the collision-less motion of vehicles in the model is a result of the second term in (A.26) in which some prediction ($v_\ell^{(\text{a})}$) of the speed of the preceding vehicle at the next time step is used. The related "anticipation" speed $v_\ell^{(\text{a})}$ at the next time step is given by formula

$$v_\ell^{(\text{a})} = \max(0, \min(v_{\ell,n}^{(\text{safe})}, v_{\ell,n}, g_{\ell,n}/\tau) - a\tau), \tag{A.34}$$

where $v_{\ell,n}^{(\text{safe})}$ is the safe speed (A.27), (A.31)–(A.33) for the preceding vehicle, $g_{\ell,n}$ is the space gap in front of the preceding vehicle. Simulations have shown that formulas (A.26), (A.27), (A.31)–(A.34) lead to collision-less vehicle motion over a wide range of parameters of the merging region of highway bottlenecks (Sects. A.6 and A.6.3) and for chosen lane changing rules (Sect. A.5).

In hypothetical steady states of traffic flow (Fig. A.1a), the safe space gap g_{safe} is determined from equation $v = v_{\text{s}}$; in accordance with Eqs. (A.26)–(A.28), at a given v in steady traffic states $v = v_\ell$ the safe speed

$$v_{\text{s}} = g_{\text{safe}}/\tau_{\text{safe}}, \tag{A.35}$$

and, therefore,

$$g_{\text{safe}} = v\tau_{\text{safe}}. \tag{A.36}$$

A.3.6 Boundary and Initial Conditions

Open boundary conditions are applied. At the beginning of the road new vehicles are generated one after another in each of the lanes of the road at time moments

$$t^{(m)} = \tau\lceil m\tau_{\text{in}}/\tau\rceil, \ m = 1, 2, \ldots. \tag{A.37}$$

In (A.37), $\tau_{\text{in}} = 1/q_{\text{in}}$, q_{in} is the flow rate in the incoming boundary flow per lane, $\lceil z\rceil$ denotes the nearest integer greater than or equal to z. A new vehicle appears on the road only if the distance from the beginning of the road ($x = x_{\text{b}}$) to the position $x = x_{\ell,n}$ of the farthest upstream vehicle on the road is not smaller than the safe distance $v_{\ell,n}\tau + d$:

$$x_{\ell,n} - x_{\text{b}} \geq v_{\ell,n}\tau + d, \tag{A.38}$$

where $n = t^{(m)}/\tau$. Otherwise, condition (A.38) is checked at time $(n+1)\tau$ that is the next one to time $t^{(m)}$ (A.37), and so on, until the condition (A.38) is satisfied. Then the next vehicle appears on the road. After this occurs, the number m in (A.37) is increased by 1.

The speed v_n and coordinate x_n of the new vehicle are

$$\begin{aligned} v_n &= v_{\ell,n}, \\ x_n &= \max(x_{\rm b}, x_{\ell,n} - \lfloor v_n \tau_{\rm in} \rfloor). \end{aligned} \tag{A.39}$$

The flow rate $q_{\rm in}$ is chosen to have the value $v_{\rm free}\tau_{\rm in}$ integer. In the initial state ($n = 0$), all vehicles have the free flow speed $v_n = v_{\rm free}$ and they are positioned at space intervals $x_{\ell,n} - x_n = v_{\rm free}\tau_{\rm in}$.

After a vehicle has reached the end of the road it is removed. Before this occurs, the farthest downstream vehicle maintains its speed and lane. For the vehicle following the farthest downstream one, the "anticipation" speed $v_\ell^{(a)}$ in (A.26) is equal to the speed of the farthest downstream vehicle.

A.4 Physical Meaning of State of Vehicle Motion

We introduce a new variable

$$A_{n+1} = (v_{n+1} - v_n)/\tau. \tag{A.40}$$

We see that this variable is the vehicle acceleration (deceleration). Correspondingly, the variable

$$\tilde{A}_{n+1} = (\tilde{v}_{n+1} - v_n)/\tau \tag{A.41}$$

is the vehicle acceleration (deceleration) without speed fluctuations ξ_n. The use of the vehicle acceleration (A.41) by the description of model formulations made in this section allows us to disclose the physics behind the parameter of the state of vehicle motion S_{n+1} in the model as follows.

In accordance with (A.14) and (A.41), the parameter of state of vehicle motion is equal to

$$S_{n+1} = \mathrm{sgn}(\tilde{A}_{n+1}). \tag{A.42}$$

Therefore, formulas (A.13) and (A.20), (A.21) can be written respectively as follows:

$$\xi_n = \begin{cases} \xi_{\rm a} & \text{if } \tilde{A}_{n+1} > 0 \\ -\xi_{\rm b} & \text{if } \tilde{A}_{n+1} < 0 \\ \xi^{(0)} & \text{if } \tilde{A}_{n+1} = 0, \end{cases} \tag{A.43}$$

$$P_0 = \begin{cases} p_0 \text{ if } \tilde{A}_n \le 0 \\ 1 \;\; \text{if } \tilde{A}_n > 0, \end{cases} \quad P_1 = \begin{cases} p_1 \text{ if } \tilde{A}_n \ge 0 \\ p_2 \text{ if } \tilde{A}_n < 0. \end{cases} \tag{A.44}$$

To explain the physics of the state of vehicle motion (A.42) in more details, we note that in empirical observations of real field traffic data it has been found that time delays of a driver can be very different in different driving situations. In particular, these time delays depend on whether the driver can accelerate or should decelerate. For example, model fluctuations (A.15) for simulations of driver's over-deceleration is applied only if the vehicle should decelerate without model fluctuations, i.e., when $S_{n+1} = -1$ (see formula (A.13)). Model fluctuations (A.17) for simulations of driver's over-acceleration are applied only if the vehicle should accelerate without model fluctuations, i.e., when $S_{n+1} = 1$ (see formula (A.13)).

A.5 Lane Changing Rules for Two-Lane Road

As in other models of lane changing on a two-lane road (e.g., [42]), in the Kerner-Klenov model a vehicle changes lane with probability p_c, if some incentive lane changing rules together with some safety conditions for lane changing are satisfied.

Incentive lane changing rules from the right lane to the left lane ($R \to L$) and from the left lane to the right lane ($L \to R$) for lane changing are chosen similar to those of the paper of Nagel et al. [42]:

$$R \to L: \quad v_n^+ \ge v_{\ell,n} + \delta_1 \quad \text{and} \quad v_n \ge v_{\ell,n}, \tag{A.45}$$

$$L \to R: \quad v_n^+ > v_{\ell,n} + \delta_1 \quad \text{or} \quad v_n^+ > v_n + \delta_1, \tag{A.46}$$

where δ_1 is constant.

Under these conditions, a vehicle changes the lane when the following safety conditions for lane changing are satisfy:

$$g_n^+ > \min(v_n\tau,\ G_n^+), \tag{A.47}$$

$$g_n^- > \min(v_n^-\tau,\ G_n^-), \tag{A.48}$$

where

$$g_n^+ = x_n^+ - x_n - d, \quad g_n^- = x_n - x_n^- - d, \tag{A.49}$$

$$G_n^+ = G(v_n, v_n^+), \quad G_n^- = G(v_n^-, v_n), \tag{A.50}$$

$G(u, w)$ is given by (A.10).

In all formulas here and below, superscripts + and − in variables, parameters, and functions denote the preceding vehicle and the trailing vehicle in the "target" (neighboring) lane, respectively (the target lane is the lane into which the vehicle

Table A.2 Parameters of lane changing often used in simulations

$\delta_1 = 1\ \mathrm{ms}^{-1}/\delta v$, $L_a = 150\ \mathrm{m}/\delta x$, $p_c = 0.2$

wants to change). In conditions $R \to L$ (A.45) and $L \to R$ (A.46), the value v_n^+ at $g_n^+ > L_a$ and the value $v_{\ell,n}$ at $g_n > L_a$ are replaced by ∞, where L_a is a constant (see Table A.2).

A.6 Models of Road Bottlenecks

A.6.1 On-, Off-Ramp, and Merge Bottlenecks

Models of road bottlenecks due to on- and off-ramps as well as a merge bottleneck, at which two road lanes are reduced to one lane, are considered in Fig. A.2.

The on-ramp bottleneck consists of two parts (Fig. A.2a):

(i) The merging region of length L_m where vehicle can merge onto the main road from the on-ramp lane.
(ii) A part of the on-ramp lane of length L_r upstream of the merging region where vehicles move in accordance with the model of Sect. A.3. The maximal speed of vehicles is $v_{\mathrm{free}} = v_{\mathrm{free\ on}}$.

At the beginning of the on-ramp lane ($x = x_{\mathrm{on}}^{(b)}$) the flow rate to the on-ramp q_{on} is given through boundary conditions that are the same as those that determine the flow rate q_{in} at the beginning of the main road (Sect. A.3.6).

The off-ramp bottleneck consists of two parts (Fig. A.2b):

(i) A merging region of length L_m where vehicle can merge from the main road onto the off-ramp lane.
(ii) A part of the off-ramp lane of length L_r downstream of the merging region where vehicles move in accordance with the model of Sect. A.3. The maximal speed of vehicles is $v_{\mathrm{free}} = v_{\mathrm{free\ off}}$.

Within a second merging region of length $L_m + L_c$, which is on the main road ($x_{\mathrm{off}}^{(s)} \le x \le x_{\mathrm{off}}^{(b)}$ in Fig. A.2b), vehicles going to the off-ramp have to change from the left lane to the right lane of the main road. The flow rate of vehicles that go to the off-ramp is given as a percentage η_{off} of the flow rate q_{in}. At the beginning of the road, with probability

$$r_{\mathrm{off}} < \eta_{\mathrm{off}}/100\%, \tag{A.51}$$

where $r_{\mathrm{off}} = \mathrm{rand}(0, 1)$, each of the vehicles becomes an "attribute", which marks a vehicle as the vehicle going to the off-ramp.

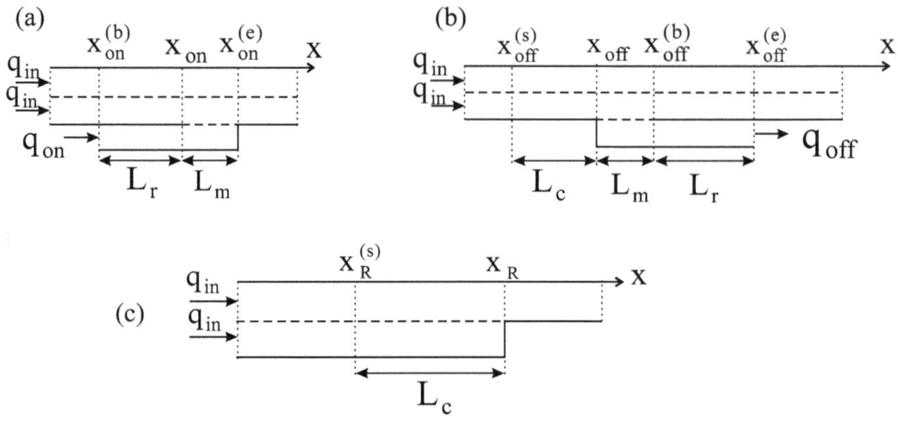

Fig. A.2 Models of on-ramp (**a**), off-ramp (**b**) and merge (**c**) bottlenecks on two-lane road

At the merge bottleneck (Fig. A.2c) within the merging region of length L_c upstream of the merge point $x = x_R$ vehicles have to change from the right lane to the left lane.

A.6.2 Moving Bottleneck

A model of a moving bottleneck is shown in Fig. A.3a. Cases when the moving bottleneck is upstream of the on- and off-ramp bottlenecks are presented in Fig. A.3b,c, respectively.

We assume that there is a slow vehicle that maximum speed v_M is smaller than the maximum speed of other vehicles: $v_M < v_{free}$. The slow vehicle that moves in the right lane (Fig. A.3) causes a moving bottleneck. If a vehicle moves at the speed $v > v_M$ in the right lane upstream of the slow vehicle, then the vehicle tries to changes from the right lane to the left lane within a merging region of the moving bottleneck of length L_M. This merging region of the moving bottleneck moves at the speed v_M of the moving bottleneck (Fig. A.3).

A.6.3 Models of Vehicle Merging at Bottlenecks

A.6.3.1 Vehicle Speed Adaptation Within Merging Region of Bottleneck

For all bottlenecks, when a vehicle is within the merging region of a bottleneck, the vehicle takes into account the space gaps to the preceding vehicles and their speeds both in the current and target lanes. Respectively, instead of formula (A.5), in (A.4)

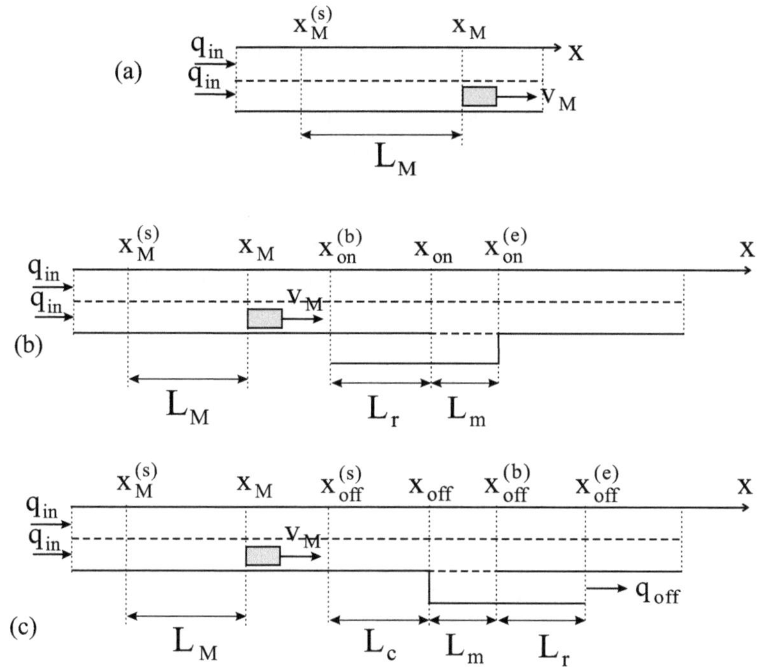

Fig. A.3 Model of moving bottlenecks on two-lane road: Moving bottleneck is outside road bottlenecks (**a**), moving bottleneck is upstream of an on-ramp bottleneck (**b**), and moving bottleneck is upstream of an off-ramp bottleneck (**c**)

for the speed $v_{c,n}$ the following formula is used:

$$v_{c,n} = \begin{cases} v_n + \Delta_n^+ & \text{at } g_n^+ \le G(v_n, \hat{v}_n^+) \\ v_n + a_n\tau & \text{at } g_n^+ > G(v_n, \hat{v}_n^+), \end{cases} \tag{A.52}$$

$$\Delta_n^+ = \max(-b_n\tau, \min(a_n\tau,\ \hat{v}_n^+ - v_n)), \tag{A.53}$$

$$\hat{v}_n^+ = \max(0, \min(v_{\text{free}},\ v_n^+ + \Delta v_r^{(2)})), \tag{A.54}$$

$\Delta v_r^{(2)}$ is constant (see Table A.3).

As in lane changing rules (Sect. A.5), superscripts $+$ and $-$ in variables, parameters, and functions denote the preceding vehicle and the trailing vehicle in the "target" (neighboring) lane, respectively. The target lane is the lane into which the vehicle wants to change.

The safe speed $v_{s,n}$ in (A.2), (A.4) for the vehicle that is the closest one to the end of the merging region is chosen in the form

$$v_{s,n} = \lfloor v^{(\text{safe})}(x_{\text{on}}^{(e)} - x_n,\ 0) \rfloor \tag{A.55}$$

Table A.3 Parameters of models of bottlenecks often used in simulations

$\lambda_{\rm b} = 0.75$ for all the bottlenecks,
$L_{\rm c} = 1.0$ or 0.7 km$/\delta x$ for off-ramp bottleneck,
$L_{\rm M} = 0.3$ km$/\delta x$ for moving bottleneck,
$v_{\rm free\ on} = 22.2\ {\rm ms}^{-1}/\delta v$, $v_{\rm free\ off} = 25\ {\rm ms}^{-1}/\delta v$,
$\Delta v_{\rm r}^{(2)} = 5\ {\rm ms}^{-1}/\delta v$ for on-ramp and moving bottlenecks,
$\Delta v_{\rm r}^{(2)} = -2.5\,{\rm ms}^{-1}/\delta v$ for off-ramp bottlenecks,
$L_{\rm r} = 1$ km$/\delta x$, $\Delta v_{\rm r}^{(1)} = 10\ {\rm ms}^{-1}/\delta v$, $L_{\rm s} = 0$,
$L_{\rm m} = 0.3$ and 0.5 km$/\delta x$
for on- and off-ramp bottlenecks, respectively

for on-ramp bottleneck and

$$v_{{\rm s},n} = \min\,(v_n^{\rm (safe)}, g_n/\tau + v_\ell^{\rm (a)}, \lfloor v^{\rm (safe)}(x_{\rm off}^{\rm (b)} - x_n - L_{\rm s},\ 0)\rfloor) \tag{A.56}$$

for off-ramp bottleneck, where $L_{\rm s}$ is a constant (see Table A.3).

A.6.3.2 Safety Conditions for Vehicle Merging

Vehicle merging at bottlenecks occurs, when safety conditions $(*)$ *or* safety conditions $(**)$ are satisfied.

Safety conditions $(*)$ are as follows:

$$\begin{aligned} g_n^+ &> \min(\hat{v}_n\tau,\ G(\hat{v}_n, v_n^+)), \\ g_n^- &> \min(v_n^-\tau,\ G(v_n^-, \hat{v}_n)), \end{aligned} \tag{A.57}$$

$$\hat{v}_n = \min(v_n^+,\ v_n + \Delta v_r^{(1)}), \tag{A.58}$$

$\Delta v_r^{(1)} > 0$ is constant (see Table A.3).

Safety conditions $(**)$ are as follows:

$$x_n^+ - x_n^- - d > g_{\rm target}^{\rm (min)}, \tag{A.59}$$

where

$$g_{\rm target}^{\rm (min)} = \lfloor \lambda_{\rm b} v_n^+ + d \rfloor, \tag{A.60}$$

$\lambda_{\rm b}$ is constant. In addition to conditions (A.59), the safety condition $(**)$ includes the condition that the vehicle should pass the midpoint

$$x_n^{\rm (m)} = \lfloor (x_n^+ + x_n^-)/2 \rfloor \tag{A.61}$$

between two neighboring vehicles in the target lane, i.e., conditions

$$\begin{aligned} &x_{n-1} < x^{(\mathrm{m})}_{n-1} \text{ and } x_n \geq x^{(\mathrm{m})}_n \\ &\text{or} \\ &x_{n-1} \geq x^{(\mathrm{m})}_{n-1} \text{ and } x_n < x^{(\mathrm{m})}_n. \end{aligned} \tag{A.62}$$

should also be satisfied.

A.6.3.3 Speed and Coordinate of Vehicle After Vehicle Merging

The vehicle speed after vehicle merging is equal to

$$v_n = \hat{v}_n. \tag{A.63}$$

Under conditions $(*)$, the vehicle coordinates x_n remains the same. Under conditions $(**)$, the vehicle coordinates x_n is equal to

$$x_n = x^{(\mathrm{m})}_n. \tag{A.64}$$

A.6.4 ACC-Vehicle Merging at On-Ramp Bottleneck

In the on-ramp lane, an ACC-vehicle moves in accordance with the model (6.16)–(6.18). The maximal speed of the ACC vehicle in the on-ramp lane is $v_{\mathrm{free}} = v_{\mathrm{free\ on}}$. The safe speed $v_{\mathrm{s},n}$ in (6.18) for the ACC-vehicle that is the closest one to the end of the merging region is the same as that for manual driving vehicles (A.55).

An ACC-vehicle merges from the on-ramp lane onto the main road, when some safety conditions $(*)$ *or* safety conditions $(**)$ are satisfied for the ACC-vehicle. Safety conditions $(*)$ for ACC-vehicles are as follows:

$$g^{+}_n > \hat{v}_n \tau, \quad g^{-}_n > v^{-}_n \tau, \tag{A.65}$$

where $\hat{v}_n$ is given by formula (A.58). Safety conditions $(**)$ are given by formulas (A.59)–(A.62), i.e., they are the same as those for manual driving vehicles. Respectively, as for manual driving vehicles, the ACC-vehicle speed and its coordinate after ACC-vehicle merging are determined by formulas (A.63) and (A.64).

A.7 Stochastic Simulation of "Strong" and "Weak" Speed Adaptation

A.7.1 Simulation of Driver Speed Adaptation Effect

We consider a vehicle speed v_n (Fig. A.4) that satisfies conditions

$$0 < v_n < v_{\rm free}. \tag{A.66}$$

If condition

$$g_n > G_n \tag{A.67}$$

is satisfied and

$$v_n + a_n\tau < \min(v_{\rm free}, v_{{\rm s},n}), \tag{A.68}$$

then from (A.2), (A.4), (A.5), (A.13), (A.14), (A.17) it follows that

$$v_{n+1} = v_n + a_n\tau, \tag{A.69}$$

i.e., with some probability vehicle accelerates (labeled by "acceleration" in Fig. A.4) with the maximum acceleration a.

If the condition

$$g_n < g_{{\rm safe},n} \tag{A.70}$$

is satisfied, then the speed is higher than the safe speed:

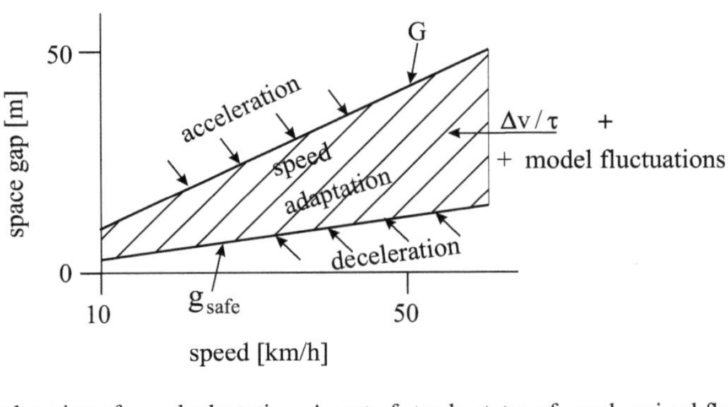

Fig. A.4 Explanation of speed adaptation. A part of steady states of synchronized flow associated with (A.66) adapted from Fig. A.1c. Δv is difference between the speed of the preceding vehicle and the vehicle speed

$$v_n > v_{s,n}. \tag{A.71}$$

In (A.70), a safe space gap $g_{\text{safe},n}$ is determined from the equation

$$v_n = v_{s,n}(g_{\text{safe},n}, v_{\ell,n}, v_\ell^{(a)}) \tag{A.72}$$

that follows from Eqs. (A.26), (A.27) at $g_n = g_{\text{safe},n}$.

Under condition (A.71), from (A.2) we find that the vehicle decelerates (labeled by "deceleration" in Fig. A.4).

A different vehicle behavior we find within a 2D-region of steady states of synchronized flow (2D-dashed region in Fig. A.4) that satisfy conditions

$$g_{\text{safe},n} < g_n \leq G_n, \tag{A.73}$$

where the safe space gap $g_{\text{safe},n}$ is found from equation (A.72). Additionally to conditions (A.73), we assume that conditions

$$| \Delta v_n | \leq a\tau, \ a_n = b_n = a, \ v_n + \Delta_n + \xi_n \leq \min(v_{\text{free}}, v_{s,n}, v_n + a\tau), \tag{A.74}$$

are satisfied. In (A.74), Δv_n is the speed difference between the speed of the preceding vehicle $v_{\ell,n}$ and the vehicle speed v_n:

$$\Delta v_n = v_{\ell,n} - v_n. \tag{A.75}$$

Under conditions (A.73) and (A.74), from (A.2), (A.4)–(A.6) with some probability the vehicle speed v_{n+1} at the next time step $n + 1$ is equal to

$$v_{n+1} = v_n + \Delta v_n + \xi_n. \tag{A.76}$$

This means that the vehicle acceleration (deceleration) is equal to

$$(\Delta v_n + \xi_n)/\tau. \tag{A.77}$$

This formula explains the speed adaptation effect in the stochastic model (labeled "speed adaptation" in Fig. A.4): Within the space gap range (A.73) the speed difference Δv_n (A.75) in (A.77) describes speed adaptation to the speed of the preceding vehicle that occurs without caring what the precise space gap to the preceding vehicle is. In addition to this deterministic effect, in the stochastic model there is a random speed change ξ_n, which leads to a stochastic behavior of speed adaptation in the model.

A.7.2 Stochastic Driver's Choice of Space Gap in Synchronized Flow

In the three-phase theory, while moving in synchronized flow a driver can arbitrary choose a space gap to the preceding vehicle within the 2D-region of synchronized flow (Fig. A.1a), i.e., under conditions (A.73). We distinguish two limit cases of driver speed adaptation under conditions (A.73) [23]:

(i) "Strong" driver speed adaptation and
(ii) "weak" driver speed adaptation.

Under strong speed adaptation, a driver chooses a large mean speed gap g_{strong} (Fig. A.5) that satisfies conditions

$$g_{\text{strong}} \leq G, \quad G - g_{\text{strong}} \ll g_{\text{strong}} - g_{\text{safe}}. \tag{A.78}$$

On contrast, under weak speed adaptation, a driver chooses a small mean speed gap g_{weak} (Fig. A.5) that satisfies conditions

$$g_{\text{weak}} \geq g_{\text{safe}}, \quad g_{\text{weak}} - g_{\text{safe}} \ll G - g_{\text{weak}}. \tag{A.79}$$

To simulate the driver's choice of a space gap within the 2D-region of synchronized flow, we use a stochastic description of driver's speed adaptation through a change in probabilities p_2 and p_1 in (A.21). Introducing a coefficient of speed adaptation ε, we write these probabilities as follows:

$$p_1 = \min(1, (1+\varepsilon)p_1^{(0)}), \quad p_2 = \min(1, (1+\varepsilon)p_2^{(0)}(v_n)), \tag{A.80}$$

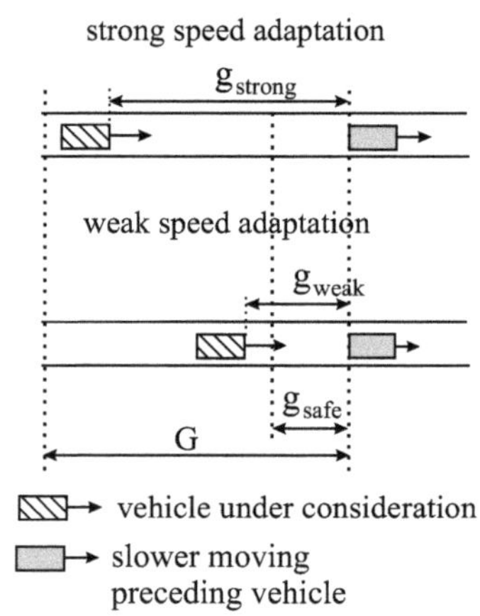

Fig. A.5 Qualitative explanation of strong driver speed adaptation and weak driver speed adaptation within 2D-region of states of synchronized flow where model parameters $p_1^{(0)}$, $p_2^{(0)}(v_n)$, and v_{21} are given in Table A.1. We have found that the larger ε, the stronger the speed adaptation in the 2D-region of synchronized flow and, therefore, the larger the mean space gap (the longer the mean time headway) between vehicles in synchronized flow.

Probability $p_2^{(0)}(v_n)$ is chosen to be an increasing speed function (Table A.1). The physical meaning of this speed dependence of probability $p_2^{(0)}$ is as follows: The larger the probability $p_2^{(0)}$, the stronger the speed adaptation within synchronized flow. The stronger this speed adaptation effect, the less the probability of moving jam emergence in synchronized flow. In empirical data, it is often observed that the smaller the speed in synchronized flow, the larger the probability of moving jam emergence. To satisfy these empirical results, the speed adaptation effect decreases with the decrease in the speed of synchronized flow.

For a qualitative explanation of this stochastic description of the driver's choice of a space gap within the 2D-region of synchronized flow, we do not take into account random fluctuations ξ_n and assume that the space gap to the preceding vehicle g_n is within this 2D-region of synchronized flow states (Fig. A.1c), i.e., the space gap satisfies conditions (A.73). Under conditions (A.73), the speed is lower than the safe speed:

$$v_n < v_{s,n}. \tag{A.81}$$

The driver's speed adaptation effect occurs, when a driver moves initially with a higher speed than the speed of the preceding vehicle

$$v_n > v_{\ell,n} > 0 \tag{A.82}$$

and the driver cannot overtake the preceding vehicle. In this case, the driver should decelerate. This vehicle deceleration adapting the vehicle speed to the speed of the preceding vehicle is described by Eq. (A.6) through the use of stochastic deceleration b_n. This stochastic deceleration depends on whether the vehicle decelerates at time step n or not (see formulas (A.19), (A.21) that define stochastic vehicle deceleration b_n).

If the vehicle does not decelerate at time step n, then according to (A.19) the stochastic deceleration b_n in Eq. (A.6) is equal to

$$b_n = a\Theta(p_1 - r_1). \tag{A.83}$$

In this case, with probability p_1, the vehicle, which did not decelerate at time step n, begins to decelerate at time step $n + 1$ trying to approach the speed of the preceding vehicle. However, from (A.83) we also see that with probability $1 - p_1$ the deceleration $b_n = 0$, that is, the vehicle does not begin to decelerate at time step $n + 1$ although in accordance with (A.82) the vehicle speed is higher than the speed of the preceding vehicle.

If the vehicle decelerates at time step n, then rather than formula (A.83), for stochastic vehicle deceleration b_n in Eq. (A.6) we should use the following formula

$$b_n = a\Theta(p_2 - r_1). \tag{A.84}$$

Thus, with probability p_2 the vehicle continues its deceleration at time step $n+1$ trying to approach the speed of the preceding vehicle; however, with probability $1-p_2$ the vehicle deceleration $b_n = 0$, that is, the vehicle *interrupts* its deceleration although in accordance with (A.82) the vehicle speed is higher than the speed of the preceding vehicle.

The synchronization space gap G and safe space gap g_{safe} in (A.12) are related to a given vehicle speed under the condition that the speed difference Δv_n (A.75) between the preceding vehicle and the vehicle under consideration satisfies condition $\Delta v_n = v_{\ell,n} - v_n = 0$. In contrast, the synchronization space gap G_n and safe space gap $g_{\text{safe},n}$ in (A.73) are dynamic variables. For example, when $\Delta v_n > 0$, then at a given speed v_n the value G_n decreases; when $\Delta v_n < 0$, then at a given speed v_n the value $g_{\text{safe},n}$ increases. In each of these cases, the dynamic 2D-region of synchronized flow states changes over time.

For this reason, the dynamic decrease in G_n or/and increase in $g_{\text{safe},n}$ can lead to a "dynamic jump" from vehicle dynamics associated with speed adaptation within the dynamic 2D-region of synchronized flow to a qualitative another vehicle behavior.

A.7.3 "Jam-Absorption" Effect

In [23], we have shown that accordingly to the three-phase theory (see Fig. 8.7 of Sect. 8.3.3), synchronized flow states that are on and above the line J in the flow–density plane are metastable states with respect to an S→J transition. Contrarily, synchronized flow states that are below line J in the flow–density plane are stable states with respect to an S→J transition. As found in [23], the probability of occurrence of either metastable synchronized flow states or stable synchronized flow states in congested traffic depends basically on the characteristics of driver speed adaptation in synchronized flow.

It turns out [23] that a driver behavior leading to metastable synchronized flow is related to weak driver speed adaptation. As mentioned in Sect. A.7.2, under weak speed adaptation, on average drivers come relatively closely to associated preceding vehicles ($g = g_{\text{weak}}$ in Fig. A.5) within the space gap range between vehicles (A.73). As a result, the average time headway between drivers is relatively small in synchronized flow. This results in states of synchronized flow that are on and above the line J (Fig. 8.7). In this case, a so-called general pattern (GP) that consists of synchronized flow and wide moving jams emerge spontaneously

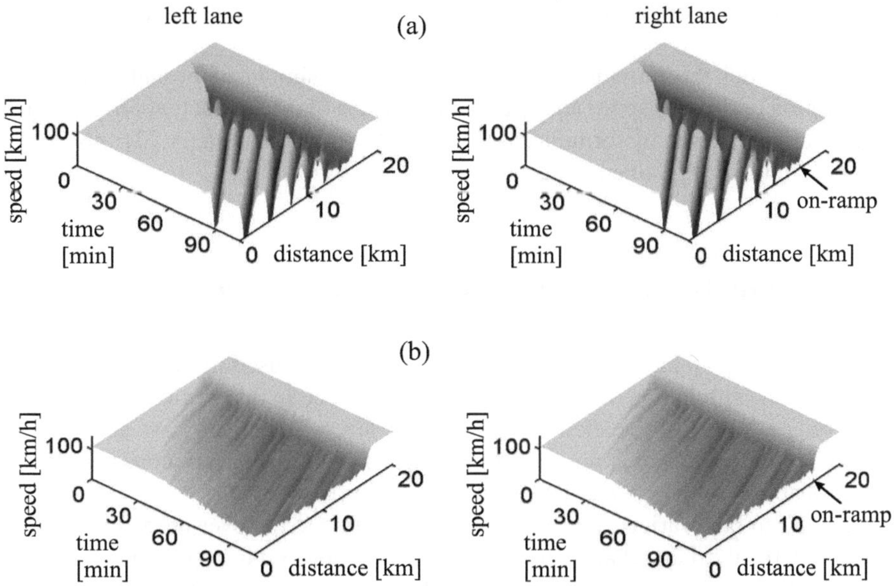

Fig. A.6 Simulations of transformation of GP (**a**) into widening synchronized flow pattern (WSP) (**b**) through driver's choice of stronger speed adaptation parameters: In (**b**), the coefficient of driver speed adaptation ε in (A.80) is considerably larger than in (**a**), while the flow rate on the main road upstream of the bottleneck and the on-ramp inflow as well as other model parameters are the same in (**a**) and (**b**). Simulations have been made on a two-lane road with an on-ramp bottleneck. Vehicle speed in space and time. Figures in left panel—left lane, right panel—right lane. Adapted from Fig. 14 of [23]

at a highway bottleneck after traffic breakdown has occurred at the bottleneck (Fig. A.6a) [19, 21].

In the opposite case of strong driver speed adaptation, a driver that approaches a slower moving preceding vehicle adapts the speed to that of the preceding vehicle quickly. In this case, while a vehicle decelerates, its space gap does not decrease considerably ($g = g_{\text{strong}}$ in Fig. A.5). This results in states of synchronized flow that are below the line J (Fig. 8.7). In this case, all wide moving jams dissolve and only synchronized flow remains in congested traffic (Fig. A.6b) [23]. The effect of moving jam dissolution can be called the "jam-absorption" effect.

Thus, the jam-absorption effect occurs, when a space gap (time headway) between vehicles in synchronized flow upstream of a moving jam are large enough and, therefore, the space gap is related to points in the flow-density plane lying below the line J (Fig. 8.7). In this case, the jam dissolves and only synchronized flow remains [23]. This jam-absorption effect in highway traffic leads to a transformation of general patterns (GPs) into synchronized flow patterns (SPs) in which no wide moving emerge spontaneously (Fig. A.6) [23].

As shown in [27, 34] this effect in city traffic leads to the occurrence of synchronized flow in over-saturated city traffic: Due to strong speed adaptation time headway between drivers increase and moving queues dissolve at some distance of traffic signal.

In the model, this moving jam-absorption effect is simulated through a stochastic description of driver's speed adaptation with the use of model fluctuations in vehicle deceleration b_n given by formula (A.19) applied under condition (A.73) only, as this has been already explained in Sect. A.7.2.

A.8 Simulation Approaches to Over-Acceleration Effect

In Sect. 5.10.2, we have already considered some ideas used in the Kerner-Klenov model for the simulation of the over-acceleration effect. In this section, due to a great importance of the over-acceleration effect in the three-phase theory, we discuss these and other mathematical approaches to simulations of the discontinuous behavior of the probability of driver over-acceleration in more details.

There can be a variety of driver maneuver leading to the over-acceleration effect (Sect. 5.10), like lane change for passing or temporary closing to the preceding vehicle. For example, if a driver believes that she/he can overtake a slow moving preceding vehicle that does not accelerate, the driver can accelerate in an initial road lane before lane change. However, it can turn out that although the driver accelerates, nevertheless she/he cannot change lane and pass. In this case, the driver must decelerate to the speed of the preceding vehicle and wait for another possibility to pass. This vehicle acceleration with the subsequent deceleration to the speed of the preceding vehicle can be repeated several times before the driver can pass.

In the stochastic model, there are several different mathematical approaches for the simulation of the over-acceleration effect. In each of these approaches, a *dynamic* speed gap appears (gray region labeled by "dynamic speed gap" in Fig. 5.19 of the main text) between free flow (line *F*) and steady states of synchronized flow of lower speeds. Within this dynamic speed gap, initially steady states of synchronized flow of higher speeds are destroyed and no long-time living synchronized flow states occur.

In the stochastic model a *discontinuous character* of over-acceleration discussed in Sect. 5.10 is modeled through the occurrence of the dynamic speed gap caused by stochastic model dynamics that destroys initially steady states of synchronized flow of higher speeds (gray region in Fig. 5.19). A competition between the speed adaptation and over-acceleration effects (labeled by "speed adaptation" and "over-acceleration" in Fig. 5.19), which occurs within a local disturbance in free flow, is associated with this dynamic speed gap.

A.8.1 *Implicit Simulation of Over-Acceleration Effect Through Driver Acceleration*

The dynamic speed gap between states of free flow and synchronized flow resulting in a discontinuous character of over-acceleration can be simulated through driver acceleration even without vehicle lane changing [29]. Under condition (A.73), the function $\xi_{\rm a}$ in (A.13) is given by formula (A.17), i.e.,

$$\xi_{\rm a} = a^{\rm (a)}\tau\Theta(p_{\rm a} - r). \qquad (A.85)$$

In accordance with (A.13), these model fluctuations are applied only if the vehicle should accelerate without fluctuations.

A.8.2 *Simulation of Over-Acceleration Effect Through Combination of Lane Changing to Faster Lane and Random Driver Acceleration*

The discontinuous character of over-acceleration can also be simulated through the use of a combination of an implicit simulation through a random driver acceleration (A.85) *and* an explicit simulation based on lane changing to a faster lane [29]. Rules for lane changing to a faster lane are given by formulas (A.45)–(A.50).

A.8.3 *"Boundary" Over-Acceleration*

In addition with the over-acceleration effect *within* a local disturbance in free flow discussed above that can be called as "bulk" over-acceleration, there can also be "boundary" over-acceleration. In contrast with bulk over-acceleration, boundary over-acceleration occurs at the downstream disturbance front within which a vehicle accelerates from a lower free flow speed to a higher one. Indeed, at the downstream disturbance front the vehicle can reduce the time delay in acceleration due to passing of the preceding vehicle, i.e., due to over-acceleration. The lower the speed and the greater the density, the smaller the over-acceleration probability, i.e., the smaller the time delay reduction due to over-acceleration.

We assume that due to "boundary" over-acceleration the mean time delay in vehicle acceleration $\tau_{\rm del}^{\rm (acc)}(v_n)$ (Sect. A.3.4), which in accordance with formula (A.22) is associated with acceleration probability $p_0(v_n)$, can be a decreasing speed function within a speed range between speeds in free and synchronized flows. The reduction in $\tau_{\rm del}^{\rm (acc)}(v_n)$ can be explained by lane changing to a faster lane at the downstream

Table A.4 Model parameters of over-acceleration often used in simulations

Models for over-acceleration of Sects. A.8.1 and A.8.2:
$p_0(v_n) = 0.575 + 0.125 \min(1, v_n/v_{01})$, $a^{(a)} = a^{(b)} = a$, $a^{(0)} = 0.2a$, $p_a = 0.17$, $v_{01} = 10\text{ms}^{-1}/\delta v$
Model for over-acceleration of Sect. A.8.3:
$p_0(v_n) = 0.575 + 0.125 \min(1, v_n/v_{01}) + {}$ $+0.15 \max(0, (v_n - v_{02})/(v_{\text{free}} - v_{02}))$, $v_{02} = 23.61\ \text{ms}^{-1}/\delta v$, $a^{(a)} = a^{(b)} = a$, $a^{(0)} = 0.2a$, $p_a = 0.17$, $v_{01} = 10\text{ms}^{-1}/\delta v$
Model for over-acceleration of Sect. A.8.4:
$p_0(v_n) = 0.575 + 0.125 \min(1, v_n/v_{01})$, $a^{(b)}(v_n) = 0.2a + {}$ $+0.8a \max(0, \min(1, (v_{22} - v_n)/\Delta v_{22}))$, $a^{(0)} = 0.2a$, $v_{22} = 12.5\ \text{ms}^{-1}/\delta v$, $\Delta v_{22} = 2.778\ \text{ms}^{-1}/\delta v$, $a^{(a)} = 0$, $p_{\text{a}} = 0$, $\lambda = 0.75$, $\Delta v^{(1)} = 2\ \text{ms}^{-1}/\delta v$, $v_{01} = 10\text{ms}^{-1}/\delta v$

front of the disturbance. To simulate this effect, we can use in (A.22) the following speed-dependence of acceleration probability

$$p_0(v) = \\ 0.575 + 0.125 \min(1, v/v_{01}) + 0.15 \max(0, (v - v_{02})/(v_{\text{free}} - v_{02})), \tag{A.86}$$

where v_{01}, v_{02} are constants, $v_{02} > v_{01}$ (see Table A.4).

A.8.4 *Explicit Simulation of Over-Acceleration Effect Through Lane Changing to Faster Lane*

Simulations show that the discontinuous character of over-acceleration can be simulated explicitly through lane changing to a faster lane *only*, i.e., when *no* other mathematical formulations for over-acceleration are used, specifically, when in (A.17) $p_{\text{a}} = 0$.

To reach this goal, we use much weaker safety conditions for lane changing in comparison with (A.47) and (A.48). In particular, when (A.47) and (A.48) are not satisfied, then under condition (A.45) or (A.46) a vehicle can nevertheless change to a faster lane with the above-mentioned given probability p_{c}, if the following safety conditions are satisfied: The space gap between two neighboring vehicles in the target lane satisfies condition

$$x_n^+ - x_n^- - d > g_{\text{target}}^{(\text{min})}, \tag{A.87}$$

where

$$g_{\text{target}}^{(\text{min})} = \lfloor \lambda v_n^+ + d \rfloor, \tag{A.88}$$

λ is constant (see Table A.4). In addition to (A.87), the vehicle should pass the midpoint $x_n^{(\text{m})}$ between two neighboring vehicles in the target lane found from the formula

$$x_n^{(\text{m})} = \lfloor (x_n^+ + x_n^-)/2 \rfloor, \tag{A.89}$$

specifically, the following conditions should be satisfied

$$\begin{aligned} &x_{n-1} < x_{n-1}^{(\text{m})} \text{ and } x_n \geq x_n^{(\text{m})} \\ &\text{or} \\ &x_{n-1} \geq x_{n-1}^{(\text{m})} \text{ and } x_n < x_n^{(\text{m})}. \end{aligned} \tag{A.90}$$

After lane changing, the coordinate of the vehicle is set to $x_n = x_n^{(\text{m})}$ and the vehicle speed v_n is set to $\hat{v}_n$:

$$\hat{v}_n = \min(v_n^+, \ v_n + \Delta v^{(1)}), \tag{A.91}$$

where $\Delta v^{(1)}$ is constant that describes the increase in the speed after lane changing.[1]

As in the Nagel-Schreckenberg (NaSch) cellular automaton (CA) model [1, 43], random fluctuations in vehicle deceleration (A.15) allows us to simulate the well-known over-deceleration effect leading to the classical traffic flow instability of Herman, Gazis, Montroll, Potts, Rothery, and Chandler [2, 6, 7, 9]. The amplitude of the random fluctuations in vehicle deceleration $a^{(\text{b})}$ in (A.15) is chosen to be a decreasing speed function:

$$a^{(\text{b})}(v_n) = 0.2a + 0.8a \max(0, \min(1, (v_{22} - v_n)/\Delta v_{22})), \tag{A.92}$$

where v_{22} and Δv_{22} are constants (Table A.4). Note that in accordance with (A.92), model fluctuations in homogeneous free flow introduced in the model of over-acceleration of Sect. A.8.2 are chosen smaller than that in synchronized flow. This is explained as follows. Traffic breakdown is associated with a competition of speed adaptation and over-acceleration. The probability of driver over-acceleration exhibits a discontinuous character. The smaller the amplitude of model fluctuations in free flow, the more accurate nucleation effects that determine the F→S transition

[1]Formulas (A.87)–(A.91) have already been used for vehicle merging at bottlenecks (Sect. A.6.3.2). However, in Sect. A.6.3.2 these formulas have been used within the merging regions of the bottlenecks *only*, whereas here for the explicit simulation of over-acceleration effect through the lane changing formulas (A.87)–(A.91) are used on the whole main road.

can be studied in free flow outside bottlenecks. Contrarily, model fluctuations in synchronized flow simulate driver time delay in vehicle deceleration, which is responsible for wide moving jam emergence in synchronized flow. To simulate this driver over-deceleration in synchronized flow, greater model fluctuations are applied.

A.9 A Markov Chain: Sequence of Numerical Calculations of Model

Because the Kerner-Klenov model consists of a number of variables and seems to be complicated, we present below the sequence of calculations of the model functions needed to reproduce results of numerical simulations of this model shown in the main text of the book. This sequence of model calculations can also be helpful for a deeper understanding of the physics behind the model. We assume that for each of the vehicles the model variables

$$v_n,\ x_n,\ A_n,\ S_n \text{ at step } n$$

have already been found with the model (A.2)–(A.7), (A.9), (A.10), (A.13)–(A.21), (A.26), (A.27), (A.31)–(A.34), (A.45)–(A.50). In the case, when the model of Sect. A.8.4 is used, to these formulas we should add formulas (A.87)–(A.91).

It should be noted that the calculation of the model variables depends on whether a vehicle is at time step n outside or within a merging region of a bottleneck (Sect. A.6.1). For this reason, we consider the sequence of calculations of the model functions separately for these two cases.

A.9.1 Vehicles Moving Outside Merging Regions of Bottlenecks

The following sequence of calculations is made in the model, to find the model variables at the next time step $n+1$:

1. Calculation of space gaps (A.7) between vehicles at time step n:

$$g_n = x_{\ell,n} - x_n - d. \tag{A.93}$$

2. Calculation of space gaps (A.49) between vehicles at time step n in the neighboring lane:

$$g_n^+ = x_n^+ - x_n - d, \quad g_n^- = x_n - x_n^- - d. \tag{A.94}$$

3. Calculation of the synchronized gaps G_n^+, G_n^- (A.50) between vehicles at time step n

$$G_n^+ = G(v_n, v_n^+), \quad G_n^- = G(v_n^-, v_n), \tag{A.95}$$

where $G(u, w)$ is given by (A.10).

4. Check of lane changing rules from the right lane to the left lane ($R \to L$) (Eq. (A.45)) and from the left lane to the right lane ($L \to R$) (Eq. (A.46)) of Sect. A.5:

$$\begin{array}{lll} R \to L: & v_n^+ \geq v_{\ell,n} + \delta_1 \quad \text{and} \quad v_n \geq v_{\ell,n}, \\ L \to R: & v_n^+ > v_{\ell,n} + \delta_1 \quad \text{or} \quad v_n^+ > v_n + \delta_1, \end{array} \tag{A.96}$$

and security conditions (A.47), (A.48), i.e., conditions

$$g_n^+ > \min(v_n\tau,\ G_n^+), \quad g_n^- > \min(v_n^-\tau,\ G_n^-). \tag{A.97}$$

When security conditions (A.97) are not satisfied, security conditions (A.87)–(A.90) for the model of Sect. A.8.4 should be proven. Under conditions (A.87)–(A.90), the vehicle speed v_n is set to $\hat{v}_n$ in accordance with (A.91). All vehicles for which lane changing rules are satisfied change their lane with probability p_c.

5. Calculation of space gaps between vehicles (A.7) at time step n *after lane changing*:

$$g_n = x_{\ell,n} - x_n - d. \tag{A.98}$$

6. Calculation of the synchronized gap between vehicles at time step n (see (A.9)):

$$G_n = G(v_n, v_{\ell,n}). \tag{A.99}$$

7. Calculation of probabilities (A.20), (A.21) of random acceleration a_n and deceleration b_n:

$$P_0 = \begin{cases} p_0 \text{ if } S_n \neq 1 \\ 1 \ \text{ if } S_n = 1, \end{cases} \quad P_1 = \begin{cases} p_1 \text{ if } S_n \neq -1 \\ p_2 \text{ if } S_n = -1. \end{cases} \tag{A.100}$$

8. Calculation of random acceleration a_n (A.18) and deceleration b_n (A.19):

$$a_n = a\Theta(P_0 - r_1),\ b_n = a\Theta(P_1 - r_1). \tag{A.101}$$

9. Calculation of vehicle acceleration (deceleration) (A.6) within the synchronization gap:

$$\Delta_n = \max(-b_n\tau, \min(a_n\tau,\ v_{\ell,n} - v_n)). \tag{A.102}$$

10. Calculation of a vehicle speed (A.5) resulting from vehicle acceleration (deceleration) of (A.102):

$$v_{c,n} = \begin{cases} v_n + \Delta_n & \text{at } g_n \leq G_n \\ v_n + a_n\tau & \text{at } g_n > G_n. \end{cases} \tag{A.103}$$

11. Calculation of a safe speed $v_n^{(\text{safe})}$ (A.27) in accordance with (A.31)–(A.33):

$$v_n^{(\text{safe})} = \lfloor v^{(\text{safe})}(g_n,\ v_{\ell,n}) \rfloor. \tag{A.104}$$

12. Calculation of the anticipation speed of the preceding vehicle (A.34):

$$v_\ell^{(\text{a})} = \max(0, \min(v_{\ell,n}^{(\text{safe})}, v_{\ell,n}, g_{\ell,n}/\tau) - a\tau). \tag{A.105}$$

13. Calculation of the safe speed (A.26) that takes into account the anticipation speed of the preceding vehicle:

$$v_{s,n} = \min\,(v_n^{(\text{safe})}, g_n/\tau + v_\ell^{(\text{a})}). \tag{A.106}$$

14. Calculation of the speed $\tilde{v}_{n+1}$ (A.4):

$$\tilde{v}_{n+1} = \min(v_{\text{free}}, v_{s,n}, v_{c,n}). \tag{A.107}$$

15. Calculation of the state of vehicle motion S_{n+1} (A.14) at step $n+1$:

$$S_{n+1} = \begin{cases} -1 & \text{if } \tilde{v}_{n+1} < v_n \\ 1 & \text{if } \tilde{v}_{n+1} > v_n \\ 0 & \text{if } \tilde{v}_{n+1} = v_n, \end{cases} \tag{A.108}$$

16. Calculation of components of random fluctuation in acceleration (deceleration) (A.15)–(A.17):

$$\xi_a = a^{(\text{a})}\tau\Theta(p_a - r),\ \xi_b = a^{(\text{b})}\tau\Theta(p_b - r), \tag{A.109}$$

$$\xi^{(0)} = a^{(0)}\tau \begin{cases} -1 & \text{if } r \leq p^{(0)} \\ 1 & \text{if } p^{(0)} < r \leq 2p^{(0)} \text{ and } v_n > 0 \\ 0 & \text{otherwise.} \end{cases} \tag{A.110}$$

17. Calculation of the resulting random fluctuation (A.13):

$$\xi_n = \begin{cases} \xi_a & \text{if } S_{n+1} = 1 \\ -\xi_b & \text{if } S_{n+1} = -1 \\ \xi^{(0)} & \text{if } S_{n+1} = 0. \end{cases} \tag{A.111}$$

18. Calculation of the vehicle speed (A.2), vehicle motion in space (A.3), and acceleration (deceleration) (A.40) at step $n + 1$:

$$v_{n+1} = \max(0, \min(v_{\text{free}}, \tilde{v}_{n+1} + \xi_n, v_n + a_{\max}\tau, v_{\text{s},n})), \tag{A.112}$$

$$x_{n+1} = x_n + v_{n+1}\tau, \tag{A.113}$$

$$A_{n+1} = (v_{n+1} - v_n)/\tau. \tag{A.114}$$

We see that the Kerner-Klenov model (A.2)–(A.7), (A.9), (A.10), (A.13)–(A.21), (A.26), (A.27), (A.31)–(A.34), (A.45)–(A.50) is a Markov chain: At time step $n + 1$, values of model variables v_{n+1}, x_{n+1}, A_{n+1}, and S_{n+1} are calculated based only on their values v_n, x_n, A_n, and S_n at step n. The same conclusion follows for the model of Sect. A.8.4 consisting of Eqs. (A.2)–(A.7), (A.9), (A.10), (A.13)–(A.21), (A.26), (A.27), (A.31)–(A.34), (A.45)–(A.50), and (A.87)–(A.91).

A.9.2 Vehicles Moving Within Merging Regions of Bottlenecks

Within a merging region of a bottleneck, rules of the vehicle motion are given by formulas (A.2)–(A.4), (A.7), (A.9), (A.10), (A.13)–(A.21), (A.26), (A.27), (A.31)–(A.34), (A.49), (A.52)–(A.64).

The following sequence of calculations is made in the model, to find the model variables at the next time step $n + 1$ for vehicles within the merging region:

1. Calculation of space gaps between vehicles (A.7) at time step n:

$$g_n = x_{\ell,n} - x_n - d. \tag{A.115}$$

2. Calculation of space gaps between vehicles (A.49) at time step n in the neighboring lane within the merging region:

$$g^+_n = x^+_n - x_n - d, \quad g^-_n = x_n - x^-_n - d. \tag{A.116}$$

3. Calculation of speed $\hat{v}_n$ in (A.58) at time step n:

$$\hat{v}_n = \min(v^+_n,\ v_n + \Delta v^{(1)}_r). \tag{A.117}$$

4. Check of safety conditions (∗) (A.57) within the merging region:

$$\begin{aligned} g^+_n &> \min(\hat{v}_n\tau,\ G(\hat{v}_n, v^+_n)), \\ g^-_n &> \min(v^-_n\tau,\ G(v^-_n, \hat{v}_n)). \end{aligned} \tag{A.118}$$

All vehicles for which safety conditions (∗) are satisfied change their lane. Under conditions (∗), the vehicle coordinates x_n remains after lane changing and according to (A.63) the vehicle speed v_n is set to:

$$v_n = \hat{v}_n. \tag{A.119}$$

5. Check of safety conditions $(**)$ (A.59)–(A.62) within the merging region:

$$x_n^+ - x_n^- - d > g_{\text{target}}^{(\min)}, \quad g_{\text{target}}^{(\min)} = \lfloor \lambda_b v_n^+ + d \rfloor, \tag{A.120}$$

$$\begin{aligned} &x_{n-1} < x_{n-1}^{(\mathrm{m})} \text{ and } x_n \geq x_n^{(\mathrm{m})} \\ &\text{or} \\ &x_{n-1} \geq x_{n-1}^{(\mathrm{m})} \text{ and } x_n < x_n^{(\mathrm{m})}, \\ &x_n^{(\mathrm{m})} = \lfloor (x_n^+ + x_n^-)/2 \rfloor. \end{aligned} \tag{A.121}$$

All vehicles for which safety conditions $(**)$ are satisfied change their lane. Under conditions $(**)$, according to (A.63), (A.64), the vehicle speed v_n and the vehicle coordinate x_n are set, respectively, to:

$$v_n = \hat{v}_n, \quad x_n = x_n^{(\mathrm{m})}. \tag{A.122}$$

6. Calculation of space gaps between vehicles at time step n *after merging*:

$$g_n = x_{\ell,n} - x_n - d. \tag{A.123}$$

7. Calculation of probabilities (A.20), (A.21) of random acceleration a_n and deceleration b_n:

$$P_0 = \begin{cases} p_0 \text{ if } S_n \neq 1 \\ 1 \;\;\text{ if } S_n = 1, \end{cases} \quad P_1 = \begin{cases} p_1 \text{ if } S_n \neq -1 \\ p_2 \text{ if } S_n = -1. \end{cases} \tag{A.124}$$

8. Calculation of random acceleration a_n (A.18) and deceleration b_n(A.19):

$$a_n = a\Theta(P_0 - r_1), \; b_n = a\Theta(P_1 - r_1). \tag{A.125}$$

9. Calculation of speed $\hat{v}_n^+$ (A.54):

$$\hat{v}_n^+ = \max(0, \min(v_{\text{free}}, \; v_n^+ + \Delta v_r^{(2)})). \tag{A.126}$$

10. Calculation of vehicle acceleration (deceleration) (A.53) within the synchronization gap

$$\Delta_n^+ = \max(-b_n\tau, \min(a_n\tau, \; \hat{v}_n^+ - v_n)). \tag{A.127}$$

11. Calculation of a vehicle speed $v_{c,n}$ in formula (A.52):

$$v_{c,n} = \begin{cases} v_n + \Delta_n^+ \text{ at } g_n^+ \leq G(v_n \hat{v}_n^+) \\ v_n + a_n\tau \text{ at } g_n^+ > G(v_n, \hat{v}_n^+). \end{cases} \tag{A.128}$$

12. Calculation of a safe speed $v_n^{(\mathrm{safe})}$ (A.27) in accordance with (A.31)–(A.33):

$$v_n^{(\mathrm{safe})} = \lfloor v^{(\mathrm{safe})}(g_n,\ v_{\ell,n}) \rfloor. \tag{A.129}$$

13. Calculation of the anticipation speed of the preceding vehicle (A.34):

$$v_\ell^{(\mathrm{a})} = \max(0, \min(v_{\ell,n}^{(\mathrm{safe})}, v_{\ell,n}, g_{\ell,n}/\tau) - a\tau). \tag{A.130}$$

14. Calculation of the safe speed (A.26) that takes into account the anticipation speed of the preceding vehicle:

$$v_{\mathrm{s},n} = \min\,(v_n^{(\mathrm{safe})}, g_n/\tau + v_\ell^{(\mathrm{a})}). \tag{A.131}$$

15. Calculation of the speed $\tilde{v}_{n+1}$ (A.4):

$$\tilde{v}_{n+1} = \min(v_{\mathrm{free}}, v_{\mathrm{s},n}, v_{\mathrm{c},n}). \tag{A.132}$$

16. Calculation of the state of vehicle motion S_{n+1} (A.14) at step $n+1$:

$$S_{n+1} = \begin{cases} -1 & \text{if } \tilde{v}_{n+1} < v_n \\ 1 & \text{if } \tilde{v}_{n+1} > v_n \\ 0 & \text{if } \tilde{v}_{n+1} = v_n, \end{cases} \tag{A.133}$$

17. Calculation of components of random fluctuation in acceleration (deceleration) (A.15)–(A.17):

$$\xi_\mathrm{a} = a^{(\mathrm{a})}\tau\Theta(p_\mathrm{a} - r),\ \xi_\mathrm{b} = a^{(\mathrm{b})}\tau\Theta(p_\mathrm{b} - r), \tag{A.134}$$

$$\xi^{(0)} = a^{(0)}\tau \begin{cases} -1 & \text{if } r \le p^{(0)} \\ 1 & \text{if } p^{(0)} < r \le 2p^{(0)} \text{ and } v_n > 0 \\ 0 & \text{otherwise.} \end{cases} \tag{A.135}$$

18. Calculation of the resulting random fluctuation (A.13):

$$\xi_n = \begin{cases} \xi_\mathrm{a} & \text{if } S_{n+1} = 1 \\ -\xi_\mathrm{b} & \text{if } S_{n+1} = -1 \\ \xi^{(0)} & \text{if } S_{n+1} = 0. \end{cases} \tag{A.136}$$

19. Calculation of the vehicle speed (A.2), vehicle motion in space (A.3), and acceleration (deceleration) (A.40) at step $n+1$:

$$v_{n+1} = \max(0, \min(v_{\text{free}}, \tilde{v}_{n+1} + \xi_n, v_n + a_{\max}\tau, v_{\text{s},n})), \tag{A.137}$$

$$x_{n+1} = x_n + v_{n+1}\tau, \tag{A.138}$$

$$A_{n+1} = (v_{n+1} - v_n)/\tau. \tag{A.139}$$

A.10 Model of Heterogeneous Traffic Flow

In this section, we extend the model of identical vehicles and drivers described above for a more realistic heterogeneous traffic flow in which drivers exhibit different behavioral characteristics and vehicles have different parameters [30]. Although in the model of heterogeneous traffic flow we assume that all vehicles move with the same update rules of vehicle motion, some model functions and parameters in these rules are different: In particular, different vehicle types exhibit different maximum speeds in free flow, different vehicle lengths, as well as different driver time delays in different driving situations. In the model, different vehicles and drivers are specified by vehicle identifier $j = 1, 2, \ldots, H$, where H is the total number of different types of vehicles ($H > 1$).

Formulas for rules of vehicle motion in traffic flow with identical vehicles and identical drivers considered above remain the same in heterogeneous traffic flow, if model parameters that are responsible for individual vehicle characteristics are specified by vehicle identifier $j = 1, 2, \ldots, H$. For example, for heterogeneous flow in Eqs. (A.2)–(A.7), (A.9), (A.10) the vehicle speed in free flow v_{free}, the vehicle length d, coefficient k in formula for the synchronization space gap G_n are

$$v_{\text{free}} = v_{\text{free}}^{(j)}, \; j = 1, 2, \ldots, H, \tag{A.140}$$

$$d = d^{(j)}, \; j = 1, 2, \ldots, H, \tag{A.141}$$

$$k = k^{(j)}, \; j = 1, 2, \ldots, H. \tag{A.142}$$

Respectively, all other model parameters that describe individual driver time delays, individual lane changing behavior, merging behavior at road bottlenecks that can be different for different vehicles are specified by vehicle identifier $j = 1, 2, \ldots, H$ as shown in examples (A.140)–(A.142).

In the model, vehicle identifier j and the related vehicle parameters are ascribed to the vehicle as its individual "attributes" when the vehicle is generated at the beginning of the road.

To show that the Kerner-Klenov microscopic stochastic three-phase traffic flow model of identical vehicles and drivers can be very easy extended for any heterogeneous traffic flow, below we limit a consideration of heterogeneous traffic flow in which there can be only three types of vehicles: "fast vehicles," "slow vehicles," and "long vehicles." The vehicle length of fast and slow vehicles is lower than the length of long vehicles. The maximum vehicle speed in free flow of fast vehicles is higher than the one for slow and long vehicles. There are also other model parameters and variables that are different for different drivers and vehicles. We choose the vehicle identifier $j = 1$ for fast vehicles, it is $j = 2$ for slow vehicles, and it is $j = 3$ for long vehicles. All model parameters and variables that are chosen in the model different for the three types of vehicles are marked by superscripts (j) where $j = 1$, 2, and 3. The percentages of fast vehicles $\eta^{(1)}$, slow vehicles $\eta^{(2)}$, and long vehicles $\eta^{(3)}$ in heterogeneous flow satisfy the obvious condition:

$$\eta^{(1)} + \eta^{(2)} + \eta^{(3)} = 100\%. \tag{A.143}$$

To take into account different driver behavioral characteristics and different vehicle parameters in heterogeneous traffic flow, the following changes have been made in the model for identical vehicles of Sects. A.3–A.8.

A.10.1 Vehicle Motion on Single-Lane Road

A.10.1.1 Steady States and Vehicle Motion

Eqs. (A.2)–(A.7), (A.9), (A.10), (A.13)–(A.21), (A.26), (A.27), (A.31)–(A.34) are also the general rules of vehicle motion in heterogeneous flow. However, in heterogeneous flow with three types of vehicles, in accordance with (A.140) the vehicle speed in free flow v_{free} in Eqs. (A.2), (A.4) is

$$v_{\text{free}} = v^{(j)}_{\text{free}}, \ j = 1, 2, 3, \tag{A.144}$$

where $v^{(1)}_{\text{free}}$, $v^{(2)}_{\text{free}}$, and $v^{(3)}_{\text{free}}$ are either constant values (Table A.5) or functions of a space gap (see Sect. A.11). In this heterogeneous flow, accordingly to (A.141) the vehicle length d is

$$d = d^{(j)}, \ j = 1, 2, 3, \tag{A.145}$$

where $d^{(1)}$, $d^{(2)}$, and $d^{(3)}$ are constant values (see Table A.5).

The space gap g_n (A.7) in heterogeneous flow has been taken in the form

$$g_n = x_{\ell,n} - x_n - d_\ell, \tag{A.146}$$

where d_ℓ is the length of the preceding vehicle that depends on the vehicle type.

The synchronization gap G_n in Eq. (A.5) is given by Eqs. (A.9), (A.10), where accordingly to (A.142) value k in (A.10) is

$$k = k^{(j)}, \ j = 1, 2, 3, \tag{A.147}$$

Table A.5 Model parameters of heterogeneous flow often used in simulations of vehicle motion in road lane. Other model parameters are the same as those in Table A.1

$v_{\rm free}^{(1)} = 33.3\ {\rm ms}^{-1}/\delta v$ (120 km/h),
$v_{\rm free}^{(2)} = v_{\rm free}^{(3)} = 25\ {\rm ms}^{-1}/\delta v$ (90 km/h),
$d^{(1)} = d^{(2)} = 7.5\ {\rm m}/\delta {\rm x}$, $d^{(3)} = 17\ {\rm m}/\delta {\rm x}$,
$k^{(1)} = k^{(2)} = 3$, $k^{(3)} = 4$,
$p_{\rm a}^{(1)} = p_{\rm a}^{(2)} = 0.17$, $p_{\rm a}^{(3)} = 0.3$,
$p_{\rm b}^{(1)} = p_{\rm b}^{(2)} = 0.1$, $p_{\rm b}^{(3)} = 0.2$,
$p_0^{(1)}(v) = 0.6 + 0.17 \min(1, v/v_{01}))$,
$p_0^{(2)}(v_n) = 1 - 1.3(1 - p_0^{(1)}(v_n))$,
$p_0^{(3)}(v_n) = 1 - 1.5(1 - p_0^{(1)}(v_n))$,
$v_{01} = 10\ {\rm ms}^{-1}/\delta v$, $a^{(\rm b)} = a$

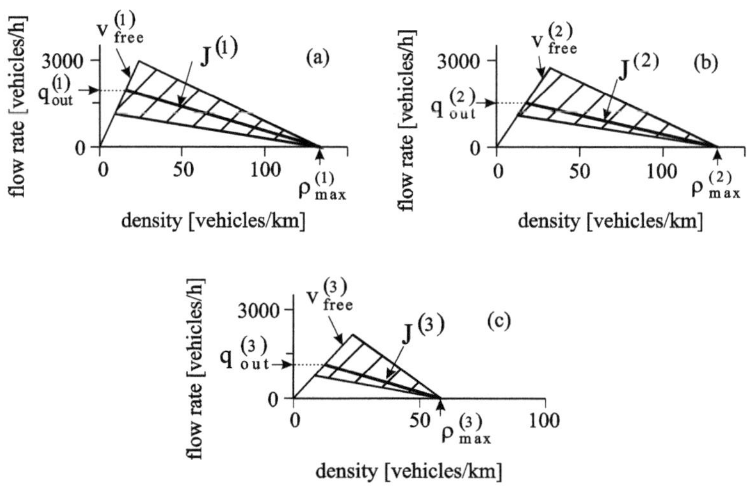

Fig. A.7 Steady-state model solutions and the line J for the following cases. (**a**) All vehicles are fast vehicles (line J is line $J^{(1)}$). (**b**) All vehicles are slow vehicles (line J is line $J^{(2)}$). (**c**) All vehicles are long vehicles (line J is line $J^{(3)}$). The flow rate in the outflow from a wide moving jam and the downstream jam front velocity are $q_{\rm out}^{(1)} = 1900$ vehicles/h and $v_{\rm g}^{(1)} = -16.2$ km/h for fast vehicles, $q_{\rm out}^{(2)} = 1510$ vehicles/h and $v_{\rm g}^{(2)} = -13$ km/h for slow vehicles, $q_{\rm out}^{(3)} = 1130$ vehicles/h and $v_{\rm g}^{(3)} = -24.5$ km/h for long vehicles, respectively. $\rho_{\rm max}^{(1)}$, $\rho_{\rm max}^{(2)}$, and $\rho_{\rm max}^{(3)}$ are, respectively, jam density for fast vehicles, slow vehicles, and long vehicles. Adapted from [30]

$k^{(1)}$, $k^{(2)}$, and $k^{(3)}$ are constants (see Table A.5).

Two-dimensional regions of steady-state model solutions in the flow–density plane for traffic flows in which either all vehicles are fast vehicles, or all vehicles are slow vehicles, or else all vehicles are long vehicles are shown in Figs. A.7a,b, and c, respectively.

A.10.1.2 Fluctuations

Random acceleration and deceleration are described by Eqs. (A.13), (A.15)–(A.21). However, in heterogeneous flow the probabilities $p_{\rm b}$ and $p_{\rm a}$ in (A.15) and (A.17) are

$$p_{\rm b} = p_{\rm b}^{(j)},\ j = 1, 2, 3, \tag{A.148}$$

$$p_{\rm a} = p_{\rm a}^{(j)},\ j = 1, 2, 3. \tag{A.149}$$

Moreover, in (A.20)

$$p_0(v) = p_0^{(j)}(v),\ j = 1, 2, 3, \tag{A.150}$$

$$p_0^{(1)}(v) > p_0^{(2)}(v) > p_0^{(3)}(v). \tag{A.151}$$

The mean time delays in vehicle acceleration

$$\tau_{\rm del}^{(\rm acc)}(v) = \tau_{\rm del}^{(\rm acc,\ \mathit{j})}(v),\ j = 1, 2, 3 \tag{A.152}$$

are

$$\tau_{\rm del}^{(\rm acc,\ \mathit{j})}(v) = \frac{\tau}{p_0^{(j)}(v)},\ j = 1, 2, 3, \tag{A.153}$$

$$\tau_{\rm del}^{(\rm acc,\ 1)}(v) < \tau_{\rm del}^{(\rm acc,\ 2)}(v) < \tau_{\rm del}^{(\rm acc,\ 3)}(v). \tag{A.154}$$

Accordingly to (A.154), we assume that fast vehicles have a shorter mean time delay than the related mean time delays for slow and long vehicles, i.e., it is assumed that fast vehicles prefer a more aggressive driving.

A.10.1.3 Safe Speed

In Eqs. (A.26), (A.27), (A.31)–(A.34) for the safe speed $v_{{\rm s},n}$ as in all other related formulae of Sect. A.3.5, the space gap g_n in heterogeneous flow is given by Eq. (A.146).

A.10.2 Lane Changing Rules in Model of Two-Lane Road

As in Sect. A.5, lane changing rules in a two-lane model of heterogeneous flow are based on incentive and security conditions. The following incentive conditions for lane changing from the right lane to the left (passing) lane ($R \rightarrow L$) and a return change from the left lane to the right lane ($L \to R$) have been used in the model of heterogeneous flow:

$$R \to L: \ v_n^+ \geq v_{\ell,n} + \delta_1 \text{ and } v_n \geq v_{\ell,n}, \tag{A.155}$$

$$L \to R: \ v_n^+ > v_{\ell,n} + \delta_2 \text{ or } v_n^+ > v_n + \delta_2, \tag{A.156}$$

where

$$\delta_1 < \delta_2 \quad \text{for fast vehicles}, \tag{A.157}$$

$$\delta_2 < \delta_1 \quad \text{for slow and long vehicles}, \tag{A.158}$$

$\delta_1 \geq 0$, $\delta_2 \geq 0$ are constants (see Table A.6).

It is assumed that if the vehicle speed in the right lane is high enough, slow and long vehicles moving in the left lane are usually forced to change to the right lane, whereas slow and long vehicles moving in the right lane retain in this lane. To simulate this effect, the following incentive conditions for slow and long vehicles have been applied. For lane changing from the left lane to the right lane ($L \to R$) for slow and long vehicles the incentive condition is (A.156) *or*

$$L \to R: \quad v_n^+ > v_{\text{free}}^{(j)} - \delta_0, \ j = 2, 3, \tag{A.159}$$

where $\delta_0 > 0$ is constant (see Table A.6). For lane changing from the right lane to the left lane ($R \to L$) for slow and long vehicles the incentive conditions are (A.155) *and*

$$R \to L: \quad v_{\ell,n} \leq v_{\text{free}}^{(j)} - \delta_0, \ j = 2, 3. \tag{A.160}$$

The security conditions for lane changing are given by the inequalities (A.47), (A.48), in which g_n^+ and g_n^- are

$$g_n^+ = x_n^+ - x_n - d^+, \quad g_n^- = x_n - x_n^- - d, \tag{A.161}$$

d^+ is the length of the preceding vehicle in the "target" (neighboring) lane, vehicle length d is given by formula (A.145).

If the incentive and security conditions are satisfied, then as in Rickert et al. [44] in this model the vehicle changes the lane with the probability $p_c < 1$ ($p_c = 1$ if (A.159) is satisfied).

Slow and long vehicles can change from the left lane to the right lane even if the security conditions (A.47), (A.48) are not satisfied. This lane changing is realized if the above incentive condition (A.159) is satisfied and the following two security conditions that are similar to Eqs. (A.87)–(A.90) are satisfied: (i) The gap between two neighboring vehicles in the right lane on the main road exceeds some value $g_{\text{target}}^{(\text{min})}$, i.e.,

$$x_n^+ - x_n^- - d^+ > g_{\text{target}}^{(\text{min})}. \tag{A.162}$$

(ii) The vehicle passes the point

$$x_n^{(\mathrm{m})} = \lfloor (x_n^+ + x_n^- + d - d^+)/2 \rfloor \tag{A.163}$$

between two vehicles in the target lane of the main road for time step n, i.e.,

$$\begin{aligned} &x_{n-1} < x_{n-1}^{(\mathrm{m})} \quad \text{and} \quad x_n \geq x_n^{(\mathrm{m})} \\ &\text{or} \\ &x_{n-1} \geq x_{n-1}^{(\mathrm{m})} \quad \text{and} \quad x_n < x_n^{(\mathrm{m})}. \end{aligned} \tag{A.164}$$

In (A.162)

$$g_{\text{target}}^{(\min)} = \lfloor \lambda v_n^+ + d \rfloor, \tag{A.165}$$

$$\lambda = \lambda^{(j)},\ j = 2, 3, \tag{A.166}$$

$\lambda^{(2)}$ and $\lambda^{(3)}$ are constants (see Table A.6), vehicle length d is given by formula (A.145). If, in accordance with (A.159), (A.162)–(A.165), a vehicle changes lane, then the vehicle coordinate in the target lane is set to $x_n = x_n^{(m)}$. These security conditions facilitate lane changing for slow and long vehicles from the left lane to the right lane when the security rules (A.47), (A.48) are not satisfied.

As known [39], two slow and/or long vehicles moving side by side in the left and right lanes hinder fast vehicles accelerating in free flow on a two-lane road. To avoid this effect, it is assumed that either a slow vehicle or a long vehicle in the left lane, which should change the lane, can move with a higher speed in free flow before the vehicle changes to the right lane. For this purpose, when the condition (A.159) is satisfied, the free flow speed v_{free} in (A.2), (A.4) for slow or long vehicles in the left lane is equal to

$$v_{\text{free}}^{(2,\ \text{left})} > v_{\text{free}}^{(2)} \tag{A.167}$$

and

$$v_{\text{free}}^{(3,\ \text{left})} > v_{\text{free}}^{(3)} \tag{A.168}$$

for slow and long vehicles, respectively; $v_{\text{free}}^{(2,\ \text{left})}$ and $v_{\text{free}}^{(3,\ \text{left})}$ are constants (see Table A.6).

Table A.6 Lane changing model parameters often used in simulations of heterogeneous flow. Other model parameters are the same as those in Table A.2

$\delta_1 = 1\,\mathrm{ms}^{-1}/\delta v$, $\delta_2 = 3.5\,\mathrm{ms}^{-1}/\delta v$ for fast vehicles,
$\delta_1 = 3.5\,\mathrm{ms}^{-1}/\delta v$, $\delta_2 = 1\,\mathrm{ms}^{-1}/\delta v$ for slow and
long vehicles, $\delta_0 = 6\,\mathrm{ms}^{-1}/\delta v$, $\lambda^{(2)} = \lambda^{(3)} = 0.8$,
$v_{\mathrm{free}}^{(2,\ \mathrm{left})} = 28.5\ \mathrm{ms}^{-1}/\delta v$, $v_{\mathrm{free}}^{(3,\ \mathrm{left})} = 27.5\ \mathrm{ms}^{-1}/\delta v$

A.10.3 Boundary, Initial Conditions, and Models of Bottlenecks

Open boundary conditions of Sect. A.3.6 are applied. However, in the related formulas the length of the preceding vehicle d should be replaced by d_ℓ. In the initial state ($n = 0$), all vehicles have a related free flow speed $v_n = v_{\mathrm{free}}^{(j)}$, $j = 1, 2, 3$ and they are positioned at space intervals $x_{\ell,n} - x_n = v_{\mathrm{free}}^{(j)}\tau_{\mathrm{in}}$, $j = 1, 2, 3$.

As above-mentioned, the vehicle identifier j and the related vehicle parameters are individual "attributes" that are ascribed to a vehicle when the vehicle is generated at the beginning of the road. We have used two different possibilities to generate vehicles of various types at the beginning of the road:

(i) Fast, slow, and long vehicles are randomly generated in the left and right lanes with the rates related to chosen values of the flow rate q_{in} and the percentages $\eta^{(1)}$, $\eta^{(2)}$, and $\eta^{(3)}$.
(ii) Fast vehicles are preferably generated in the left lane whereas slow and long vehicles are preferably generated in the right lane.

In case (ii), only $\max(0,\ 2\eta^{(1)} - 100\%)$ of fast vehicles are randomly generated in the right lane, whereas only $\max(0,\ 2\eta^{(2)} + 2\eta^{(3)} - 100\%)$ of slow and long vehicles are randomly generated in the left lane.

In heterogeneous flow, the models of bottlenecks are described by formulas of Sect. A.6. However, in safety conditions (A.57) the values g_n^+ and g_n^- are given by Eq. (A.161). Formulas (A.59) and (A.61) should be respectively rewritten as

$$x_n^+ - x_n^- - d^+ > g_{\mathrm{target}}^{(\mathrm{min})} \tag{A.169}$$

and

$$x_n^{(\mathrm{m})} = \lfloor (x_n^+ + x_n^- + d - d^+)/2 \rfloor, \tag{A.170}$$

where d and d^+ are given by formula (A.145).

A.11 Realistic Heterogeneous Traffic Flow

A.11.1 Dependence of Free Flow Speed on Space Gap

To simulate the dependence of the speed in free flow on the vehicle density, we assume that the free flow speed $v_{\rm free}(g_n)$ depends on the space gap between the vehicle and the preceding vehicle. In simulations, we have used either formula

$$v_{\rm free}(g_n) = \lfloor \max(v_{\rm free}^{(\rm min)}, v_{\rm free}^{(\rm max)}(1 + c_1 d/(g_n + d))^{-1}) \rfloor, \tag{A.171}$$

or formula

$$v_{\rm free}(g_n) = \lfloor \max(v_{\rm free}^{(\rm min)}, v_{\rm free}^{(\rm max)}(1 - \kappa d/(d + g_n))) \rfloor, \tag{A.172}$$

where $v_{\rm free}^{(\rm min)}$, $v_{\rm free}^{(\rm max)}$, κ, and c_1 are constants that can be different for passenger and long vehicles. We have already used formula (A.172) in Sect. 12.4 of the main text (see formula (12.18) and Fig. 12.5).

Clearly that for heterogeneous traffic flow consisting of drivers and vehicles with different characteristics model parameters (Sect. A.10) functions $v_{\rm free}(g_n)$ can also be chosen very different for different vehicles (drivers). For example, for heterogeneous traffic flow with different types $j = 1, 2, 3, \ldots$ of vehicles (or drivers), formula (A.144) should be replaced by

$$v_{\rm free}(g_n) = v_{\rm free}^{(j)}(g_n),\ j = 1, 2, 3, \ldots, \tag{A.173}$$

where both functions $v_{\rm free}^{(j)}(g_n)$ and model parameters in these functions can be chosen differently for different types $j = 1, 2, 3, \ldots$ of vehicles (or drivers).

A.11.2 Simulations of Traffic Patterns on Realistic Three-Lane Highway

The model of heterogeneous traffic (Sects. A.10 and A.11.1) has been used for simulations and analysis of real field traffic data that has been measured on a three-lane highway in Germany (Fig. A.8a). There have been two sources of real field traffic data: (i) 1-min average data measured by road detectors and (ii) single-vehicle data measured by TomTom probe vehicles (Figs. A.8b and A.9). One of the objectives of these simulations has been a study of a quality of traffic service "danger warning" messages [38]. In this section, we present the further development of the Kerner-Klenov model made for these simulations.

Additionally to different driver characteristics and vehicle parameters, in realistic heterogeneous traffic flow there are often more than two-lanes (in one direction)

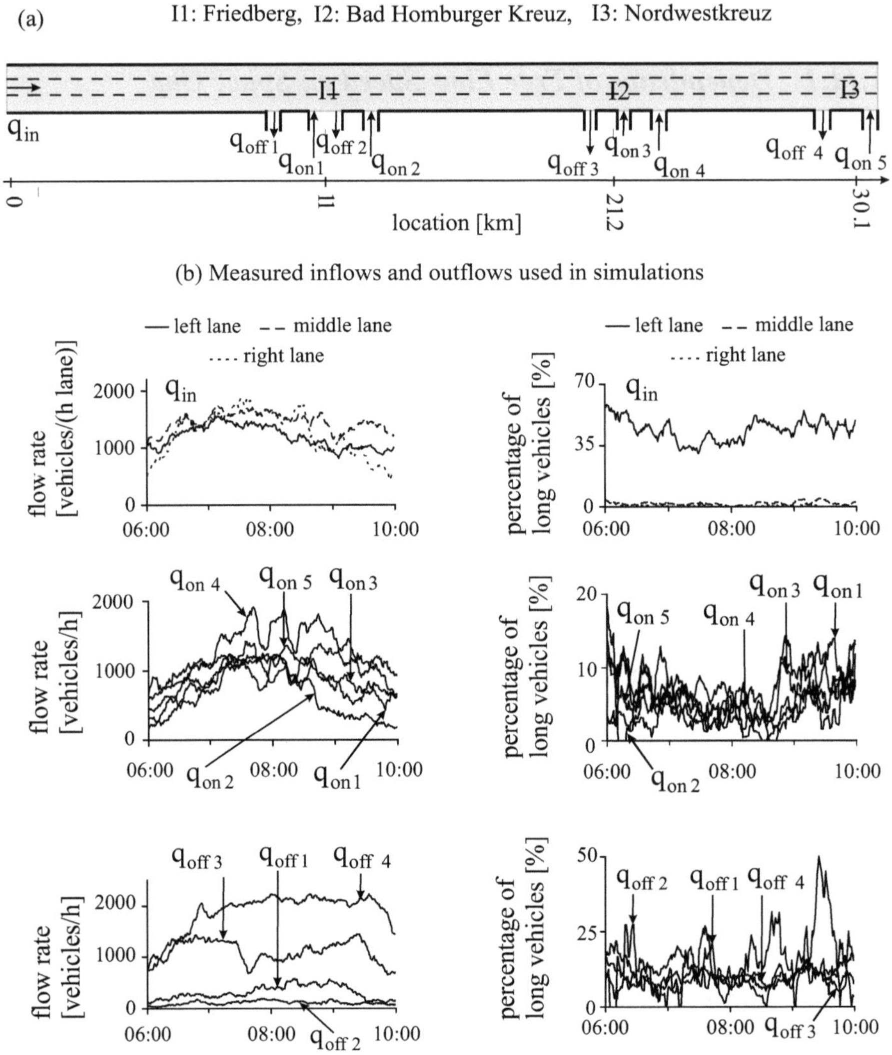

Fig. A.8 Empirical inflows and outflows used in simulations: (**a**) Simplified schema of three-lane road section of A5-South highway in Germany on which both probe vehicle data and detector data presented in Fig. A.9 were measured on December 10, 2009 (**a**). (**b**) Measured time-functions of inflows and outflows (*left*) with percentages of long vehicles (*right*) used in simulations. In (**a**), locations of on-ramps $x_{\text{on}\,i}, i = 1, 2, \ldots, 5$ are 11.085, 11.789, 21.522, 21.86, 30.19 km, respectively; locations of off-ramps $x_{\text{off}\,y}, y = 1, 2, 3, 4$ are 10.674, 11.366, 18.612, 27.768 km, respectively. Adapted from [38]

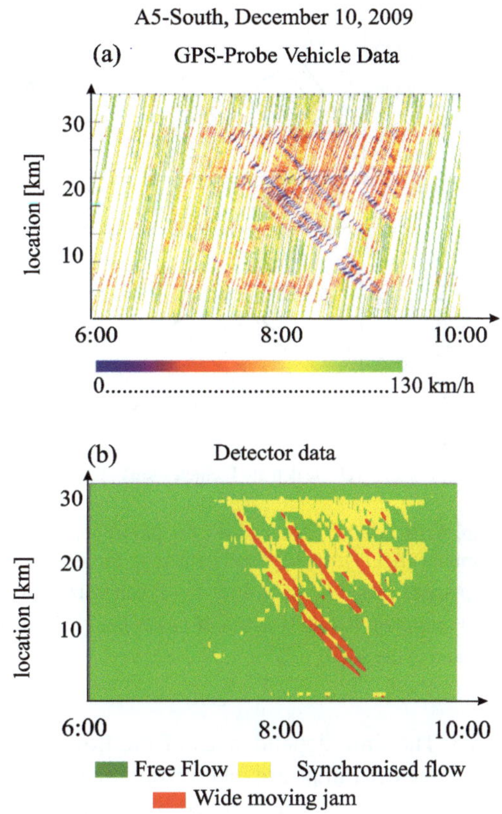

Fig. A.9 Overviews of TomTom GPS-probe vehicle data (**a**) and road detector data (**b**) measured on December 10, 2009 on A5-South highway in Germany (Fig. A.8) within a congested pattern: (**a**) Vehicle trajectories in space and time. (**b**) The three traffic phases reconstructed with ASDA and FOTO models from raw road detector data. Adapted from [38]

on many real highways. Therefore, in this section we extend the model for this realistic traffic flow. However, for the simplicity of the consideration, we limit by a heterogeneous traffic flow consisting of only two types of vehicles, passenger (usual) and long vehicles, which move on a three-lane road. In the model, free flow speed of passenger and long vehicles depend on the space gap as shown in Fig. A.10.

As in real traffic, in a model of three-lane road with on-ramp and off-ramp bottlenecks (these bottlenecks are within road intersections denoted by I1-I3 in Fig. A.8a) vehicles appear at the upstream road boundary $x = 0$ of the main road and from the on-ramps; vehicles leave the main road to the off-ramps and at the downstream road boundary $x = 30.5$ km. In this open traffic process, all traffic variables on the main road within road locations $0 < x < 30.5$ km result *only* from traffic modeling at the following boundary conditions:

(i) The time-dependence of the flow rate at the upstream road boundary $x = 0$ is taken from measurements at the detectors located at $x = 0$ (Fig. A.8).
(ii) The time-dependences of the on-ramp inflow rates are taken from measurements of these flow rates at the detectors located at the associated on-ramps (Fig. A.8).

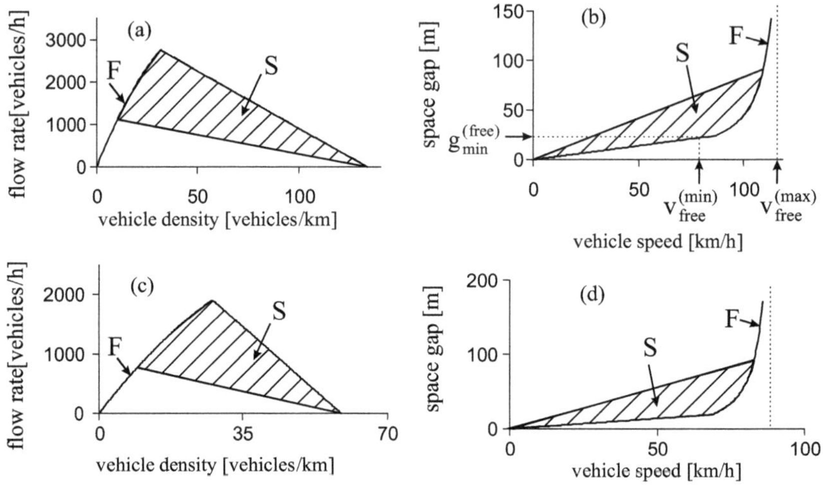

Fig. A.10 Model steady states for passenger vehicles (**a**, **b**) and long vehicles (**c**, **d**) related to formula (A.171): (**a**, **c**) Steady states in the flow-density plane. (**b**, **d**) Steady states in the space-gap-speed plane. F—free flow. S—synchronized flow. In (**b**), $v_{\text{free}}^{(\text{max})}$ and $v_{\text{free}}^{(\text{min})}$ are the maximum and minimum free flow speed of passenger vehicles, $g_{\text{min}}^{(\text{free})}$ is the minimum space gap in free flow related to the speed $v_{\text{free}}^{(\text{min})}$

(iii) The time-dependences of the flow rates of vehicles leaving the main road to the off-ramps are taken from measurements of these flow rates at the detectors located at the associated off-ramps (Fig. A.8).
(iv) At the downstream model boundary $x = 30.5$ km, free flow conditions for vehicles leaving the road section are given.

A.11.3 Update Rules of Vehicle Motion in Road Lane

Update rules of vehicle motion in the model are given by Eqs. (A.2)–(A.7). However, as explained in Sect. A.11.1, in Eqs. (A.2) and (A.4) the free flow speed v_{free} depends on the space gap g_n. Respectively, the model of vehicle motion in road lane used in numerical simulations reads as follows:

$$v_{n+1} = \max(0, \min(v_{\text{free}}(g_n), \tilde{v}_{n+1} + \xi_n, v_n + a\tau, v_{\text{s},n})), \tag{A.174}$$

$$x_{n+1} = x_n + v_{n+1}\tau, \tag{A.175}$$

$$\tilde{v}_{n+1} = \min(v_{\text{free}}(g_n), v_{\text{c},n}, v_{\text{s},n}), \tag{A.176}$$

$$v_{c,n} = \begin{cases} v_n + \Delta_n & \text{at } g_n \leq G_n \\ v_n + a_n\tau & \text{at } g_n > G_n, \end{cases} \tag{A.177}$$

Table A.7 Model parameters often used for simulations of traffic flow on three-lane road

Parameters for vehicle motion in road lane:
$\tau = \tau_{\text{safe}} = 1$ s, $d = 7.5$ m$/\delta x$ for personal vehicles,
$d = 17$ m$/\delta x$ for long vehicles, $\delta x = 0.01$ m,
$v_{\text{free}}^{(\max)} = 33.33$ ms$^{-1}/\delta v$,
$v_{\text{free, long}}^{(\max)} = 25$ ms$^{-1}/\delta v$,
$v_{\text{free}}^{(\min)} = 15$ ms$^{-1}/\delta v$,
$c_1 = 1.4$ for personal vehicles, $c_1 = 0.6$ for long vehicles,
$b = 1$ ms$^{-2}/\delta a$, $\delta v = 0.01$ ms^{-1},
$\delta a = 0.01$ ms^{-2}, $k = 3$, $p_1 = 0.3$, $p_{\text{b}} = 0.15$, $p_{\text{a}} = 0.15$,
$a^{(\text{b})}(v_n) = 0.2a+$
$+0.8a \max(0, \min(1, (v_{22} - v_n)/\Delta v_{22}))$,
$v_{22} = 12.5$ ms$^{-1}/\delta v$, $\Delta v_{22} = 2.778$ ms$^{-1}/\delta v$,
$p^{(0)} = 0.005$, $p_2(v_n) = 0.48 + 0.32\Theta(v_n - v_{21})$,
$p_0(v_n) = 0.52 + 0.23 \min(1, v_n/v_{01})$ for personal vehicles and
$p_0(v_n) = 0.33 + 0.37 \min(1, v_n/v_{01})$ for long vehicles,
$v_{01} = 10$ ms$^{-1}/\delta v$, $v_{21} = 15$ ms$^{-1}/\delta v$, $a = 0.5$ ms$^{-2}/\delta a$,
$a^{(0)} = 0.2a$, $a^{(\text{a})} = 0$
Lane changing parameters (model of over-acceleration):
$\delta_1 = 1$ ms$^{-1}/\delta v$ and $\delta_2 = 3.5$ ms$^{-1}/\delta v$ for passenger vehicles,
$\delta_1 = 3.5$ ms$^{-1}/\delta v$ and $\delta_2 = 1$ ms$^{-1}/\delta v$ for long vehicles,
$\delta_0 = 6$ ms$^{-1}/\delta v$,
$L_{\text{a}} = 150$ m$/\delta x$,
$p_{\text{c}} = 0.2$, $\lambda = 0.8$, $\Delta v^{(1)} = 2$ ms$^{-1}/\delta v$

where

$$\Delta_n = \max(-b_n\tau, \min(a_n\tau,\ v_{\ell,n} - v_n)). \tag{A.178}$$

Model functions ξ_n, G_n, a_n, b_n, $v_{\text{s},n}$ as well as their physics are the same as those in formulas (A.9), (A.10), (A.13)–(A.21), (A.26), (A.27), (A.31)–(A.34) of Sects. A.3.1–A.3.5 for two-lane road; the probability p_0 in (A.20) is chosen different for passenger and long vehicles (see Table A.7).

A.11.4 Lane Changing Rules on Three-Lane Road

By the formulating of lane changing rules on three-lane road, we have used formulas (A.47), (A.48), (A.155)–(A.165) for lane changing in heterogeneous traffic flow of Sect. A.10.2.

In particular, lane changing rules from the right lane to the middle lane and from the middle lane to the left lane are described by (A.155). Lane changing rules from the middle lane to the right lane and from the left lane to the middle lane are described by (A.156). In (A.155) and (A.156), $\delta_1 > 0$ and $\delta_2 > 0$ are constants, $\delta_1 < \delta_2$ for personal vehicles and $\delta_1 > \delta_2$ for long vehicles (Table A.7).

When lane changing rules (A.155), (A.156) are satisfied, a vehicle changes the lane with probability p_c, if in addition one of the following sets of safety conditions for lane changing are satisfied. The first set of the safety conditions for lane changing is given by formulas (A.47), (A.48), and (A.161).

When safety conditions (A.47), (A.48) are not satisfied, then a vehicle can nevertheless change to a faster lane with the above-mentioned given probability p_c, if the second set of safety conditions given by formulas (A.162)–(A.165) is satisfied. For the model of heterogeneous traffic flow on three-lane road, in formulas (A.161), (A.162)–(A.165) we assume that

$$\lambda = \lambda^{(j)}, \ j = 1, 3, \tag{A.179}$$

$\lambda^{(1)}$ (for passenger vehicles) and $\lambda^{(3)}$ (for long vehicles) are constants, vehicle length d is given by formula

$$d = d^{(j)}, \ j = 1, 3, \tag{A.180}$$

where $d^{(1)}$ and $d^{(3)}$ are constant values. These and other model parameters in formulas (A.47), (A.48), (A.155)–(A.165) for the model of heterogeneous traffic flow on three-lane road are given in Table A.7.

It is assumed that if the vehicle speed in the right lane is high enough, long vehicles moving first in the middle lane should usually change to the right lane, whereas long vehicles moving in the right lane keep the lane. To simulate this effect, the following lane changing rules that are similar to rules (A.159), (A.160) of Sect. A.10.2 have been added to the incentive criteria (A.155) and (A.156). A long vehicle changes from the left lane to the middle lane or from the middle lane to the right lane even if conditions (A.156) are not satisfied but condition

$$v_n^+ > v_{\text{free, long}}^{(\text{max})} - \delta_0 \tag{A.181}$$

is satisfied. Here, $v_{\text{free, long}}^{(\text{max})}$ is the maximum speed for long vehicles, $\delta_0 > 0$ is constant (see Table A.7). A long vehicle changes from the right lane to the middle lane or from the middle lane to the right lane only if the related condition

$$v_{\ell,n} \leq v_{\text{free, long}}^{(\text{max})} - \delta_0. \tag{A.182}$$

is satisfied together with (A.155). When formulas (A.181), (A.182) and security rules (A.47), (A.48), (A.161) or security rules (A.162)–(A.165) are satisfied, the vehicle changes the lane with probability $p_c = 1$.

A.11.5 Models of On- and Off-Ramp Bottlenecks on Three-Lane Road

There are five on-ramps and four off-ramps on the three-lane freeway section under consideration, which can act as effectual bottlenecks (Fig. A.8a). In the model of an on ramp bottleneck, we assume that within a road section of length $L_{\rm d}$ upstream of an on-ramp and within the on-ramp merging region of length $L_{\rm m}$ (Fig. A.11a) personal vehicles moving in the right lane on the main road try to change to the middle lane, and the vehicles moving in the middle lane try to change to the left lane.

Similarly, in the model of an off-ramp bottleneck we assume that within a road section of length $L_{\rm d}$ upstream of an off-ramp and within the off-ramp merging region of length $L_{\rm m}$ (Fig. A.11b) the vehicles that are not going to the off-ramp try to change from the right lane to the middle lane and from the middle lane to the left lane.

To simulate this effect, within the region

$$x_{\rm on} - L_{\rm d} \le x_n \le x_{\rm on} + L_{\rm m} \qquad \text{(A.183)}$$

for the on-ramp bottleneck and within the region

$$x_{\rm off} - L_{\rm d} \le x_n \le x_{\rm off} + L_{\rm m} \qquad \text{(A.184)}$$

for the off-ramp bottleneck and when conditions (A.155), (A.47), (A.48) are satisfied, the probability of lane changing is set to $p_{\rm c} = 1$.

Vehicle merging occurs within the merging region $L_{\rm m}$ of the on-ramp bottleneck (Fig. A.11a) and within the merging region $L_{\rm c} + L_{\rm m}$ of the off-ramp bottleneck (Fig. A.11b) when the following safety conditions (∗) *or* (∗∗) are satisfied. The safety conditions (∗) are as follows:

$$g_n^+ > \hat{v}_n \tau \text{ and } g_n^- > \min(\lambda_{\rm b,1} v_n^- \tau,\ G(v_n^-, \hat{v}_n)), \qquad \text{(A.185)}$$

or

$$g_n^+ > \min(\lambda_{\rm b,1} \hat{v}_n \tau,\ G(\hat{v}_n, v_n^+)) \text{ and } g_n^- > v_n^- \tau, \qquad \text{(A.186)}$$

$$\hat{v}_n = \min(v_n^+,\ v_n + \Delta v_r^{(1)}), \qquad \text{(A.187)}$$

$\Delta v_r^{(1)} > 0$ is constant, $\lambda_{\rm b,1} = c_{\rm b}\lambda_{\rm b}$, $c_{\rm b}$ is constant (see Table A.8), $\lambda_{\rm b}$ can be constant for some bottlenecks, for other bottlenecks $\lambda_{\rm b}$ can be chosen as a function of the average space gap $g_{{\rm avr},n}^{\rm (tag)}$ between $N_{\rm avr}$ subsequent vehicles ahead in the target lane:

$$g_{{\rm avr},n}^{\rm (tag)} = \frac{1}{N_{\rm avr}} \sum_{w=1}^{N_{\rm avr}} g_{w,n}^{\rm (tag)}, \qquad \text{(A.188)}$$

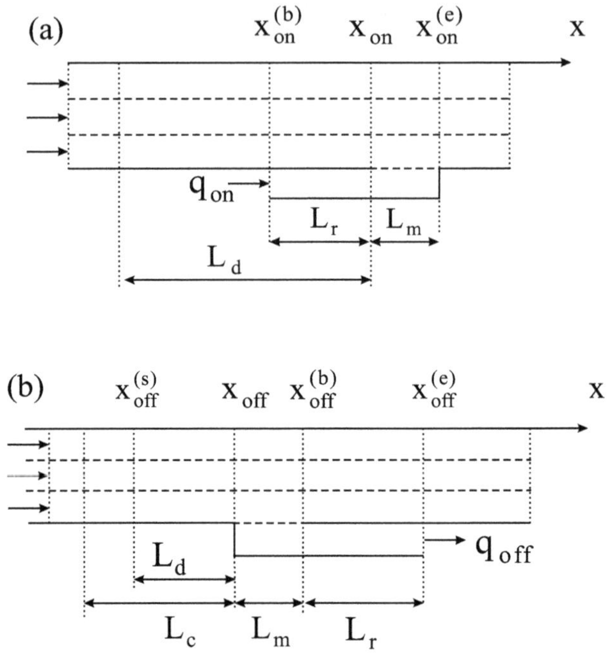

Fig. A.11 Models of on-ramp (**a**) and off-ramp (**b**) bottlenecks on three-lane road

where $g^{(\mathrm{tag})}_{w,n}$ is a space gap in front of the vehicle with index w in the target lane, the initial number $w = 1$ corresponds to the trailing vehicle in the target lane, the vehicle with index w is the preceding vehicle for the vehicle with index $w-1$, N_{avr} is a model parameter. A function $\lambda_{\mathrm{b}}(g^{(\mathrm{tag})}_{\mathrm{avr,n}})$ has been chosen as follows:

$$\lambda_{\mathrm{b}}(g^{(\mathrm{tag})}_{\mathrm{avr,n}}) = \lambda^{(0)}_{\mathrm{b}} + (\lambda^{(1)}_{\mathrm{b}} - \lambda^{(0)}_{\mathrm{b}}) \max(0, \min(1, (g^{(\mathrm{tag})}_{\mathrm{avr,n}} - g_1)/\Delta g_1), \tag{A.189}$$

where $\lambda^{(0)}_{\mathrm{b}} < \lambda^{(1)}_{\mathrm{b}}$, g_1, Δg_1 are constants (see Table A.8).

The safety condition $(**)$ is given by formula (A.169) in which

$$g^{(\mathrm{min})}_{\mathrm{target}} = \lfloor \lambda_{\mathrm{b}} v^{+}_{n} + d \rfloor, \tag{A.190}$$

λ_{b} is given by (A.189); in addition, the vehicle should pass the midpoint $x^{(\mathrm{m})}_{n}$ (A.170) between two neighboring vehicles in the target lane, i.e., conditions (A.62) should be satisfied.

The vehicle speed after vehicle merging is given by formula (A.63). Under the rule $(*)$ the vehicle coordinate x_n does not change, under the rule $(**)$ the vehicle coordinate x_n is given by formula (A.64). Vehicle speed adaptation before vehicle merging is determined by Eqs. (A.52)–(A.54), $\Delta v^{(2)}_{r}$ is constant (see Table A.8).

Table A.8 Parameters of bottleneck models often used for traffic simulations on three-lane road

$\lambda_b = 0.75$ for all bottlenecks except with
off-ramps at the middle and the downstream bottlenecks,
for the latter bottlenecks λ_b is given by formulas (A.189)
where $g_1 = 20$ m/δx, $\Delta g_1 = 20$ m/δx,
$\lambda_b^{(0)} = 0.72$, $\lambda_b^{(1)} = 0.8$ for the off-ramp at the middle bottleneck,
$\lambda_b^{(0)} = 0.65$, $\lambda_b^{(1)} = 0.8$ for the off-ramp at the downstream bottleneck,
$\lambda_b = 0.6$ for long vehicles at all bottlenecks,
$\lambda_b = 0.65$ for all off-ramp lanes,
$c_b = 0.5$, $N_{avr} = 10$,
L_c for passenger vehicles:
for the off-ramp at the downstream bottleneck:
$L_c = 0.3$ km/δx in the middle lane, $L_c = 0.6$ km/δx in the left lane,
for the off-ramp at the middle and the upstream bottlenecks:
$L_c = 0.5$ km/δx in the middle lane, $L_c = 1.0$ km/δx in the left lane,
for long vehicles and for all bottlenecks:
$L_c = 1.0$ km/δx in the middle lane,
$L_c = 2.0$ km/δx in the left lane,
$v_{free\ off} = 23.6$ ms$^{-1}/\delta v$, $L_d = 0.5$ km/δx,
$\Delta v_r^{(2)} = 5$ ms$^{-1}/\delta v$ for on- ramp bottlenecks,
$\Delta v_r^{(2)} = 10$ ms$^{-1}/\delta v$ for off-ramp bottlenecks, $\Delta v_r^{(1)} = 10$ ms$^{-1}/\delta v$,
$L_s = 0.1$ km/δx in the left lane, $L_s = 0.05$ km/δx in the middle lane,
$L_s = 0$ in the right lane, $L_r = 0.2$ km/δx for on- ramps and
$L_r = 0.1$ km/δx for off- ramps,
values L_m are taken from road infrastructure
for different on-ramps $L_{m,\ i}$, $i = 1, 2, \ldots, 5$
are, respectively, 281, 300, 300, 415, 250 m and for different off-ramps
$L_{m,\ y}$, $y = 1, 2, 3, 4$ are 210, 273, 2150, 1500 m related to Fig. A.8a

The same rules for vehicle merging are used in models of the on-ramp bottleneck (Fig. A.11a) and the off-ramp bottleneck (Fig. A.11b), i.e., when a vehicle merges from the on-ramp onto the main road or a vehicle leaves the main road to the off-ramp.

The safe speed $v_{s,n}$ in (A.174), (A.176) for the vehicle that is the closest one to the end of the merging region is chosen in accordance with Eqs. (A.55), (A.56).

A.11.6 Some Results of Simulations

None of the flow rates (or vehicle speeds or else other traffic variables) measured by the detectors on the main road within the simulation road section $0 < x < 30.5$ km are used in simulations. Moreover, we use in the boundary conditions of simulations the flow rates measured separately for personal and long vehicles (Fig. A.8b). In other words, we simulate the *self-development of traffic* on the main three-lane road section with the same locations of on- and off-ramps and the same associated flow rates to the on-ramps and flow rates to the off-ramps as those measured in real traffic within the road locations $0 < x < 30.5$ km, i.e., between the detectors located at $x = 0$ and $x = 30$ km on the main road of the section of the freeway A5-South. Note that initial conditions in simulations are taken as described in Sect. A.3.6.

Simulations with this model show the following results [38]:

1. As in the real data used, in all simulations traffic breakdown at a bottleneck is always an F→S transition. The F→S transition exhibits the nucleation nature. We have found that both in real field data and in the simulations there is the metastability of free flow with respect to the F→S transition at highway bottlenecks.
2. Wide moving jams do not emerge spontaneously in free flow. The jams emerge spontaneously in synchronized flow only (S→J transitions).
3. Through an appropriated choice of these bottleneck parameters we have been able to simulate a spatiotemporal congested pattern (Fig. A.12a) that is very close both in time and space to the empirical pattern reconstructed from detector data through the use of the ASDA and FOTO models (Fig. A.12b). In accordance with the three-phase theory [19, 21], such a congested pattern is an example of an expanded congested traffic pattern (EP).
4. However, we should mention that the time instant of traffic breakdown at a bottleneck (F→S transition) and the spatiotemporal distribution of traffic variables within the emergent synchronized flow are random characteristics, which can be different in different simulation realizations (runs). These different realizations have been made at the same inflow and outflow rates and other boundary conditions, however, at different initial values of the function rand() in the model. Due to a random time delay of traffic breakdown, we have found that in realizations 1, 2, and 3 shown in Figs. A.12a and A.13a,b, time instances of traffic breakdown at the bottlenecks are indeed different from each other.
5. The time instant of the emergence of a wide moving jam in synchronized flow is also random value, which can be different in different simulation realizations (runs). Moreover, in contrast with traffic breakdown that occurs in different simulation realizations at almost the same location in a neighborhood of a bottleneck, locations of the emergence and dissolution of wide moving jams are random values. There are considerable differences in time instants and locations of S→J transitions in different simulation realizations. In particular, some of wide moving jams that appear in some of the realizations do not occur in other ones (compare realization 1 (Fig. A.12a) with realizations 2 and 3 (Fig. A.13)). In addition, we should mention that the jam dissolution is also a random pattern characteristic that is different in different simulation realizations.

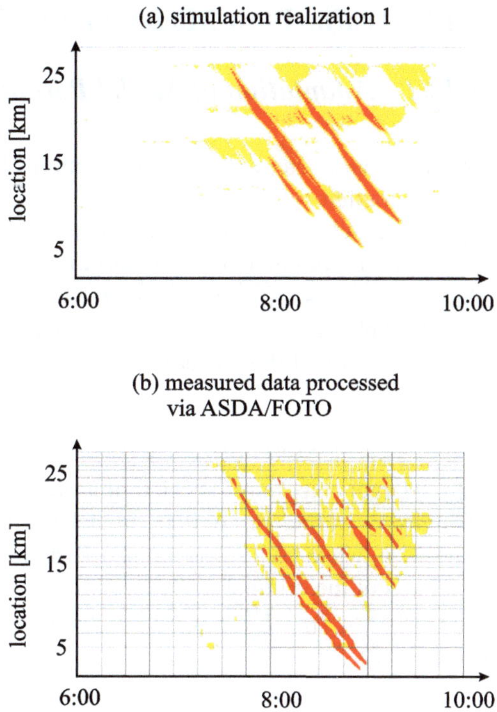

Fig. A.12 Comparison of realization 1 of microscopic simulations (**a**) with real measured data processed via the ASDA and FOTO methods (**b**). Figure (**b**) is adapted from Fig. A.9b. Adapted from [38]

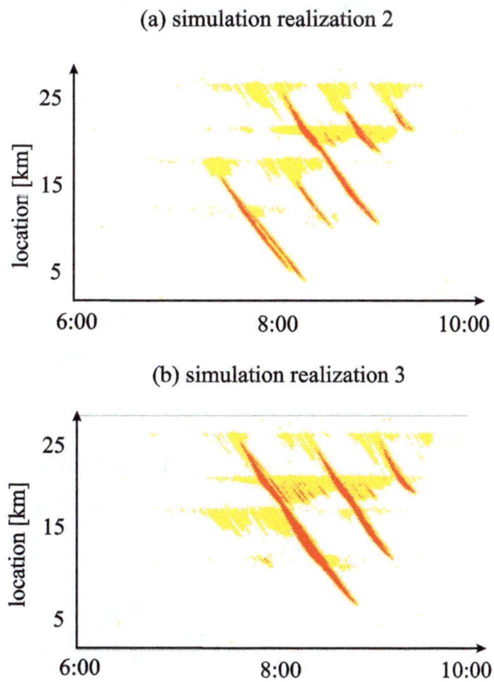

Fig. A.13 Two different realizations 2 and 3 of microscopic simulations. Adapted from [38]

A.12 Traffic Flow Model for City Traffic

A.12.1 Adaptation of Model Parameters for City Traffic

Empirical parameters of city traffic patterns [5] have been used for the adaption of model parameters for city traffic [25]. In particular, in city traffic we should ensure a larger vehicle acceleration from a standstill in a queue at the signal. This larger acceleration is required to satisfy an empirical value of the lost time during the green phase $\delta t = T_{\rm G} - T_{\rm G}^{\rm (eff)} \approx$ 3–4 s [4, 5] (Sect. 9.2.2). We have made the following model development. When the speed difference Δv_n (A.75) is large enough and/or the acceleration of the preceding vehicle $A_{\ell,n}$ is large enough satisfying condition

$$\Delta v_n + A_{\ell,n}\tau \geq \Delta v_{\rm a}, \tag{A.191}$$

then a larger maximum acceleration $a_{\rm max} = k_{\rm a}a$ is used than that in the model of highway traffic. In (A.191), model parameter $k_{\rm a} > 1$; $k_{\rm a}$, $\Delta v_{\rm a}$ are constants. Otherwise, when (A.191) is not satisfied, the maximum acceleration remains to be equal to a of the original model used for highway traffic (Sect. A.3). As a result, the time lost $\delta t = T_{\rm G} - T_{\rm G}^{\rm (eff)} \approx 3.2$ s found in the model satisfies empirical values [4, 5] (Sect. 9.2.2).

A.12.2 Rules of Vehicle Motion

In the Kerner-Klenov model for city traffic on a single-lane road, the vehicle speed v_{n+1}, coordinate x_{n+1}, and acceleration A_{n+1} at time step $n+1$ are found from equations:

$$v_{n+1} = \max(0, \min(v_{\rm free}, \tilde{v}_{n+1} + \xi_n, v_n + a_{\rm max}\tau, v_{{\rm s},n})), \tag{A.192}$$

$$x_{n+1} = x_n + v_{n+1}\tau, \tag{A.193}$$

$$A_{n+1} = (v_{n+1} - v_n)/\tau, \tag{A.194}$$

$$\tilde{v}_{n+1} = \min(v_{\rm free}, v_{{\rm s},n}, v_{{\rm c},n}), \tag{A.195}$$

$$v_{{\rm c},n} = \begin{cases} v_{{\rm c},n}^{(1)} & \text{at } \Delta v_n + A_{\ell,n}\tau < \Delta v_{\rm a} \\ v_{{\rm c},n}^{(2)} & \text{at } \Delta v_n + A_{\ell,n}\tau \geq \Delta v_{\rm a}, \end{cases} \tag{A.196}$$

$$v_{{\rm c},n}^{(1)} = \begin{cases} v_n + \Delta_n^{(1)} & \text{at } g_n \leq G_n \\ v_n + a_n\tau & \text{at } g_n > G_n, \end{cases} \tag{A.197}$$

Table A.9 Model parameters often used in simulations of vehicle motion in city traffic

$\tau_{\rm safe} = \tau = 1$, $d = 7.5\ {\rm m}/\delta x$,
$\delta x = 0.01$ m, $\delta v = 0.01\ {\rm ms}^{-1}$, $\delta a = 0.01\ {\rm ms}^{-2}$,
$v_{\rm free} = 15.278\ {\rm ms}^{-1}/\delta v$, (55 km/h), $b = 1\ {\rm ms}^{-2}/\delta a$, $a = 0.5\ {\rm ms}^{-2}/\delta a$,
$\Delta v_{\rm a} = 2\ {\rm ms}^{-1}/\delta v$, $k_{\rm a} = 4$, $\gamma = 1$,
$k = 3$, $p_1 = 0.35$, $p_b = 0.1$, $p_a = 0.03$, $p^{(0)} = 0.005$,
$p_0(v_n) = 0.667 + 0.083 \min(1, v_n/v_{01})$,
$p_2(v_n) = 0.48 + 0.32\Theta(v_n - v_{21})$,
$v_{01} = 6\ {\rm ms}^{-1}/\delta v$, $v_{21} = 7\ {\rm ms}^{-1}/\delta v$,
$a^{(0)} = 0.2a$, $a^{(\rm a)} = a$,
$a^{(\rm b)}(v_n) = 0.2a + 0.8a \max(0, \min(1, (v_{22} - v_n)/\Delta v_{22}))$,
$v_{22} = 7\ {\rm ms}^{-1}/\delta v$, $\Delta v_{22} = 2\ {\rm ms}^{-1}/\delta v$,
$\delta_1 = 2\ {\rm ms}^{-1}/\delta v$, $L_{\rm a} = 80\ {\rm m}/\delta x$, $p_{\rm c} = 0.1$

$$\Delta_n^{(1)} = \max(-b_n\tau, \min(a_n\tau,\ v_{\ell,n} - v_n)), \tag{A.198}$$

$$v_{{\rm c},n}^{(2)} = v_n + \Delta_n^{(2)}, \tag{A.199}$$

$$\Delta_n^{(2)} = k_{\rm a} a_n \tau \max(0, \min(1,\ \gamma(g_n - v_n\tau))), \tag{A.200}$$

$$a_{\max} = \begin{cases} a & \text{at } \Delta v_n + A_{\ell,n}\tau < \Delta v_{\rm a} \\ k_{\rm a} a & \text{at } \Delta v_n + A_{\ell,n}\tau \geq \Delta v_{\rm a}, \end{cases} \tag{A.201}$$

where γ is constant (see Table A.9).

In (A.192)–(A.201), functions and variables a_n, b_n, ξ_n, G_n, $v_{{\rm s},n}$ are given by associated formulas of Sects. A.3.1–A.3.5. We can use the same rules for lane changing and models of on- and off-ramp bottlenecks as those presented in Sects. A.5 and A.6 for highway traffic.

When a heterogeneous city traffic should be modeling, then the model (A.192)–(A.201) presented for city traffic of identical vehicles and drivers is extended for any heterogeneous city traffic as that has been made in Sects. A.10 and A.11 for highway traffic. The physical features of the model for city traffic are also the same as those considered above for highway traffic. For the model of city traffic the same approaches of simulations of over-acceleration effect of Sect. A.8 can be applied. As the model for highway traffic (Sect. A.9), the model of city traffic is a Markov chain.

To use the space-gap dependence of the free flow speed $v_{\rm free}(g_n)$ (Sect. A.11) in the model of city traffic, Eqs. (A.192) and (A.195) should be changed accordingly to Eqs. (A.174) and (A.176), respectively.

A.12.3 Reduction of Three-Phase Model to Two-Phase Model

In Sect. 9.6, a comparison of simulations of the breakdown of a green wave (GW) in city traffic with three-phase and two-phase traffic flow models has been made. For this comparison, we have used a two-phase traffic flow model following from a reduction of the Kerner-Klenov stochastic microscopic three-phase model presented above in Sect. A.12.2. Here, we consider the derivation of the two-phase traffic flow model used in Sect. 9.6 for simulations of traffic at traffic signal on a single-lane road in city traffic.

We remove the 2D-region of synchronized flow states (dashed region S in Figs. 9.5a,b) and the speed adaptation effect within these states as well as the over-acceleration effect. This has been done through the use of

$$G_n = p_{\mathrm{a}} = p_1 = p_2 = 0 \tag{A.202}$$

in the three-phase model (A.13)–(A.21), (A.26), (A.27), (A.31)–(A.34), (A.192)–(A.201). As a result, the two-phase stochastic microscopic model for city traffic reads as follows:

$$v_{n+1} = \max(0, \min(v_{\mathrm{free}}, \tilde{v}_{n+1} + \xi_n, v_n + a_{\max}\tau, v_{\mathrm{s},n})), \tag{A.203}$$

$$x_{n+1} = x_n + v_{n+1}\tau, \tag{A.204}$$

$$A_{n+1} = (v_{n+1} - v_n)/\tau, \tag{A.205}$$

$$\tilde{v}_{n+1} = \min(v_{\mathrm{free}}, v_{\mathrm{s},n}, v_{\mathrm{c},n}), \tag{A.206}$$

$$v_{\mathrm{c},n} = \begin{cases} v_n + a_n\tau & \text{at } \Delta v_n + A_{\ell,n}\tau < \Delta v_{\mathrm{a}}, \\ v^{(2)}_{\mathrm{c},n} & \text{at } \Delta v_n + A_{\ell,n}\tau \ge \Delta v_{\mathrm{a}}, \end{cases} \tag{A.207}$$

$v^{(2)}_{\mathrm{c},n}$ is determined by (A.199), (A.200), $a_{\max}$ is given by (A.201),

$$\xi_n = \begin{cases} -\xi_{\mathrm{b}} & \text{if } S_{n+1} = -1 \\ \xi^{(0)} & \text{if } S_{n+1} = 0. \end{cases} \tag{A.208}$$

In (A.203)–(A.208), functions and variables a_n, S_{n+1}, ξ_{b}, $\xi^{(0)}$, and $v_{\mathrm{s},n}$ are given by associated formulas of Sects. A.3.1–A.3.5. Boundary and initial conditions as well as other model parameters remain the same in both three-phase and two-phase models (Sect. A.3.6 and Table A.9). Steady states of the resulting two-phase traffic flow model are related to a fundamental diagram (see Fig. 9.5c,d). Characteristics of a moving queue in the Kerner-Klenov stochastic microscopic three-phase model and the two-phase model are identical (line *J* shown in Fig. 9.5a,c).

References

1. R. Barlović, L. Santen, A. Schadschneider, M. Schreckenberg, Eur. Phys. J. B **5**, 793–800 (1998)
2. R.E. Chandler, R. Herman, E.W. Montroll, Oper. Res. **6**, 165–184 (1958)
3. D. Chowdhury, L. Santen, A. Schadschneider, Phys. Rep. **329**, 199 (2000)
4. F. Dion, H. Rakha, Y.S. Kang, Transp. Rec. B **38**, 99–122 (2004)
5. N.H. Gartner, Ch. Stamatiadis, in *Encyclopedia of Complexity and System Science*, ed. by R.A. Meyers (Springer, Berlin, 2009), pp. 9470–9500
6. D.C. Gazis, R. Herman, R.B. Potts, Oper. Res. **7**, 499–505 (1959)
7. D.C. Gazis, R. Herman, R.W. Rothery, Oper. Res. **9**, 545–567 (1961)
8. P.G. Gipps, Trans. Res. B **15**, 105–111 (1981)
9. R. Herman, E.W. Montroll, R.B. Potts, R.W. Rothery, Oper. Res. **7**, 86–106 (1959)
10. B.S. Kerner, Phys. Rev. Lett. **81**, 3797–3400 (1998)
11. B.S. Kerner, in *Proceedings of the 3rd Symposium on Highway Capacity and Level of Service*, ed. by R. Rysgaard, Vol. 2 (Road Directorate, Ministry of Transport, Denmark, 1998), pp. 621–642
12. B.S. Kerner, Trans. Res. Rec. **1678**, 160–167 (1999)
13. B.S. Kerner, in *Transportation and Traffic Theory*, ed. by A. Ceder (Elsevier Science, Amsterdam, 1999), pp. 147–171
14. B.S. Kerner, Phys. World **12**, 25–30 (August 1999)
15. B.S. Kerner, J. Phys. A Math. Gen. **33**, L221–L228 (2000)
16. B.S. Kerner, in *Traffic and Granular Flow '99: Social, Traffic and Granular Dynamics*, ed. by D. Helbing, H.J. Herrmann, M. Schreckenberg, D.E. Wolf (Springer, Heidelberg, Berlin, 2000), pp. 253–284
17. B.S. Kerner, Transp. Res. Rec. **1710**, 136–144 (2000)
18. B.S. Kerner, Netw. Spat. Econ. **1**, 35–76 (2001)
19. B.S. Kerner, *The Physics of Traffic* (Springer, Berlin, New York, 2004)
20. B.S. Kerner, J. Phys. A Math. Theor. **41**, 215101 (2008)
21. B.S. Kerner, *Introduction to Modern Traffic Flow Theory and Control* (Springer, Heidelberg, Dordrecht, London, New York, 2009)
22. B.S. Kerner, J. Phys. A Math. Theor. **44**, 092001 (2011)
23. B.S. Kerner, Phys. Rev. E **85**, 036110 (2012)
24. B.S. Kerner, Europhys. Lett. **102**, 28010 (2013)
25. B.S. Kerner, Physica A **397**, 76–110 (2014)
26. B.S. Kerner, Phys. Rev. E **92**, 062827 (2015)
27. B.S. Kerner, P. Hemmerle, M. Koller, G. Hermanns, S.L. Klenov, H. Rehborn, M. Schreckenberg, Phys. Rev. E **90**, 032810 (2014)
28. B.S. Kerner, S.L. Klenov, J. Phys. A Math. Gen. **35**, L31–L43 (2002)
29. B.S. Kerner, S.L. Klenov, Phys. Rev. E **68**, 036130 (2003)
30. B.S. Kerner, S.L. Klenov. J. Phys. A Math. Gen. **37**, 8753–8788 (2004)
31. B.S. Kerner, S.L. Klenov, J. Phys. A Math. Gen. **39**, 1775–1809 (2006); 7605
32. B.S. Kerner, S.L. Klenov, Phys. Rev. E **80**, 056101 (2009)
33. B.S. Kerner, S.L. Klenov, J. Phys. A Math. Theor. **43**, 425101 (2010)
34. B.S. Kerner, S.L. Klenov, G. Hermanns, P. Hemmerle, H. Rehborn, M. Schreckenberg, Phys. Rev. E **88**, 054801 (2013)
35. B.S. Kerner, S.L. Klenov, A. Hiller, J. Phys. A Math. Gen. **39**, 2001–2020 (2006)
36. B.S. Kerner, S.L. Klenov, A. Hiller, Non. Dyn. **49**, 525–553 (2007)
37. B.S. Kerner, S.L. Klenov, A. Hiller, H. Rehborn, Phys. Rev. E **73**, 046107 (2006)
38. B.S. Kerner, H. Rehborn, R.-P. Schäfer, S. L. Klenov, J. Palmer, S. Lorkowski, N. Witte, Physica A **392** 221–251 (2013)
39. W. Knospe, L. Santen, A. Schadschneider, M. Schreckenberg, Physica A **265**, 614–633 (1999)

40. S. Krauß, PhD thesis, DRL-Forschungsbericht 98-08 (1998) (available from www.zaik.de/~paper)
41. S. Krauß, P. Wagner, C. Gawron, Phys. Rev. E **55**, 5597–5602 (1997)
42. K. Nagel, D.E. Wolf, P. Wagner, P. Simon, Phys. Rev. E **58**, 1425–1437 (1998)
43. K. Nagel, M. Schreckenberg, J. Phys. (France) I **2**, 2221–2229 (1992)
44. M. Rickert, K. Nagel, M. Schreckenberg, A. Latour, Physica A **231**, 534 (1996)
45. M. Takayasu, H. Takayasu, Fractals **1**, 860–866 (1993)

Appendix B
Kerner-Klenov-Schreckenberg-Wolf (KKSW) Cellular Automaton (CA) Three-Phase Model

B.1 Motivation

As explained in Sect. A.1, one of the most complex requirements for the Kerner-Klenov microscopic stochastic three-phase traffic flow model considered in Appendix A is that the model should simulate a variety of different driver characteristics independent of each other. In particular, there are many different model parameters that allows us the choice of different values of driver time delays in acceleration and deceleration made independent of each other in a diverse variety of driving situations. This explains why the Kerner-Klenov model is a complex one and it has many model parameters.

In 1992, Nagel and Schreckenberg [16] introduced a cellular automaton (CA) traffic flow model. The Nagel-Schreckenberg (NaSch) CA traffic flow model belongs to two-phase traffic flow models of the GM model class. One of the main novel achievements of the NaSch CA model is as follows: The model incorporates the classical traffic flow instability of Herman, Gazis, Montroll, Potts, Rothery, and Chandler [2, 4–8] through the use of model fluctuations. The classical work of Nagel and Schreckenberg [16] has inspired us to develop a three-phase CA traffic flow model that additionally to the classical traffic flow instability incorporates the metastability of free flow with respect to an F→S transition as introduced in the three-phase theory [9, 10].

The result was the Kerner-Klenov-Wolf (KKW) CA three-phase traffic flow model developed in 2002 [14]. The KKW CA model incorporates the Nagel-Schreckenberg approach for the mathematical description of driver reaction time through model fluctuations *and* some rules of vehicle motion introduced in the Kerner-Klenov microscopic stochastic three-phase traffic flow model (Appendix A) that are required to simulate the metastability of free flow with respect to the F→S transition.

As mentioned, in the Kerner-Klenov microscopic stochastic three-phase traffic flow model (Appendix A) there are many different model parameters that allows us the choice of different values of driver time delays in acceleration and deceleration

B.S. Kerner, *Breakdown in Traffic Networks*, DOI 10.1007/978-3-662-54473-0

made independent of each other in a diverse variety of driving situations. Contrarily to the Kerner-Klenov microscopic stochastic three-phase traffic flow model, the main motivation for the development of the Kerner-Klenov-Wolf (KKW) CA three-phase model has been to find a model with a minimum possible number of model parameters. Nevertheless, the model should be able to simulate a 2Z-characteristic for phase transition of the three-phase theory and spatiotemporal features of traffic patterns postulated in the three-phase theory. Through the minimization of the number of model parameters in the model, we have understood that in the KKW CA model, in contrast with the Kerner-Klenov stochastic model, some of the different driver time delays in acceleration and deceleration that have different physical meaning depend on the same model parameter.

To mitigate these disadvantages of the KKW CA model, we have developed two other three-phase CA traffic flow models: The Kerner-Klenov-Schreckenberg (KKS) CA model [11] and the Kerner-Klenov-Schreckenberg-Wolf (KKSW) CA model [12, 13]. The main advantage of the three-phase CA models is that due to a small number of model parameters the CA models are much easier to understand in comparison with the Kerner-Klenov microscopic stochastic three-phase traffic flow model.

In the book, we have presented simulation results of the KKSW CA model. For this reason, below we consider rules of vehicle motion in the KKSW CA model.

B.2 Rules of Vehicle Motion in KKSW CA Model

In the KKSW CA model, discrete (and dimensionless) values of space coordinate and speed are used, which are measured respectively in values δx and δv, and time is measured in values of time step τ. As in the Kerner-Klenov model (Appendix A), time step τ is equal to 1 s. In the KKSW CA model, a discretization cell $\delta x = 1.5$ m and discretization speed interval $\delta v = 1.5$ m/s are considerably larger than those in the Kerner-Klenov model, in which $\delta x = 0.01$ m and $\delta v = 0.01$ m/s. For this reason, at least some of the microscopic traffic phenomena can be simulated with the use of the Kerner-Klenov model in more details. However, traffic simulations with the KKSW CA model can be considerably quicker than that with the Kerner-Klenov model of Appendix A. Therefore, the KKSW CA model can be more suitable for on-line simulations of large scale traffic networks.

In the KKSW CA model, the following subsequent calculations of rules of vehicle motion are made:

(a) "over-acceleration through lane changing to faster lane," lane changing (for passing) occurs with probability p_c when the intention (B.1)–(B.6) and safety conditions (B.7) are satisfied:

$$R \to L: v_n^+ \geq v_{\ell,n} + \delta_1 \text{ and } v_n \geq v_{\ell,n}, \tag{B.1}$$

$$L \to R: v_n^+ \geq v_{\ell,n} + \delta_2 \text{ or } v_n^+ \geq v_n + \delta_2, \tag{B.2}$$

$$R \to L, \text{ trucks}: v_n^+ \geq v_{\ell,n} + \delta_1 \text{ and } v_n \geq v_{\ell,n} \tag{B.3}$$

$$\text{and} \quad v_{\ell,n} < v_{\text{free, long}}^{(\text{max})} - \delta_0, \tag{B.4}$$

$$L \to R, \text{ trucks}: v_n^+ \geq v_{\ell,n} + \delta_2 \text{ or } v_n^+ \geq v_n + \delta_2 \tag{B.5}$$

$$\text{or} \quad v_n^+ \geq v_{\text{free, long}}^{(\text{max})} - \delta_0, \tag{B.6}$$

$$g_n^+ \geq \min(v_n,\ g_c) \text{ and } g_n^- \geq \min(v_n^-,\ g_c), \tag{B.7}$$

(b) "comparison of vehicle space gap with the synchronization space gap":

$$\begin{gathered} \text{if } g_n \leq G(v_n) \\ \text{then follow rules (c), (d) and skip rule (e),} \end{gathered} \tag{B.8}$$

$$\begin{gathered} \text{if } g_n > G(v_n) \\ \text{then skip rules (c), (d) and follow rule (e),} \end{gathered} \tag{B.9}$$

(c) "speed adaptation within the synchronization space gap":

$$v_{n+1} = v_n + \text{sgn}(v_{\ell,n} - v_n), \tag{B.10}$$

(d) "over-acceleration through random acceleration within the synchronization space gap"

$$\begin{gathered} \text{if } v_n \geq v_{\ell,n}, \text{ then with probability } p_a \\ v_{n+1} = \min(v_{n+1} + 1,\ v_{\text{free}}), \end{gathered} \tag{B.11}$$

(e) "acceleration":

$$v_{n+1} = \min(v_n + 1,\ v_{\text{free}}), \tag{B.12}$$

(f) "deceleration" associated with a safe speed $v_{s,n} = g_n$:

$$v_{n+1} = \min(v_{n+1},\ g_n), \tag{B.13}$$

(g) "randomization" is given by formula:

$$\text{with probability } p, \quad v_{n+1} = \max(v_{n+1} - 1,\ 0), \tag{B.14}$$

(h) "motion" is described by formula:

$$x_{n+1} = x_n + v_{n+1}. \tag{B.15}$$

Rule (a) "over-acceleration through lane changing to faster lane" is applied, when

$$r_1 < p_c, \tag{B.16}$$

$r_1 = \text{rand}()$ is a random value distributed uniformly between 0 and 1. Rule (d) "over-acceleration through random acceleration within the synchronization space gap" is applied, when

$$r < p_a, \tag{B.17}$$

where $r = \text{rand}()$ is a random value distributed uniformly between 0 and 1. Rule (g) "randomization" is applied, when

$$p_a \le r < p_a + p, \tag{B.18}$$

where $p_a + p \le 1$. Probability of over-acceleration p_a in (B.11) is chosen as the increasing speed function:

$$p_a(v_n) = p_{a,1} + p_{a,2} \max(0, \min(1, \ (v_n - v^{(\text{syn})})/\Delta v^{(\text{syn})})), \tag{B.19}$$

where $p_{a,1}$, $p_{a,2}$, $v^{(\text{syn})}$ and $\Delta v^{(\text{syn})}$ are constants (Table B.1).

In (B.1)–(B.15), $n = 0, 1, 2, \ldots$ is the number of time steps; x_n and v_n are the coordinate and speed of the vehicle; v_{free} is the maximum speed for passenger vehicles; $v^{(\text{max})}_{\text{free, long}}$ is the maximum speed of trucks; the subscript ℓ marks variables related to the preceding vehicle;

$$g_n = x_{\ell,n} - x_n - d_\ell \tag{B.20}$$

is a space gap; the length of the preceding vehicle d_ℓ depends on whether the preceding vehicle is a passenger vehicle or a truck;

$$G(v_n) = kv_n \tag{B.21}$$

is a synchronization space gap; $R \to L$ and $L \to R$ denote, respectively, lane changing from the right lane to the left lane and from the left lane to the right lane[1];

[1]Main features of lane changing rules (B.1)–(B.6) are taken from Nagel et al. [15]. In particular, in these rules overtaking in the right lane is allowed. This can be seen from conditions (B.2). Additionally, drivers tend to return to the right lane after passing a slower moving vehicle even if they can travel at the same speed in either lane. This is governed by the condition that the speed

Table B.1 Model parameters of the KKSW model often used in simulations

Parameters for vehicle motion in road lane:
$\delta x = 1.5$ m, time step is 1 s,
$d = 5$ (7.5 m) for passenger vehicles,
$d = d^{(\mathrm{truck})} = 12$ (18 m) for trucks,
$v_{\mathrm{free}} = 25$ (135 km/h) for passenger vehicles,
$v_{\mathrm{free}} = v^{(\max)}_{\mathrm{free,\ long}} = 16$ (86.4 km/h) for trucks,
$v^{(\max)}_{\mathrm{free,\ L}} = 14$ (75.6 km/h),
$p^{(2)}_0 = 0.5$ for passenger and $p^{(2)}_0 = 0.7$ for trucks,
$p^{(2)}_2 = 0.35$, $p_3 = 0.01$,
$p_{\mathrm{a},1} = 0.07$, $p_{\mathrm{a},2} = 0.08$, $v^{(\mathrm{syn})} = 14$ (75.6 km/h),
$\Delta v^{(\mathrm{syn})} = 3$ (16.2 km/h),
$v_{\mathrm{pinch}} = 8$ (43.2 km/h), $k_1 = 3$, $k_2 = 2$
Parameters for lane changing:
$p_{\mathrm{c}} = 0.07$, for passenger vehicles: $\delta_1 = 1$ (5.4 km/h)
and $\delta_2 = 3$ (16.2 km/h);
for trucks: $\delta_1 = 3$ (16.2 km/h),
$\delta_2 = 1$ (5.4 km/h), $\delta_0 = 4$ (21.6 km/h);
$g_{\mathrm{c}} = 16$ (24 m), $L_{\mathrm{a}} = 100$ (150 m);
if for trucks condition (B.6) is satisfied,
then $p_{\mathrm{c}} = 1$ and $g_{\mathrm{c}} = 10$ (15 m)

in lane changing rules (B.1)–(B.7), superscripts "+" and "−" denote variables and functions related, respectively, to the preceding and trailing vehicles in the "target" lane between which the vehicle appears if it changes to this target lane; in particular,

$$g^{+}_{n} = x^{+}_{n} - x_n - d^{+} \tag{B.22}$$

is the space gap between the vehicle and the preceding vehicle that has the coordinate x^{+}_{n} in the target lane, d^{+} is the length of the preceding vehicle in the target lane,

$$g^{-}_{n} = x_n - x^{-}_{n} - d \tag{B.23}$$

is the space gap between the vehicle and the trailing vehicle that has the coordinate x^{-}_{n} in the target lane, d is vehicle length; the speed v^{+}_{n} or speed $v_{\ell,n}$ in (B.1)–(B.6) is set to ∞ if the space gap g^{+}_{n} (B.22) or space gap g_n (B.20) exceeds a given look-

of the trailing vehicle in the right lane is set to ∞ when the space gap to the trailing vehicle in the right lane exceeds L_{a}; as a result, condition (B.2) is satisfied.

ahead distance L_a. In (B.1)–(B.21), δ_1, δ_2, δ_0, g_c are model parameters (Table B.1). In (B.14), (B.18), we use for probability p formula

$$p = \begin{cases} p_2 \text{ for } v_{n+1} > v_n \\ p_3 \text{ for } v_{n+1} \leq v_n. \end{cases} \tag{B.24}$$

The importance of formula (B.24) is as follows. This formula allows us to consider a time delay in vehicle acceleration at the downstream front of synchronized flow independent of over-deceleration: The time delay in vehicle acceleration depends on probability p_2, whereas the effect of over-deceleration depends on probability p_3. In the KKSW CA model, we assume different values of time delays in vehicle acceleration at the downstream front of synchronized flow and at the downstream front of a wide moving jam. For this reason, probability p_2 is given by formula

$$p_2(v_n) = \begin{cases} p_0^{(2)} \text{ for } v_n = 0 \\ p_1^{(2)} \text{ for } v_n > 0. \end{cases} \tag{B.25}$$

In the KKSW CA model, in formula (B.25) probability $p_1^{(2)}$ is chosen to provide a time delay in vehicle acceleration only if the vehicle does not accelerate at previous time step n:

$$p_1^{(2)} = \begin{cases} p_2^{(2)} & \text{for } v_n \leq v_{n-1} \\ 0 & \text{for } v_n > v_{n-1}, \end{cases} \tag{B.26}$$

In (B.24)–(B.26), p_3, $p_0^{(2)}$, and $p_2^{(2)}$ are constants (Table B.1). Steady states of the KKSW CA model are shown in Fig. B.1a,b.

To describe the pinch effect resulting in spontaneous wide moving jam emergence in synchronized flow (S→J transition) [9], we also assume that in (B.21) [14]

$$k(v_n) = \begin{cases} k_1 \text{ for } v_n > v_{\text{pinch}} \\ k_2 \text{ for } v_n \leq v_{\text{pinch}}, \end{cases} \tag{B.27}$$

where v_{pinch}, k_1, and k_2 are constants ($k_1 > k_2 \geq 1$) (Table B.1). The model parameter v_{pinch} defines a range of speeds in synchronized flow $0 < v \leq v_{\text{pinch}}$ (Fig. B.1b) within which wide moving jams occur spontaneously with a larger probability. Parameters of the KKSW CA model that are often used in simulations are presented in Table B.1.

In some simulations presented in the main text of the book, we have used the KKSW CA model for heterogeneous traffic flow. As in the Kerner-Klenov stochastic microscopic three-phase traffic flow model (Sect. A.10), formulas for rules of vehicle motion of the KKSW CA model for traffic flow with identical vehicles and identical drivers remain the same in heterogeneous traffic flow, if model parameters that are responsible for individual vehicle characteristics are chosen

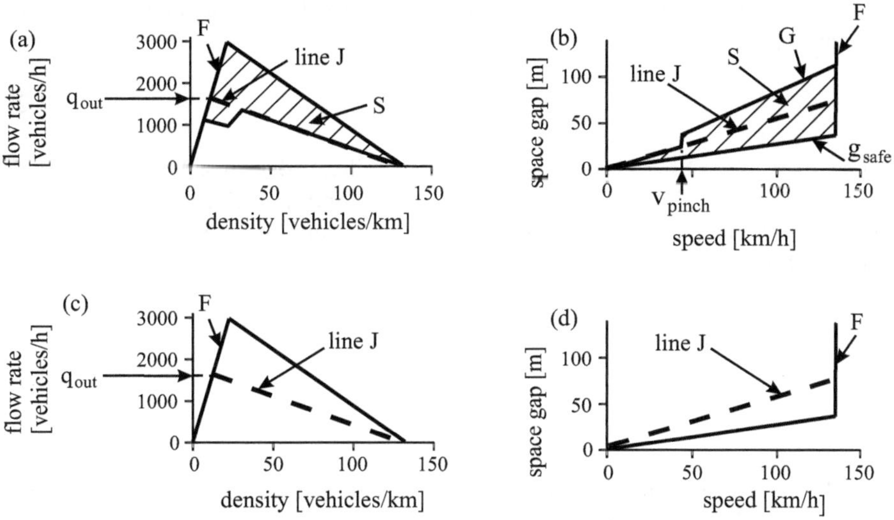

Fig. B.1 Steady states of the KKSW CA model (**a, b**) and the Nagel-Schreckenberg CA model (**c, d**) and in the flow–density (**a, c**) and space-gap–speed plane (**b, d**). F—free flow, S—synchronized flow (hatched 2D-regions in (**a, b**)); line J is the line J that represents the downstream front of a wide moving jam (Sect. 4.5.1); q_{out} is the flow rate in free flow related to the outflow from the wide moving jam; G is the synchronization space gap between vehicles; $g_{\text{safe}} = v$ is the safe space gap between vehicles moving at the speed v

individually for each of the vehicle (or driver) types. In particular, in heterogeneous flow consisting of passenger vehicles and trucks, the length of the preceding vehicle d_ℓ depends on whether the preceding vehicle is a passenger vehicle or a truck. We choose $\delta_1 < \delta_2$ for passenger vehicles and $\delta_1 > \delta_2$ for trucks. If a truck is in the left lane and condition (B.6) is satisfied, the vehicle decreases its maximum speed to the value $v^{(\text{max})}_{\text{free, L}} < v^{(\text{max})}_{\text{free, long}}$, where $v_{\text{free}} = v^{(\text{max})}_{\text{free, long}}$ is the maximum speed of trucks, and changes to the right lane with probability $p_c = 1$, when safety conditions (B.7) are satisfied. We assume that the truck's length is considerably longer than that for a passenger vehicle; the maximum speed of trucks $v_{\text{free}} = v^{(\text{max})}_{\text{free, long}}$ is considerably lower than that for passenger vehicles.

In traffic flow simulations with the KKSW CA model presented in this book, we have used the following boundary and initial conditions. Open boundary conditions are applied. At the beginning of the road new vehicles are generated one after another in each of the lanes of the road at time moments

$$n^{(m)} = \lceil m\tau_{\text{in}} \rceil, \ m = 1, 2, \ldots . \tag{B.28}$$

In (B.28), $\tau_{\text{in}} = 1/q_{\text{in}}$, q_{in} is the flow rate in the incoming boundary flow per lane, $\lceil z \rceil$ denotes the nearest integer greater than or equal to z. A new vehicle appears on the road only if the distance from the beginning of the road ($x = x_b$) to the position

$x = x_{\ell,n}$ of the farthest upstream vehicle on the road is not smaller than the safe distance $v_{\ell,n} + d_{\ell}$:

$$x_{\ell,n} - x_{\mathrm{b}} \geq v_{\ell,n} + d_{\ell}, \tag{B.29}$$

where $n = n^{(m)}$. Otherwise, condition (B.29) is checked at time $(n + 1)$ that is the next one to time $n^{(m)}$ (B.28), and so on, until the condition (B.29) is satisfied. Then the next vehicle appears on the road. After this occurs, the number m in (B.28) is increased by 1.

The speed v_n and coordinate x_n of the new vehicle are

$$\begin{aligned} v_n &= v_{\ell,n}, \\ x_n &= \max(x_{\mathrm{b}}, x_{\ell,n} - \lfloor v_n \tau_{\mathrm{in}} \rfloor). \end{aligned} \tag{B.30}$$

The flow rate q_{in} is chosen to havc the value $v_{\mathrm{free}}\tau_{\mathrm{in}}$ integer. In the initial state ($n = 0$), all vehicles have the maximum speed $v_n = v_{\mathrm{free}}$ and they are positioned at space intervals $x_{\ell,n} - x_n = v_{\mathrm{free}}\tau_{\mathrm{in}}$.

After a vehicle has reached the end of the road it is removed. Before this occurs, the farthest downstream vehicle maintains its speed and lane.

B.3 Models of Bottlenecks for KKSW CA Model

B.3.1 On- and Off-Ramp Bottlenecks

Models of road bottlenecks due to on- and off-ramps as well as a merge bottleneck, at which two road lanes are reduced to one lane, are shown in Fig. A.2.

The on-ramp bottleneck consists of two parts (Fig. A.2a):

(i) The merging region of length L_{m} where vehicle can merge onto the main road from the on-ramp lane.
(ii) A part of the on-ramp lane of length L_{r} upstream of the merging region where vehicles move in a road lane in accordance with the KKSW model (B.8)–(B.15), (B.17)–(B.27). The maximal speed of vehicles is $v_{\mathrm{free}} = v_{\mathrm{free\ on}}$.

At the beginning of the on-ramp lane ($x = x^{(\mathrm{b})}_{\mathrm{on}}$) the flow rate to the on-ramp q_{on} is given as the flow rate q_{in}.

The off-ramp bottleneck consists of two parts (Fig. A.2b):

(i) A merging region of length L_{m} where vehicle can merge from the main road onto the off-ramp lane.
(ii) A part of the off-ramp lane of length L_{r} downstream of the merging region where vehicles move in road lane in accordance with the KKSW model (B.8)–(B.15), (B.17)–(B.27). The maximal speed of vehicles is $v_{\mathrm{free}} = v_{\mathrm{free\ off}}$.

Within a second merging region of length $L_{\mathrm{m}}+L_{\mathrm{c}}$, which is on the main road ($x_{\mathrm{off}}^{(\mathrm{s})} \le x \le x_{\mathrm{off}}^{(\mathrm{b})}$ in Fig. A.2b), vehicles going to the off-ramp have to change from the left lane to the right lane of the main road. The flow rate of vehicles that go to the off-ramp is given as a percentage η_{off} of the flow rate q_{in}. For this purpose, at the beginning of the road a vehicle is given an "attribute," which marks this vehicle as a vehicle going to the off-ramp, if condition

$$r_{\mathrm{off}} < \eta_{\mathrm{off}}/100\% \tag{B.31}$$

is satisfied, where $r_{\mathrm{off}} = \mathrm{rand}(0, 1)$.

A model of a moving bottleneck is shown in Fig. A.3a. Cases when the moving bottleneck is upstream of the on- and off-ramp bottlenecks are presented in Fig. A.3b,c, respectively. We assume that a slow vehicle that moves in the right lane at the speed v_{M}, where $v_{\mathrm{M}} < v_{\mathrm{free}}$, is the vehicle that causes a moving bottleneck. If a vehicle moves initially in the right lane upstream of the slow vehicle, then within the moving merging region L_{M} (Fig. A.3) the vehicle tries to changes from the right lane to the left lane.

B.3.2 Vehicle Motion Rules in Merging Region of Bottlenecks

Vehicle motion rules in merging region are the same for on-ramp, off-ramp, and moving bottlenecks. These vehicle motion rules are as follows:

(a) "conditions for vehicle merging" is given by formula:

$$g_n^+ \ge \min(\hat{v}_n,\ g_{\mathrm{c}}^{(\mathrm{m})}) \quad \text{and} \quad g_n^- \ge \min(v_n^-,\ g_{\mathrm{c}}^{(\mathrm{m})}), \tag{B.32}$$

where

$$\hat{v}_n = \min(v_n + 1,\ v_n^+), \tag{B.33}$$

(b) "comparison of vehicle space gap with the synchronization space gap":

$$\begin{gathered} \text{if } g_n^+ \le G(v_n) \\ \text{then follow rules (c), (d) and skip rule (e),} \end{gathered} \tag{B.34}$$

$$\begin{gathered} \text{if } g_n^+ > G(v_n) \\ \text{then skip rules (c), (d) and follow rule (e),} \end{gathered} \tag{B.35}$$

(c) "speed adaptation within the synchronization space gap" is given by formula:

$$v_{n+1} = v_n + \mathrm{sgn}(\hat{v}_n^+ - v_n), \tag{B.36}$$

(d) "over-acceleration through random acceleration within the synchronization space gap" is given by formula

$$\text{if } v_n \geq \hat{v}_n^+, \text{ then with probability } p_{\mathrm{a}},$$
$$v_{n+1} = \min(v_{n+1} + 1, \ v_{\mathrm{free}}), \tag{B.37}$$

(e) "acceleration" is given by formula (B.12):

$$v_{n+1} = \min(v_n + 1, \ v_{\mathrm{free}}), \tag{B.38}$$

(f) "deceleration" associated with a safe speed $v_{\mathrm{s},n}$:

$$v_{n+1} = \min(v_{n+1}, \ v_{\mathrm{s},n}), \tag{B.39}$$

(g) "randomization" is given by formula (B.14):

$$\text{with probability } p, \quad v_{n+1} = \max(v_{n+1} - 1, \ 0), \tag{B.40}$$

(h) "motion" is described by formula (B.15):

$$x_{n+1} = x_n + v_{n+1}. \tag{B.41}$$

In (B.36), $\hat{v}_n^+$ is given by:

$$\hat{v}_n^+ = \max(0, \ \min(v_n^+ + \Delta v_{\mathrm{r}}^{(2)}, \ v_{\mathrm{free}})). \tag{B.42}$$

After merging, the vehicle co-ordinate does not change and the vehicle speed is set to

$$v_n = \hat{v}_n. \tag{B.43}$$

Rule (d) "over-acceleration through random acceleration within the synchronization space gap" (B.37) is applied, when

$$r < p_{\mathrm{a}}, \tag{B.44}$$

where $r = \mathrm{rand}()$ is a random value distributed uniformly between 0 and 1. Rule (g) "randomization" (B.40) is applied, when

$$p_{\mathrm{a}} \leq r < p_{\mathrm{a}} + p, \tag{B.45}$$

Table B.2 Parameters of merging regions of on-ramp, off-ramp and moving bottleneck in KKSW CA model often used in simulations

$g_{\mathrm{c}}^{(\mathrm{m})} = 10$ (15 m) for merging region
of the on-ramp, the off-ramp
and the moving bottleneck,
$\Delta v_{\mathrm{r}}^{(2)} = -1$ (−5.4 km/h) for the off-ramp
$\Delta v_{\mathrm{r}}^{(2)} = 5$ (27 km/h) for the on-ramp
and the moving bottleneck,
length of the merging region of the on-ramp
$L_{\mathrm{m}} = 200$ (300 m),
length of on-ramp lane upstream of the merging region
$L_{\mathrm{r}} = 200$ (300 m),
length of the merging region of the off-ramp
$L_{\mathrm{m}} = 400$ (600 m),
length of the merging region upstream of the off-ramp
$L_{\mathrm{c}} = 300$ (450 m),
length of off-ramp lane downstream of the merging region
$L_{\mathrm{r}} = 600$ (900 m),
length of the merging region of the moving bottleneck
$L_{\mathrm{M}} = 300$ (450 m),
maximum speed in the on-ramp lane
$v_{\mathrm{free\ on}} = 17$ (91.8 km/h),
and in the off-ramp lane $v_{\mathrm{free\ off}} = 17$ (91.8 km/h)

where $p_{\mathrm{a}} + p \leq 1$. In (B.32)–(B.42), v_n^- is the speed of the following vehicle in the target lane before merging, v_n^+ is the speed of the preceding vehicle in the target lane before merging, g_n^- (B.23) is the space gap to the following vehicle in the target lane before merging, g_n^+ (B.22) is the space gap to the preceding vehicle in the target lane before merging. Values p_{a} in (B.37), $G(v_n)$ in (B.35), and p in (B.40) are given by formulas of the KKSW CA model (B.19), (B.21), and (B.24), respectively. $\Delta v_{\mathrm{r}}^{(2)}$, $g_{\mathrm{c}}^{(\mathrm{m})}$ are constants (Table B.2).

Safe speed $v_{\mathrm{s},n}$ in rule (f) "deceleration" (B.39) for different bottlenecks are described by the following formulas. For the on-ramp bottleneck

$$v_{\mathrm{s},n} = \min(g_n,\ g_{\mathrm{on,\ n}}), \tag{B.46}$$

where $g_{\mathrm{on,\ n}} = x_{\mathrm{on}} + L_{\mathrm{m}} - x_n$ is a space gap to the downstream end of the on-ramp merging region. For the off-ramp bottleneck

$$v_{\mathrm{s},n} = \min(g_n,\ g_{\mathrm{off,\ n}}), \tag{B.47}$$

where $g_{\mathrm{off,\ n}} = x_{\mathrm{off}} + L_{\mathrm{m}} - x_n$ is a space gap to the downstream end of the off-ramp merging region. For the moving bottleneck

$$v_{\mathrm{s},n} = g_n. \tag{B.48}$$

In (B.46)–(B.48), g_n is the space gap given by formula (B.20).

Model parameters used in all simulations shown in the main text of the book are presented in Tables B.1 and B.2.

B.4 Comparison of KKSW CA Model with Nagel-Schreckenberg CA Model

To see qualitative difference between the KKSW CA model and the classical Nagel-Schreckenberg (NaSch) CA model [1, 3, 16, 17], we use the following methodology: We remove in the KKSW CA model rules of vehicle motion in road lane, which incorporate hypotheses of the three-phase theory:

(i) We remove the synchronization space gap G and the rule (b) described by Eqs. (B.8), (B.9) from the KKSW CA model. Then, steady states of a new CA model lie on the fundamental diagram (Fig. B.1c,d).
(ii) We remove the rule (c) "speed adaptation within the synchronization space gap" described by Eq. (B.10).
(iii) We remove the rule (d) "over-acceleration through random acceleration within the synchronization space gap" described by Eq. (B.11).

The remaining model rules of vehicle motion in road lane of the new CA model are as follows:

(a) "acceleration":

$$v_{n+1} = \min(v_n + 1,\ v_{\text{free}}), \tag{B.49}$$

(b) "deceleration":

$$v_{n+1} = \min(v_{n+1},\ g_n), \tag{B.50}$$

(c) "randomization" is given by formula:

$$\text{with probability } p, \quad v_{n+1} = \max(v_{n+1} - 1,\ 0), \tag{B.51}$$

(d) "motion" is described by formula:

$$x_{n+1} = x_n + v_{n+1}. \tag{B.52}$$

We see that the resulting CA traffic flow model (B.49)–(B.52) is the NaSch CA model with the slow-to-start rule [1, 3, 16, 17].

References

1. R. Barlović, L. Santen, A. Schadschneider, M. Schreckenberg, Eur. Phys. J. B **5**, 793–800 (1998)
2. R.E. Chandler, R. Herman, E.W. Montroll, Oper. Res. **6**, 165–184 (1958)
3. D. Chowdhury, L. Santen, A. Schadschneider, Phys. Rep. **329**, 199 (2000)
4. D.C. Gazis, *Traffic Theory* (Springer, Berlin, 2002)
5. D.C. Gazis, R. Herman, Trans. Sci. **26**, 223 (1992)
6. D.C. Gazis, R. Herman, R.B. Potts, Oper. Res. **7**, 499–505 (1959)
7. D.C. Gazis, R. Herman, R.W. Rothery, Oper. Res. **9**, 545–567 (1961)
8. R. Herman, E.W. Montroll, R.B. Potts, R.W. Rothery, Oper. Res. **7**, 86–106 (1959)
9. B.S. Kerner, *The Physics of Traffic* (Springer, Berlin, Heidelberg, New York, 2004)
10. B.S. Kerner, *Introduction to Modern Traffic Flow Theory and Control* (Springer, Heidelberg, Dordrecht, London, New York, 2009)
11. B.S. Kerner, S.L. Klenov, M. Schreckenberg, Phys. Rev. E **84**, 046110 (2011)
12. B.S. Kerner, S.L. Klenov, G. Hermanns, M. Schreckenberg, Physica A **392**, 4083–4105 (2013)
13. B.S. Kerner, S.L. Klenov, M. Schreckenberg, Phys. Rev. E **89**, 052807 (2014)
14. B.S. Kerner, S.L. Klenov, D.E. Wolf, J. Phys. A Math. Gen. **35**, 9971–10013 (2002)
15. K. Nagel, D.E. Wolf, P. Wagner, P. Simon. Phys. Rev. E **58**, 1425–1437 (1998)
16. K. Nagel, M. Schreckenberg, J. Phys. (France) I **2**, 2221–2229 (1992)
17. A. Schadschneider, D. Chowdhury, K. Nishinari, *Stochastic Transport in Complex Systems* (Elsevier Science, New York, 2011)

Appendix C
Dynamic Traffic Assignment Based on Wardrop's UE with Step-by-Step Method

The objective of this Appendix C is a comparison of the application of congested pattern control method used in Sect. 13.3.2 of the main text of the book (Figs. 13.2 and 13.3) with a step-by-step method for the searching of the Wardrop's UE (13.3).

To find solutions for dynamic traffic assignment with the use of the Wardrop's UE for two-route network (Fig. 12.3a) at the total network inflow rate $Q = 7000$ vehicles/h as that used in Figs. 13.2 and 13.3, we have applied dynamic traffic assignment of the flow rates q_1 and q_2 in (13.3) based on simulated route travel times T_1 (route 1) and T_2 (route 2) with the use of a well-known step-by-step iteration method [1] (see Sects. 20.2–6 and 20.2–7 of [1]). Eq. (13.3) can be rewritten as follows:

$$f(q_1) = 0, \tag{C.1}$$

where

$$f(q_1) = T_1(q_1, q_{\rm on}^{(1)}) - T_2(q^{(o)} - q_1, q_{\rm on}^{(2)}), \tag{C.2}$$

time unit is [s]. Solutions of Eq. (C.1) can be found based on the following iteration procedure:

$$q_1^{(n+1)} = q_1^{(n)} - 3600\delta f(q_1^{(n)}) \text{ vehicles/h}, \tag{C.3}$$

where $n = 1, 2, \ldots,$ is the index of n-th time interval (we have used time interval 10 min), $\delta = 0.0002$ [s^{-2}]. For calculation of values $f(q_1^{(n)})$ (C.2) at the end of n-th time interval, travel times $T_1(q_1^{(n)}, q_{\rm on}^{(1)})$ and $T_2(q_0 - q_1^{(n)}, q_{\rm on}^{(2)})$ for each of the vehicles have been found from simulations and averaged (to avoid fluctuations) over 20 s intervals.

One of the simulation realizations made with this iteration procedure is shown in Fig. C.1. In this simulation realization, after delay time $T^{(B)} = 40$ min, traffic breakdown (labeled by F→S transition in Fig. C.1a) has occurred at bottleneck 1.

B.S. Kerner, *Breakdown in Traffic Networks*, DOI 10.1007/978-3-662-54473-0

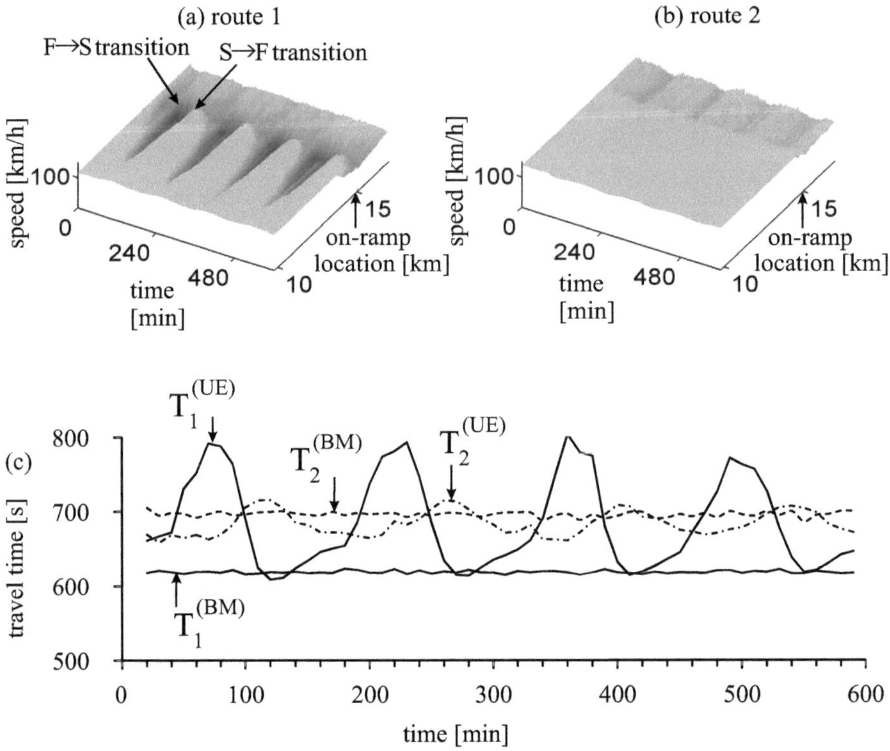

Fig. C.1 Simulations of dynamic traffic assignment of the flow rates q_1 and q_2 in (13.3) with the use of the Wardrop's UE for two-route network (Fig. 12.3a) based on step-by-step method (C.3). (**a**, **b**) Speed in space and time on routes 1 (**a**) and 2 (**b**) for a simulation realization. (**c**) Travel times for route 1 and 2 as time-functions: $T_1 = T_1^{(UE)}$, $T_2 = T_2^{(UE)}$ for the Wardrop's UE (13.3). As in Fig. 13.2, travel times $T_1 = T_1^{(BM)}$, $T_2 = T_2^{(BM)}$ are related to application of the network throughput maximization approach (BM principle for "zero breakdown probability" (10.8)) for which $q_1 = 3560$ vehicles/h, $q_2 = 2340$ vehicles/h. $Q = 7000$ vehicles/h

After the breakdown has occurred, a congested pattern has begun to develop on route 1 leading to increase in the travel time $T_1 = T_1^{(UE)}$ (Fig. C.1c). In accordance with the iteration method (C.3), the flow rate q_1 decreases and, respectively, the flow rate q_2 increases. As a result, over time congested pattern has dissolved (labeled by S→F transition in Fig. C.1a) and free flow has returned at bottleneck 1. Then, after another time delay $T^{(B)} = 74$ min, traffic breakdown has occurred at bottleneck 1 once more resulting in the formation of a second congested pattern, and so on.

Thus, we can confirm the common result of the application of the Wardrop's UE for dynamic traffic assignment of Sect. 13.3.2: Even when the total network inflow rate is still smaller than the network capacity, the application of the Wardrop's UE for dynamic traffic assignment is a random process of the congested pattern emergence with the subsequent dissolution of the congested pattern.

The term *random process* used above is explained by random time delays for traffic breakdown. For this reason, in different simulation realizations we have found different sequences of the congested pattern emergence and dissolution. Clearly, this random process leads to large osculations of travel times $T_1 = T_1^{(\mathrm{UE})}$ and $T_2 = T_2^{(\mathrm{UE})}$ (Fig. C.1c), whereas travel times $T_1^{(\mathrm{BM})}$ and $T_2^{(\mathrm{BM})}$ associated with the application of the network throughput maximization approach are almost time-independent.

We can see that when the congested pattern control approach is used for dynamic traffic assignment with the use of the Wardrop's UE (Fig. 13.2a,b), then a congested pattern propagates less than 2 km upstream of the bottleneck. Contrarily, when a step-by-step method (C.3) is used for dynamic traffic assignment with the use of the Wardrop's UE (Fig. C.1a,b), then a congested pattern propagates about 4 km upstream of the bottleneck. It should be noted that we have tried to find the optimal model parameters of the step-by-step method (C.3) at which the upstream propagation of traffic congestion exhibits a minimum value.

Therefore, we can assume that the performance of the congested pattern control approach for dynamic traffic assignment is better as that with standard methods based on a travel time change due to congestion. The reason of this result is as follows. When the congested pattern control approach for dynamic traffic assignment is applied (Sect. 13.3.2), feedback control of traffic congestion starts almost without time delay after the breakdown has been measured by a road detector. Contrarily, when the step-by-step method (C.3) for dynamic traffic assignment is applied, feedback control of traffic congestion starts only when travel time $T_1^{(\mathrm{UE})}$ begins to increase due to the congestion propagation. Therefore, in the case of the application of standard methods based on a travel time change due to congestion there is a time delay between the occurrence of traffic breakdown at a network bottleneck and the beginning of feedback control. This time delay results in a longer congestion propagation upstream of the bottleneck.

Reference

1. G.A. Korn, T.M. Korn, *Mathematical handbook for scientist and engineers* (McGraw-Hill Book Company, New York/San Francisco/Toronto/London/Sydney, 1968)

Glossary

ANCONA On-Ramp Metering **A**utomatic o**n**-ramp **c**ontrol of c**on**gested p**a**tterns (ANCONA) is feedback on-ramp metering based on a congested pattern control approach. In the congested pattern control approach, congestion is allowed to set in at a bottleneck. The basic idea is the congestion dissolution or the maintenance of congestion conditions at the bottleneck to the minimum possible level (when the dissolution of congestion is not possible); in particular, a congested pattern should not propagate upstream.

Automatic Driving Vehicle Based on Three-Phase Theory An automatic driving vehicle based on the three-phase theory is the automatic driving vehicle for which there is a speed range within which at each vehicle speed that is larger than zero there are the infinite number of the operating points of the automatic driving vehicle. Automatic driving is also called as self-driving or automated driving, or else autonomous driving.

Bottleneck Traffic breakdown leading to the onset of traffic congestion in a traffic network occurs mostly at a network bottleneck. The bottleneck can be a result of road works, on- and off-ramps, a decrease in the number of road lanes, road curves and road gradients, traffic signal, etc. The probability of traffic breakdown at the bottleneck is considerably larger than that is outside the bottleneck.

Breakdown Minimization (BM) Principle The BM principle states that the optimum of a traffic or transportation network with N bottlenecks is reached, when dynamic traffic assignment, optimization and/or control are performed in the network in such a way that the probability $P_{\rm net}$ for the occurrence of traffic breakdown in at least one of the network bottlenecks during a given time interval $T_{\rm ob}$ for observing traffic flow reaches the minimum possible value. The BM principle is equivalent to the maximization of the probability that during the time interval $T_{\rm ob}$ traffic breakdown occurs at none of the network bottlenecks.

Catch Effect The catch effect is an induced traffic breakdown at a bottleneck caused by the propagation of a synchronized flow pattern (SP) to the bottleneck. The

B.S. Kerner, *Breakdown in Traffic Networks*, DOI 10.1007/978-3-662-54473-0

term *catch effect* emphasizes that the initial SP is usually caught at the bottleneck as a result of the induced traffic breakdown. The catch effect can occur only, if free flow conditions have been at the bottleneck before the SP reaches it.

Characteristic Parameters of Wide Moving Jam Propagation Characteristic parameters of the propagation of a wide moving jam are average traffic variables and jam parameters that do not depend on traffic variables in traffic flow upstream of the wide moving jam. The characteristic parameters are the flow rate, density, and the average speed in free flow formed in the jam outflow, the mean velocity of the downstream jam front as well as the jam density; the characteristic parameters are the same for different wide moving jams. However, the characteristic parameters can depend considerably on traffic parameters (percentage of long vehicles (trucks), weather, road conditions, etc.).

Congested Traffic Congested traffic is defined as a state of traffic in which the average speed is *lower* than the minimum average speed that is still possible in free flow.

Congested Traffic Pattern A congested traffic pattern (congested pattern for short) is a spatiotemporal traffic pattern within which there is congested traffic.

Congested Pattern Control Approach In congested pattern control approach, *no control* of traffic flow at a network bottleneck is realized as long as free flow is realized at the bottleneck. This means that the occurrence of random traffic breakdown is permitted to occur at the bottleneck: Congestion at the bottleneck is allowed to set in. The main idea of this approach is as follows. Only after traffic breakdown has occurred, traffic control starts. In this approach, two control scenarios are possible: (i) Due to the congested pattern control approach, congestion dissolves and free flow recovers at the bottleneck. In this case, the congested pattern control approach can also be called "control of traffic breakdown at bottleneck". (ii) Due to the congested pattern control approach, traffic congestion at the bottleneck is maintained at a minimum possible level, in particular, traffic congestion does not propagate upstream of the bottleneck, i.e., congested pattern is localized at the bottleneck.

Constrain "Alternative Network Routes (Paths)" By any application of the breakdown minimization (BM) principle there is a constrain "alternative network routes (paths)". The constrain "alternative routes" prevents the use of too long routes in dynamic traffic assignment with the use of the BM principle.

Criteria [S] and [J] for Traffic Phases Criteria (definitions) [S] and [J] for traffic phases are spatiotemporal macroscopic criteria for traffic phases in congested traffic that define the synchronized flow and wide moving jam traffic phases in congested traffic, respectively.

Double Z-Characteristic (2Z-Characteristic) for Phase Transitions A double Z-characteristic (2Z-characteristic) for phase transitions is a macroscopic characteristic of phase transitions in traffic flow. The 2Z-characteristic consists of an

Z-characteristic for traffic breakdown together with an Z-characteristic for S→J and J→S transitions.

Dynamic Traffic Assignment Dynamic traffic assignment is traffic assignment in a traffic or transportation network considering the temporal dimension of the assignment problem.

Effective Location of Bottleneck The effective location of bottleneck is a road location in a neighborhood of a highway bottleneck at which traffic breakdown (F→S transition) is observed with the largest probability. In real field traffic data, the effective bottleneck location is the road location at which a permanent speed disturbance is localized at the bottleneck. The average speed within the permanent local speed disturbance at the bottleneck is smaller than the speed in free flow outside the permanent local speed disturbance.

After traffic breakdown has occurred at the bottleneck, a congested pattern emerges at the bottleneck. The effective location of the bottleneck can be different from a road location in a neighborhood of the bottleneck at which the downstream front of the congested pattern is fixed at the bottleneck. Moreover, both effective location of the bottleneck and the location at which the downstream front of the congested pattern is fixed at the bottleneck are probabilistic values in real traffic.

Empirical Fundamental of Transportation Science The nucleation nature of traffic breakdown (F→S transition) at a highway bottleneck is an empirical fundamental of transportation science.

Empirical Traffic Breakdown Empirical traffic breakdown (F→S transition) at a highway bottleneck can occur in a metastable state of free flow with respect to the F→S transition. In the metastable state of free flow, traffic breakdown occurs only if a nucleus required for the breakdown appears. Empirical traffic breakdown (F→S transition) at the bottleneck can be either spontaneous or induced.

Expanded Traffic Congested Pattern (EP) An expanded traffic congested pattern (EP) (expanded pattern for short) is a congested traffic pattern whose synchronized flow affects at least two adjacent highway bottlenecks. An EP can be either an SP or an GP. The term *expanded pattern* emphasizes only that rather than the SP or GP occurs at an isolated bottleneck, synchronized flow of the SP or GP affects two or more adjacent highway bottlenecks.

Empirical Traffic Data Empirical traffic data is a synonym of real field traffic data. This data is solely obtained through measurements in real world traffic based on road detectors, video-cameras, global navigation satellite system (GNSS) in vehicles, and other methods for the measurement of traffic variables (vehicle position, vehicle speed, flow rate, density, time headway between vehicles, etc.).

Free Traffic Flow Free traffic flow (free flow for short) is usually observed, when the vehicle density in traffic is small enough. The flow rate increases in free flow with the increase in the vehicle density, whereas the average vehicle speed is a decreasing density function. In accordance with real field traffic data, there is a limit in the increase in the vehicle density in free flow: In the associated limit state of free flow, the flow rate and density reach their maximum values while the average speed reaches a minimum value that is still possible in free flow.

Front of Traffic Pattern A front of a traffic pattern is either a moving or motionless region within which one or several of the traffic variables change abruptly in space (and in time, when the front is a moving one). There are downstream front and upstream fronts of the traffic pattern. The downstream pattern front separates the pattern from other traffic states downstream. The upstream pattern front separates the pattern from other traffic states upstream.

F→S transition In all known observations, traffic breakdown at a highway bottleneck is a phase transition from the free flow phase to synchronized flow phase called an F→S transition. The terms *traffic breakdown* and an *F→S transition* are synonyms related to the same phenomenon of the onset of congestion in free flow.

Fundamental Diagram of Traffic Flow The fundamental diagram of traffic flow is a relationship between the flow rate and density in vehicle traffic. Because the flow rate is the product of the average speed and density, the fundamental diagram is associated with a relationship between these traffic variables. In accordance with an obvious result of traffic measurements, on average the speed decreases when the density increases. Thus, in the flow–density plane, the fundamental diagram should pass through the origin (when the density is zero so is the flow rate) and should have at least one maximum. The fundamental diagram gives also a connection between the average space gap (net distance) between vehicles and the average speed.

General Congested Traffic Pattern (GP) A general congested traffic pattern (GP) (general pattern for short) is a spatiotemporal congested pattern within which congested traffic consists of the synchronized flow and wide moving jam phases.

Highway Capacity of Free Flow at Bottleneck Any flow rate in free flow at a highway bottleneck at which traffic breakdown (F→S transition) can be induced at a highway bottleneck is equal to one of the highway capacities of free flow at the bottleneck. There are the infinite number of such flow rates: *At any time instant*, there are the infinite number of highway capacities of free flow at the bottleneck that are between a minimum highway capacity and a maximum highway capacity. The existence of an infinite number of highway capacities at any time instant means that highway capacity is stochastic.

Incommensurability of Classical Traffic Theories with Three-Phase Traffic Theory The three-phase theory has been introduced for the explanation of the empirical free flow metastability with respect to an F→S transition at a highway bottleneck. This metastability of free flow has no sense for the classical traffic and transportation theories. In accordance with the classical theory by Kuhn about scientific revolutions, this shows the incommensurability of the three-phase theory and the classical traffic-flow theories. The term "incommensurable" has been introduced by Kuhn to explain a paradigm shift in a scientific field.

Induced Traffic Breakdown (Induced F→S Transition) at Bottleneck An induced traffic breakdown (induced F→S transition) in metastable free flow at a bottleneck is traffic breakdown that is induced at the bottleneck by the propagation of a moving spatiotemporal congested traffic pattern. This congested pattern has occurred *earlier* than the time instant of traffic breakdown at the bottleneck and at a *different* road location (for example at another bottleneck) than the bottleneck location. When this congested pattern reaches the bottleneck, the pattern induces traffic breakdown at the bottleneck.

The empirical induced F→S transition at the bottleneck proves the empirical nucleation nature of traffic breakdown at the bottleneck. The empirical nucleation nature of traffic breakdown at a highway bottleneck can be considered the empirical fundamental of transportation science.

Line *J* The line *J* is a characteristic line in the flow–density plane representing a steadily propagation of the downstream front of a wide moving jam. The slope of the line *J* is determined by the mean velocity of the downstream jam front. All infinite number of states of traffic flow in the flow–density plane that lie on the line *J* are threshold states for wide moving jam existence and emergence. The line *J* separates all steady states of synchronized flow in the flow–density plane into two qualitatively different classes: (i) States below the line *J* are associated with synchronized flow in which no wide moving jam can emerge spontaneously and persist. (ii) States on and above the line *J* are associated with metastable synchronized flow in which a wide moving jam can emerge and exist.

Maximum Highway Capacity A maximum highway capacity is equal to a critical flow rate at a highway bottleneck that separates metastable states of free flow with respect to traffic breakdown (F→S transition) at the bottleneck from unstable states of free flow: When the flow rate is equal to or larger than the maximum highway capacity, free flow is unstable with respect to traffic breakdown at the bottleneck. The maximum highway capacity is a stochastic value.

Maximum Signal Capacity A maximum signal capacity is equal to a critical average arrival flow rate at traffic signal that separates metastable states of under-saturated traffic with respect to traffic breakdown at traffic signal from unstable states of under-saturated traffic at the signal: When the flow rate is equal to or larger than the maximum signal capacity, under-saturated traffic is unstable with respect to traffic breakdown at the signal. The maximum signal capacity is a stochastic value.

Metastable Free Flow with Respect to Traffic Breakdown (F→S Transition) Metastable free flow with respect to traffic breakdown (F→S transition) at a highway bottleneck is free flow in which small enough local speed (and density) disturbances decay over time, i.e., the small local disturbances cause no traffic breakdown. However, if a large enough local disturbance appears in metastable free flow at the bottleneck, the F→S transition does occur. In the latter case, the local disturbance is called a nucleus for the F→S transition.

Metastable Synchronized Flow with Respect to Wide Moving Jam Emergence (S→J Transition) Metastable synchronized flow with respect to wide moving jam emergence (S→J transition) is synchronized flow in which a small enough local speed disturbance, within which the speed is smaller than the synchronized flow speed outside the disturbance, decays over time, i.e., the small disturbance causes no wide moving jam emergence. However, if a large enough local disturbance appears in metastable synchronized flow, the S→J transition does occur. In the latter case, the local disturbance is called a nucleus for the S→J transition.

Microscopic Criterion for Wide Moving Jam Within wide moving jams, there are regions in which traffic flow is interrupted: The inflow into the jam has no influence on the jam outflow. The flow interruption effect is realized when the time interval of the flow interruption is considerably longer than the mean time delay in vehicle acceleration at the downstream front of a wide moving jam. The flow interruption effect determines the microscopic criterion for the wide moving jam phase. If in measured data of congested traffic the wide moving jam phase has been identified through the use of the microscopic criterion for the wide moving jam phase, then with certainty all remaining congested states in the data set are related to the synchronized flow phase. For this reason, the microscopic criterion for the wide moving jam phase can be considered a microscopic criterion for traffic phases in congested traffic.

Minimum Highway Capacity A minimum highway capacity is equal to a critical flow rate at a highway bottleneck that separates stable states of free flow with respect to traffic breakdown (F→S transition) at the bottleneck from metastable states of free flow at the bottleneck: When the flow rate is smaller than the minimum highway capacity, free flow is stable with respect to traffic breakdown at the bottleneck: No traffic breakdown can occur at the bottleneck. This is independent of the amplitude of a time-limited local disturbance in free flow at the bottleneck. The minimum highway capacity is a stochastic value.

Minimum Signal Capacity A minimum signal capacity is equal to a critical average arrival flow rate at traffic signal that separates stable states of under-saturated traffic with respect to traffic breakdown at traffic signal from metastable states of under-saturated traffic at the signal: When the flow rate is smaller than the minimum signal capacity, under-saturated traffic is stable with respect to traffic breakdown at the signal. This is independent of the amplitude of a time-limited local disturbance in under-saturated traffic at signal. The minimum signal capacity is a stochastic value.

Moving Jam A moving traffic jam (moving jam for short) is a congested traffic pattern of great vehicle density spatially limited by two jam fronts. Within the downstream jam front vehicles accelerate escaping from the jam; within the upstream jam front, vehicles slow down approaching the jam. The jam as a whole localized structure propagates upstream in traffic flow. Within the jam (between the jam fronts) vehicle density is great and speed is very low (sometimes as low as zero).

Network Throughput Maximization Approach A network throughput maximization approach is an approach to dynamic traffic assignment and control in a traffic or transportation network. The network throughput maximization approach is an application of the BM principle that maximizes the network throughput while keeping free flow conditions in the whole network. The network throughput maximization approach maintains condition $P_{\text{net}} = 0$, where P_{net} is the probability of traffic breakdown in the network. Therefore, the network throughput maximization approach is also called the BM principle for "zero breakdown probability". Through the maximization of the network throughput with the network throughput maximization approach no traffic breakdown can occur in the traffic network.

Network Capacity A network capacity denoted by C_{net} is a measure (or "metric") of a traffic or transportation network. Under a steady-state analysis of the network, when the network throughput maximization approach is applied for dynamic traffic assignment in the network, as long as the total network inflow rate is smaller than the network capacity traffic breakdown cannot occur in the network. The network capacity can considerably depend on the amplitude of fluctuations (flow rate, the vehicle speed, and/or the vehicle density) in free flow at network bottlenecks. The network capacity is a stochastic value.

Nucleation Nature of S→F Instability There can be many speed peaks (local disturbances of the speed increase in synchronized flow) with random amplitudes that occur randomly over time at the downstream front of synchronized flow at a highway bottleneck. Only when a large enough speed peak appears, the S→F instability occurs. Speed peaks of smaller amplitude cause dissolving speed waves of the increase in the speed (dissolving acceleration waves) in synchronized flow: All these waves dissolve over time while propagating upstream within synchronized flow. As a result, the synchronized flow persists at the bottleneck. Thus, the S→F instability exhibits the nucleation nature.

Nucleation Nature of Traffic Breakdown (F→S Transition) The empirical nucleation nature of the F→S transition is as follows. Traffic breakdown at a highway bottleneck is a phase transition from free flow to synchronized flow (F→S transition). The F→S transition occurs in a metastable state of free flow with respect to this transition at a highway bottleneck. There can be many different local speed disturbances of the speed decrease in metastable free flow at the bottleneck. Small enough local speed disturbances in metastable free flow at the bottleneck decay. However, large enough local speed disturbances in metastable free flow lead to the F→S transition at the bottleneck. A local speed disturbance occurring in the metastable free flow that leads to the F→S transition is called a nucleus for the F→S transition. The nucleation nature of the F→S transition is explained by the nucleation feature of the S→F instability: The S→F instability governs traffic breakdown (F→S transition) at a highway bottleneck.

Nucleus for Moving Jam Emergence A nucleus for wide moving jam emergence in metastable synchronized flow is a local speed (density) disturbance in the synchronized flow that occurrence causes wide moving jam emergence in synchronized flow. Wide moving jam emergence in synchronized flow is an S→J transition. The average speed within the nucleus for moving jam emergence is lower than the average speed in synchronized flow outside the nucleus. A growing nucleus for wide moving jam emergence that propagates upstream in metastable synchronized flow is also called a growing narrow moving jam.

There can be local speed (density) disturbances with different disturbance amplitudes that can be nuclei for wide moving jam emergence in metastable synchronized flow. A local speed (density) disturbance with the minimum possible disturbance amplitude that occurrence causes moving jam emergence in metastable synchronized flow is called a critical nucleus for moving jam emergence. The amplitude of local speed (density) disturbance that is the critical nucleus for moving jam emergence depends on traffic variables (average speed and density) in metastable synchronized flow.

Nucleus for Traffic Breakdown A nucleus for traffic breakdown (F→S transition) in metastable free flow is a local speed (density) disturbance in free flow that occurrence causes traffic breakdown at a highway bottleneck. The average speed within the nucleus for traffic breakdown is lower than the average speed in free flow outside the nucleus.

There can be local speed (density) disturbances with different disturbance amplitudes that can be nuclei for traffic breakdown in metastable free flow at the bottleneck. A local speed (density) disturbance with the minimum possible disturbance amplitude (critical disturbance) that occurrence causes traffic breakdown in metastable free flow at the bottleneck is called a critical nucleus for traffic breakdown. Thus, a critical disturbance with respect to the F→S transition in free flow is a synonym of a critical nucleus for traffic breakdown. The amplitude of local speed (density) disturbance that is the critical nucleus for traffic breakdown depends on the flow rate in metastable free flow at the bottleneck.

In real field traffic data, a nucleus for empirical spontaneous traffic breakdown at the bottleneck occurs through an interaction of one of the waves in free flow with a permanent speed disturbance localized at a highway bottleneck. When the wave reaches the location of the disturbance at the bottleneck (effective bottleneck location), spontaneous traffic breakdown, i.e., phase transition from free flow to synchronized flow (F→S transition) occurs.

Operating Point of Automatic Driving Vehicle An operating point of an automatic driving vehicle is a hypothetical steady state of the vehicle that follows the preceding vehicle moving at a time-independent speed. In the operating point, both the acceleration (deceleration) of the automatic driving vehicle and the speed difference between the speed of the preceding vehicle and the speed of the automatic driving vehicle are equal to zero, i.e., the automatic driving vehicle moves at the time-independent speed of the preceding vehicle.

In the classical model of an automatic driving vehicle, at each given speed of the vehicle that is larger than zero, there is only one operating point of the automatic driving vehicle. The operating point corresponds to a time headway that is equal to the desired time headway. Contrarily, for an automatic driving vehicle based on three-phase theory there is a speed range within which at each vehicle speed that is larger than zero there are the infinite number of the operating points. This means that for the automatic driving vehicle based on the three-phase theory there is no fixed time headway to which the automatic driving vehicle should necessarily be steered.

Over-Acceleration Effect An over-acceleration effect is driver maneuver leading to a higher speed from an initial car-following at a lower speed, i.e., when the vehicle has initially been within a synchronization space gap. The over-acceleration effect is an opposite effect to the speed adaptation effect. In particular, following a slow moving preceding vehicle on a multi-lane road, a driver searches for the opportunity to accelerate and to pass. In this case, the over-acceleration effect is vehicle acceleration for passing, i.e., lane changing to a faster lane.

Over-Deceleration Effect An over-deceleration effect is as follows. If a vehicle begins to decelerate unexpectedly, then the following vehicle starts deceleration with a time delay. When this time delay is long enough, the driver decelerates stronger than it is needed to avoid collisions. As a result, the speed becomes lower than the speed of the preceding vehicle. If this over-deceleration effect is realized for the following drivers, a wave of vehicle speed reduction appears and increases in amplitude over time. The over-deceleration effect is the reason for the classical traffic flow instability that leads to moving jam emergence in traffic flow.

Permanent Local Speed Disturbance in Free Flow at Highway Bottleneck At a great enough flow rate at a highway bottleneck, the bottleneck introduces a permanent local speed disturbance in free flow. The disturbance is localized in a neighborhood of the bottleneck. On average the speed is lower and density is greater within this disturbance than these traffic variables are in free flow outside the disturbance. The existence of the permanent local speed disturbance at the bottleneck explains why the probability of traffic breakdown (F→S transition) is considerably larger at the bottleneck than outside the bottleneck. A road location in a neighborhood of the bottleneck at which the permanent local speed disturbance is localized is called the effective location of the bottleneck.

Probability of Traffic Breakdown in Traffic or Transportation Network The probability of traffic breakdown in a traffic or transportation network P_{net} is the probability that traffic breakdown occurs during a given time interval for observing traffic flow in at least at one of the network bottleneck (probability of traffic breakdown in the network, for short).

S→F Instability An S→F instability is an instability of synchronized flow. Due to the S→F instability an initial speed disturbance of the increase in the speed within synchronized flow (S) at the bottleneck transforms into a growing speed wave of the increase in the speed (growing acceleration wave) that propagates upstream within synchronized flow and leads to free flow (F) at the bottleneck: The S→F instability results in an S→F transition. The S→F instability is caused by driver over-acceleration. The S→F instability occurs due to a finite time delay in driver over-acceleration.

S→J Transition An S→J transition is a phase transition from the synchronized flow phase to wide moving jam phase.

Safe Space Gap If the space gap of the vehicle to the preceding vehicle is smaller than a safe space gap, the vehicle decelerates. The safe space gap depends on the vehicle speed. The upper boundary of the 2D-region of steady states of synchronized flow is associated with the safe space gap .

Saturation Flow Rate The maximum flow rate in the outflow from a vehicle queue at traffic signal during the green signal phase determines the saturation flow rate when the following conditions are satisfied: (i) The vehicles are in a standstill within the queue, (ii) vehicles escape freely from the queue at its downstream front, and (iii) the vehicles can accelerate to the maximum vehicle speed permitted in city traffic downstream of the signal. In empirical observations, the saturation flow rate is a characteristic flow rate that does not depend on distances between vehicles that are in a standstill in the queue, on the queue length, and on the arrival flow rate upstream of the queue. However, the saturation flow rate can depend considerably on traffic parameters (percentage of long vehicles (trucks), weather, road conditions, etc.).

Sequence of F→S→F Transitions A sequence of F→S→F transitions occurring at a highway bottleneck determines a time delay of traffic breakdown at the bottleneck. When an F→S transition begins to develop at the bottleneck, i.e., the upstream front of synchronized flow begins to propagate upstream from the bottleneck, an S→F instability can randomly occur. The S→F instability is realized when in the emergent synchronized flow at the bottleneck a local speed disturbance of the increase in the speed appears within which driver over-acceleration becomes on averaged stronger than driver speed adaptation. Due to the S→F instability, free flow appears at the bottleneck. As a result, the downstream front of the synchronized flow departs upstream from the bottleneck. In its turn, this results in the dissolution of the synchronized flow, i.e., in the interruption of the formation of a congested pattern at the bottleneck. We call this effect as the sequence of F→S→F transitions. The sequence of F→S→F transitions interrupts the formation of the congested pattern at the bottleneck.

Sources of Nuclei for Traffic Breakdown at Bottleneck Rather than the nature of traffic breakdown, the terms *empirical spontaneous* and *empirical induced* traffic breakdowns at a highway bottleneck distinguish different *sources* of a nucleus that occurrence leads to traffic breakdown. The source of a nucleus for empirical spontaneous traffic breakdown at the bottleneck is usually associated with an interaction of one of the waves in free flow with a permanent speed disturbance localized at the bottleneck. The source of a nucleus for empirical induced traffic breakdown at the bottleneck is usually a congested traffic pattern that propagates to the bottleneck location. Therefore, the source of a nucleus for traffic breakdown

(F→S transition) is the solely difference between empirical spontaneous and induced traffic breakdowns at the bottleneck.

Spatiotemporal Analysis of Real Field Traffic Data To understand spatiotemporal phenomena in vehicular traffic, a spatiotemporal analysis of real field (empirical) traffic data measured in vehicular traffic is needed. To perform the spatiotemporal analysis of real field traffic data, traffic variables (the flow rate, vehicle speed, vehicle density, space gap between vehicles, time headway (net time gap) between vehicles, etc.) should be measured simultaneously both in space and time.

In the book, the term *real field traffic data* is a synonym of the term *empirical traffic data*. Respectively, the term *spatiotemporal analysis of real field traffic data* is a synonym of the term *spatiotemporal analysis of empirical traffic data*.

Spatiotemporal Phenomena in Vehicular Traffic Traffic occurs in space and time. A diverse variety of spatiotemporal interactions of vehicles occurring in different driving situations results in the spatiotemporal character of real traffic phenomena. An important example of spatiotemporal traffic phenomena is traffic breakdown at a highway bottleneck.

Spatiotemporal Traffic Pattern A spatiotemporal traffic pattern (traffic pattern for short) is a specific distribution of traffic flow variables in space and time. To find real features of the spatiotemporal traffic pattern, a simultaneous measurements of traffic flow variables in space and time should be available. Examples of the traffic flow variables are the flow rate, speed, density, occupancy, space gaps, and time headway between vehicles.

Speed Adaptation Effect An speed adaptation effect is driver speed adaptation of the vehicle speed to the speed of the preceding vehicle occurring usually when the vehicle cannot pass a slower moving preceding vehicle. The speed adaptation effect occurs within a synchronization space gap without caring, what the precise space gap is, as long as the space gap is not smaller than a safe space gap.

Spontaneous Traffic Breakdown (Spontaneous F→S Transition) at Bottleneck If before traffic breakdown occurs at a highway bottleneck there is free flow at the bottleneck as well as upstream and downstream in a neighborhood of the bottleneck, then traffic breakdown is called spontaneous traffic breakdown (spontaneous F→S transition) at the bottleneck. In empirical data, a spontaneous traffic breakdown occurs, if a nucleus required for traffic breakdown appears spontaneously at the bottleneck.

Spontaneous Wide Moving Jam Emergence in Synchronized Flow Spontaneous wide moving jam emergence in metastable synchronized flow is a phase transition from the synchronized flow phase to the wide moving jam phase occurring due to the growth of a nucleus required for moving jam emergence in synchronized flow. There can be various sources for the occurrence of a nucleus required for moving jam emergence in metastable synchronized flow: unexpected braking of a vehicle in synchronized flow, lane changing on the main road leading to time headway reduction (and as a result, to local speed reduction), fluctuations in flow rates, vehicle merging from other roads, etc.

Stable Free Flow with Respect to Traffic Breakdown Stable free flow with respect to traffic breakdown at a highway bottleneck is free flow in which any time-limited speed (density) disturbances decay over time, i.e., no traffic breakdown (no F→S transition) can occur. Stable free flow with respect to traffic breakdown is realized when the flow rate in free flow at the bottleneck is smaller than a minimum highway capacity.

Stable Synchronized Flow with Respect to Wide Moving Jam Emergence Stable synchronized flow with respect to wide moving jam emergence is synchronized flow in which no wide moving jams can emerge or persist over time (no S→J transition can occur). Synchronized flow states that lie below the line J in the flow–density plane are stable synchronized flow states with respect to wide moving jam emergence.

Standard Dynamic Traffic Assignment Standard dynamic traffic assignment is related to approaches to dynamic traffic assignment with the use of the Wardrop' UE or SO whose objective is the minimization of route travel times and/or other travel costs in a traffic and transportation network. The term "standard" emphasizes that the approaches to dynamic traffic assignment are related to the state-of-the-art in traffic and transportation research.

Steady States of Traffic Flow Steady states of traffic flow are hypothetical states of homogeneous (in time and space) traffic flow of identical vehicles (and identical drivers) in which all vehicles move with the same time-independent speed and have the same space gaps.

Stochastic Highway Capacity Stochastic highway capacity has been explained in item *highway capacity of free flow at bottleneck*.

Synchronized Flow Traffic Pattern (SP) A synchronized flow traffic pattern (SP) is a spatiotemporal congested pattern in which congested traffic consists only of the synchronized flow traffic phase.

Synchronized Flow Traffic Phase In the three-phase theory, the following definition (criterion) [S] of the synchronized flow phase (synchronized flow for short) in congested traffic is made. In contrast to the wide moving jam phase, the downstream front of the synchronized flow phase does *not* maintain the mean velocity of the downstream front. In particular, the downstream front of synchronized flow is often fixed at a highway bottleneck. In other words, synchronized flow does not exhibit the characteristic jam feature [J].

Synchronization Space Gap In the three-phase theory, a synchronization space gap is a characteristic of the driver speed adaptation effect. If the space gap of the vehicle to the preceding vehicle is greater than the synchronization space gap, the vehicle accelerates. If the space gap of the vehicle moving on a single-lane road is equal to or smaller than the synchronization space gap, the vehicle adapts its speed to the speed of the preceding vehicle without caring about the space gap as long as it is not smaller than a safe space gap. This means that if the preceding vehicle moves at a time-independent speed, and if we neglect fluctuations, the vehicle moves at the same time-independent speed (i.e., the vehicle acceleration is equal to zero) with an arbitrary space gap within a space gap range between the synchronization space gap and safe space gap.

The lower boundary of the 2D-region of steady states of synchronized flow in the flow–density plane is associated with the synchronization space gap.

Theoretical Fundamentals of Transportation Science The three-phase theory is the theoretical fundamental of transportation science that explains the empirical nucleation nature of traffic breakdown. The BM principle is the theoretical fundamental of transportation science that permits to maximize the network throughput preventing traffic breakdown in the network.

Threshold Flow Rate for Traffic Breakdown at Highway Bottleneck A threshold flow rate for traffic breakdown (F→S transition) at a highway bottleneck is the flow rate that separates metastable states of free flow at the bottleneck into two different flow rate ranges: (i) When the flow rate in free flow at the bottleneck is smaller than the threshold flow rate, in the metastable free flow the probability of spontaneous traffic breakdown during a given time interval for observing traffic flow is equal to zero; this means that only induced traffic breakdown is possible in this metastable free flow. (ii) When the flow rate in free flow at the bottleneck is equal to or larger than the threshold flow rate, in the metastable free flow either spontaneous traffic breakdown or induced traffic breakdown is possible at the bottleneck.

Three-Phase Traffic Flow Model A three-phase traffic flow model is a mathematical traffic flow model in the framework of the three-phase theory. The basic requirement for a three-phase traffic flow model is that the model should show a double Z-characteristic (2Z-characteristic) for phase transitions in traffic flow.

Three-Phase Traffic Theory In the three-phase traffic theory (three-phase theory for short), besides the free flow phase there are two phases in congested traffic, the synchronized flow and wide moving jam phases. The synchronized flow and wide moving jam phases in congested traffic are defined through the use of empirical spatiotemporal criteria (definitions) [S] and [J], respectively. The three-phase theory is a qualitative theory that explains traffic breakdown and resulting congested traffic patterns based on the criteria [S] and [J] for traffic phases as well as on hypotheses of this theory.

Time Delay of Traffic Breakdown at Highway Bottleneck There is a time delay of traffic breakdown (F→S transition) at a highway bottleneck. The time delay is a random value. A random time delay of traffic breakdown at the bottleneck is caused by a sequence of F→S→F transitions occurring at the bottleneck: The S→F transition in this sequence prevents traffic breakdown at the bottleneck. The S→F transition is determined by the S→F instability at the bottleneck. Because the S→F instability exhibits the nucleation nature, the S→F instability does not occur, if no nucleus required for the S→F instability appears in synchronized flow at the bottleneck. In this case, traffic breakdown does occur at the bottleneck. The occurrence of the nucleus required for the S→F instability in synchronized flow at the bottleneck is a random event. This explains that there is a random time delay of traffic breakdown at the bottleneck. This explains also a result of the three-phase theory that the S→F instability governs traffic breakdown (F→S transition) at the bottleneck. For this reason, traffic breakdown at the bottleneck can be called a time-delayed traffic breakdown at the bottleneck.

Time Delay of Traffic Breakdown at Traffic Signal Traffic breakdown at the signal is a transition from under-saturated traffic to over-saturated traffic at the signal. There is a time delay of traffic breakdown at traffic signal. The time delay is a random value. The time delay of traffic breakdown at traffic signal is associated with the metastability of under-saturated traffic at the signal with respect to the transition to over-saturated traffic at the signal. For this reasons, traffic breakdown at the signal in city traffic can be called a time-delayed transition from under-saturated traffic to over-saturated traffic at the signal.

Time Headway A time headway is a net time gap between two vehicles following each other.

Traffic Assignment Traffic assignment is a procedure used to find the network link flow rates from the original-destination demand.

Traffic Breakdown at Highway Bottleneck Traffic breakdown at a highway bottleneck is the phenomenon of the onset of congestion in an initial free flow at the bottleneck. In all observations, traffic breakdown is an F→S transition at the bottleneck. The terms *F→S transition*, *breakdown phenomenon*, *traffic breakdown*, and *speed breakdown* are synonyms related to the same effect: the onset of congestion in free flow at the bottleneck.

Traffic Flow Model A traffic flow model is devoted to the explanation and simulation of traffic flow phenomena, which are observed in measured data of real traffic flow, and to the prediction of new traffic flow phenomena that could be found in real traffic flow.

Traffic Parameters Traffic parameters are those parameters of the dynamic spatiotemporal process "vehicular traffic," which can influence on traffic variables and traffic patterns. Examples of traffic parameters are characteristics of traffic network infrastructure (e.g., types and parameters of network bottlenecks and their locations), weather (whether the day is sunny or rainy or else foggy, dry or wet road, or even ice and snow on road), the percentage of long vehicles (trucks), day time, working day or week-end, other road conditions, and vehicle technology.

Traffic Pattern A traffic pattern is a spatiotemporal structure in traffic that exhibits some specific features of the pattern. The traffic pattern is separated by the upstream and downstream pattern fronts from other traffic patterns and traffic states.

Traffic State A traffic state is a state of traffic that exhibits some specific features that distinguish this traffic state from other traffic states. There can be different traffic states related to the same traffic phase. In particular, there can be different free flow states within the free flow phase as well as there can be different synchronized flow states within the synchronized flow phase. For example, there can be homogeneous synchronized flow states that are nearly homogeneous in space and time as well as non-homogeneous synchronized flow states within the pinch region of synchronized flow of an GP within which narrow moving jams emerge and propagate upstream while growing in the jam amplitude.

Traffic Variables A distribution of traffic variables in space and time determines features of spatiotemporal traffic patterns. Examples of traffic variables are the vehicle speed, the time headway between vehicles, the space gap between vehicles, the flow rate, the vehicle density, and the vehicle occupancy.

Two-Dimensional (2D) States of Synchronized Flow A hypothesis of the three-phase theory assumes that in synchronized flow there are the infinite number of space gaps to the preceding vehicle moving at a time-independent speed at which a vehicle can move with this time-independent speed, while following the preceding vehicle. Thus, hypothetical steady states of synchronized flow cover a two-dimensional (2D) region in the flow–density plane, i.e., there is no desired space gap in hypothetical steady states of synchronized flow. Each of the steady states describes a hypothetical non-disturbed and noiseless vehicle motion in synchronized flow.

Wardrop's Equilibria The Wardrop's user equilibrium (UE) and system optimum (SO) are also called the Wardrop's equilibria or the Wardrop's principles.

Wardrop's System Optimum (SO) The Wardrop's SO states that the average travel times of all vehicles in a network is a minimum, which implies that aggregate vehicle-hours spent in travel is also minimum. In other words, in accordance with Wardrop's SO, the network-wide travel time should be a minimum.

Wardrop's User Equilibrium (UE) The Wardrop's UE states that traffic on a network distributes itself in such a way that the travel times on all routes used from any origin to any destination are equal, while all unused routes have equal or greater travel times.

Waves in Free Flow and Spontaneous Traffic Breakdown at Bottleneck In real field traffic data, a nucleus for empirical spontaneous traffic breakdown at a highway bottleneck occurs usually through an interaction of one of the waves in free flow with a permanent speed disturbance localized at the bottleneck. When the wave reaches the location of the permanent speed disturbance at the bottleneck (effective bottleneck location), spontaneous traffic breakdown, i.e., spontaneous phase transition from free flow to synchronized flow (F→S transition) occurs. The waves appear often due to slow vehicles moving in free flow. These waves propagate with the average speed of slow vehicles in free flow. Within a wave, the total flow rate is larger and the speed averaged across the highway is smaller than outside the wave. The waves in free flow exhibit a two-dimensional (2D) asymmetric spatiotemporal structure whose characteristics are different in different highway lanes. A wave that amplitude is large enough can become a nucleus for the breakdown while the wave propagates through the bottleneck; small enough waves in free flow do not become nuclei for traffic breakdown.

Wide Moving Jam Traffic Phase In the three-phase theory, the following definition (criterion) [J] of the wide moving jam traffic phase (wide moving jam for short) in congested traffic is made. A wide moving jam is a moving jam that maintains the mean velocity of the downstream jam front, even when the jam propagates through any other traffic states or highway bottlenecks. This is the characteristic feature [J] of the wide moving jam phase.

Z-Characteristic for S→J and J→S Transitions A Z-characteristic for S→J and J→S phase transitions is a macroscopic characteristic for phase transitions between the synchronized flow and wide moving jam traffic phases in congested traffic. The Z-characteristic for S→J and J→S phase transitions consists of states of synchronized flow (S), the average (macroscopic) value of the speed within the critical nucleus for the S→J transition, and states related to a wide moving jam (J).

Z-Characteristic for Traffic Breakdown A Z-characteristic of traffic breakdown is a macroscopic characteristic of traffic breakdown at a highway bottleneck. The Z-characteristic is built by metastable states of the traffic phases F and S together with a branch for the average speed within critical nuclei for traffic breakdown (F→S transition) at the bottleneck.

Index

B.S. Kerner, *Breakdown in Traffic Networks*, DOI 10.1007/978-3-662-54473-0

MIX
Papier aus verantwortungsvollen Quellen
Paper from responsible sources
FSC® C105338

If you have any concerns about our products,
you can contact us on
ProductSafety@springernature.com

In case Publisher is established outside the EU,
the EU authorized representative is:
Springer Nature Customer Service Center GmbH
Europaplatz 3, 69115 Heidelberg, Germany

Printed by Libri Plureos GmbH
in Hamburg, Germany